SUPERCOLLIDER 1

SUPERCOLLIDER 1

Edited by
Michael McAshan

Superconducting Supercollider Laboratory
Dallas, Texas

Plenum Press • New York and London

Library of Congress Cataloging in Publication Data

International Industrial Symposium on the Supercollider (1989: New Orleans, La.)
  Supercollider 1 / edited by Michael McAshan.
      p.   cm.
  "Proceedings of the International Industrial Symposium on the Supercollider, held
February 8–10, 1989, in New Orleans, Louisiana."
  Includes bibliographical references.
  ISBN 0-306-43365-6
  1. Superconducting Super Collider—Congresses. 2. Superconducting magnets—
Congresses. I. McAshan, Michael. II. Title. III. Title: Supercollider one.
QC787.P7I57   1989   89-16388
539.7′3—dc20   CIP

Proceedings of the International Industrial Symposium on The Supercollider,
held February 8–10, 1989, in New Orleans, Louisiana

PREFACE

IISSC '89 was a tremendous success. A total of 635 people attended this educational forum which was dedicated to further the understanding of the design, construction and operation of the Superconducting Supercollider (SSC). A total of 110 presentations and addresses were given. The topics discussed covered all aspects of the SSC including:

Magnet Technology	Cryogenics
Conventional Facilities	Technical Systems
Detectors	Related Accelerator Technology
Superconducting Wire/Cable	

Approximately 38% of the presentations addressed superconducting magnet technology, 16% were devoted to detector technology, 10% addressed superconducting wire/cable, and the balance was equally split between the remaining topics.

A special award was presented to Professor M. Tigner for his meritorious contribution to the Superconducting Supercollider (SSC). The award was presented on behalf of the IISSC Board of Directors.

Keynote speakers included:

Gerald Bachy, CERN
Joe Barton, Representative from Texas, 6th Disctrict
Ed Bingler, Exec. Director, Texas National Research Laboratory Commission
James Decker, Deputy Director, Office of Energy Research, (DOE)
Helen Edwards, Fermi National Accelerator Laboratory
M. G. D. Gilchriese, SSC Central Design Group
Robert Hunter, Director, Office of Energy Research, (DOE)
Leon Lederman, Director, Fermi National Accelerator Laboratory
Roy Schwitters, Director, SSC Laboratory
Alvin Trivelpiece, Director, Oak Ridge National Laboratory
Gus Voss, DESY

Highlights of the symposium included two panel sessions. The first panel discussed the growing role of industry in accelerator technology. The second panel addressed the congressional perspective on SSC.

<u>Industrial Panel</u>	<u>Congressional Panel</u>
J. R. Faulkner, Varian-Continental	Joe Barton (R), Texas, 6th Dist.
A. J. Favale, Grumman	Jim Chapman (R), Texas, 1st Dist.
R. Jacobsen, Maxwell Labs, Inc.	Mike Espy (D), Mississippi, 2nd Dist.
B. Prichard, Jr., SAIC	Jimmy Hayes (D), Louisiana, 7th Dist.
J. M. Rawls, General Atomics	Carl Pursell (R), Michigan, 2nd Dist.

The 1989 International Industrial Symposium on the Supercollider was produced by the IISSC Corporation. The 1989 IISSC Board of Directors was comprised of a consortium from industry, universities, national laboratories, and government.

<u>Member</u>	<u>Affiliation</u>
Mr. Charles Anderson	Air Products and Chemicals
Mr. Owen Anglum	Armco, Inc.
Mr. Robert Baldi (*)	General Dynamics
Dr. David Berley	National Science Foundation
Dr. Dietrich Bonmann	ABB Technology Company
Dr. Wendell Chen	University of Texas
Mr. Michael Davis	Martin Marietta
Dr. Robert Diebold	U. S. Department of Energy
Ms. Rene Donaldson(*)	SSC Design Group
Mr. Paul Gilbert	Parsons, Brinckerhoff
Dr. Leonard Goldman	Bechtel National, Inc.
Dr. Eric Gregory	Supercon, Inc.
Mr. Kim Hoag (*)	Ralph M. Parsons Company
Mr. Andy Jarabak	Westinghouse Electric Corp.
Dr. Thomas Kirk	SSC Central Design Group
Dr. David Larbalestier	University of Wisconsin
Dr. Peter Limon	SSC Central Design Group
Dr. Paul Mantsch (*)	Fermi National Accelerator Laboratory
Dr. Michael McAshan	SSC Central Design Group
Mr. Paul Reardon	Science Applications International Corp.
Mr. E. Parke Rohrer (*)	Brookhaven National Laboratory
Dr. Giuseppe Scarfi	Ansaldo Componenti S.p.A.
Mr. Clyde Taylor	Lawrence Berkeley Laboratory
Mr. Kuniyasu Toga	Hitachi Ltd.

(*) Denotes Board of Directors who also served as IISSC corporate officers.

Combined, there were 24 members on the board of directors for IISSC '89. Thirteen of the members represented industry, three of which were from foreign countries. Six of the members represented universities and three the national laboratories. The remaining two board members represented the Department of Energy and the National Science Foundation. In addition, conference management was headed by Ms. Pamela E. Patterson.

The following 46 companies, organizations, societies and agencies provided support to help produce IISSC '89:

ABB Technology Company
Air Products and Chemicals, Inc.
American Society of Civil Engineers
American Society of Mechanical Engineers
Ansaldo S.p.A.
Armco, Inc.
ASM International
BBC Brown Boveri AG

Bechtel National, Inc.
Brookhaven National Laboratory
Brown & Root, Inc.
CRS Sirrine
Daniel, Mann, Johnson and Mendenhall
Division of Particles and Fields, American Physical Society
EG&G, Inc.
Fermi National Accelerator Laboratory
Fluor Daniel
General Atomics
General Dynamics Space Systems Division
Grumman Corporation
Harza Engineering Company
Hitachi, Ltd.
Institute of Electrical and Electronics Engineers
Intermagnetics General Corporation
International Association of Bridge, Structural and Ornamental Iron Workers
Lawrence Berkeley Laboratory
Lester B. Knight & Associates, Inc.
Martin Marietta Corporation
Morrison Knudsen
National Science Foundation
National Society of Professional Engineers
Parsons Brickerhoff Quade & Douglas, Inc.
Science Applications International Corporation
Sheet Metal Workers' International Association
Society of Automotive Engineers
Stone & Webster Engineering Corporation
STV/Seelye Stevenson Value & Knecht
Supercon Inc.
Sverdrup Corporation
The Ralph M. Parsons Company
The University of Texas at Arlington
UNISTRUT Corporation
U. S. Department of Energy
University of Wisconsin
Universities Research Association—SSC Central Design Group
Westinghouse Electric Corporation

Those listed provided either financial support, volunteer assistance, or both. In total, $36,500 was raised to provide startup and operating funds for IISSC. Of this total, the U. S. Department of Energy provided $20,000 in the form of a grant.

The breakdown of the 635 people who registered for IISSC '89 is as follows: approximately 10% of the participants represented 11 foreign countries, and the domestic participants represented 37 states. Combined, these people represent a total of 278 organizations, 197 of which are industrial firms. The balance of the organizations represented are principally from universities, national laboratories and government.

Planning for IISSC '90 is already under way. The Board of Directors has added five new members:

<table>
<tr><td><u>Member</u></td><td><u>Affiliation</u></td></tr>
<tr><td>Ed Bingler</td><td>Texas National Research Laboratory Commission</td></tr>
<tr><td>Bob Marsh</td><td>Teledyne Wah Chang</td></tr>
<tr><td>Skip Porter</td><td>Houston Area Research Council</td></tr>
<tr><td>Carl Rosner</td><td>Intermagnetics General Corporation</td></tr>
<tr><td>Sven Svendsen</td><td>Daniel, Mann, Johnson & Mendenhall</td></tr>
</table>

The Board of Directors also elected Mike Davis of Martin Marietta Astronautics to be the chairman for IISSC '90.

Robert W. Baldi
Chairman
IISSC '89

ACKNOWLEDGMENTS

The editor is greatly indebted to several members of the staff of the SSC Central Design Group who carried the largest part of the burden of preparing these proceedings for publication. In particular, we would like to thank Ms. Valerie Kelly. Her powers of organization, her prodigious capacity for work, and her enthusiasm have made this project a pleasure.

In addition we are grateful to Annie Calinog for preparing many of the manuscripts, to Nathalie Guyol for updating the database, to Kate Metropolis for her powerful editing skills, to Darlene Moretti for taking the time between other responsibilities to proof, and to Stephen Sporn for handling the xeroxing and mailings.

CONTENTS

1. GROWING ROLE OF INDUSTRY IN ACCELERATOR TECHNOLOGY

Moderator: P. Reardon
Science Applications International Corporation

2. PARALLEL TECHNICAL SESSIONS

2A. MAGNET TECHNOLOGY

Chairman: R. Coombes
SSC Central Design Group

2B. CONVENTIONAL FACILITIES AND TECHNICAL SYSTEMS

Chairman: R. Robbins
Sverdrup Corporation

2C. ADVANCED PARTICLE DETECTORS

Chairman: M.G.D. Gilchriese
SSC Central Design Group

3. PLENARY TECHNICAL SESSION

Chairman: R. Diebold
U.S. Department of Energy

4. PARALLEL TECHNICAL SESSIONS II

4A. INDUSTRIAL OPPORTUNITIES ON THE SSC

Chairman: K. W. Chen
The University of Texas at Arlington

4B. MAGNET TECHNOLOGY II

Chairman: P. Mantsch
Fermi National Accelerator Laboratory

ATTENDEES

1 Growing Role of Industry in Accelerator Technology

Moderator:

P. Reardon
Science Applications International Corporation

HIGH PERFORMANCE PROTON ACCELERATORS†

Anthony J. Favale

Grumman Space Systems
Bethpage, New York 11714-3588

The theme* for the talks given in this session is that,
given accelerator performance requirements, industry can bring
to completion major accelerator subsystems. This includes the
capability to analyze the requirements by use of appropriate
accelerator and other codes, perform a conceptual design for
approval by the customer, accomplish the engineering using
appropriate codes and requirements, procure or build the
subcomponent parts and ultimately install, commission and test
the completed subsystem in almost a turn-key fashion.

In concert with this theme this paper briefly outlines how
Grumman, over the past four years, has evolved from a company
that designed and fabricated a Radio Frequency Quadrupole (RFQ)
accelerator from the Los Alamos National Laboratory (LANL)
physics and specifications to a company who, as prime
contractor, is designing, fabricating, assembling and
commissioning the U.S. Army Strategic Defense Command's (USA
SDC) Continuous Wave Deuterium Demonstrator (CWDD) accelerator
as a turn-key operation. In the case of the RFQ, LANL
scientists performed the physics analysis, established the
specifications supported Grumman on the mechanical design,
conducted the RFQ tuning and tested the RFQ at their
laboratory. For the CWDD Program Grumman has the
responsibility for the physics and engineering designs,
assembly, testing and commissioning albeit with the support of
consultants from LANL, Lawrence Berkeley Laboratory (LBL) and
Brookhaven National Laboratory. In addition, Culham Laboratory
and LANL are team members on CWDD. LANL scientists have
reviewed the physics design as well as a USA SDC review board.

I must mention that in this evolution we have benefited
greatly and are indepted to accelerator physicists and
engineers at the LANL, BNL, LBL & MIT. Without their help via

---

† Presented at International Industrial Symposium on the
  Super collider 2/8/89 Session II: Growing Role of Industry
  in Accelerator Technology.

* As presented by the Session Moderator Paul Reardon

tech transfer and or consultations, we would not be here today reporting on this work.

By high performance we mean high brightness, i.e., high current and low emittance systems in which transmission is high and emittance growth is kept low. A brief description of the RFQ, RGDTL and CWDD projects will be discussed.

<u>RFQ</u>

In 1985, Grumman was awarded a contract by LANL to design and fabricate a space qualified RFQ for the Strategic Defense Initiative Office's (SDIO) Beam Experiments Aboard a Rocket (BEAR) Program. The BEAR project is to be the first space flight for SDIO's Neutral Particle Beam program. LANL scientists performed the physics analysis and established the requirements for the RFQ. Grumman, with considerable support from LANL, designed the RFQ and fabricated an engineering model. LANL fabricated a cold model to test the RF features of the design. The unique features of this design are that it has no movable vanes and no, by virtue of the electroformed assembly, residual stress. This helps ensure that it stays tuned after being subjected to launch loads. The RFQ is fabricated by machining four vane sections, mounting them in an assembly fixture, tuning the RFQ while in this fixture then immersing the whole assembly, vanes and fixture, in an electroforming tank to produce a monolithic structure which is its own vacuum enclosure with integral RF seals.

Figure 1 shows the RFQ in its assembly fixture being tuned just prior to electroforming at GAR Electroforming, Danbury, CT. This test utilizing the bead pull apparatus shown in figure one is repeated after the electroforming operation to verify no changes have taken place in the RF characteristics of the RFQ. The bead pull apparatus, associated electronics and software are Grumman's. This system is an excellent example of technology transfer from LANL to Grumman. The Grumman system is modelled after LANL's and was built with their support. Figure 2 shows the completed RFQ. Figures 3 and 4 give the RFQ features and achievements. Grumman has built a second BEAR RFQ for the purpose of conducting physics studies both at LANL and at Grumman.

<u>RGDTL</u>

In 1987, Grumman won a competition to fabricate a RGDTL for the Accelerator Test Stand, ATS, at LANL. LANL did the physics calculations and the mechanical and thermal design. Modifications to the design were made in two areas. The material was changed from copper plated steel to copper plated aluminum and the thermal design was changed to comply with space tracability. Figure 5 shows the completed RGDTL. Final assembly was done by LANL personnel. Specifications and achieved performance are shown in Figure 6.

<u>CWDD</u>

In 1988 Grumman won a competition to design, fabricate, assemble, test and commission as prime contractor the CWDD accelerator for the USA SDC. Grumman has as its team members

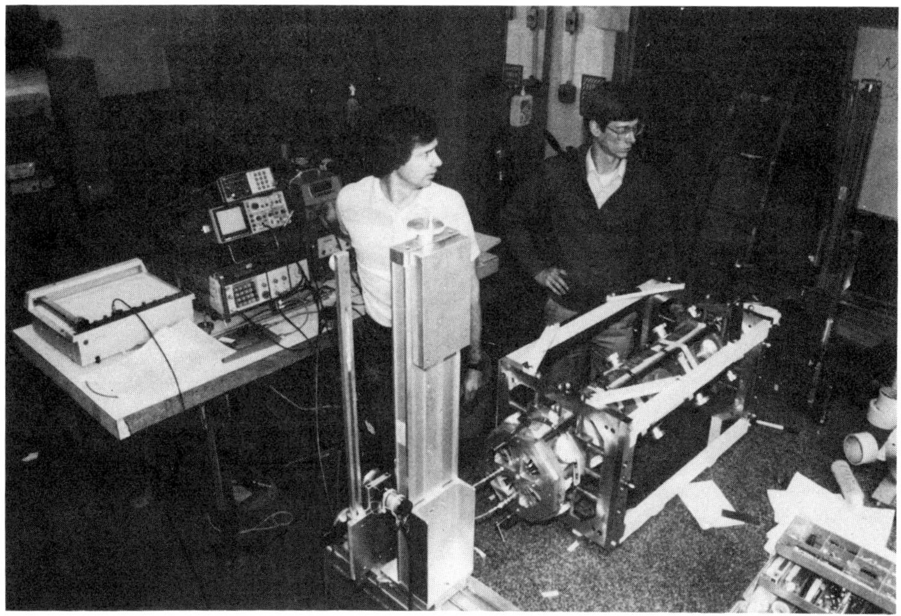

Figure 1.   Tuning of the RFQ by LANL and Grumman Scientists prior to electroforming at GAR

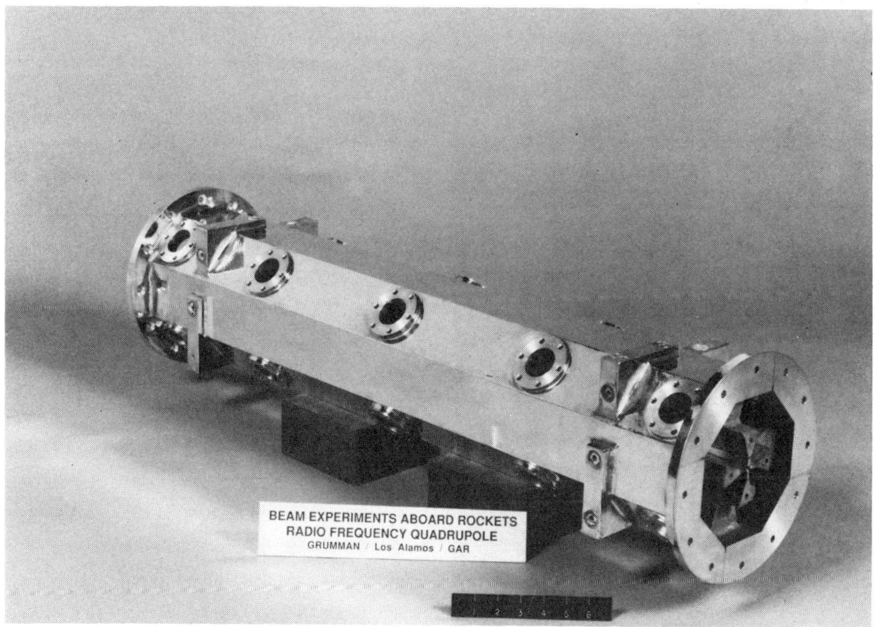

Figure 2.   The completed RFQ showing the four electroformed joints.

PARAMETER	REQUIRED	ACHIEVED
BEAM CURRENT	26 mA	28 mA
OUTPUT EMITTANCE (rms norm)	.01 π cm-mrad	.01 π cm-mrad
RF FIELDS	≤5% Dipole ≤5% Tilt	<1% Dipole <1% Tilt
CAVITY QUALITY (Q)	≥5000	6330
PEAK SURFACE FIELDS	1.8 Kp	3.0 Kp (sparker)
WEIGHT	≤150 lbs	147 lbs
THERMAL LOADS	60° F Rise	>60° F Rise

STRUCTURAL LOADS
  Pressure
  Launch / Separation
	17 psid
	5 g's lateral
	50 g's axial
		11 psid (tested)
		Exceeded ⎤ Trivial
		Not Tested ⎦

Random Vibration	AFGL ARIES Spec	Exceeded Statically Passed Shake Test

Figure 3.   BEAR RFQ features

ENERGY	1 MeV
FREQUENCY	425 MHz
LENGTH	1 Meter
COPPER POWER	70 kW
BEAM POWER	26 kW
DUTY FACTOR	.025%

BRIGHTNESS
  RFQ Output
  Neutralized
	280 A/(π cm mrad)^2
	120 A/(π cm mrad)^2

POWER EFFICIENT RFQ
FLIGHT QUALIFIED DESIGN
NO LONGITUDINAL RF JOINTS
STABLE - MONOLITHIC STRUCTURE
INTEGRAL VACUUM VESSEL

Figure 4.   BEAR RFQ Achievements

Figure 5.   The completed RGDTL

RAMP GRADIENT DTL - CONFIRMED PERFORMANCE

OPERATIONAL PARAMETERS	DESIGN SPECIFICATION	RGDTL PERFORMANCE CONFIRMED AT LANL'S "ATS" TEST FACILITY
• BEAM PARTICLE	H^-	H^-
• BEAM CURRENT	100 MA	80 MA (LIMIT OF THE TEST STAND INJECTOR AND RFQ)
• BEAM EMITTANCE GROWTH	(NOT SPECIFIED)	0.028% (NO GROWTH)
• BEAM TRANSMISSION	(NOT SPECIFIED)	97% OR GREATER (AS MEASURED)
• CAVITY FREQUENCY	425 MHZ	425 MHZ
• CAVITY RF DRIVE POWER (2 LOOPS)	1000 KW	824 KW (450 KW INTO CAVITY WALL 374 KW INTO BEAM)
• POWER EFFICIENCY	(NOT SPECIFIED)	45%
• INPUT ENERGY	2.07 MEV	2.07 MEV
• INITIAL ACCEL. GRADIENT	2.0 MV/METER	2.0 MV/METER
• FINAL ACCEL. GRADIENT	4.4 MV/METER	4.4 MV/METER
• OUTPUT ENERGY	6.67 MEV	6.67 MEV

Figure 6.   RGDTL confirmed performance

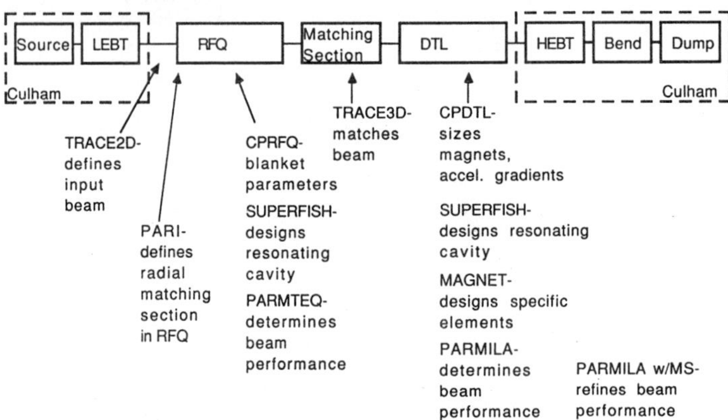

Figure 7.   Physics Design codes used for CWDD program

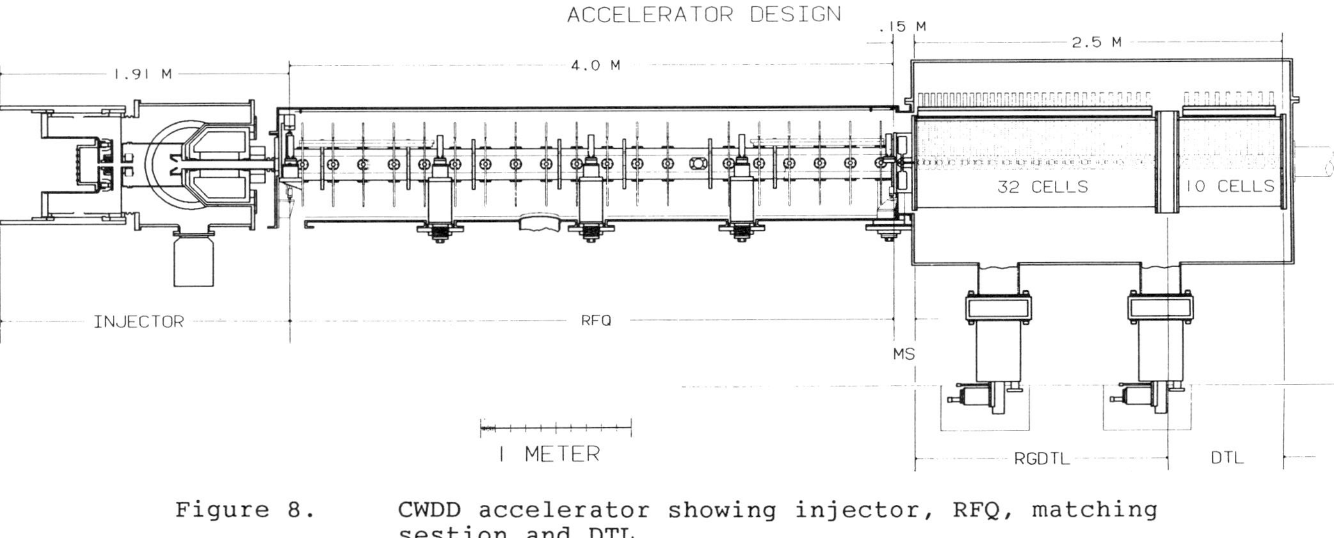

Figure 8.   CWDD accelerator showing injector, RFQ, matching sestion and DTL

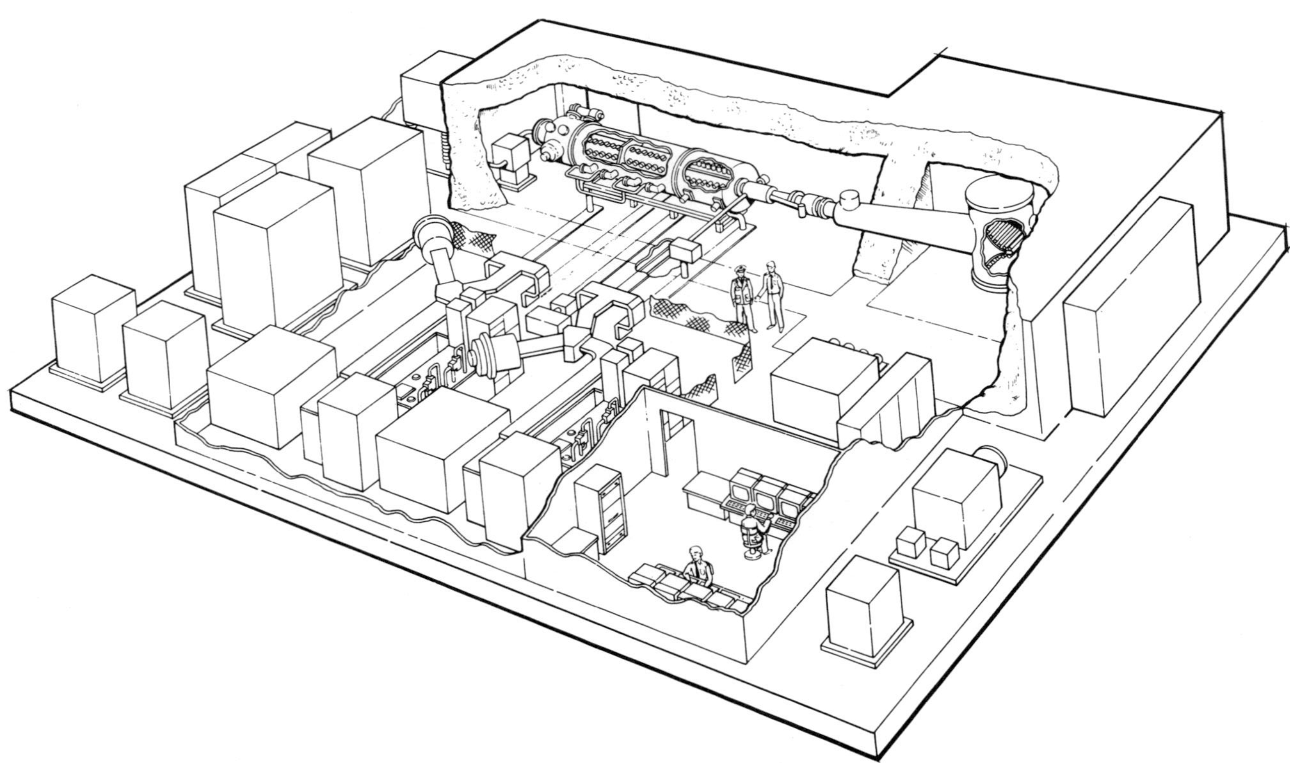

Figure 9.   Sketch of CWDD accelerator and facility

the Culham Laboratory and LANL. This accelerator operates on
CW mode, uses deuterium gas and the accelerator major
components are run at cryogenic temperatures. On satisfactory
commissioning, this accelerator will be turned over to the USA
SDC as a turn-key device. The detailed design and performance
features of this device are classified hence will not be
covered herein. Figure 7 delineates the physics design codes
used by Grumman physicists in designing the RFQ, matching
section and DTL. These codes were acquired from LANL via
technology transfer and assistance in their utilization was
provided by LANL physicists as well as consultants from BNL and
LBL. The final design was reviewed by these consultants and
the AT division of LANL.

Figures 8 and 9 show a schematic of the CWDD device and an
artists conception of the accelerator and its facility
respectively.

In addition to the above projects Grumman scientists and
engineers are involved in two other major accelerator programs;
namely, the Ground Test Accelerator, GTA at LANL and the
Compact Synchrotron X-Ray Light Source, SXLS, at BNL. In both
projects Grumman competed and won the industrial support
contractor role. For the GTA, Grumman has 26 engineers and
scientists working side by side with their LANL counterparts
supporting the design. This is a six year program through
commissioning of the machine. For the SXLS we will soon have
as many as 7 engineers and scientists working with their BNL
counterparts designing, assembling and commissioning this
device. In both of the above programs "Technology Transfer" to
industry is a stated objective of the program.

In summary, we believe that U.S. Industry can supply
accelerator components or turn-key devices given specification
by other industries, national laboratories or government
agencies as shown in this paper. The primary technology for
accelerators as well as new innovations resides in the national
laboratories and at several universities. Their support to
industry via technology transfer programs and or consultation
is imperative to a growing U.S. Accelerator Industry.

PARTICIPATION OF SAIC IN THE

LLUMC PROTON SYNCHROTRON PROJECT

Ben A. Prichard, Jr.

Science Applications International Corporation
227 Wall Street
Princeton, N.J.

ABSTRACT

Loma Linda University Medical Center (LLUMC) is
constructing a 250 MeV proton synchrotron and associated
treatment facilities for the purpose of the control of
cancer through particle beam irradiation or proton therapy.
The synchrotron and beam transport line are being developed
by Fermi National Accelerator Laboratory (FNAL) with
participation by Science Applications International
Corporation (SAIC) as an industrial partner for technology
transfer. SAIC is supporting the project in three ways: 1)
by accomplishing specific tasks in the design and
development of the facility; 2) by participating directly
with LLUMC and FNAL in areas of technology transfer; and 3)
by being directly responsible for the installation,
commissioning, and early operation of the facility.

INTRODUCTION

The use of protons in cancer therapy was first proposed
in 1946 by Robert R. Wilson.[1] With the exception of highly
specialized treatment utilizing accelerators designed for
high energy or nuclear physics experiments, the LLUMC device
will be the first facility using this major modality for
cancer treatment in a facility specifically designed for
that purpose.

The recent availability and widespread use of Computed
Tomography devices (CAT scanners) and Magnetic Resonance
Imaging devices (MRI) has stimulated the interest of
segments of the medical community in exploiting the promise
of proton therapy. The ability of the new imaging devices
to precisely locate tumor volumes makes proton therapy and
its inherent precision useable as a practical treatment
tool.

Loma Linda University Medical Center (LLUMC), as a leader
in advancing the frontiers of medical technology, has

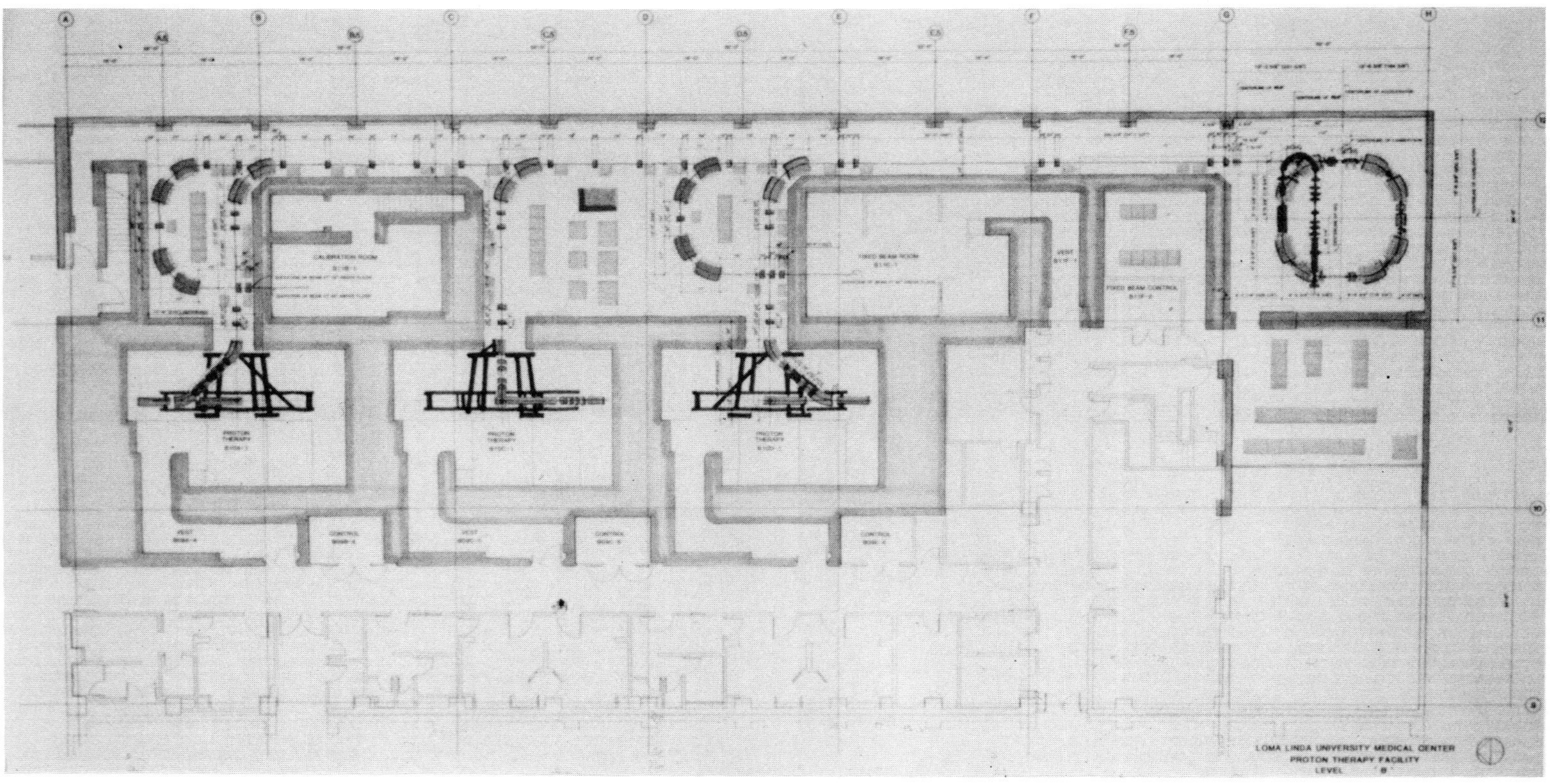

Fig. 1 Floor Plan for Proton Treatment Center at Loma Linda University Medical Center.

undertaken to develop a 250 MeV proton synchrotron as a
hospital based device strictly for the purpose of proton
therapy.

LLUMC has contracted with Fermi National Accelerator
Laboratory[2] (FNAL) with the approval of the Department of
Energy (DoE), to design and prototype a 250 MeV proton
synchrotron with the performance parameters listed in Table
1.

Table 1.   LLUMC Proton Synchrotron
Performance Specifications

Accelerated Particle	proton
Peak Energy	70 to 250 MeV
Peak Intensity	1×10^{11} ppp
Acceleration Cycle	1.1 seconds
Acceleration Time	0.5 seconds
Flattop (Extraction time)	0.1 seconds to ~1.0 seconds
Reset Time	0.5 seconds

Furthermore, LLUMC has contracted with Science
Applications International Corporation (SAIC) to participate
in the project as an industrial partner for technology
transfer. The intent of LLUMC is to make this new
technology available to the rest of the medical community
through SAIC.

The LLUMC facility plan is shown in Fig. 1. The main
proton therapy floor houses the accelerator, a beam
transport system, three gantry treatment rooms, a fixed beam
treatment room, and a calibration room. The general layout
of the accelerator and beam transport lines is shown in Fig.
2. The gantry treatment rooms allow for treatment of a
stationary patient from any angle about the gantry axis of
rotation (0°-360°). Treatment in the fixed beam room
requires movement of the patient for treatment from multiple
angles or ports.

In order to accomplish technology transfer, SAIC is
participating in the project in three modes which are: 1)
performing specific tasks in the design and development of
the project; 2) participating with LLUMC and FNAL in areas
of technology transfer; and, 3) direct responsibility in
installation, commissioning, and operation of the
accelerator as it is moved from FNAL to LLUMC. These three
modes are discussed in more detail below.

SUPPORT TASKS

Many of the skills needed for project tasks to be
performed exist within SAIC. These capabilities have been
selected from in a manner to complement those of FNAL and
LLUMC to benefit the overall project. In general, the
engineering design skills of SAIC coupled with an
understanding of the physics of particle accelerators and
beam transport systems have been used to supplement the

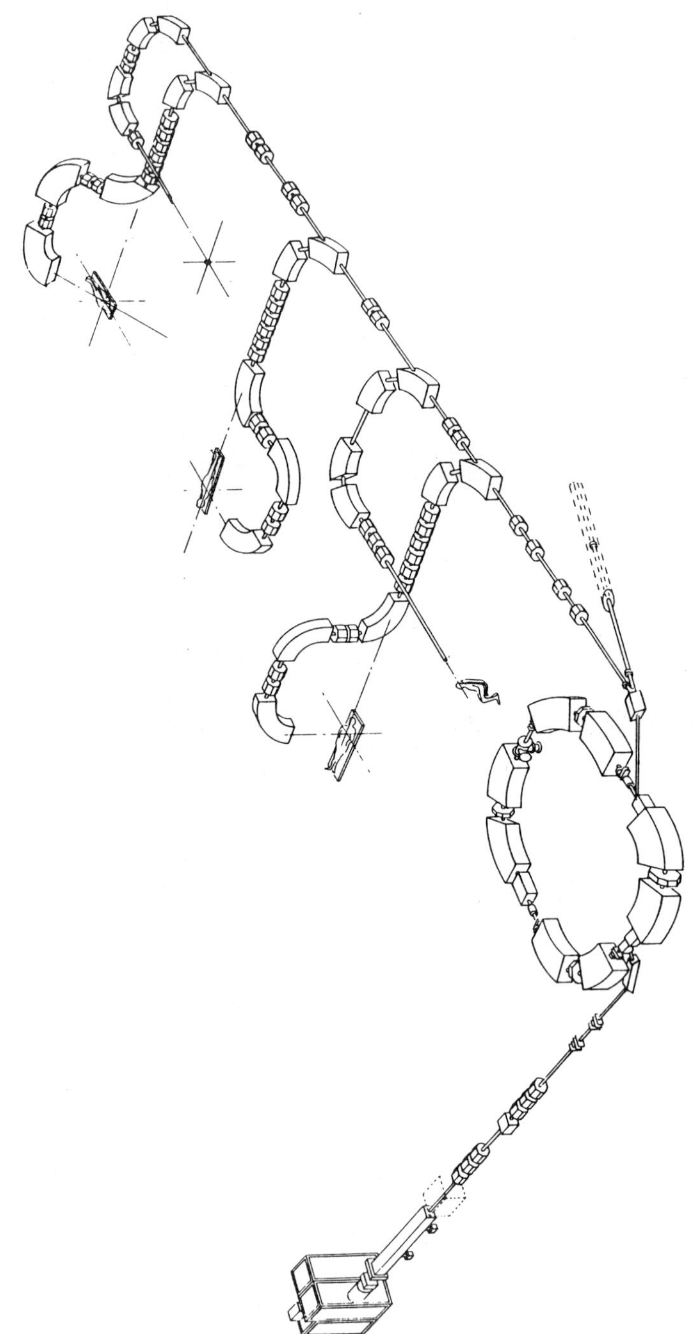

Fig. 2 A schematic diagram of the proton accelerator and delivery system for the Loma Linda University Medical Center Proton Therapy Facility.

talents of FNAL which has great strength in accelerator
physics but whose engineering resources were less available
to the project because of other priorities in the
laboratory. Similarly, LLUMC has looked to SAIC to
supplement its hardware design needs complementing its
medical and radiology expertise.

In particular, SAIC has supported FNAL by the design,
specification, procurement, and vendor liaison of the
accelerator and beam transport magnet power supplies.
Additionally, SAIC has designed and fabricated the unique
pulsed injection septum and kicker power supplies shown in
Fig. 3. The specifications of these supplies are given in
Table 2. SAIC is also designing and fabricating an SCR
switch assembly for switching power between the various
treatment room beam transport magnets as well as designing
various elements of the power distribution system and
analyzing power distribution system interaction with the
various pulse loads.

SAIC is supporting LLUMC by the mechanical design of the
three treatment gantries, nozzle system components and
patient support components. The gantries shown in Fig. 4
are 36 foot diameter, 190,000 pound devices which rotate the
beam transport system such that a patient can be treated
from any angle about the axis of rotation. The size of the
gantry is determined by the length of the treatment nozzle
and the bend radius of the transport magnets for 250 MeV
protons.

SAIC performed the engineering design of the gantries and
produced a build-to-print procurement package for LLUMC. In

Fig. 3 SAIC injection kicker power supply and
pulsed septum magnet power supply.

                 Table 2.   Specifications for Pulsed Injection
                        Septum and Kicker Power Supplies.

Injection System
  Output Current   28,000 A
  Output Voltage   2,800 V
  Pulse Length   120 microseconds
  Pulse Flattop   50 microseconds
  Pulse Repetition Rate   1 Hz
  Switching Element   Thyristor

Injection Kicker
  Output Voltage   65 kV
  Full-Time   <50 nanoseconds
  Repetition Rate   1 Hz
  Switching Element   Thyratron

addition, SAIC administered the fabricator selection process
and is responsible as LLUMC's technical representative to
oversee its fabrication and installation.

   SAIC has designed the nozzle raster deflection magnets
and raster control power supplies. These magnets control
the lateral location of beam as it is scanned over the tumor
volume. SAIC has performed numerous other design tasks in
support of both LLUMC and FNAL and in so doing has become an
integrated partner in the project.

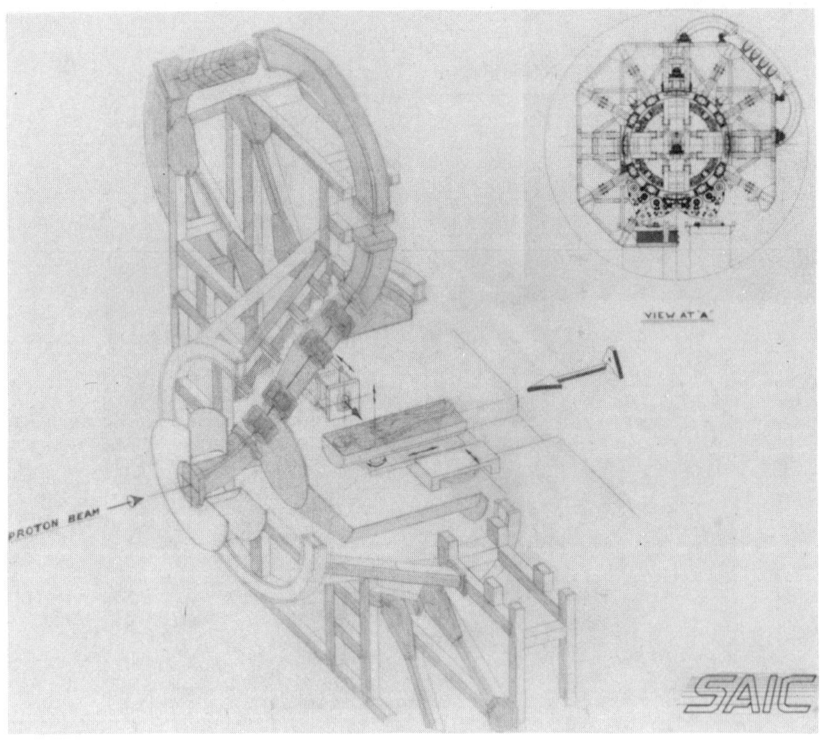

Fig. 4 Artist's rendition of LLUMC gantry.

TECHNOLOGY TRANSFER

   In order to facilitate the transfer of technology and
insure its availability for subsequent medical applications,
SAIC has located key individuals at both FNAL and LLUMC.
These individuals represent a mixture of physics, mechanical
engineering, electrical engineering, and software skills
relevant to the needs of the project. The group located at
FNAL is participating in the commissioning and documentation
of the accelerator system and will be responsible for its
disassembly, moving to LLUMC, assembly, installation,
recommissioning, and early operation at Loma Linda. A
photograph of the prototype accelerator is shown in Fig. 5.
The group at FNAL, after moving from Illinois to California,
will remain at LLUMC through approximately the first year of
operation.

   A similar group of SAIC employees is already located at
LLUMC working with employees of the hospital on activities
related to the application and control of the treatments.
They will continue to be located at Loma Linda during the
first year of operation of the device.

INSTALLATION, COMMISSIONING, AND OPERATION

   The tasks of installation, commissioning, and early
operation accomplish multiple purposes. SAIC's role in
installation and commissioning of the accelerator at LLUMC

Fig. 5 A photograph of the LLUMC synchrotron
at FNAL.

reduces the need for FNAL to supply personnel for that purpose which would be a taxing requirement on the laboratory given their other responsibilities. Similarly, LLUMC does not have the personnel to accomplish this task and would be required to hire such personnel.

Early operation of the device by SAIC provides the necessary incremental personnel needed during the debugging phase of operation without creating the need for the hospital to staff up during this phase and cut back later. Also during this period, SAIC will be training hospital staff to operate the accelerator facility while at the same time experiencing on the job training for training others in future facilities. Developing this capability is important in considering the provision of this technology to other medical institutions. At the completion of this phase, SAIC will have experienced all the relevant aspects of technology transfer and the ability to provide this new and important technology on a commercial basis.

SUMMARY

SAIC, along with LLUMC and FNAL, is privileged to participate in one of the model programs of technology transfer. Motivated by the promise of proton therapy, LLUMC has taken the lead in transferring the technology available in the national laboratories (primarily FNAL) to a commercial source (SAIC) for the purpose of advancing the frontiers of cancer therapy and making that capability available to other medical institutions. In addition to the major role played by FNAL, Lawrence Berkeley Laboratory (LBL) and the Harvard Cyclotron Laboratory (HCL) have been contributors as well.

The transfer of technology and the successful partnership of the national laboratories and industry demonstrated in the LLUMC proton therapy project can serve as a model of similar enterprises in the future.

REFERENCES

1. R. R. Wilson, _Radiology_ 47:487 (1946).

2. F. Cole, P. V. Livdahl, F. Mills and L. Teng. Design and Application of a Proton Therapy Accelerator _in_: "Proceedings of the 1987 IEEE Particle Accelerator Conference,"E. R. Lundstrom and L. S. Taylor, ed., IEEE Service Center, Piscataway (1987).

HIGH POWER NEUTRAL BEAM SYSTEMS

John M. Rawls

General Atomics
P. O. Box 85608
San Diego, CA 92138

ABSTRACT

Spurred by the requirement to supply megawatts of power to heat magnetically confined plasmas to temperatures of interest for fusion research, a new class of low energy, high power accelerators termed neutral beam injectors has been developed. Industry has played an important role in building upon technology advances at the national laboratories to engineer neutral beam injectors to meet the needs of specific users. A brief retrospective of the field is presented, with emphasis upon one particular application, that of DIII-D, a large tokamak at General Atomics. In this instance, the role of industry has been especially extensive because the user/system integrator is itself an industrial concern.

Accelerator technology has played a significant role in the ongoing research and development program to harness the energy from thermonuclear reactions to useful purposes. In inertial confinement fusion, one of the two major branches of the program, the principle technical challenge has been to demonstrate a cost-effective driver to compress pellets of fusion fuel. This has led to the development of high current linacs, such as the PBFA series at Sandia, and to innovative heavy ion acceleration techniques, an example being the Megalac experiment at LBL. In magnetic confinement fusion, the other major branch of fusion R&D, two issues predominate (Figure 1):

<u>Confinement</u> Confine the plasma energy in the magnetic bottle well enough that the overall energy balance of the process is favorable.

<u>Heating</u> Raise the temperature of the ions to levels ($\gtrsim 10$ keV) corresponding to a large fusion cross section.

Progress toward the first objective has been made by designing more effective magnetic bottles, the most successful being tokamaks whose magnetic geometry has been carefully chosen to avoid plasma instabilities that result in loss of confinement. The greatest laboratory success in meeting the second objective has been through the injection into the torus of energetic atoms produced by a low energy, high current accelerator termed a neutral beam injector.

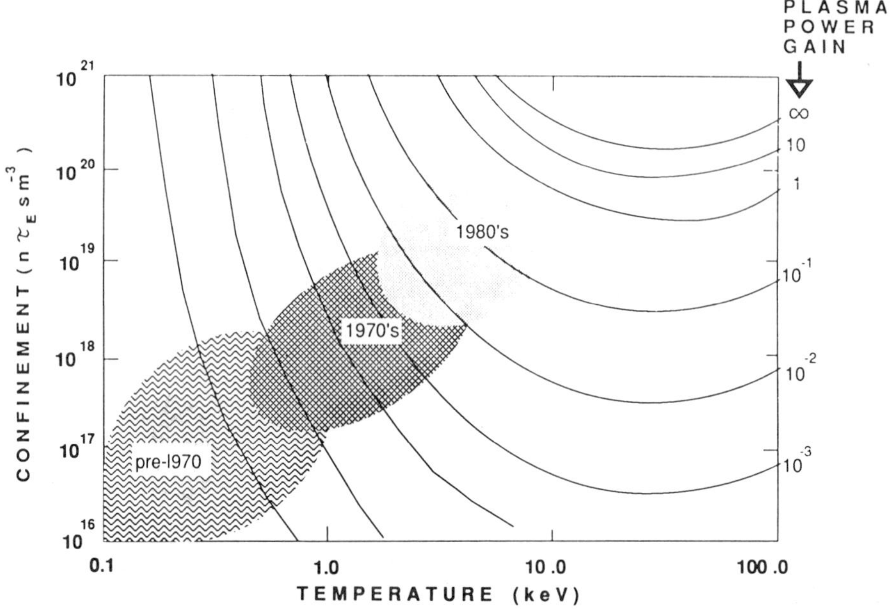

Figure 1.   Progress toward fusion conditions

The development of neutral beam injectors suitable for this purpose is a process that has spanned nearly two decades and involved many individual and institutional contributors. [Ref. 1.]   A brief presentation can of necessity only scratch the surface of this wide field.   To be in keeping with the thrust of this conference, I will concentrate on what has transpired at General Atomics (GA), which has long been the world's most active industrial participant in fusion research.   GA has been at the forefront of the confinement issue by virtue of experimental results on a series of devices that have pointed the way to optimal shaping of the magnetic geometry.   The configuration of the major research device

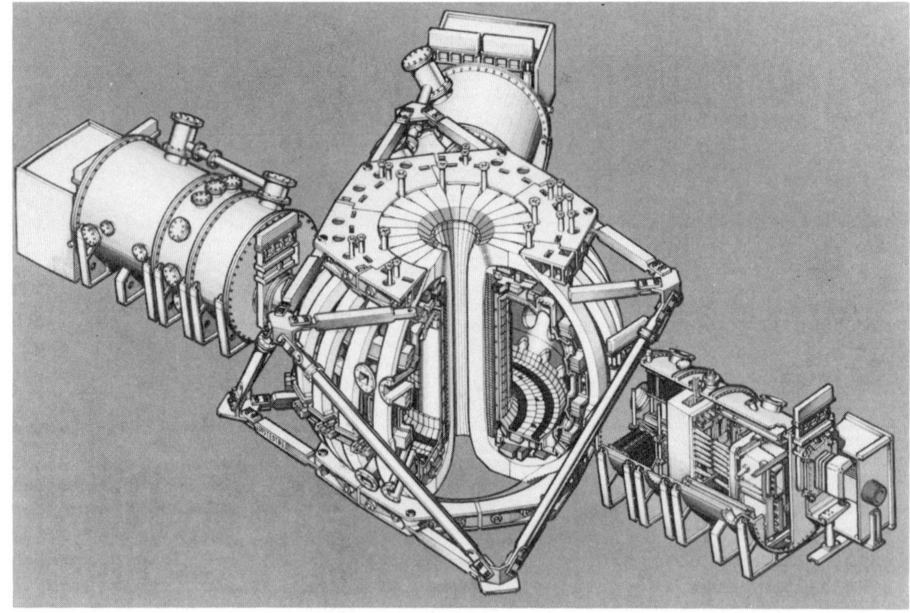

Figure 2.   Isometric of DIII-D, with one of four
neutral beamlines removed.

at GA (DIII-D, Figure 2), that of a dee-shaped cross section torus, is presently the favored approach for a magnetic fusion reactor. The development and implementation of neutral beam injection to heat plasmas in this device is representative of a program-wide success story.

The requirements on a neutral beam injector to raise the temperature of a tokamak plasma by approximately 1 keV are summarized in Table 1. The species injected into the plasma is a neutral isotope of hydrogen - neutral to enable the beam to propagate across the confining magnetic field, the choice of isotope being dependent on the particulars of the experiment. The energy is chosen so that the atomic processes involved (stripping and charge exchange) produce a distribution of beam-generated hot ions that will give rise to a satisfactory heating source term. For the plasma line densities in question this corresponds to tens of kilovolts. The energy requirement is on the order of 1 MW, implying beam currents of tens of amperes. To achieve an equilibrium thermal state of the plasma, the beam must be on for several plasma energy confinement times, leading to a pulse length requirement typically about 0.5 seconds. These requirements, especially those for the current, were well outside the state-of-the-art at the time the tokamaks themselves were coming into operation (@ 1975-1980).

DIII-D, being a large device with the capability of confining plasmas at near reactor parameters [Ref.2], entails requirements at the upper end of the range shown in Table 1. The configuration adopted has four neutral beam injectors, each consisting of two 80 keV, 80 A proton sources rated at 5 second pulse length [Ref. 3]. Such a system corresponds to a total of 14 MW of power delivered to the torus.

A schematic of the neutral beam injectors for DIII-D is shown in Figure 3; for simplicity, only a single ion source is illustrated. Positive ions are extracted from a plasma source, accelerated electrostatically to 80 keV, passed through a gas cell to convert a significant fraction of the beam into neutral atoms, sent through a magnetic field that sweeps the remaining charged portion of the beam away to a dump, and transports the beam through a valve into the torus.

The principal technical hurdles faced in implementing such a system are summarized in Table 2, together with the type of solutions found. The high current necessitated the engineering of large, uniform plasma sources with extraction accomplished via multi-aperture grid structures. The extraction geometry of the sources in use on DIII-D (Figure 4 is a photograph of the prototype, developed at LBL) is rectangular, 12 cm by 48 cm. Achieving the required multi-second pulse lengths necessitated hollow, water-cooled grid structures. The high current and long pulse nature of the operation, in combination with the gas load requirement to neutralize the beam, leads to a gas throughput much larger than the fusion plasma can tolerate. As a consequence, high power neutral beam injectors must be sized to accommodate many square meters of cryopanels. Figure 5, which shows the three major subassemblies comprising the DIII-D neutral beam injectors, provides a feeling for the scale of the undertaking. Beam stops also present a problem, because the power density exceeds those encountered in all but a few applications. It has been possible to utilize a number of ideas having their origin in the nuclear and rocket fields to deal with this. In the final analysis, reliability has turned out to be the problem requiring the greatest effort. Progress in this area emerged from close cooperation over a period of a decade between the developer and user communities and from innovations in the industrial sector in manufacturing engineering.

Table 1

TECHNOLOGY REQUIREMENTS FOR BEAM HEATING OF TOKAMAKS

Parameter	Requirement	Rationale	
Charge State	Neutral	Cross B-Fields	1970 State of the Art
Energy	20 to 100 KeV	Penetrate Plasma	
Pulse Length	~1 sec	Confinement Time	
Power	~1 MW	Significant Heating	0.01 MW (at 1 sec)
Current	10 to 100 Amps	Power/Energy	0.1 Amp (at 1 sec)

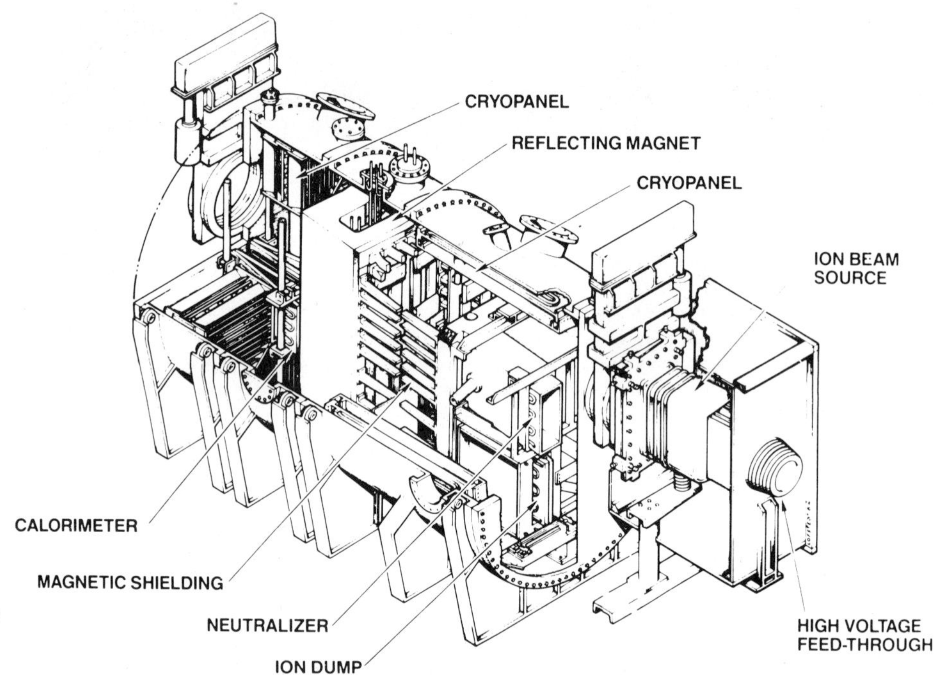

Figure 3.   DIII-D neutral beam injector.

Table 2

TECHNOLOGY HURDLES AND SOLUTIONS

Problem	Solution	Technology Base
High Current	Large, Uniform Plasmas; Multi-Aperture Grids	Space Thrusters
Long Pulse	Internally-Cooled Grids	
Gas Throughput	10^6 L/S Cryopumps	Space Simulators
Beam Stops	~10 kW/cm^2 Techniques	Fission Reactors
Reliability	Developer/User/Industry Collaboration	

Figure 4. The first DIII-D ion source, produced at LBL.

Figure 5.   The three major subassemblies comprising
the DIII-D neutral beam injectors.

Figure 6 represents in summary fashion how the DIII-D neutral beam
program has evolved from a national lab-dominated effort, through a
period in which the technical advances were engineered through lab-
industry collaboration to meet the needs of the particular application,
to the present stage where industry is fully self-sufficient.

The program has been a complete success. The neutral beam injectors
in operation at General Atomics, shown in Figure 7, are real workhorses
in the fusion research effort.   They perform at or near specification
with sufficient reliability and repeatability to support a broad experimental agenda. Indeed, the overall availability for neutral beam injection experiments exceeds 90% [Ref. 4].

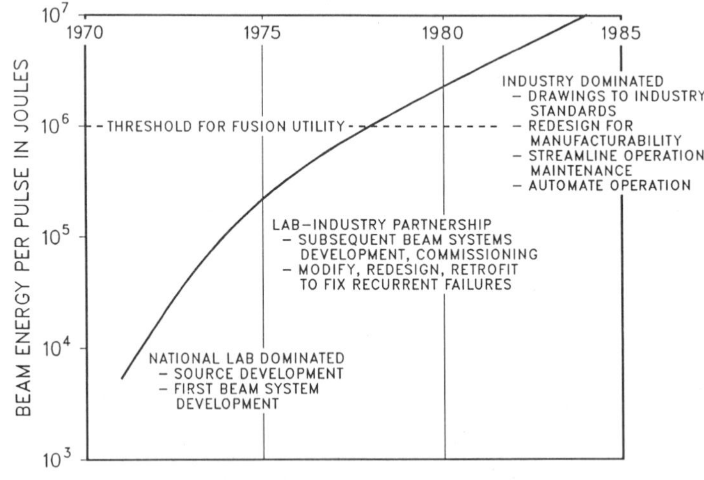

Figure 6.   Timeline to high performance,
reliable neutral beam systems

Figure 7. DIII-D with neutral beam injectors installed.

REFERENCES

1. M. M. Menon, "Neutral Beam Heating Applications and Development,"
    Proceedings of the IEEE, vol. 69, p. 1012 (1981).

2. J. L. Luxon and J. G. Davis, "Big Dee - A Flexible Facility
    Operating Near Breakeven Conditions," Fusion Technology, vol. 8,
    p. 441 (1985).

3. A. P. Colleraine, et al., "The DIII-D Long Pulse Neutral Beam
    System," in Proceedings of 11th Symposium on Fusion Engineering,
    1985, p. 1278.

4. J. Kim, et al., "Performance of the DIII-D Neutral Beam Injection
    System," Proceedings of the 12th Symposium on Fusion Engineering,
    1987.

2 Parallel Technical Sessions

Magnet Technology

Chairman:

R. Coombes
SSC Central Design Group

PERFORMANCE OF FULL-LENGTH SCC MODEL DIPOLES:

RESULTS FROM 1988 TESTS*

J. Tompkins, M. Chapman, J. Cortella, A. Desportes, A. Devred,
J. Kaugerts, T. Kirk, K. Mirk, R. Schermer, and J. Turner

SSC Central Design Group
c/o Lawrence Berkeley Laboratory
One Cyclotron Road, MS 90-4040
Berkeley, CA 94720 USA

J.G. Cottingham, P. Dahl, M. Garber, G. Ganetis, A. Ghosh,
C. Goodzeit, A. Greene, J. Herrera, S. Kahn, E. Kelly, G. Morgan,
A. Prodell, E.P. Rohrer, W. Sampson, R. Shutt, P. Thompson,
P. Wanderer, and E. Willen

Brookhaven National Laboratory, Upton, NY 11973 USA

M. Bleadon, B.C. Brown, R. Hanft, M. Kuchnir, M. Lamm, P. Mantsch,
P.O. Mazur, D. Orris, J. Peoples, J. Strait, and G. Tool

Fermi National Accelerator Laboratory, Batavia, IL 60510 USA

S. Caspi, W. Gilbert, R. Meuser, C. Peters, J. Rechen, J. Royet,
R. Scanlan, C. Taylor, and J. Zbasnik

Lawrence Berkeley Laboratory, Berkeley, CA 94720 USA

Abstract

Over the past year, magnets that meet the SSC field and quench training criteria have
been developed as a result of the detailed understanding of magnet performance made
possible by tests on a series of model magnets. The SSC dipole magnet design is a $\cos\theta$
style coil with a 4-cm aperture and a magnetic length of 16.6 m. The design operating field
is 6.6 T at a current of 6.5 kVA. Design, fabrication, and testing of full-length model
magnets has been a cooperative effort among the SSC Central Design group and three major
national laboratories: BNL, FNAL, and LBL.

I. INTRODUCTION

The Super Collider will require nearly 8,000 superconducting dipole magnets to form
the 83.1-m rings in which beams of protons will be circulated. The design parameters of the
dipole magnets, as described in the Conceptual Design Report,[1] include an operating field
strength of 6.6 T at a current of 6.5 kVA and an operating temperature of 4.35 K.

---

* This work is supported by the U.S. Department of Energy.

The full-scale model dipole magnets have a $\cos\theta$ style coil with a 4-cm aperture and a magnetic length of 16.6 m. The dipole inner assembly, known as the cold mass, is shown in Figure 1. The superconducting coil is restrained by stainless steel collars ,which provide both the precise alignment necessary to maintain high field quality and the primary clamping force against conductor motion. Iron yoke laminations are located outside of the collars to augment the magnetic field by roughly 20 percent. The dipole inner assembly is completed by a stainless steel skin (0.1875 inches thick) that forms the boundary of the region in which the 4.35 K helium circulates. In some of the magnets tested in 1988, the region between the iron yoke and stainless steel collars (nominally a 0.010-inch gap in older designs), was shimmed to ensure direct contact. In this manner, the collared coil assembly is restrained in its axial motion with respect to the iron yoke, and the yoke laminations and outer skin become structural elements that contribute to the effective stiffness and clamping ability of the collars. (This is discussed in more detail below.)

The design parameters were chosen to optimize cost, field strength, superconductor properties, and operational characteristics such as the machine aperture (in which the proton orbits are stable) and control and corrections schemes. The $\cos\theta$ style coil uses super-conductor economically while maintaining the high field quality necessary for a storage ring. Typically, the magnetic field quality is described in terms of the strength of coefficients in a multi-pole expansion of the field.* In a perfect dipole magnet, only the lowest-order (dipole) turn would exist. Symmetry considerations dictate that certain terms should vanish. In general, the field must be uniform to a few parts in 10^4 of the central field value (with this requirement becoming more stringent for higher-order multipoles).

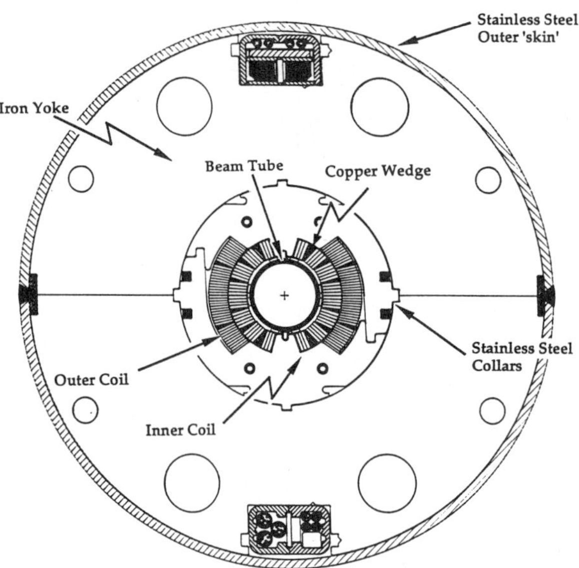

Figure 1. Cross sectional view of a model SSC dipole cold mass with prominent features labelled.

*The multi-pole expansion used is:

$$B_x + iB_y = B_0 \sum_{n=0}^{\infty} (b_n + ia_n)(x + iy)^n,$$

where x & y are the usual space coordinates, and B_0 is the dipole field strength. The b_ns represent the normal multipole components, while the a_ns are the skew components.

Evidence for the success of the present design in meeting the desired field quality limits[2] can be seen in measurements of short (1.8 -m) SSC model dipole magnets. The field is measured in the uniform, two-dimensional cross section to determine the geometric multipoles, which depend on coil design and conductor placement. Data from several short magnets are shown for both normal and skew multipole components up to $n = 8$ (18-pole in our nomenclature) in Figure 2. The upper and lower limits acceptable for each term are shown as bars on the figure. The data clearly fall well within the specified tolerances.

The goal of the magnet R&D program is to develop a design suitable for industrial production of full-length model dipole magnets that meet all system requirements. The first long magnets tested (late in 1986 and early in 1987) did not achieve this goal; they did not routinely reach the operating current and they exhibited excessive training.[3] While these early model dipoles did not fully realize the goals of the R&D program, they provided sufficient data to identify for us which features of the design needed improvement and to establish the need for expanded and refined diagnostic instrumentation.

The primary focus of the magnet R&D program in 1988 was to solve the training problem. Improvements in magnet design, along with specific design variations, were incorporated into the series of magnets to be tested. The diagnostic instrumentation was also greatly expanded to provide more detailed information on quench origins and a more reliable and complete picture of mechanical behavior. Specific mechanical issues included improvement in the design of the stainless steel collars and in the treatment of the coil ends, as well as variations of the design of the interface between the collared coil assembly and iron yokes. The instrumentation expansion included a large increase in the number of voltage taps used for quench localization and major improvements in the design of mechanical sensors (strain gauge assemblies) used to measure the stress/strain state of magnet components.

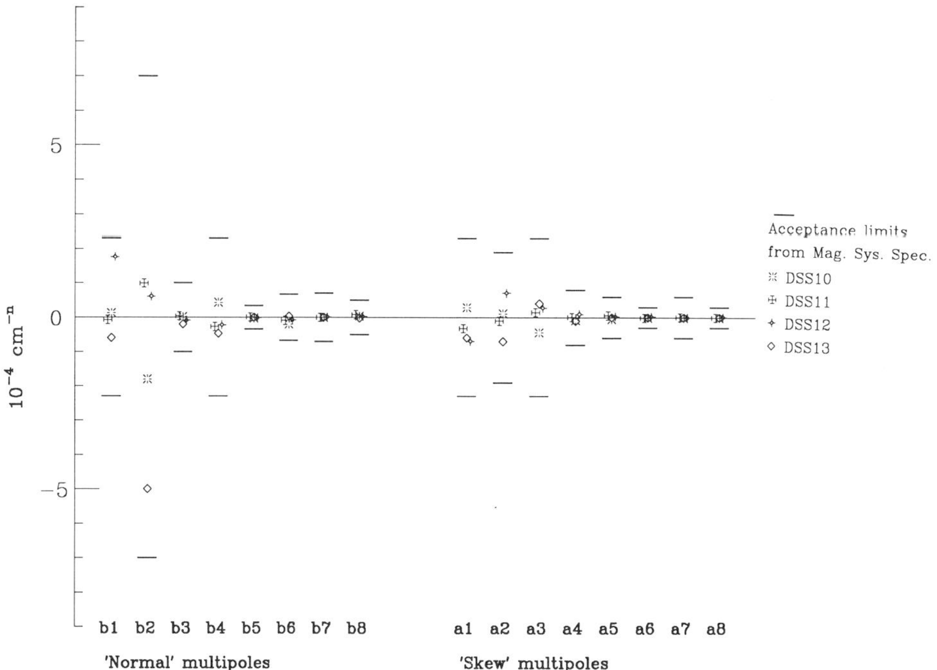

Figure 2.   Geometric multipoles measured in 1.8-m model dipoles; horizontal bars indicate SSC acceptance limits.

II. ASPECTS OF MECHANICAL DESIGN

The stainless steel collared long magnets (BNL design) represent the main thrust of the magnet R&D program. An alternative design, based on aluminum collars (LBL design), had been developed only in short (1-m) models prior to 1988, when the first full-length magnet was tested. Thus the long magnet test program reflects the development of the stainless steel collar design, and most of this discussion addresses only that design. In this section three aspects of the magnet mechanical design are examined. These discussions, although in general applicable to both magnet designs, are specific to the stainless steel collar models and are based on results from tests of early models.

A. Coil Clamping–The Collars

Increasing the initial (zero current) clamping force and maintaining it during magnet excitation had become a design focus in 1987. Collars come in left-right pairs and mesh together in top and bottom sets. The left-right pairs, originally held together by pins, were spot-welded together to provide greater rigidity. (See Figure 1.) The increased strength of the spot-welded collars provides better mechanical restraint against the forces on the coils during excitation. Spot-welded collars were first used in a long magnet in mid-1987 and are now incorporated into all magnets.

The initial clamping pressure is determined during the collaring process by the azimuthal thickness of the coils and by the thickness of shims placed between the collar pole faces and the coil. (Maximum pressure limits are set to maintain the integrity of the insulation.) The upper and lower collar pack assemblies are pushed together by collar presses until slots (keyways) on both sides are aligned and a metal bar (called a key) can be inserted. In collars used in long magnets prior to mid-1988, the keys and keyways were square in cross section (called straight keys). By the third long magnet of 1988, following developments in shorter model magnets,[4] the collar design was modified to incorporate tapered keys and keyways.[5]

The assembly procedure followed with tapered keys differs from that with straight keys and obtains a given collared coil pressure with less overpressure during collaring, lessening the possibility of damage to the coil insulation. In the collaring procedure for square keys, the collaring force is vertical, and significant overpressure must be applied to allow alignment of the keyways and insertion of the keys. The clamping force obtained is significantly lower than the maximum collaring force. For tapered keys the procedure is quite different: a vertical force sufficient to get the keyways nearly aligned is applied, and then the keys are driven into the keyways by a horizontal force, the taper forcing the keys and keyways into alignment. In this manner, it is possible to obtain a desired clamping pressure with a minimum of overpressure.

Intimately coupled to the design issue of coil clamping is the consideration of coil size variation. Variation in the azimuthmal size (arc length in the collars) of the coil translates into variations in the clamping pressure. Shim sizes are determined by the requirement of bringing the low point to a minimally acceptable level.[6] This consideration directly interacts with the collaring: shimming to the low point must not overpressure the high point.

The variation in coil size in the magnets tested in 1988 was typically ±0.004 inches, nearly a factor of two greater than design requirements. The variation in coil size was determined to be caused by the coil molding tooling. However, until recently, funds were not available to improve the tooling. Coils made for long magnets to be tested this year have greatly improved uniformity. Further improvements are under way at BNL and entirely new, more robust coil fabrication tooling is being developed at FNAL.[7]

B. <u>Coil End Treatment</u>

The treatment of the coil ends, where the cable must bend up and over the beam pipe to return along the opposite side, was also enhanced. Analyses of 1987 test results determined that quench origins were predominantly in or near the end regions of the coil.* (During excitation, the Lorentz force exerts several thousand pounds of force on the coil ends.) A weak coil end structure, coupled with relatively weak end plates holding the coils, was thought to contribute to end quenching.

The first attempt to strengthen the end region was to use an alumina-loaded epoxy to fill in the the spaces between coil windings, after the molding process, to form a hardened unit. Subsequent revisions to the design used epoxy-laden inserts between coil turns and fiberglass layers on the inner and outer surfaces, added as part of the molding process. The result is a smooth coil end of uniform radial thickness. Short (1.8-m) magnets tested with hardened ends showed improved quench performance[8] and this procedure was incorporated into the fabrication of long magnets. The ends of magnets DD0010 and DD0012 were filled with epoxy, while the later magnets, DD0014 and DD0015, were constructed with the revised end treatment .

The older design, including DD0010, had "thin" (0.75-inch) stainless steel end plates, split at the middle. These thin end plates deflect when the coil is excited, allowing some coil motion. To further constrain the coil ends, magnets after DD0012 were constructed with thicker (1.5-inch stainless steel), solid end plates. The thick end plates show very little deflection during excitation and prevent motion of the coil end. The added stiffness of the end plate, however, results in a greater buildup of force between the coil end and the plate during excitation.

C. <u>Collared Coil–Yoke Lamination Interface</u>

A third, and perhaps the most important, design issue to be examined in the 1988 magnet test series was the treatment of the collared coil-iron yoke lamination interface. In the original design, there was a 0.010-inch clearance between the collared coil assembly and the yokes. This clearance was to facilitate centering of the coil within the iron to obtain optimal field characteristics. The clearance, however, led to a somewhat poorly defined mechanical interface, which could result in erratic motion between the two components: the coil could both slip and stick as it was elongated by the Lorentz force during excitations. Motion of the collared coil assembly within the yoke could cause premature quenching; it could push the end of the coil against the end plate or generate heat along its length.

In addition to the evidence for quenching near the ends, the 1987 test progam showed evidence of inelastic "stick-slip" motion of the collared coil assembly within the yoke blocks. The first magnet tested in 1988, DD0010, was of the older design and clearly exhibited the same mechanical behavior, dubbed "ratcheting," during its test history. Following each quench to successively higher currents, it was observed that the coil moved outward during excitation but did not return completely to its initial position. This ratcheting was detected as a buildup of the quiescent force on the end plates and by direct measurement of the change in the quiescent position of the coil ends.

The other stainless steel collar magnets designed for 1988 addressed this design issue in two ways: two magnets (DD0012 and DD0014) were constructed with shims placed between the collars and the yoke to prevent motion, while one magnet (DD0015) was built with a "slide plane" to allow free, elastic motion of the coil within the yokes.

---

*The analysis, based on the pressure transducer data, could only localize the quench origin to roughly ± 1 m along the axis of the magnet; the actual extent of the coil end region is less than 20 cm.

The effectiveness of the shimming technique was first tested in short magnets, and improved quench performance was observed.[4] Magnets constrained in axial motion of the collared coil assembly by the shimming technique had increased stiffness in the radial direction. The shims directly coupled the collared coil assembly to the iron yokes and thereby to the outer stainless steel skin. This tightly coupled package of collars-yoke-skin results in a stronger clamping of the coil against the Lorentz force: during excitation, the coils are more tightly held and hence the restraint against quench-inducing motion is increased.

The design of the current (1989) R&D magnets follows the axially constrained coil assembly line of development: the new magnets have no clearance between the collars and yoke (a so-called "line fit"). An additional clamping force is obtained when the outer stainless steel skin is welded around the assembly and weld shrinkage applies additional compression.

III. IMPROVEMENTS IN INSTRUMENTATION

The results from magnet tests in 1987 illuminated certain design features for study and improvement. The 1987 test results also clearly demonstrated the need for improved instrumentation to determine where quenches originated and to understand what aspects of magnet mechanics caused quenching below the cable short sample limit. Quench origins can be localized by detecting the resistive voltage associated with the presence of a normal zone: an elaborate system of voltage taps was inserted in the coils to trace quench evolution. The mechanical state of the magnet assembly was determined with an improved and expanded set of strain gauge transducers. These topics are discussed in greater detail elsewhere in these proceedings;[9,10] we present only a brief overview below.

A.   Voltage Taps

A quench is detected by sensing the resistive voltage generated in the magnet when a region of the conductor goes out of the superconducting state. This voltage is detected by attaching sense wires to various points on the coil wires (voltage taps) to measure the voltage difference between taps. The coil in an SSC dipole magnet consists of four parts, which are wound separately and then joined: two inner (upper and lower) and two outer (upper and lower) quarter-coils. In magnets prior to DD0010, the voltage tap information was typically limited to the voltage across an entire quarter coil. Beginning with magnet DD0010, the turns of the inner coils were divided into four parts by connecting four voltage tap wires to them, separating the more complicated end regions of the coil from the body, resulting in 128 additional channels of voltage information. The quench origin is determined by finding the first segment of coil to develop voltage.

By analyzing the rate of voltage development in the initial coil segment and the time at which voltage is seen in adjacent coil segments, the quench origin can be determined to roughly ±10 cm along a given turn.[9] With this resolution, it is possible to determine a particular site or mechanical feature associated with the quench origin and thus to understand and improve magnet performance.

B.   Strain Gauge Assemblies

Premature quenching (quenches below the cable short sample limit) is believed to be associated with inelastic motion of the conductor. Magnets tested in 1988 had specific design modifications to control the motion of the conductor. Using improved strain gauge assemblies, we are able to determine the mechanical response of key system components, such as the collar clamping pressure and the force of the coil on the end plates, and to correlate the measurements with magnet performance.

Specially designed collar packs incorporate strain gauges to measure the collar-coil pressure.[10,11] These gauge collar packs were mounted at the point where the inner coils were smallest in azimuthal length, corresponding to the lowest prestress. The strain gauge assemblies were carefully designed to give accurate, reproducible measurements of the average stress at the collar-coil interface.

Forces at the coil ends were measured by another type of strain gauge assembly.[11] Threaded bolts, in which pairs of strain gauges were mounted, connected the end plates to the coil end. Four of these assemblies, called "bullets" because of their shape, transmitted and measured the Lorentz force between the coil and the end plates at each end of the magnet.

IV. THE 1988 TEST PROGRAM

In 1988, five full-length model SSC dipole magnets were tested. Four of these were constructed with stainless steel collars (the BNL design), representing a progression of design improvements as well as variations in the approach to a specific design issue; the fifth magnet was the first aluminum-collared (LBL design) long magnet to be tested.

To place the magnets in perspective from both chronological and design evolution viewpoints, a genealogical chart of the long magnets is shown in Figure 3. Time and design evolve from left to right; distinct design variants are separated vertically. Only a few of the design elements, those considered most likely to affect quench performance, have been included.

The design issue of axial restraint (for the BNL design magnets) is separated into three classes of implementation: in the earlier magnets, including DD0010, the collared coil yoke interface was not well defined. Full axial constraint describes magnets (DD0012 and DD0014) in which shims were inserted between the collars and yokes to lock them together. Magnet DD0015, on the other hand, was axially free: a slide plane was introduced between collars and yokes to permit free, elastic motion. The treatment of the collars also evolved with time: from straight keys (DD0010 and DD0012) to tapered keys (DD0014 and DD0015). The details of the coil cross section and changes in the coil end treatment are also indicated on the figure.

By the end of 1988, a clearly identified mainstream design had emerged. A mature version, incorporating axially constrained coils and the various collar and end treatment improvements (as well as improvements in the coil fabrication procedure), has been employed in the next series of long magnets, DD0016–DD0018.

Table 1 presents a summary of design characteristics and other properties of these five magnets.

A.   Quench Performance

The quench performance for the five magnets tested in 1988 is presented in Figure 4 where the quench currents are displayed. The dashed lines represent the short sample current, calculated using the measured $I_c(4.2\ \text{K}, 5\ \text{T})$ of the inner coil conductor, for each magnet. The variation in short sample current results from differing superconductor J_c as well as variations in the cable copper-to-superconductor ratio. Two of the magnets, DD0012 and DD0014, were "thermally cycled": the first series of tests was followed by a warmup to room temperature and then a second cooldown and test period. The thermal cycles are indicated by vertical dashed lines.

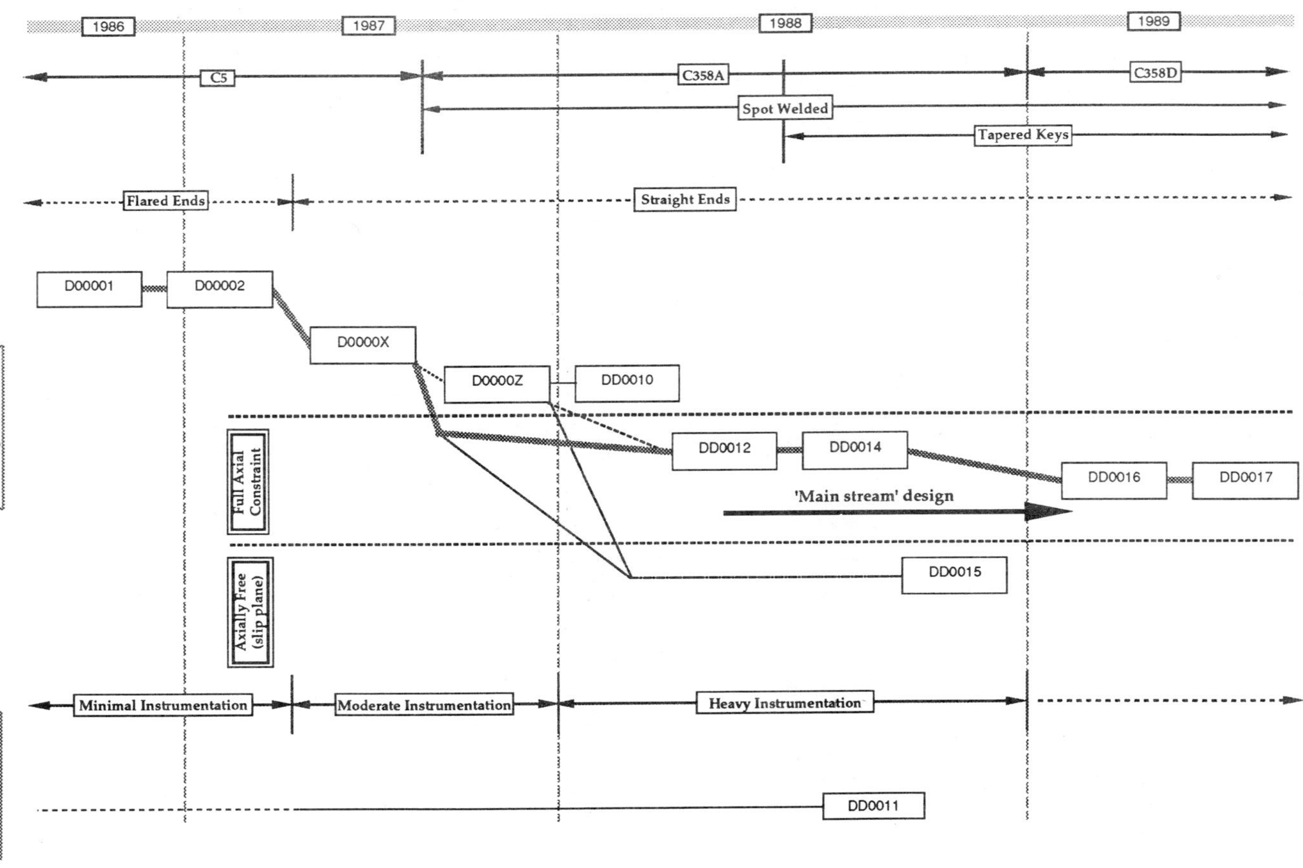

Figure 3.   A 'genealogical' chart of the SSC model dipoles; the main stream design path is indicated by heavy lines.

TABLE 1. MAGNET DESIGN CHARACTERISTICS

Magnet	Collars	Collar/Yoke* Interface	Coil End Treatment	End Plates	Inner Coil Cu:Sc	Inner Coil I_{SS} (4.35K)	Inner Coil Pre-Stress (Cold)
DD0010	S.S. C358A St. keys	? Nom. 0.010" clearance	Filled w/Al. loaded epoxy	3/4" S.S. split	1.4:1	6840 amps	4050 psi
DD0012	S.S. C358A St. keys	Constrained w/shims	Filled w/Al. loaded epoxy	$1\frac{1}{2}$" S.S. solid	1.6:1	6450 amps	4990 psi
DD0014	S.S. C358A tapered keys	Constrained w/shims	Molded inserts & fiberglass 'shoes'	$1\frac{1}{2}$" S.S. solid	1.24:1	6900 amps	4830 psi
DD0015	S.S. C358A tapered keys	Free 'DU' slide	Molded inserts & fiberglass 'shoes'	N.A.**	1.24:1	6910 amps	5360 psi
DD0011	Al. NC9 tapered keys	? constrained warm	G-10 spacer with Al. clamps	1" S.S. solid***	1.25:1	7170 amps	****

*? $\Rightarrow$ not well defined when cold; stick-slip motion possible and ratchetting observed.

**DD0015 end plate was backed off from, and not in contact with, coil end.

***LBL end plate design differs from that used in the other magnets in table.

****Measurements not directly comparable due gauge effects.

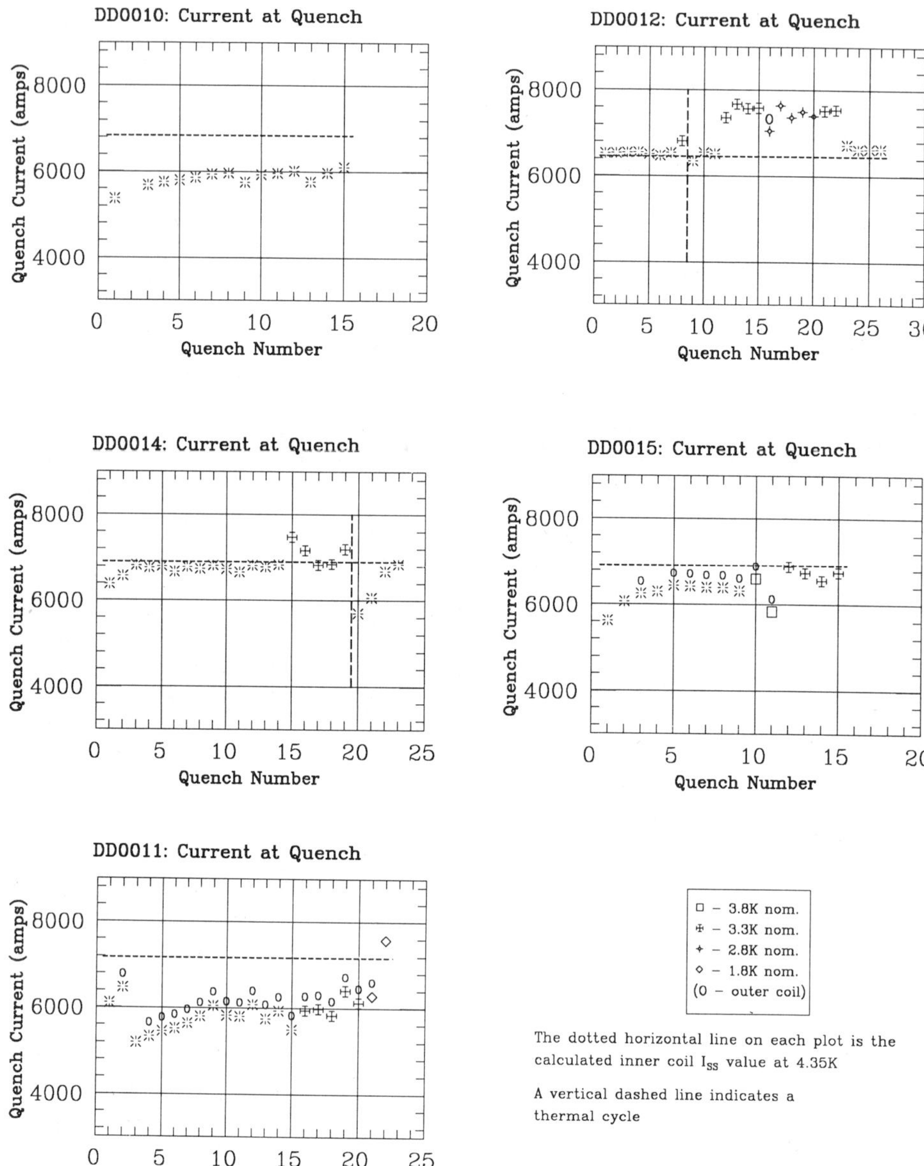

Figure 4. Summary of quench currents obtained in tests of SSC full length model dipoles during 1988. The horizontal dashed lines indicate the predicted short sample current; vertical dashed lines indicate a thermal cycle.

The axially constrained magnets, DD0012 and DD0014, exhibit quench performance that is clearly superior both to the other magnets and to previous long magnets. In the initial tests at 4.35 K, DD0012 exceeded the SSC operating current of 6500 A without a training quench, while DD0014 has only one quench below that value. Both perform near or above their predicted short sample limits. Magnet DD0010 displayed slow training at currents significantly below its short sample limit, as was characteristic of its predecessor, DD000Z,[12] and other magnets of the old design. Magnet DD0015 appeared to plateau at currents well below its short sample prediction.

Finally, magnet DD0011—the first long dipole magnet with aluminum collars—showed erratic training behavior at relatively low currents. The quench performance of each magnet is examined in more detail below.

DD0010

Magnet DD0010 was both the last long magnet of the "old" design (in which the collar-yoke interface was not well defined) and the first long magnet to have expanded diagnostic instrumentation. During the final stages of assembly at FNAL, the cold mass was heated during the application of strain gauges to the stainless steel outer skin. Subsequent measurements indicated an apparent drop in inner coil prestress of about 1000 psi. Fifteen spontaneous quenches were recorded during its first test cycle.*

The voltage tap data were used to determine quench origins. In DD0010, all quenches occurred on turn 13 (counting from the midplane) of the inner coils, both upper and lower. The axial location of the quenches was distributed along the length of the magnet, with two clusters at locations roughly 350 and 440 inches from the feed end. Subsequent disassembly revealed that the copper wedge between turns 13 and 14 (refer to Figure 1) was not bonded to the adjacent cable. The combination of diagnostic measurements—prestress monitoring and quench origin determination—and observations during disassembly presented a reasonably complete explanation for the magnet performance.

DD0012

Magnet DD0012 was the first axially constrained long magnet having shims between the collared coil and yoke assemblies to prevent relative motion. It also was the first long magnet to have thick (1.5-inch stainless steel) end plates to further constrain the coil ends. DD0012 went directly to its short sample limit on its first quench and displayed very good quench behavior at 4.35 K. At 3.2 K, DD0012 exhibited some training before reaching currents near the calculated short sample value of about 7500 A. Lowering the temperature further did not increase the quench current obtained: a mechanical limit had been reached.

The 4.35 K quenches, although very close in current, fell into two classes depending upon whether they exhibited short sample behavior. Short sample behavior is distinguished by several characteristics: the current reached, the location and repeatability of the quench origin, and strong correlation with temperatures are among the most important criteria used. The voltage tap data indicated that the short sample quenches of DD0012 occurred on turn 16 (the high field turn) near the warmer return end of the magnet in essentially identical patterns of voltage development. The correlation of quench current with temperature is shown in Figure 5. The slope dI_q/dT is in good agreement with an expected value of about 1070 A/K.

---

* Following a second cooldown, a short was detected between the quench-protection heaters and the lower outer coil to ground, and further testing was cancelled. The magnet was returned to BNL, disassembled, inspected, and repaired. Modifications were made to the quench-protection heater design.

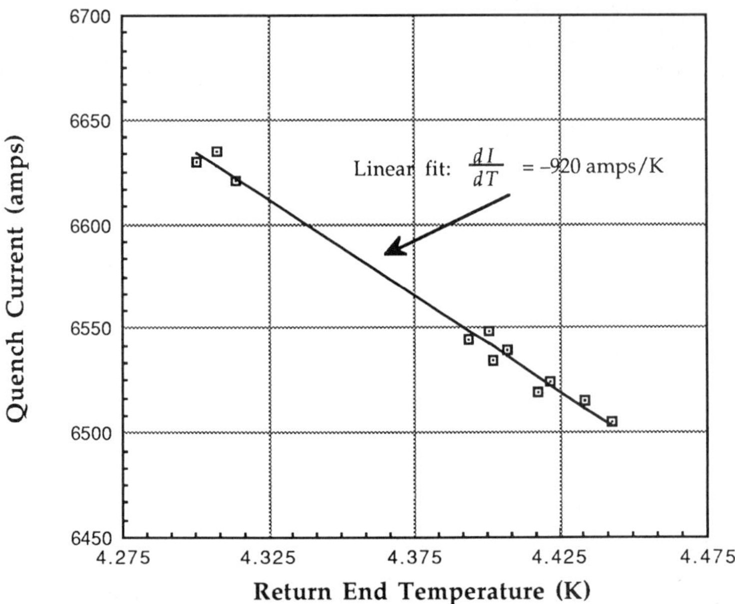

Figure 5.   A plot of DD0012 quench currents versus return end temperature for quenches identified as short sample quenches. The line is the result of a linear fit to the data which yields $\frac{dI}{dt}$ = –920 amps/K.

The non-short sample quenches were dominantly in a region of the coil called the ramp splice. In both halves of the magnet, the inner and outer coils are joined on the innermost (pole turns) at the feed end. This joint occurs in the straight section of the magnet just before the end. Turn 16 of the inner coil is bent radially outward to the outer coil radius while turn 20 of the outer coil is brought around the end to the azimuthal position of inner turn 16. The solder joint and ramp of the inner cable to the outer radius are held inside a G-10 box. The collars in this region must have a notch cut out to hold the splice assembly and thus are weaker than the regular collars. The voltage tap data provided the resolution necessary to identify this relatively short 10-inch length of conductor as the source of quenches. The ramp splice subsequently has been improved, and a revised ramp splice assembly has been included in recent R&D magnets (DD0016 and after.)

<u>DD0014</u>

In addition to having its coil axially constrained with shims in the same manner as DD0012, magnet DD0014 was the first long magnet with tapered-key collars and coil ends strengthened during the molding process. It exceeded the SSC operating current on its second quench and then reached a plateau near its short sample value. The plateau shows considerably more variation than DD0012, and close inspection of the data reveals that 9 of the 14 quenches at 4.35 K are in the ramp splice regions. The first quench, a training quench, occurred on turn 16 of the upper inner coil in the return end region. The remaining quenches were near the return end of turn 16 of the upper inner coil.

The temperature was lowered to 3.3 K and further quench tests were performed. The first quench occurred at 7478 A, but subsequent quenches were lower in current. All of the 3.3 K quenches were in the upper inner coil ramp splice region.

A thermal cycle was followed by rapidly cooling the magnet to 4.35 K; the usual limitation placed on the temperature differential across the magnet of <125 K* was removed. This took less than 13 hours; a normal cooldown typically takes about 36 hours. It then took two training quenches (at currents lower than previously observed) before DD0014 was back to its pre-thermal cycle level. All quenches in the second test cycle were in the lower inner coil; the first quench was on turn 16 in the body of the magnet; the remainder were in the ramp splice segment.

<u>DD0015</u>

DD0015 was the last of the BNL design, stainless steel collared magnets to be tested in 1988. Unlike DD0012 and DD0014, DD0015 was an axially free magnet in which the collared coil assembly was free to move with respect to the yoke laminations. The end plates were not in contact with the coil ends. Due to lack of cable while DD0015 was being fabricated, the lower outer coil was re-used from a previously tested magnet, DD000Z.[12] Other design characteristics included tapered-key collars and coil ends strengthened in the molding procedure.

Three of the first four quenches at 4.3 K were in the lower inner coil ramp splice segment; the remainder were in the lower outer coil, which did not have voltage taps, so precise quench location was not available. Following the fourth quench, the quench current reached a plateau at about 6400 A. The expected short sample limit for the inner coils was about 6910 A (upper inner); for the lower outer coils, the value expected was essentially the same.

A plot of the lower outer coil quench currents versus temperature, shown in Figure 6, reveals a strong correlation at nearly the expected slope. This is highly suggestive of conductor- (as opposed to mechanically) dominated behavior. A crude longitudinal position analysis using pressure transducer data indicated that all but one of the lower outer coil quenches occurred in a localized region near the return end. It was also noted that magnet DD000Z, the source of the lower outer coil, had failed with a short to ground in the upper inner coil *near the return end.*[12] The observed behavior is consistent with a loss of J_c for the lower outer coil conductor. (However, the highest current reached in DD000Z testing was about 6300 A, below the DD0015 plateau value.)

The temperature was lowered to 3.8 K, and two additional lower outer coil quenches occurred. The first quench was at 6600 A; the second quench was at 5840 A. The temperature was then lowered to 3.3 K, and four quenches were recorded. These last quenches all occurred in the lower inner coil ramp splice region.

The quench behavior of DD0015, when looked at in detail, is encouraging: quenching was associated with either the ramp splice (as in other magnets) or the outer coil. There were no other inner coil quenches. Freedom of the collared coil assembly to move during excitation and removal of the constraining end plate do not appear to have resulted in inner coil end quenching.

<u>DD0011</u>

Magnet DD0011 was the first long magnet with aluminum collars (LBL design) to be tested. The coil cross section (NC9) differs slightly from that of the stainless steel (C358A) design, as do other aspects of the mechanical assembly. Specifically, both the coil winding

---

* The temperature is monitored in interconnect regions where the magnet is attached to the test stand. Temperature sensors mounted in the interconnect regions measure the helium temperature before it enters the magnet (feed end) and after it has traversed the length of the magnet (return end).

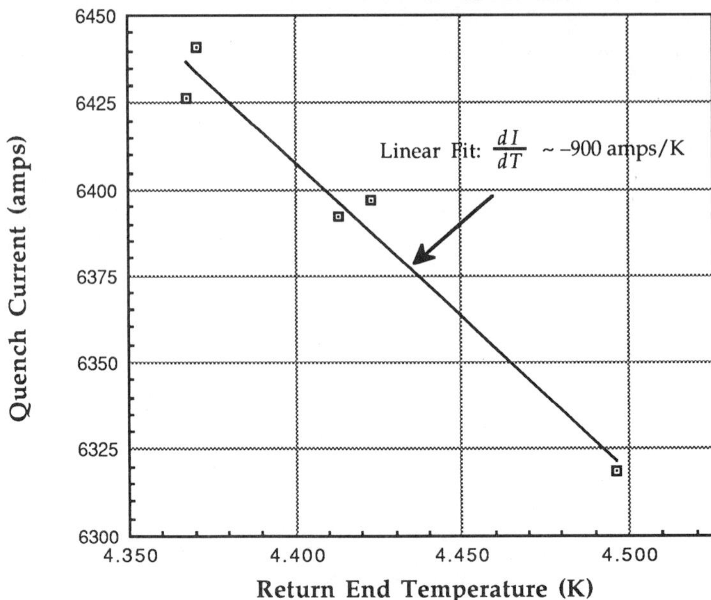

Figure 6.   A plot of lower outer coil plateau quench currents in DD0015 versus return end temperature; a linear fit yields $\frac{dI}{dt}$ = ~ –900 amps/K.

and mechanical assembly procedures for the end regions differ from the BNL design. Not unexpectedly, the first NC9 long coil winding and fabrication encountered some difficulties, and several shorts had to be repaired.

The quench performance of DD0011 was erratic, reminiscent of the first long stainless steel collared dipole, DD0001.[3] Its first quench was at 6107 A in the upper inner coil; the second was at 6464 A, in the upper outer coil. The second quench was followed by a quench at 5192 A (a precipitous drop in current) in the upper outer coil, and subsequent quenches remained in the upper outer coil showing slow, erratic training. A total of 15 quenches were recorded at 4.35 K. Lowering the temperature to 3.3 K did not appreciably improve quench behavior.

At this time in the test program, the ability to operate in helium II became available on the DD0011 test stand. Tests of 1-m magnets at LBL had shown promising improvements of quench behavior at 4.35 K by conditioning magnets at lower temperatures.[14] DD0011 was run at 1.8 K, and its initial quench was in the upper outer coil at 6273 A. The next quench was in the lower inner coil at 7579 A. Post-quench tests revealed that a ground fault had developed. Opening and disassembly revealed that the short occurred at the feed end of the magnet, in the vicinity of one of the shorts that had been repaired during assembly.

B.   <u>Prestress Comparison</u>

An important variable in magnet performance is the clamping pressure exerted by the collars on the coil. Two aspects of the clamping are of significance: the quiescent (zero current) value of the pressure or prestress and the rate at which this pressure decreases as a function of magnet current. The combination of these two factors determines the clamping force at high currents, an important factor in determining quench performance.

Figure 7 displays the average inner coil stress versus I^2 for the four magnets with stainless steel collars tested in 1988. (Note: the operating current of 6.5 kA corresponds to a value of 42 on the I^2 scale.) The correlation of stress values with performance is quite clear:

- DD0010, which performed well below specifications, had low prestress and fell rapidly to low values with I^2.

- DD0012 and DD0014, which performed quite well, started with acceptable prestress values and maintained reasonable clamping pressure at the operating current.

- DD0015, whose performance was dominated by outer coil limitations, had sufficiently high prestress to maintain an acceptable level at the operating current.

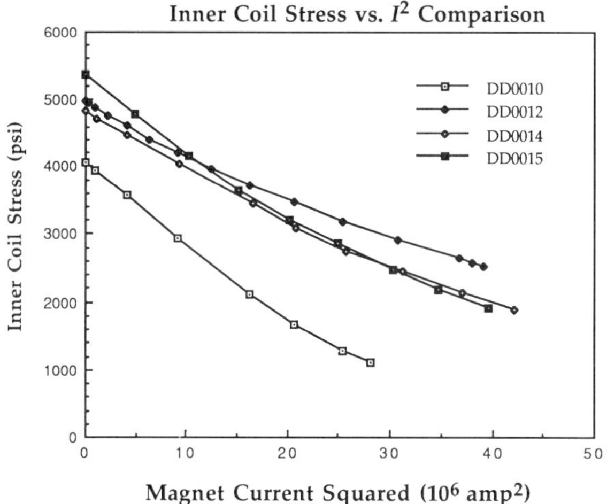

Figure 7.   Average inner coil azimuthal stress versus I^2 for the four full length SSC model dipoles with stainless steel collars tested in 1988.

The shimmed magnets, DD0012 and DD0014, show a slower decrease in pressure with I^2 than either DD0010 (older-style collar-yoke interface) or DD0015 (axially free). The rate of stress change with I^2 reflects the greater effective stiffness of the shimmed magnets in which the yoke and stainless steel skin of the cold mass are contributors. The fact that DD0015 performed reasonably well, given the limitations discussed above, indicates that the value of stress obtained at the operating current is among the most important variables.

C.   Ramp-Splice Quenches

The voltage tap data permitted identification of the ramp splice region, a length of cable less than 30 cm long, as the source of a large number of quenches. As noted above, the collars in this region are weakened by a notch cut out to accommodate the ramp splice.

In the axially constrained magnets, shims provide additional support in this region, and quenching does not occur until relatively high currents; in unshimmed magnets, ramp splice quenches can occur at lower currents. The fact that quenches at the ends of older magnets were largely correlated with the feed end is strongly suggestive of a ramp splice origin.

The support of the ramp splice by the collars in this region has been revised. New magnets will incorporate design changes to the ramp splice assembly to prevent motion and better constrain the cable.

VIII. SUMMARY

The SSC magnet R&D program made significant advances in 1988. A clearly defined set of mainstream magnet design parameters has been determined. The magnetic field quality of the present design falls well within the limits set for accelerator operation. The problem of training in the long magnet is rapidly being solved. The two magnets representative of the mainstream design, DD0012 and DD0014, reached operating current with one or zero training quenches.

Advances in instrumentation have resulted in greatly improved diagnostic information. Quench origins have been determined to within 10 cm. The ramp splice region has been clearly identified as a source of quenches at higher currents and is being revised. The end regions of the coils have been strengthened and are associated with only a few quenches. Quenching in the body of the magnet—the region of uniform two-dimensional cross section—has largely been eliminated by controlling the collar clamping force.

Great improvements have been made in the system of strain gauge sensors that measure the mechanical state of the magnet. Repeatable, reliable coil stress and end force measurements are now available. The coil stress measurements correlate well with magnet performance.

The next series of long magnets—DD0016, DD0017, and DD0018—incorporate the features of the mainstream design which evolved in 1988:

- an axially constrained collared coil assembly,

- revised ramp splice,

- strengthened ends, and

- modifications to the coil cross section (wedge shape),

and are made using improved coil molding fixtures that make more uniform coils. We expect in the near future, with some further minor tuning, to have resolved most of the major issues affecting magnet quench performance.

ACKNOWLEDGMENTS

We would like to thank the technical staff at the Fermilab Magnet Test Facility, whose efforts made these tests possible; the staffs at BNL and LBL who contributed to the design and fabrication of the magnets; and the members of the SSC Central Design Group who participated in the testing and data analysis.

We would also like to thank Annie Calinog, Nathalie Guyol, Valerie Kelly, and Darlene Moretti for their excellent work in preparing this manuscript and Kate Metropolis for invaluable editorial assistance.

REFERENCES

[1] J. D. Jackson, ed., Conceptual Design of the Superconducting Super Collider, SSC-SR-1020, March 1986; revised, September 1988.

[2] SSC Specification Number, SSC-Mag-D-1010, Dec. 15, 1988; also, A. Chao & J. Peterson, internal SSC/CDG memo to T. Kirk and M. Tigner (unpublished).

[3] J. Strait, et al., "Tests of Prototype SSC Magnets," Proceedings of the 12th Particle Accelerator Conference, Washington, D.C. (1987). Also, Fermilab technical memo TM-1451 and SSC note, SSC-N-321.

[4] P. Wanderer, et al., "Test Results from Recent 1.8-m SSC Model Dipoles," presented at the Applied Superconducting Conference, San Francisco, California, Aug. 22-25, 1988. *IEEE Trans. Magn.* **25, No. 2** (1989): 1451

[5] C. Peters, et al., "Use of Tapered Key Collars in Dipole Models for the SSC," *IEEE Trans. Magn.* **24** (1988): 820.

[6] M. Anerella, E. Willen, private communication.

[7] J. Carson, et al., "SSC Dipole Coil Production Tooling," presented at the First International Industrialization of the SSC Conference, New Orleans, Louisiana, Feb. 8-10, 1989, in these proceedings.

[8] P. Wanderer, et al., "Test Results from 1.8-m SSC Model Dipoles," *IEEE Trans. Magn.* **24** (1988): 816.

[9] A. Devred, et al., "Quench Start Localization in Full-length SSC R&D Dipoles," presented at the First International Industrialization of the SSC Conference, New Orleans, Louisiana, Feb. 8-10, 1989, in these proceedings.

[10] M. Anerella, et al., "Recent Measurements of Coil Stresses in SSC Model Magnets Using Beam Type Strain Gauge Transducers," [IISSC ref., in these proceedings].

[11] C. L. Goodzeit, et al., "Measurement of Internal Forces in Superconducting Accelerator Magnets with Strain Gauge Transducers," presented at the 1988 Applied Superconductivity Conference, San Francisco, California, Aug. 1988. *IEEE Trans. Magn.* **25, No. 2** (1989): 1451.

[12] R. Coombes, ed., "SSC 17-Meter Dipole Magnet DD000Z Test Results and Investigation of Coil Failure," SSC-SR-1034, March 1988.

[13] W. S. Gilbert and W. V. Hassenzahl, "1.8 K Conditioning (Non-Quench Training) of a Model SSC Dipole," Lawrence Berkeley Laboratory, *IEEE Trans. Magn.* **23** (1987): 1229.

SSC DIPOLE COIL PRODUCTION TOOLING

J. A. Carson, E. J. Barczak, R. C. Bossert, J. S. Brandt
and G. A. Smith

Fermi National Accelerator Laboratory
Technical Support Section
Batavia, IL 60510

ABSTRACT

Superconducting Super Collider dipole coils must be produced to high
precision to ensure uniform prestress and even conductor distribution within
the collared coil assembly. Tooling is being prepared at Fermilab for the
production of high precision 1M and 16.6M SSC dipole coils suitable for mass
production. The design and construction methods builds on the Tevatron
tooling and production experience. Details of the design and construction
methods and measured coil uniformity of 1M coils will be presented.

INTRODUCTION

The Superconducting Super Collider will require approximately 8,000
dipole magnets 16.6M in length. Each magnet will have two inner and two
outer coil assemblies. Each coil assembly must be made to exacting
tolerances. Of primary concern is the cross sectional uniformity of
individual conductor placement within the magnet assembly along its full
length. In application, the coils are paired and assembled using laminated
stainless steel clamping collars. The collars serve to compress or prestress
the coils so as to resist the magnetic forces which act on the coil when
energized. Redistribution of conductor placement within the coil from its
specified design results in undesirable multipole fields. Variations in
prestress which contribute to conductor placement errors have other
undesirable effects. Areas of low prestress can allow conductor motion
resulting in coil heating and subsequent quenching of the magnet below its
design load. Areas of overly high prestress hasten the breakdown of the
insulation system which can result in turn to turn coil shorting and magnet
failure (Figure 1).

The coils are composite structures and as such each constiuent is a
possible contributor to coil conductor placement and size variations. The
term size is always tied to a specific load. The prestress for the SSC will

range from approximately 8 KPSI to 12 KPSI*. The constiuents are the
superconducting cable, the superconducting strands within the cable, the
insulation; kapton and B-stage epoxy impregnated fiberglass tape (Figure 2),
the epoxy in the fiberglass tape and the copper wedges as well as the
application of the insulation onto the cable and wedges. All of these items
require stringent quality control. In any event there is little the coil
fabricator can do about them except follow criteria for accepting or
rejecting their use in fabricating the coil. The burden on the coil
fabricator is to assemble these components in a way which produces coils as
good as the constiuents allow without adding variations. This burden is
manifested in the tooling and procedures used in making the coils.

The coil production sequence for making coils is straight forward. The
steps are not complex. What is important is the design and quality of the
tooling and rigid conformance to procedures. Figure 3 illustrates the
process.

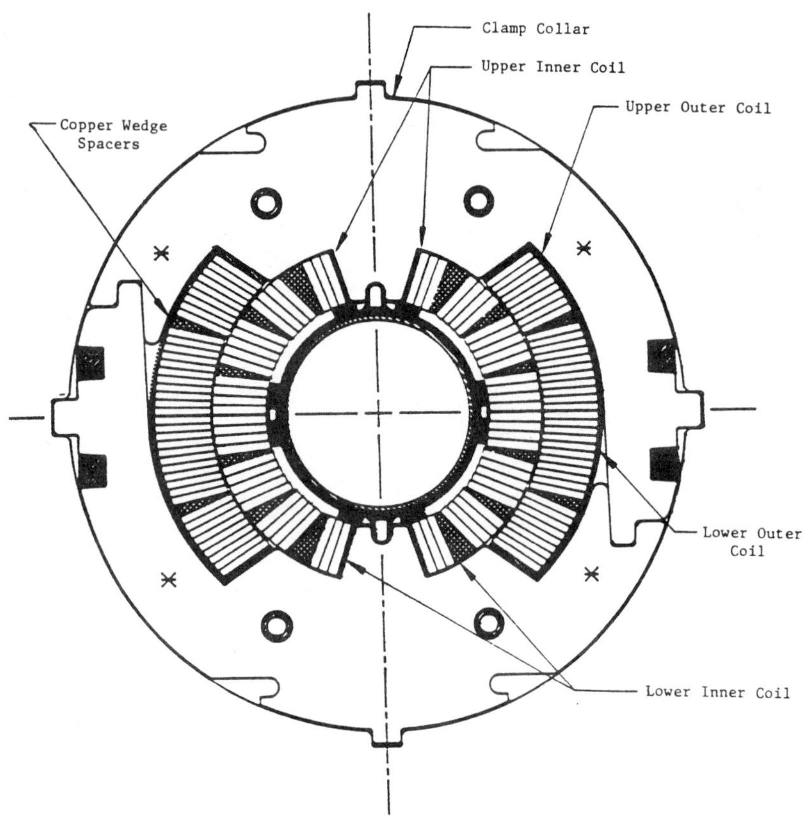

Figure 1. SSC Dipole - Collared Coil Cross Section

*The prestress range will be established from SSC Research and Development
 program.

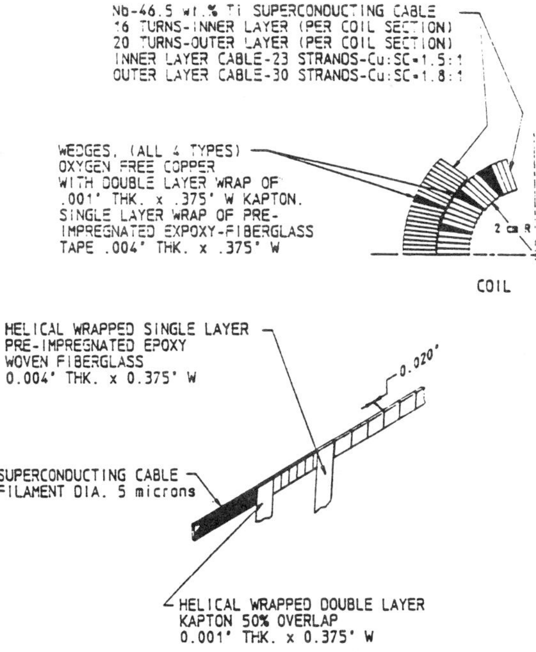

Figure 2. Coil Components

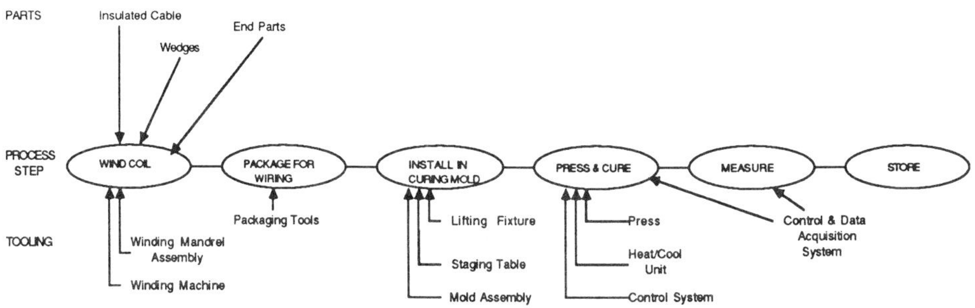

Figure 3. Coil Production Process Sequence

LAMINATED TOOLING

The Technical Support Group at Fermilab has designed and is in the process of assembling 16.6M tooling for winding and curing SSC C358D cross section coils. One meter tooling of like design has been completed and is starting to be used to produce a series of 1M coil for purposes of coil-end development[1] and the evaluation of the tooling itself. The tooling under construction is an outgrowth of our experience in the production of more than 1,000 magnets which are now in operation at Fermilab's accelerator, the Tevatron.

The tooling used in forming the coil structure is a closed cavity mold with the coil winding mandrel assembly being the male portion and the forming mold being the female portion. The insulated cable along with wedges and end components are wound onto the mandrel at constant tension, secured, and fitted to the mold cavity. The cavity formed is a long semi-cylinder of dimensions equivalent to the "as collared" coil size. During the development of the Tevatron tooling, we found it was very difficult to produce long tooling which satisfied the uniform cross section criteria. Our most frustrating problem was trying to achieve the required dimensional accuracy of the complex tooling shapes, a combination of radii and angles. Short tooling, (1 foot) could be produced, however long tooling with the same dimensional requirements exceeded convention machining capabilities. Compounding the problem was the requirement for many sets of identical tooling necessary to achieve the required coil production throughput. Besides the cross section problems, the tooling had to be massive so as not to distort the mold cavity during the high pressure molding process. This resulted in mold forms with high longitudinal stiffness. The stiffness imposed stringent flatness and straightness tolerances on both the tooling and the presses which closed the mold. Our solution was to use stacks of steel laminations for the tooling. The laminations could be stamped to high precision with piece parts of nearly identical size. With laminations we could create multiple tooling which was identical. Since the tooling was now a stack of laminations, longitudinal stiffness became nil, therefore, the tooling could easily comply to press platten irregularities.

The SSC tooling Fermilab is producing uses the lamination theme. Conventional machining methods are limited to precision items which have simple, easily machined geometries such as rectangular bars. Tooling components which involve radii and angles that are not practical for laminating are made using computer controlled electrical discharge wire cutting machining (EDM)[2] methods.

Figure 4 illustrates the outer coil winding mandrel assembly. The winding key is an EDM part. The curved portion of the mandrel is made of stacked laminations. A conventionally machined strongback carries the laminations. Figure 5 illustrates the same mandrel with the coil wound on it, packaged and ready to install in the form mold. The sizing bars are used to define the azimuthal cavity and are ground-bar stock. The sheet metal retainer serves to retain the coil to the mandrel and provide a smooth surface for the outer radius of the coil. Figure 6 illustrates the mold assembly. The body of the mold is stacked laminations tied together with full length tie rods. Full length tubes are installed in six places to provide for heating and cooling the mold. The tubes, low-carbon steel, are nominally .010" smaller in diameter than the lamination hole. Once installed, the tubes are expanded using ball swagging. Ball swagging is a process where a hard steel ball is forced through the tube using hydraulic pressure. As the ball progresses, the tube yields. The ball is sized to give intimate contact of the tube and the lamination providing a good thermo connection. The tubes are terminated at manifolds on each end for connection to a fluid heating/cooling system. Spring loaded roller assemblies are bolted to the "T" ways on the sides of the laminations and allow transport of the mold into the press following a guide rail. The ground steel bars fastened to the mold top are provided as an adjustable interface with the upper press platten. These bars can be shimmed to fine tune the azimuthal cavity size if required. The laminations are 1050 spheroidize annealed steel throughout. We would have preferred laminations produced by the fine-blanking[3] process for reasons of accuracy and edge finish but because of schedule demands opted for compound die stamped laminations with shaved edges at critical areas. All other steel components are 4150 HT alloy steel.

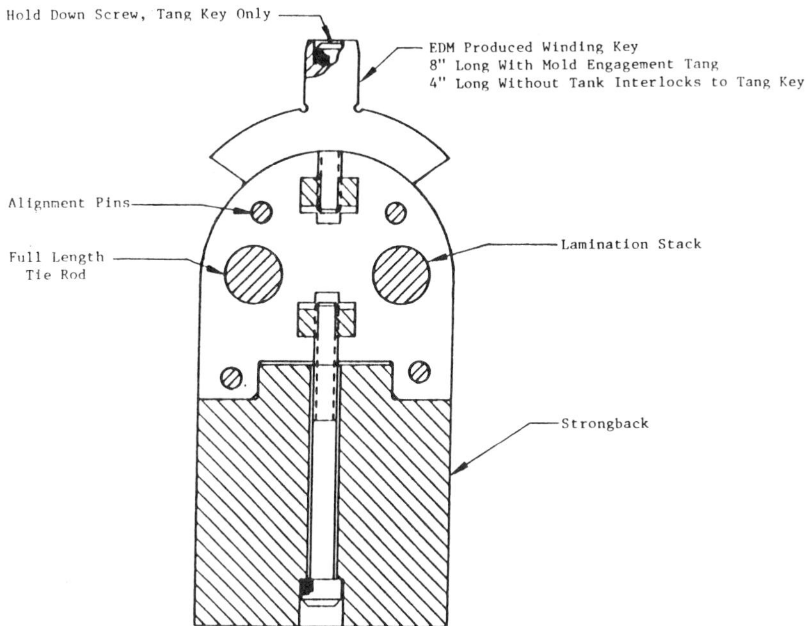

Figure 4.   Outer Winding Mandrel and Winding Key Assembly - Length = 56'

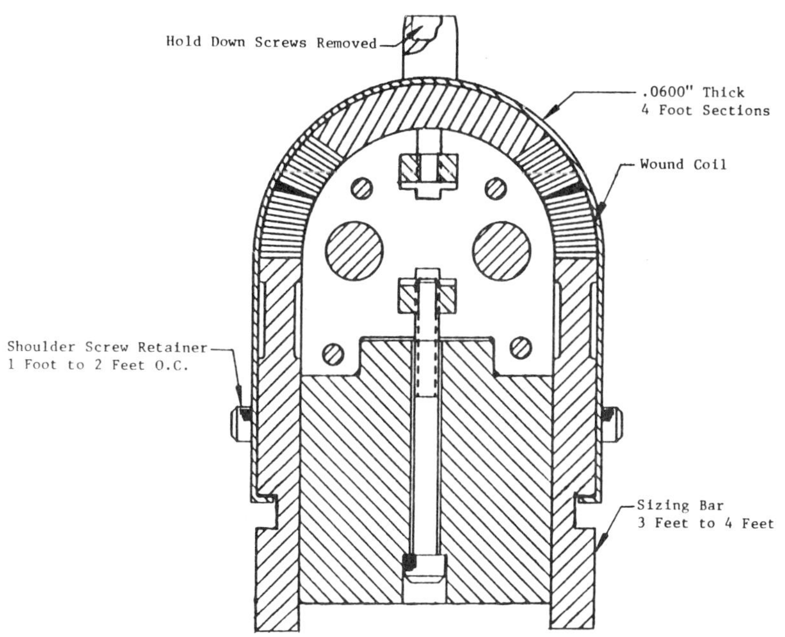

Figure 5.   Packaged Outer Coil Ready to Install in Mold

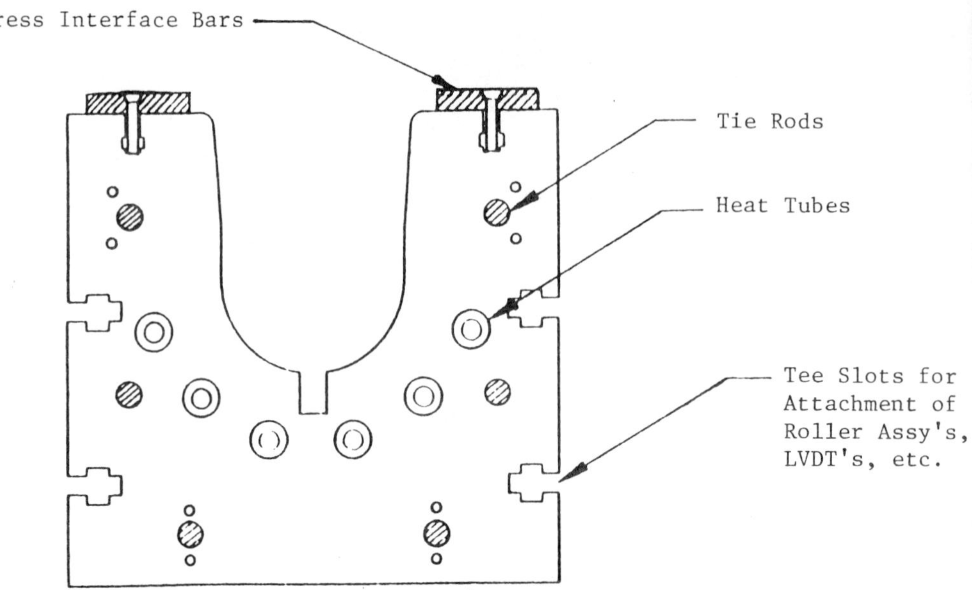

Figure 6. Laminated Outer Coil Mold Assembly - Length = 56'

        Assembly of the laminated tooling has its own unique problems.
Regardless of the accuracy of the lamination there is bound to be asymmetries
in the lamination features. These asymmetries preclude flipping the
lamination because the overlap of asymmetric features would effectively
change the critical mold feature dimensions. To guard against accidental
flipping, witness notches are provided in the stamping die. Because the
laminations are not perfectly parallel, stacking can result in fanning of the
stack. To avoid this, the coil material is measured before stamping and
equal quantities of laminations are stamped with opposite sheet taper.
Laminations of opposite but equal taper are then stacked alternately to avoid
fanning. Before stacking can start the laminations must be deburred and
cleaned. The stacking is done in stages. First a stack of approximately 6'
length is made. Stacking pressure is approximately 300 psi. During this
initial stack up, corrections are made to keep end to end parallelism to
within .015" over 10". This is accomplished by the removal and replacements
of a few laminations with appropriate taper. If the stack calls for a series
of notched laminations at some interval (molds), it is accomplished at this
point. Once the stack is deemed acceptable it is unstacked and reassembled
in the same order with alignment pins installed. The alignment pins, 2" long
roll pins, assure lamination to lamination alignment without the need for
elaborate and extremely precise stacking surfaces. The completed 6'
assemblies are then joined, using the roll pins, to form an uninterrupted
stack of laminations. End plates are added and tie rods inserted to retain
the stack. The coils have no straightness requirement, therefore, the
tooling need be only straight enough to affect engagement. It should be
noted that nothing in the tooling assembly process is allowed which would
compromise the integrity of the laminated tooling's main attributes, i.e.
uniformity and compliance. For this reason welding and or secondary
machining of the assembly is not allowed. The design tolerance for the
components allows a closed cavity mold accuracy of: radial uniformity of $\geq$
.0015" and azimuthal uniformity of $\geq$ .002" along the full length.

APPLICATION

The process steps for utilization of the tooling is depicted in Figure 7.
The tooling is designed to work in a two-stage press; i.e. the press must
provide independent loading against the mandrel and the sizing bars. The
intent here is to define the radial cavity before loading the coil
azimuthally. The purpose is to achieve better conductor placement by not
allowing the cable to cable radial interface to lockup through friction
before radial placement is achieved. Figure 8 shows the applications of the
loads during molding process.

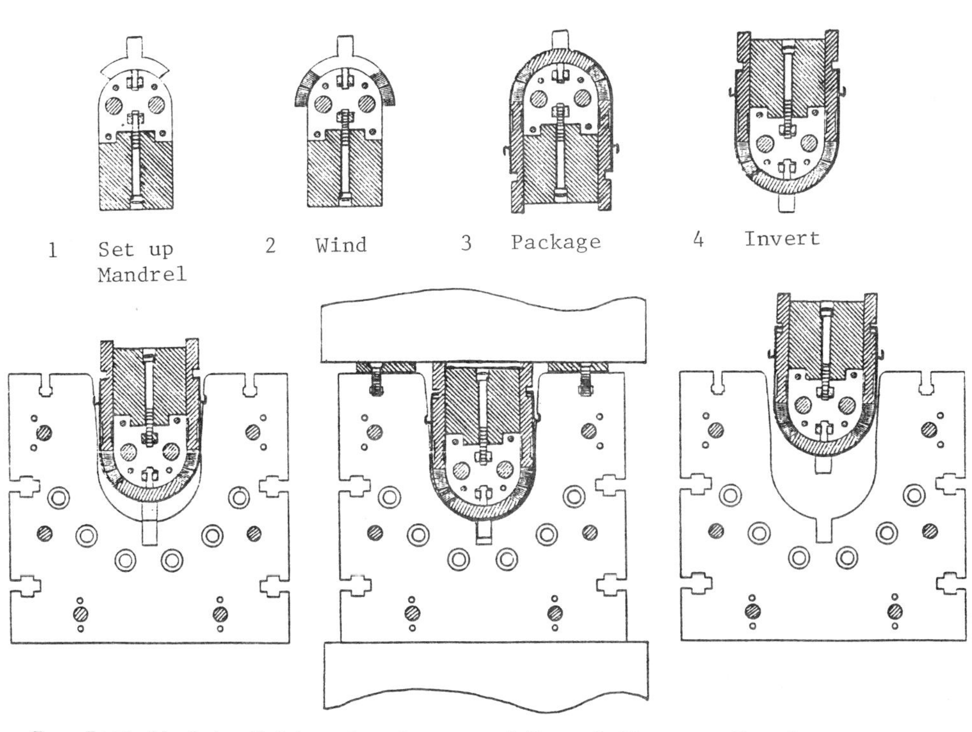

Figure 7.   Coil Fabrication Process

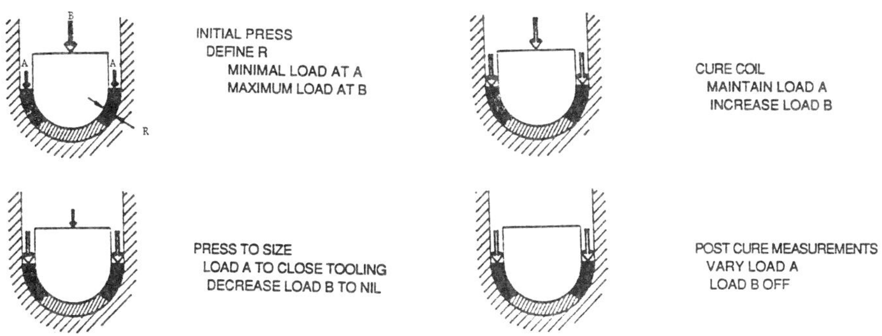

Figure 8.   Coil Molding Cycle

The curing press under construction is shown in Figure 9. The press capacity is 3,333 lbs/in against the mandrel and 18,889 lbs/in against both sizing bars. The geometric design constraints on the press require that the plattens' longitudinal flatness be $\leq$ .001" between hydraulic cylinders and $\leq$ .001" at any transverse section under full load. The press will be fitted with load cells and LVDT's at 18" increments on both sides of the coil to enable measurement of stress and strain of the coil in the press during and after the mold cycle. Outputs will be recorded by a computer operated data acquisition system. The hydraulic and heating/cooling system will be controlled by the same computer.[4]

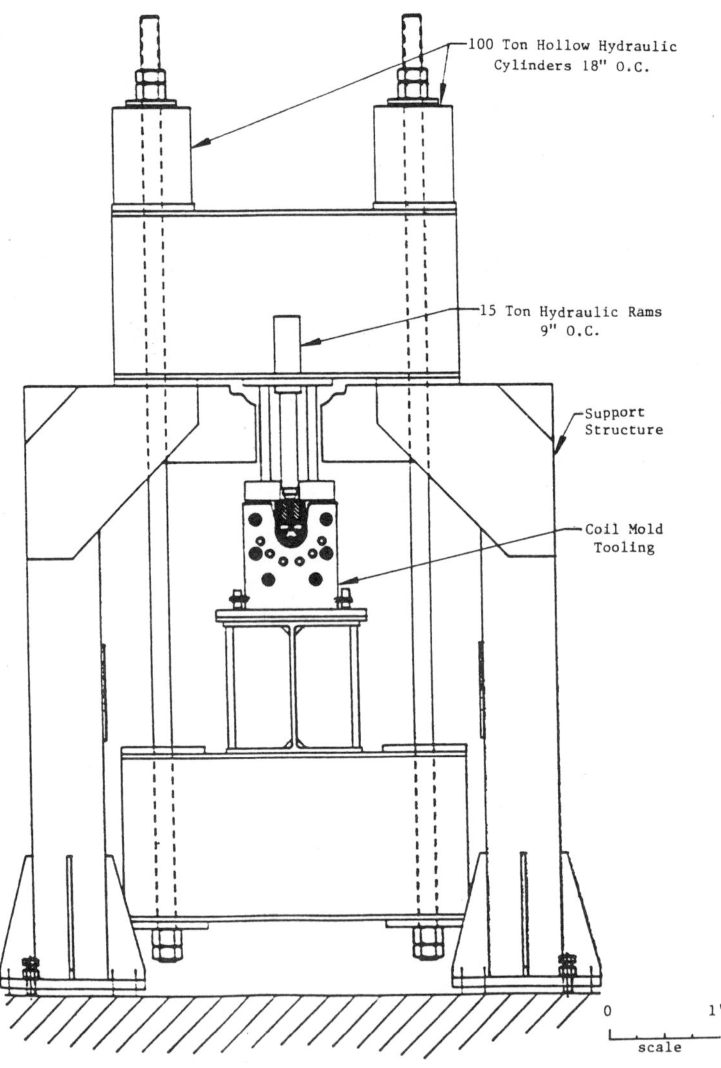

Figure 9. Coil Molding Press - Length of Press = 60' + 60' for Staging Table

PROGRESS TO DATE

   The 1M tooling has been completed and is currently in use. The long
16.6M tooling is under construction with completion scheduled for March 1989.
The 1M tooling utilizes existing Tevatron coil presses modified to provide
the required two-stage loading. We have made four test coils of the SSC NC9
cross section. Coil measurements are being compiled. At Fermilab another
magnet program is in progress to produce a series of low beta quadrupole
magnets for a new detector installation. Tooling used in the quadrupole
program is identical in design, function and materials as the SSC tooling,
the only exception being coil cross section and cable and insulation type.
Figure 10 plots the measurements of twelve (12) coils produced from this
tooling.

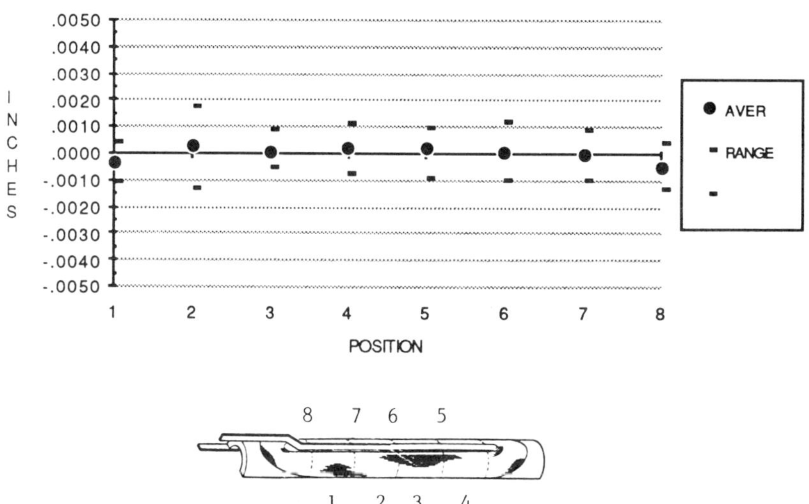

Figure 10. Size Distribution of 12 Coils at 6,857 PSI

REFERENCES

1. R. C. Bossert, J. S. Brandt, J. A. Carson, H. J. Fulton, G. C. Lee, and
    J. M. Cook, "Analytical Solutions to SSC Coil End Design", these
    proceedings.
2. Tool and Manufacturing Engineer's Handbook, SME, 3rd edition, 13-1.
3. Tool and Manufacturing Engineer's Handbook, SME, 3rd edition, 15-17.
4. C. E. Dickey, "Coil Measurement Data Acquisition and Curing Press Control
    System for SSC Dipole Magnet Coils", these proceedings.

INSPECTION AND TEST PLANNING FOR PRODUCTION SSC DIPOLE MAGNETS

Benjamin Wiant

Westinghouse Electric Corporation
Orlando, Florida

Bruce Brown

Fermi National Accelerator Laboratory
Batavia, Illinois

ABSTRACT

B. Wiant for the Westinghouse SSC Program Team. It is
important that the inspections and tests performed during the
manufacturing cycle of the SSC dipole magnets ensure
performance, reliability and the lowest possible price to the
customer. The degree of inspection and the level of testing can
have a direct economical effect on the entire SSC magnet
program. Therefore, careful planning for inspection and test
must be done to ensure magnets that can be readily produce at
reasonable costs. More importantly, the magnets must meet design
requirements. How inspection planning helps achieve these goals
is the subject of this paper.

INTRODUCTION

All manufactured products need some degree of inspection
and test to ensure they will meet their performance requirements
as well as perform reliably during their specified operating
life. To optimize the scope of inspection and test during
manufacture the process of inspection planning is utilized. The
extent of inspection and test for products is an important
factor in keeping the life cycle costs of the product at a
minimum. Excessive inspection and inadequate inspection are both
costly with respect to a company's profit, reputation and
potential future business as a quality supplier. Therefore,
comprehensive inspection plans properly implemented can
contribute significantly to the present and future capability of
companies to compete.

INSPECTION PLANNING

Inspection planning for the SSC dipole magnets is no

different than for any other manufactured product. It is a
systematic process where inspection points and methods are
determined and how they fit and support the manufacturing cycle.
Inspection planning is technology, and like all technology it
also has expanded with time, therefore the factors to be
considered and the methods available change with time. Quality
control has moved from an era of inspecting product for
attributes during manufacturing, to controlling manufacturing
processes and more recently to product design and process
improvement. Therefore, the mind set of inspection today is to
monitor both manufacturing process parameters and the product to
determine whether a process has changed. A number of factors
must still be considered when planning for inspection. Typically
they are as follows:

1) What to inspect - key processes must be identified,
   measured and controlled. It is then that the product is
   inspected, but only to verify that the process has not
   changed.
2) When to inspect - when/where during the manufacturing
   process should the inspection be done?
3) How to inspect - methods, procedures or devices?
4) Who/What should inspect - operator, inspector or
   automated equipment?
5) Where to inspect - on the assembly line, off line, a
   special facility, at the vendor, or at the customer's
   site?
6) Inspection tools - whether they are manual or automated
   tools, a measurement should be consistent regardless of
   who or what does it. Calibration and selection of tools
   is critical.
7) Sampling plans - how many are to be inspected. Are
   certain sampling plans required?
8) Work environment - both product and personnel need to be
   considered when designing a work area; lighting,
   temperature, ventilation, humidity, access to tools and
   instructions are major considerations.
9) Records - what to record, how it will be recorded and
   how it will be retrieved. To do this cost effectively,
   suitable data management technologies must be used
   including data base managers, graphics and statistical
   analysis packages.
10) Standards - what standards are required in the industry
    and by the customer?
11) Feedback system - a mechanism must be in place to
    evaluate results of inspections in order to take the
    needed steps to correct any adverse findings. This
    evaluation may be done by people, and/or machines.

To formulate a comprehensive and workable inspection plan
all of these factors should be addressed.

WHY PLAN? - A QUESTION NO LONGER ASKED?

Inspection planning is only part of the much larger process
of quality planning. The goal of quality planning is to supply
customers with products that meet requirements and that have the
minimum life cycle cost to them and yourself. Results of good
quality planning show up everyday in profits, customer
satisfaction and market share. In today's global market it is a

given that to survive you must be a quality supplier.

KEY PROCESS IDENTIFICATION

The initial step in the inspection planning process is to assure you know and understand the product's requirements and function. Once this step is completed the design of the product can begin. Part of the design process is manufacturing and inspection planning. As the design progresses a manufacturing process map can be developed as seen in Figure 1. This map can then be used as a tool in identifying the key processes in the manufacturing cycle. Key processes are all processes which materially affect the deliverables to the customer and whose output is highly dependent on the control of certain process parameters. Generally, these key processes can be controlled to produce satisfactory output only by keeping these process parameters within prescribed limits. Three tools used to characterize the key processes are Pareto analysis, cause and effect diagrams and design of experiments. There are numerous texts and published articles which deal with these topics in depth.

What is a process parameter? It is any factor (such as feed rate, temperature, speed, pressure, etc.) which can cause variability in the output of the process. Some parameters are allowed to remain uncontrolled but those which can influence the success or failure of the process must be kept within control limits.

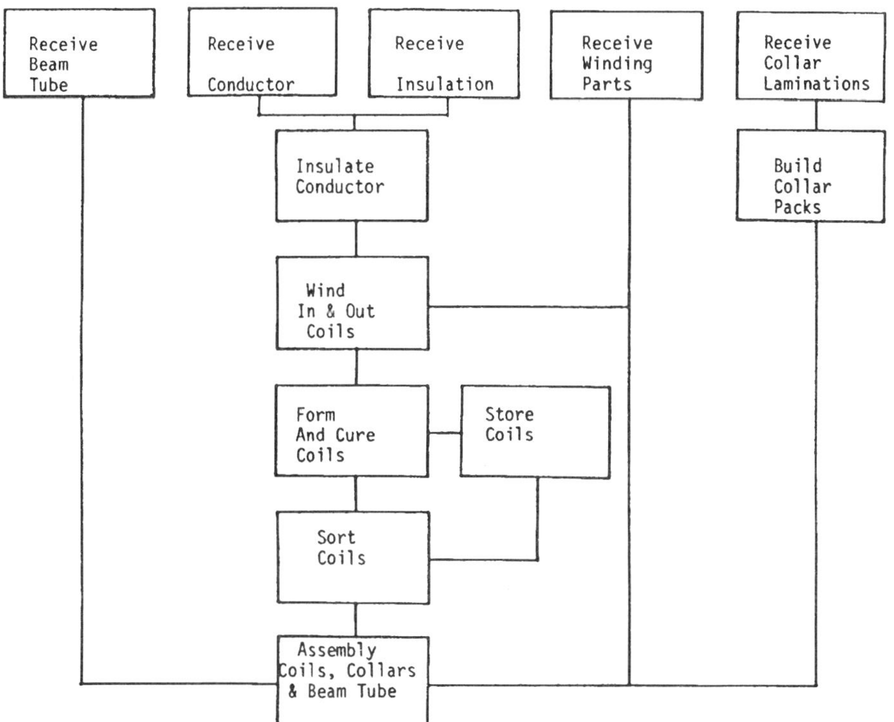

Figure 1. Manufacturing Process Map for SSC dipole magnets coil collaring.

Multifunctional quality teams should be formed for the life cycle of the product to perform the tasks associated with process characterization and process performance. The task of identifying key processes is important in inspection planning. It allows us to concentrate on the more important processes, which in turn assures total quality at minimal quality costs. In addition three other effects are realized:

1) During the review of the design it verifies what dimensions and tolerances are critical.
2) During production it emphasizes control of key processes.
3) During vendor selection it provides guidance in qualifying only vendors capable of meeting process requirements.

As a result of performing this planning process and defining the key processes, we determine what measurements will be made. We also determine how the measurements will be taken, where in the process they will be made, the extent and frequency of measurements, the tools and equipment used, who will take the measurement and if the measurement should be accomplished automatically. The method to store and retrieve data is also established.

INSPECTION PLAN DEVELOPMENT

Inspection plans should be written for all manufactured items as well as purchased items. When preparing an inspection plan the manufacturing flow diagram showing all the basic steps to produce the product is required. In conjunction with the flow diagram there are some basic principles to follow when developing the inspection points.

1) Identify the key processes and construct process models as seen in Figure 2.
2) Identify the variables to monitor for each process.
3) Identify inspection points to verify process is still in control. Checking for process parameter variability rather than product attributes.
4) Do each inspection as early in the manufacturing cycle as possible.
5) Inspect only once each step.
6) Do the inspection the same way each time regardless of the inspector. This ties in with planning for the selection of the measuring equipment.
7) Start at the end of the manufacturing flow diagram to identify each parameter in the finished product that requires inspection. Then decide at what point during manufacturing each parameter could first be inspected and mark this on the chart. Then, continue to work backwards through the manufacturing cycle, doing the same at each point along the way. When finished, a clear pattern of inspection points will be defined on the flow chart. Now, starting at the beginning of the manufacturing cycle, detail the inspection points and decide whether each inspection is indicative of the process at that point. If it isn't the inspection should be reevaluated and moved to the correct process in the flow diagram.

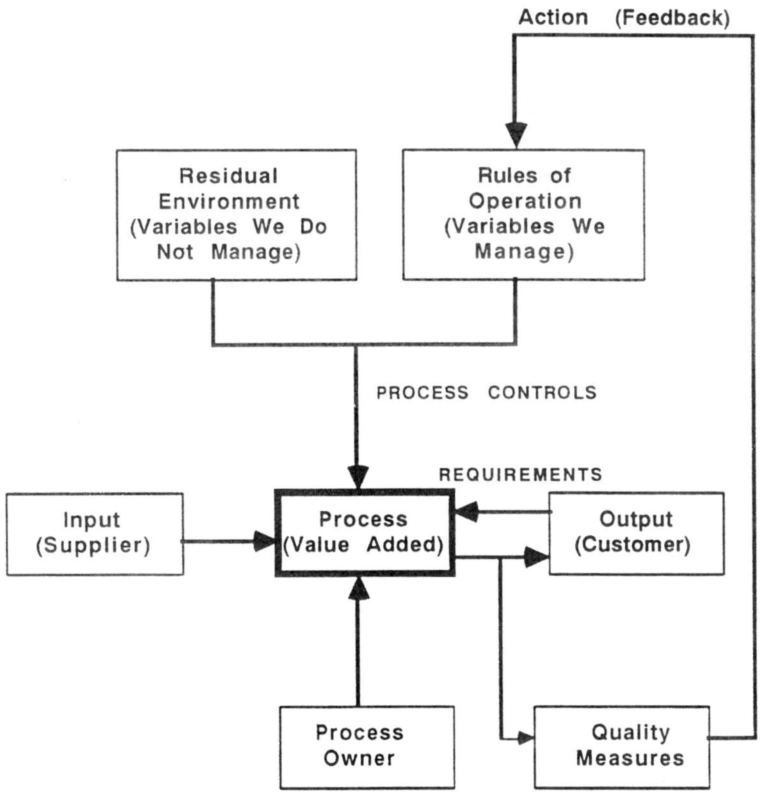

Figure 2. Process model used in characterizing individual
processes.

8) Write the detailed inspection plan and implement the
plan as written.

When qualifying vendors who supply materials and components
for the end product it is important that both the vendor and the
buyer understand the requirements of the order. Each vendor is
responsible for monitoring his processes to ensure they stay in
control. The buyer must audit the vendor's process controls and
product to ensure continued product integrity.

SUMMARY

If inspection and test are to be done both correctly and
economically, an inspection plan is required. Inspection
planning must be done in the design phase of a program before
production begins. Through management of the manufacturing
processes, corrective action can be taken as process variations
and product non-conformances are detected. Then subsequent
changes to the inspection plan can be made.

It is hard work to develop and implement an inspection plan
but well worth the effort in reduced quality costs and providing
a quality product to the customer.

ELECTRICAL MEASUREMENTS DURING MAGNET CONSTRUCTION

George Sintchak, George Ganetis and
Garry Cottingham

Brookhaven National Laboratory*
Accelerator Development Department
Upton, New York 11973

ABSTRACT

Throughout the construction phase of the cold mass for SSC magnets,
electrical tests are made to determine that no faults in the coil structure
have developed. These tests include ones designed to measure turn-to-turn
voltage hold-off, hypot tests to ground, coil resistance, and
instrumentation checks. These various tests will be described and the test
parameters that are used will be covered.

Electrical tests are performed throughout the magnet assembly and
fabrication process to verify that coil integrity and insulation quality of
the various components and sub-assemblies are within nominal limits. These
tests are also required to certify each dipole for SSC acceptance before it
is installed in the cryostat and leaves the factory for final installation.
The following series of tests, which are conducted at room temperature, are
listed below.

- Resistance
- Inductance and "Q"
- Insulation
- Impulse
- Ratiometer

<u>Resistance Tests</u>

Resistance measurements are performed using a one ampere (usually)
precision constant current power supply and measuring the resultant voltage
drop across the element under test. This is analogous to a four-wire
ohmmeter. This test is easy to perform and is uncomplicated. The maximum
output voltage required from the power supply is less than ten volts.

The main coils are connected in series as per the final wiring
configuration, and, with the one ampere current flowing in the coils,
the voltage drop across each coil is measured and recorded. The DC
resistance test will usually indicate a turn-to-turn short. However,

---

* Work performed under the auspices of the U.S. Department of Energy.

the coil resistance will vary somewhat with changes in room (or coil)
temperature. Therefore, difference voltages and temperatures are compared
with previous readings. Also, the resistance of the two inner (or outer)
coils should track each other closely. A voltage drop change (or
difference) of 80 - 90 mVolts is usually an indication of a shorted turn.

Refer to Figure 1 for typical voltage drop DC measurements of SSC dipole
coils with one ampere current flowing in the coils. Please note that the
voltage drop per turn for inner and outer coils is different because of a
slight difference in the superconductor cable composition.

This resistance test of the main coils is repeated frequently throughout
the assembly process, and in particular, before and after the collaring
operation, impulse testing, iron yoke installation, shell welding, Helium
leak test, and for final testing.

- INNER COIL HAS 16 TURNS.

- OUTER COIL HAS 20 TURNS.

- TOTAL WINDING = (2 x 16) + (2 x 20) = 72 TURNS.

- INNER COIL VOLTAGE DROP IS APPROX. 82 mV/TURN
        82 mV x 16 TURNS - <u>1.312 VOLTS</u>

- OUTER COIL VOLTAGE DROP IS APPROX. 95 mV/TURN
        95 mV x 20 TURNS = <u>1.900 VOLTS</u>

- TOTAL MAGNET WINDING = (2 x 1.312) + 2(1.90) = 6.424 VOLTS

- CABLE VOLTAGE DROP IS 60-75 μV/INCH
        (DEPENDS ON CABLE COMPOSITION & TEMPERATURE)

Figure 1. Typical DC Measurements
SSC Dipole Coils

<u>Inductance and "Q" Test</u>

The inductance and "Q" measurement provides another low voltage test
on individual coils that will check for turn-to-turn shorts. This test is
particularly sensitive to "soft" shorts. Q is the quality factor of the
coil and is defined as the ratio of inductive reactance divided by the
effective resistance of the coil. The effective resistance includes the DC
resistance of the coil plus all the other resistive and eddy current losses
due to the core (if any) material and manner in which the coil is wound.
The inductance and "Q" measurement is done using two test frequencies. 1
kHz is used for individual open coils in an air medium, and 120 Hz is used
when the dipole coils are in their final wired configuration within the
yoked iron core. See Figure 2 for typical values of inductance and "Q" for
various coil configurations. Turn-to-turn shorts will show a reduction of
10 - 20 % in the value of "Q".

<u>Insulation Tests</u>

There are several insulation tests done to insure there are no shorts or excessive leakage currents between various components and sub-assemblies within the magnet dipole assembly. The high voltage, or hypot, leakage tests are done at a voltage level that exceeds what the magnet may experience during operation. In general, this test voltage is determined by doubling the expected voltage, and adding 1000 volts. The maximum test voltage required is 5000 volts, and the short circuit current should be limited to 2 mA to avoid damaging any magnet components should a flashover occur.

Simple low precision ohmmeter tests are done before making any high voltage test to be sure that the resistance between the components under test is greater than 20 megohms. The insulation hypot test requirements are as follows:

- Main coils to all other components and ground at 5 kV.
- Main lower coils to upper coils (midplane) at 3 kV.
- Trim coils to all other components and ground at 5 kV.
- Quench protection heaters to other components and
      ground at 5 kV.
- Ground is defined as the collar/yoke/shell and beam tube.
- Leakage current should be less than 50 μA after one minute of the
      applied test voltage.

Typical Values, Series Mode, f = <u>1 KH</u>$_z$
<u>Uncollared</u> Coils - No Metal

	L	Q
Inner Coil	2.35 mHy	10.25
Outer Coil	5.29 mHy	17.2

Typical Values, Series Mode, f = <u>120 H</u>$_z$
Collared & Yoked, Magnet Winding in Iron

L	Q
49.75 mHy	2.89

Data taken with a General Radio 1657 Bridge.

Figure 2.   Inductance & "Q" Measurements
SSC Dipole Magnet Coils

<u>Impulse Test</u>

The impulse test is a high voltage test that checks the turn-to-turn voltage hold-off insulation integrity. This simulates the conditions that may occur during a quench. The coil winding insulation is stressed by discharging a capacitor that delivers a 2 kV pulse to produce (approximately) a 50 volt per turn voltage drop. See Figure 3 for a simplified diagram of the High Voltage Impulse Generator that is used for the impulse test. The resulting damped oscillation is stored on a digital

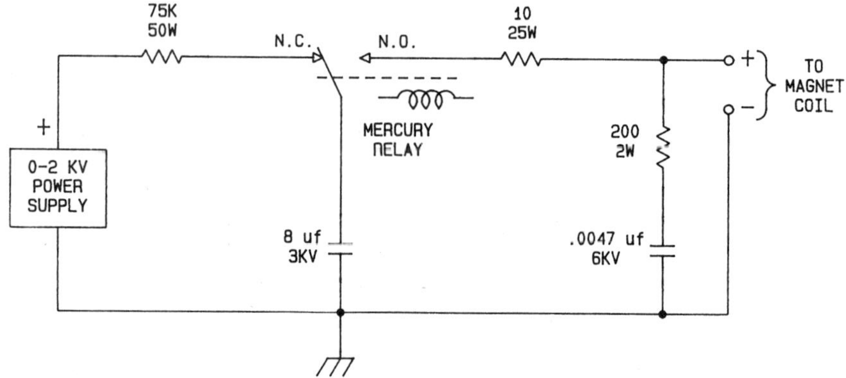

Fig. 3.   High Voltage Impulse Generator

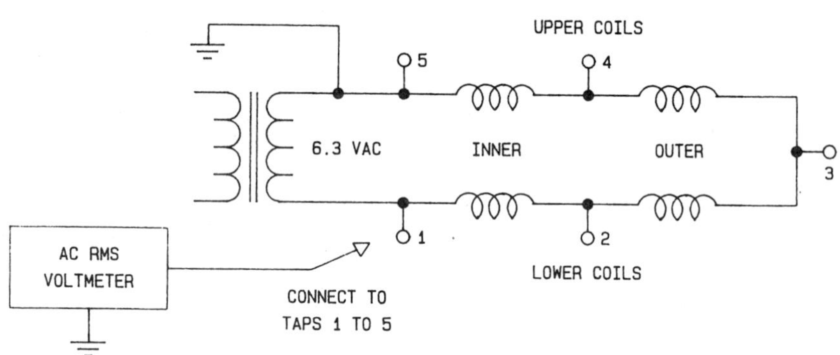

Fig. 4.   Magnet Ratiometer

storage oscilloscope and photographed or plotted. Any change in the waveform will indicate a turn-to-turn breakdown of insulation. This test is usually done before and after the collaring operation, and should not be repeated any more than necessary to avoid stressing the insulation. As mentioned before, a resistance test of the main coils is made before and after this test.

<u>Ratiometer Test</u>

The ratiometer test checks the turns ratio of the coils in a dipole magnet. By comparing turn ratios, it is possible to check if a coil has a shorted turn. The advantage of this method is that it is insensitive to temperature and may be used in a liquid helium environment. The test is performed with coils in iron (yoked) at 60 Hz with approximately one ampere AC current flowing in the magnet. The coils are connected in series as per the final configuration to form an autotransformer as shown in Figure 4. An AC RMS voltmeter is used to measure the voltage developed across the four individual coils using the voltage taps.

The inner upper and lower coils should develop 0.1885 of the total excitation voltage measured. The outer upper and lower coils should develop 0.3115 of the total excitation voltage. If the computed ratio is different by more than +/-0.007 from the given ratio, it indicates there may be a shorted turn in the coil in question. Because the magnet coils are not closely coupled as in a transformer, the inner and outer ratio numbers were determined experimentally using a known good magnet at room temperature and at liquid helium temperature.

In conclusion, some of the tests may seem redundant, but, in many circumstances a cross check or different method of testing is very helpful when interpreting test results. Proper safety procedures should be followed when making any electrical tests and the equipment should be properly grounded. Following the final electrical testing, warm magnetic measurements are done to check magnetic field quality.

QUENCH START LOCALIZATION IN FULL-LENGTH SSC R&D DIPOLES

A. Devred, M. Chapman, J. Cortella, A. Desportes, J. Kaugerts, T. Kirk,
K. Mirk, R. Schermer, J. C. Tompkins, and J. Turner

SSC Central Design Group[*]
c/o Lawrence Berkeley Laboratory
One Cyclotron Road, MS 90-4040
Berkeley, CA 94720 USA

J. G. Cottingham, P. Dahl, M. Garber, G. Ganetis, A. Ghosh, C. Goodzeit,
A. Greene, J. Herrera, S. Kahn, E. Kelly, G. Morgan, A. Prodell,
E. P. Rohrer, W. Sampson, R. Shutt, P. Thompson, P. Wanderer,
and E. Willen

Brookhaven National Laboratory
Upton, NY 11973 USA

M. Bleadon, B. C. Brown, R. Hanft, M. Kuchnir, M. Lamm, P. Mantsch,
P. O. Mazur, D. Orris, J. Peoples, J. Strait, and G. Tool

Fermi National Accelerator Laboratory[*]
Batavia, IL 60510 USA

S. Caspi, W. Gilbert, R. Meuser, C. Peters, J. Rechen, J. Royet, R. Scanlan,
C. Taylor, and J. Zbasnik

Lawrence Berkeley Laboratory[†]
Berkeley, CA 94720 USA

ABSTRACT

Full-length SSC R&D dipole magnets instrumented with four voltage taps on each
turn of the inner quarter coils have been tested. These voltage taps enable (1) accurate
location of the point at which the quenches start and (2) detailed studies of quench
development in the coil. Attention here is focused on localizing the quench source.
After recalling the basic mechanism of a quench (why it occurs and how it propagates),
the method of quench origin analysis is described: the quench propagation velocity on
the turn where the quench occurs is calculated, and the quench location is then verified
by reiterating the analysis on the adjacent turns. Last, the velocity value, which appears
to be higher than previously measured, is discussed.

---

[*] Operated by the Universities Research Association, Inc., for the United States
Department of Energy.

[†] Work supported by the Office of Energy Research, Office of High-Energy Physics,
High-Energy Physics Division, Department of Energy under Contract No. DE-AC03-
76SF00098.

I.   INTRODUCTION

I.1.   What Is A Quench?

The superconducting state only exists when materials are maintained below a temperature called the *critical temperature*. This critical temperature depends on many parameters, including the amount of current carried by the conductors and the magnetic field to which the conductors are subjected. (For a given turn of a dipole coil, the current and the magnetic field are, of course, correlated.) A three-dimensional surface, called the *critical surface*, is defined by the boundaries of critical temperature, critical magnetic field, and critical current, as shown in Figure 1. The critical surface is the very boundary between superconducting and normal resistive states.

A dipole is normally operated at conditions corresponding to a point located underneath the critical surface, where the entire coil is superconducting. Let us assume that, starting from this operating point, we ramp up the current. In ramping up the current (and thus the magnetic field) we get closer and closer to the critical surface, and soon, somewhere in the coil, we cross it. Crossing the critical surface means that, somewhere in the coil, a small volume of conductor switches to the normal resistive state. When switching to the normal resistive state, this small volume dissipates power by the Joule effect. The dissipated power then overheats the small volume of conductor and also, by thermal diffusion or another mechanism of heat-transfer, the region surrounding this small volume. If the Joule heating is sufficient, the surrounding region can then, in turn, reach the critical temperature, switch to the normal resistive state, and dissipate power. Under certain conditions, the normal region, in which the conductors have switched to the normal resistive state, can continue to grow in this manner until the coil is no longer superconducting. This process is called a *quench*.

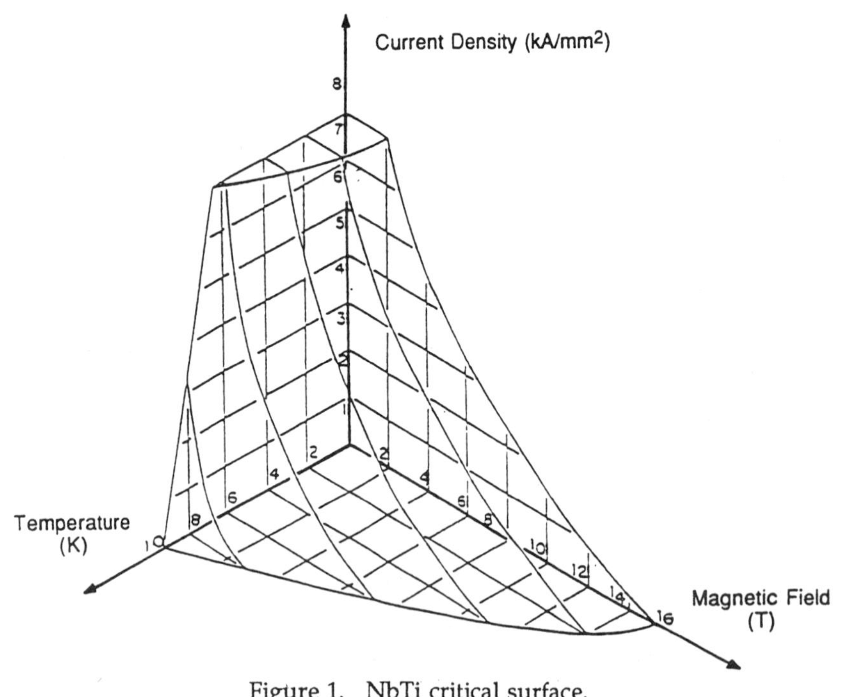

Figure 1.   NbTi critical surface.

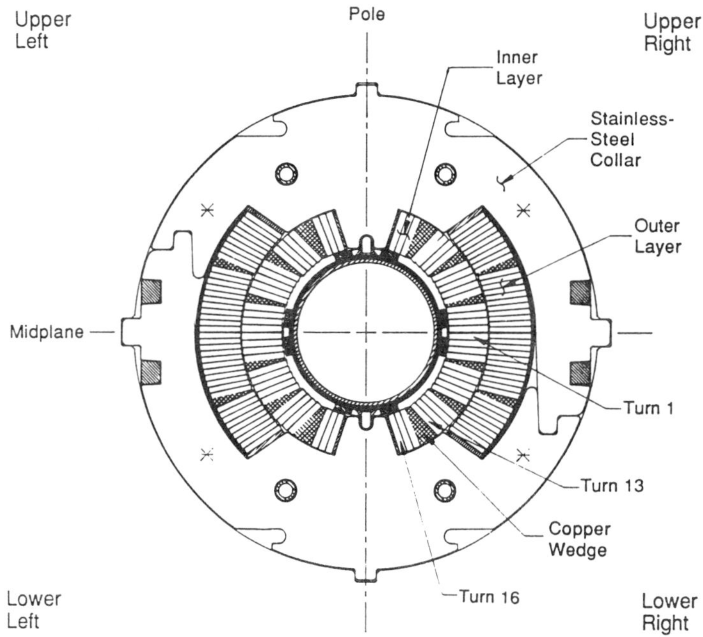

Figure 2. SSC coil cross section (C358).

I.2. Quench Origins

Quench origins can be classified in two categories: *training quenches* and *plateau quenches*. In a dipole, the turn where the conductor sees the maximum magnetic field is the pole turn of the inner coil, as shown in Figure 2. Let $B/I = f_0(I)$ designate the transfer function of this particular turn, and T_0 designate the operating temperature of the magnet. At a given temperature, the maximum value of current that can circulate in the coil is defined by the intersection, in the plane $T = T_0$, of the transfer function $B/I = f_0(I)$ with the critical surface. This maximum value is called the *short-sample current*, I_{ss}. When ramping up the current from a point located underneath the critical surface, quenches can occur at currents lower than I_{ss}. Such quenches are called *training quenches*. Quenches can also occur when the short-sample current is reached; those are called *plateau quenches*.

Plateau quenches are a consequence of the intrinsic characteristics of the conductor, and nothing can be done about them, except by improving the conductor. Training quenches are of a different nature. Since they occur at a current, I, lower than $I_{ss}(T_0)$, somewhere in the coil a volume of conductor sees a temperature increase, $T_0 + \Delta T$, such that $I = I_{ss}(T_0 + \Delta T)$. Thus, the main difference between the plateau and the training quenches is that in plateau quenches the critical surface is crossed because of an increase in current, in training quenches because of an increase in temperature. The local heating that can induce a training quench has mainly one cause in non-impregnated coils: the motion and friction of a conductor (or length of conductor) under the Lorentz force. During the quench, conductor that is not properly restrained mechanically will move. It is then possible for the conductor to lodge in a more secure position, so that in a subsequent test a higher current can be achieved. Eventually, when all the conductor is well positioned, the magnet will reach the short-sample limit. This improvement of the magnet performance is called *training*. Of course, the goal is to build magnets that will not exhibit quenches until currents are reached that are well above the operating current of the machine.

I.3. <u>Quench Propagation Velocity</u>

The coil in a SSC dipole magnet consists of four separately wound parts that are joined during assembly: two inner (upper and lower) and two outer (upper and lower) quarter coils. The inner quarter coils have 16 turns; the outer quarter coils have 20 turns. The turns are counted starting from the midplane of the coil. Figure 2 shows a cross section of a Brookhaven design dipole coil. In this design three copper wedges are inserted in the inner quarter coils: one between turns 4 and 5, one between turns 9 and 10, and the third between turns 13 and 14. The outer quarter coils contain only one wedge, located between turns 11 and 12. In this geometrical configuration, the quench can propagate in three directions: 1) axially (or longitudinally) along the conductors, 2) azimuthally (or transversely) to conductor in the same layer, through the insulation between conductors, and 3) radially, from one layer of conductors to another, through the insulation between the two layers. Thus, the development of a quench in a dipole coil is really a three-dimensional problem.

Most investigations have concentrated on longitudinal propagations. In the early 1960s it was discovered[1,2] that the propagation of a quench along a conductor occurred with a constant velocity, called the *propagation velocity*. Since their discovery, these propagation velocities have been the object of numerous papers; most of them can be found in reference 3. Very few studies have been performed on transverse propagations. They are usually described as following a mechanism similar to the longitudinal propagation, except that the thermal conductivity along the conductor is replaced by the transverse thermal conductivity along the given direction.[4]

II. DETERMINING THE LOCATION OF A QUENCH

II.1. <u>The Technique</u>

When prototype SSC dipole magnets are tested, the current is ramped up from zero until a quench occurs. Two parameters are carefully recorded during the test: the temperature of the helium (assumed to be equal to that of the coil, with some allowance for phase lag) and the current at quench. The quench current is then compared with the estimated short-sample current at the given temperature, and the nature of the quench is identified. If it is a training quench, the next step is to determine where the quench started. Knowing where the quench originated can be very helpful to the magnet builders, because it pinpoints the mechanically weak areas of the coil that need improvement.

The question is how to accurately locate the quench origin. Because the main manifestation of a quench is the development of a resistive zone, it is logical to instrument the coil with voltage taps. To perform any clever analysis, we have to be able to discriminate among the three possible axes of propagation, so each turn of the coil is instrumented with four voltage taps, as shown in Figure 3. These four voltage taps break the turn into four sections, along which the propagation is only longitudinal. During testing, voltages across all sections of all turns are permanently recorded. As long as the coil is entirely superconducting, of course, they are nil. Once a quench has occured, they sequentially exhibit resistive voltages as the quench propagates. By determining which section shows the first resistive voltage, then estimating the propagation velocity along this particular section, we are able to locate the quench origin to within 10 cm.

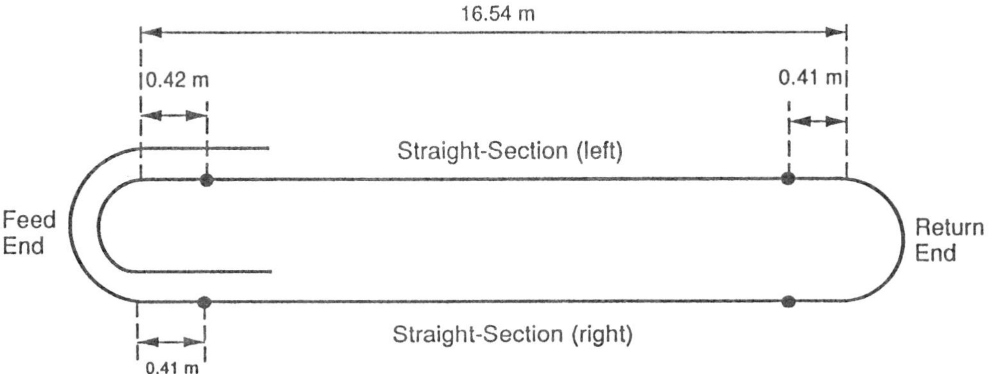

Figure 3.   Tap location on turn 13 of lower inner quarter coil. (not to scale)

Of the full-length SSC dipole prototypes tested in 1988, two were equipped with four voltage taps per turn of the inner quarter coils (128 taps total): magnet DD0010 and magnet DD0012. The features of these magnets and their test results have been presented elsewhere.[5,6,7] Descriptions of the magnet test set up can also be found in references 8 and 9. As shown in Figure 3, the voltage taps were located about 40 cm inside the body of the magnet. This defines two *end sections*, about 1 meter long, and two *straight sections*, about 15.75 meters long. In our jargon, the *feed end* is the magnet end where the power leads are located, and the *return end* is the opposite end. *Right* and *left* sides are defined by facing the magnet at the feed end. We will now discuss an example of quench localization on magnet DD0010.

II.2.   An Analysis Example

To illustrate our method of analysis, we have chosen a typical quench of long magnet DD0010. The characteristics of this quench are presented in Table I.

The first rising voltage shows that the quench occured in the right-hand straight section of the lower inner quarter coil on turn 13. As shown in Figure 2, turn 13 is located next to the copper wedge closest to the pole. Figure 4 shows a plot of voltages across this straight section and across the two corresponding end sections of turn 13. The quench starts somewhere in the straight section and propagates in both directions (with two fronts), toward the two end taps. After 71 milliseconds it hits the return end tap, and the voltage across the return end section takes off. In the meantime, the quench continues to propagate towards the feed end tap (with only one front) and reaches it after 84 milliseconds. At that time, the voltage across the feed end section starts to rise; the entire straight section has switched to the normal resistive state.

Table I. Selected Quench Characteristics

Temperatures			
Feed End	4.32 K	Return End	4.46 K
Pressures			
Feed End	3.99×10^5 Pa	Return End	3.98×10^5 Pa
Current at quench	5,680 A		
Estimated short-sample[*]	6,717 A		

[*] Using the return-end temperature.

First of all, we want to point out that these signals, which are unfiltered and presented as recorded, are of very good quality. They reveal what is going on in amazing detail. For example, let us follow the quench when it arrives in the feed end section. This section can be geometrically divided into three parts: a straight segment of about 40 cm; a curved section of a few centimeters, where the conductor makes a U-turn; and another straight segment, on the left-hand side of the turn. From the magnetic point of view, the two straight segments are similar and see the same magnetic field as the straight section. The curved section, on the other hand, is outside the iron yoke and sees a much lower magnetic field (by a factor of 10%). We then expect to see three phases as the quench propagates along the end section: a fast propagation mode very similar to the one on the

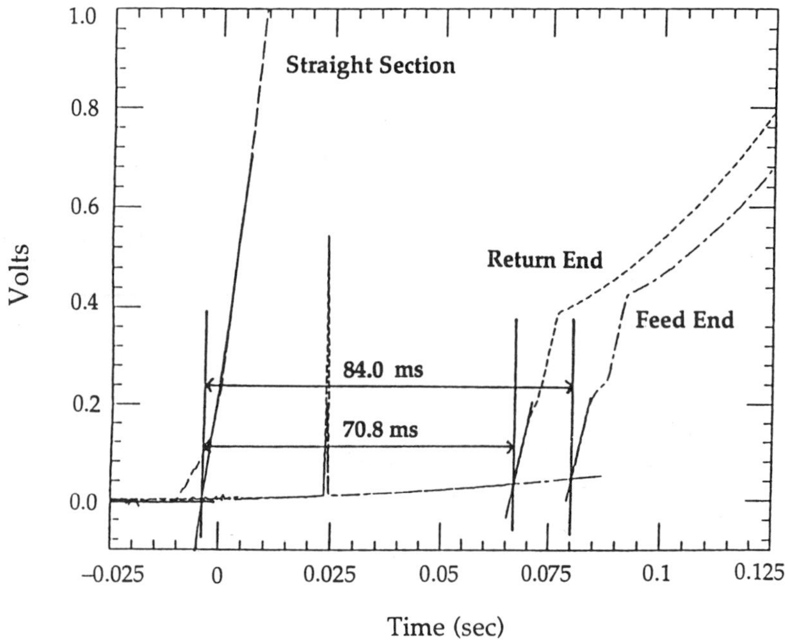

Figure 4. Quench localization on turn 13.

straight section, followed by a slow-down when the quench enters the low-field curved section, and then a recovery of the fast propagation mode once the quench is on the other side of the turn. This is exactly what the traces of Figure 4 show. Also, once the entire end section has switched to the normal resistive state, the voltage continues to grow, although less rapidly, due to the heating of the conductor by the Joule effect, which continuously increases the copper's resistivity.

Determination of the quench start location is rather straightforward. Let us assume that the quench propagation velocity, v, is constant along the straight section (this assumption will be verified later). If t_1 is the time difference between the starts of the straight section and the return end section voltages, and t_2 is the time difference between the starts of the straight-section and the feed-end voltages, v is given by

$$v = \frac{L}{(t_1 + t_2)} \tag{1}$$

where L is the length of the straight section.

An estimation of the quench start location, L_{qs}, from the return end tap is then given by

$$L_{qs} = v\, t_1 \quad . \tag{2}$$

In our example, we have

$$L = 15.72 \text{ m} \quad .$$

It yields

$$v = 101.6 \text{ ms}^{-1} \qquad\qquad L_{qs} = 7.2 \text{ m} \quad .$$

This method for calculating the quench propagation velocity is called the *time-of-flight* technique.

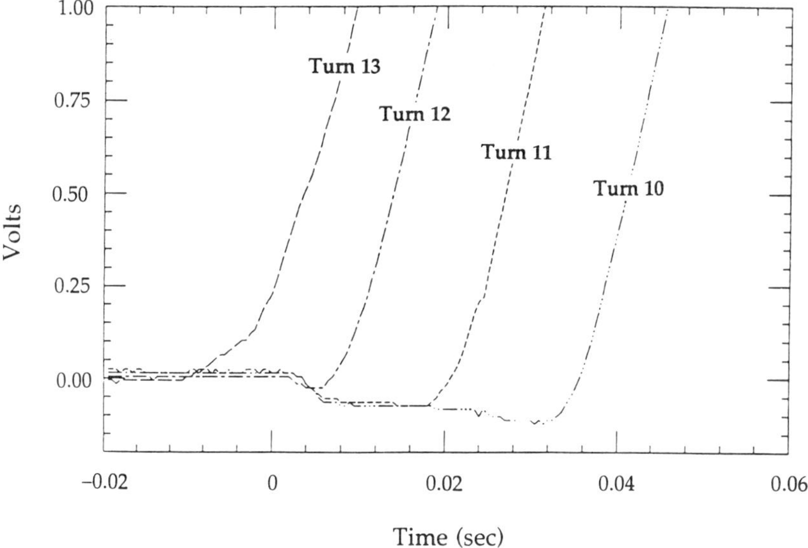

Figure 5.   Transverse propagation.

II.3. <u>Accuracy of the Time-of-Flight Technique</u>

How accurate is this estimation of the quench origin? We do have some means to evaluate it. As we described earlier, while the quench is propagating along the conductor, it is also propagating transversely, from turn to turn, through the insulation between conductors. After a while, we expect to see the quench reach turn 12, then turn 11, and so on. Figure 5 shows a plot of the voltages across the right-hand straight sections of turns 13, 12, 11, and 10. They exhibit exactly the expected behavior. Of course, the quench is also propagating in the other direction, toward turn 14, but the copper wedge between turns 13 and 14 interposes a thermal resistance, which delays the propagation and makes the signals more difficult to interpret. This is also the case for the wedge located between turns 10 and 9.

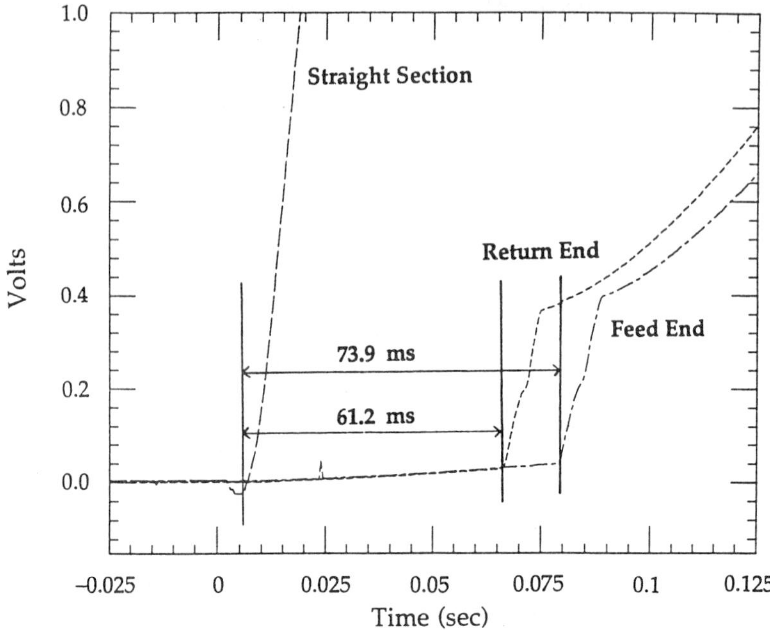

Figure 6. Quench localization on turn 12.

To propagate transversely, the quench will follow the shortest thermal path between conductors. Since the conductor insulation and the coil prestress have been made as uniform as possible along the coil, the shortest path is a straight line perpendicular to the conductors. The quench will thus start to propagate on adjacent turns at the same longitudinal location as its origin on turn 13. Verification of the quench start location on turn 13 can then be made by reiterating the quench start analysis on turns 12, 11, and 10. For example, a plot of voltages across the turn 12 right-hand straight section and across the two corresponding end sections is given in Figure 6. These traces look very similar to that of turn 13, and the propagation velocity can be calculated using the same time-of-flight technique. The results of the quench start analyses on turns 13 through 10 are presented in Table II.

Table II. Quench Start Analyses on Different Turns of Magnet DD0010

Turn Number	Propagation Velocity	Quench Start Location[*]
13	102 ms^{-1}	7.17 m
12	117 ms^{-1}	7.12 m
11	127 ms^{-1}	7.18 m
10	136 ms^{-1}	7.07 m

[*] The origin of the axis is the turn 13 right-hand return-end tap.

The results in Table II agree within 5 cm on turns 13, 12 and 11. The error bar grows to 10 cm with turn 10, which takes off more than 40 ms after turn 13. We consider this conservative number of 10 centimeters to be the accuracy with which we are able to longitudinally locate the quench origin. These 10 centimeters have, of course, to be compared with the 15.75 meter-long straight sections.

III. THE PUZZLE OF THE VELOCITIES

Magnet DD0010 was the first full-length SSC dipole to be equipped with so many voltage taps. One of the most unexpected results of the analysis described above was the values of the quench propagation velocities that were obtained. They are much higher than the velocities previously measured and reported in the literature,[3] for both conductor samples in specific laboratory tests and actual coils. They are even higher than the velocities measured previously on short SSC model dipoles (a 4.5-m model[10] and a 1.8-m model[11]), where they did not exceed 30 ms^{-1}, even for currents close to short sample. How can we be certain that the time-of-flight technique is reliable?

There is another method for estimating the propagation velocity: using the slope of the voltage traces when they start to rise. The beginning of the rise of any voltage trace, across any section of the coil, corresponds to the early moments of the quench propagation along the given section when no significant heating by the Joule effect has occurred that would affect the conductor's resistivity. As long as the heating is negligible, the derivative of the voltage U across the section can then be related to the current I, considered as constant, and to the propagation velocity v by

$$\frac{dU}{dt} = \frac{\rho I}{A} \, n v \quad , \tag{3}$$

where ρ is the resisitivity per unit length of conductor, A is the conductor area, and n is the number of propagation fronts along the given section at the time of the slope calculation.

Once ρ and A are known, it is then easy to deduce the velocity from the voltage slope. This method of calculation is called the *derivation* technique.

The characteristics of the conductor are summarized in Table III. Of course, the resistivity value to be introduced in Formula (3) has to take into account the magnetic field effect. At the given current, the field is calculated from the transfer function $B/I = f(I)$ of the given turn. Of course, the magnetic fields do vary across the cable. For turn 13, the peak field is estimated to be given by

$$B / I = 0.9128 \; 10^{-3} + \frac{0.7505}{I} \tag{4}$$

Table III. Selected DD0010 Inner Layer Cable Characteristics

Number of strands	23
Strand diameter	0.808 mm
Filament diameter	6 μm
Copper to niobium-titanium ratio	1.41
Copper RRR (between 10 and 295 K)	64
Critical current (at 5 T and 4.22 K)	12,980 A

Table IV. Turn 13 Velocities (Derivation Technique)

Section	Number of fronts	Voltage slope	Velocity
Straight right	2	84.5 Vs^{-1}	92.7 ms^{-1}
Return end	1	42.6 Vs^{-1}	93.3 ms^{-1}
Feed end	1	43.0 Vs^{-1}	92.7 ms^{-1}

Table IV shows the velocities calculated by the derivation technique at the beginning of the rise of the voltages across the turn 13 right-hand side straight section and the two corresponding ends. We first notice that the values appear to be very close to each other. This justifies, *a posteriori*, our assumption that the quench was propagating longitudinally with a constant velocity. Second, the constant value estimated by the derivation technique appears to be in fairly good agreement with the velocity estimated by the time-of-flight technique (the 10% discrepancy can easily be attributed to the uncertainties of the conductor characteristics). Thus, these two independent methods lead to the same results. The 100 ms^{-1} are then fully credible.

Another enigma appears when one considers these 100 ms^{-1} in terms of existing theories of quench propagation. In the classical description of longitudinal propagation, the "engine" of the quench is located close to the transition front, where a fraction of the power generated in the normal zone is transmitted by conduction along the conductor copper to the superconducting zone, which, in turn, heats up and goes into transition. In this model, the helium surrounding the conductor can only slow down the propagation by absorbing part of the dissipated power. An upper limit, v_0, of the velocity can then be calculated by considering a single bare conductor behaving adiabatically. Using formula (29) of reference 3, with the quench characteristics of Table I and the conductor characteristics of Table III, yields

$$v_0 = 27.6 \text{ ms}^{-1}$$

The actual propagation velocity is then almost four times higher than the upper limit predicted by the classical model. This means that another mechanism, different from the simple Fourier conduction along the copper conductor, must be speeding up the propagation. What could this be? It is too early for us to draw any conclusions; we only want to mention an interesting conjecture that can be found in reference 12.

IV. CONCLUSION

The above results demonstrate indisputably the successful use of voltage taps to test SSC R&D dipoles. Without them, we would never have been able to locate to within 10 cm the quench origin, as well azimuthally (which turn of the coil) as longitudinally (where on the turn). Nor would the unexpectedly fast quench propagation velocities of 100 ms^{-1} have come to light. Of course, magnet DD0010 has been quenched more than once. For each quench we have been able to repeat these analyses and to obtain similar, equally accurate results. The particular example we have presented here is then reproducible. In fact, the 15 quenches of magnet DD0010 were all training quenches and all occured on turn 13, either in the upper inner or the lower inner quarter coil, in either the right-hand or the left-hand straight section. The longitudinal locations also varied from quench to quench. These data revealed a global misbehavior of turn 13. When the magnet was subsequently disassembled, it was found that the copper wedges next to the turn were loose (a possible consequence of a handling error during final preparation prior to testing). The analysis of voltage tap data has given us considerable information about quench source, which has significantly strengthened the SSC dipole design.

V. ACKNOWLEDGMENT

We are grateful to Katharine Metropolis for insightful comments on the manuscript.

REFERENCES

[1] W. R. Cherry and J. R. Gittleman, "Thermal and electrodynamic aspects of the superconductive transition process," *Solid State Electron*, **1**, p. 287, 1960.

[2] R. F. Broom and E. H. Rhoderick, "Thermal propagation of a normal region in a thin superconducting film and its application to a new type of bistable element," *Br. J. Appl. Phys.*, **11**, p. 292, 1960.

[3] A. Devred, "General formulas for the adiabatic propagation velocity of the normal zone," *IEEE Trans. Magn.*, **MAG-25 No. 2**, p. 1698, 1989.

[4] N. K. Wilson, *Superconducting Magnets*, Oxford, Clarendon Press, p. 204, 1983.

[5] J. Peoples, et al., "Status of the SSC superconducting magnet program," *IEEE Trans. Magn.*, **MAG-25 No. 2**, p. 1444, 1989.

[6] J. Strait, et al., "Test of full scale SSC R&D dipole magnets," *IEEE Trans. Magn.*, **MAG-25 No. 2**, p. 1455, 1989.

[7] J. Tompkins, et al., "Performance of full length SSC model dipoles: results from 1988 tests," presented at the First International Industrialization of the SSC Conference, New Orleans, Louisiana, February 8-10, 1989, in these Proceedings.

[8] J. Strait, et al., "Fermilab R&D test facility for SSC magnets," presented at the First International Industrialization of the SSC Conference, New Orleans, Louisiana, February 8-10, 1989, in these Proceedings.

[9] P. Mazur and T. Paterson, "A cryogenic test stand for full length SSC magnets with superfluid capability," presented at the First International Industrialization of the SSC Conference, New Orleans, Louisiana, February 8-10, 1989, in these Proceedings.

[10] G. Ganetis and A. Prodell, "Results from heater-induced quenches of a 4.5 m reference design D dipole for the SSC," *IEEE Trans. Magn.*, **MAG-23 No. 2**, p. 495, 1987.

[11] W. Hassenzahl, "Heater-induced quenches in SSC model dipoles," *IEEE Trans. Magn.*, **MAG-23 No. 2**, p. 934, 1987.

[12] C. Luongo, et al., "Thermal-hydraulic simulation of helium expulsion from a cable-in-conduit conductor," *IEEE Trans. Magn.*, **MAG-25 No. 2**, p. 1589, 1989.

A VERY LARGE SUPERCONDUCTING SOLENOID

R.W. Fast, J.H. Grimson, R.D. Kephart, H.J. Krebs,
M.E. Stone, E.D. Theriot, and R.H. Wands

Fermi National Accelerator Laboratory
P.O. Box 500
Batavia, IL 60510

ABSTRACT

A detector utilizing a superconducting solenoid is being discussed for the
Superconducting Super Collider (SSC). A useful field volume of 8 m diameter
x 16 m length at 1.5-2 T is required. The magnet will have a stored energy
of ~1.5 GJ at 2T. All particle physics calorimetry will be inside the bore of
the solenoid such that there is no need for the coil and cryostat to be "thin"
in radiation lengths. An iron yoke will reduce the excitation required and will
provide muon identification and a redundant momentum measurement of the
muons. We have developed a conceptual design to meet these requirements.
The magnet will use a copper-stabilized Nb-Ti conductor sized for a cryostable
pool boiling heat flux of ~0.025 W/cm^2. A thermosiphon from a storage
vessel above the cryostat will be used to prevent bubble stagnation in the
liquid helium bath. The operating current, current density, coil subdivision
and dump resistor have been chosen to guarantee that the coil will be
undamaged should a quench occur. The axial electromagnetic force will be
reacted by metallic support links; the stainless steel coil case will support the
radial force. The 5000 metric tons of calorimetry will be supported from the
iron yoke through a trussed cylindrical shell structure separate from the
cryostat. The coil and case, radiation shield and stainless vacuum vessel
would be fabricated and cryogenically tested as two 8-m sections. These
would be lowered into the underground experimental hall and installed into the
iron flux return yoke to provide the required 16-m length.

INTRODUCTION

The SSC will have six interaction regions located approximately 50
meters below grade. A large superconducting solenoid is being considered as a
part of a detector facility for one of the interaction halls. Figure 1
schematically shows such an SSC detector. The calorimetry and central
tracking chambers, which will be located inside the solenoid, require a field
volume 8 m in diameter by 16 m long. The field in the bore is required to
be 1.5-2.0 T; the field in the iron flux return yoke will be ~1.5 T. Locating
the calorimetry internal to the solenoid eliminates the need to have a "thin"
coil and cryostat in terms of radiation lengths. Because thickness is not a
constraint, several coil types are possible. The design presented here is a
cryostable pool boiling coil.

The required field volume will be provided by two solenoids, each 9.5 m in diameter and 8 m in length, connected in series electrically with a common power supply. Each 8-m unit will contain four identical 2-m long liquid helium/coil modules. The modules are mechanically connected and share a common vacuum vessel. Helium is supplied by thermosiphons to each module from a storage vessel located on top of the iron. Thus each module would have its own helium supply and return   piping.

The calorimetry and flux return iron will be split at the longitudinal center and the halves can be moved independently; therefore, we will use two 8-m support structures to support the 5000 metric tons of calorimetry and central tracking chamber. There are notches in the iron for hangers for the supports and for instrumentation cables. The support structures are immediately inside the solenoids but are completely independent of the cryostat and vacuum shell. The ends of each structure will be supported from the iron. The support structures, consisting of concentric stainless steel cylinders with a trussed web, have a radial thickness of 25 cm and are expected to have a deflection of ~3 cm when loaded with the calorimetry. The calorimeter will rest on rails and can be inserted from either end.

QUENCH PROTECTION

Four 2-m modules are connected in series. The modules could in principle have separate gas cooled current leads. However, to reduce the overall heat load on the magnet, we assume that the modules in each 8-m section will be connected with superconducting bus in the liquid helium filled interconnecting pipes. It is essential that the coil survive quenches without damage and to this end a low current density and an external fast dump resistor were chosen. This resistor was chosen to be 0.1 Ω to limit the discharge voltage to 500 V across the terminals. The eddy current heating in both the coil and helium vessel during a fast dump is expected to be ~1300 W for an 8-m assembly. This might be sufficient to quench the coil and so the discharge will be through the fast dump resistor only if a normal zone is detected. The highest temperature reached in the coil due to a quench will be 100 K. A discharge for any other reason will be through a 0.02 Ω slow dump resistor with the power supply reversed to maintain a constant discharge voltage. The routine charge or discharge time, with 100 V across the terminals, would be 100 min and the eddy current heat load would be 26 W per 8-m assembly. Figure 2 is the electrical schematic.

CONDUCTOR AND COIL SPECIFICATIONS, WINDING AND ASSEMBLY SCHEME

The superconductor will be copper stabilized Nb-Ti wire which has a short sample current of 10 kA at 3 T and 4.5 K. This wire will be soldered into additional copper stabilizer. For the purpose of this study we assume the conductor will be 26 x 18 mm and will have a copper to superconductor area ratio of 146 to 1. Here we have chosen the stabilizer area to insure safe discharge of the coil in the event a normal region in the coil should develop. An operating current of 5000 A results in a current density of 1.56 x 10^5 A/cm^2 in the superconductor and 1068 A/cm^2 overall. The surface heat flux, with 25% wetting and a RRR of 100, is 0.025 W/cm^2.

There are seven conductor layers having a total of 651 turns in each 2-m coil module. Since the forces are radially outward, the coil will be wound starting with the outside layer, using the outer shell and flat annular heads as a coil form. Insulation will consist of G-10 buttons between turns, slotted G-10 sheets between layers and slotted G-10 and Kapton adjacent to the helium vessel. The packing factor will be 74% with 13% being helium space.

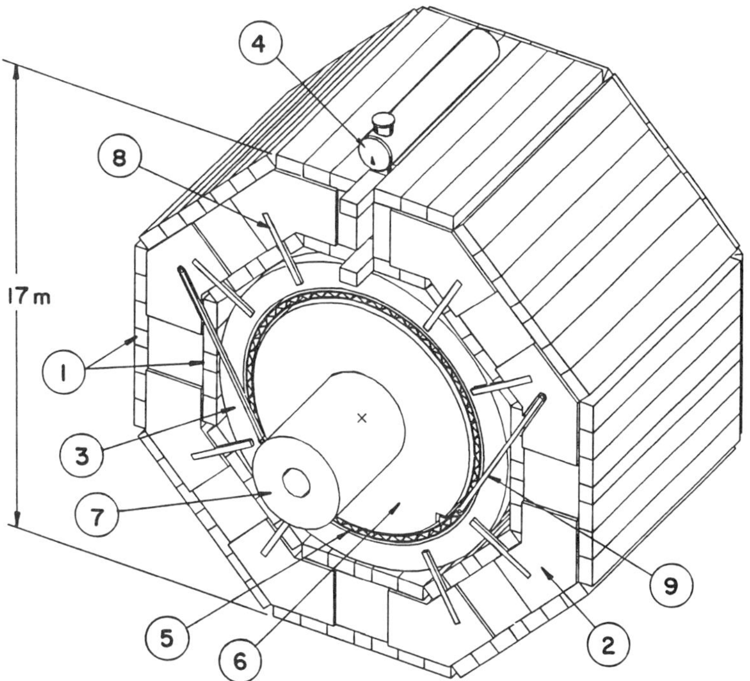

Fig. 1.   One-half SSC Detector: 1, muon tracking; 2, flux return iron; 3, coil assembly; 4, 5000 L helium dewar; 5, calorimeter support structure; 6, calorimeter; 7, central tracking chambers; 8, coil supports; 9, support structure hangers.

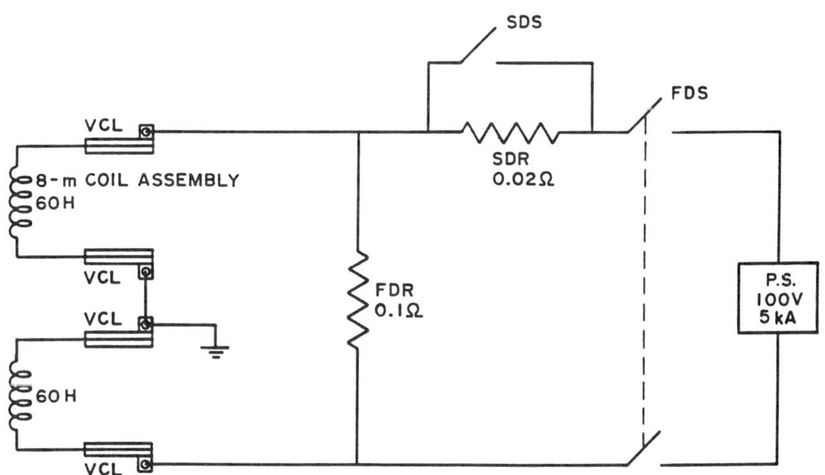

Fig. 2.   Electrical Schematic: FDR, fast dump resistor; FDS, fast dump switch (NC, open to initiate a fast dump); SDR, slow dump resistor; SDS, slow dump switch (NC, open to initiate a slow dump); VCL, vapor cooled leads; P.S., power supply.

Each 2-m coil module will be designed in accordance with Section VIII of the ASME Boiler and Pressure Vessel Code for an internal pressure differential of 0.8 MPa (116 psia) and will be adequately relieved for a loss of insulating vacuum or a quench of the coil. The axial electromagnetic force will be transmitted through the shells to the supports, and the radial force will be resisted by both the conductor and outer shell.

Winding will be done with the coil axis vertical. The winding fixture will provide the radial preload and a series of compression bars at one end of each coil can be individually adjusted to provide a circumferentially uniform axial preload. These axial compression bars are adjustable only during assembly. After winding, the inner shell of the helium vessel will be welded to the coil form. The four 2-m He vessels are bolted together. Each is located inside an inner and outer vacuum jacket and integral nitrogen shield. The entire 8-m cold mass is supported with respect to the magnet iron by separate metallic radial and axial supports. Figure 3 is a cross section of the coil, cryostat and vacuum vessel.

SUPPORTS

The axial body force on the 8-m cold mass is dependent on the geometry of the end wall. The supports are designed for an axial or radial misalignment, or the equivalent due to iron non-homogeneity, of 2.5 cm. The force constants are 8.8 MN/m (5 x 10^4 lb/in) for radial and 149 MN/m (8.5 x 10^5 lb/in) for axial misalignment. The supports to the cold mass have been chosen to be metallic and to have both a forced flow liquid nitrogen and a thermosiphon liquid helium intercept.

The supports for the cold mass are assumed to be separated function with different elements to react the axial and radial forces. We initially favored a combined function support because by properly adjusting the angle of the support with the axial direction, the force due to thermal contraction could be eliminated, but, since these supports are at an angle to the axial direction, they generate a large radial buckling force while reacting the axial body decentering forces. This disadvantage does not occur with a separated function support. An additional advantage is that both elements of the separated function supports can be much longer which greatly reduces the heat load.

The warm ends of all supports will be attached near the ends of the vacuum vessels to avoid transmitting the loads and weight through the shells. Both the axial and radial supports are located on the outer diameter of the helium vessel. The axial supports will have their cold ends connected near the inside bolted flange on the outboard 2-m module; whereas the cold ends of the radial supports will be attached near either end of the 8-m assembly.

REFRIGERATION AND CRYOGENICS

The thermal shield cooling tubes and the support intercepts are fed with forced flow subcooled liquid nitrogen at an average temperature of 83 K. The inner and outer shields and each set of supports will have independent heat intercept circuits. The expected heat load to the nitrogen system could be as much as 5 kW for the complete 16-m magnet.

A refrigerator at the surface supplies helium to two 5000-L storage dewars on top of the iron. A cold compressor will be used if needed to maintain the storage dewar at about 30 kPa and 4.5 K. Each dewar has supply and return lines to the four modules in one 8-m assembly. These lines are sized such that the return helium is less than 1% gas by weight and the supply lines are carefully insulated to ensure the maximum liquid fraction

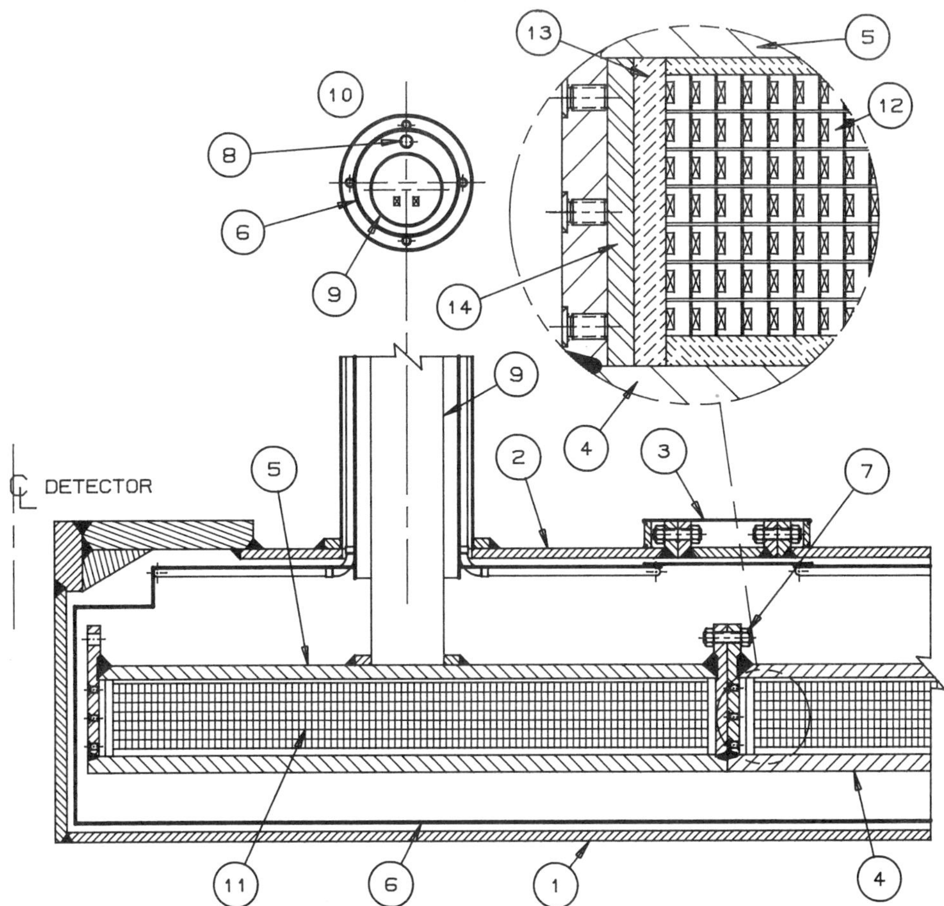

Fig. 3. Axial cross section of coil, helium and vacuum vessels: 1, inner and 2, outer, vacuum shells; 3, assembly joint (if required); 4, inner and 5, outer helium vessel shell; 6, radiation shield; 7, coil module attachment; 8, liquid helium supply pipe; 9, helium return/vent pipe; 10, chimney to storage dewar; 11, coil winding; 12, conductor; 13, G-10 insulation; 14, axial preload bar.

entering the bottom of the magnet. The steady state thermosiphon flow will be about 25 g/s to each module. Separate helium circuits, which will have a return flow of less than 7% by weight, will intercept the supports. The heat load from the separated function support system might be as high as 100 W depending upon the design. The total steady state heat load for the magnet system could be 410 W (230 W plus 36 L/h for lead flows). The 4.5 K refrigerator will have a capacity of 1600 to 1800 W.

In order to facilitate cooldown, a separate refrigerator would probably be employed. It would consist of a GHe/LN_2 heat exchanger and a turbo-expander which could provide 400 g/s of 55 K helium gas. This would be adequate to cool down the total cold mass (both 8-m sections) of 1000 tons in approximately two and a half weeks.

CONCLUSIONS

Preliminary work has indicated that the size and field requirements of this magnet are a reasonable extrapolation of existing superconducting magnet technology. The magnet design must be very conservative to guarantee high reliability, since the time to fix even a small problem could have a major impact on the SSC physics program. More detailed study must be done to optimize parameters both in terms of reliability and cost-effectiveness. Because of the large size of this magnet, fabrication techniques must be examined very closely and the necessity of on-site construction must be rigorously evaluated. However, at this point, we see nothing that would preclude the construction of such a magnet.

ACKNOWLEDGEMENTS

This work was supported by the U.S. Department of Energy, Contract numbers DE-AC02-76CH03000 and DE-AC02-89ER40486.

Conventional Facilities and Technical Systems

Chairman:

R. Robbins
Sverdrup Corporation

TUNNELING TECHNOLOGIES FOR THE COLLIDER RING TUNNELS

Peter Frobenius

Bechtel Civil, Inc.

San Francisco, California

ABSTRACT

The Texas site chosen for the Superconducting Super Collider has
been studied, and it has been determined that proven, conventional
technology and accepted engineering practice are suitable for
constructing the collider tunnels. The Texas National Research
Laboratory Commission report recommended that two types of tunneling
machines be used for construction of the tunnels: a conventional hard
rock tunnel boring machine (TBM) for the Austin chalk and a double
shielded, rotary TBM for the Taylor marl. Since the tunneling machines
usually set the pace for the project, efficient planning, operation,
and coordination of the tunneling system components will be critical to
the schedule and cost of the project. During design, tunneling rate
prediction should be refined by focusing on the development of an
effective tunneling system and evaluating its capacity to meet or
exceed the required schedules.

INTRODUCTION

The Texas site was recently selected for construction of the
Superconducting Super Collider (SSC) project, and preliminary design of
the SSC is to begin in the near future. The proposed construction of
53 miles of collider ring tunnels, along with the construction of six
underground collision halls, puts this work into the category of major
underground projects such as rapid transit systems.

The purpose of this paper is to summarize geotechnical
investigations and tunnel concepts that have been developed for
collider tunnel design and construction. It also outlines the future
design and construction planning needed to complete the collider
tunnels within the overall schedule and budget constraints of the SSC
project.

The geologic conditions of undisturbed soft rock favor extensive
use of tunneling machines. The major advantage of TBMs is the
tunneling speed. Other advantages are elimination of overbreak and
early, simple installation of tunnel support.

Disadvantages include factors that can lead to loss of time and investment due to improperly applied tunnel machine technology. Among these factors is the sensitivity of tunneling machines to unexpected changes in geologic conditions. Others are the potential for excessive downtime of tunneling machine backup due to inability to keep up with muck removal and material supply at the required pace.

The Texas National Research Laboratory Commission (TNRLC) report[1] recommended two different tunneling machines for Austin chalk and Taylor marl. These machines and their backups will be discussed, and tunnel support requirements for the different ground conditions will be outlined. To illustrate the work required, a case history of a similar tunnel project will be reviewed.

In addition to the TNRLC report, reports of the SSC Conceptual Design Group[2] and U.S. Department of Energy[3] were reviewed in preparation of this paper.

BACKGROUND

Collider Ring Layout

The 10-foot finished diameter collider ring tunnel is approximately 53 miles in circumference. A plan view of the ring (Figure 1) indicates its major features: (a) upper and lower arc sections and two diametrically opposed clusters (near and far clusters) of straight tunnel lengths containing the injector complex and six experimental or

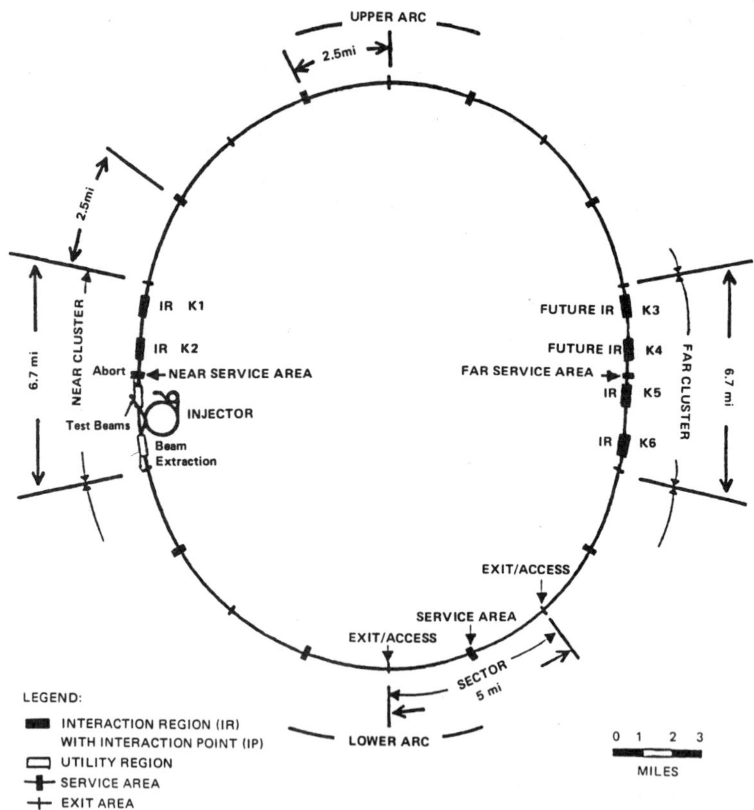

Figure 1 Typical Layout of the SSC

interaction regions, where future experiments will be conducted; (b) ten service areas comprising the facilities for cooling and energizing the superconducting magnet system of the collider; and (c) ten access areas which subdivide the ring into four sectors, each approximately 5 miles long, in the upper and lower arcs. Both service and access areas have access shafts from the surface to the tunnels.

Geologic Conditions

Site conditions are favorable for the construction of the SSC collider tunnels using proven, conventional technology and accepted engineering practice. This determination is based on the site investigations and the documented successful completion of tunnels and deep excavations in the same geologic units in the region. Figure 2 shows that 70 percent of the tunneling for the ring will be in Austin chalk and 30 percent in Taylor marl. The Austin chalk is characterized as a low strength rock with an average unconfined compressive strength of 2,000 psi. Unweathered claystone of the Taylor marl displays even lower strength than chalk. The groundwater table is below the tunnel alignment; thus, groundwater is not a problem. Both rock types are suitable for excavation by a tunnel boring machine (TBM).

The TNRLC report recommends that the collider tunnels be excavated using rock tunneling machines and that alcoves, chambers, and the short sections of connector tunnel needed around the main collider ring be excavated using road headers. The following discussion is based on findings of the TNRLC report. Additional investigations will be conducted to finalize tunneling methods.

Tunneling in Austin Chalk

For tunneling in the Austin chalk, the TNRLC report recommends the use of 12-foot-diameter conventional TBMs with domed rotary cutterheads and disc cutters. The relative softness of the material will allow high penetration rates, and since the chalk is not abrasive, low cutting tool wear should result in fewer cutting tool changes and greater machine availability. The report predicts average advance rates of 200 feet per day.

The report assumes that in the Austin chalk, ground support will be provided during excavation only when required in localized fracture areas or areas of lower cover. Support, where required, will consist of rock bolts or welded wire-mesh fabric with or without shotcrete. In heavily fractured areas, partial circumference steel ribs can be set initially, and a cast-in-place concrete lining installed after completion of tunneling. Such supported lengths in the SCC are not expected to exceed 5 percent of the tunnel length in the chalk. Although it is estimated that most of the tunnel in Austin chalk can remain unsupported during the operational phase of the collider project, the exposed tunnel walls above the invert should be covered with shotcrete to control dust and possible seepage.

Tunneling in Taylor Marl

Most of the Taylor marl should have adequate stand-up time; slickensided joint and fracture areas, however, will need immediate support. The key to maintaining stability of the tunnel in marl is to preserve the original moisture content of the marl as much as possible.

To preserve the original moisture content of the marl, the TNRLC report recommends that double shielded, rotary TBMs equipped with drag

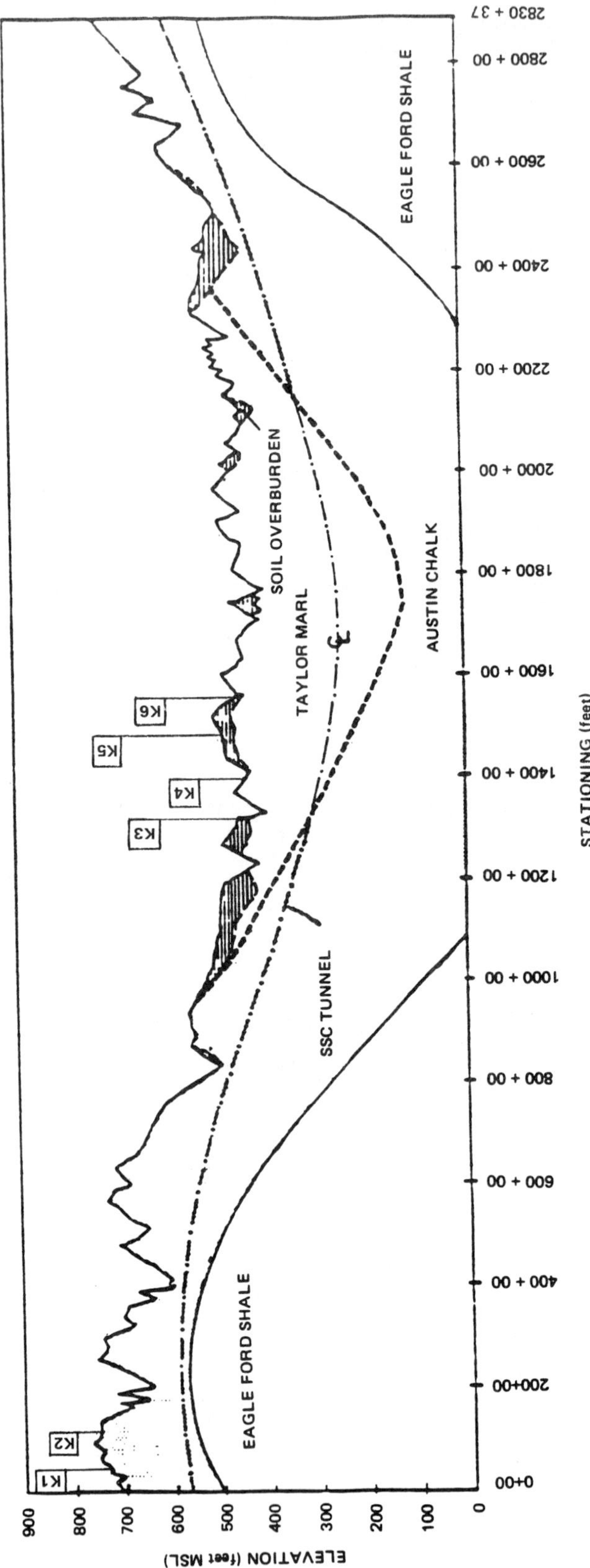

Figure 2 Geologic Profile Along Collider Ring

bits be used in the marl. The report assumes that the following steps
will take place. A permanent/final precast segmented concrete liner
will be installed directly behind the machine as it advances. Segments
will be moved to the face on special cars and placed within the tail
skirt of the TBM by an erector ring. After installation of the liner
and emergence from the tail skirt, the annular space between erected
segments and excavated tunnel opening will be filled by grout
injection. The rotary TBM is expected to advance at an average rate of
130 feet per day. Typical tunnel support for chalk and marl is shown
in Figure 3.

DEVELOPMENT OF THE TUNNELING SYSTEMS

Design and Construction Planning Objectives

As the TNRLC report indicates, the Texas site conditions favor
extensive use of TBMs. Advantages of fast tunnel progress and early
project occupancy can be derived from TBM use.

In the following discussion, the construction planning and design
steps are outlined that will contribute to the proper application of
tunneling machine technology and will avoid potential disadvantages of
tunnel machine use leading to loss of time and investment.

Cost and Schedule Considerations

Completion of a tunneling job within budget and schedule depends on
the achieved average tunnel progress rate. The major portion of tunnel

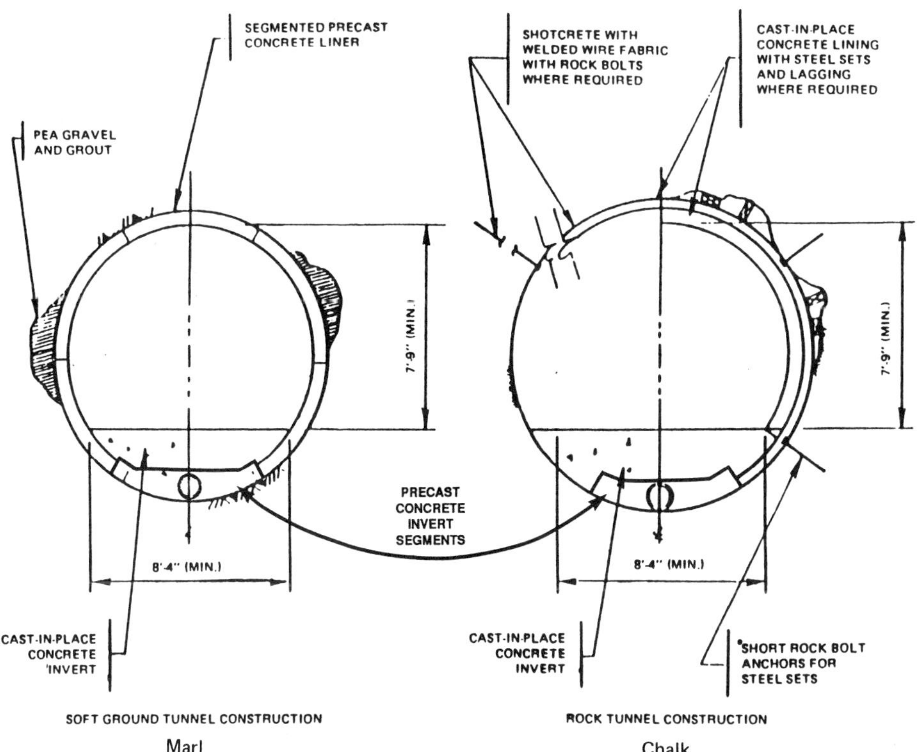

Figure 3 Typical Tunnel Support Sections

cost consists of labor costs, power costs, and operation and maintenance cost of equipment. These costs are practically constant on a daily basis. Therefore, a fast advance rate is crucial in keeping a low cost per foot of tunnel[4].

For the SSC project, most major milestones and constraints concerning underground permanent equipment installation will depend on the projected advance rates and completion rates of the collider tunnel sections. Therefore, the advantages of early occupancy and reduced construction financing costs are directly dependent on tunneling schedule.

<u>Tunneling Rate Prediction</u>

Tunneling rate prediction should be refined during preliminary and final design by focusing on the development of an effective tunneling system and evaluating its capacity to meet or exceed the required construction schedules.

Average tunnel progress rate is a function of rock strength and composition, cutter type, cutterhead penetration rate, and tunnel system utilization (defined as the percentage of actual TBM excavation time out of total available working hours).

<u>Rock Strength and Composition</u>. It appears that the geology of the site is excellent for excavation of the collider tunnels. The strength characteristics described earlier classify chalk and marl as soft rocks in terms of tunnel boring characteristics. In addition, the geology is not complicated by faults, discontinuities, or ground water. There is a good chance that exploration will continue to produce straightforward data that favor a smooth tunneling operation.

<u>Cutter Types</u>. Cutterheads of rock tunneling machines may be equipped with several different types of cutters (Figure 4):

    o   Soft to medium rock – drag bits, disc cutters
    o   Medium to hard rock – disc cutters
    o   Hard to very hard rock – rolling cutters

Drag cutters work by the undercutting principle in which the rock is undercut by the rotating drag cutters mounted on the cutting wheel. The required thrust is applied tangentially in the plane of the cutting wheel by the torque of the cutting wheel. The remaining uncut rock portion is broken by a light wedging action produced by a wedge shape of the cutting tips of drag cutters. The drag bit requires less forward thrust than any other cutter type. This can be an advantage in soft rock in reducing machine cost and in avoiding the high gripping forces on weak tunnel walls required to resist the high thrust forces for other types of cutters. Drag bits are cost efficient in soft rock formations, but cutter life is reduced rapidly as rock becomes harder and more abrasive.

Rolling cutters and disc cutters operate by attacking the rock face with a wedge-shaped cutting edge which causes the brittle rock to shatter and disintegrate around the cutting tip. The forward thrust required to press the rotating cutting wheel against the rock face is produced by hydraulic cylinders supported by the stationary machine body.

98

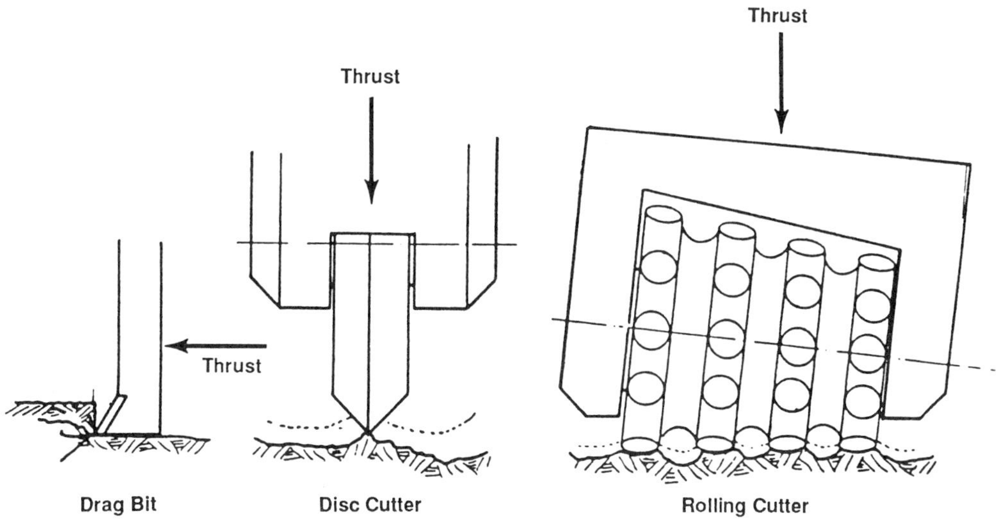

Figure 4   Types of Cutters

The disc cutter can cut a very wide range of rock, from weak to 30,000 psi, at economical rates. At the low end of the scale, soft plastic material will clog the cutter and render it ineffective. The disc cutter requires higher thrust loads than the drag bit, but not as high as the rolling bits in rock of equivalent hardness. It will break out larger chips than the rolling bit and is, therefore, a more efficient cutter.

The rolling cutter cuts the hardest rock, using tungsten carbide inserts or buttons. Very high thrusts and machine weight and rigidity are required[5].

Cutterhead Penetration. The achieved penetration rate of a cutterhead equipped with disc or rolling cutters is a function of rock strength, cutterhead thrust against the face, and running speed of the individual cutters across the rock tunnel face. Rock strength has to be accepted as found. Cutter thrust and cutter running speed can be selected within limits. Increased thrust and running speed increase rock penetration rate. However, an offsetting cost in increased cutter wear by abrasion and cutter bearing wear has to be paid because wear and abrasion are direct functions of cutter thrust and speed.

Figure 5 slows the theoretical penetration rates that can be achieved in soft rock using 20,000 lb/cutter thrust and 250 ft/min cutter speed. The figure shows that the theoretical penetration rates increase in a hyperbolic function as the tunnel diameter decreases. The figure also shows achieved penetration rates and average tunneling rates for selected projects and confirms that maximum penetration rates predicted by the TNRLC report are feasible. Average tunneling rates predicted by TNRLC are also feasible as long as average tunneling machine utilization can be maintained at a rate of 50 percent. That means that the machine has to be boring 50 percent of the available time. Similar penetration predictions can be developed for cutterheads equipped with drag cutters.

Tunneling System Utilization. Two specific tunneling systems are proposed by the TNRLC report for the collider tunnels. One uses a conventional rock tunneling machine with side grippers. The other uses a double shielded tunneling machine.

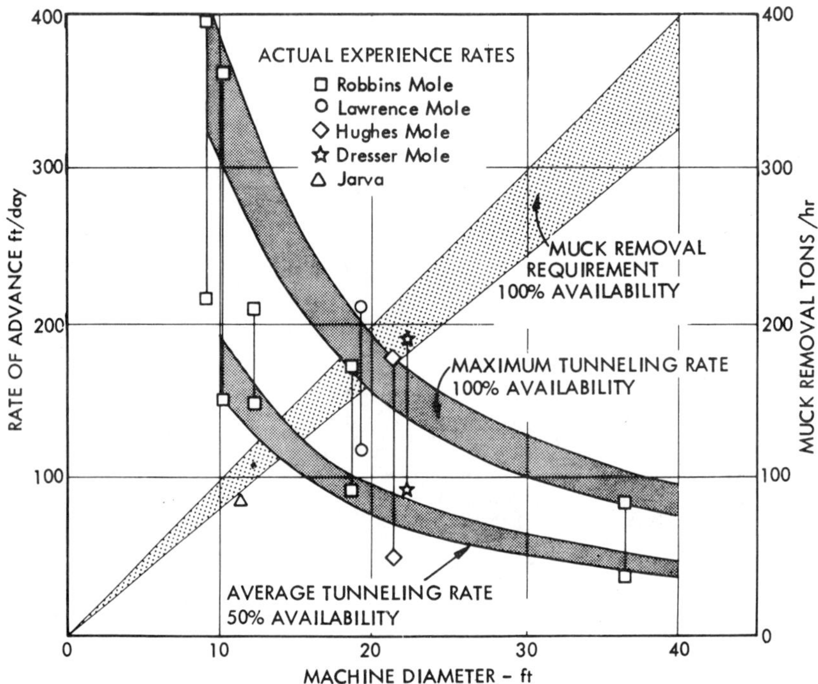

Figure 5   Progress Rates and Muck Removal Requirements of Tunneling Machine in Soft Rock

Each tunneling system incorporates equipment to perform the following interrelated activities: excavation, ground support, and materials handling. Material handling includes transport of muck spoil from the heading to the disposal site and transport of construction materials and personnel to the heading and return of personnel to the surface.

Figure 6 shows the tunnel system component alternatives that can be considered for incorporation in the tunneling system. Under ideal conditions, the selected TBM sets the pace for all the tunnel system components following in the tunnel train. Conversely, the TBM can be seriously delayed by failure of one of the many interrelated components of the system.

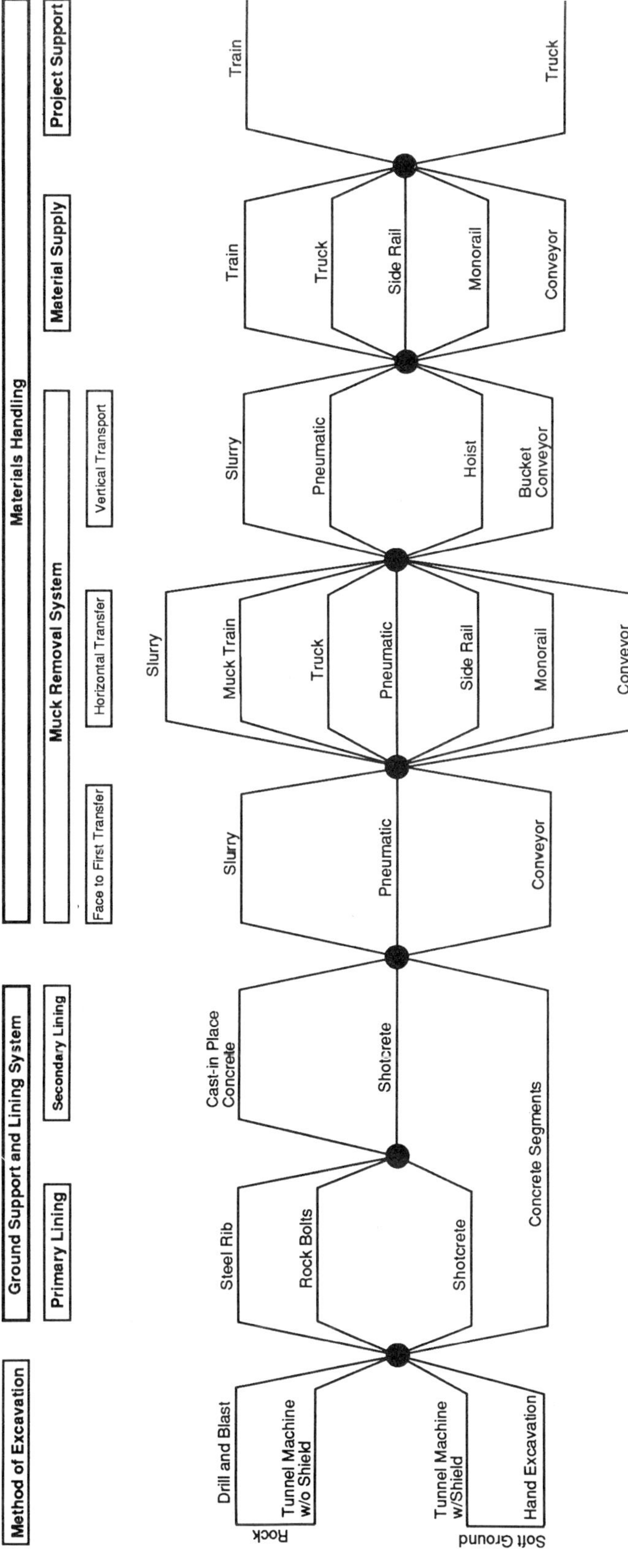

Figure 6 Tunnel System Component Alternatives

<u>Recommended Tunneling Systems</u>

The component specifications for the two tunneling systems proposed for cutting the collider tunnels are described in the following.

<u>Conventional Rock Tunneling Machine</u>. In Austin chalk, where only a minimum ground support consisting of rock bolts and shotcrete is anticipated, a conventional rock tunneling machine of 12-foot diameter with side grippers is proposed to be used (Figure 7). The finger canopy extending backward from the cutterhead is considered adequate for roof support and protection. The tunnel side walls and floor in chalk will be firm enough to provide adequate support for the side gripper pads of the TBM during the advance cycle of the cutterhead. Installations on the rear of the TBM include rock bolt drill rigs and a ring beam erector. On the tunnel train, platforms are provided from which shotcreting will be conducted. The operating cycle by which the machine moves forward is described in Figure 8.

<u>Double Shielded TBM</u>. Taylor marl may need immediate support after excavation in slickensided joint and fracture areas. To provide this support and to preserve the moisture content of the marl, the TNRLC report recommends the installation of a precast segmented lining directly behind the tunneling machine. This installation has to be performed under the protection of a tunnel machine shield with the help of a segment erector. Therefore, the TNRLC report recommends using a double shielded TBM of 12-foot diameter equipped with drag cutters. A TBM incorporating the requirements outlined by the TNRLC report is shown in Figure 9. It is equipped with telescoping double shields. Installations at the rear of the TBM include a precast lining segment erector for erection of precast lining under the protection of the tail shield. The operating and ring erection cycle by which this machine moves forward is shown in Figure 10.

While this machine offers the advantage of roof protection during face excavation and lining erection, several problems may be encountered when driving through areas of more or less cohesionless materials or through squeezing rock.

o   Large blocky rock fragments can become lodged in the openings of the cutterhead and block cutterhead rotation.

o   The reciprocating motion of the tail shield, during the advance cycle, tends to form a hard cake of compacted muck in the annular space between the two telescoping shields. When it builds up, this hard cake must be removed; otherwise, telescoping action between the two shields is restricted.

o   Double shielded machines are more restricted in their maneuverability than open, conventional TBMs. Therefore, cutterheads on shielded TBMs are designed to cut a slightly larger size tunnel than the outside diameter of the shield. This provides clearance for steering both shields and for retracting the tail shield forward. In squeezing rock, the double shielded TBM is vulnerable. The clearance can close up as the rock moves inward, stopping progress of the machine.

These problem areas will need careful study and resolution during the tunnel design.

    <u>Tunnel Train</u>. Both tunneling machines require a tunnel train
several hundred feet in length to be towed behind the TBM. The tunnel
train consists of a series of 20-foot-long platforms running on the
tunnel rail. The train contains power supply and hydraulic power
packs, transformers, provisions for ventilation, dust suppression, and
compressed air supply. Muck is transferred to the shafts by diesel
locomotives pulling muck cars. Tunnel lining materials are supplied by
flat cars and/or concrete haul units. A gantry conveyor transports
excavated spoil back along the tunnel train to dump the muck through a
dump chute into rail cars, which are pulled one by one into loading
position under the chute.

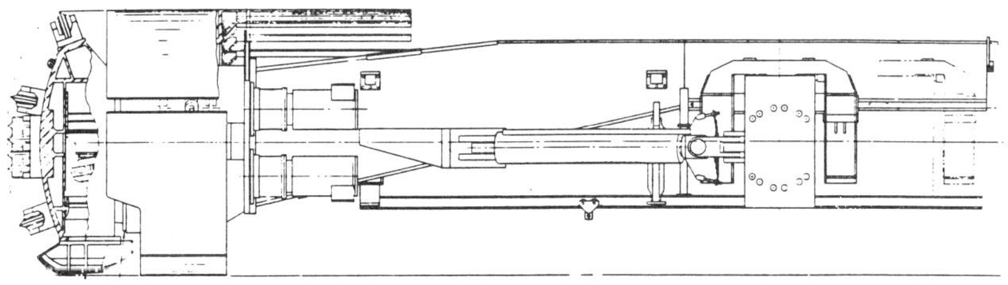

Figure 7   Conventional Rock Tunneling Machine

<u>Tunnel System Backup</u>

    Having discussed the proposed tunneling systems, it is now useful
to review the tunnel system backup that can be provided from the shafts
and the length of tunnel sections that have to be serviced by tunnel
muck trains. Improper selection of tunnel section and shaft facilities
will have the same effect as tunnel system component failure and will
contribute to tunnel system downtime and low TBM utilization.

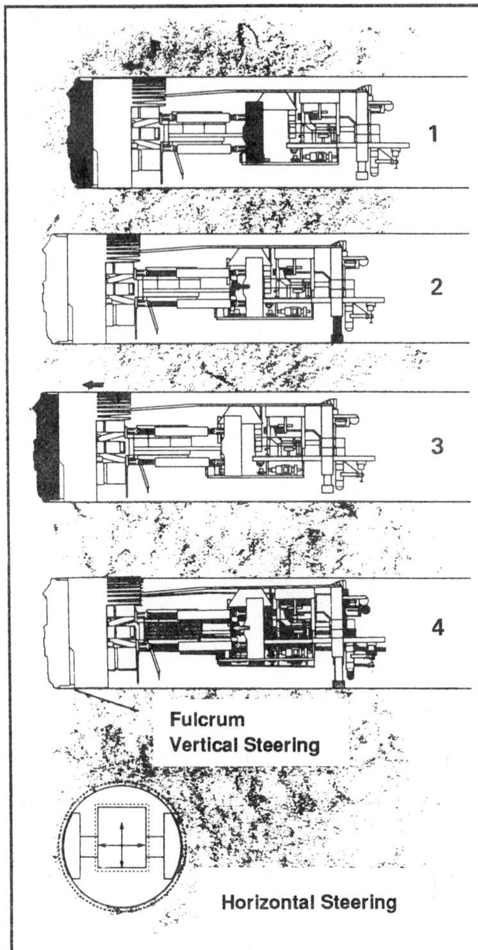

1. BORING CYCLE: With side grippers firmly planted on the tunnel wall, hydraulic cylinders advance the rotating TBM cutterhead into the rock.

2. RETRACTION: When the cylinders reach their full stroke, the cutterhead is stopped; rear support legs are lowered; grippers are retracted and thrust cylinders retract, pulling side gripper assembly forward.

3. RESET: Side grippers are planted against the tunnel wall; rear support legs are raised; and the cutterhead is restarted to begin new boring cycle.

4. STEERING: The boring direction is changed by shifting the rear of the machine up, down, or to one side. This movement causes a slight shift of the cutting head in the opposite direction, resulting in a gentle curve and the establishment of a new heading.

Steering is accomplished with the gripper shoes fully gripped on the tunnel wall. Machine attitude may be set at the start of each boring cycle, and altered during the boring cycle without resetting the gripper shoes.

Figure 8   Hard Rock TBM Operating Cycle

Dividing the collider ring tunnels into ten segments appears reasonable. Each segment will include approximately 28,000 linear feet of main collider tunnel. TBMs can tunnel this footage without major overhaul. Typically, a tunnel segment runs between refrigeration shaft facilities, with one intermediate vent/access shaft. Both shaft facilities will be used for tunnel access and material removal as described in the TNRLC report. (During construction and operation, these shafts will also provide ventilation intake and exhaust as well as emergency exit.) At a maximum hourly advance rate of 16 feet per hour, four tunneling machine excavation cycles may be completed per hour, each filling one muck train. Using track turnouts, it appears that four full trains and four empty trains can be accommodated in the tunnels and shafts each hour.

<u>Performance Record of Tunneling Systems</u>

The description of the factors that are important for tunnel system utilization indicates that efficient planning, operation, and coordination of the components of the tunneling system are of prime importance. As stated previously, the tunneling machine should set the pace for the materials handling and ground support installation. Capacity, size, and arrangement of the tunnel equipment should allow for uninterrupted transfer of outgoing muck and incoming materials.

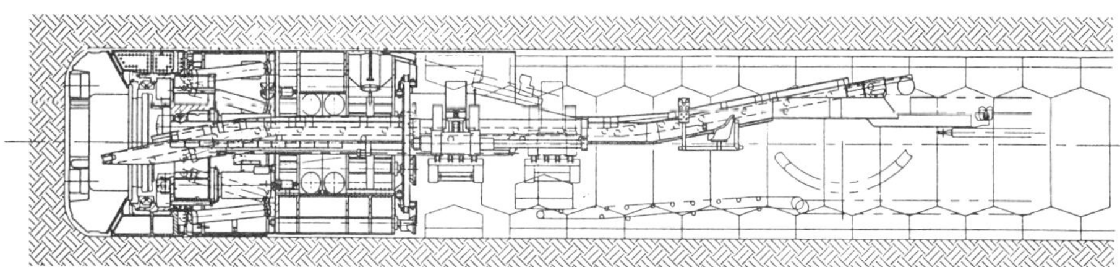

Figure 9   Double Shield Tunnel Machine

During work with tunneling machines on the Bay Area Rapid Transit (BART) project in the 1960s, performance and utilization of five tunneling machines working on different contracts were recorded[6].

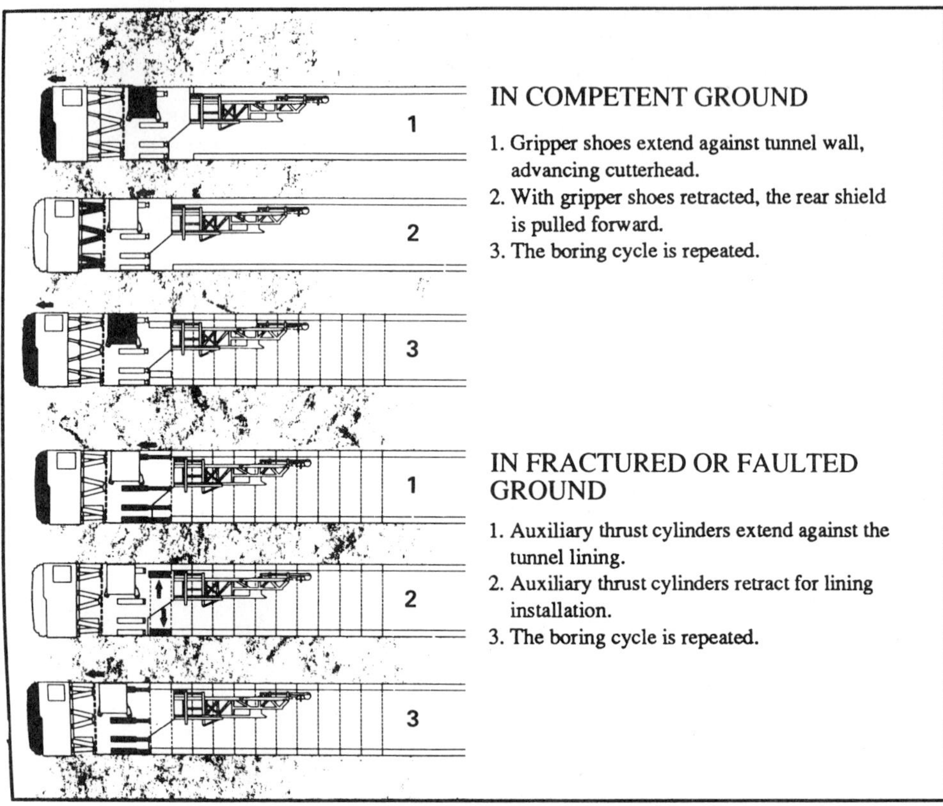

Figure 10   Double Shield Operating Cycle

A review of this and other past jobs shows that it is difficult to achieve perfect performance records. Even on successful tunneling jobs, the actual drilling time of the tunneling machine is usually less than 50 percent of working time, and unscheduled delays and breakdowns can use up as much as 30 to 60 percent of available production time. A record of experience with a rock tunneling machine in Australia, singled out for its recent date and clear description of system performance and delays, is shown in Figure 11. It shows typical delays that can occur.

Typically, tunnel system performance can be broken down into the following percentages of available working time:

	Percentage
Machine drilling	30 – 50
Regripping and repositioning of machine	1 – 5
Scheduled maintenance of tunnel machine and tunneling system	10 – 15
Unscheduled downtime of tunneling system	59 – 30
Total	100

106

WEEK INTERVAL / NO. SHIFTS / SUPPORT TYPE/SPACING	ADVANCE (m)	SHIFT AV PENETR RATE (m/hr)	BORING B1 (%)	SYSTEM UTILIZATION (%)	DL 1 MECH-SURFACE (%)	DL 2 MECH-U/GROUND (%)	DL 4 ELECTRICAL (%)	DL 5 DERAIL (%)	DL 6 SURVEY-GEOLOGY (%)	DL 7 % RAKE B2 (B1%)	DL 9 SERVICES (%)	DL 10 INDUST DISPUTE (%)	DL 11 MAINT (%)	DL 3 CLEAN MUCK (%)	CL 8 GROUND (%)	UNACCOUNTED (%)	COMMENTS
53 6 Oct - 10 Oct 86 9300.7 - 9440.4 15 R.B/Sets 1.2m/0.6m	139.7	9.31 3.9	29.7 2 38hr	34.3		20.3	1.4	6.5	-	12.1 (41)	6.2	6.3		3.2	9.7 RB12		DL2 TUNNEL TRAIN SEPARATED FROM TBM 3 TIMES DL5 6 No DERAILMENTS
54 13 Oct - 17 Oct 86 9440.4 - 9504.1 15 Sets 1.2m/0.6m	63.7	4.25 3.4	15.5 1 24hr	18.4		19.7	6.5	6.2	-	2.8 (18)	3.1	6.3		0.6	34.8	16	DL2 WINDER PROBLEMS BLOCK HOIST REPAIR SHEAR PINS, SHUNTING LOCO. PUMPS DL8 PACKING/JACKING RINGS. GROUND SUPPORT 1 SHIFT
55 20 Oct - 24 Oct 86 9504.1 - 9658.5 15 R.B./Sets 1.2m	154.4	10.29 4.1	31.3 2 5hr	36.9		8.6	6.9	3.8	0.1	16.3 (52)	10.5	6.5		1.2	6.8 RB23	24	
56 27 Oct - 31 Oct 86 9658.5 - 9745.3 15 Sets 1.2m/0.6m	86.8	5.79 3.6	19.9 1 59hr	23.5		27.1	3.6	3.9	0.1	8.1 (40)	6.8	7.4		7.1	10.6	18	DL2 TBM HYDRAULICS STEERING CLUTCH No 1 MOTOR. LEAD UP RAMPS
57 5 Nov - 7 Nov 86 9745.3 - 9820.8 9 Sets 1.2m	75.5	8.39 4.8	22.0 1 76hr	25.6		11.2	3.3	17.5	-	16.5	8.0	8.0		4.7	2.8	24	DL5 11No DERAILMENTS
58 10 Nov - 14 Nov 86 9820.8 - 9997.9 15	177.1	11.81 5.0	29.6 2 37hr	35.1		8.6	5.4	4.1	0.1	19.6 (66)	8.4	6.5		4.3	7.5	04	
59 17 Nov - 21 Nov 86 9997.9 - 10079.5 15	81.6	5.44 2.1	32.8 2 62hr	38.3		6.9	2.6	0.6	-	13.2 (40)	7.4	10.7		4.9	13.3	21	DL8 PACK & ERECT SETS
60 25 Nov - 26 Nov 86 10079.5 - 10107.8 5	28.3	5.66	37.0	41.0		26.1	1.4	-	-	13.0 (35)	3.9	7.8		4.5	1.1	12	DL2 DIXON CONVEYOR

Figure 11   Sample Summary of Tunneling Performance and Delays

English Channel Crossing Case History

Tunneling machine experience for the English Channel Crossing project illustrates the importance of selection and realistic evaluation of the tunneling system. Because of the similarity of ground conditions and the impact of tunnel advance rates on project schedules and cost, the experience of this project is relevant to our approach to the SSC project. The 31-mile underground link comprises two rail tunnels and a central service tunnel to be driven in the chalk/marl beds beneath the English Channel between the rail terminals at Folkstone and Calais (Figure 12). Tunnel construction represents a large percentage of the channel crossing project, and project schedule and cost are very dependent on tunneling performance. Seaward tunnel drives are just starting now.

The French team plans to use five TBMs, three seaward and two landward. The first TBM, designed by Robbins, is a fully shielded machine 18 feet 4 inches in diameter. The similarities to SSC tunnel requirements are: (a) tunneling in chalk/marl, (b) high-volume material handling requirements associated with projected high tunnel advance rates in soft chalk rock, and (c) the fact that the TBMs combine rock boring features and ground support features just like the shielded TBMs recommended for Taylor marl on the SSC project. An important difference is the fact that the machine at the English Channel has to be sealed against groundwater pressure of 10 atmospheres.

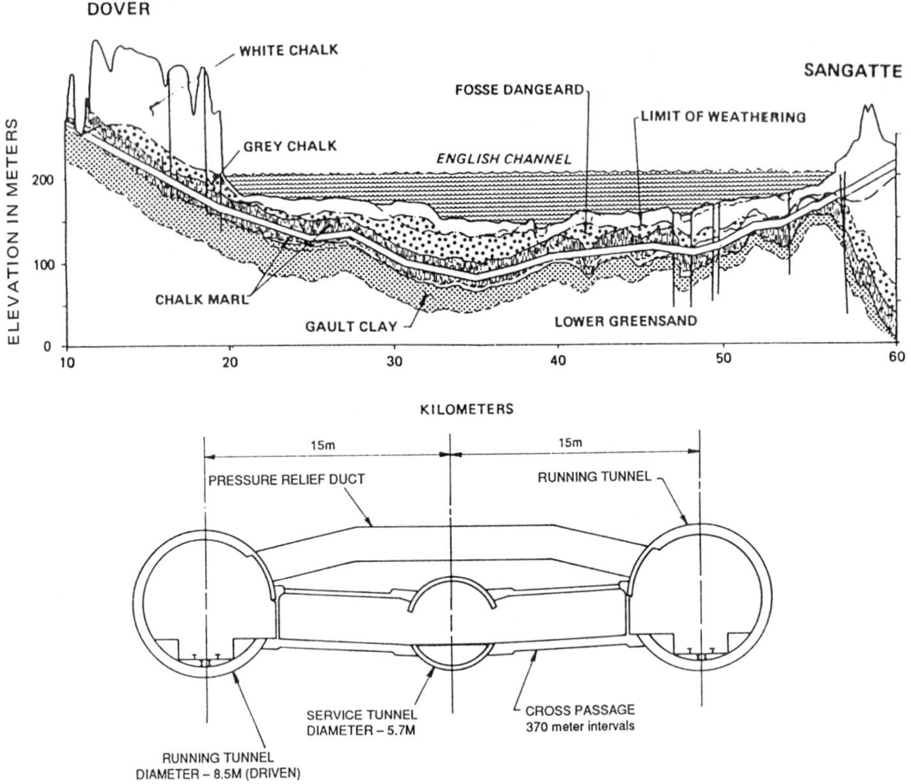

Figure 12   English Channel Crossing

Rock boring features of the TBM are identified in Figure 13 as follows:

1. A cutterhead is equipped with drag pick cutters. Disc cutters can be installed in place of the drag picks in the same cutter housing.

2. Grill bars on the face and at the periphery of the cutterhead prevent ingestion of caving rock blocks.

3. Grippers provide thrust reaction during the boring so segments can be placed while the shield jacks are retracted.

4. An extensible telescoping cutterhead provides for boring advance during segment placing.

Ground support features (Figure 13) of the TBM include:

A. A complete full circle shield.

B. An articulation joint, which facilitates steering, sealed to withstand 10 bar hydrostatic pressure.

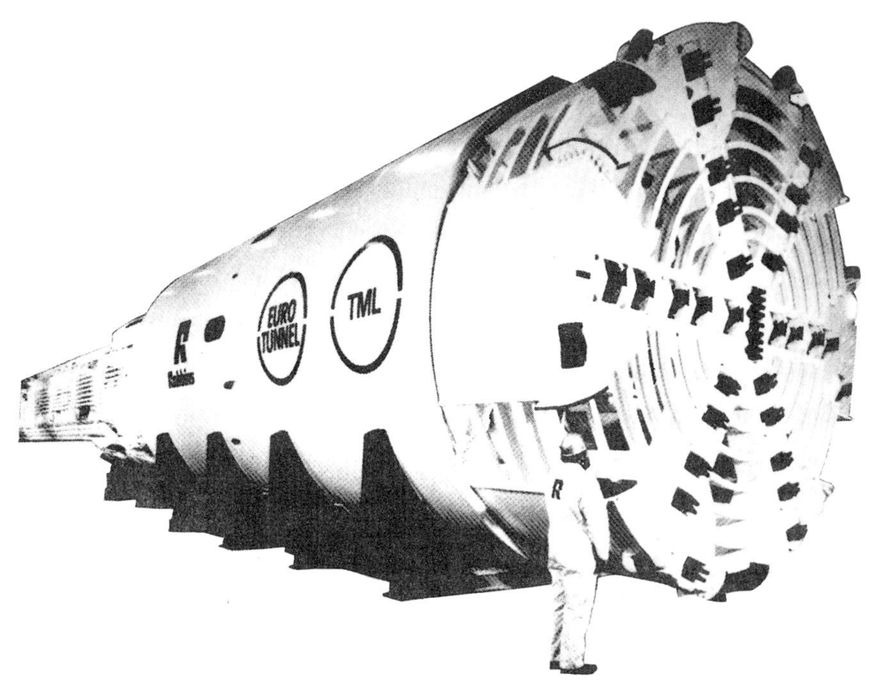
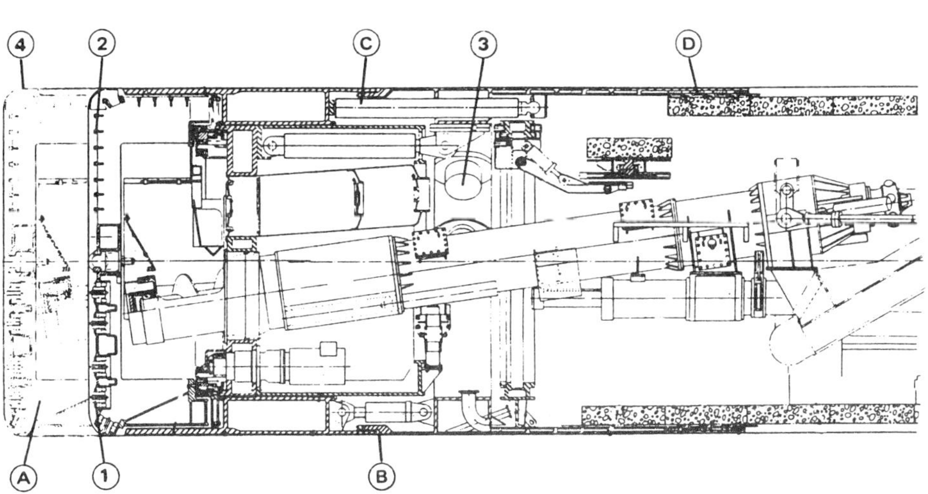

Figure 13   Shield Machine for Soft Rock and High Water Pressure

C. High-thrust shield jacks bearing against the erected ring
    segment, which provide the thrust to overcome shield friction
    in squeezing conditions, and counteract the 10 bar water
    pressure against the front of the shield when boring in the
    closed mode.

D. Continuous extruded backfill grout pumps, which fill the cavity
    between the segments and the rock as it is exposed by the
    forward progress of the shield.

This TBM started the seaward service tunnel in March 1988. Another
TBM designed by the British firms James Howden/Decon started tunnel
driving on the British seaward service tunnel in the spring of 1988.

Currently, channel tunnel contractors and engineers have to
demonstrate that they can achieve and sustain the required tunneling
rates to meet the project completion dates. Initial shakedown
difficulties are being experienced on both sides of the channel, and
all project participants are working to resolve such difficulties.
Feedback from the experience on this project will provide valuable
input to SSC design studies.

CONCLUSION

The planning and efficient execution of tunneling for the 53 miles of
collider tunnels will be one of the major challenges of the SSC
project. Rock conditions are ideal for use of modern tunneling
machines. The tunneling rates predicted by the TNRLC report can be
achieved through careful design and planning of the tunneling
operation, with the selection of suitable rock cutters, cutterheads,
tunneling systems, and backup.

ACKNOWLEDGMENTS

The help of the Robbins Company, in particular Messrs. O. Askilsrud
and E. R. Kennedy, in providing information and photos concerning
Robbins Company rock TBMs is deeply appreciated.

REFERENCES

1. Dallas-Fort Worth Texas National Research Laboratory Commission,
    Superconducting Super Collider.
2. SSC Central Design Group, SSC Conceptual Design, Conventional
    Facilities (March 1986).
3. U.S. Department of Energy, Draft Environmental Impact Statement,
    Superconducting Super Collider (August 1988).
4. R. C. Robbins, "Economic Factors in Tunnel Boring," Paper delivered
    at Tuncon 70, Johannesburg, S.A.
5. J. G. Thon, "Tunnel Boring Machines," Section in: Tunnel
    Engineering Handbook, Van Nostrand Reinhold Company, p. 236 ff.
    (1982).
6. E. Peterson and P. Frobenius, "Soft Ground Tunneling Technology on
    the BART Project," Civil Engineering (October 1971).
7. P. W. Hunter, "Excavation of a Major Tunnel by Double Shielded TBM
    Through Mixed Ground of Basalt and Clayey Soils," RETC Proceedings,
    Volume 1 (1987).
8. R. J. Robbins, "Tunnel Boring Machines for High Water Inflows,"
    Proceedings International Tunneling Association Congress, Madrid,
    Spain (1988).

CONSTRUCTION TECHNIQUES USED AT CERN FOR THE LEP PROJECT

C. Laughton

SSC Central Design Group[*]
c/o Lawrence Berkeley Laboratory
Berkeley, CA 94720

INTRODUCTION

The Large Electron Positron (LEP) machine was selected by the European Committee for Future Accelerators (ECFA) in the 1970's as the machine best suited to achieving the next step forward in high-energy physics research. LEP is a natural forerunner to the Superconducting Super Collider, having a similar layout and operating strategy.

The LEP accelerator is located adjacent to existing European Laboratory for Particle Physics (CERN) facilities in the Leman Basin astride the French-Swiss border near Geneva. The site allows CERN laboratory accelerators to be used for the pre-acceleration and injection of the electrons and positrons into the LEP ring, with the construction of the main tunnel and experimental halls in competent rock structures, avoiding overlying moraine or till deposits and deep excavation under the water table.

The LEP machine evolved considerably from conception to construction as cost and risk factors were taken into account in fixing the final site dimensions and orientation.[1] A balance was struck between construction costs and operational power consumption, which is reduced with an increase in the ring's radius of curvature.

This paper describes the siting and execution of the LEP underground works.

DESCRIPTION OF WORKS

Machine Layout

The LEP tunnel and experimental halls are sited underground in the bedrock of the Leman plain and in the foothills of the Jura mountains.[2] The machine ring is housed in a 26.6-km quasi-circular tunnel around which are sited eight potential interaction points (Fig. 1). At four of these points are constructed experimental zones, at Points 2, 4, 6 and 8, each incorporating machine services, personnel access and a detector cavern. At Point 1 is sited a machine hall, around which are placed inclined injection tunnels to allow particle transfer down from the Super Proton Synchrotron (SPS) system. At Points 2 and 6 additional galleries running parallel to the main tunnel are constructed to house klystron and power supply equipment.

---

[*]Operated by the Universities Research Association, Inc., for the U.S. Department of Energy.

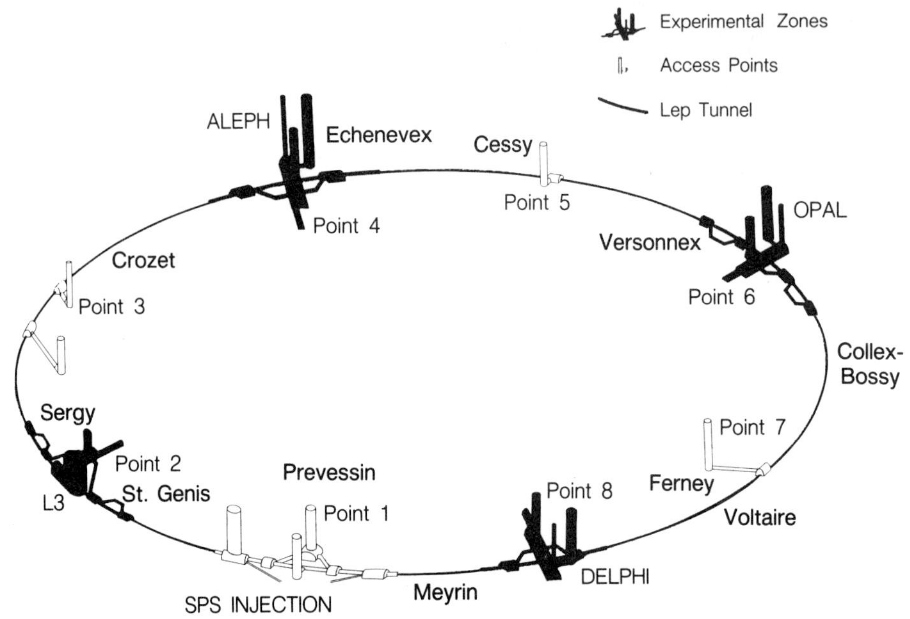

Fig. 1.   The LEP accelerator underground works.

Most of the auxilliary machine equipment (cooling towers, ventilation plant, electrical substations etc.) has been located on the surface, either above or adjacent to the vertical access shaft collars.

The LEP pre-injector has been integrated into the CERN accelerator complex and is situated on the existing laboratory site (Fig. 2).

<u>Geology Outline</u>

The tunnel lies in two very distinct geological structures (Fig. 3). In the Jura Cretaceous limestone and limestone-marl strata are encountered; these are underlain by a series of poor quality Triassic rocks. The bedding of this anticlinal formation is relatively uniform but faulting, associated with the folding movements, is encountered, particularly in the lower portions of the fold. Karstically enlarged zones exist within the limestone bedding and were intersected during the site investigation phase, at shallow depth.

In the Leman basin Quarternary moraine deposits overlie the Tertiary bedrock, in which the majority of the LEP structures are housed.[3] The bedrock, referred to as the molasse, consists of marl and sandstone subhorizontal bedding, varying in thickness from 0.1 to 2.0 m. The rock matrix is relatively indurated and unsaturated and is basically a "good excavation medium" despite the occasional presence of weaker marl strata liable to alteration if not protected from air and water.

The glacial moraine deposits were excavated during the shaft sinks; these deposits are neither consistent with depth nor horizon. Gravel and clay fractions are present, and both superficial and deep water tables exist around the site.

<u>Siting Constraints</u>

As stated in the introduction, the location of the LEP main tunnel was changed from its original alignment (Fig. 4). Firstly the circumference was reduced from 30.6 to 26.6 km while maintaining the orientation with respect to the existing SPS tunnel, to allow particle transfer. This change had the advantage of reducing the tunnel length

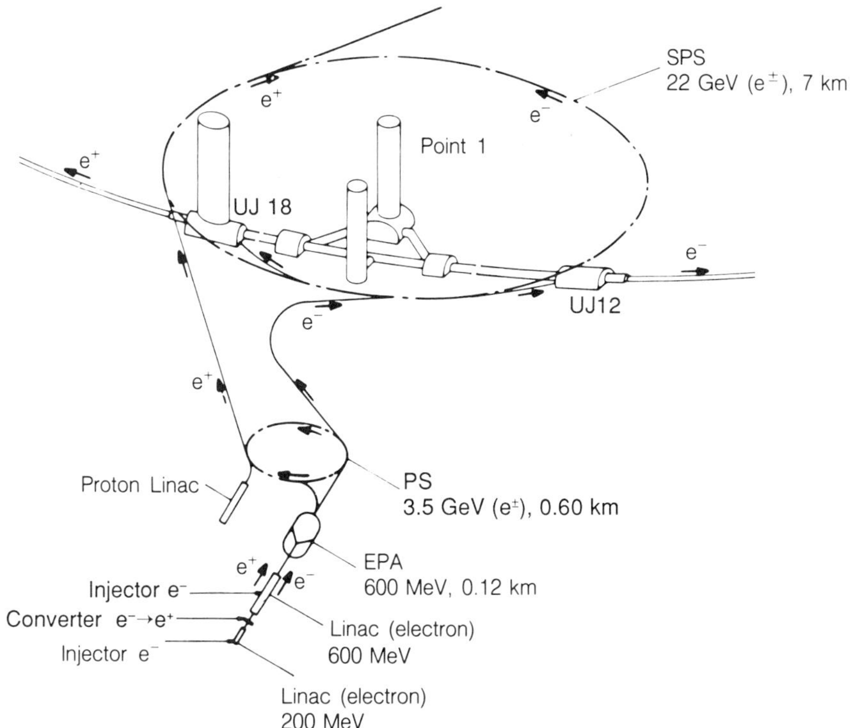

Fig. 2. Schematic representation of the LEP injector chain.

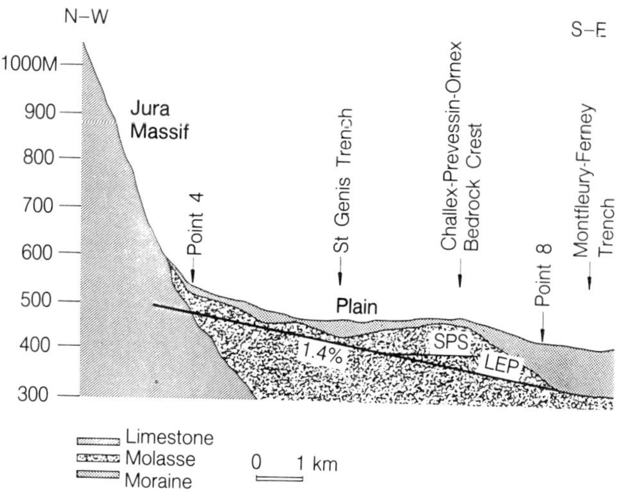

Fig. 3. Simplified geological section.

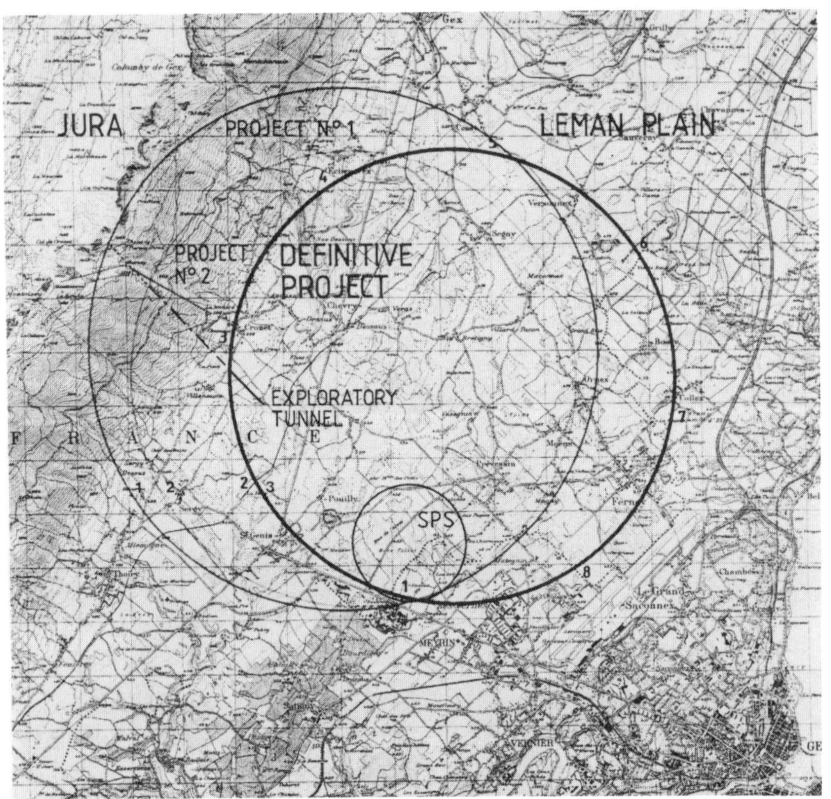

Fig. 4.   Plan showing the positions of the three LEP ring alignments.

in the formations of the Jura from 12 to 8 km, hence avoiding the anhydrites and salts of the Triassic rocks found at depth under the mountain range. Although the tunnel in this position could have been constructed without entering the Triassic rocks, long-term stability, costs and time scheduling were still in doubt in the Jura formation owing to the presence of water under pressures of up to 600 m head. This led to a further modification of the ring layout. The second move was made prior to the call for tenders, at the end of 1981, allowing the ring to be rotated away from the Jura massif. This left only 3 km in the mainly limestone strata of the Cretaceous and one major fault zone to traverse. The ring plane was also inclined at an angle of 1.43 percent, dipping to the SE away from the Jura; this ensured that the tunnel and experimental cavern excavation stayed within the plain bedrock and reduced the water pressure head in the Jura to a minimum. Liasion to the SPS ring was again maintained.

The final adjustment of the LEP machine and the addition of access tunnels at the secondary points, notably 3 and 7, also ensured that a minimal amount of interference was generated with existing surface structures. Despite a local population density of over 100 inhabitants per square kilometer, in the project region, it was possible to place the surface areas of LEP in lowly populated zones.

At all points measures were taken to limit the environmental impact. No relocation was necessary and only five houses were encountered within 100 m of the site access shafts. At four points, where housing was close by, top-soil mounds and tree planting was practiced to give building screening. At Point 4 the entire surface platform was lowered by up to 12 m with respect to ground level and the cooling tower facility moved 500 m, thus reducing the site's visual nuisance value considerably.

UNDERGROUND CONSTRUCTION

Contractual Division of Works

The underground works were carried out under four contracts (Fig. 5). Three contracts, for excavation of the underground works in the Plain bedrock, were awarded to Eurolep—a consortium of five contractors (Fougerolle [F], Astaldi [I], Phillip Holzmann [W.G.], Entrecanales y Tavova [Sp], and Rothpletz Lienhard [CH]) who used Tunnel Boring Machine (TBM) and roadheader for excavation of the sandstone-marl strata.

The fourth contract was awarded to GLLC (Locher [CH], Chantier Modernes [F], Intrafor Cofor [F], Baresel [W.G.], Wayss-Freytag [W.G.]) who undertook excavation of the main tunnel in the Jura and contact zone between the molasse and Jura formations, using conventional drill and blast techniques. A change in excavation and support techniques was considered necessary owing to the different ground conditions expected within the two rock formations.

In the Plain, CERN had already gained considerable knowledge of the in-situ ground conditions, having constructed the 8 km SPS tunnel excavated in the same rock strata. Here, once the bedrock topography had been established by borehole and refraction surveying and the rock characteristics confirmed, the underground structures were designed to allow excavation by mechanical equipment and temporary support by precast linings, passive rock bolting and reinforced shotcrete.

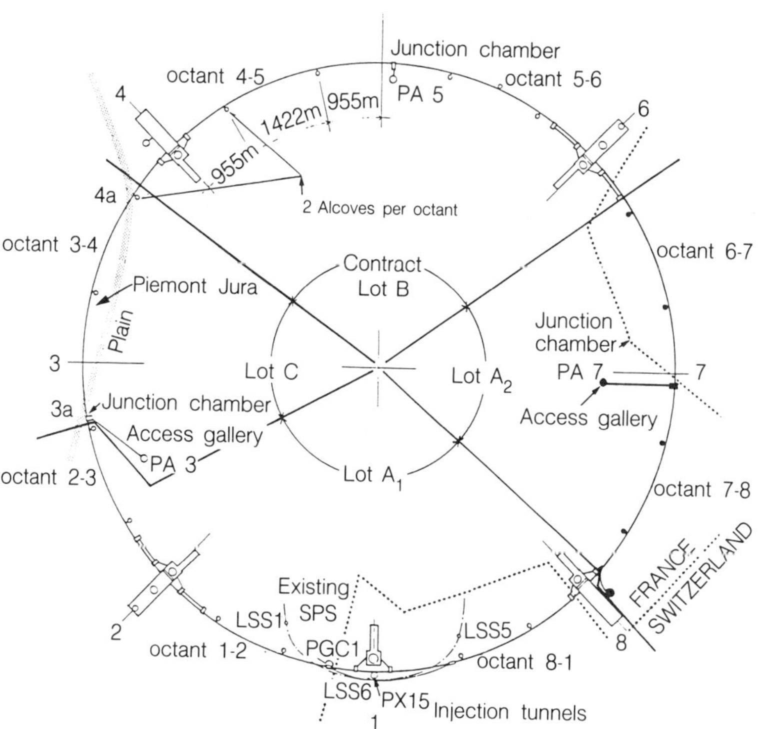

Fig. 5. Division of the underground works for contractual purposes.

In the Jura massif a more detailed site investigation campaign was undertaken, including the driving of a TBM pilot gallery through the Tertiary-Secondary unconformity. Here drill and blast was selected for the main tunnel. The site investigation showed the majority of the strata to be relatively competent but evidence of karst development was found in fractured zones and poor quality, permeable rock was encountered, particularly in the vicinity of the Allondon fault to the North of the section. The use of a TBM was excluded at the tender stage as the ground treatment operations, to be performed at depths of up to 180 m below the water table, would have been severely hampered by the presence of a full-face machine.

<u>The Main Tunnel</u>

The Plain main tunnel has a finished internal diameter of 3.8 m and an excavated diameter of 4.5 m, thus allowing placement of the temporary rock support system, drainage or waterproofing complex and final, cast-in-situ lining. To accommodate errors in survey alignment and those caused by operational deviation of the TBMs a displacement of the axis of the excavated tunnel from the theoretical position of ±8 cm was allowed while still guaranteeing a minimum internal radial clearance of 1.82 m (see Fig. 6). The inner side of the tunnel cross-section is reserved for the movement of personnel and equipment by monorail and locotractor. The outside houses the machine components and services, notably the magnet chain, power and cooling supplies.

In the molasse, three TBM's were used. Originally two prototype double shield machines were employed but a third "open" TBM was introduced to accelerate the excavation after five months of strike had put back the works scheduling. The two double-shield TBMs used a five-piece precast concrete segmental shell, inserted and contact-grouted against the rock as soon as practicable after excavation to provide temporary rock support (see Fig. 6). The open type TBM had a partial "finger" shield in the roof section and used a system of 2 swellex rock bolts (1.8-m long and variable longitudinal spacing 0.60 to 1.20 m) pinned-back roof arching and wire mesh. The bolts and arching were placed as the shield advanced and shotcrete applied 15 to 30 m behind the face (see Fig. 7).

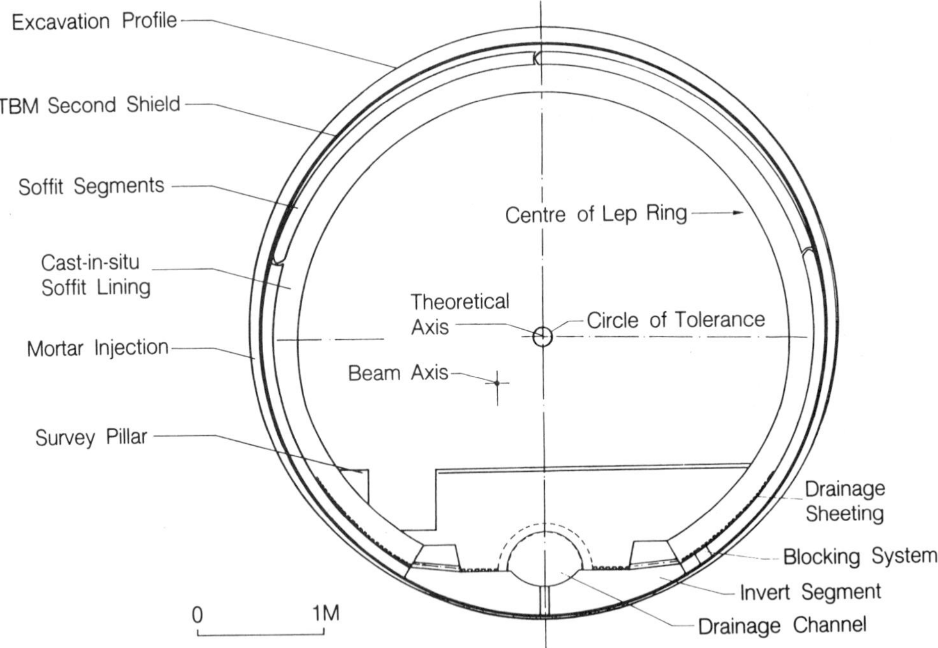

Fig. 6.   Cross-section of the main tunnel showing the drainage and secondary lining (double shield TBM temporary support).

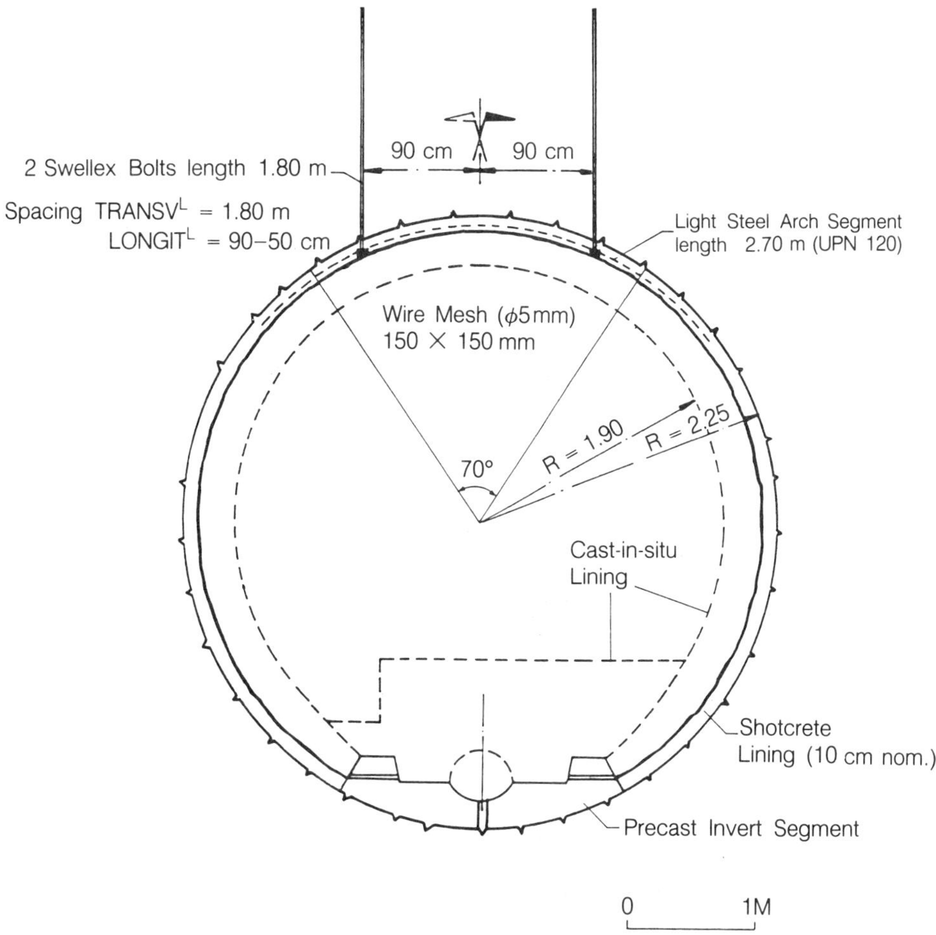

Fig. 7.   Temporary support of the main tunnel by rock bolt and shotcrete (open TBM).

For all the molasse excavations a systematic approach was adopted for rock support to ensure safe access for personnel and medium-term stability of the openings. The rock structure contains very weak marl bands, often fractured and susceptible to alteration; their occurrence was not predictable sufficiently in advance of the excavation to allow "tailor-made" support to be selected, so a continuous support system was adopted to give the necessary "worst case" protection.

All the underground works at CERN are given a secondary lining to ensure the long-term stability and water-tightness of the structures. The importance of these qualities to the success of the laboratory operation cannot be over-emphasized. CERN, which at present has a permanent staff of over 3000, is dedicated to the production and collision of sub-nucleic particles so the operational efficiency of the machine has a direct effect upon the overall "productivity" of the laboratory. Any intervention for reasons of machine or detector instability or water ingress would result in an accelerator shut-down leaving experimental runs, which nowadays may involve hundreds of physicists and permanent staff, at a standstill.

To achieve a high degree of efficiency in the production and collision of the subnucleic particles care is taken to create a laboratory environment underground. This results in the use of waterproofing membranes and heavy foundation works in detector and machine regions and the use of a sophisticated survey network. The specifications of the LEP ring and experimental zones are significantly more severe than those associated with more familiar underground works such as transport and service tunnels or hydro-electric facilities in terms of precision and required operational environment. The strict tolerances and high level of finish, which may be considered "luxury" in most tunneling situations, have ensured that the underground accelerator and detector systems at CERN have operated with only scheduled maintenance and modification shut-downs since their commissioning.

In the Plain a second, "cast-in-situ" lining was poured after the placement of waterproofing, drainage and ducting systems. In the molasse a drainage system, placed between the temporary and cast-in-situ lining, ensured that water seepage into the tunnel was minimal. A continuous soffit membrane was applied locally where sandstone strata impregnated with hydrocarbon oils were encountered, but at no time were fluids under pressure taken into account in the final lining design (see Fig. 8).

All excavation in the Jura was carried out using drill and blast techniques. For the majority of the length full rounds of 3.5 m were drilled out with ground support being provided by rock bolt and light shotcrete cover, applied upon excavation.

Towards the end of the Jura tunnel drive, in proximity to the major fault zone in the North of the contract section, a large water inflow was encountered and the excavation and support techniques were consequently modified. The water inflow, loaded with suspended sand-silt deposits under pressures of up to 15 bars, reached nearly 200 l/sec during the spring melt and stopped the tunnel drive for a total period of some five months. The inflow occurred locally some fifteen to twenty meters behind the tunnel face, initially manifesting itself as a flow of limited importance and low pressure but increasing rapidly as the enlarged fracture network fill became increasingly "flushed out" by the passage of water.

Similar conditions were envisaged for the remaining length of the tunnel so once the large inflow had been sealed-off by internal reinforcement (steel arch, lagging and concrete) shallow (<5 m) chemical resin grouting and cement consolidation at depth (<10 m), the decision was made to erect and concrete-in a cast-iron ring directly upon excavation to profile (Fig. 9). The tunnel blasts were also modified and the pull

Fig. 8.   Enlarged section of the main tunnel prior to final lining (soffit shutter in background) in a hydrocarbon rich zone.

Fig. 9.   Cast-iron tubbing in the karstic zone of the Jura formation.

reduced to ensure there was a minimal stand-up time and low vibration level. Although this modified excavation and support system reduced considerably the overall production rate it guaranteed against the recurrence of the large water inflow and enabled a re-scheduling of the waterproofing and final lining operations to be undertaken with confidence. The superposing of these latter operations has allowed the project commissioning date to be maintained for mid-1989.

To give long-term stability and above all, watertightness along the main tunnel in the Jura formations a complete waterproof sleeve and an independent secondary lining, capable of withstanding the full hydrostatic pressure head, were placed. This lining will eliminate long-term draw-down of the groundwater tables and provide a dry environment for the installation and operation of the machine.

<u>Experimental Zones</u>

The main caverns and auxiliary chambers were sited exclusively in the molasse and were almost entirely excavated by roadheaders. A D9 ripper and some explosives were used in the body of the hall but all profiles were formed using mobile head equipment and New Austrian Tunneling Method (NATM) temporary support mechanisms, shotcrete and rock bolting, installed systematically (Fig. 10).

In all cases the horseshoe caverns were developed in passes of 4 to 5-m height from the crown downwards, access being given by axially aligned shafts (Fig. 11).

The contractor used 3 and 6-meter passive rock bolts fully bonded by cement mortar and reinforced shotcrete 10 to 25 cm thick to provide support to the rock opening for up to 2 years prior to the placement of the final cast-in-situ lining. The majority of the rock strata sandstones and sandstone-marl showed little sign of overstressing and failure but the weak marl bands, which in some cases had uniaxial compressive strengths of below 2 MPa, failed locally, rupturing the shotcrete lining. Reinforcement of these zones, by additional rock bolting and occasionally the further application of shotcrete, was undertaken following NATM principles.

The main caverns and chambers received a final cast-in-situ lining incorporating a waterproofing and drainage complex (Fig. 12). In the main cavern steel arching was also placed in the 50-cm lining to give anchorage for the overhead crane system (Fig. 13).

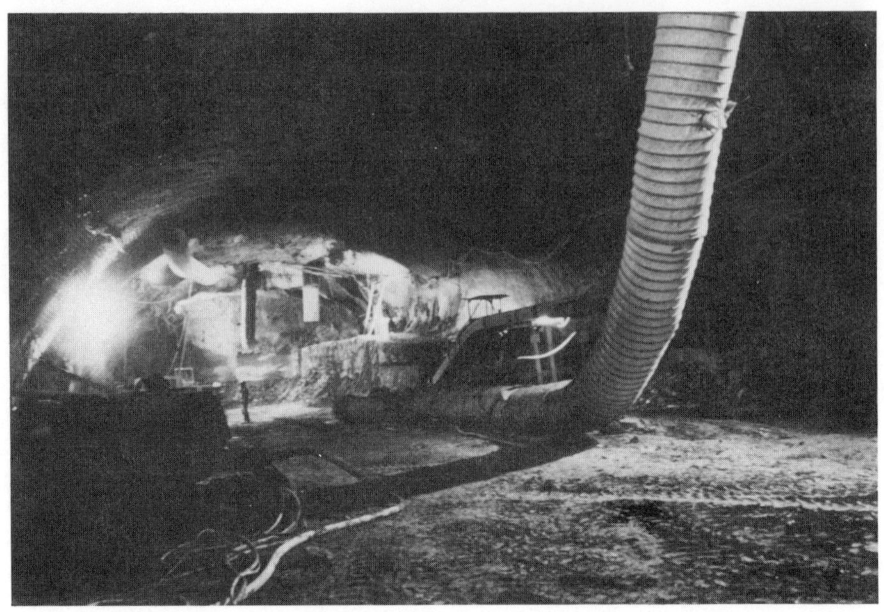

Fig. 10. Development of the cavern crown showing the West Falia Lunen Buffel WAV170 in the background.

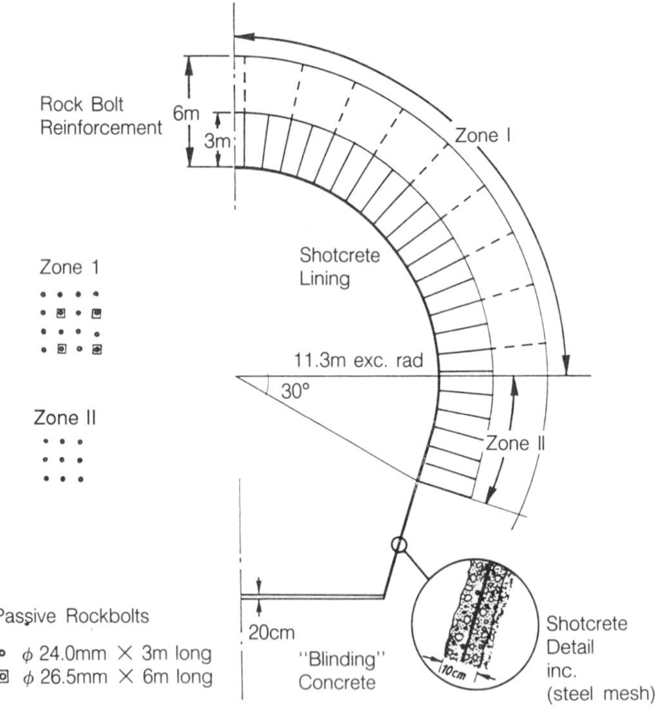

Fig. 11. Cavern temporary rock support mechanisms.

Fig. 12. The Point 2 experimental cavern (UX25) lining showing the soffit shutter and waterproof membrane.

Fig. 13. The fully lined UX25 cavern with the detector cradle in place.

An "umbrella type" waterproofing and drainage sheet system was placed in the roof of the caverns to ensure the absence of water in the detector zone; this was particularly important owing to infiltration of water from the overburden water tables via the shafts. Once in the footwalls only drainage sheeting was used to guide the water to the floor drains and sump.

<u>Access Shafts</u>

Eighteen permanent shafts connect the tunnel and experimental zones with the surface; they were excavated by backhoe and the use of explosives in the bedrocks. Diaphragm walling and/or freezing was used to support the ground during sinking in the saturated moraines and protect the natural groundwater tables (Fig. 14). Stage excavation, with the placement of cast-in-situ concrete rings was carried out in dry overburden. In the bedrocks reinforced shotcrete provided adequate support except in fractured zones where spot rockbolting was installed.

Three types of shafts have been constructed to give electrical and fluid supply to the main tunnel and permit the transportation of personnel and equipment to and from the experimental zones. The permanent shafts have a final slipform lining varying in thickness from 40 cm for the 5-m shafts up to 100 cm for the 23-m diameter shaft (Fig. 15). All the linings have a waterproof membrane backing and incorporate modular insets used to fix the pre-assembled shaft infrastructure units.

Eight machine shafts (7 and 9 m finished diameter) are spaced around the ring and give i) a free cross-section area for vertical lifts of components and materials (9-m shafts only), ii) an area for the electrical and fluid ducting installations, iii) a fire resistant zone in which are housed the lift and stairwells. A further machine shaft, of 14-m diameter, has been used to lower the main tunnel magnets to machine level.

Fig. 14. Excavation in the frozen water-bearing moraines with the ring shutter anchored above the working level.

Fig. 15. Slipform shutter advancing up the large experimental shaft at Point 2.

At points 4, 6 and 8 two shafts of 5 and 10-m diameter provide fire resistant personnel access and a free passage for detector components, respectively. At Point 2 a 23-m diameter shaft was excavated to allow handling and installation of particularly large and heavy detector equipment.

CLOSING REMARKS

The bulk of the civil engineering work was completed early in 1988 with the exception of the Jura tunnel, where a considerable delay had accumulated owing to problems encountered during the tunnel drive. In general the in-situ ground conditions have conformed to those expected from the site investigation studies, undertaken prior to final design of the underground structures. The excavation and support techniques were well adapted to the ground conditions and very little on-site intervention was required to modify support systems during excavation. By far the most important factor affecting ground stablility in the molasse proved to be the rapid protection of the excavated surface to ensure that minimal alteration was initiated in the marl strata by prolonged (>24 hrs) exposure to air.

In the Jura the rapid application of a comparatively heavy temporary support system became necessary in the karstically affected limestone strata. This technique guaranteed the medium-term stability of the rock structure and groundwater regimes until placement of the final lining. Once personnel became familiar with the new driving work, tunnel lining activities were undertaken in parallel to allow machine installation to proceed on schedule (Fig. 16).

Fig. 16. Monorail transport used for installation works in the main tunnel.

REFERENCES

1. H. Laporte, Le LEP, Grand Projet souterrain. Tunnels et ouvrages souterrains. No. 63—May–June, 1984, pp. 115–122.

2. C. Laughton, CERN Accelerator Project. World Tunnelling and Subsurface Excavation. Vol. 1, No. 2, June 1988, pp. 114–125.

3. G. Amberger, La molasse du bassin genevois: La recherche scientifique dans les hautes écoles et universités suisses, 1986, pp. 5–8.

CRITICAL ISSUES CONCERNING TUNNEL VENTILATION

IN THE SUPER CONDUCTING SUPER COLLIDER RING TUNNEL

Philip E. Egilsrud

Sverdrup Corporation
801 North Eleventh Street
St. Louis, MO 63101

ABSTRACT

In this paper, the author discusses the problem of designing a ventilation system for the SSC Collider Ring Tunnel. The history of ventilation of tunnels is briefly reviewed and the types of systems used for ventilation of rail, transit and highway tunnels are described in detail.

Next, the ventilation system prepared for the Collider Ring Tunnel in the RTK conceptual conventional design report is described.

The author questions this design from a safety and practical standpoint. He offers several other solutions which provided a more powerful system for emergencies. One supplies fresh outside air uniformly along the tunnel length. All suggested systems increase costs over the RTK proposed system. The possible problem of troublesome condensation forming on equipment in the tunnel and tunnel walls in hot, humid weather is also a major concern.

INTRODUCTION

It is not my intention in this report to solve the problem of ventilation of the 53-mile Collider Ring Tunnel or to make any specific recommendation regarding a system design.

Rather, in this document, I will apply my background and experiences in designing many ventilation systems for different types of tunnels to this problem of tunnel ventilation for the Collider Ring Tunnel.

This paper will be organized as follows:

I will describe some of these systems by type and outline their design, advantages and disadvantages.

Next, I will review the ventilation system called for in the conceptual design report for the conventional facilities and in the SSC Safety Review document.

Last, I will express my thoughts regarding this proposed ventilation
system design and express my concerns about the proposed approach.
Other methods, as well as consideration to the design, will be discussed.

General Discussion of Ventilation Tunnels

The Super Conducting Super Collider (SSC) Ring Tunnel is a very
unique tunnel compared to a tunnel built for other purposes.

Control of the environment in various other types of tunnels has
been a concern even on the earliest of these structures. Mine tunnels
were probably the first tunnels of any significant length. Control of
the environment, particularly the air quality in mines, was an early
concern, and even today, we read of mine accidents where lack of air in
safe quantities and quality is the cause of death or impairment of mine
personnel's activities. The U.S. Bureau of Mines has currently, and in
the past years, established criteria for mine ventilation based on its
mine research and investigations of mine disasters.

Early tunnels for transportation were built to carry roads and
railroads through or under natural barriers such as hills, mountains,
rivers and other bodies of water. Several steam engine driven trains
had crews who died of asphyxiation in these early unventilated facilities. This problem mandated that electric locomotives be used in the
longer tunnels. The first long tunnels were through the Alps to provide
year round dependable transportation between the North and South of
Europe. Ventilation during construction of these bores was a critical
consideration because of high temperatures inside the mountains and the
dust and the fumes resulting from the tunneling methods. Many
tunnelers died because of poor air and silicosis due to inadequate
ventilation. Electric locomotives to pull the trains through these
tunnels became universal in order to ease the control of the environment in the tunnels. This became the answer for the Cascade Mountain
tunnels is the Western US, the Hudson, East River Tunnels in New York
City and also the New York Central Trains coming into Grand Station in
New York. This solution was very satisfactory and still serves in many
areas. Later, the use of diesel locomotives made modifications
necessary to ventilation systems of many long railroad tunnels, mainly
to provide cooling air for the radiators of multiple engines coupled
together at the head end or middle of long freight trains.

In rail and transit tunnels, the piston effect of the moving trains
moves enormous quantities of air in and out of the tunnels through
ventilation shafts, provided for that purpose. Reversible fans and
dampers in these shafts provide fire emergency capability to move air
through the tunnels when the train is stopped and standing in that
section of the system. Air quantities and pressures are much larger,
in this case, than suggested for the SSC conventional design of the
Collider Ring Tunnel.

The first ventilated highway tunnel was the Holland Tunnel under the
Hudson River in New York. The feasibility of providing a safe environment for motorists using this proposed tunnel was controversial from the
very beginning. Common sense, backed by extensive research on carbon
monoxide (CO) emitted by motor vehicles of that day, plus careful
measurement of the tolerance of humans to CO emissions, as well as tests
concerning the size and length and airflow in ducts, saved the day, for
a successful solution was found. Many other highway tunnels have been
built since the Holland tunnel, most with similar ventilation systems.
Anti-air pollution devices, now common on most US passenger cars, has

126

greatly reduced the ventilation levels required to keep the CO and other vehicle emitted contaminants at a tolerable level. As a result, ventilation for fire and other emergencies has become the chief concern for establishing maximum ventilation capability for highway tunnels.

Other long tunnels, such as for carrying utilities, (i.e., water, sewerage, electrical and communication cables), require special considerations for environmental control. There must be a safe environment when personnel enter these tunnels for maintenance.

Removal of internally generated heat is often a reason for ventilation of many tunnels (see Fig. 1).

<u>The Following is a Description of Methods Employed to Ventilate Highway Tunnels by Mechanical Ventilation</u>

The most appropriate mechanical ventilation systems for tunnels are longitudinal ventilation, semi-transverse ventilation, and full transverse ventilation.

Longitudinal ventilation is defined as any system, where the air is introduced to or removed from the tunnel roadway, at a limited number of points, thus creating a longitudinal flow of air within the roadway. The injection-type longitudinal system has been frequently used in rail tunnels; however, it has also found application in vehicular tunnels. Air is injected into the tunnel roadway at one end of the tunnel where it mixes with the air brought in by piston effect of the incoming traffic (see Fig. 2a).

This system is most effective where traffic is unidirectional. The air velocity within the roadway is uniform throughout the tunnel, and the concentration of contaminants increases from zero at the entering

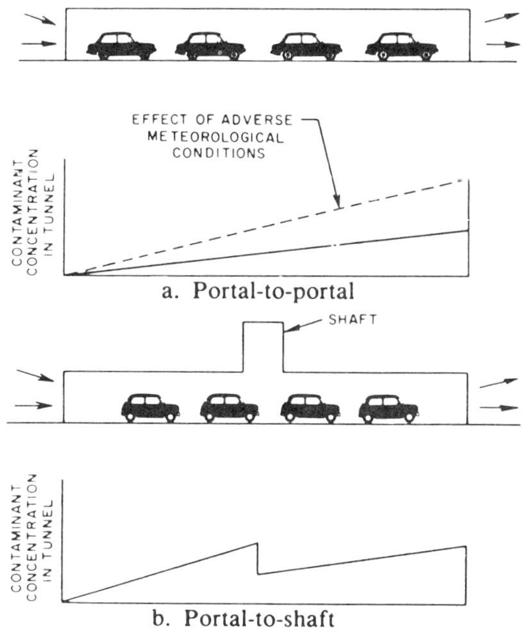

Fig. 1. Natural Ventilation vs Contaminant Concentration in Tunnel.

portal to a maximum at the exiting portal. Adverse external atmospheric
conditions can reduce the effectiveness of this system. The contaminant
level increases at the exit portal as the air flow decreases or the
tunnel length increases.

The longitudinal system with a shaft (see Fig. 2b) is similar to the
naturally ventilated system with a shaft, except that it provides a
positive stack effect. Bi-directional traffic in a tunnel ventilated in
this matter will show a peak contaminant concentration at the shaft
location. This system is applicable only for bi-directional tunnels;
for unidirectional tunnels, the contaminant levels become unbalanced.

The longitudinal system has two shafts located near the center of
the tunnel, one for exhausting and one for supplying (see Fig. 2c). It
will provide a reduction of contaminant concentration in the second half
of the tunnel. A portion of the air flowing in the roadway is replaced
in the interaction taking place at the shafts. Adverse wind conditions
can cause a reduction of flow and a subsequent rise in contaminant
concentration in the second half of the tunnel, along with short
circuiting of the fan flow.

In a growing number of tunnels, a longitudinal ventilation system is
achieved with booster fans mounted at the tunnel ceiling (see Fig. 2d).
Such a system eliminated the need for space to house ventilation fans in
the building; however, it may require a tunnel of greater height or
width to fit the booster fans.

The standard longitudinal ventilation systems (excluding the
booster fan system), with either supply or exhaust at a limited number
of locations within the tunnel, are the most economical systems because
they require the least number of fans, place the least operating burden

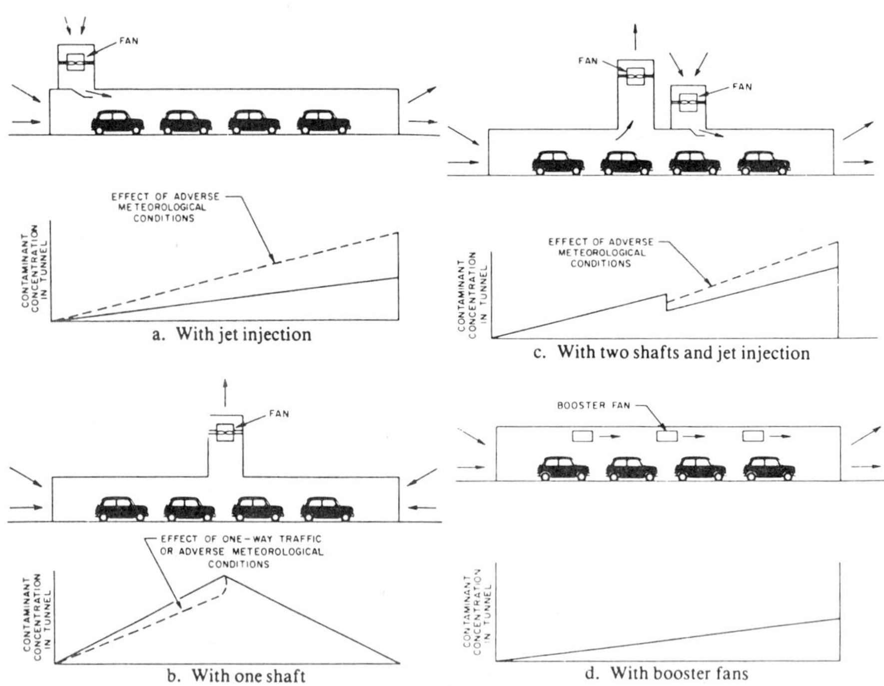

Fig. 2. Longitudinal Ventilation.

on these fans, and do not require separate air ducts. As the length of the tunnel increases, however, the disadvantages of these systems become apparent, such as excessive air velocities in the roadway and smoke being drawn the entire length of the roadway during an emergency. To alleviate these problems, consider uniform air distribution.

Semi-transverse Ventilation

Uniform distribution or collection of air through the length of a tunnel is this system's chief characteristic. The supply air version of the system (see Fig. 3a) produces a uniform level of carbon monoxide throughout the tunnel because the air and the vehicle exhaust gases enter the roadway area at the same rate. In a tunnel with unidirectional traffic, an additional air flow will be created within the roadway area.

This system, because of the fan-induced flow, will not be adversely affected by atmospheric conditions. The air will be transported the length of the tunnel in a duct fitted with periodic supply outlets. The best location for the introduction of air to the roadway is at exhaust pipe level of the vehicles to permit immediate dilution of the exhaust gases. To accomplish the air distribution, an adequate pressure differential must be generated between the duct and the roadway to counteract piston effect and atmospheric winds.

During a fire within the tunnel, the air supplied provides dilution of the smoke. To aid in fire fighting efforts and in emergency egress, the fresh air should enter the tunnel through the portals to create a respirable environment for these activities. Therefore, the fans in a supply semi-transverse system should be reversible and ceiling supply should be considered. With a ceiling supply system and reversible fans, the smoke will be drawn upward.

The exhaust semi-transverse system (see Fig. 3b) in a unidirectional tunnel produces a maximum contaminant concentration at the exiting portal. In a bi-directional tunnel, a zone of zero fresh air is created near the center of the tunnel, producing the maximum level of contaminants. A combination supply and exhaust system (see Fig. 3c) is applicable only in a unidirectional tunnel where the air entering the traffic stream is exhausted in the first half, and air is supplied in the second half to be exhausted through the exit portal.

The accepted semi-transverse system is the supply type, which is the only one not affected by adverse meteorological conditions or opposing traffic. Semi-transverse systems are used in tunnels up to about 1,000 m (3,280 ft) at which point the tunnel air velocities near the portals become excessive.

Full transverse ventilation is used in longer tunnels. A full exhaust duct is added to a supply type semi-transverse system which achieves a uniform distribution of supply air and a uniform collection of vitiated air (see Fig. 4). This system was developed for the Holland Tunnel in 1927. With the arrangement a uniform pressure will occur throughout the roadway and no longitudinal air flow will occur except that generated by traffic piston effect, which tends to reduce contaminant levels. An adequate pressure differential between the ducts and the roadway is required to assure proper air distribution under all ventilation conditions.

Enclosed Vehicular Facilities

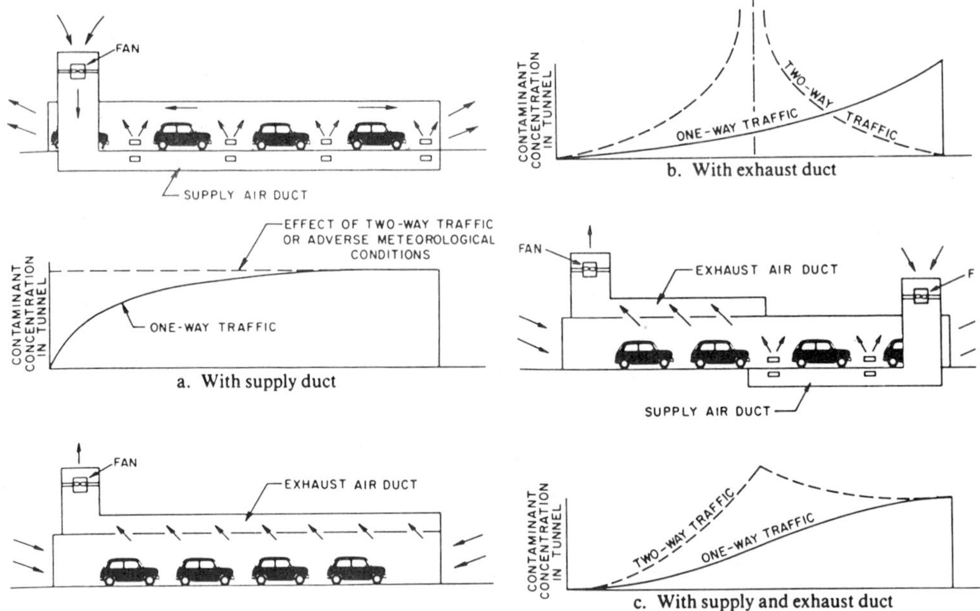

Fig. 3.   Semitransverse Ventilation.

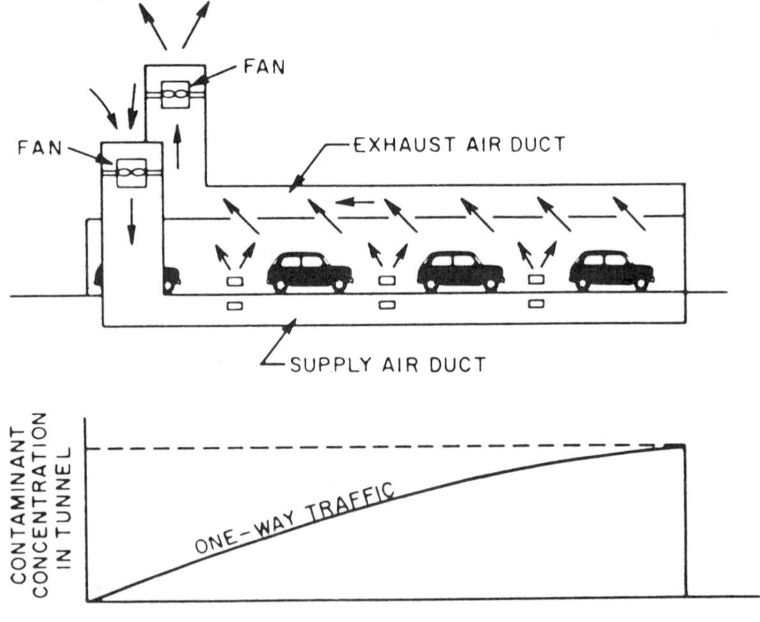

Fig. 4.   Full Transverse Ventilation.

130

The desirable location of the supply air inlets for rapid dilution of exhaust gases is at the level of the vehicle emission with the exhaust outlets in the ceiling. This location was established by full scale tests conducted by the U.S. Bureau of Mines. The air distribution can be one-sided or two-sided.

Others

There are many variations and combinations of the systems described. A combined system for a unidirectional tunnel approximately 425 m (1,394 ft) long is shown in Fig. 5. Section 3 utilizes a full transverse system because of the upgrade roadway; Section 2 utilizes a semi-transverse supply with a longitudinal exhaust, and the remainder of the tunnel (Section 1) is a semi-transverse supply system. Such a system is not recommended for a long tunnel.

Emergency Conditions

The existence of an emergency condition in any vehicular tunnel, particularly one generating smoke and heat, such as during a fire, can lead to a disaster if proper steps are not taken in the design and operation of the tunnel ventilation system. The primary objective in such a situation is the rapid removal of smoke and heat from the tunnel to permit the safe departure of the vehicle occupant and the entry of fire-fighting personnel to the scene of the fire. This can be accomplished most effectively by removing the smoke and heat at a high level within the tunnel and supplying fresh air at a low level either through a designated supply system or through the tunnel portals. A typical full transverse ventilation system will provide the ideal emergency ventilation system, i.e., a high level (ceiling) exhaust and a low level positive supply (roadway). A semi-transverse ventilation system which is configured to utilize the overhead system to exhaust thus causing fresh air to enter the roadway area through the portals will also present an effective emergency ventilation system. Prior to the design of a tunnel, the criteria for emergency conditions must be established including the minimum level of ventilation required to create the environment described previously.

Conventional Facilities Ventilation Design

The conventional facilities design of ventilation for the Collider Ring 53-mile tunnel calls for ventilation systems in which axial fans are installed in each service area and each exit area shaft. A fan at each exit shaft, supplies air to the tunnel and a fan at each service

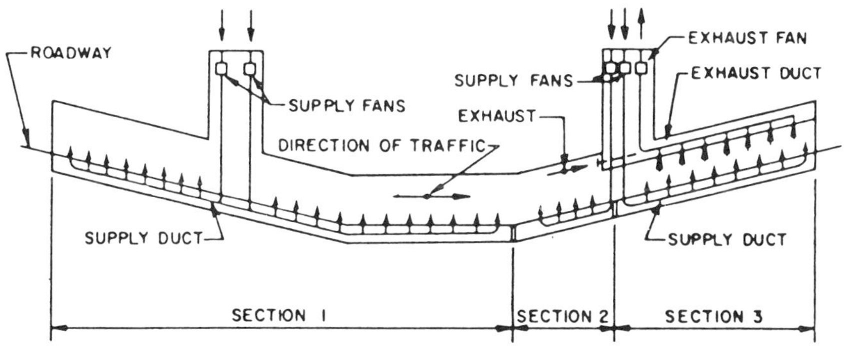

Fig. 5. Combined Ventilation System.

area, exhausts air from the tunnels. Each fan will have a capacity of 10,000 cubic feet pair per minute at 2 inches of water static pressure for exhaust and 3 inches for supply. The assumption is that equal flow will be achieved because of the symmetrical arrangement of the tunnel between each shaft. The air balance between each segment is apparently not controlled or assured in any manner. The assumed 5,000 cfm travelling in each direction from a supply fan in an exit shaft causes an air velocity in the tunnel cross section of approximately 88 feet per minute. This means with 2-1/2 miles between supply and exhaust fans, it would take fresh air approximately 2-1/2 hours to travel from the supply to the exhaust point. These velocities and fan pressures are very low and could be diverted, or reversed by small pressure differences. In case of fire, the system would be very weak and slow in pushing in oxygen laden air, and removing the products of combustion from the tunnel. Also, the tunnel being the duct in this case, would have heat and/or smoke in a large portion of its length for a long time, depending, of course, where the fire is located relative to the service areas and exit shafts.

The cycle of operation calls for 14 days of round-the-clock operation for experiments, during which time, the tunnel has no personnel present, nor is any ventilation provided.

During this operating period, the facility shuts down operation for two days while repairs, maintenance, and cleaning takes place. Personnel will enter the 54-mile ring at various service areas or exit shafts to work on whatever tasks are necessary. Transportation is available on electric powered vehicles called transporters.

Fire Mode Ventilation

A fire in any tunnel is a serious situation. This has been the subject of many research projects and investigations in the past. Fire protection experts will tell you ventilation is the key to controlling a fire. Using ventilation by whatever means, clears out heat and smoke enabling fire fighters to do their job. In a tunnel, this problem relates to the length of the tunnel, with the general rule, that the longer the tunnel the more serious the problem.

For highway tunnels, ASHRAE has a standard of 100 cfm per lane foot of ventilation air for a fire emergency mode. On the basis of an envelope, 13-ft wide by 18-ft high and 1-ft deep, this provides 22 air changes per hour. The conventional SSC design provided for a 0.4 of an air change per hour.

As discussed previously in this paper, highway tunnels having a length more than 2,000 feet or more, are equipped with one or two parallel ducts, depending upon whether semi or fully transverse ventilation is provided. In each case, fresh outside air is supplied, under the normal mode, uniformly through the tunnel's length. During a fire, ventilation rates are increased, sometimes significantly to remove heat and provide access and safety. Only shorter highway tunnels have no parallel duct, therefore, the roadway cell or bore is used to move the air in and out of the tunnel and, in most cases, some arrangement using fans is provided to exhaust heat and smoke during a fire.

The danger of a fire in the Collider Ring Tunnel of the SSC is more remote than either rail, transit or highway tunnels. However, I have read that during tests, the magnets have burned. There are dozens of these magnets in any 2-1/2-mile segment of the Collider tunnel. The air changes with 5,000 cfm for the 2-1/2 miles between shafts provides 0.4 air changes per hour, while the ASHRAE and other standards require 22 or more. As can be seen, the proposed system falls very short of this ASHRAE standard. Of course, these standards relate to tunnels where larger vehicles carrying fuel, burnable, and sometimes hazardous cargoes, could produce a large BTU output.

It seems that the present design, offers little protection in an emergency. It is true that few personnel are ever in the tunnel, and these personnel could be well trained, and carry breathing apparatus and other protection on their transporter, but a stronger ventilation capability for emergencies is worth considering.

Air Distribution - Balance

A 10,000 cfm supply of air will be provided at each exit shaft. Apparently, the air flow is expected to divide equally at the Collider tunnel due to the symmetry of the layout, with 5,000 cfm travelling in each direction, towards each service area with a 10,000 cfm exhaust fan. This distance is 2-1/2 miles away in each direction.

Maintaining this equal balance of airflow direction is not accomplished by positive means. Therefore, restrictions to airflow in one side, which might occur when work is being done with transporters, personnel and other equipment, blocks the cross-sectional area of the tunnel at that point. This could cause increased resistance to airflow, and consequent reduction of airflow down that segment of the Collider ring where the air is needed most.

Variable and controlled artificial restrictions to airflow could be provided in each Collider tunnel branch at the exit shafts, and assure a balanced airflow between each segment.

A FEW ALTERNATIVE SCHEMES WHICH MIGHT BE CONSIDERED

Longitudinal Approach

One of the least costly enhancements of the capability of the ventilation system would be to significantly increase the capacity of the fans in the service areas and exit shafts. This could be an independent set of fans with reversible air flow capability, which would bypass the normal fans, filters and heaters to move air in either direction from one shaft to the other. This flow would be for 2-1/2 miles in a direction, chosen, based on the location of the fire or other emergency to provide the safest environment for personnel who might be in the tunnel. This could provide for a quick and safe entry for fire fighters and those who must enter for repairs, observation or other reasons.

An example would be to provide 9 air changes per hour or a fan capacity of about 112,500 cfm at each shaft. Velocity in the Collider tunnel would be 1,980 feet per minute or 22.8 miles per hour. For this scheme to work, means should be devised to shut off the tunnel in the direction of the next segments in order to concentrate the airflow in the affected segment.

Capacity control could be employed for controlling this airflow to suit conditions. Added cost over the currently proposed RTK conceptual design would be the additional fans, perhaps 2 per shaft at 51,000 cfm each, for a total of 40 plus ductwork controls, and means of closure of the Collider tunnel segments at each shaft.

This scheme would provide the ability to move air, heat and smoke in and out of a 2-1/2-mile tunnel segment much more rapidly. The exposed equipment in the tunnel would be less likely to be damaged by the heat and smoke of a fire or because of its quicker removal, and the cooling effect of the greater quantity of fresh air moving through. For this example of a ventilation rate of 112,500 cfm, providing 9 air changes per hour, 6.6 minutes would be required for air to travel the 2-1/2-mile segment between shafts. This 9 air changes per hour was an arbitrary choice in the author's part based on a velocity in the tunnel approximately of 2,000 ft per minute, which seems reasonable. Greater or lesser quantities of airflow could be considered, if this scheme is to be adopted.

A Semi-Transverse Approach

An air-duct installed to run parallel to the Collider Ring Tunnel would provide many advantages over the longitudinal ventilation methods considered above. Fresh air could be introduced into the Collider tunnel uniformly along its length, providing fresh air from a separate source, not from a distant source travelling along the tunnel itself.

The area under the floor could be the location for this duct, or a duct at the ceiling could be installed. This latter solution is probably difficult.

At each shaft, service and exit supply fans with filters and heaters would supply air to the duct servicing one half of a 2-1/2-mile segment. A bulkhead in the duct at the midpoint would divide the system.

The fresh air introduced into the tunnel uniformly along its length, would travel in the tunnel returning to either shaft, where exhaust fans would remove it from the system.

Each duct would be equipped with two fans or four supply fans per shaft, one set of exhaust fans could serve to remove air from two one-half segments. Assuming an 8 square foot duct could be provided under the tunnel floor, a velocity of say 4,000 feet per minute in the duct, at the shaft, would allow 32,000 cfm. One-half of a 2-1/2-mile segment could provide about 5 air changes per hour at the maximum capacity of the fans. Capacity control could reduce this air change rate to suit a particular need or condition.

This would provide, in the author's opinion, a much safer environment in the tunnel, and would not be very expensive, if the duct could be easily provided by use of the under floor space.

A further refinement would be the capability to reverse this system to exhaust (by reversing capability of the fans), and by means of dampers, concentrate this exhaust capacity where smoke or heat is present, or, when and where helium or nitrogen is leaking. Normal exhaust fans at the shafts would reverse to supply air at those points in this mode.

<u>Longitudinal Using Exhaust Fans Along the Tunnel</u>

This alternative would require exhaust fans installed in shafts to the surface above, spaced along the Collider tunnel segments at about 2,000 ft intervals. In an emergency, the fan nearest the emergency area would be energized. This would exhaust air, plus smoke, heat and gases, which are present in the area of the tunnel near the operating fan. Fans would be 30,000 to 50,000 cfm causing a velocity of 600 to 900 feet per minute in each direction toward the fan. Additional supply fans would be required at each service area and exit, to make-up this air being removed from the tunnel.

<u>Conditioning Air Supplied to the Collider Tunnel</u>

The RTK conceptual design indicates that air supplied to the tunnel will be filtered and where necessary, tempered by electric or natural gas-fired air heaters. Not mentioned are those times in the summer when the Texas heat and humidity will be present.

Introduction of humid air into this underground tunnel, whose walls and installed equipment could be cold, may cause condensation on walls and sensitive equipment.

The incoming air could be dried by use of desiccant or refrigeration type de-humidifying equipment. Both these methods are expensive and energy consuming.

SUMMARY

In this paper, I have expressed the following concerns about the proposed RTK Conceptual design ventilation method for the Collider Ring Tunnel.

o   Low airflow - 5,000 cfm per 2-1/2-mile segment. 0.4 air changes per hour - 2-1/2 hours for the air to travel from supply to exhaust.

o   Low air pressure, which might be overcome by other events such as gas leaks, fire, or obstructure.

o   Lack of positive balancing methods to insure that 5,000 cfm flows in both directions at an exit shaft supply point.

o   No provisions to de-humidify air during hot humid weather.

o   Lack of a fire mode capability to remove excessive amounts of heat and smoke in an effective manner.

<u>Alternative Ventilation Solutions to Consider</u>

o   Beef up the existing method by installing larger fans to provide a larger flow and consequent change of air in the 2-1/2-mile long segments. Install blocking doors or inflatable stoppers to control flow.

o   Install a parallel duct, probably under the floor of the
     tunnel. Supply air from each service shaft and each exit shaft
     to this duct. Each such system would supply air to 1/2 a
     segment or 1-1/4 miles of tunnel. Air would be uniformly
     delivered to each unit of length of the tunnel. Exhaust would
     be back through the tunnel to exhaust fans at each shaft.

o   Install exhaust fan units in the ceiling of the tunnel spaced
     at 2,000 ft. Vent would be to the surface. Capacity to
     exhaust would be sufficient to remove heat, smoke and gases at
     a localized point. Larger supply fans in each shaft would
     make up the air being exhausted. Fire detectors or other
     controls would activate the exhaust fan nearest the problem.

---

1   The above text was adapted from the 1982 ASHRAE Guide, Chapter on
    Enclosed Ventilation Facilities.

IMPACT OF RADIATION SHIELDING REQUIREMENTS

ON CONVENTIONAL CONSTRUCTION FOR THE SSC

George J. Doddy

Daniel, Mann, Johnson & Mendenhall
1900 M Street, NW Suite 800
Washington, D.C.

1. INTRODUCTION

The purpose of the conventional facilities in an accelerator is to support the technical components of the accelerator and this is particulary true in the case of the radiation shielding. Because of the vast parameters of the SSC, the requirement for shielding protection will have a corresponding impact in the design and construction of the conventional facilities.   The intent of this paper, therefore, is to review the sources of radiation in lay terms and to address the areas where protection is required in general terms.

Basically, the SSC is a complex of five cascaded accelerators in which protons are accelerated from rest to 20 TeV and consists of the 410 feet long 600 MeV Linear Accelerator (LINAC), the 820 feet circumference 8 GeV Low Energy Booster (LEB), the 6200 feet circumference 100 GeV Medium Energy Booster (MEB), the 19,700 feet circumference 1000 GeV (TeV) High Energy Booster (HEB) and the 53 mile circumference 20 TeV Collider.

The first four accelerators comprise the injector which introduces two 1 TeV proton beams in opposite directions into the collider which then accelerates and stores these proton beams at 20 TeV in preparation for the collisions in the interaction regions of the expermental areas with a total energy level of 40 TeV.   The accelerators which contain the magnets, the interaction regions which contain the detectors and the beam aborts systems which contain the beam dumps are the sources of radiation and are located in tunnels and underground enclosures which must be designed to accommodate the respective sources and types of radiation. See Figure 1-1

The LINAC, LEB and MEB utilize room temperature magnets whereas the HEB and the COLLIDER utilize superconducting magnets.

The underground cnolosures for the injector, the collider and the experimental areas require access from grade for the rf systems, power supplies, cryogenics, controls, utilities, personnel access and equipment through service areas.   Ventilation and additional personnel access is provided in exit areas located midway between the service areas.   The experimental areas which contain the interaction regions are located in two clusters-one near the injector and one diametrically opposite in the far ring.   Access to all of the underground facilities requires special

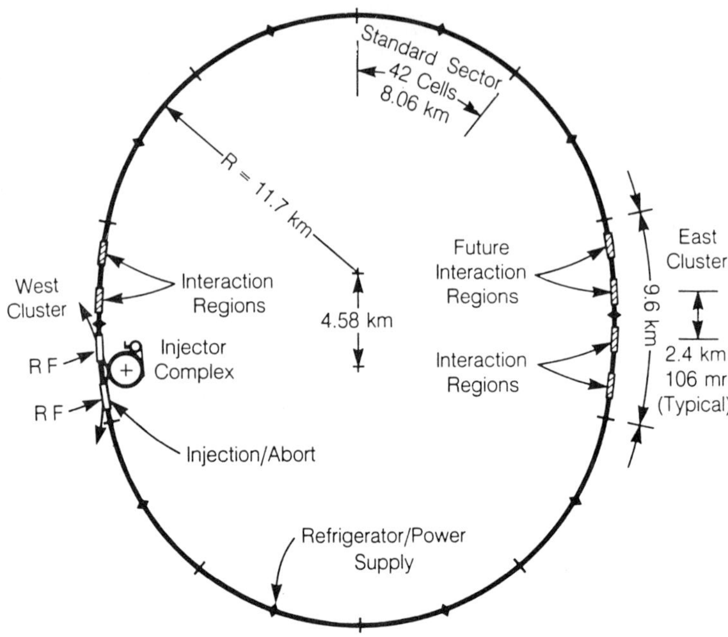

Fig. 1-1. SSC complex.

attention to provide the appropriate shielding and geometry for protection against radiation.

The LINAC, LEB, MEB, AND HEB comprise the injector which in essence is similar to the Tevatron at Fermilab and the SPS at CERN where comparable radiation conditions to those which will be experienced in the SSC have been observed and recorded for years.

The campus which consists of the central laboratory, industrial buildings, emergency buildings and treatment facilities will not be addressed in this paper whose purpose is to discuss only those components of the SSC which are impacted by radiation shielding requirements.

2.   TYPES OF RADIATION IN THE SSC

Radiation is a form of energy and when radiation encounters matter its energy is transferred to that matter. Generally, this energy is transformed to heat just as a microwave oven transforms its energy into heating the food.   In fact radiation is a natural phenomenon that has been around from the beginning, bathing all life as it evolved with low levels of radiation.

Accelerated beam particles are a form of radiation and as Jones states in "Radiation Safety in the SSC" there are two distinctly differently aspects of radiation involved with the SSC; prompt radiation and induced radioactivity.

The prompt radiation is that produced by ionizing particles when the beam interacts with something. When the beam is turned off, this radiation goes off and the effect can be compared with light going out of a light bulb when it is turned off.   It is this radiation that requires the accelerator structure to buried or shielded with earth berms.

The induced radioactivity refers to radiation from radioactive isotopes produced by protons which have come into contact with matter. This aspect persists after the beam is turned off and can be compared to the faint glow that remains in the dark after a television set or certain fluorescent lamps have been turned off.

The SSC will produce different types radiation in certain locations and the requirements for shielding will have an impact on the design of the conventional facilities. As Jackson states in "SSC Environmental Radiation Shielding" there are five types of radiation in the SSC: hadrons; photons and electrons; muons; residual radioactivity; and synchrotron radiation.

a. Hadrons

Initially, the massive and strongly interacting particles that are produced by the collision of protons in the beams are hadrons. Some particles decay rapidly while other collide with matter and are transformed into lower energy particles such as electrons, protons, and neutrons that make up all the common materials around us. Hadrons are stopped within meters depending on the primary energy, the type of radiation and the density of the absorbing materials. Radiation in the transverse direction is dominated by hadrons which will have an impact on the design of the conventional facilities.

b. Photons and Electrons

High energy photons and electrons accompany the hadrons. As a consequence, the degradation via electron magnetic interactions of the photons and electrons, called the electromagnetic cascade, occurs over distances of approximately a few meters in solid materials. Photons and electrons will have no significant impact on the design of the conventional facilities.

c. Muons

The characteristics of muons require special attention. The high energy muons follow the direction of the beam and are not easily absorbed at high energies. Furthermore, muons are significant in the High Energy Booster, the Collider, the IR Halls and the Beam Dumps. These muons require several kilometers in earth before experiencing decay and travel tangentially to the circumference of the main ring. Muons will have an impact on the design of the conventional facilities.

d. Residual Radioactivity

Radioactive nuclei are produced where the beams interact, where they are directed into a target, and where unused beams are absorbed. Both air and water can become radioactive if the beam passes through it. While radioactive products decay rapidly in light absorbers, such as oxygen and helium, those nuclei produced in heavy materials such as copper, iron, or lead can have long life times. However, the artificially induced radioactivity in the SSC is mostly very low level. Residual radioactivity will have an impact on the design of the conventional facilities.

e. Synchrotron Radiation

While normally associated with circular electron accelerators, synchrotron radiation is a factor in the design of the cryogenic cooling systems for superconducting magnets because of the high energy. However, it is not presently a concern in the design of the personnel radiation shielding.

f. _Tritium_

As stated by Toohig in the "SSC Safety Review Document" tritium, which is produced by the interaction of hadrons and liquid helium in the superconducting magnets at various beam loss points, is calculated to be negligible. However, tritium will have an impact on the design of the conventional facilities for the beam absorbers.

3. RADIOLOGICAL PROTECTION

L. Jones in "Radiation Safety of the Superconducting Super Collider" of April 1986 states the following:

> "The U.S. Department of Energy (D.O.E.) has established radiation safety standards for individual members of the general public due to operation of a D.O.E. facility as less than 500 mrem (millirem) in any one year (D.O.E. order 5480.1). Furthermore, the D.O.E. guidance for reducing radiation exposure to "as low as reasonably achievable" (ALARA) specifies that new facilities be designed such that anticipated exposures will be less than 20% of the maximum allowed dose equivalent. Hence, the SSC will be shielded to limit exposure to any member of the general public to less than 100 mrem per year under worst-case accident conditions.
>
> Sustained operation of a D.O.E. facility should not result in off-site continuous exposure of more that 100 mrem per year (D.O.E. order 5480.1A). In addition, D.O.E. has specified a reference value of 25 mrem per year as a level above which special D.O.E. approval would be required. The SSC design will be based on 10 mrem per year as the maximum exposure to the general public resulting from the routine operation of the facility."

a. _Standards_

Department of Energy (DOE)
Environmental Protection, Agency (EPA)
Internal Commission on Radiological Protection of the Environment (ICRP)
National Council on Radiation Protection Measurements (NCRP)
National Fire Codes
DOE/EP - 0023, "A Guide For Environmental Radiological Surveillance at U.S. DOE Installation, 1981
DOE N5480.1 - 1986
Environmental Protection, Safety & Health Program
NCRP Report No. 88 - 1986 "Radiation Alarm Access Control System"
NCRP Report No. 59 - 1980
CFR 87a,b - 1987

b. Limits and Action Levels

EXPOSURE PATHWAY	AGENCY & AUTHORITY	ANNUAL DOSE EQUIVALENT (mrem/yr)
EXTERNAL RADIATION		
	DOE Occasional exposure to general population	500
	DOE Prolonged exposure to general population	100
	DOE Action level	25
	Design goal at boundary	10
AIRBOURNE RELEASES		
	EPA Clean Air Act	25
	DOE Reporting level	12.5
DRINKING WATER (Applied to groundwater)	EPA Safe Drinking Water Act	4

Radiation shielding is therefore designed to both attenuate and absorb radiation. The amount of both is a function of the beam source as well as the material and density of the shielding. The most commonly used matters are earth, concrete, steel, graphite and lead. The design of all radiation shielding and environmental radiation protection is an integral part of the accelerator design.

4. RADIATION DESIGN GOALS

Although the SSC is vast in scale, the handling of shielding and mitigation measures are similar to those at Fermilab, Brookhaven National Laboratory and CERN. The experience gained at these laboratories provides evidence that no new radiological problems are posed and that the SSC can be operated safely based on prudent design. The following components are described in the CDR and in the "SSC Safety Review Document" edited by T. Toohig.

a. Injector

In particular, the injector complex of the SSC is comparable to the Tevatron at Fermilab and the SPS at CERN. In fact, the SSC injector is less of a problem because of the absence of a fixed-target program where major shielding and monitoring problems must be addressed.

The Linac and LEB are concrete enclosures with 8 feet of earth shielding; the MEB a is concrete enclosure with 11 feet of earth shielding; and the HEB is a concrete enclosure with 13 feet of earth shielding. The earth density for this shielding is assumed at $1.8 g/cm^3$. See Figures 4-1, 4-2, 4-3,

b. Beam Injection

The relationship of the HEB and the Collider is influenced by the requirement that allows personnel access to the HEB while the collider is operating and vice versa. Therefore, the plane of the HEB as defined in the CDR lies above and inside the Collider ring with a minimum separation between the Collider and HEB rings of 30 feet horizontally and 23 feet vertically. The placement of the rings at different elevations as well as

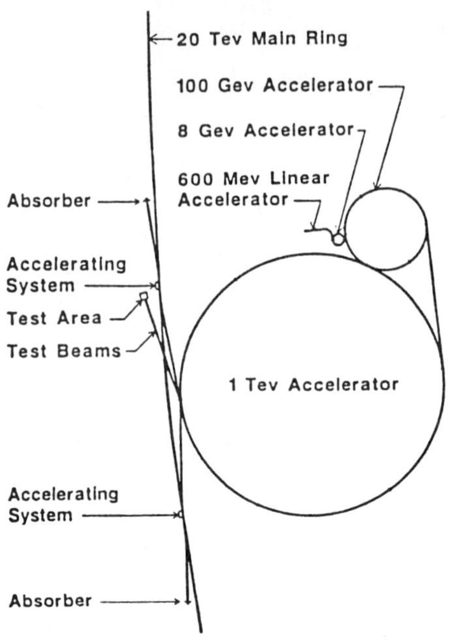

Fig. 4-1.   Injector complex.

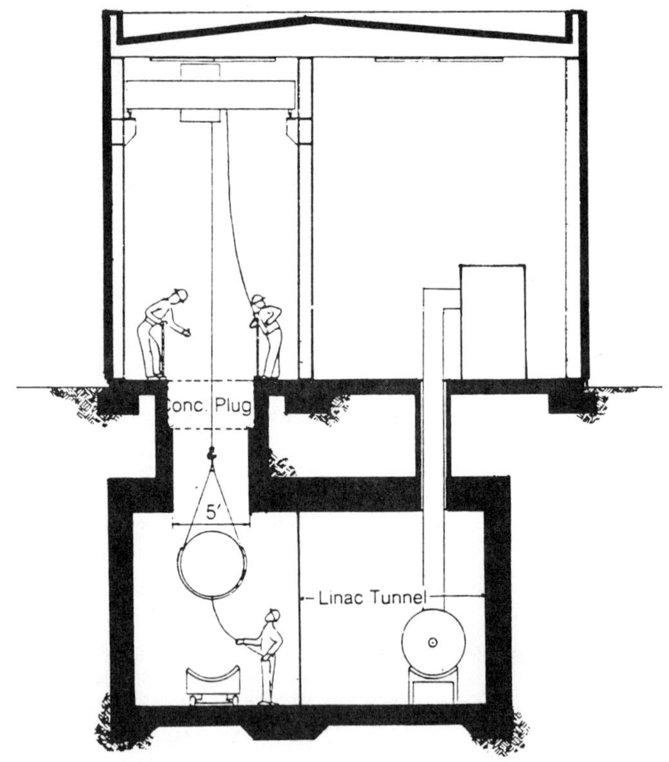

Fig. 4-2.   Linac and gallery.

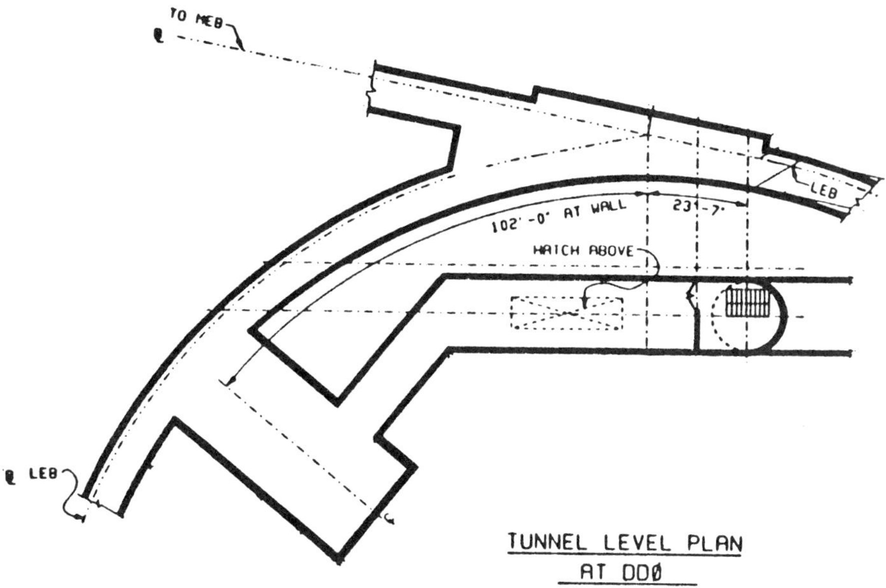

Fig. 4-3. Access labyrinth from tunnel.

the separation by earth shielding prevents the muons of either ring from being a hazard. See Figure 4-4.

Special attention is required at the beam scrapers since the induced radiation is relatively high and extra shielding will be required at those locations.

c. Collider

In normal operation, the beams are confined in the vacuum chambers of the magnet rings. Radiation then occurs only at the interaction region, at the scrapers and at the primary beam absorbers. A very small amount of radiation, which is absorbed within the walls of the tunnel, is produced as a uniform halo around the ring resulting from a collision of the beam protons with the residual gas molecules.

Should a problem occur, beam position and beam log monitors will almost instantly trigger a system of special magnets to begin controlled ejection of the beam into the beam absorbers. This protection system coupled with the earth shielding of 30 feet minimum will protect the public.

The forward muon emission will extend along a tangent for a distance up to two kilometers with a 200 meter zone extending radially outward in a plane from the tunnel. Because muons are far less interactive than hadrons, this zone demands less restrictive personnel access. The shielding will be designed to accommodate the extremely rare scenario of a total beam loss anywhere on the periphery with a frequency of once a year. See Figure 4-5

d. Experimental Areas

Although six experimental halls are proposed, only four of these halls will be included in the design at this time. These halls are located in two clusters in the Collider. See Figures 4-6, 4-7, and 4-8.

The source of radiation for the experimental areas occurs in the Collision Halls where the proton beams collide within the detectors. Hadron shielding

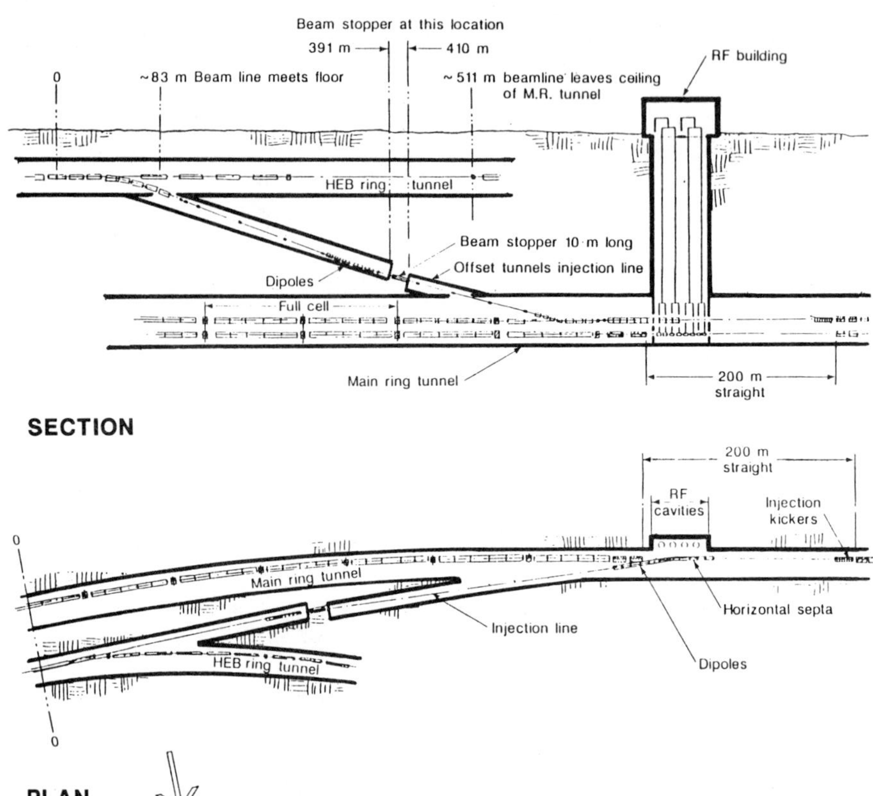

Fig. 4-4. Injection tunnels from the HEB to the collider.

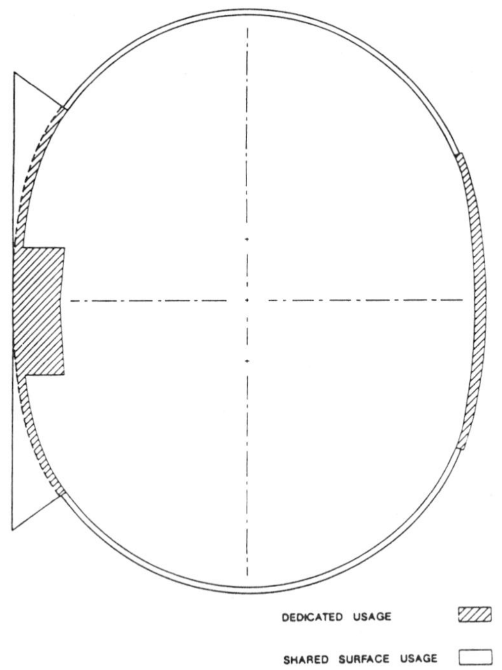

Fig. 4-5. Land areas from the I.S.P.

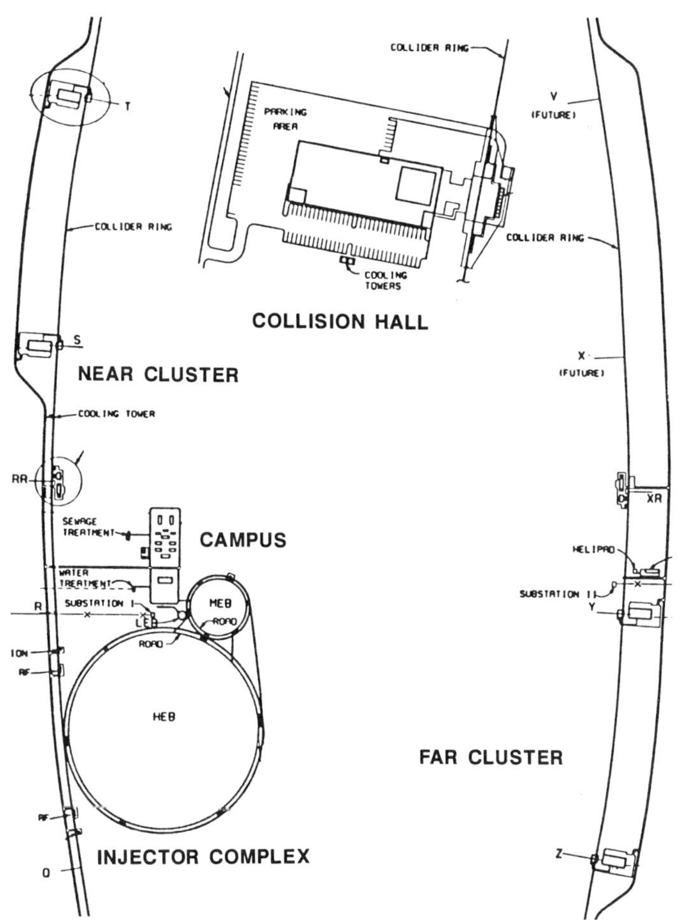

Fig. 4-6. Experimental areas.

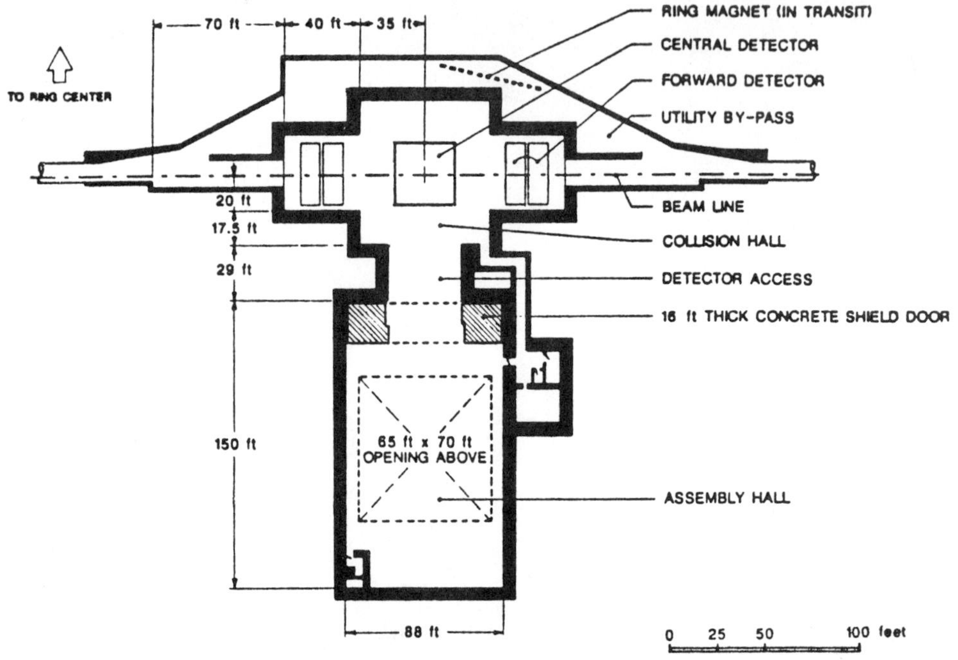

Fig. 4-7. Experimental hall — plan at beam level.

is required transversely to allow activity to occur in the Assembly Hall while the beam is running. One possibility is to provide sliding concrete doors 16 feet thick between the Collision Halls and the Assembly Halls and to provide a labyrinth to allow controlled personnel access while the shielding doors are closed and the beam is not running.

e. <u>Beam Absorbers</u>

At the end of a machine cycle, the beam run is terminated by ejecting the beams from the machine and into beam absorbers, one for each ring.

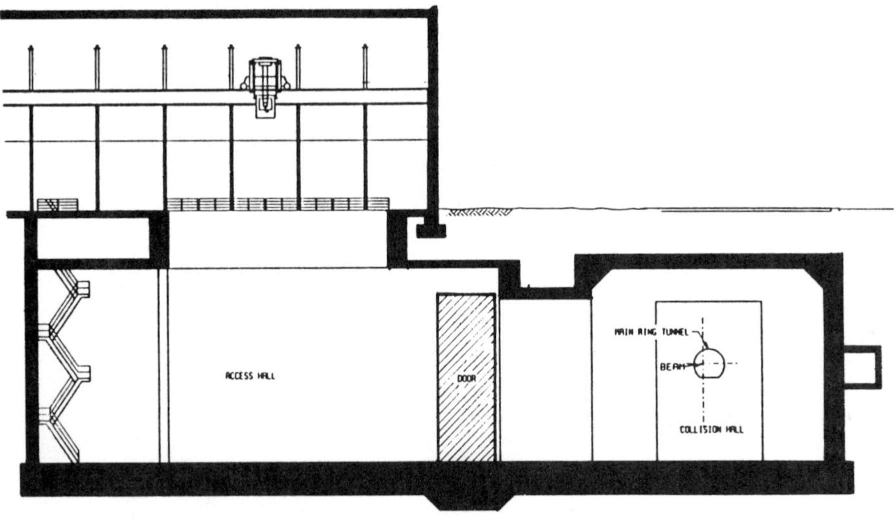

Fig. 4-8. Experimental hall — section.

The absorbers are designed so that contamination levels are minimized. Absorption of the prompt muons will be provided in a restricted zone 30 feet wide and 3.25 miles long down stream from the absorbers with an assumed soil density of 2.24g/cm^3. See Figure 4-9, 4-10, 4-11, 4-12. The test beam in the injector is treated similarly to the beam absorber. See Figure 4-1.

f. Access Control Systems

Except for some residual activation in the injection and extraction areas, accelerator related radiation disappears when the machine is shut down - much like the disappearance of the picture when a television set is turned off. For this reason, primary emphasis in accelerator radiation safety is placed not only on ensuring that personnel be excluded from beam enclosures during the operation of the beam but also that the beam be excluded from areas occupied by personnel. The process of ensuring that personnel are excluded from the machine during operation begins with a sequential search of the affected beam enclosures. Because approximately 60 miles of underground enclosures are involved, these enclosures will be divided into interlock segments based on the service areas at the mid point of each sector. Each sector will be monitored and a positive signal will be required from the search sequence before a beam can be introduced into the facility. To be certain that no beam enters the enclosures during the highest level of interlock, two independent devices are required-either of which alone when disarmed is capable of preventing the beam from entering an accessed enclosure. This requires orientation and separation of adjacent accelerator enclosures so that when either or both of the critical devices is off, the beam from one facility cannot enter another. Since access control and critical signal devices involve life safety, they will be connected to the Main Control Center whose power will be backed up by an uninterruptable power supply. See Figure 1-1.

g. Monitoring Systems

In addition to the operational monitoring of personnel, a general environmental and monitoring system will be installed throughout the site. Included in this system will be the monitoring of radiation, fire and facility safety with appropriate sensors and transducers at strategic locations throughout the site. This monitoring system will be connected to

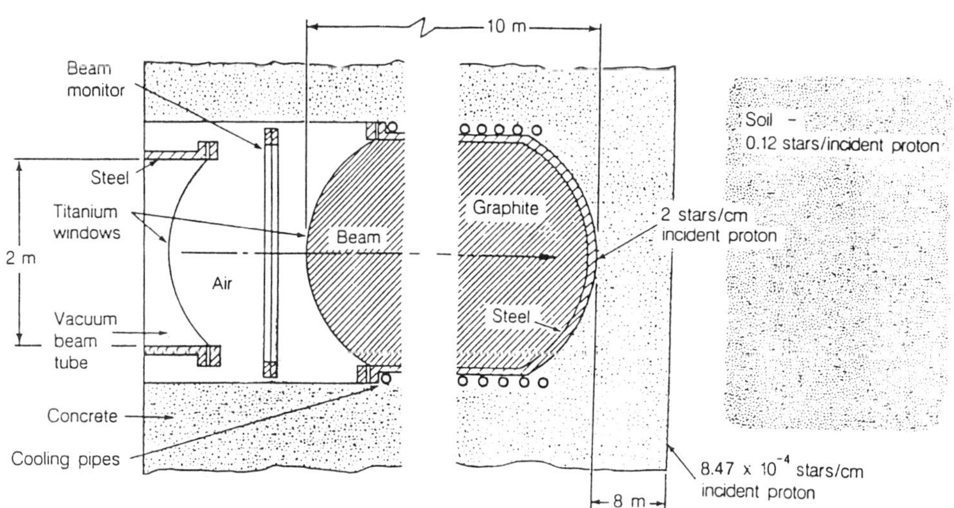

Fig. 4-9. Collider ring beam abort dump — longitudinal section.

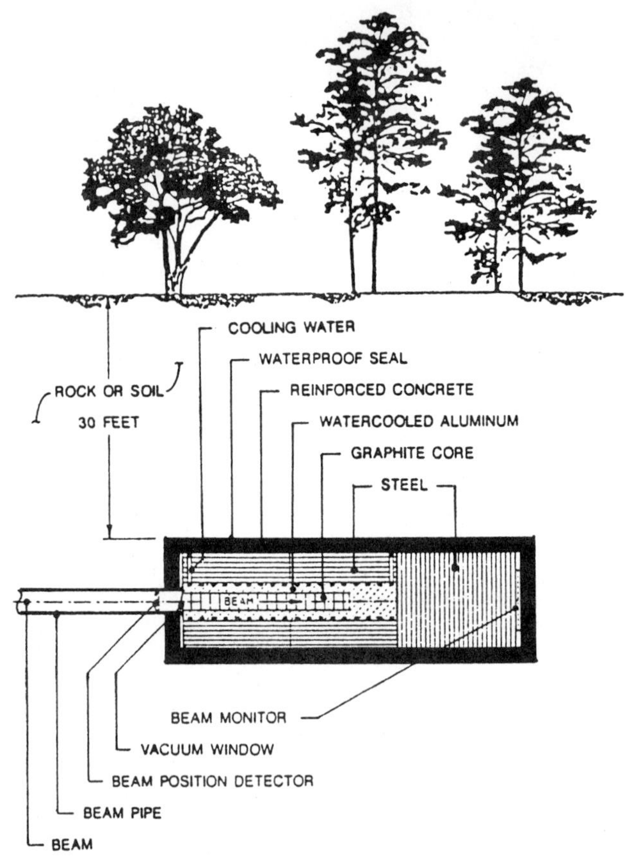

Fig. 4-10. SSC beam absorber — example.

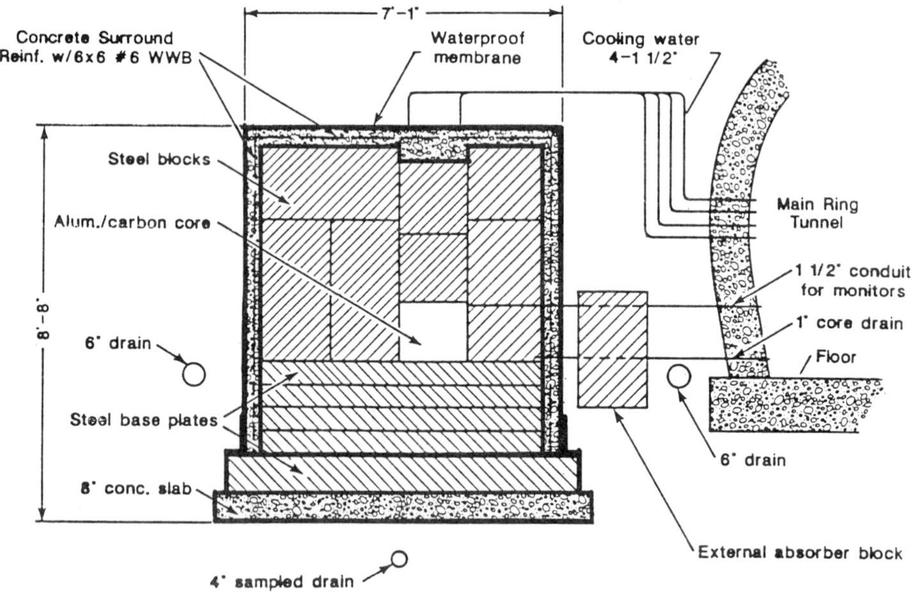

Fig. 4-11. Water management at beam dump.

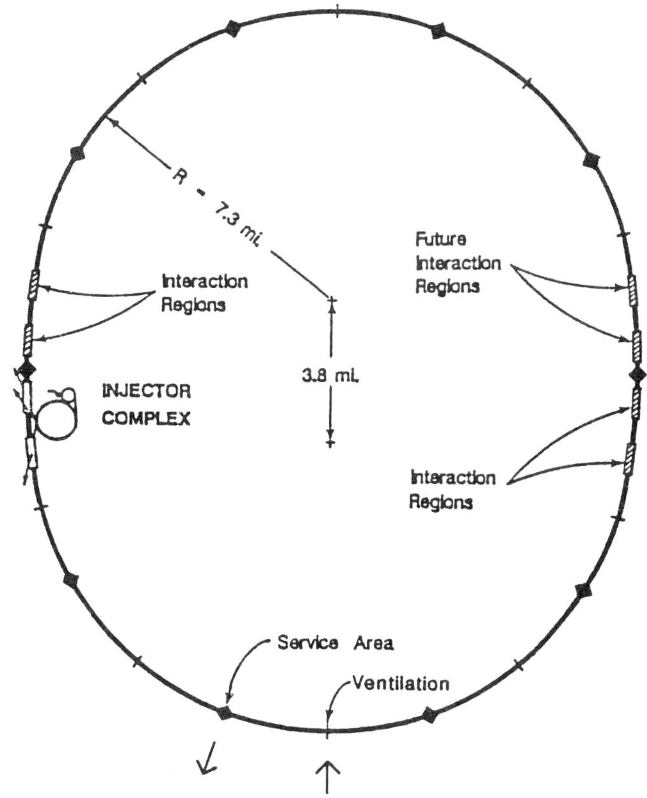

Fig. 4-12. SSC ventilation concept.

the laboratory operations center with personnel on duty 24 hours a day. Access to radiation areas will be monitored and controlled through the accelerator control system. Fire and security will be monitored at the laboratory emergency center. Included in the environmental monitoring will be the sampling of off-site water, air, effluent and soil.

h. Cooling Water Systems

Water that is used to cool magnets beam dumps and targets becomes measurably radioactive. Protection is afforded by keeping monitoring regularly.

The cooling water will be circulated away from the radiation sources. In the case of the beam dumps (absorber) where the experience of Fermilab will be applied to the design of these systems for the SSC. See Figure 4-11

i. Ventilation System

The accelerator tunnels are not occupied during operation of the accelerator. However, access to the tunnels is required for installation, modification and maintenance. Air becomes activated when exposed to radiation and the prime sources for this activation are the beam dumps, targets, and interaction points. Protection is afforded by utilizing Filters and by controlling ventilation. Consequently the ventilation system is designed for operation during these periods. Ventilation fans will be provided for the tunnel with exhaust shafts at the service areas and supply shafts at the exit points located midway between the service areas. This criteria assumes a seven man crew to replace magnets. See Figure 4-12, 4-13

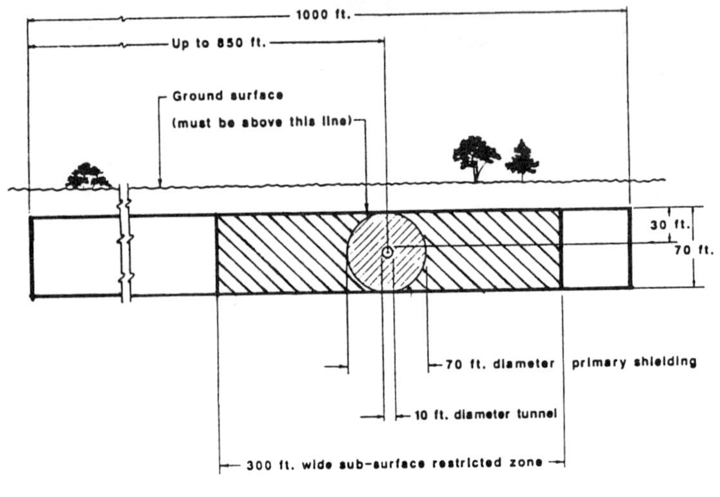

SUBSURFACE AREA REQUIRED FOR COLLIDER RING WITH 30FT COVER

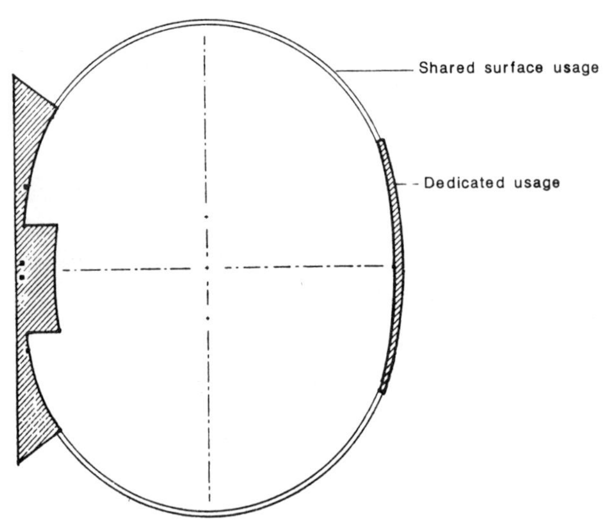

LAND REQUIRED FOR THE SSC

Fig. 4-13.

j. Ground Water Management

In the primary beam absorbers where a small fraction of the radionuclides that are soluble with water may reach the water outside the dump, it is necessary to insure that the site ground water is protected from concentration of these nuclei. The experience gained at Fermilab will be utilized to avoid high activation of the ground water by leaching or by mixing the ground water with cooling water. In addition to the water proof seal around the dump, a network of under drains to remove any water along with monitoring drains and monitoring wells will be provided to confirm the freedom from radioactivity. See Figure 4-11

k. Site Boundary Requirements

The SSC describes an oval approximately 16 miles by 20 miles. The 53 mile circumference collider ring requires an exclusive use of a subsurface area of 70 feet high by 1000 feet wide to provide the necessary radiation protection during operation. If the collider ring is near the surface, the arcs will require a 1000 feet wide land zone totalling approximately 3800 acres.

The beam absorber and test beam areas are located outside the main ring and require approximately 4600 acres. In order to avoid interference with test beams and absorbers, the injector is located inside the main ring and requires about 1700 acres. Provision is also made for a possible electron-proton option and requires about another 1700 acres.

The two experimental clusters totalling 4000 acres provides space for up to ten halls although only four will be constructed initially. The campus area accounts for another 350 acres. See Figure 4-13

1. Baseline Radiation

Issues of radiation and radioactivation are important to the SSC and baseline information on the existing background levels of radioactivity in the air, soil and ground water will be gathered in the vicinity of the site. The three existing sources of individual exposure to radiation are naturally occuring background sources, industrial and agricultural source introduced by man, and radiation used in medical diagnosis and treatment. The unit of measure for biological dose of radiation in humans is the rem. Even now, cosmic radiation from outer space is penetrating the earth's atmosphere and is also passing through us. In addition, we are exposed to radioactive substances in the earth's crust, building materials and foods. The main source of naturally occuring background radiation is radon, a gaseous radionuclide produced by radioactive decay of radium in the ground. The presence and intensity of radon varies with location and could be as much as 200 mrem/yr whereas the dose from cosmic rays varies with altitude and could be as much as 100 mrem/yr.

Of interest is the fact that cosmic radiation contributes 35 to 90 mrem/yr and that this radiation increases with altitude because cosmic radiation is absorbed by the atmosphere. Therefore, persons in high mountains or flying receive more radiation than those at sea level. Of greater personal interest is the fact that we all have small amounts of radiation in our bodies that is transferrable to others.

Industrial and agricultural source contribute about 5 to 1 mrem/yr. The largest source of man made radiation is medical X-rays of 3.9 mrem/yr and nuclear medical procedures of 14 mrem/yr.

The SSC designers have set the goal for the public of 10 mrem/yr - well below the legal limit of 100 mrem/yr and substantially below the background radiation which is estimated at 364 to 524 mrem/yr. Figure 4-14

5. SUMMARY

Although the SSC will be the largest and most powerful device to be constructed for the purpose of conducting basic research, it is different from existing accelerators only in scale. The types of radiation that are produced by the SSC accelerators is similar to that in Fermilab and CERN and therefore not only are well understood but also can be readily controlled. Years of experience with prompt and induced radiation in accelerators, interaction regions and primary absorbers enable the SSC designers to address specific radiation sources with an inventory of data and computer programs for analysis and application.

Except for areas where the beam interacts, particularly in the beam dumps, the residual radioactivity is negligible and the prompt radiation will disappear when the machine is turned off. The shielding in the special areas will respond to the radiation requirements and this together with the access control and the monitoring systems will make the SSC safe for the operators as well as the general public who, by design, will not be exposed to more than 10 mrem/yr. This exposure is considerably less than the estimated 364 to 524 mrem/yr that nearly everyone receives from natural and technological radiation that constitute the background sources of radiation.

From the experience gained in working with other existing accelerators, the impact of radiation shielding requirements on the Conventional Facilities is well understood. It then becomes a matter of working closely with the staff of the management and operations contractor to incorporate that experience into the design and construction of the SSC.

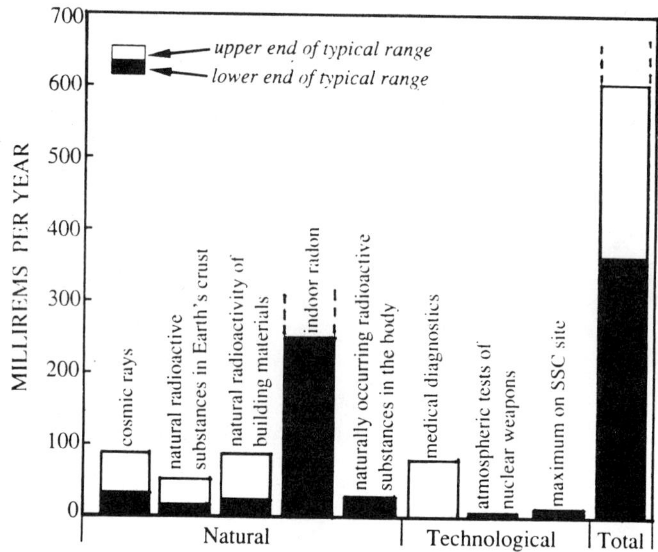

Fig. 4-14. Sources of public radiation exposure (Continental U.S.)

6. BIBLIOGRAPHY

Baker, "Environmental Report for Calendar Year 1987" Fermilab 88/40, 1104.100, UC-41.
Coulson, Freeman, and Toohig, "A Guide to Understanding the Radiation Environment of the Superconducting Super Collider (SSC), SSC-147.
Environmental Impact Statement, Superconducting Super Collider, DOE/EIS 0138D-DRAFT-August 1988, DOE/EIS 0138-FINAL-December, 1988.
Fasso, et. al. "Radiation Problems in the Design of the Large Electron-Positron Collider (LEP) CERN 84-02.
Gilchriese, "Radiation Effects at the SSC", SSC-SR-1035.
Groom, Editor "Radiation Levels in the SSC Interaction Regions "Task Force Report, SSC-SR-1033.
Invitation For Site Proposals for the Superconducting Super Collider, DOE/ER-0315 April, 1987.

Jackson, Editor "SSC Environmental Radiation Shielding", SSC-SR-1026.

Jones, "Radiation Safety of the Superconducting Super Collider" SSC-54.

Metropolis, Editor "An Introduction to Radiation Protection for the Superconducting Super Collider," SSC-SR-1027.

Reference Design Study, Superconducting Supercollider, May, 1984.

"Report of the DOE Review Committee on the Conceptual Design of the Superconducting Super Collider" DOE/ER-0267, May 1986.

"Report of the Task Force on Radioactivation," SSC-SR-1029.

SSC Site Evaluations, "A Report by the SSC Site Task Force", DOE/ER-0392 November, 1988.

Superconducting Super Collider Conceptual Design Appendix C - Conventional Facilities, SSC Central Design Group, March, 1986.

Superconducting Super C321ider Conceptual Design SSC Central Design Group, March, 1986.

Texas National Research Laboratory Commission, Superconducting Super Collider, Dallas-Fort Worth SSC Site Report September, 1987.

Toohig, Editor "SSC Safety Review Document", SSC-SR-1037, November 1988.

Toohig, Editor "Workshop on Radiological Aspects of SSC Operations Task Force Report", SSC-SR-1031.

SYSTEMS ENGINEERING AND INTEGRATION FOR THE SSC

D. J. Laintz, B. Crosby, M. Davis, D. Eben, J. Gliozzi, E. Kientz,
J. Knafelc, J. Phelps, M. Rider, and K. Shearer

Martin Marietta Astronautics, Denver, Colorado 80201

I. INTRODUCTION

Experience in high technology projects of scale and scope similar to the SSC, leads us to propose utilization of a Systems Engineering & Integration (SE&I) process tailored specifically to the SSC project.

We begin by giving an overview of SE&I . This overview includes the purpose, history, definition, and a discussion of when to use it. We then proceeded to give a description of a formalized SE&I process. We discuss tailoring the process to a project and close by recommending an early commitment to an SE&I methodology for the SSC.

II. OVERVIEW OF SYSTEMS ENGINEERING AND INTEGRATION

SE&I objectives are intended to organize, sequence and optimize system acquisition and program goals. Although programs differ in underlying requirements, there are disciplined, logical methods that can be employed to benefit the system analysis, design, build, test and activation process.

Purpose

The purpose, of utilizing a disciplined SE&I process is to establish order and accomplish specific objectives. These objectives include, but are not limited to:
* Ensuring that system definition and design, reflects requirements for all system elements: equipment, software, personnel, facilities, and procedural data.
* Integrating technical efforts of the design team specialists to produce a balanced design.
* Providing a comprehensive indentured framework of system requirements for use as performance, design, interface, and test criteria.
* Providing source data for development of technical plans and contract work statements.
* Providing a systems framework for logistic analysis, logistic trade studies, and logistic documentation such as technical manuals.
* Ensuring that Life Cycle Cost considerations and requirements are fully considered at all phases of the design process.

These objectives optimize system requirements and program goals by emphasizing:
  - coherent, integrated systems development
  - least risk, including risk assessment and risk mitigation
  - lowest cost, including cost avoidance and Life Cycle Cost assessment
  - schedule compliance, including program master schedule and re-plan re-schedule activities

History

The formalization of the SE&I process started after World War II. It was driven primarily by increasing technical complexity of large Department of Defense (DOD) projects. During this period, some large program failures occurred which possibly could have been avoided, or at least mitigated, through the use of disciplined SE&I process. Variations of SE&I techniques were evolving as a result of different contractors and different government agencies, i.e., Air Force, Army. In an effort to standardize SE&I techniques, MIL-STD-499 was drafted and approved in 1969. MIL-STD-499 has served as the guide for SE&I on many large, technically complex projects. Some of these projects, which Martin Marietta participated in, include:

- Viking Mars Lander
- Shuttle Launch Complex
- Skylab
- Federal Aviation Air Traffic Control (FAATC) upgrade
- Peacekeeper (MX) weapon system
- Midgetman (Small Missile) Weapon System

Utilization of SE&I is now standard practice on DOD and NASA projects. Multiple contractors have successfully completed projects employing the techniques of formalized SE&I. This approach has also proven to be valuable in integrating the efforts of major contractors working on the same project.

Definition

MIL-STD-499 defines SE&I as: "both a technical process and a management process. This process applies both scientific and engineering efforts to:

- Transform an operational need into a description of system performance parameters and a system configuration through the use of an iterative process of definition, synthesis, analysis, design, test and evaluation;
- Integrate related technical parameters and ensure compatibility of all physical, functional and program interfaces in a manner that optimizes the total system definition and design;
- Integrate reliability, maintainability, safety, human factors and other such disciplines into the total engineering effort to meet the cost, schedule, and technical performance objectives."

When to use SE&I

The methodologies of SE&I are utilized informally or formally on all projects. Scale and complexity of the project will dictate application technique. Formal application of a defined, disciplined SE&I process adds the greatest value to a project when some of the following characteristics are present:

- Large systems or subsystems with inherent advanced or advancing technology;
- Contributing participants are spread throughout the country and world;
- Large and diverse design teams;
- Several related pieces of interfacing hardware and equipment;
- Hardware and interfacing software systems are being concurrently developed;
- Complex operational and logistics support requirements;
- Schedules for subsystems development, assembly and test that are interdependent and critical.

Experience has shown that projects that contain all or some of the characteristics have benefited by utilizing a formalized SE&I process. <u>The SSC project contains all of the above characteristics</u>.

III. THE SE&I PROCESS

The SE&I process is applied iteratively through the project/design phases. Program and product element descriptions become more detailed with each application. The final product is an integrated operational system that includes documentation of all system elements.

Figure 1 depicts the SE&I process overlaid on the various phases of a project. As the project advances from the initiation and analysis phase through the operations phase, the design matures. To support the maturing designs, a systems engineering process is applied and iterated through each project/design phase. Significant importance is placed on having available experienced personnel resources that include all of the speciality engineering skills (identified on figure 1 as Engineering Integration) applicable to the project. These personnel support the activities of the iterative systems engineering process, by applying their special engineering skills and automated tools to support the SE&I process.

The current status of the SSC program phase can be tracked on figure 1. The program initiation and analysis phase is complete. This effort included the decisions and recommendations made by HEPAP at Snowmass in 1983, and the development of the critical design parameters including a definition of the operations concept. The program has since transitioned into the conceptual phase. The CDG conceptual design report was published in March 1986, and the dipole magnet R&D program is being conducted. The dipole magnet program is currently transitioning into the next project phase of requirements definition and technology demonstration. This is consistent with the magnet industrialization program that is underway. Other project elements are still in the conceptual phase.

Engineering Documentation

As the project/design phases mature and evolve, the products of the systems engineering process are retained as program documentation. This documentation can be structured as a rational data base that facilitates a project wide exchange of data and identifies configuration and status of project elements. The documentation that is generated during the application of the iterative Systems Engineering process evolves along a parallel path with the various project/design phases. Using the products of the iterative Systems Engineering process, experienced personnel resources are again applied to develop program supporting documentation.

The types of personnel required, changes as the project/design phases evolve, for example, during the early SSC program phases of initiation and conceptual development, personnel required were primarily Scientists and Physicists. These people develop Critical Design Parameters and the Conceptual Design Report and their supporting analysis and studies. These products are retained as program documentation and are used as baseline data as the program phase advances. During the next phase of the project, personnel skill requirements expand. These additional skills must include all of the experienced specialty engineering resources as well as trained, experienced systems engineers. These systems engineers utilize the documentation that evolves from the iterative process to generate the program specifications. These specifications include, but are not limited to; the Top Level Systems Specification, Development Specifications, Product Specifications, Process and Materials Specifications, Drawing Trees, Specification Trees, and Design Drawings. Figure 2 is a sample specification tree.

As the project/design phases evolve and the systems engineering process is applied, the engineering documentation expands and becomes more complete and accurate. The data base contains program baselines and configuration. Specifications and designs are maintained and continually statused against approved baselines. This documentation and data base is utilized in achieving the objectives stated in the purpose above.

Management Documentation

Since the SE&I process has the flexibility to be adapted to project specific objectives, the methods, controls and procedures that will be utilized need to be defined. A Systems Engineering Management Plan (SEMP) is developed to serve as the governing plan. Prior to starting the formalized SE&I process, project management and the SE&I contractor(s) agree on methodology and document the approaches that will be utilized. Specialty engineering plans are also identified in the SEMP.

PROGRAM PHASES
PROGRAM INITIATION & ANALYSIS
• Identify Program Opportunities & Purpose
• Determine Program Goals & Critical Design Parameters
• Define Operations Concept
CONCEPTUAL
• Define System Concepts
• Select Concept to Develop
• Conceptual Design
REQUIREMENTS DEFINITION & TECHNOLOGY DEMONSTRATION
• New Technology Demonstration
• System & Equipment Configuration
• Preliminary Design
DESIGN DETAIL
SYSTEM DESIGN PARAMETERS
SYSTEM LEVEL DESIGN
CONFIGURATION DESIGN
SYSTEMS ENGINERING PROCESS
Hardware Engineering
Parts & Materials
Test Engineering
Software Engineering
ENGINEERING
Logistics Engineering
Assembly Engineering
Human Engineering
Radiation Hardness
Systems Engineering
Program Goals
System Design Parameters
Program Opportunity & Purpose
Criteria Design Parameters
Trade Studies
Requirements Allocation & Validation
System Concept
Concept Model
Functional Flow Analysis
Concept Evaluation
Requirements Analysis
Trade Studies
Requirements Allocation & Validation
Preliminary Design
Configuration Engineering Model
Functional Flow Analysis
Model Test & Evaluation
Continued Requirements Analysis
Allocated Requirements
Allocated Requirements
Allocated Requirements
ENGINEERING DOCUMENTATION
SYSTEMS ENGINEERING
CRITICAL DESIGN PARAMETERS • FUNCTIONAL FLOWS • TRADE STUDIES •
FUNCTIONAL REQUIREMENTS • LOGISTICS REQUIREMENTS •
Critical Design Parameters
Conceptual Design Report
SPECIFICATION TREE
TOP-
MANAGEMENT DOCUMENTATION
SYSTEMS ENGINEERING
TECHNICAL
ORGANIZATION •
MASTER SCHEDULE •
SYSTEMS
INTERFACE CONTROL •
ENG DRAWING CONTROL •
SPECIALTY
RELIABILITY •
MAINTAINABILITY •

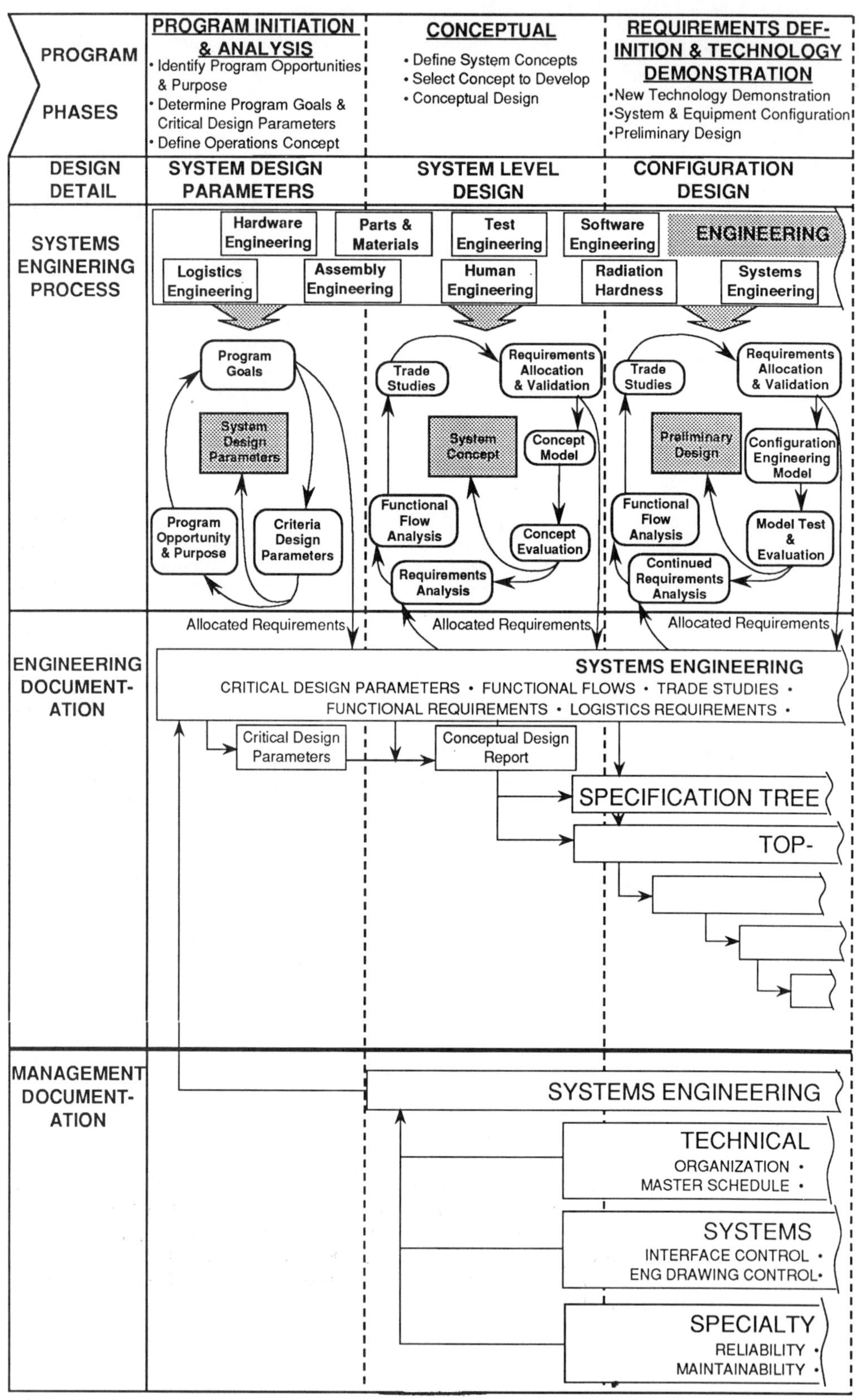

Figure 1

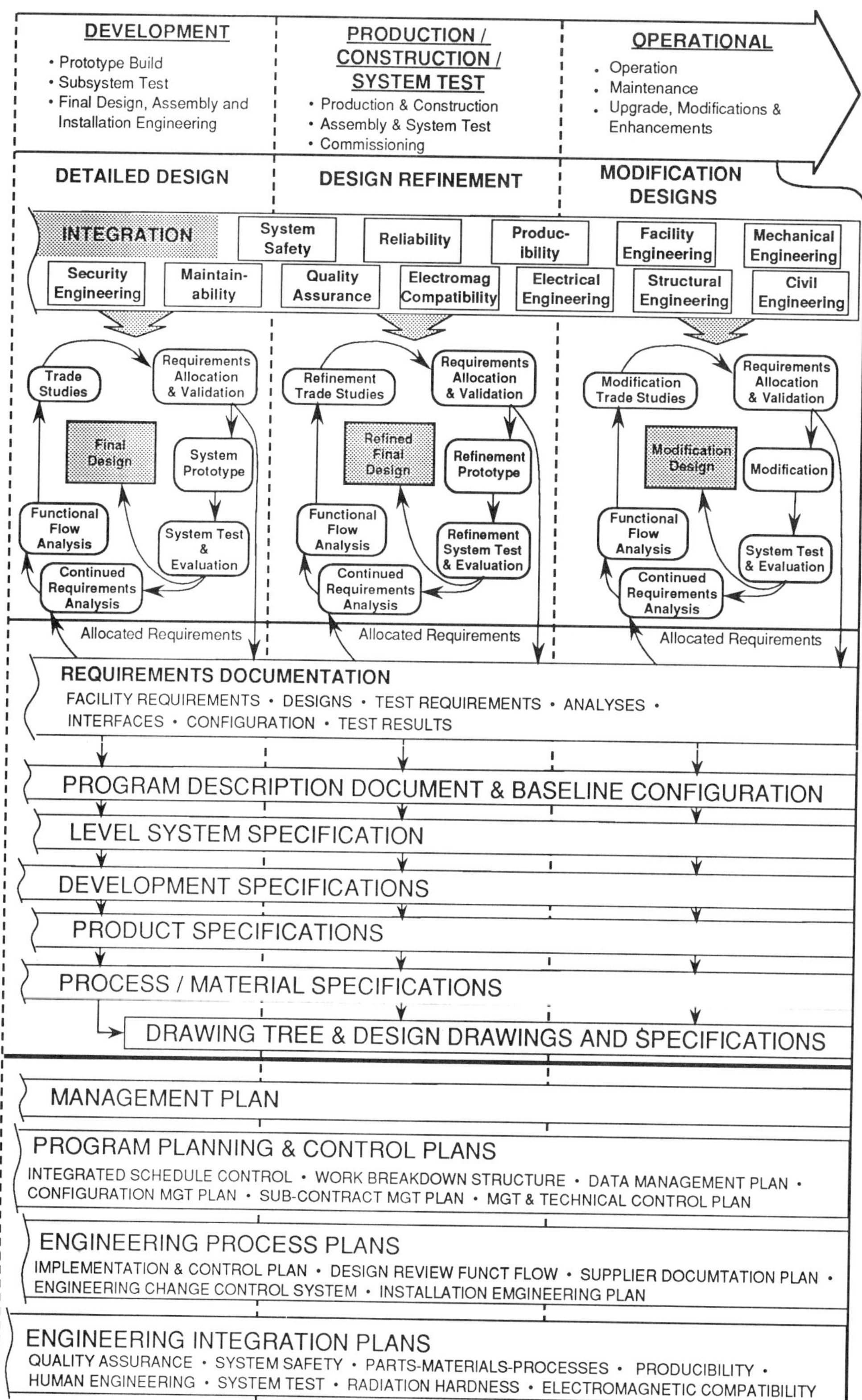

Figure 1 (continued)

Superconducting Super Collider System Specification
(Top-Level System Specification)

(Sub-Systems Specifications)

Site Support
- Roads, Parking & Landscaping
- Water Treatment Plant
- Sewage Treatment Plant
- Central Office Laboratory
- Heavy Works Building
- Shops Buildings
- Warehouse
- Emergency Service Building
- Vehicle Service Building
- Entrance Kiosk
- Utilities Distribution
- Electrical Substations
- Refridgerator / Power Supply Buildings

LINAC
- Main Enclosure
- Support System Specification
- Ion Source
- RFQ
- DTL
- SCL
- RF Power Systems
- Transfer Line
- Instrumentation & Controls
- Safety Systems

Low Energy Booster
- Tunnel
- Surface Buildings
- Support Systems Specifications
- Magnets
- Vacuum System
- Power Supplies
- RF Systems
- Injection System
- Extraction System
- Abort System
- Instrumentation
- Controls
- Safety Systems

Medium Energy Booster
- Tunnel
- Surface Buildings
- Support Systems Specifications
- Magnets
- Vacuum System
- Power Supplies
- RF Systems
- Injection System
- Extraction System
- Abort System
- Instrumentation
- Controls
- Safety Systems

High Energy Booster
(Incl Test Beam Area)
- Tunnel
- Surface Buildings
- Support Systems Specifications
- Magnets (SC & Conv.)
- Cryogenics
- Vacuum System
- Power Supplies
- RF Systems
- Injection System
- Extraction System
- Abort System
- Instrumentation
- Controls
- Safety Systems

Main Ring
- Tunnel
- Surface Buildings
- Support Systems Specifications
- Magnets (SC & Conv.)
- Cryogenics
- Vacuum System
- Power Supplies
- RF Systems
- Beam Feedback System
- Injection System
- Abort System
- Instrumentation
- Controls
- Safety Systems

Detector
- Facility
- Support System Interfaces

SUPERCONDUCTING SUPER COLLIDER
System Specification Tree

SYSTEM SPEC
SUB SYSTEM SPEC
SUB SYSTEM SPEC
DVLMNT SPEC
DVLMNT SPEC
DVLMNT SPEC
BUILD SPEC
BUILD SPEC
SPECIFICATION FLOWDOWN

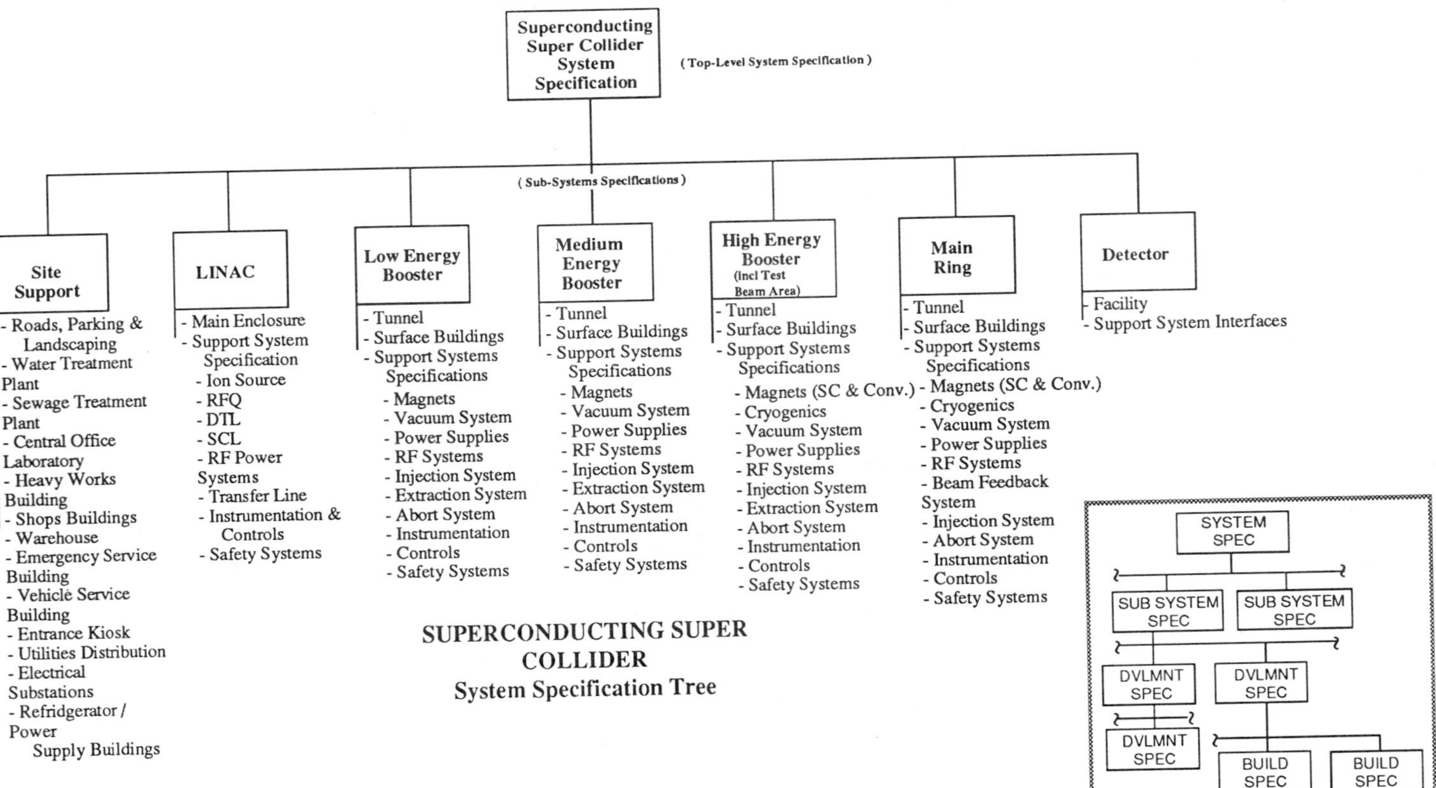

Figure 2

THE SEMP

The SEMP's principal role is to assure control and identify the overall engineering process. This includes defining the fundamental relationships required for engineering integration. The SEMP contains specialty plans that describe activities in specific critical areas. Combined with these plans, the SEMP serves as the program's top level technical management master plan.

The SEMP is a program- and contractor-specific document identifying the organizational configuration, functions and responsibilities, management techniques, analyses, simulations, technical performance measurement (TPM) parameters, and schedules that will be initiated, tracked, or employed on the program. SEMPs may use the standard format defined in MIL-STD-499, "Engineering Management," or, if accepted by the Project/Program Manager, any contractor-proposed format which provides all necessary information is acceptable. The standard format in MIL-STD-499 has three parts, as discussed below.

Part I, Technical Program Planning and Control, identifies organizational responsibilities and authority for the contractor's systems engineering management, including control of subcontracted engineering, verification, configuration management, and document management, as well as plans and schedules for design and technical program reviews.

Part II, System Engineering Process, describes the process used in defining and allocating requirements and their documentation for the program. This part also explains the contractor's intended design strategy at each development level, and the trade-off results which may trigger iterations of the design process.

Part III, Engineering Specialty Integration, defines how the engineering specialties of reliability, maintainability, human engineering, safety, logistics, producibility, and other areas are integrated into the mainstream design effort.

The SEMP can also specify the models and simulations that will be employed in defining system requirements, optimizing system configuration, and performing verification.

Specialty Plans

Additional plans for engineering specialities are referenced in Part III of the SEMP. The need for separate specialty plans is a key part of this specification tailoring effort. Where separate plans are not prepared, often much of the basic work is described in the overall SEMP. Acquisition programs include plans for the following specialty disciplines integrated under the systems engineering umbrella:
- Technical performance measurement
- Producibility
- Maintainability
- Quality
- Human engineering
- Safety
- Logistics support analysis
- Reliability
- Production engineering
- Contamination and corrosion control
- Parts, materials, and processes control
- Electromagnetic control
- Radiation hardening
- Commissioning Decommissioning
- Test planning analysis
- Packaging, handling, storage, and transportation

Where engineering specialty plans are required, they must clearly indicate how the specialty contributes to and benefits from the general systems engineering process and documentation. These detailed plans also provide the initial basis for the development and review of program cost estimates and schedules. Although each specialty plan will probably have its own data item description, all plans should, at a minimum, contain the following system engineering interface information:
- Objective - Purpose of including the specialty and the scope of its role within the systems engineering process

- Activity Definition - Summary description of all tasks required to fulfill the specified function, the content of required inputs from systems engineering, and the expected products to be provided to systems engineering.
- Responsibilities - Definition of all organizations supporting (or supported by) the activity, which tasks they are responsible for, and their lines of authority, with particular emphasis on the division of analytical tasks between the systems engineering organization and the specialty organization.
- Schedules - Timing and sequence of all engineering tasks related to major milestones for system development and design, and to specific inputs from supporting organizations in the systems engineering process, including critical path assessments.
- Resource Definition - Identification of specific hardware, software, personnel, and facilities required to complete the engineering tasks according to the schedule and to provide the required support, from this discipline's point of view, to the overall systems engineering process.

IV. SUMMARY

History has demonstrated that projects of the scale and technical complexity of SSC can be successfully completed, utilizing a formalized SE&I approach. Industry has experienced personnel resources, familiar with the SE&I process, available in sufficient numbers and skills to apply SE&I techniques to the SSC project. History has also shown that applying the SE&I process during the early project/design phases.yields the most value added.

Conclusions

The SE&I process can:

- Be tailored to specific project objectives.
- Add the most value when implemented early. Utilizing the iterative process to define, develop, and document requirements.
- Enhance key project goals of:
  -- System performance with least risk
  -- Reduced Life Cycle Cost
  -- Schedule performance
- Be responsive to the highly visible funding and contracting environment

Recommendations

It is our recommendation that the SSC project organization take the necessary actions to assure that a formalized Systems Engineering and Integration approach be utilized on the SSC.

Early management commitment and development of a Systems Engineering Management Plan tailored to the SSC project is essential to early application of a formalized process to the project.

References

1. MIL-STD-499 Engineering Management (USAF)
2. Systems Engineering Management Guide (Defense Systems Management College) 2nd Edition, Dec 1986

Advanced Particle Detectors

Chairman:

M. G. D. Gilchriese
SSC Central Design Group

AN INTEGRATED 3D DESIGN, MODELING AND ANALYSIS RESOURCE FOR SSC DETECTOR SYSTEMS

N.J. DiGiacomo, T. Adams, M. K. Anderson, M. Davis, B. Easom, J. Gliozzi, W. M. Hale, J. Hupp, K. Killian, M. Krohn, R. Leitch, M. Lajczok, L. Mason, J. Mitchell, J. Pohlen and T. Wright

Martin Marietta Astronautics, Denver, Colorado 80201

I. INTRODUCTION

Integrated computer aided engineering and design (CAE/CAD) is having a significant impact on the way design, modeling and analysis is performed, from system concept exploration and definition through final design and integration. Experience with integrated CAE/CAD in high technology projects of scale and scope similar to SSC detectors leads us to propose an integrated computer-based design, modeling and analysis resource aimed specifically at SSC detector system development. The resource architecture emphasizes "value-added" contact with data and efficient design, modeling and analysis of components, sub-systems or systems with fidelity appropriate to the task.

We begin with a general examination of the design, modeling and analysis cycle in high technology projects, emphasizing the transition from the classical "islands of automation" to the integrated CAE/CAD-based approach. We follow this with a discussion of "lessons learned" from various attempts to design and implement integrated CAE/CAD systems in scientific and engineering organizations. We then consider the requirements for design, modeling and analysis during SSC detector development, and describe an appropriate resource architecture. We close with a report on the status of the resource and present some results that are indicative of its performance.

II. THE DESIGN, MODELING AND ANALYSIS CYCLE

The development of a complex device or system proceeds through a number of distinct, well-defined phases. Figure 1 depicts this evolution as a flow, starting with concept exploration and ending with system commissioning. The success of such a project can be viewed as the convergence of the flow; careful, thorough concept exploration and requirements definition assure the expeditious assembly, test and commissioning of the system. Conversely, the process can well diverge if the early phases of the project are passed over too quickly.

A constant feature of each project phase is the design, modeling and analysis cycle. The cycle is repeated continuously as the project proceeds, albeit with requirements and goals that are functions of the phase. Concept exploration, for example, requires fast turnaround analysis in order to explore a broad range of design possibilities, but can tolerate rather lower fidelity. The final design, however, demands thorough, high fidelity modeling and analysis.

The success of a project, then, can depend strongly on the availability of the design, modeling and analysis tools appropriate to the task. It is important to stress the words AVAILABLE and APPROPRIATE. Sophisticated analysis packages are of little use in concept exploration if they are too cumbersome to use quickly. Likewise, simple, easy to use modeling packages may be inadequate to an engineer performing a detailed part design.

The Classical Approach: "Islands of Automation"

Individual design, modeling or analysis packages can be viewed as "Islands of Automation". Sophisticated analyses can be done by computer, but data transfer, manipulation and translation between programs must be done by hand. The fidelity of modeling and analysis is generally limited by the need for extensive manual setup, gridding and meshing. The analyses performed are quite independent of each other, and turnaround times between iterations are long compared to other timescales in the project, such as drawing preparation and technical or management milestones. In essence, the ability to exercise design and engineering judgment and to explore the space of design possibilities is limited by the time-consuming tedium of handling the data flow between the "Islands".

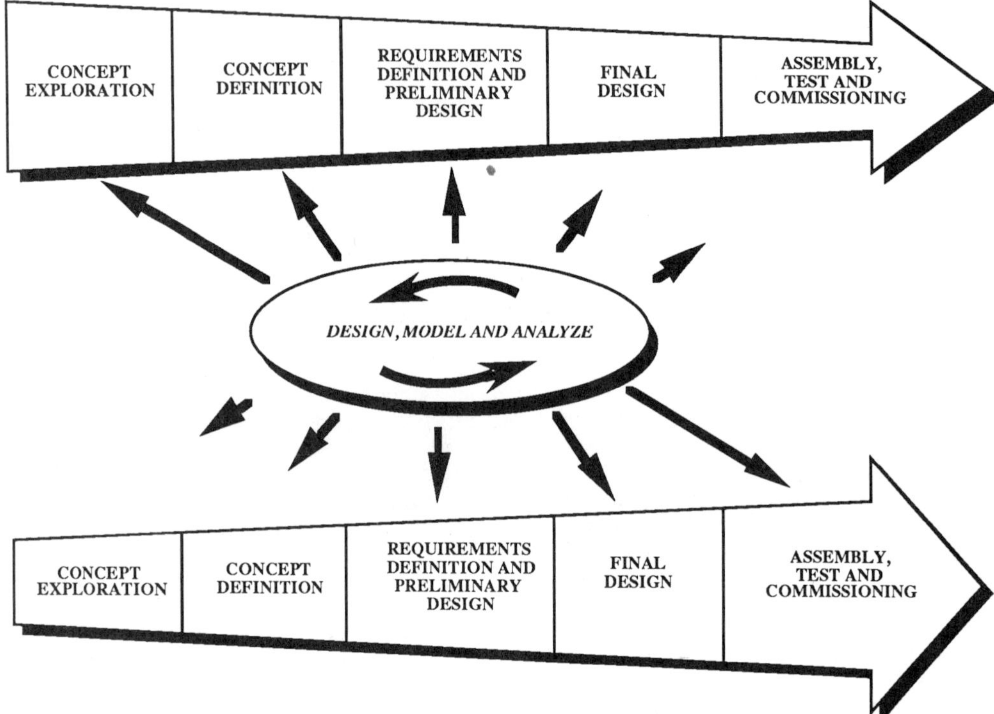

Figure 1. The project flow, with phases indicated. The design, modeling and analysis cycle reoccurs throughout the project lifetime, with requirements and goals that depend on the phase. Investment in design, modeling and analysis tools that allow thorough concept exploration and definition helps smooth the subsequent final design, assembly, test and commissioning (upper arrow). Conversely, passing through the concept and requirements definition phases too quickly is a primary cause of cost and schedule problems during assembly, test and commissioning (bottom arrow).

Integrated CAE/CAD

Integrated CAE/CAD provides designers and engineers in all phases of the project with linked design, modeling and analysis tools that can be tailored to the task at hand. The essential emphasis is on "value-added" contact with data. One only intervenes in the flow of data between packages to contribute engineering or design judgment. The user does not participate in routine data transfer, translation and manipulation. Interactive graphics are used extensively to input design, define modeling and analysis parameters and conditions and to evaluate results. The system is optimized to provide fast turnaround and to allow parallel analyses. The fidelity of modeling and analysis is determined by the task requirements and not by the tedium of setting up new grids or meshes, and errors due to manual data manipulation are eliminated.

Concept Exploration and Definition

Groups (i.e. potential collaborations) develop a conceptual detector system design for a given set of physics requirements. For each design, one must:
- Evaluate the viability of different sensor/component detector technologies within the context of the particular system design.
- Simulate the overall response of the detector system
- Determine the engineering feasibility and implications of the design.

Groups should be able to perform the above quickly and easily enough to allow the exploration of the maximum fraction of "detector possibility space". In addition, the Laboratory must have the means to compare and evaluate competing designs.

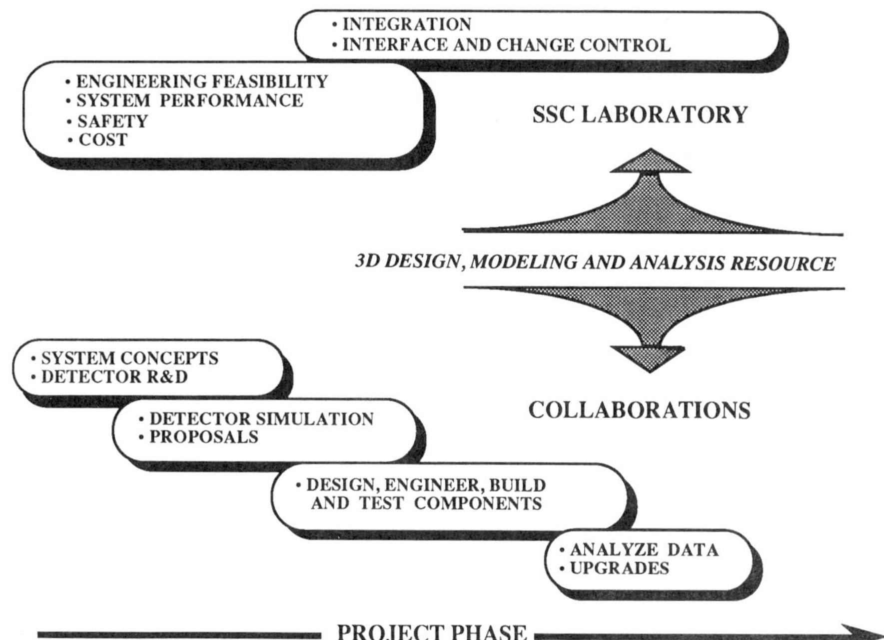

Figure 2. SSC laboratory and collaboration tasks that could utilize 3D design, modeling and analysis capability, shown versus project phase.

Requirements Development and Preliminary Design

At this stage, collaborations have rather well-defined physics goals and preliminary detector designs. One must:
- Carefully determine the performance (i.e. physics response) and "buildability" of each detector system.
- Define the interfaces among components and sub-systems, and establish an interface control methodology.
- Generate preliminary drawing packages of the system and its components for distribution to collaborating institutions to allow evaluation of task magnitude and to constrain prototype development to overall system requirements.
- Feed back progress in prototype R&D into overall system design and analyze impacts on system performance, other detector components and cost.

Design, Modeling and Analysis of "One of a Kind" Devices

Investments in extensive prototyping, or more recently, in design, modeling and analysis tools is easily justified in "production" high technology by distributing the expenditure over many units. It is much harder to quantify the return on such investments for "one of a kind" devices such as planetary spacecraft or SSC detectors. The real return is probably best measured by:

- The performance enhancements that result from broader and more thorough concept exploration and definition. In essence, a larger fraction of "possibility space" is explored, resulting in a better device.
- The improvements in the assembly, test and commissioning phases of the project, such as fewer change orders and better schedule performance.

III. IMPLEMENTATION OF INTEGRATED CAE/CAD

There are a number of "lessons learned" from various attempts at designing and implementing integrated CAE/CAD systems within scientific or engineering organizations, including Martin Marietta Astronautics [1]. These include the following:

- CAE/CAD capability consists, first and foremost, of people. One does not simply buy software and hardware tools. One must train and maintain a team of designers, engineers, analysts and managers.
- Integration of individual design, modeling and analysis packages into a useful, coherent tool requires a significant amount of work by systems professionals. The integration cannot be done solely by the users or by those responsible for the component packages. In addition, optimal use of the CAE/CAD tools requires that one establish, and follow, procedures and formats to track and catalog designs, drawings, specifications, etc.
- The individual design, modeling and analysis skills of the project team become less separable when integrated CAE/CAD is employed. An interdisciplinary approach seems to work best.
- Project management must adapt, both programmatically and organizationally, to the interdisciplinary environment. A major contributor to successful implementation of CAE/CAD in technical organizations is management commitment. In addition, management must believe that the up-front investment in CAE/CAD tools and an emphasis on more modeling and analysis in the early stages of a project will lead to a better design and faster commissioning.
- Careful hardware and software evaluation is critical, because one is building a SYSTEM, not a series of stand-alone codes. Again, systems professionals must assist users or specialists in making these choices.
- The often conflicting requirements of resource distribution and centralization must be satisfied. The resource capabilities must be distributed to users wherever they are, but the main database must be controlled and code modifications, revisions and updates implemented by a central authority. One approach that seems to work is to have a fraction of the central facility staff rotate in from the distributed user facilities. This allows for good user familiarity with, and direct and timely user input to the central facility.

IV. AN INTEGRATED CAE/CAD RESOURCE FOR SSC DETECTORS

We now consider integrated CAE/CAD in the context of SSC (or other large collider) detector system development. Figure 2 shows our project flow arrow along with tasks that the SSC laboratory and the collaborating institutions must perform. We propose an integrated CAE/CAD resource centered at the SSC Laboratory, with distributed access to collaborating institutions. As such, the resource is available to the collaborations that will develop the detector system concepts and components, and to the Laboratory administration and staff that must evaluate, select and then integrate the detectors. We examine below the requirements for design, modeling and analysis during each phase of detector system development. We then propose a candidate integrated CAE/CAD resource architecture.

Final Design

For each approved and funded detector system, one must:
- Efficiently evolve the detailed design, in an iterative manner, based on component development, interface control, interference checks and engineering modeling and analysis.
- Generate component drawings and manufacturing sequences from a central design database.
- Control quality and uniformity of components and parts through approved parts lists and a uniform cataloging system.

Assembly, Test and Commissioning

In this phase, components built at collaborating institutions, industry and the Laboratory are assembled, integrated and tested at the Laboratory. One must:
- Control changes and interfaces from a central authority.
- Redesign, reanalyze and recalculate detector response on a short timescale to adapt the detector system to changes in component design or performance, or to changes in physics requirements.
- Perform simulations of detector performance for data analysis purposes, from the highest fidelity model of the detector system.
- Generate assembly drawings and sequences.

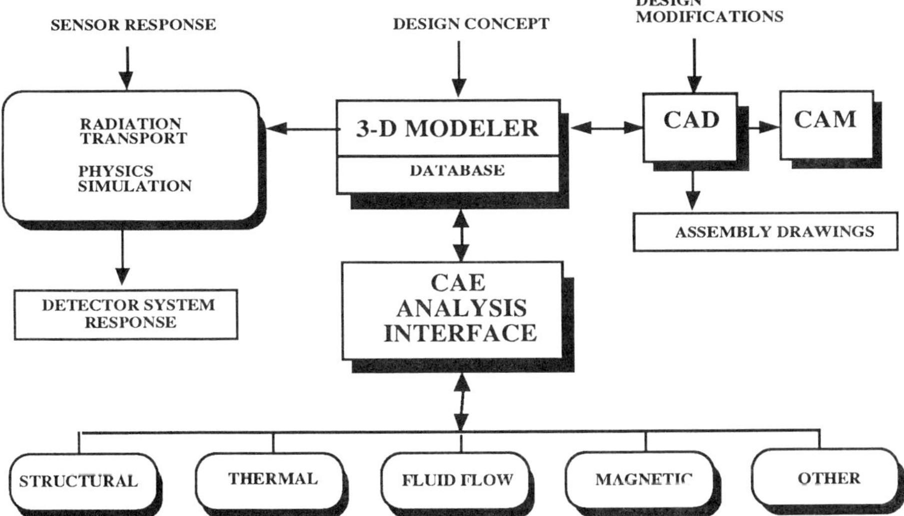

Figure 3. Candidate architecture of an integrated CAE/CAD resource for detector system development. The elements of the resource and the data flow between packages are described in Section III of the text.

Resource Architecture

Figure 3 is a schematic representation of an integrated CAE/CAD resource or "tool kit" for SSC detector development. The resource architecture, based on existing systems at Martin Marietta, is designed to satisfy the requirements discussed above. The schematic nature of Figure 3 must be emphasized; as noted earlier, integrating the design, modeling and analysis packages is a complex task for systems professionals. A graphic example of the actual complexity can be seen in Figure 4, where the thermophysics part of the "demonstration" integrated CAE/CAD resource (discussed in Section IV) is shown.

An example of code integration is shown in Figure 4, where the thermophysics capability of the resource is presented. A design created in GEOMOD is gridded in SUPERTAB for mechanical analysis by NASTRAN. SUPERTAB's gridding is translated for both TRASYS and SINDA by TSIG [7] and the thermal analysis is performed. The temperature profiles are then passed back to NASTRAN via SUPERTAB for the mechanical analysis, with iteration occuring as necessary.

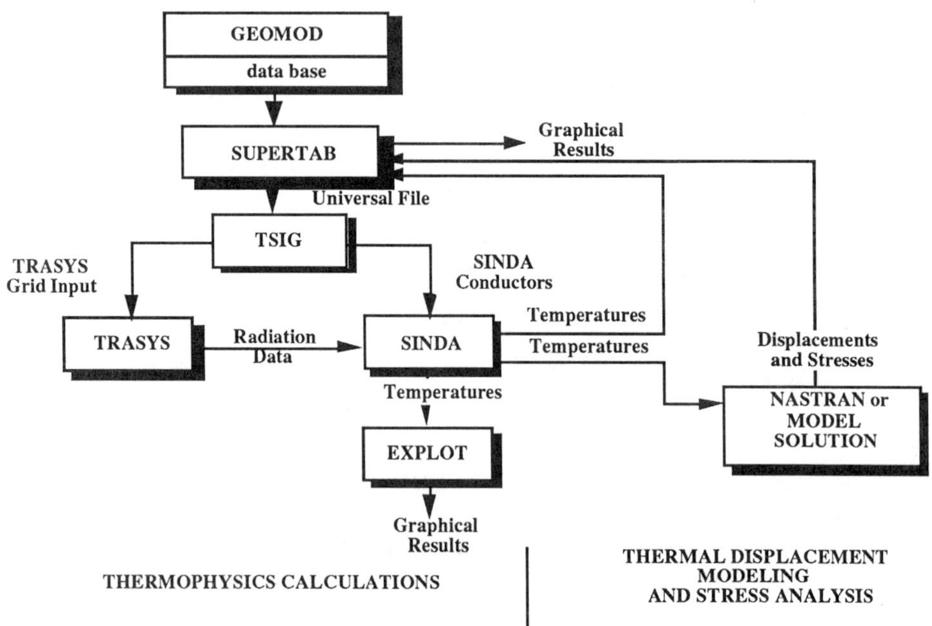

Figure 4. The thermophysics section of the integrated CAE/CAD resource. GEOMOD is a 3D modeler, and SUPERTAB is the CAE interface that generates finite element meshes. TRAYSYS abd SINDA are thermal radiation transport and thermal analyzer codes, with EXPLOT and TSIG as interface packages. NASTRAN is one of the suite of stress and dynamics codes available within the resource.

Figures 5-7 are representative results from our design, modeling and analysis recource applied to model detector. They are meant to indicate the rapid, efficient process by which one moves from design to a 3D model to interactive analysis and engineering evaluation. The true utility of the resource's graphic capability is not apparent in Figures 5-8, as they are necessarily black and white; color is particularly important in making full use of contour plots such as Figure 7.

Figure 5 shows cuts through the model detector. The model is based on the DØ liquid argon calorimeter [9] and a cryostable superconducting coil similar to one proposed for the CDF detector [10]. Figure 6 depicts the GEOMOD-resident design calorimeter modules supported by a stainless steel strap and beam structure (the cradle) that attaches to the external coil support (not shown). Figure 7 shows the axial (i.e. along the beam axis) stresses on the junction under gravity, cryogenically loaded and in a 2 Tesla magnetic field. The indicated stresses are due to torque on the strap generated by the calorimeter modules. The analysis employed MSC/NASTRAN, TRASYS/SINDA and POISSON.

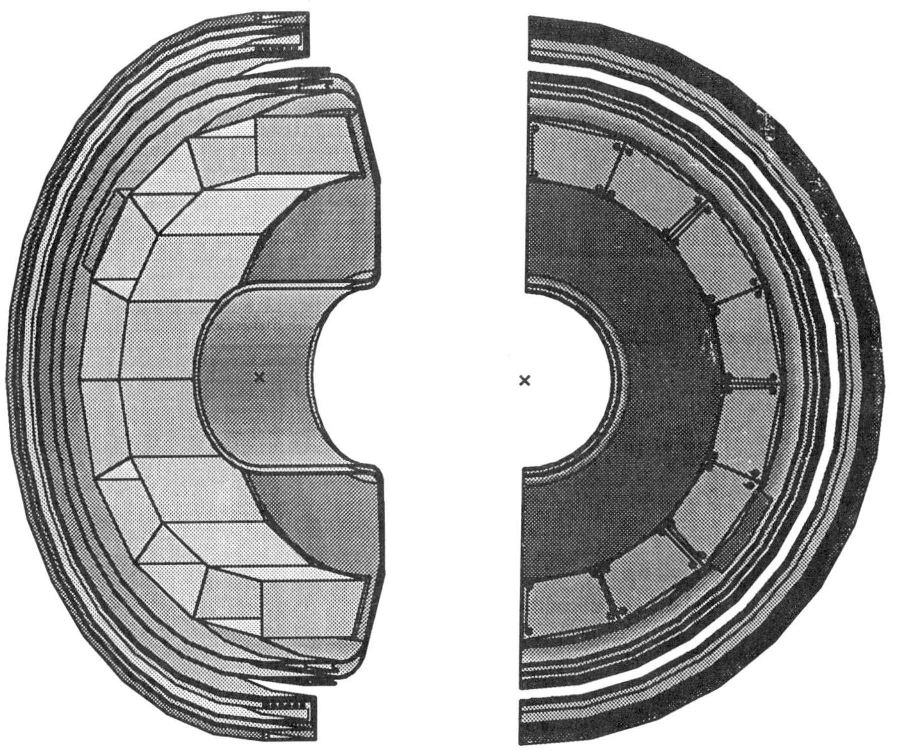

Figure 5. GEOMOD sections of the model detector, showing the cryostat, outer calorimeter modules, and the superconducting coil.

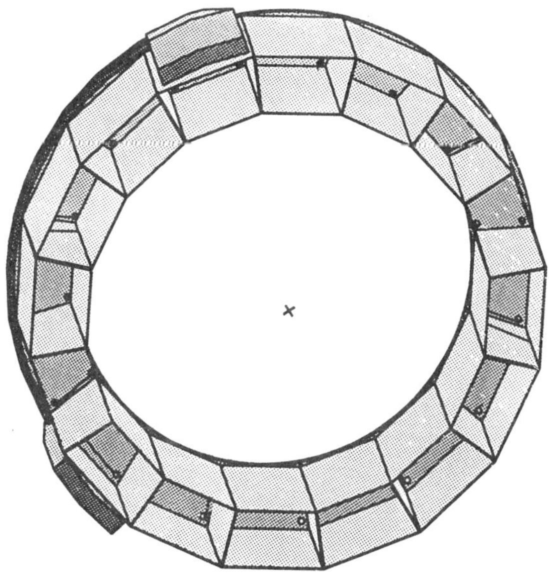

Figure 6. A GEOMOD detail of the calorimeter modules supported by a stainless steel strap and beam.

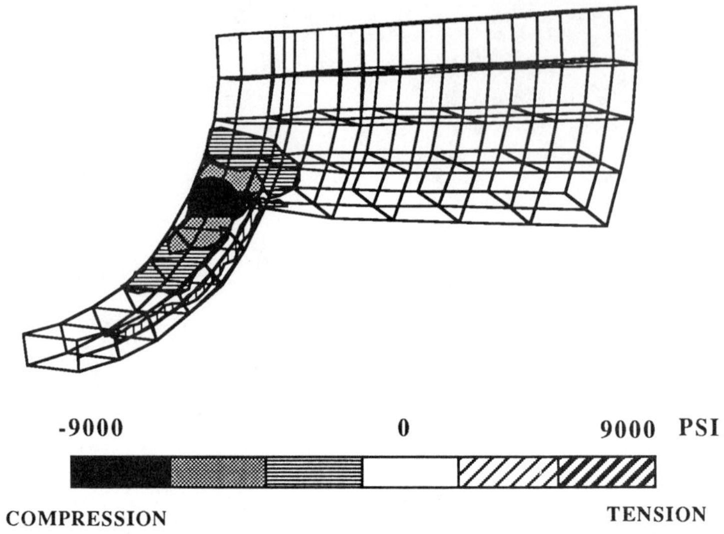

Figure 7. A stress contour plot of the intersection of the support strap and beam, showing the axial stresses on the junction.   The compression/tension scale is shown at the bottom of the picture.   The analysis was performed with gravity, a 2 Tesla magnetic field, and under cryogenic load.

CONCLUSIONS

The complexity, cost and performance requirements of SSC detectors make it imperative that the design, modeling and analysis cycle be made as efficient, and cost effective as possible. To this end, and based on similar challenges with "one-of-a-kind" high technology in the aerospace industry, we are developing an integrated CAE/CAD resource for SSC detector system development.

ACKNOWLEDGEMENTS

We wish to thank Gil Gilchriese for bringing this subject to our attention, Howard Gordon for DØ liquid argon calorimeter documentation and Ron Fast for the design study on a cryostable superconducting coil for CDF.

REFERENCES

1. Project Challenge CAE/CAD/CAM Initiative, Martin Marietta Astronautics Group, Denver CO. (1987)
2. MSC/NASTRAN, MacNeal-Schwendler Corp., Los Angeles, CA, (1987).
3. ABAQUS, Hibbitt, Karlsson and Sorensen Inc., Providence, RI, (1987).
4. ANSYS, Swanson Analysis Systems, Inc. Houston, PA, (1987).
5. I-DEAS, Structural Dynamics Research Corp., Milford. OH, (1987).
6. POISSON, Los Alamos National Laboratory, Group AT-6, Los Alamos, NM, (1987).
7. SINDA and TRASYS, Martin Marietta Astronautics Group, Denver CO., (1987).
8. CADDS 4X, Computervision Corp., Bedford, MA, (1987).
9. DØ Revised Design Report, DØ Note 512, Fermilab, (1985).
10. R. Fast et al., "Design Report for a Cryostable 3m Diameter Superconducting Solenoid for the Fermilab CDF Facility", TM-1075, Fermilab, (1981).

SILICON PIN DIODE HYBRID ARRAYS FOR CHARGED PARTICLE DETECTION: BUILDING BLOCKS FOR VERTEX DETECTORS AT THE SSC*

Stephen Gaalema and Gordon Kramer**
Hughes Aircraft Company, Carlsbad, CA 92009

Stephen L. Shapiro and William Dunwoodie
Stanford Linear Accelerator Center, Stanford University, Stanford, CA 94309

John F. Arens and Garrett Jernigan
Space Sciences Laboratory, University of California, Berkeley, CA 94720

ABSTRACT

Two-dimensional arrays of solid state detectors have long been used in visible and infrared systems. Hybrid arrays with separately optimized detector and readout substrates have been extensively developed for infrared sensors. The characteristics and use of these infrared readout chips with silicon PIN diode arrays produced by MICRON SEMICONDUCTOR for detecting high-energy particles are reported. Some of these arrays have been produced in formats as large as 512×512 pixels; others have been radiation hardened to total dose levels beyond 1 Mrad. Data generation rates of 380 megasamples/second have been achieved. Analog and digital signal transmission and processing techniques have also been developed to accept and reduce these high data rates.

INTRODUCTION

High-resolution vertex and tracking devices at the SSC will be important for the reconstruction of complex events, including secondary vertices close to the primary interaction point. The main properties required of these devices are: fast response time, fine spatial resolution, multiple particle resolution, and radiation hardness.

A resolution of better than 10 μm and the ability to distinguish the many particles of a jet can be achieved with two-dimensional pixel devices. Due to the three-dimensional nature of the coordinate information provided, these provide efficient track finding with a minimum number of layers in the high-multiplicity environment of the SSC.

The SSC beam crossing period will be 15 ns, and interesting events will occur every 100 to 10,000 crossings. Hit times must be recorded to an accuracy of one crossing period, and the readout time of the detectors must be less than the mean interval between interesting events to avoid large dead time losses.

---

* Work supported by the Department of Energy, contract DE–AC03–76SF00515.

** Permanent address: Electro-Optical & Data Systems Group, Hughes Aircraft Company, El Segundo, CA 90245.

The expected combination of high luminosity and large particle multiplicity will produce a high-radiation environment in the vicinity of the collision point. For a proton-proton total cross section of approximately 100 mbarn, the SSC will yield approximately 10^8 interactions per second when operating at its design luminosity of $10^{33}/cm^2/sec$. At a distance of 5 cm from the beam, the absorbed radiation dose will be approximately 1 Mrad per year.

New electronic detector systems usually require at least several years for development. Examples are CCDs, infrared detector arrays and x-ray bolometers. The schedule for the SSC construction is commensurate with the schedule for new vertex detector development if the latest existing technology is used as a base.

DEVICE CANDIDATES

Candidates for vertex/tracking devices at the SSC include: wire chambers, scintillating fiber detectors, silicon microstrip detectors, double-sided microstrip detectors, silicon drift chambers; and pixel devices in the form of CCD arrays, monolithic silicon arrays and hybrid microdiode arrays. It is generally recognized that in terms of space, momentum and two-track resolution, wire chambers at present are inferior to silicon devices. It is also generally recognized that pixel devices are preferable to strip devices or drift devices, as far as two-track resolution is concerned.

Pixel devices—in particular, silicon diode arrays—are a natural choice for vertex detectors. These devices provide three-dimensional coordinate information with a spatial resolution of a few microns. However, as we move farther from the collision point to the tracking detector, many of the other devices mentioned above become attractive candidates for a variety of reasons.

A vertex detector based on the use of a pixel device would provide efficient track-finding with a minimum number of layers in the high-multiplicity environment of the SSC due to the three-dimensional nature of the coordinate information provided. The absence of ambiguities in coordinate matching which are always present in nonpixel devices allows the number of detector layers to be minimized, thereby reducing the size and cost of the vertex detector.

The only pixel devices presently in use in high energy physics experiments today are CCDs.[1] These devices are inappropriate to the SSC environment for a variety of reasons, including: (i) a small signal size of about 1000 electrons, which necessitates their being cooled, renders them extremely susceptible to degradation caused by radiation damage, and requires costly and delicate readout electronics; (ii) a varying transfer efficiency across their faces; (iii) no off gate during readout, resulting in a time/space ambiguity; and (iv) a serial readout resulting in a readout time of about 10 msec for a device with 10^5 pixels.

A number of researchers are presently working on the problem of fully integrated pixel detectors in monolithic arrays.[2,3] The goal is to fabricate the readout electronics on the same high-resistivity silicon as the detector diodes. Success in these efforts will result in detectors having the minimum thickness possible. However, there will be a loss of flexibility. A change in either the detector specification or the readout specification will result in not only a circuit change, but also a re-analysis of the production process, since they are now intimately coupled. There will be difficulty in production, since both the detector and the readout electronics must be fabricated in high-resistivity facilities which can handle this new process. There are far fewer of these facilities worldwide than there are those which can produce more conventional circuits.

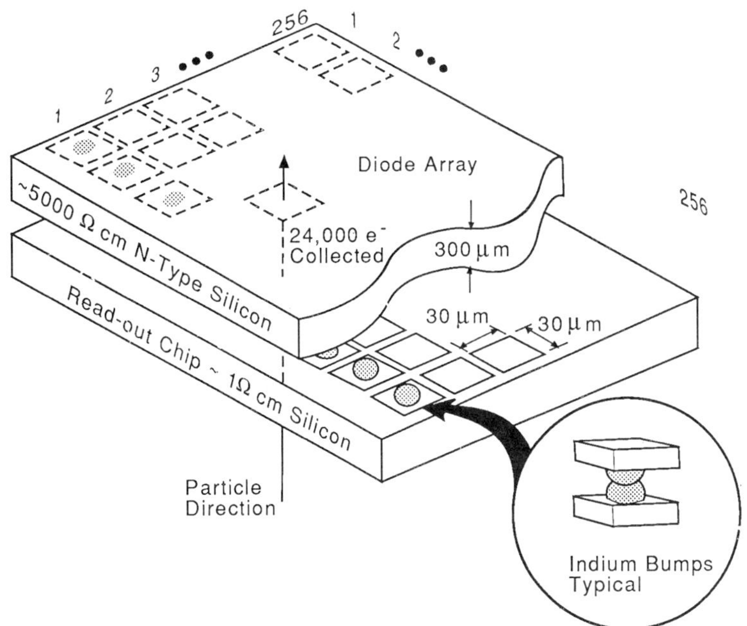

Fig. 1   Schematic representation of a Silicon PIN Diode hybrid detector.

The architecture which we find to be well suited for charged particle detection at the SSC is that of a hybrid.[4] The charged particle detector and the readout electronics are constructed as two separate silicon chips, each optimized for its specific function. The two chips, indium bump bonded together, then provide the basic building block for the construction of a detector array.

SILICON HYBRID ARRAY EXPERIENCE

The choice of the hybrid design (*viz.*, one in which each diode of the detector array is bonded to an independent amplifier readout circuit on a mating VLSI chip via an array of aligned indium metal bumps that cold weld under pressure to form ohmic contact), allows for additional flexibility in the selection of detector and readout electronics.[5] For instance, a change in the leakage current specification of the detector array will not affect the readout electronics, nor will a change in the VLSI chip oxide thickness to accommodate a radiation hardness specification affect the detector array. Figure 1 is a schematic representation of a silicon PIN diode array hybrid. Figure 2 is a microphotograph of an array of indium bumps prior to the bonding process. The bumps shown are approximately 15 μm in diameter. A small (10 × 64 pixel) hybrid mounted in a 68-pin leadless carrier is shown in Fig. 3. The readout chip in this device, the Hughes CRC–198, was originally designed to mate with a long-wave infrared detector chip which for optical reasons required 120 μm square pixels. The success of this chip lead to a series of larger arrays with smaller pixels (for 3–4 μm infrared platinum silicide Schottky barrier detectors) as shown in Table I. A photograph of the 256 × 256 pixel CRC–365 hybrid, which has become fairly mature, is shown in Fig. 4. Figure 5 is a histogram of the infrared sensitivity of each pixel of a 256 × 256 hybrid array. Of note is the nearly identical electronic response of the 65,536 pixels of the array and the notation that only three pixels in the array are dead. Currently being developed are 512 × 512 (see Fig. 6) and larger arrays. The circuit used in the array unit cell of all of the chips listed in

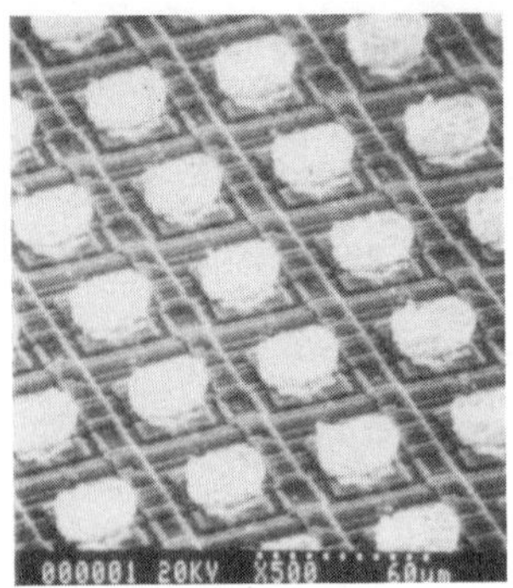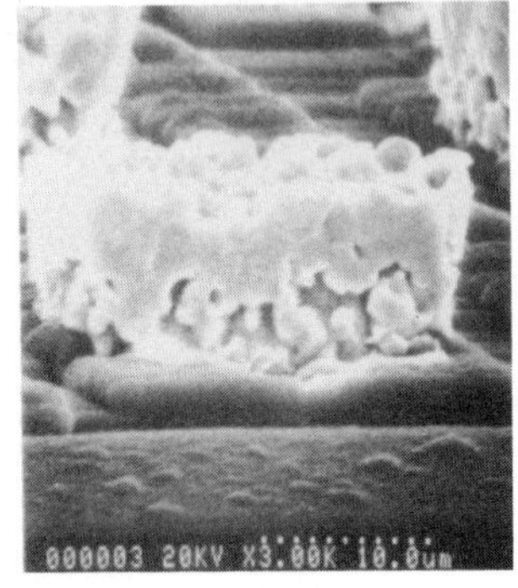

Fig. 2   A microphotograph of an array of indium bumps prior to the bonding process.

Table I is called a source follower per detector (pixel) or SFD with direct readout (DRO) multiplexing as opposed to CCD multiplexing.

Figure 7 is a schematic diagram of the MOSFET circuit of the 10×64 readout array. The diagram has been divided into its several functional portions. The section replicated for each pixel contains four MOSFETs. Signal charge is generated by the detector diode, and is fed to the gate of the signal MOSFET, where it stays until a readout is made. The pixel selection circuit indicates how a sequence of address lines can select an individual pixel by turning on the gates of the V_{DD} bias MOSFET and the enable gate of the reset MOSFET. The U_{reset} signal allows the gate of the signal MOSFET to be reset to the V_{reset} level for any pixel that is enabled by the reset MOSFET. All of the signal MOSFETs in a column of the array are connected to a readout MOSFET in a source follower configuration which provides power for driving an external circuit.

PIN DIODE HYBRID ARRAYS

The existing infrared readout chips can be mated to PIN diode arrays to produce hybrids with a number of advantages as high-energy particle detectors.

The present readout chips allow random access to any pixel, which then operates as an independent detector. By virtue of its geometry alone, each pixel detector (PIN diode)

Fig. 3   A photograph of a 10×64 readout chip bump bonded to a Silicon PIN diode array.

Fig. 4   A photograph of a 256 × 256 silicon hybrid.

provides about 3,000 times less sensitivity to diode leakage current than a microstrip detector, which will increase the radiation hardness to neutrons. To complement this increase in the radiation hardness of the detector diode, one of the present readout chips (the 10×64 array) has been fabricated in a technology which is radiation hard at cryogenic temperatures to 1 Mrad of Cobalt 60 gamma rays.

The hybrid operates at room temperature, eliminating the need for a cryostat and easing the problems of installation, alignment and accessibility for power and control cabling.

The detector thickness itself can now be optimized. The PIN diode array can be fabricated to give the requisite signal (80 e/μm of silicon), while the VLSI chip can be thinned to less than 100 μm if necessary. This feature is important for minimizing multiple-scattering effects and also for limiting the number of adjacent pixels turned on by tracks incident at large angles to the surface.

It should be noted that the diode anodes are indium bump bonded to the silicon readout chip, leaving the cathode available to be etched into a pattern of pads or strips. Signals picked up on the cathode pads would be available for fast triggering or for reducing

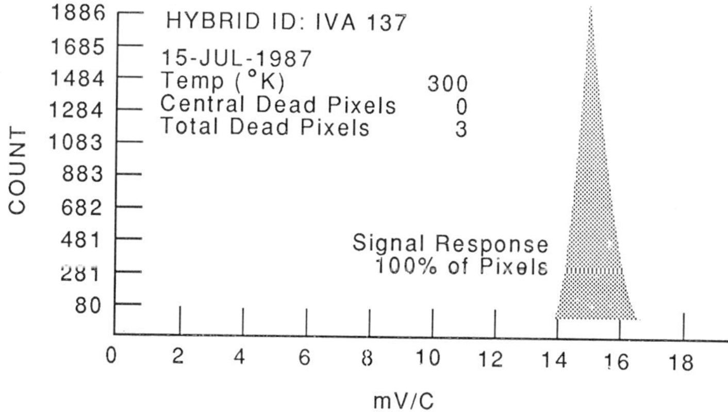

Fig. 5   A histogram of the infrared sensitivity of each pixel of a 256 × 256 hybrid array.

Fig. 6   A photograph of a 512 × 512 silicon infrared hybrid.

the readout time by using the random access nature of the readout electronics. System considerations will determine if the cathode signals are useful.

To address the problem of data which may accumulate during the readout cycle, gating circuitry can be designed to inhibit the flow of charge between the detector and the readout chip after an event is tagged. Alternatively, an input buffer can be added which leaves the detector alive but which does not alter the data being read out. These features do not exist on the present readout chips.

To address the problem of data being recorded from various beam crossings, a design must be adopted which allows the data and the time of arrival to be correlated. Present chips do not have this feature. They do appear to have the ability to provide a "fast indication that a particle has been recorded by the detector. However, we have not yet explicitly studied this feature in the laboratory. This ability, important if confirmed, can

Table I

Design	Array Size	Pixel Size (μm)	Year
CRC–171	10 × 32	120 × 120	1980
CRC–198	10 × 64	120 × 120	1981
CRC–228	58 × 62	75 × 75	1983
CRC–234	128 × 128	50 × 50	1984
CRC–304	256 × 256	30 × 30	1985
CRC–350	256 × 256	30 × 30	1986
CRC–352	244 × 400	24 × 24	1986
CRC–365	256 × 256	30 × 30	1986
CRC–389A	244 × 400	24 × 24	1987
CRC–389B	512 × 512	20 × 20	1987
CRC–412	488 × 640	20 × 20	1988

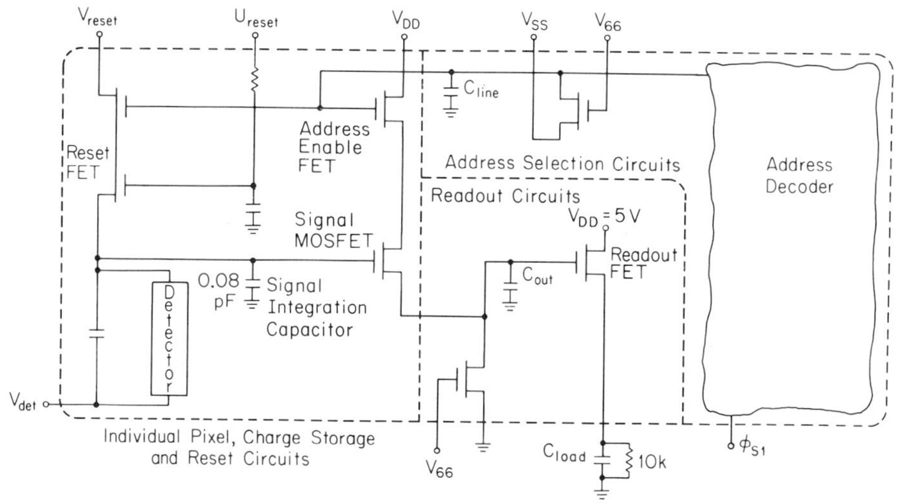

Fig. 7   Schematic drawing of the readout electronics of the 10 × 64 readout array.

be used in designs to strobe shift registers which record time information while retaining the small pixel size of our present chips.

In the present generation of readout chips, it should be noted that the power used during the write cycle is essentially zero. Power is used only when reading, and cooling needs thus are lessened.

DESCRIPTION OF PRESENT HARDWARE

There are two readout chip geometries currently being built into detector arrays. Figure 8 is a photograph of the two mating PIN diode arrays on a production wafer.[6] Shown are a number of 10×64 arrays with 120 μm square pixels and a number of 256×256 arrays with 30 μm square pixels. The diode arrays are identically processed and differ only in that they must mate with their respective readout chips. The readout chips are

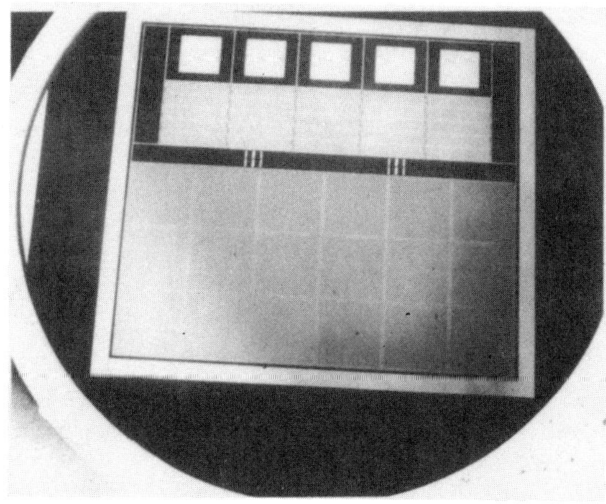

Fig. 8   A photograph of 256 × 256 and 10 × 64 PIN detectors on a wafer. The five square structures at the top of the wafer are test diodes. The wafers have been manufactured by MICRON SEMICONDUCTOR, LTD., Sussex, England.

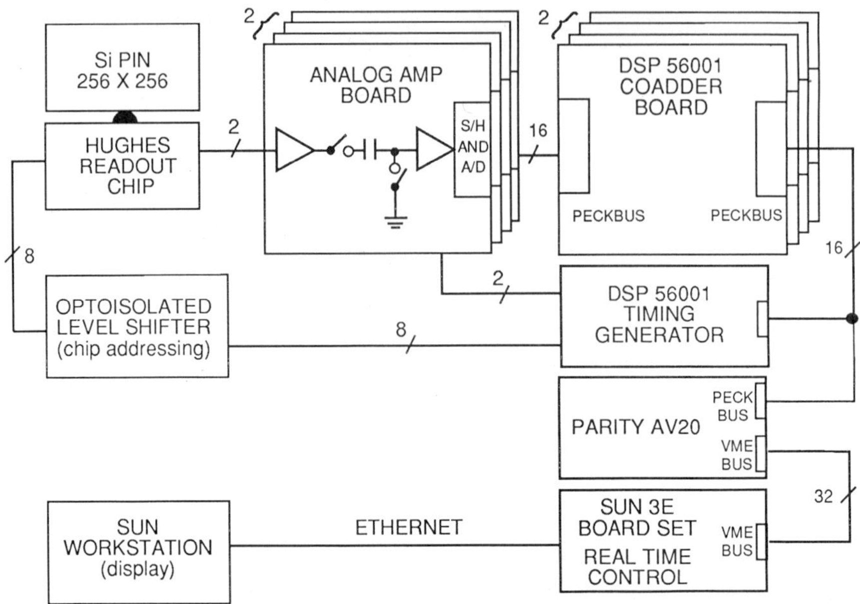

Fig. 9   A block diagram of the high energy physics data acquisition electronics.

similar, but have some differences. For instance, the 10 × 64 array has been optimized to be radiation hard to 1 Mrad at 10° K. It has a random access architecture in that a unique setting of its address lines will select one and only one row of pixels. The 256 × 256 array, on the other hand, is random access in that the pixels are addressed via row and column shift registers. This feature makes addressing a given pixel more complicated but allows easy implementation of a sparse scan algorithm. This particular readout chip has been optimized to collect electrons but is bipolar at the signal levels we expect.

Figure 9 is a block diagram of the high energy physics data acquisition system. A Sun Microsystems Sun–3/110LC–4 workstation controls a system housing a Sun 3/E CPU, a Motorola 68020 bus converter board, amplifiers, ADCs, digital signal processors and a clock generator. The digital signal processor is the Motorola DSP56001. This device acquires data at a rate of 10 MIPS, processes it and passes it via the MC68020 to the Sun 3/E. The bus converter board, a Parity systems AV20 dual-port processor, interfaces the local analog bus (PECKBUS) to the VME bus.

The noise performance of the 10 × 64 readout chip at 10° K has been measured to be 45 electrons rms referred back to the input. This is within a factor of two of the expected Johnson noise of the gate of the input signal MOSFET. At room temperature, the expected noise signal should be about 200 electrons rms, taking into account the square root of the temperature ratio in the Johnson noise component. However, the measurement has not yet been made. If the noise performance is as expected, detectors thinner than the traditional 300-μm-thick devices currently being fabricated would offer more than adequate signal-to-noise performance.

RESEARCH PROGRESS

Using Detector Development funds from the Department of Energy, a contract was made with Micron Semiconductor to design and fabricate two silicon PIN diode arrays.

The array specifications were chosen to allow the detector chips to mate with two Hughes Aircraft Co. readout chips. The arrays fabricated are a 10×64 array with 120 μm square pixels and a 256×256 array with 30 μm square pixels. The first silicon 10×64 hybrid arrays were completed in December 1988, and the silicon 256×256 hybrid arrays are being built. The data discussed in this paper were taken with the 10×64 array.

Additional Detector Development funds were used to begin the fabrication of the dedicated high energy physics data acquisition system described above. The amplifier/ADC/DSP56001 boards and the clock generator in the present system, however, are circuits remaining from the infrared data acquisition system.[7] These need to be redesigned to high energy physics criteria. Figure 10 is a photograph of the data acquisition electronics for the 256×256 array.

A data acquisition and display software package based on the infrared system has been written. The operating system is UNIX, the DSP has been programmed in assembly language, and various control functions are written in Magic/L, an interactive language

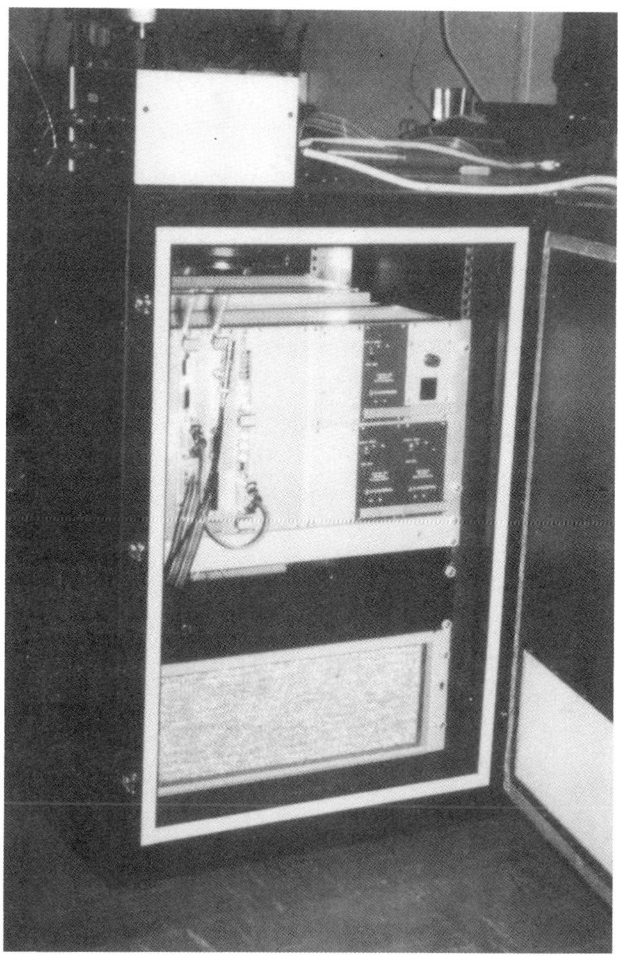

Fig. 10   A photograph of our present data acquisition system for the 256×256 array.

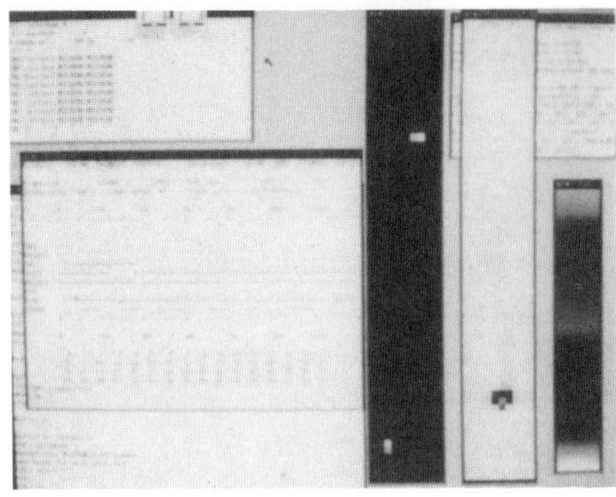

Fig. 11   A photograph of the SUN display screen. Included is a display of the timing and control pattern, the pulse height scale, and two event windows each showing a 10 × 64 representation of the hybrid.

derived from Forth. The Parity AV20 dual port processor has been programmed in C, and the DSP will be reprogrammed in C. Figure 11 is a photograph of a SUN display during data acquisition. The large window on the left displays the timing pattern used to read out the hybrid. The small window on the right shows the pulse height scale, which is a color scale from blue at the top to white at the bottom. The two rectangular windows just right of the center represent the pixel array and contain two events; the one on the right has detected an alpha particle, while the event on the left has two alphas and a 59.5 KeV x-ray.

Figure 12 is a photo of the hybrid placed under an Americium 241 alpha source. Americium 241 is also a source of 59.5 KeV and 26.3 KeV x-rays. The mounting fixture is a low temperature Dewar for tests to be performed later, which will characterize the hybrid's performance as a function of temperature.

Figure 13 is a scope trace of an output channel. The faint traces are the signals from the detected alpha particles. Correlated double-sampling electronics described

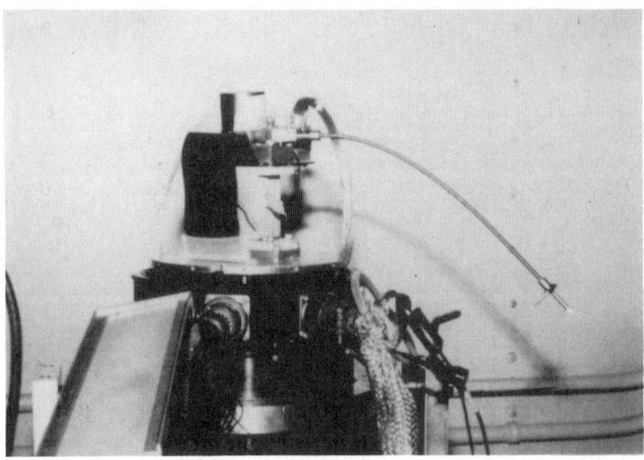

Fig. 12   A photograph of the hybrid mounted in the Dewar, under an Americium source.

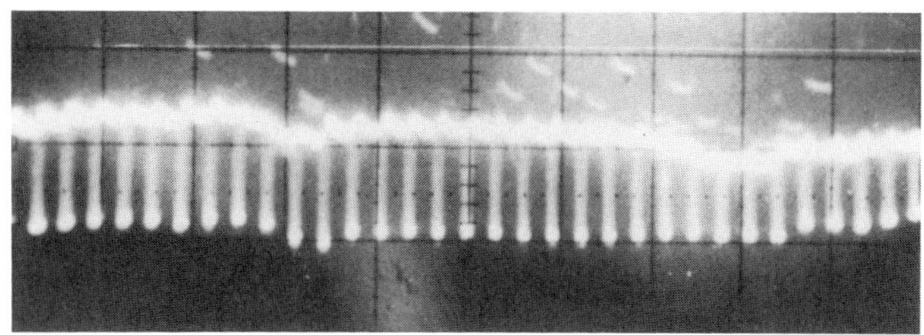

Fig. 13   A photograph of a scope trace of an output channel (one row) of the 10×64 detector.

elsewhere[8] is needed to reduce the 1/f noise and provide the excellent signal-to-noise ratio performance observed.

At the present time, a system exists which can read with either of two detector geometries: 10×64 or 256×256. The program of testing and characterizing these devices has just begun. The first hybrid arrays have just been manufactured. Preliminary data demonstrate the detection of alpha particles and 60 KeV x-rays with a high signal-to-noise ratio. However, the readout system has not been optimized. The timing pattern and bias voltages have not been swept through their respective ranges, and environmental noise has not yet been removed. The preliminary data are encouraging. The detection of a 60 KeV x-ray means that we have detected a signal having approximately 16,700 electrons. This is less than the 24,000 electrons generated by a minimum ionizing particle traversing 300 μm of silicon.

Table II is a summary of current device characteristics.

LOW-MASS, HIGH-DATA-RATE PACKAGING

Packaging techniques have also been developed for the infrared application of these chips that should be very useful for the SSC. A beryllium module pedestal is shown in Fig. 14 that holds several 10×64 hybrids used on the Airborn Optical Adjunct (AOA) Program.[9] The beryllium structure that holds up to 48 of these modules in a focal plane (Fig. 15) is over 16 inches long. This focal plane generates over 380 million analog data

Table II   Summary of Device Parameters

Array dimension	10×64	256×256
Pixel size	120 μm	30 μm
Detector material	Germanium Silicon	Silicon
Number of readout channels	10	2
Power during "write" cycle	0 mW	0 mW
Power during read cycle	10 mW	2 mW
Present clock speed	1 MHz	2 MHz
Theoretical clock speed	10 MHz	10 MHz
Readout mode	Random Access	Random Access
Processing power	20 MIPS/channel	20 MIPS/channel
Radiation hardness	1 Mrad	?
Noise at room temperature	$< 300\ e^-$ rms	$< 300\ e^-$ rms

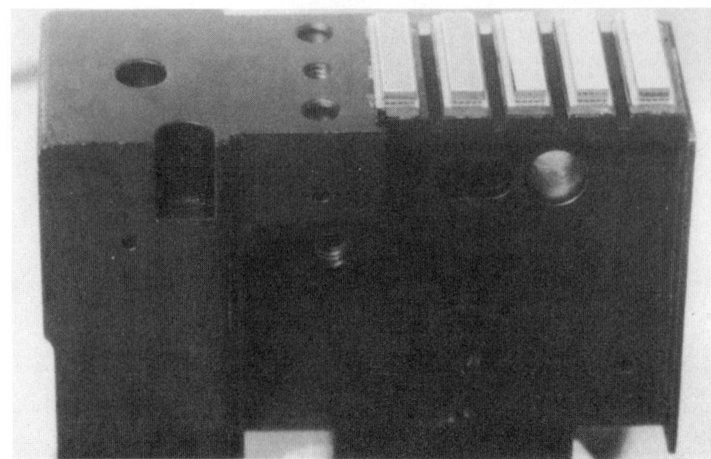

Fig. 14   A photograph of a beryllium module pedestal holding several 10 × 64 hybrid arrays.

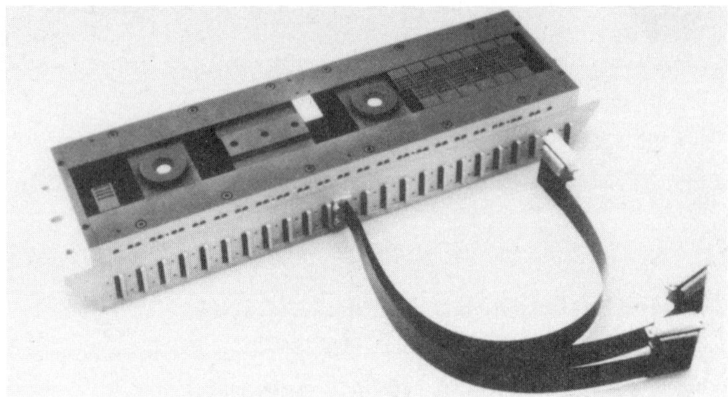

Fig. 15   A photograph of the AOA focal plane.

samples per second. The data is then converted from analog to digital information and processed at a rate of 15 billion operations per second by the data processor shown in Fig. 16.

CONCLUSION

Silicon PIN diode pixel detectors appear to be the prime candidate for an SSC vertex detector. Pixel detectors should have the best performance in terms of resolution, signal-to-noise ratio, radiation hardness and unambiguous high-speed response. An architecture to achieve correlation between the pixel containing the data and the time of the beam crossing using a combination of on-chip and off-chip signal processing appears feasible to implement advanced radiation-hardened silicon CMOS technology. More study is required to ensure that the power that must be dissipated to achieve full operating speed is compatible with the maintenance of PIN diode operating temperatures. The produciblity of large hybrid diode array/readout chips of up to 512 × 512 pixels has been demonstrated.

Very preliminary data from our present 10 × 64 silicon hybrid arrays demonstrate the detection of alpha particles and 60 KeV x-rays.

Fig. 16   Sensor signal processor for the AOA program.

REFERENCES

1. C. Damerell; Experiment NA32 at the CERN SPS; Vertex Detectors; Rutherford Appleton Laboratory Report RAL 86–077 (July 1986).

2. S. Parker, "A Proposed VLSI Pixel Device for Particle Detection," submitted to the International Workshop on Silicon Pixel Detectors for Particles and X-rays, Leuvan, Belgium (June 1988).

3. G. Vanstraelen, I. Debusschere, C. Claeys, and G. Declerck, "Fully Integrated CMOS Pixel Detector for High Energy Particles," submitted to the International Workshop on Silicon Pixel Detectors for Particles and X-Rays, Leuvan, Belgium (June 1988).

4. S. Shapiro and T. Walker, "The Microdiode Array—A New Hybrid Detector," SLD–New Detector Note No. 122 (October 15, 1984).

5. S. Gaalema, "Low Noise Random–Access Readout Technique for Large PIN Detector Arrays," *IEEE Trans. on Nucl. Sci.* NS–32 1:417 (February 1985).

6. Fabricated by MICRON Semiconductor Ltd., Sussex, England.

7. J. F. Arens, J. G. Jernigan, M. Peck, C. A. Dobson, E. Kilk, J. Lacy, and S. Gaalema, "10 Micrometer Infrared Camera" *Applied Optics,* 26:18 (1987).

8. S. Shapiro, W. Dunwoodie, J. Arens, J. Gernigan, and S. Gaalema, "Silicon PIN Diode Array Hybrids for Charged Particle Detection," SLAC–PUB–4701 (December 1988), submitted to *Nucl. Inst. and Meth.*

9. The Airborn Optical Adjunct Program, *Aviation Week and Space Tech.* (November 1988).

DEVELOPMENT OF NEW SCINTILLATING MATERIALS FOR THE SSC

Charles R. Hurlbut

Bicron Corporation

12345 Kinsman Road, Newbury, OH, 44065

ABSTRACT

Development of organic scintillators for the SSC is concentrated in the
two areas of scintillating optical fibers and resistance to radiation damage.
While good progress has been made in adapting small-scale fiber making
methods to the conventional plastic scintillator technology, much work
remains in optimizing light output and transmission in small fibers and in
mass production techniques. New solvent and fluor systems are being
investigated for liquid fiber optics, and novel polymers are being tried for
greater rad-hardness. The greatest challenge is the radiation damage issue.
Systematic research is building a base for postulating damage mechanisms and
how scintillators should be utilized in high intensity radiation fields.

INTRODUCTION

Basic research has the habit of initiating the development of new technologies. This is certainly true in the area of scintillating materials for
high energy physics research. After a relatively quiescent period of at
least ten years a new wave of scintillator development began quietly in the
late seventies as Sam Borenstein at Brookhaven began efforts to fabricate
clad optical fibers made from commercial plastic scintillator (Ref. 1).
During the 1980's a number of particle accelerators worldwide were upgraded
to higher energies and luminosities. With these, the problem of radiation
damage to detector components became an issue of concern. The Superconducting Super Collider represents the next generation of such accelerators, and
the accompanying demands for new types of detectors has spurred a major
resurgence in scintillator research. In this regard, the subjects of particular interest are the following:

1. Fast timing: to handle event rates 1000 times higher than previously
   known.

2. Radiation hardness: to operate in correspondingly high dose fields for
   long periods.

3. Suitable to the particular physics and geometries of the SSC.

4. Manufacturable inexpensively and in large, high precision quantities.

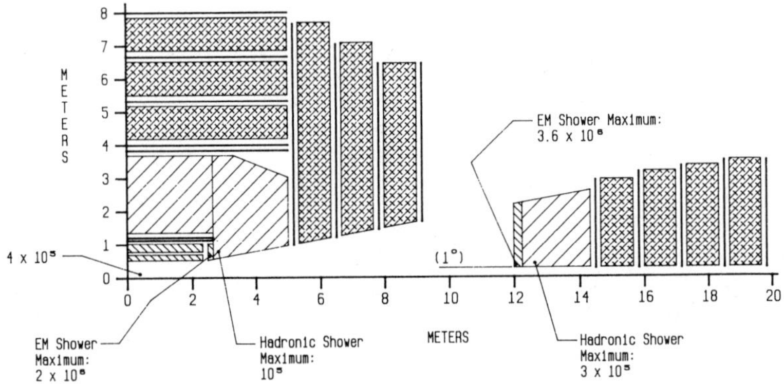

Fig. 1 Section of a solenoid type detector showing estimated
annual radiation dose values in rads

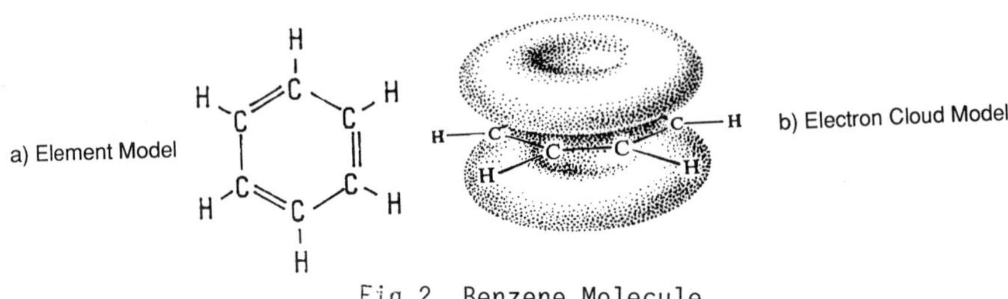

Fig 2. Benzene Molecule

OVERVIEW OF SCINTILLATORS

<u>General Description</u>

Scintillators are materials which generate light from energy deposited in them. Their most common forms are solid crystals, solid plastics, and organic liquids, and they are normally made by chemical processes. They are used extensively to detect nuclear radiations of almost every type known ranging from the more commonly known gamma rays and beta particles to the elusive neutrino. The brightness of the light, or scintillations, generated in them is proportional to the amount of energy left in them from their interaction with radioactivity. In most instances, as in particle physics, the energy is deposited as the particle or ray passes through the scintillator. The scintillations, which are extremely faint and unable to be seen by even a dark-adapted human eye, are isotropic bursts of many individual photons of light. Therefore the surfaces of a scintillator are prepared by polishing or application of reflector coatings to direct a maximum number of the photons to an electronic photosensor which is optically coupled to the scintillator surface. Important aspects of scintillators are their atomic composition, density, light output, optical clarity, spectra of their emitted light, scintillation timing properties, and mechanical properties.

<u>Scintillators Considered for the SSC</u>

Scintillator research has encompassed inorganic crystals and organic liquids and plastics. The inorganic materials of interest have all been characterized by decay times of ten mano-seconds or less. One of these, cesium fluoride, with a single decay component of about 3 nanoseconds (ns), is extremely hygroscopic and therefore eliminated for extensive use in the SSC. Pure cesium iodide has two decay components the shorter of which is about 6 ns (Ref. 2). This crystal scintillator is severely is damaged by radiation doses of as little as 1000 rads and is consequently unsuitable in the SSC. Barium fluoride is extremely radiation-hard (Ref. 3) and has a fast scintillation decay component of 0.6 ns. The emission spectrum of this output peaks at 220 nm, and this presents difficulties of finding suitable photosensors. Also, it has a long decay emission which accounts for 90% of its total light output and which is at a longer wavelength and therefore contaminates the acquisition of useful data. Because of its great radiation hardness, however, there continue to be investigations into making barium fluoride useful in SSC detection applications.

One of the persistent limitations of inorganic crystal scintillators is the narrow range of available sizes and shapes. The largest crystals grown are about 80 cm in diameter, and some can be heat forged to form meter long bars. These large sizes are the exception rather than the rule, and all suffer from brittleness and poor mechanical strength. Therefore a major task confronting their use in the SSC is the development of useful physical shapes and overcoming the accompanying light collection difficulties.

Glass scintillators, a special class of inorganic fluorescent materials, has been the subject of extensive work in fabricating arrays of fibers with coherent imaging capability. Bundles of 25 micron diameter fibers have been made which exhibit excellent tracking for microvertex detection. The glasses are relatively resistant to radiation damage showing little degradation from a megarad dose and would therefore be well suited for detectors located near the main beam line. Their primary drawback is that about half of their scintillation light is emitted in a very long decay time mode, and this would cause serious pulse pile-up during an experimental run.

Organic scintillators are of considerably greater interest for the SSC. They have fast decay times typically of about 3 ns, and are rich in hydrogen

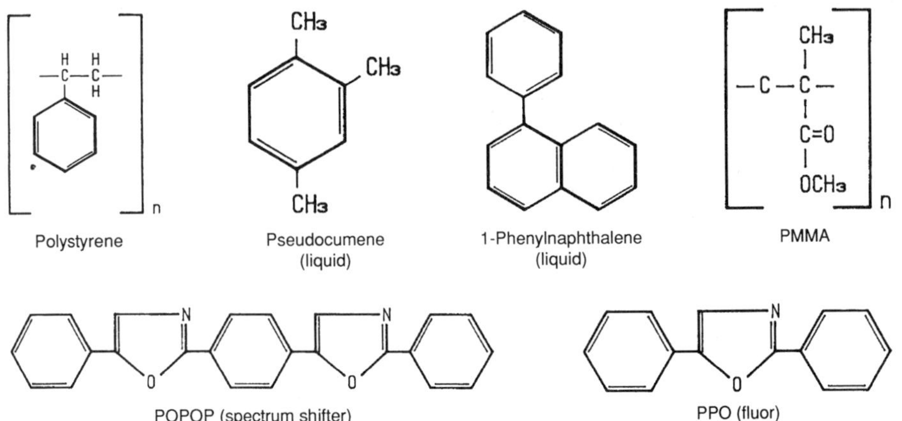

Fig. 3 Organic Scintillator Chemicals

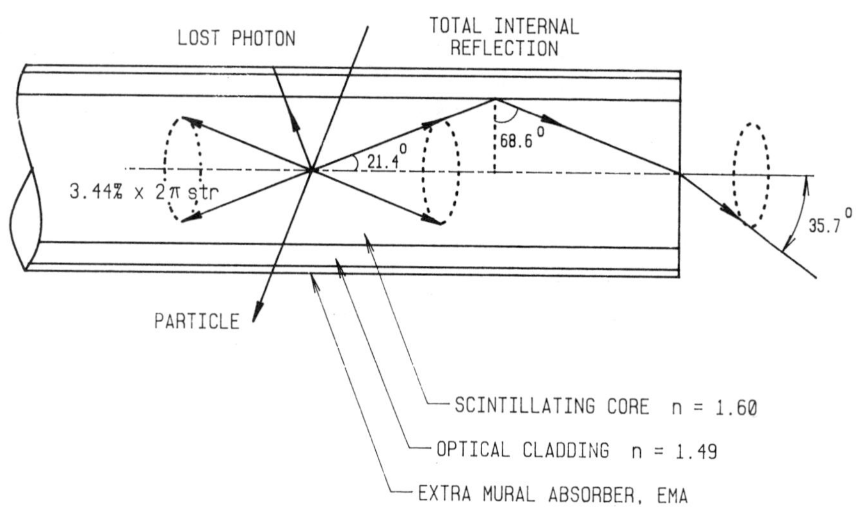

OPTICAL CLADDING THICKNESS: >5λ (APPROX. 3 MICRONS)
TYPICALLY 3-5% x O.D.

Fig. 4 Parameters of an Individual Scintillating Fiber
(Core Base: Polystyrene)
(Cladding: Polymehtylmethacryate)

which makes them excellent sensors of the neutron components of the showers
generated in hadron calorimeters (this, in turn, contributes greatly to
achieving "compensation" in which the relative electron/hadron response is
approximately equal to one, e/h = 1). They possess a degree of radiation-
hardness suitable to some of the SSC detector needs, and this has the potential of being extended. In addition, they are relatively inexpensive and can
be manufactured in a large variety of shapes and sizes such as optical fibers
and large sheets. They constitute a class of sensor materials high on the
list of those most likely to be extensively used in the several large detector systems to be deployed around the accelerator ring. In this paper we
review organic scintillators, their scintillation mechanism and composition,
and the present status of research to improve and develop them to the needs
of SSC physics.

OVERVIEW OF SSC DETECTORS

General Detector Description

    A brief review of the general kinds of detection devices that might be
incorporated in a large SSC detector system will serve as a backdrop to the
ensuing discussion of scintillators. Figure 1 is a longitudinal section of a
quadrant of a detector similar to the one discussed in the papers presented
by M.D. Gilchriese and H.H. Williams elsewhere in these proceedings. There
will be several such devices, each differing from the others in one or more
respects, constructed along the 83 km (51 mi.) circumference of the SSC accelerator ring.

    At the SSC target luminosity of 10^{33} cm^{-2} sec^{-1} and crossing rate for
the ellipsoidal proton "bunches" of one every 16 ns, there will be 10^8
collision events per second. Great numbers of subatomic particles and gamma
rays generated by each collision would be spewed out in a spherical cloud.
The intensity of this radiation will be lowest in the regions at 90° to the
proton beam line at the collision point and increasing with reducing scatter
angle. The vast majority of collision products would be showered in narrow
cones of less than 10° half-angle along the directions of the colliding
proton beams.

    An assemblage of tracking and calorimetry detectors resembling a sphere
surrounds the collision point with conical apertures along the beam lines to
allow these extremely intense fluxes to escape the central region. The
vertex detectors are not shown but would be located in a cylindrical array
very close to the collision point. It is in these components, the vertex
detectors, the trackers and, particularly, the calorimeters, that organic
scintillators are likely to play a major role. Analogous sets of detectors
are constructed farther out along the main beam line in the "forward regions"
to cover the conical openings down to within a central cone of one degree or
less. All parts of the large detector, including the massive magnetized iron
covers, play a roll in producing information about the myriad of energetic
processes occurring within.

Radiation Exposure

    The radiation exposure and consequent damage will vary widely throughout
the detector system in proportion to the distribution of the collision
products. The resulting degradation of the various detector components is a
major concern and has been an important topic of SSC related detector
research. A summary of anticipated radiation exposure levels and general
formulae for estimating the dose rates of specific areas and conditions in
the SSC detectors is available from the SSC office in Berkeley (Ref. 4). In
general, dose rates will fall off in proportion to the distance squared along

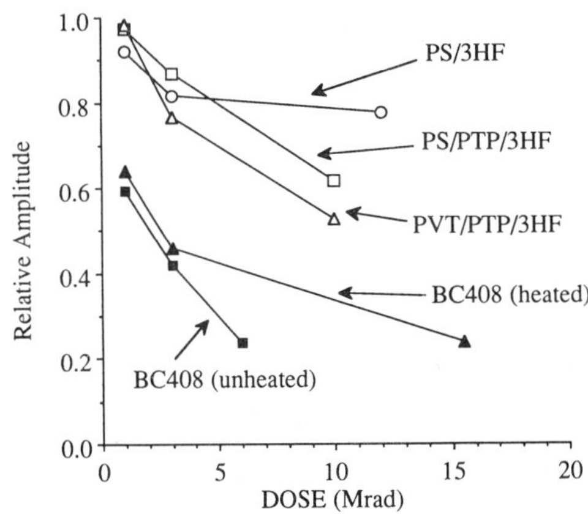

Fig. 5(a) Loss of light output from irradiation in air

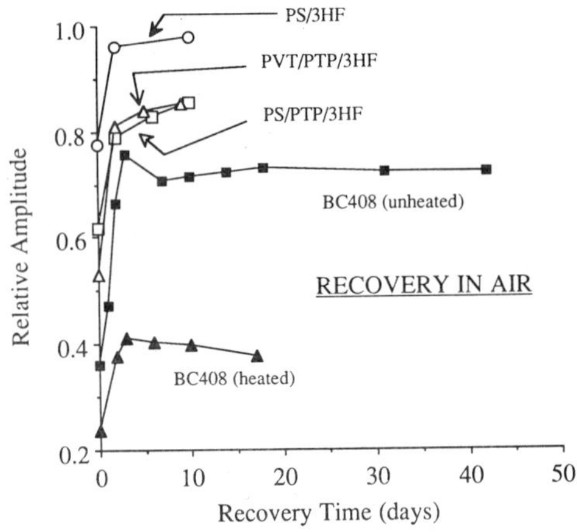

Fig. 5(b) Recovery of light output after irradiation is ceased

a straight line from the collision point and will be extremely high at small
scatter angles from the collision beam line. This will be perturbed as the
particles and gammas interact with the various light and heavy solid
materials associated with the detectors, superconducting magnet (if present),
and structural components. The shower maxima occurring about two inches
behind the front face of the EM calorimeters are of particular note. Some
examples of estimated dose rate levels are presented here for reference in
the later discussion:

a) In the central cavity where tracking
   devices are located, from charged
   particles emanating from primary p-p 4×10^3 Gy/yr.
   collisions, radius = 10 cm.

b) At the entrance face of an electro-
   magnetic calorimeter from incident 2×10^3 Gy/yr.
   gammas, approx. 2 m radius on a 10° cone:

c) At the maximum of the EM shower, about
   6 cm into the calorimeter, for (b): 2×10^4 Gy/yr.

d) At the maximum of the hadron initiated
   cascades in a hadronic calorimeter, 1×10^3 Gy/yr.
   same geometry as in (b):

e) At the maximum of the EM shower in a
   forward region calorimeter, r = 20 m 2×10^6 Gy/yr.
   on a 0.5° cone:

The need for continued research into the mechanisms and prevention of
radiation damage, in electronics as well as scintillators, is well demonstrated.

REVIEW OF ORGANIC SCINTILLATION PROCESS

The fluorescence mechanism occurs only in "aromatic" organic materials
containing the benzene ring, the atomic structure of which is shown in Fig. 2
(a). This is a planar structure in which some of the electrons available for
creating the bonds that hold the molecule together are not localized between
the adjacent atoms. Rather, the molecule is thought of as having rigid
sigma-bonds between the carbon atoms in its six membered ring and clouds of
more loosely held pi-electrons on either side as rendered schematically in
Fig. 2 (b). These pi-electrons are easily excited by the energy deposited in
the material during the passage of a high energy particle. The resultant
excited electronic states decay releasing their energy by either thermal
molecular vibrations or fluorescence.

Organic scintillators are solutions in which selected fluorescent compounds are dissolved in an appropriate solvent. In the case of plastic
scintillators the solvent is a polymer such as polystyrene and the solution
is mechanically rigid. A number of such molecules are shown in Fig. 3. All
are characterized by the presence of benzene and similar ring structures.
Scintillators made with the nonaromatic polymethylmethacrylate (PMMA), of
Plexiglas (Ref. 5) fame, have light yields of at best 30% of fully aromatic
scintillators, and that is achieved only after the plastic is heavily doped
with an aromatic compound such as naphthalene.

Conventional organic scintillators are composed typically of 98%
solvent, 1 - 2% primary fluor, and about .01% spectrum shifter. The pure
solvent would scintillate weakly at UV wavelengths with a peak at about 320
nm at which it is moderately opaque and with a relatively long decay time.

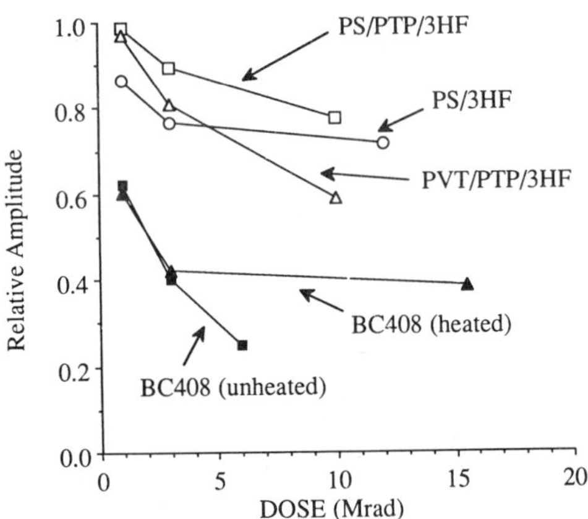

Fig. 6(a) Loss of light output from irradiation in argon gas

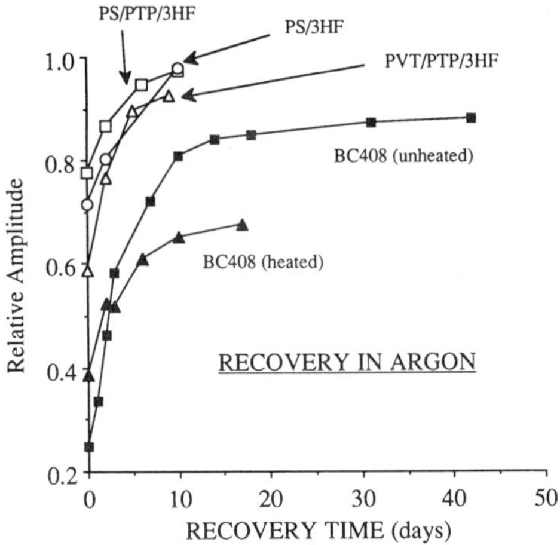

Fig. 6(b) Recovery of light output after irradiation is ceased

The primary fluor such as PPO in Fig. 3 has a light absorption spectrum well placed to efficiently absorb this emission and re-emit it at a longer wavelength typically with a peak at 370 nm. However, at the stated concentration of 10 to 20 grams per liter the radiative energy transfer is superceded by a non-radiative mechanism based on dipole-dipole interaction known as Forster transfer (Ref. 6), and the short decay time of the primary fluor dominates. A spectrum shifter is used to absorb and remit the near-UV scintillation into the visible where the bulk clarity of the solution is high and the sensitivity of most photosensors are optimized.

The choice of solvent and additives and their concentrations is made to optimize the scintillator properties for such parameters as light output, emission spectrum, bulk clarity, decay time, radiation hardness, atomic composition, physical properties, and application conditions.

PLASTIC SCINTILLATING FIBERS

The operation of a single scintillating optical fiber is illustrated in Fig. 4. This example consists of a plastic scintillator core surrounded by a thin cladding of nonscintillating plastic. As the scintillation light cascades along the fiber, the electromagnetic wave penetrates into the cladding to a depth of several wavelengths. This sets the minimum cladding thickness at about 3 microns and requires that it be transparent. The EMA outer layer is opaque in order to eliminate optical crosstalk between adjacent fibers in an imaging bundle or tracking assembly. It does not function in the optical transmission of the fiber and is omitted when not needed. The EMA is commonly used with glass fiber optics and is likely to be of more limited use with plastic fibers.

The prime function of the cladding is to protect the smooth surface of the core which must be maintained defect-free and highly reflective. Its refractive index, n, must be lower than that of the core, providing a stepped index surface. The critical angle for total internal reflection = $\sin^{-1}$ ($n_{cladding}/n_{core}$) and is 68.4° in this example. The fraction of the total light transmitted in each direction along the fiber (the trapping efficiency) is indicated by the cone with the half-angle equal to the complement of the critical angle. (The angle calculation, $0.5[1-n_{cl}/n_{co}]$, yields a value lower than the true trapping fraction because it does not account for off-axis scintillations. The actual trapping efficiency is about 4% instead of the 3.44% indicated.) The trapping efficiency can be increased by selecting core and cladding materials with greater differences in their refractive indices. A polystyrene core clad with material of n=1.4 would have a trapping efficiency near 7%, a target well worth pursuing.

The fabricating of fused fiber bundles and ribbons where the individual fibers are always quite small, 200 to 25 microns, requires special scintillator formulae to reduce crosstalk. Over 90% of the light in a fiber can escape into adjacent fibers as shown in Fig. 4. This poses little crosstalk problem since it is outside of the trapping cone of all parallel fibers in the bundle. It is possible, however, that these escaping photons are ultraviolet from the primary scintillator, having passed through a sufficiently thin layer of plastic to have missed a spectrum shifter molecule. With decreasing fiber size the fraction of unshifted escaping photons increases exponentially to be shifted isotropically in the adjacent fibers causing degraded position resolution. Efforts to offset this by correspondingly increasing shifter concentrations in the plastic are met with reductions in bulk clarity. A second approach is to use one-step primary scintillators that emit at longer wavelengths where good clarity is possible; in order to achieve efficient Forster transfer from the plastic base, the compound would be incorporated at the high 1% to 2% concentrations and must

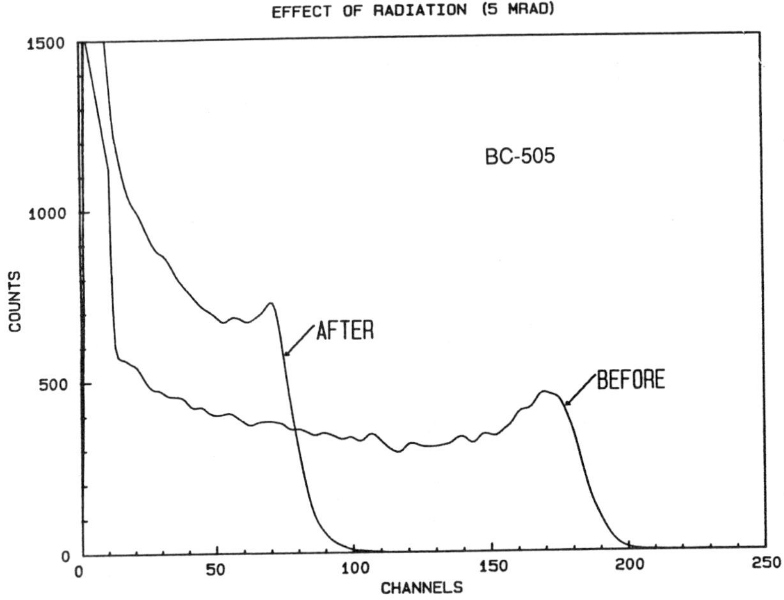

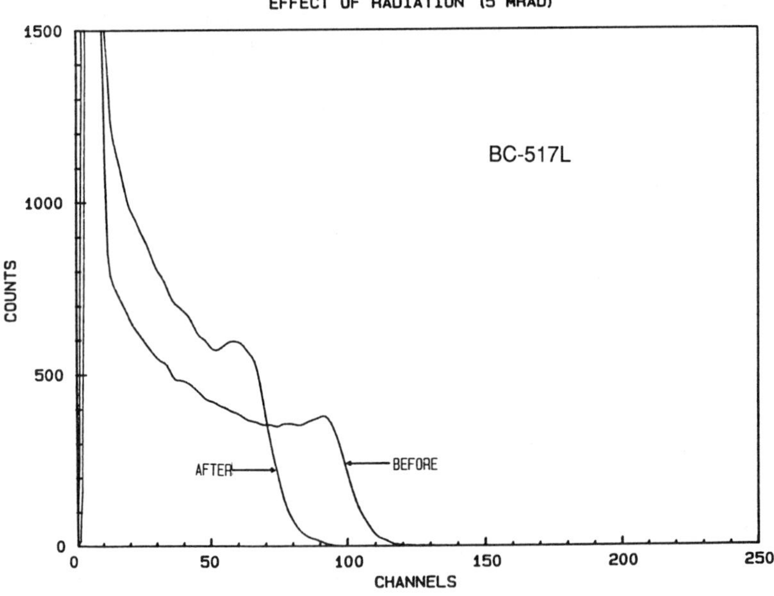

Fig. 7 Scintillation efficiency reduction due to radiation damage for two commercial scintillators -BC-505(top) and BC-517L(bottom)

not have the self-absorption common to nearly all known scintillation
compounds. Efforts to achieve this have so far met with limited success.

RECENT ORGANIC SCINTILLATOR RESEARCH

Plastic Fiber Manufacture

The manufacture of scintillating plastic fiber has been underway
commercially in Europe and Japan for several years. In the United States,
however, activities in this regard have been carried out largely by a few
individual research groups to supply their own needs, making their fiber in
joint projects with a few small companies.

Bicron Corporation, is the sole commercial manufacturer of aromatic
based plastic scintillators in North America, having, until recently,
supplied only plastic sheets and bulk liquid scintillators to the high energy
physics community. During the last twelve months we have conducted an
intensive development program in plastic scintillating fiber technology with
the assistance of an SBIR Phase I research grant. This work has involved
technical collaborations with a number of research groups at universities and
national laboratories investigating both plastics and liquids. An independent commercial capability now exists that can produce both round and square
fibers ranging in size from 0.2 mm to 3 mm. For the most common 1 mm diameter, size tolerances of +/- 1% and unfiltered light attenuation lengths exceeding 2 meters have been achieved.

Manufacturing plastic scintillating optical fiber must be done on two
scales reflecting the needs of physics research. The first level is for small
scale batches of several kilometers or less and must have the flexibility to
vary the fiber size and formula. The fibers are made by pulling from a
larger preform, and batches up to about 2 km of 1 mm diameter fiber can be
made by a single machine in a day. It is also necessary to perform some
secondary detector assembly work to support the diverse research needs. This
approach can be scaled up by duplicating the equipment to supply hundreds of
kilometers over periods of several months. All the fibers made to date have
come from this process.

Producing the million or more meters of 1 mm fiber for the envisioned
full-scale SSC detectors will require mass production methods similar to
those employed for making optical telecommunications fiber. This would be an
engineering task chartered to implement the results of present and ongoing
research of fundamental materials.

Liquid Scintillator Fibers

Liquid scintillators encapsulated in narrow nonscintillating plastic or
glass tubes can potentially be substituted for the plastic fibers. Their
particular use would be in the calorimeters of the forward regions where the
very high radiation doses (10^6 - 10^8 Gy/yr.) are likely to seriously damage
the most rad-hard plastics. Liquids are five to ten times more resistant to
radiation damage than the plastics and have the further advantage of being
able to be circulated in their tubular systems and therefore periodically
exchanged for fresh liquid.

The optics of tubular liquid systems are identical to those for solid
fibers requiring that the tubing be of lower refractive index than the
liquid. Two high refractive index solvents, 1-methylnaphthalene and 1-
phenylnaphthalene (Fig. 3) both with n above 1.6, have been studied at Notre
Dame University and show good promise for this use being among the most
efficient of scintillator solvents (Ref. 7).

<u>The Radiation Damage Problem</u>

This is the major problem faced by all the detector designs and possible sensor materials, and a sizeable portion of SSC detector research is aimed at its solution.

The symptoms of radiation damage in plastic and liquid scintillators are reduction in intrinsic light output and reduction in transparency. In an effort to develop rad-hard scintillators we began by studying the damage symptoms and kinetics in hopes of determining the mechanisms. In collaboration with the group headed by Dr. Stan Majewski at the University of Florida in Gainesville we have irradiated several dozens of standard sized scintillator samples with a ^{60}Co gamma source at a dose rate of 0.08 Mrad (800 Gy) per hour. The variables studied include the following:

o Type of solvent: three plastics and several liquids
o Atmosphere: air and inert gas
o Types and concentrations of scintillation fluors
o Stabilizing additives such as anti-oxidants.
o Temperature: 21°C (room temp.) and 50°C.

The effect of damage on the light output of several plastic scintillators is shown in Fig. 5 (a). The samples had been allowed to become saturated with air prior to and during irradiation. This is a common condition for scintillators which have sat in air for as little as one month. BC 408 is a standard blue emitting (425 nm) plastic used extensively in physics re search. Not indicated is the fact that none of the samples exhibited damage symptoms at a dose of 0.5 Mrad. At one Mrad there is a pale yellow discoloration in the BC 408 samples. This color becomes more intense with increasing dose. Fig. 5 (b) shows that once the irradiating is stopped the plastics recover some or all of their original light output. The analogues of these two processes for plastics saturated in the inert gas Argon are shown in Figs. 6 (a) and 6 (b). While the loss of light output is the same for the two conditions, the recovery is better in the inert gas atmosphere. This latter phenomenon was consistently observed whenever we made such comparisons.

In both sets of graphs the scintillators exhibiting the least damage and most complete recovery contain the fluor 3HF (3-hydroxyflavone) which has its emission maximum in the green at about 540 nm. The loss in scintillation efficiency was found to be consistent with the increase of discoloration in the plastic. With increasing dose the induced color extends further into the longer wavelengths of the visible spectrum. In addition to its longer emission wavelength, 3HF also has a large Stokes shift in which the absorption and emission maxima are separated by about 180 nm. (The more common fluors have Stokes shifts of about 60 nm.) Consequently the 3HF can shift light directly from the UV region to beyond the blue spectral region where radiation induced yellow color is strongly absorbent.

Of particular note is the recovery phenomenon in the plastics. In the region of maximum EM shower in the central detector the dose rate would be about 2.3 Gy/hour, much less than the 800 Gy/hr. dose rates of our tests. The recovery kinetics might cancel the slower damage kinetics of the SSC. This is particularly likely if the scintillation output is at longer wavelengths. There is preliminary confirmation of this actually occurring even in blue scintillators, and further studies are planned.

Polysiloxanes (silicone rubbers) with aromatic appendages on the polymer chains are moderately efficient scintillation bases and exhibit extreme resistance to yellowing from irradiation. They require a dose of 0.22 MGy to be discolored to the extent as is polystyrene by a dose of 0.1 MGy. However,

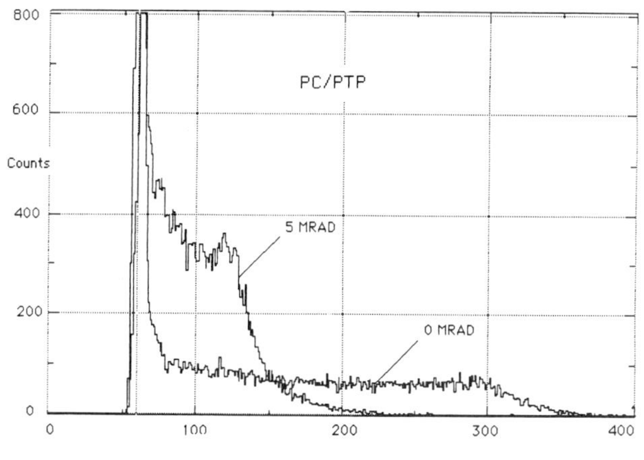

Fig 8(a) Light output loss due to radiation damage
          for a UV-emitting scintillator

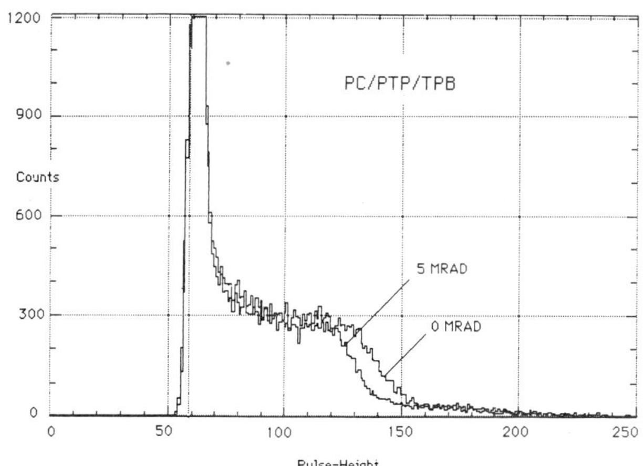

Fig 8.(b) Light output loss due to radiation damage
           for a blue-emitting scintillator

scintillators made with these have exhibited yellowing comparable to the above plastics but with much slower recovery properties. Nevertheless, continued research with this polymer is worthwhile.

Liquid scintillators also suffer from radiation damage as is demonstrated in Fig. 7 for two commercial products also used extensively in physics research. The plots show the typical Compton spectra from gamma excitation. The two liquids differ in the kinds of solvents and fluors in their formulae, and the upper one is clearly the more easily damaged of the two. The scintillation loss is comparable to that of the plastics but less severe at the lower doses, being first noticed at 1 Mrad or about twice the level at which it is first apparent in plastics. However, it differs significantly from the plastics in that there is no observed discoloration of the liquid even at a 5 Mrad dose where the reduction in scintillation efficiency is similar to that in plastics. It further differs in that the liquids exhibit no recovery of light output after the irradiating is stopped.

The results of further investigations with liquids are shown in Figures 8 and 9. Fig. 8 (a) demonstrates the extreme reduction of scintillation output for a solution of the UV emitting fluor p-terphenyl in pseudocumene. Fig. 8 (b) indicates how the percentage of reduction is minimized with the addition of the spectrum shifter tetraphenylbutadiene (TPB) which emits at about 440 nm with a sky blue fluorescence. There is a direct correspondence of this data to the light transmission measurements for the same liquids exhibited in Figures 9 (a) and 9 (b). The upper edge of the TPB absorption spectrum renders the solution (9 (b)) opaque at wavelengths below 400 nm. With this blockage omitted we can see in Fig.9 (a) the radiation damage induced absorption band.

A number of general statements can be made about the nature of radiation damage.

1. The basic damage mechanism is the breaking of chemical bonds creating chemically active free-radicals and altering the optical properties of the matrix. Since the solvents make up about 99% of the scintillator, their response to radiation dominates.

2. The plastics polystyrene and polyvinyltoluene are far more easily colored than the liquids indicating preferential breakage of the polymer chain bonds. This mechanism may also disproportionately direct the damaging energies to the polymer rather than to the fluors.

3. Recovery occurs in the non-fluid matrices in which the broken bonds, both in the plastics and fluors, can reconnect like new.

4. Oxygen can readily bond with the free radicals causing irreversible damage and also increase discoloration.

5. The solvents which are resistant to discoloration, the liquids and polysiloxanes, do not protect the fluors and do not provide for recombination of the original chemical bonds.

SUMMARY

We now have considerable understanding of the radiation damage phenomena in plastic and liquid scintillators and can implement specific chemical research to increase their rad-hardness. There are already many regions in the SSC detectors where even the existing commercial products will give excellent service. It is very likely that such traditional products can be significantly improved in their rad-hardness by modifying their formulas and correctly installing them in the SSC detectors.

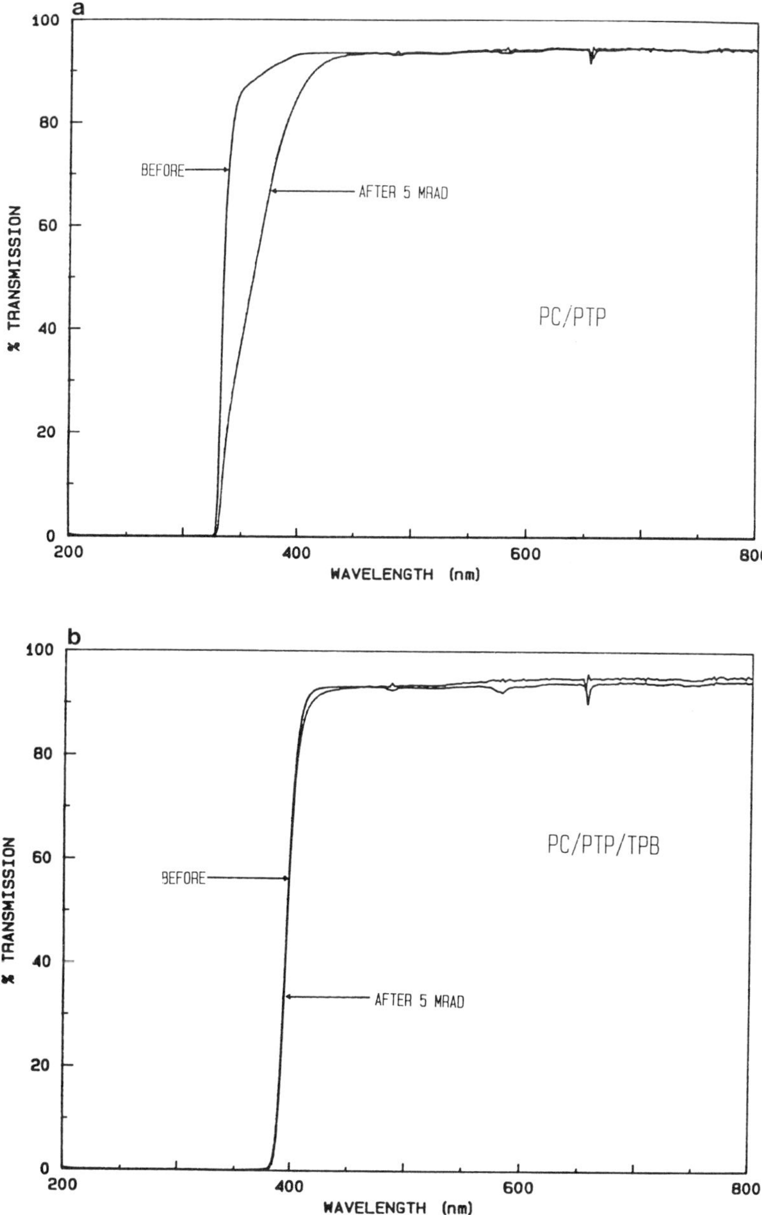

Fig. 9 Radiation damage effects on transmission for the liquid
scintillators in 8 (a) and 8 (b)

References

1. S.R. Borenstein et al, Physica Scripta 23 (1981) 550.

2. Bicron Corporation, Newbury, Ohio, internal report, Oct.13 1988.

3. S. Majewski & M. Bentley, Nucl. Instr. and Meth. A260 (1987) 373.

4. Radiation Levels in the SSC Interaction Regions, ed. D.E. Groom, Task
   Force Report, SSC Central Design Group, LBL, Berkeley, California, 1988.

5. Trademark of Rohm and Haas Company, Philadelphia, Pennsylvania

6. T. Forster, Ann. Physik 2 (1948) 55.

7. R. Ruchti et al, "Development of New Scintillating Fiber Detectors for
   High Energy Physics Applications," Univ. of Notre Dame publication UND-
   HEP-11-29-88, November 1988.

DEVELOPMENT OF RADHARD VLSI ELECTRONICS FOR SSC CALORIMETERS

John W. Dawson and Lawrence J. Nodulman

High Energy Physics Division
Argonne National Laboratory
Argonne, IL 60

ABSTRACT

A new program of development of integrated electronics for liquid
argon calorimeters in the SSC detector environment is being started at
Argonne National Laboratory. Scientists from Brookhaven National
Laboratory and Vanderbilt University together with an industrial
participant are expected to collaborate in this work. Interaction rates,
segmentation, and the radiation environment dictate that front-end
electronics of SSC calorimeters must be implemented in the form of highly
integrated, radhard, analog, low-noise, VLSI custom monolithic devices.
Important considerations are power dissipation, choice of functions
integrated on the front-end chips, and cabling requirements. An
extensive level of expertise in radhard electronics exists within the
industrial community, and a primary objective of this work is to bring
that expertise to bear on the problems of SSC detector design. Radiation
hardness measurements and requirements as well as calorimeter design will
be primarily the responsibility of Argonne scientists and our Brookhaven
and Vanderbilt colleagues. Radhard VLSI design and fabrication will be
primarily the industrial participant's responsibility. The rapid-cycling
synchrotron at Argonne will be used for radiation damage studies
involving response to neutrons and charged particles, while damage from
gammas will be investigated at Brookhaven.

INTRODUCTION AND OVERVIEW

At design luminosity the beam-beam interactions in the SSC will
create a radiation environment which may damage electronics associated
with the detectors. The very large number of readout segments required
by the physics in conjunction with the geometrical constraints of the
detectors, will require that the electronics be situated close to the
interaction region. It is important, therefore, that the nature of this
radiation environment be accurately projected and the effects on the
detector electronics be simulated in test measurements. The constraints
that the radiation fields place on the detectors can then be taken into
account in the design.

The Central Design Group has undertaken studies of radiation levels in the SSC interaction regions and the results of these studies are summarized in a detailed report[1]. The predicted radiation levels require that detector electronics be radiation hard in order that it may operated satisfactorily for a reasonable lifetime. In general, builders of previous high energy physics detectors have not been forced to deal with the problems of radiation hardness in the electronics. Accordingly, the difficulties posed by the SSC bring a new dimension to detector design.

Fortunately, there is a segment of the semiconductor industry that has been developing radiation hard electronics for military and space applications for something over 20 years. A wealth of experience, expertise, and understanding of damage mechanisms exists in this sector of the electronics industry. It is essential that these capabilities be brought to bear on the problems associated with detector design for the SSC.

We have undertaken a program of development of radiation-hard integrated electronics for SSC calorimeters. Specifically this electronics is intended to be optimized for a liquid argon or warm liquid drifting calorimeters[2,3]. This work will also provide some information applicable to the electronics for calorimeters of other types and to tracking electronics.

We have made the following assumptions:

a. An adequate level of expertise in radhard electronics exists in the United States at this time so that the development effort does not require fundamental research into radiation damage mechanisms or other basic aspects of radhard VLSI design. Specifically, extensive capabilities have developed over the last twenty years in the industrial community in response to military and space needs. This expertise is available, and accordingly, the work should be approached as an engineering project, with industrial participation as a basic part of the effort.

b. It is important that SSC detector development activities proceed promptly so that there will be confidence in the projections of cost and feasibility at the time when detector construction must begin.

c. Interaction rates, segmentation, and the radiation environment dictate that front-end electronics must be implemented in the form of highly integrated, radhard, low-noise, analog, VLSI custom monolithic devices. Important engineering considerations come from the power dissipation, the choice of functions integrated on the front-end chips, and the cabling requirements.

The guidelines for this development effort are as follows:

a. The development efforts will be focused on the calorimeter electronics. Argonne has long experience in calorimeter design, construction, and use as evidenced by the calorimeters for the HRS, CDF, ZEUS, and other detectors.

b. The work will focus upon the electronics needed for liquid argon or warm liquid drifting calorimeters as stated above.

c. This project will include extensive radiation damage studies as well as VLSI device and process development. The two areas of work will proceed jointly, in close coordination. Radiation damage exposures and evaluation will be primarily Argonne's responsibility, while radhard device design and fabrication will be primarily the industrial participant's responsibility.

d. No effort will be expended on developing non-radhard devices or designs. An adequate level of radhard expertise exists so that radhard devices will be used at the outset.

e. We will proceed by characterizing the radiation environment in which the electronics must operate, selecting a currently operating radhard process capable of yielding devices which are adequately radhard, and proceeding with a radhard design in that process.

f. The rapid-cycling synchrotron of the Intense Pulsed Neutron Source (IPNS) at Argonne produces 400 MeV protons and a neutron fluence of 10^{12}/sec/sq-cm from the spallation target. This Argonne facility together with a gamma source such as the source at Brookhaven will be used for the radiation damage studies. These studies will be done both at 25C and at liquid argon temperatures. Small prototype calorimeter sections will be operated in the radiation environment.

g. A senior participant or consultant who is an expert in radiation damage will be brought into the project.

OBJECTIVES AND GOALS

The objective of this development effort is to design a prototype VLSI version of the SSC calorimeter electronics using a specific radhard process. A phase 2 follow-on project would involve a prototype production run of a number of wafers. Developing this design will require that radiation damage studies have been conducted on devices from the designated process, that devices from the designated process have been run in prototype calorimeter sections in appropriate radiation environments, and that the radhard electronics design is compatible with the calorimeter physics requirements. As discussed above, technologies exist that meet our radiation hardness requirements for neutron and charged particle total dose. The calculations indicate that the gamma-ray levels experienced by the electronics in very forward locations poses a very difficult challenge that cannot be met with current CMOS technology.

a. **Goals for Argonne and Brookhaven**

Argonne's contribution will be to provide calorimeter expertise, to conduct radiation damage tests relating to neutrons and charged particles, to conduct the radiation damage tests relating to the prototype calorimeter section, to provide structure and guidance to the development effort, and to provide the interface to the CDG and DOE. The goal for Brookhaven personnel is to produce a detailed set of specifications for the analog calorimeter electronics, participate in the testing, and then to work in conjunction with the industrial participant to design and develop this analog electronics for the VLSI implementation.

b. **Goals for Industrial Participant**

The Industrial Participant must bring expertise in the design and production of radhard electronics. It is essential that the Industrial Participant have a radhard process that is capable of producing VLSI

devices that can operate in the SSC calorimeter environment. We believe
that a successful program requires that radiation damage studies begin
immediately with existing devices produced in the radhard process that we
envision ultimately being used for the final production of radhard VLSI
devices. It will be convenient if the Industrial Participant has a gamma
source suitable for the gamma radiation damage studies.

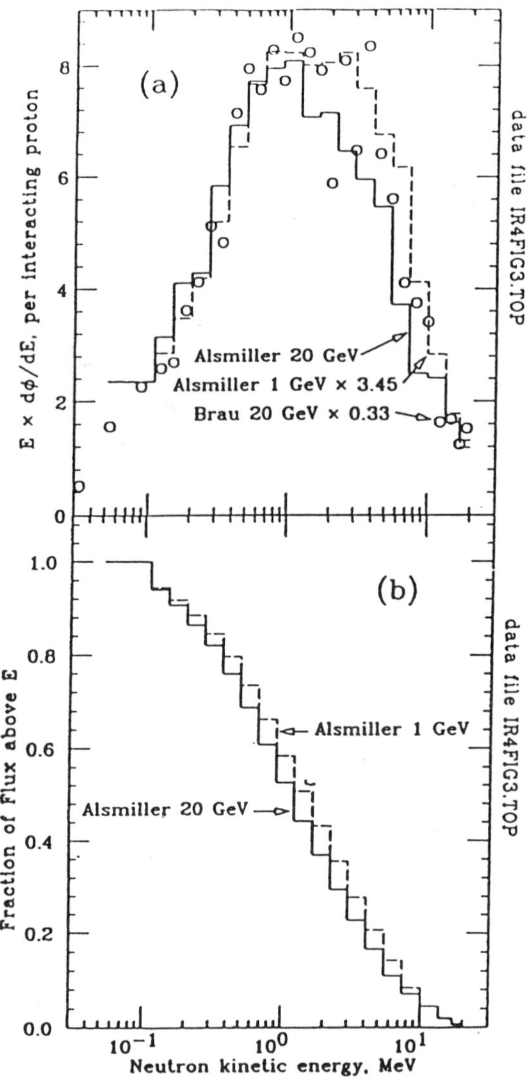

Fig. 1. The spectra of albedo neutrons from a
uranium/scintillator calorimeter. Figure
taken from SSC Report SSC-SR-1033, "Radiation Levels in the SSC Interaction Region,"
Task Force Report, D. E. Groom, Editor
(June 10, 1988).

c. Goals for Consultant

As described above, we will have at least one senior participant/-
consultant who is an expert in radiation damage to electronic devices
involved with the development effort from the beginning. This consultant
will provide direction to the research effort and participate in
evaluating progress.

RADIATION ENVIRONMENT IN THE SSC CALORIMETER

The radiation levels in a typical SSC detector have been examined by
a task force convened by the Central Design Group and the results are
summarized in a report SSC-SR-1033, edited by D. E. Groom. The major
damage results from the electromagnetic and hadronic cascades that are
initiated in the detector calorimeters by secondary particles produced in
the 40-TeV pp collisions.

For a typical detector in which a central magnetic field volume
containing tracking chambers is surrounded by dense calorimeter, the
neutron gas created in the calorimeter leaks back into the tracking
volume. The calculated neutron spectrum of this albedo, shown in Fig. 1,
peaks near 1 MeV and is similar to the neutron spectrum at IPNS, the
neutron facility at Argonne, which is discussed in detail below.

The number of albedo neutrons, as well as the total number of
neutrons at cascade maximum in a uranium-scintillator calorimeter stack,
are reasonably represented by

$$2.5 \left(\frac{p}{1\ \text{GeV/c}}\right)^{0.5} \text{n/hadron} \quad \text{and} \quad 18 \left(\frac{p}{1\ \text{GeV/c}}\right)^{0.67} \text{n/hadrons} .$$

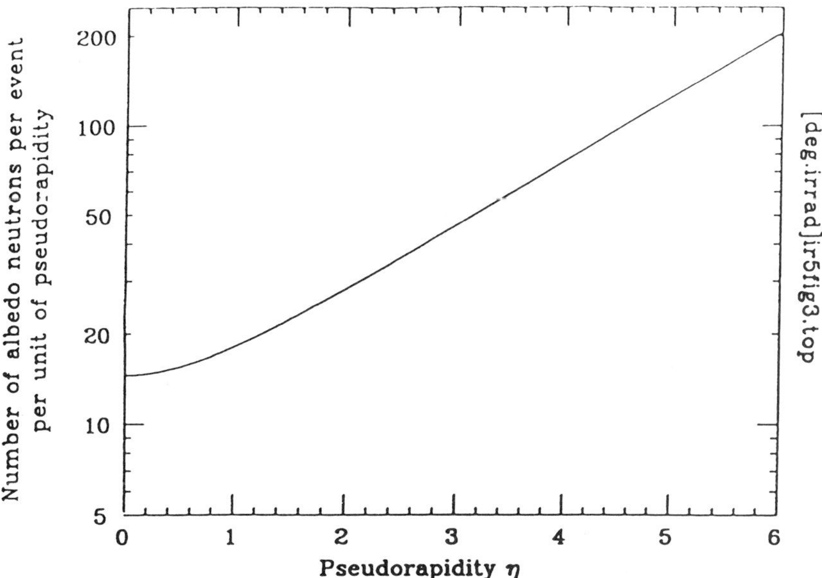

Fig. 2. The number of albedo neutrons per event per unit pseudorapidity as a function of pseudorapidity for a fine-
sampling U-scintillator calorimeter. Normal incidence
is assumed.

respectively, for an incident hadron of momentum p. Folding these
expressions with reasonable assumptions about the rapidity distribution
of secondary particles from the 40-TeV pp collisions gives the number of
neutrons to be expected in different parts of the detector. The results
are shown in Fig. 2 for the albedo neutrons and in Fig. 3 for a uranium-
scintillator calorimeter. The assumptions used for the latter correspond
to 10^8 events/sec for 10^7 operating sec per year or 10^{15} events per
year. For a central calorimeter in which the neutron flux maximum is
assumed to be at a radius of 2 m, the flux varies from 4×10^{11} to 10^{14}
neutrons/cm^2/yr. For a forward detector, the flux can approach 10^{16}
neutrons/cm^2/yr. Taking into account the reflections, the albedo
neutrons inside a spherical cavity of 2 m radius have a fluence of 2 -
4^{12}/cm^2/yr. This value will scale inversely as the square of the cavity
radius.

One concludes that, if implemented at very forward angles, the
calorimeter electronics must be able to withstand a total neutron flux,
over the lifetime of the experiment, of up to 10^{17} neutrons/cm^2. The
neutron flux has a typical energy of 1 MeV.

Similar estimates have been made for the gamma-ray dose with the
result shown in Fig. 4. The dose over a 10-year detector life will be
about 10 mrad for a 2m radius central calorimeter at (eta) = 2.5 and can
get as high as 100 Mrad at very forward angles. The dose from charged
hadrons is more an order of magnitude less as shown in Fig. 5.

We do not presently have data relating to the effect of beam loss on
the radiation environment. This effect may require an additional safety
margin. For the purposes of this work we will address only the radiation
environment resulting from beam-beam interactions.

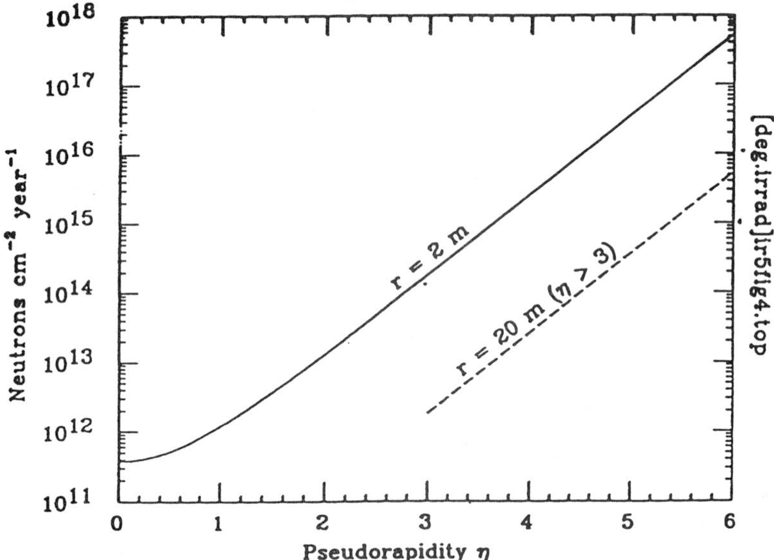

Fig. 3. The maximum neutron fluence for a uranium/scin-
tillator calorimeter. The full curve shows the
result assuming the maximum occurs at a radius
of 200 cm. Also shown is the result for a radius
of 20 m, typical of forward detectors, for
rapidity > 3.

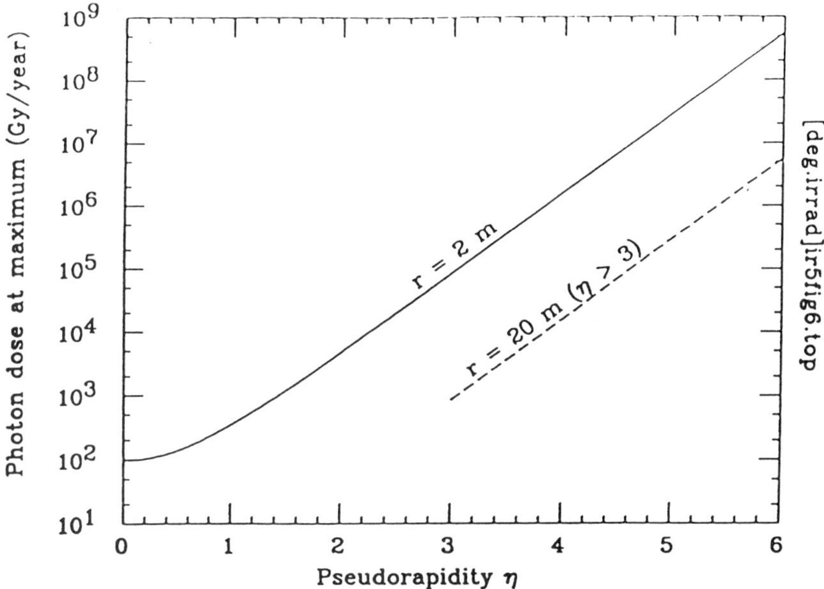

Fig. 4. The maximum dose from incident photons. The full curve assumes the maximum occurs at 200 cm. The other curve is calculated for 20 m, typical of forward detectors.

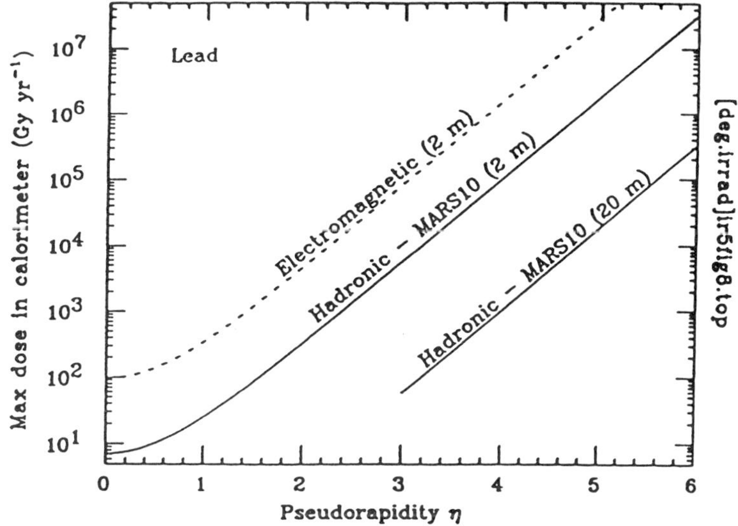

Fig. 5. The maximum hadronic dose as a function of pseudo-rapidity for a lead sphere, assuming that the maximum dose occurs at the indicated radius. The electromagnetic maximum dose is copied from Fig. G for comparison purposes.

a.   Assumptions Relating to Calorimeter Geometry

We assume that the calorimeter is configured as a barrel calorimeter and forward calorimeters with dimensions set forth in Fig. 1 of Ref. 4. For this calorimeter geometry, shown in Fig. 6, the barrel calorimeter will experience doses from a pseudorapidity of 0 to 3, and the forward calorimeters will experience doses from a pseudorapidity of 3 to some maximum. We will assume the forward calorimeter covers a pseudorapidity range of 3 to 4. The radiation environment can be easily scaled to other dimensions and configurations.

b.   Neutron, Gamma, and Charged Particle Fluences

Given the pseudorapidity coverage of the barrel and forward calorimeters, one can then use the data given in SSC-SR-1033 to determine the nature of the radiation environment. Some of the data has been tabulated for lead and some for uranium, and of course the geometrical assumptions are oversimplified. This process allows one to determine values for the doses that will be experienced by the calorimeter. If one assumes an exposure of 10 years, and uses the data given in Figs. 5-4, 5-6, and 5-8 of SSC-SR-1033, one obtains the following approximate maximum total doses.

Table 1. Approximate Maximum Total Doses

	Hadronic	Gamma	Neutrons
Barrel Cal.	2.0 Mrad	20 Mrad	$5e^{14}$/sq-cm
Forward Cal.(3-4)	1.4 Mrad	14 Mrad	$5e^{14}$/sq-cm

(Although the calculations above refer to values at shower maximum. we use the inside radius of the calorim

These total doses are growing v ry rapidly with increasing pseudorapidity in the forward calori eters as one would expect. In the barrel calorimeter, the high dose oc urs at the inner radius just past the central tracking where the radiu. is 4 meters and pseudorapidity is 3.0. This geometry must be modified to reduce the dose.

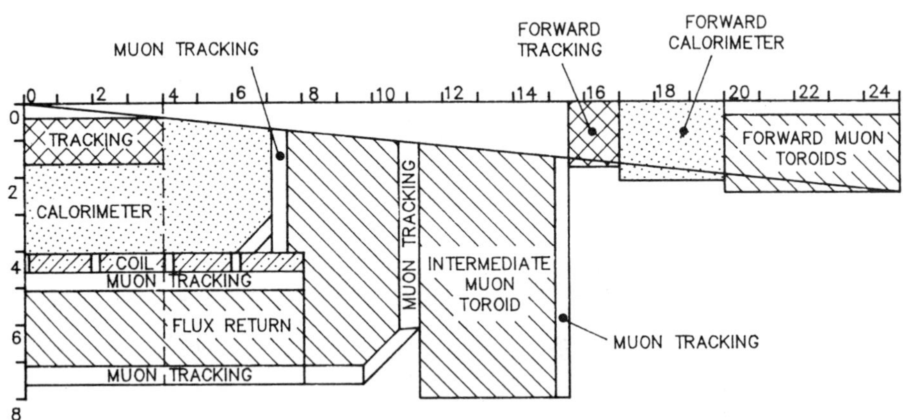

Fig. 6. The large solenoid detector.

RADIATION HARDNESS REQUIREMENTS FOR CALORIMETER ELECTRONICS

The time, segmentation, and hermeticity constraints placed upon the SSC calorimeter will require that the electronics be dispersed throughout the volume of the calorimeter[5]. Accordingly, this work is based on the assumption that the electronics is more or less uniformly distributed throughout the volume of the calorimeter. Therefore, the radiation environment of the calorimeter volume is the radiation environment within which the electronics must function.

An early goal of this project will be to determine, together with the radiation consultant, detailed radiation hardness requirements for the calorimeter electronics and to assess the ability of potential industrial participants to meet these requirements.

Due to the nature of the buried channel, JFETS are insensitive to total dose effects. JFETS have been successfully tested to 10 Mrads with almost insignificant changes to pinchoff voltages and Idss. Gate leakage (Igs) increases by an order of magnitude from less than 10 pA typical to approximately 100 pA. For neutron levels above 10^{14} neutrons/cm^2, N-channel JFETS are a must in order to minimize performance degradation.

Radiation effects on noise performance of JFETS have not been very well researched. After gamma exposures to a total dose of 10 Mrad, device noise for NJFETS typically increases from 10 nV/root hertz to 50 nV/root hertz at f = 10 Hz. While JFETS show promise for the SSC environment, JFET integration into a CMOS technology is not straightforward. In addition, pre-rad noise levels for a JFET in a CMOS technology will be significantly higher than those in a bipolar technology due to side effects of the processing steps.

This problem of gamma total dose in the forward calorimeters is important and must be carefully studied. We intend that in the initial plans for damage studies this problem be given priority because its solution could have profound implications for the solution chosen for the rest of the detector.

TESTS IN RADIATION ENVIRONMENT AT LIQUID ARGON TEMPERATURE AND 25C

Radiation damage studies are an integral part of this work. We propose that all radiation damage studies be undertaken both at liquid argon temperature and at 25C, and that no non-radhard devices be included in this work, except where needed for comparison purposes. The objectives of this damage testing program are to demonstrate that the devices we intend to use for the calorimeter electronics can operate satisfactorily over a reasonable lifetime and to provide confidence that the engineering and physics feasibility of the calorimeter plus the electronics are clearly established in the radiation environment.

Argonne operates the Intense Pulsed Neutron Source (IPNS) for neutron scattering studies[6]. This facility uses a 400-MeV proton beam incident on a heavy metal target. The spallation neutrons are moderated, and the resulting radiation field is viewed by many ports equipped with neutron spectrometers. There is also an alternative IPNS facility for doing radiation damage studies in which the sample is placed close to a high Z moderator.

There are few other particles accompanying the neutrons. The
secondary proton flux is 0.4% of the neutron flux with E > 0.1 MeV. An
upper limit to the gamma-ray flux is 15% of the total dose in rads.

The proton synchrotron typically operates at 30 Hz with an average
beam current of 15 ua so that the neutron flux at the peak of the
spectrum is about 10^{12} per cm^2 per sec.

There are several beam holes that can be used for irradiating
samples with diameters between 5 cm and 10 cm. The integral neutron
fluxes in three such tubes are given in Table 2.

Table 2. Integral Neutron Fluxes per Incident Proton

| Neutron Energy | REF | | NSF |
	Vertical Tube	Horizontal Tube	Horizontal Tube
Total	311	203	194
Thermal	2.4	1.7	44
> 0.1 MeV	199	122	55
> 1 MeV	66	36	13

The neutron flux with energy > 0.1 MeV for the REF vertical tube. The
flux varies by less than a factor of two over a cylindrical volume of 10
cm diameter by 10 cm high. This vertical tube is equipped with a
cryostat so that experiments can be done at low temperature.

In summary, the facility provides a good capability of studying
neutron irradiation of small samples, as well as 450 MeV protons for
investigating damage by charged particles. The neutron spectrum peaks
near 1 MeV but by choice of irradiation site, some control over the
spectrum is possible. Peak irradiation levels up to 10^{12} neutrons/
cm^2/sec can be achieved, and long-term lower level irradiations can also
be done on a parasitic basis.

Gamma irradiation facilities are not available at Argonne.
Brookhaven has gamma sources that are suitable for these damage
studies. We assume that the dependence on dose rate will be negligible,
but of course this must be clearly understood before the damage studies
begin.

The specific goals of the damage studies and the specific initial
program to be undertaken will be specified in a planning meeting between
personnel from Argonne, Brookhaven, and the industrial participant with
the participation of the consultant. It is important that there be
periodic reviews of the results of the damage testing program, and that
these results are available to the VLSI design program, which we envision
as taking place concurrently.

a. **Device Damage Studies**

We plan that, as soon as the details of the damage testing program
can be decided damage studies on discrete devices will begin. These
damage tests would be conducted at liquid argon temperature and 25C,
would measure noise, threshold voltage, transconductance, etc., and would
be carried out separately for charged particle, neutron, and gamma
fluences. A large enough selection of devices would be subjected to
these test to give confidence in the statistics and in wafer-to-wafer
variation in the process.

We intend to include in these tests, representative JFET devices
which would reasonably allow us to extrapolate radiation damage results
to the use of JFET's in appropriate processes.

b.   Prototype Calorimeter Operation in Radiation Environment

We will operate and take data from a small prototype calorimeter
section operating with radhard electronics in a radiation environment.
This prototype calorimeter section would be uranium/liquid argon, and can
operate with discrete devices created in the radhard process operating in
the cryogenic volume. Naturally, it will not be possible to implement
this prototype with the final VLSI device which results from the design
phase of this work, but it should be possible to implement several
channels with discrete devices from the process, or with radhard devices
where the results can reasonably be extrapolated.

Our objectives in this part of the study are to examine noise levels
of the electronics in the real calorimeter environment, and to examine
degradation of the data as a function of dose. Every effort will be made
to simulate the SSC calorimeter. For example, the front-end capacitance
will be padded to appropriate values. Naturally, it is difficult to make
hard statements about such things as calorimeter resolution from tests of
this sort, but we intend to provide an experimental basis for confidence
in the overall performance of the radhard VLSI design.

VLSI ELECTRONICS DEVELOPMENT

a.   Assumptions on Calorimeter Readout, Architecture, and Layout

We assume in this work that the calorimeter is organized as a barrel
calorimeter and forward calorimeters, and that the electronics is
dispersed throughout the volume of the calorimeter. The electronics is
implemented in the form of radhard VLSI chips, produced in a radhard low
noise CMOS process, probably in a CMOS technology, and is bonded to the
electrodes using an advanced technology. At the minimum, each VLSI chip
must have a multiplicity of charge sensitive amplifiers, circuitry to
store the output of the charge sensitive amplifiers, the ability to
multiplex the stored analog voltages on an output bus, calibration
capability, and a fast clear[7]. In addition, the VLSI devices must
provide an output that is a representation of the sum of the outputs of
all the charge sensitive amplifiers to be used in formulating the first
level trigger[8].

b.   Specifications of VLSI Device

In this work, we prefer to think of the VLSI radhard device specs in
a very general sense.   One of the first milestones is the completion and
approval of conceptual specifications for the VLSI device. These
conceptual specifications must be generated by a team consisting of
Argonne and Brookhaven electronics and calorimeter people, radhard
electronics experts from the industrial participant, and the consultant
who has been retained for this work. These conceptual specifications
would form the basis for the radhard design which could be begun by the
industrial participant at that time, subject to review as the radiation
damage studies progressed.

c. Proposed Radhard Process

The considerations driving process selection are primarily analog
device performance and packing density. Devices should have the lowest
noise levels and highest transconductances possible per microwatt of
power consumption. This implies leading edge device technology is
required with the thinnest and highest quality gate oxides attainable.
The requirement also exists for a floating (2 terminal) capacitor with
maximum capacitance density (thin dielectric). This is driven by the
potential need for large operational amplifier compensation capacitors.
Lastly, the tightest interconnect and device isolation photolithography
is required to maximize the area available for device gates and
capacitors.

The major effects of a total ionizing dose on MOS devices are
positive charge buildup in the oxide layer and interface state generation
at the oxide-silicon interface[9,10]. Positive charge buildup and
interface state generation both give rise to unwanted threshold voltage
shifts. Interface state generation also reduces the carrier channel
mobility. The generation of interface states is generally believed to be
completely suppressed at 75K, however, positive charge buildup is
worsened. Recent investigations have concluded that interface states may
be generated at cryogenic temperatures, however, they are in effect
"frozen out" at the low temperature and thus cause no device
degradation. The frozen states become active upon warming to room
temperature. A large percentage of the holes created by ionizing
radiation in an oxide are trapped at low temperatures, leading to larger
threshold voltage shifts. This situation is in contrast to room
temperature irradiations where only a small fraction of the generated
holes actually become trapped. The holes which escape initial
recombination with generated electrons (which are quickly swept out of
the oxide even at low temperatures) are trapped very near their point of
origin due to the extremely small low temperature hole mobility in
oxide. Hole movement is thought to be significant at low temperatures
only when under the influence of an electric field greater than around
two megavolts/cm.

Threshold voltage shift at liquid argon temperatures is a strong
function of device biasing during irradiation due to two field-dependent
effects. First, at low fields a portion of the electron-hole pairs
generated by radiation in the oxide recombine by geminate recombination
(recombination which occurs before the generated electron-hole pairs are
thermalized and become mobile in the oxide) and thus neutralize a portion
of the trapped holes. Electrons escaping recombination are swept out
while remaining holes are "frozen in". As the field is increased, a
larger fraction of electrons escape recombination because they are swept
out more quickly and have a smaller probability of recombining, leaving
an increasingly larger amount of uniformly distributed positive charge.
Therefore, the threshold voltage shift at a given dose increases with
increasing applied voltage. When the applied voltage is increased to a
point where an oxide field of around two megavolts/cm is achieved, holes
become mobile in the oxide and a fraction of them will be transported out
of the oxide. This gives rise to less positive charge in the oxide and a
decrease in threshold voltage shift with increasing applied bias.

To summarize, the combined result of these two effects is that the
threshold voltage shift first increases with increasing bias due to an
increasing geminate recombination escape probability, and then the threshold voltage shift eventually starts to decrease with further increases
in bias due to the initiation of hole transport out of the oxide.

CMOS processes hardened to total dose at room temperature are
usually not hard at liquid argon temperatures. Any room temperature
hardening technique which depends on the contribution of negative charge
due to interface state generation to compensate positive trapped hole
charge will be ineffective at low temperatures. Hardening of the gate
oxide consists primarily of using the thinnest high quality oxide
allowable to minimize the charge generation volume. Room temperature
hardened oxide processes do not seem to display increased hardness over
non-hardened oxides at liquid argon temperatures. In addition, the pre-
radiation interface state density can be minimized for low noise
performance since interface state generation is not relied on for
negative compensating charge during low temperature radiations.

d. **Proposed Radhard Design**

The design of a radiation hardened circuit requires a radhard
technology. Then circuit techniques are used to compensate the
parametric degradations of devices and elements, or to avoid them
completely. There are three items to be concerned with. The first is
burn-out/survival, second is device/element degradation, and the third is
radiation-induced ac/transient behavioral errors.

Burn-out is possible if the incident transient ionizing dose rate is
very high, which is probably not applicable to the SSC with the possible
exception of a beam spill. All the devices then become shorted, and the
only resistance between power supplies is bulk silicon. The major cause
of burn-out is the interconnect fused open. Survival is achieved through
the use of resistors in series with supplies, either on chip or off chip,
to limit the short circuit current. Internal resistors must be able to
absorb the energy and the interconnect must be designed with this in
mind.

Circuit designs must avoid dependence on those device/element
parameters that degrade by radiation. Techniques such as using
device/element ratio matching, cascoding, and threshold voltage tolerant
circuits are the prevalent design methods. Parametric magnitude
dependent circuit should be avoided if possible.

Transient radiation induced errors can be minimized if PN junctions
are photocurrent compensated. Closed-loop design should be preferred
over open-loop system. Closed-loop circuit tends to suppress the induced
errors by the loop-gain. Differential mode amplification is more
tolerant to this induced aberrations mainly due to common mode rejection.

In some cases, circuits can use self-calibration techniques to
adjust to the effects of radiation. The penalties are complexity and
real estate.

PLAN OF RESEARCH

a. **Milestones of This Research and Development Work**

1. Appointment of outside consultant.

2. Agreement on radiation hardness requirements and preliminary
    determination of suitability of potential industrial
    participant's radhard process.

3. Specification of program for initial radiation damage studies.

4. Completion of administrative procedures necessary to secure participation of an Industrial Participant.

5. Completion and approval of conceptual specifications for VLSI device.

6. Damage testing program activities, with progress reviews.

7. Radhard VLSI design work, with progress reviews.

8. Completion of damage testing program.

9. Completion of radhard VLSI design.

Accomplishment of the above set of milestones is expected to require approximately one year and bring us to readiness to begin producing wafers. Actual production and testing of such wafers would be carried out as a follow-on phase 2 project.

Work supported by the U.S. Department of Energy, Division of High Energy Physics, under contract W-31-109-ENG.

REFERENCES

1. D. E. Groom, Editor, "Radiation Levels in the SSC Interaction Regions," SSC Central Design Group, Task Force Report, SSC-SR-1033, (June 10, 1988).
2. Richard Wigmans, "Calorimetry at the SSC," Invited talk at the Workshop on Experiments, Detectors, and Experimental Areas for the Supercollider, Berkeley (July 7-17, 1987).
3. C. Baltay, J. Huston, and B. G. Pope, "Calorimetry for SSC Detectors," Proc. of the Summer Study on the Physics of the Superconducting Super Collider, Snowmass, CO, p. 355 (1986).
4. G. G. Hanson, et al, "Report of the Large Solenoid Detector Group," Proceedings of the Workshop on Experiments, Detectors, and Experimental Areas for the Supercollider, Berkeley (July 7-17, 1987).
5. V. Radeka and S. Rescia, "Speed and Noise Limits in Ionization Chamber Calorimeters," Nucl. Instr. and Methods A265, 228 (1988).
6. R. C. Birtcher, T. H. Blewitt, M. A. Kirk, T. L. Scott, B. S. Brown, and L. R. Greenwood, "Neutron Irradiation Facilities at the Intense Pulsed Neutron Source," Journal of Nucl. Materials 108-109, 3 (1982).
7. H. H. Williams, "Detectors for the SSC: Summary Report," Proc. of the Summer Study on the Physics of the Superconducting Super Collider, Snowmass, CO, p. 327 (1986).
8. T. J. Devlin et al., "Electronics, Triggering, and Data Acquisition for the SSC," Proc. of the Summer Study on the Physics of the Superconducting Super Collider, Snowmass, CO, p. 439 (1986).
9. J. Srour, "Basic Mechanisms of Radiation Effects of Electronic Materials, Devices, and Integrated Circuits," IEEE 1982 Annual Conference on Nuclear and Space Radiation Effects.
10. J. E. Grover, "Basic Radiation Effects in Electronics Technology," IEEE Tutorial Short Course on Radiation Effects (1984).ffects in Electronics Technology," IEEE Tutorial Short Course on Radiation Effects (1984).

Superconducting Wire and Cable

Chairman:

C. Taylor
Lawrence Berkeley Laboratory

DEVELOPMENT OF SUPERCONDUCTING STRAND AND CABLE WITH IMPROVED PROPERTIES FOR USE IN SSC MAGNETS*

Ronald M. Scanlan

Lawrence Berkeley Laboratory
1 Cyclotron Road
Berkeley, CA 94720

ABSTRACT

The critical current requirement for the NbTi superconductor strand was set at 2750 A/mm^2 (5 T, 4.2 K) in the SSC Conceptual Design, compared with a value of 1800 A/mm^2 which was specified for the strand used in the Tevatron dipoles. In addition, a filament diameter of 5 μm, instead of the 9 μm diameter used in the Tevatron, was chosen to reduce field distortion at injection. In order to meet the requirements for field homogeneity, the dimensional requirements for both strand and cable were also tightened. The technical solutions employed to achieve these improved properties and the resulting specifications will be discussed.

DISCUSSION

The present SSC dipole design is a descendent of a design originally developed at LBL in late 1983/early 1984 as the LBL proposal for the SSC dipole. This design drew upon earlier dipoles built at FNAL for the Tevatron and at DESY for the HERA accelerator, together with an extrapolation of key parameters, where such extrapolation appeared reasonable in light of new technical developments. An example of this process are the choices of copper to superconductor ratio and number of strands in the cable. Discussions with the strand manufacturers and actual coil winding experience at LBL indicated that a Cu/SC ratio of 1.0 was possible to fabricate, but presented problems for reliable large scale manufacturing and coil winding. On the other hand, a Cu/SC ratio of 1.8 was considered as quite conservative, since the Tevatron magnets and their successors (CBA, HERA, and UNK) had all demonstrated the relative ease in working with this ratio. Hence, a ratio of 1.3 was chosen as a value which would extend the state of the art, yet which would meet the requirements for strand fabricability and coil winding.

The choice of number of strands and strand diameter resulted from similar considerations. Fabricability of a 23-strand cable had been demonstrated by the Tevatron cable; the main change introduced into the SSC design was a larger strand diameter (.0318") as compared with .0268" diam., and the above mentioned decrease in the Cu/SC ratio. The larger strand diameter is beneficial in achieving a high overall current density and a high operating current, at a price of somewhat increased difficulty in coil winding. Another important cable parameter is keystone angle, and the choice here is a trade-off between the ideal (large) value

---

* This work is supported by the Office of Energy Research, Office of High Energy and Nuclear Physics, High Energy Physics Division, Dept. of Energy under Contract No. DE-AC03-76SF00098.

that is associated with a small bore dipole and a small (or zero) value that yields minimum critical current degradation during cabling. Trial cable was made by SSC with the ideal keystone angle of 2.05° for the inner cable and 1.57° for the outer. Both cables showed a large degradation as indicated by many broken filaments, so the decision was made to reduce the angle until acceptable cable was achieved. The values found acceptable are 1.6° for the inner cable and 1.2° for the outer. It is interesting to note that several years later, designers of the RHIC dipoles elected to modify the SSC outer cable to produce a fully keystoned design with a 1.6° angle. The excessive critical current degradation experienced with these cables forced them to return to the 1.2° angle chosen for the SSC outer cable.

The most important parameter to choose in terms of the ultimate performance of the magnet is the critical current density. The experience base (again, the Tevatron dipoles) indicated that a minimum value of 1800 A/mm^2 at 5T, 4.2 K could be achieved in production. However, new R&D results from Baoji Institute (China)[1] indicated that, with proper processing, values in excess of 3000 A/mm^2 at 5T, 4.2 K are possible. Since this magnet was being proposed for a machine to be built in the 1990's, it was reasonable to assume that an aggressive R&D program aimed at improved critical current density could be completed in time to provide the necessary conductor. Accordingly, a value of 2400 A/mm^2 was chosen as a design value and an R&D program was initiated in order to demonstrate that this value could be achieved in production-size billets (10" diam.). In June, 1984, this value was exceeded in a production-scale billet[2] funded by LBL, and processed by Intermagnetics General Corp. (IGC) following a heat-treatment schedule proposed by D. Larbalestier (U. Wisconsin).

The selection of parameters for the outer layer cable follows logically from the inner layer cable parameters, since we want to match the operating current and quench protection in the two layers. Accordingly, the Cu/SC ratio can be increased (this also reduces the cost of the cable), and the strand diameter reduced in order to increase the flexibility of the cable. The Cu/SC ratio was increased to 2.0, and a 30-strand cable was proposed, although a 30-strand cable had not yet been made in a long length.

These parameters remained basically unchanged during the next 18 months, while a series of 1-m models at LBL and 4.5 m magnets at BNL were being built. One minor change was in the outer layer Cu/SC ratio, from 2.0 to 1.8, in order to more closely match the inner and outer cable critical currents. During this period, a total of 9 billets 10" in diam. and 5 billets 12" in diam. were produced for the SSC R&D program and all met the minimum 2400 A/mm^2 critical current value.

In parallel with this effort, R&D work was begun in October, 1984, to explore the possibility of producing a fine filament NbTi conductor. The first R&D contract[3] was placed with Supercon, Inc. for the fabrication of a 12" diam. billet to produce a conductor with 4000 filaments, with a minimum J_c value of 2400 A/mm^2. In November, 1984, proposals were solicited from the other U.S. superconductor manufacturers and they were asked for their recommendations with respect to filament size and critical current density goals for the fine filament R&D program. These proposals and the scope of the fine filament R&D program were discussed at a Superconductor Working Group meeting at LBL in January, 1985. Following this meeting, and after additional funding had been obtained, the fine filament program was expanded to include work at IGC and Oxford Superconducting Technology (OST) on double extrusion approaches, and at LBL on hydrostatic extrusion. Results on the first Supercon billet were obtained in July, 1985, and they showed a value of J_c (5T) = 2985 A/mm^2 for a filament size of 4 μm.[4] Based on these results, the target for the SSC fine filament R&D program was changed to a value of J_c = 2750 A/mm^2 at 5T with a filament size of 2.5-5 μm. Progress toward this goal continued to look promising, so that the following values were selected for the SSC Conceptual Design Report[5,6] (March, 1986): J_c (5T, 4.2 K) = 2750 A/mm^2; filament diam. = 5 μm. After much discussion and a careful study of the results for the LBL 1-m dipoles and the BNL 4.5 m dipoles, an operating field point of 6.6 T at 4.35 K was also selected.

Using these values, a series of margins can be calculated. Three definitions of margins which are typically used are:

1) Critical current margin at a constant operating field
2) Critical current margin along the magnet load line
3) Temperature margin (operating vs. critical), at the operating field

The first two margins are calculated in SSC-MAG-160 by C. Taylor;[7] the third margin can easily be calculated using the analytical expression from G. Morgan (BNL SSC-MD-84), and has been obtained for several cases by C. Quigg and M. Tigner[8].

The important point to note is that these "margins" defined in this manner include a number of real effects which will act to limit magnet performance. These are also listed in SSC-MAG-160. Two of these factors will be discussed here, since they are important in the development of a working SSC cable specification.

In the process of converting strands with a critical current density of 2750 A/mm^2 to cable, a certain amount of degradation occurs. In the past, this has been as high as 15-20%; however, as wire quality and cable manufacturing have improved, this has decreased. The difference in strand critical current vs. cable critical current also depends on the method of measurement and corrections applied to the measurements.[9] Using these test methods, in which cable self-field effects are included, but wire self-field effects are not included, and for the applied field perpendicular to the wide face of the cable, the recent degradation values are typically less than 5%. In the present version of the SSC cable specification SSC-MAG-M-4142, Rev. 2, a value of 5% was used for the maximum allowable degradation in cabling, for both the inner and outer cable.

In order to specify a minimum critical current, another factor must be taken into account; during the manufacturing process, the Cu/SC ratio can vary from one end of the wire to the other. In order to maintain an acceptable yield, the Cu/SC ratio is permitted to vary $\pm$ 0.1; this is taken into account in converting the J_c value into a minimum I_c value for the cable.

The following is a step by step procedure used to convert from the Conceptual Design Report J_c value to the present SSC cable minimum I_c values:

1) Find the Jc value at the high field point in the dipole:
$$J_c(7T, 4.2\ K) = 1650\ A/mm^2\ \text{Inner Cable}$$
$$J_c(5.6T, 4.2\ K) = 2420\ A/mm^2\ \text{Outer Cable}$$
These values are obtained by a linear interpolation between 2750 A/mm^2 at 5T and 1100 A/mm^2 at 8T.
2) Find the minimum I_c value based on these J_c values and the maximum allowable Cu/SC ratio:
For inner cable, nominal 1.5/1 ratio,
   minimum $I_c(7T, 4.2\ K) = 1650$ A/mm^2 • minimum area
   of superconductor = 7479 A.
For outer cable, nominal 1.8/1 ratio,
   minimum $I_c(5.6T, 4.2\ K) = 2420$ A/mm^2 • minimum area
   of superconductor = 8249 A.
3) Find the minimum I_c value with an allowance of 5% for cabling degradation:
   Inner cable minimum $I_c(7T, 4.2\ K) = 7479A$ • 0.95 = 7105 A.
   Outer cable minimum $I_c(5.6T, 4.2\ K) = 7837$ A

An obvious question at this point is, "What are the prospects for increasing the dipole magnet margins?". With respect to critical current density, the recent production experience indicates that the wire manufacturers can meet the 2750 A/mm^2 value, but without a large margin; the values have been ranging from 2750 to 2950 A/mm^2. Indeed, the SSC spec.

value is already the highest in the world, with HERA being approx. 2500 A/mm^2, RHIC is 2600 A/mm^2, and LHC is equivalent to 2450 A/mm^2. Also, we have observed an increase in strand stiffness with increased J_c, and this in turn makes cabling more difficult. We are continuing the push toward higher critical current densities in the SSC R&D program, but it is premature to extrapolate those results to production quantities at this time. With respect to cable degradation, the values allowed for the SSC cable are again the lowest in the world; RHIC allows 10% and HERA allows 15%. Further significant improvement in this area seems unlikely at this time, unless we decide to change the degree to which the cables are keystoned. The most significant parameter for increasing the operating margin appears to be the operating temperature, but that requires a discussion of trade-offs that are beyond the scope of this paper. The most obvious change in the conductors which could add to the present margin is a return to a copper to superconductor ratio of 1.3 in the inner layer cable.

REFERENCES

1. L. Chengren, W. Xiao-Zu, Z. Nong, IEEE Trans. on Magnetics, MAG-19, 284, 1983.

2. D. C. Larbalestier, A. W. West, W. Starch, W. Warnes, P. Lee, W. K. McDonald, P. O'Larey, K. Hemachalam, B. Zeitlin, R. Scanlan, and C. Taylor, IEEE Trans. on Magnetics, MAG 21, 1985.

3. LBL P.O. #7456406, October, 1984.

4. E. Gregory, T. S. Kreilick, and J. Wong, "Fine Filamentary Materials for Accelerator Dipoles and Quadrupoles", in proceedings of the ICFA Workshop, P. Dahl, Ed. BNL Report 52006, pp. 85-88, 1986.

5. SSC Conceptual Design Report, SSC-SR-2020, March, 1986.

6. R. M. Scanlan, J. Royet, and R. Hannaford, "Evaluation of Various Fabrication Techiques for Fabrication of Fine Filament", NBT, Superconductors, IEEE Trans. on Magnetics, MAG-23, pp. 1719-1723, 1987.

7. C. E. Taylor, "Design "Margin" and the Effect of Different Copper-to-Superconductor Ratios on B_0 of SSC Dipoles", SSC-MAG-160, September, 1987.

8. C. Quigg and M. Tigner, SSC Central Design Group Memo, May, 1988.

9. M. Garber and W. B. Sampson, SSC-N-488, March, 1988, and SSC.Cable Specification SSC-MAG-M-4142.

SUPERCONDUCTING WIRE AND CABLE FOR THE SSC - PROGRESS AT

INTERMAGNETICS GENERAL CORPORATION TOWARDS PRODUCTION

H. Kanithi, F. Krahula, M. Erdmann, R. Schaedler and B. Zeitlin

IGC Advanced Superconductors Inc.
1875 Thomaston Ave.
Waterbury, CT , 06704

ABSTRACT

The requirements of the SSC are examined from the point of view of a wire and cable manufacturer. The present status of the conductor development at IGC is described. Since the SSC will require a large volume production at the most economical price, billets measuring 12" in diameter have been used for prototype wire production. Both inner and outer grades of wire have been processed with NbTi filament diameters of 5 to 9 μm. Considerations of the production scale up that will have to be made to meet the demands of the SSC, are also discussed.

A novel conductor design which promises to have greatly increased flux pinning capability and consequently can sustain significantly higher current densities than can conventional materials, is mentioned.

INTRODUCTION

The SSC will require considerably more superconducting cable than any previous single machine. In addition the wire and cable is relatively sophisticated as it contains many fine filaments and both wire and cable have to be made to relatively close tolerances. Since the early 1980's, the conductor design and manufacturing techniques have improved to the point where the SSC specifications can now be met by several manufacturers. These specifications are: 6 μm diameter filaments, ~4000 for the outer wire and ~7000 for the inner wire. The outer conductor must carry 2750 A/mm^2 at 5 T and the inner 1600 A/mm^2 at 7 T. The tolerance on the wire diameter is ±0.0001", on the cable thickness, ±0.00025" and on the cable width, ±0.001". For economical production of large quantities of wire and to meet the specifications (only 10% < 10,000ft) it is necessary to make the wires in long piece lengths. Also, because of problems that can be encountered at the injection field of the accelerator, it is necessary to have low magnetization in the low field region.

In this paper, in addition to some of the technical aspects, we will address some of the considerations that have to be taken into account due to the considerable scale-up that will be required to meet the SSC requirements.

WIRE AND CABLE REQUIREMENTS.

For the dipoles and quadrupoles alone, 5.4 million lbs. of cable will be required. This is equivalent to 2.7 billion ft. of wire or 96 million ft. of cable. To make this amount of conductor, it will be necessary to make 6765 12" diameter billets and the maximum rate of production will be 1810 billets, (1.4 million lbs.),of wire per year. In addition to a significant scale up of present production facilities, this additional work will also require the revamping of the existing quality control and assurance procedures.

REVIEW OF IGC-ASI CONDUCTOR EXPERIENCE.

IGC made the bulk of the conductor for the Fermilab Tevatron and, although the current density in this material was low, by 1984 the state of the art was 2600 - 2800 A/mm^2 at 5 T. in large diameter filaments [1]. At the present time [2] the J_c has been increased to 2900 - 3000 A/mm^2 in much finer filaments. This increase has been accomplished by :

Improvements in the NbTi alloy [1].
A better understanding of the heat treatments [3,4].
Elimination of intermetallics [5].
A realization of the importance of local area ratio and the importance of the relative strength of the constituents in the composite [2].

After the importance of the homogeneity of the alloy was realized, only "hi-ho", or equivalent, alloys were used in the work and, to eliminate the formation of the intermetallics, a diffusion barrier layer was introduced around all filaments. Throughout the work the latest heat treatment schedules available in the literature and also developed internally, were employed.

At first an attempt was made to make a 2.5 μm diameter filament conductor as this was still being considered in the specification. To accomplish this, two first stage 10" diameter billets, each containing 19 filaments, were extruded [6]. These were drawn to two different restack sizes and assembled into two second stage 10" diameter billets, one for the inner cable with 36,613 (19 x 1927) filaments and the other with 20,634 (19 x 1086) filaments for the outer cable, Figure 1. In addition sample quantities of the 19 element restack were processed to fine wire in order to explore the potential J_c that could be obtained in the subelement alone. In parallel, a monofilament material was made by cold cladding and a 10" diameter billet containing 10,500 restack rods was processed to a size where the filaments were 5 μm. in diameter.

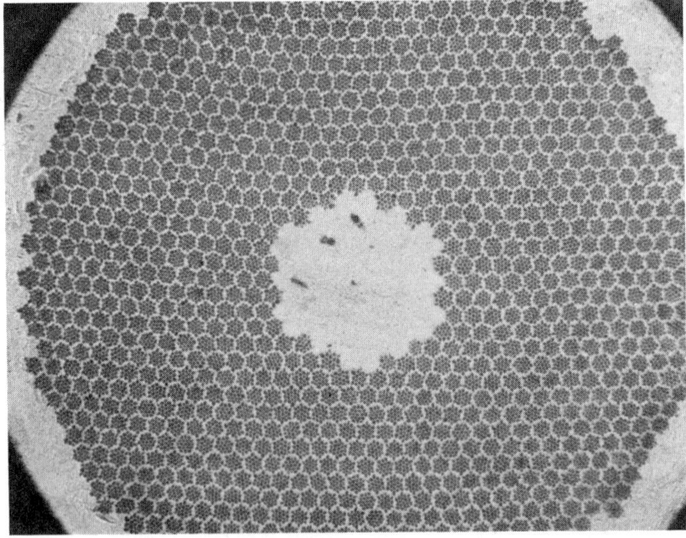

Figure 1. Cross section of an outer strand with 20634 2.5 μm. diameter filaments.

I_c of all the wires was measured at magnetic fields from 5 to 8 T, at a sensitivity of 10^{-12} Ωcm. and the "n" values were calculated. The outer grade preproduction samples with 2.7 μm diameter filaments, had a 5 T J_c of 2500 A/mm^2 and an "n" value of 22. Inner grade material with a similar filament size had a 5 T J_c of 2300 A/mm^2 and an "n" value of 20. Despite the fact that there was minimum compound formation, the filaments were still "sausaged" and the J_c of the conductor was "extrinsically" limited. The subelement, however, drawn down to a size where the filaments were 15 μm in diameter, showed a J_c of 3400 A/mm^2 and an "n" of 84. The filament quality was found to be superior to that of the restacked material, showing the importance of geometrical configuration, and filament size.

Although some difficulties were encountered in the fabrication of the restacked material, a sufficient amount of inner grade conductor was delivered and made into a full length cable. The potential feasibility of a 2.7 μm diameter filament conductor was established, although this material showed a somewhat lower J_c than the 6 μm. diameter filamentary material.

About this time, however, the specification for the J_c at 5 T was raised from 2400 A/mm^2 to 2750 A/mm^2 and the filament diameter increased from 2.5 μm. to 6 μm. The J_c of the 6 μm. diameter filament material, was 2700 A/mm^2 and an "n" value of 35. Since the filaments were more uniformly distributed throughout the matrix and more evenly constrained by their neighbors, they "sausaged" less. Also the overall ductility and drawability of the monofilament restack conductors far exceeded the multifilament restacked material. Several pieces of the former conductor were processed to final size in single lengths weighing more than 45 kg. each.

In more recent work to produce conductor for the SSC, the billet size has been increased from the 10" diameter to a 12" diameter for more economical production. The monofilament restack has been adopted, and another key design parameter, the local area ratio, (LAR) has been investigated and the filament diameter varied from 4.8 to 9 μm. The LAR is defined as the the volumetric copper/NbTi fraction in the immediate vicinity of the filaments. In this work, however, a double extrusion approach was employed. In place of a cold clad monofilament, an extruded monofilament billet was fabricated to produce the restack element. Appropriate numbers of copper clad monofilament rods were stacked in 12" diameter billets and these processed to finished wire in the conventional manner.

Figure 2 shows a typical filament distribution and geometry for billet 5251-1000 which is an inner conductor with nominally 7250 6 μm diameter filaments and a Cu/Sc ratio of 1.5.

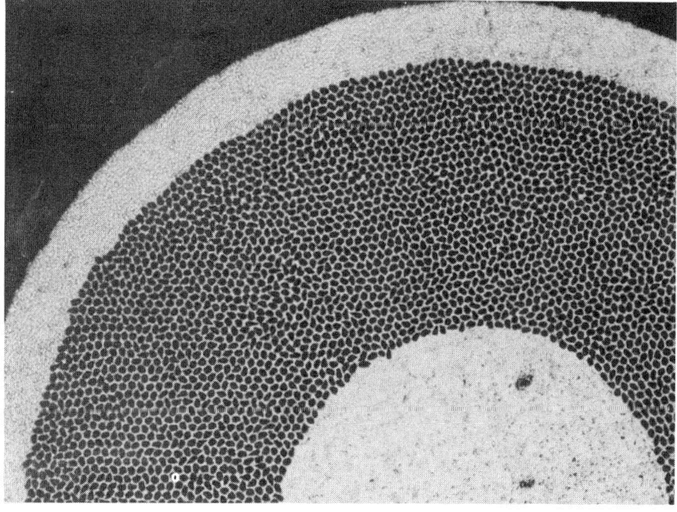

Figure 2. Cross section of an inner strand with 6 μm. diameter filaments.

A similar billet, 5251-1001, which is an outer conductor with nominally 6470, 4.8 μm diameter filaments and a Cu/Sc ratio of 1.8, was made. The piece lengths for these two materials are compared in Figure 3. The average piece length for the 6 μm, filament inner grade was 7600 ft. while it was 5200 ft. for the 4.8 outer grade material.

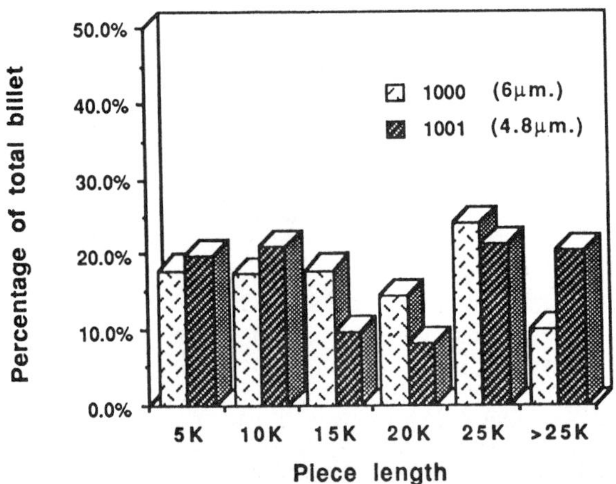

Figure 3. Piece length comparison between 6 μm. and 4.8 μm. diameter filamentary strands.

The strand from the inner grade billet 1000, exhibited slightly higher J_c and higher "n" values than the outer grade billet 1001 at all the fields tested (Figure 4). From the total cold work point of view [5], one would have expected a higher J_c for the outer grade conductor. However, the lower "n" values in the finer filaments of outer grade wire seem to counteract any possible J_c enhancement due to extra cold work.

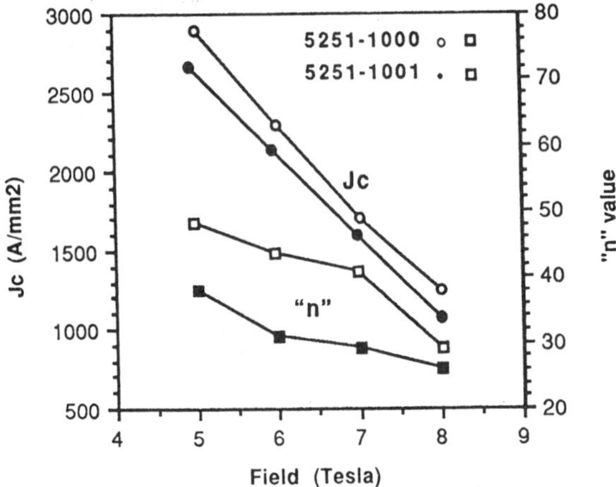

Figure 4. Current Density (solid line) and n value (dashed line) as a function of magnetic field for billets 5251-1000 and 5251-1001.

The critical current characteristics of cables made from these wires were measured at BNL. All the cables met the I_c specification and the degradation was around 4.0% for the 6 μm. diameter inner material and 6.5% for the 4.8 μm. outer diameter cable. These low degradation numbers prove that the NbTi filaments are ductile and their microstructures close to being optimum. If the filament quality is not well controlled , I_c degradation of 12 to 15 % is not uncommon in such flat cables.

SUMMARY OF IGC'S CONDUCTOR DEVELOPMENT STATUS

By using the improved alloy, diffusion barriers and by paying attention to the geometry of the filament array, we have explored the feasibility of making both inner and outer cables from strand with filament sizes ranging from 2.5 to 9 µm. in diameter.

The process developed employs an extruded monoelement restack rod, assembled in a 12"restack billet. Cabling has been carried out on the Dour machine [7] but in addition IGC has arranged to purchase a U.S. made machine from AFA in Garfield, New Jersey .

There has been a steady improvement in J_c over the past 4 years (Fig. 5) and J_c values approaching 3000 A/mm^2 and "n" values ~ 50 at 5 T have been achieved in 6 µm. diameter filamentary material. Such material has also been successfully cabled with low degradation.

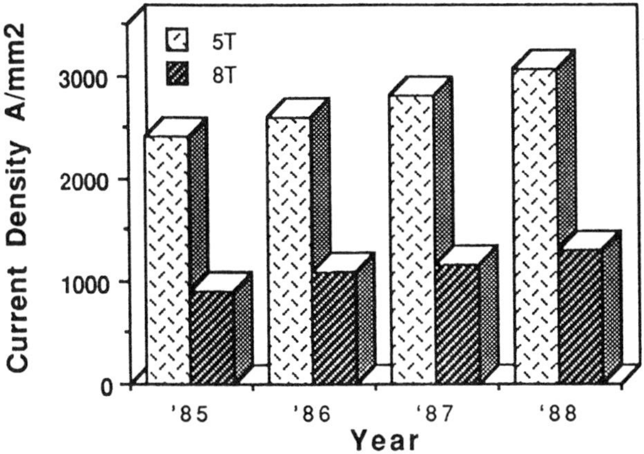

Figure 5. IGC J_c Performance.

Challenges still exist, however, before the 6 µm. diameter filamentary cable can be produced economically in the quantities required for the SSC . There is need to improve the consistency of wire and cable production. In particular, it is important to be able to reliably produce the longest piece lengths repeatedly and also to improve the J_c margins on the conductor for the inner cable.

The total number of billets which all the manufacturers have been asked to make are very small and more billets must be made to take us further up the learning curve and enable us to obtain statistically meaningful data from production quantities of material. Such orders will also enable us to introduce process improvements, increase non destructive testing, improve quality control and assurance procedures, and train our production personnel in the intricacies of fine filamentary conductor processing. In summary, we need volume orders so that we can take the process out of the development phase into the production mode.

The goal of our work on the 6 µm. diameter filamentary material is to minimize cost and maximize performance. If, however, 2.5 µm. diameter filamentary materials are needed, triple extrusions will probably be required and this will ensure that further development work will be required to develop new billet assembly techniques and refine the processing parameters.

CONSIDERATIONS FOR PRODUCTION SCALE-UP

While IGC produces more superconducting material than any other company, we obviously must make changes to our plant and our overall organization to meet the needs of the SSC. To prepare for these changes, we have engaged consulting engineers to plan for SSC plant expansion in a phased approach. Expansion space requirements are modest and relatively easy to implement as IGC has contiguous expansion space available.

The SSC would help justify a relatively modest capital expansion and modernization, which would benefit all customers and strengthen us in the world market. The tooling and equipment required specifically for the SSC will consist of cablers and twisters only. All additional equipment would be considered as a capital investment in the superconducting business on the part of IGC-ASI.

Summarizing, we conclude that IGC can handle easily the SSC wire and cable needs without jeopardizing the needs of our current customers.

INCREASED CURRENT DENSITY IN NbTi

The superconducting wire producers have been constantly required to increase the current density in commercially available materials. While we have made significant progress in recent years, Figure 5, it is reasonable to expect that, in the short term, i.e. the next 1-2 years, further progress will at least enable us to widen the margins with which we can meet the present SSC requirements.

In the longer term, we at IGC feel that a very significant improvement in J_c in NbTi is possible by the use of a novel flux pinning approach, which is the subject of a recently issued patent [8]. Figure 6. shows schematically the concepts involved. A uniform and

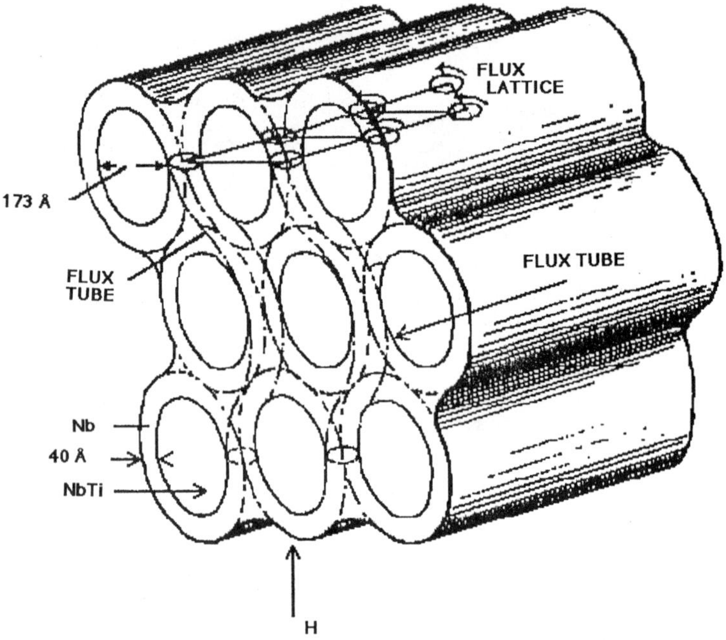

Figure 6. A uniform array of pinning sites created by Nb clad NbTi rods.

periodic layered pinning structure is created , one possible way of doing this is to clad NbTi with Nb, which at moderate fields provides these pinning layers. The spacing of the layers can be adjusted to correlate with the coherence length of the filament core and the field. This approach, offers the potential of J_c's as high as 6000 A/mm^2 at 5 T. If such J_c's can be achieved in practical wires which can be economically fabricated, the whole field of applied superconductivity would be significantly affected.

SUMMARY

We have traced the development of high J_c fine filament conductors at IGC, to the point where conductors with 6 μm. diameter filaments meeting the SSC specifications for both inner and outer cable can now be produced.

It is necessary, however to make large numbers of billets in the near future in order to install the QC and QA procedures , gather statistical data and transfer the operations from development into a fully fledged production mode.

The capital and manpower needs have been alluded to in general terms and the conclusion reached that IGC is ready to meet the SSC challenge and can gear up to meet the needs relatively easily with minimal capital and manpower requirements. Small short term J_c improvements can be expected but the problems of changing the filament diameter to 2.5 μm., while still maintaining the present high J_c's, should not be underestimated.

Mention is made of a recently issued IGC patent which offers the possibility of significantly increasing J_c in NbTi by a new fabricating technique which produces new types of pinning sites.

ACKNOWLEDGEMENTS

Thanks are due to the DOE, Lawrence Berkeley Laboratory (LBL), and the Small Business Administration Innovative Research Organization (SBIR), for continued financial support in this area. Also the NbTi workshops, organized by DOE headquarters, LBL, the Central Design Group (CDG) and the University of Wisconsin have proved to be a very significant factor in the rapid developments that have been made in the SSC conductors, as have the contributions made by many individuals from all these organizations. We are also deeply in debt to both the Brookhaven National Laboratory (BNL) and the National Institute of Standards and Testing (NIST) for helping in the testing and evaluation of our materials.

REFERENCES

1. D.C. Larbalestier, A.W. West. W. Starch, W. Warnes, P. Lee, W.K. McDonald, P. O'Larey, K. Hemachalam, B. Zeitlin, R. Scanlan and C. Taylor, "High critical current densities in industrial composites made from high homogeneity Nb 46.5 Ti", IEEE Trans., MAG-21, Vol. 2, pp. 269-272, 1985.

2. H.C. Kanithi, C.G. King, B.A. Zeitlin, "Fine filament NbTi Conductors for the SSC", IEEE Trans., vol 25, 2, pp. 1922-1925, 1989.

3. Li Chengren and D.C. Larbalestier, "Development of high critical current densities in Niobium 46.5 wt.% titanium", Cryogenics, 27, 4, 171-177, 1987.

4. D.C. Larbalestier and P. Lee, NbTi Workshop Asilomar, January 1989.

5. H. Kanithi, "Expectations and limitations of J_c in practical NbTi conductors", Adv. in Cryo. Eng., A.F, Clark and R. P. Reed, eds. Plenum Press, New York 1988, vol. 34, pp. 951 958.

6. C. King, K. Hemachalam and B. Zeitlin, "Prototype fabrication of ultrafine filament NbTi conductor for the SSC". IEEE Trans., MAG-23, 2, pp. 1351-1354, 1987.

7. J.M. Royet, R. Armer, R. Hannaford and R.M. Scanlan, "An Industrial Cabling Machine for the SSC", Paper III D-8, International Industrial Symposium on the Super Collider, Feb. 8-10, New Orleans, 1989.

8. United States Patent 4,803,310, B.A.Zeitlin, M.S. Walker and L.R. Motowidlo, "Superconductors having controlled laminar pinning centers and method of manufacturing same" issued Feb. 7, 1989.

S. Hong, D. Geschwindner, A. Mantone, W. Marancik, S. Zalek
and R. Zhou

Oxford Superconducting Technology
600 Milik Street, P.O. Box 429
Carteret, NJ 07008

INTRODUCTION

The charged particle beam in the accelerator is kept on its orbit by powerful dipole magnets. The dipole magnets for large accelerators are often made of superconductors for technical and economical reasons. Based on the technical success of the Tevatron, the proton synchrotron at the Fermi Accelerator Laboratory, the dipoles for the Superconducting Super Collider (SSC) are of high performance superconducting magnets. However, the SSC requires quite advanced superconducting composites compared to the Tevatron composites for economical reasons. The requirement of critical current density (J_c) is high. The Tevatron required ~1800 A/mm^2 at 5 T and the SSC requires ~3000 A/mm^2 at 5 T. This enhanced J_c requirement necessitates the finer filament diameter of superconductor in the composite to minimize the sextapole field errors due to the magnetization of filaments at the low injection field. For the industrial fabrication of this composite, the superconducting composite manufacturers in the U.S.A. have been developing this composite for the past few years with close collaboration with national laboratories and universities. In this paper, we will review the progress for SSC superconducting composites developed at Oxford Superconducting Technology (OST).

CRITICAL CURRENT DENSITY

As shown in Fig. 1, improvement of J_c over the past decade, we have demonstrated that the J_c of Nb 46.5 w/o Ti superconductor can be as high as 3800 A/mm^2 at 5 T in a very fine simple composite. However, a somewhat lower J_c is expected in the composite of SSC wire diameter due to the limit of coldwork and self-field effect. The fine filament and a large number of filaments add complexity to the fabricability of high performance composites although some of the difficulties have been resolved in the past few years. It is likely that production quantities of SSC composites can be supplied at slightly over 3000 A/mm^2 which provides a comfortable margin in performance.

EARLY GENERATION OF FINE FILAMENT COMPOSITES

OST introduced a fine filament NbTi composite for ac application in

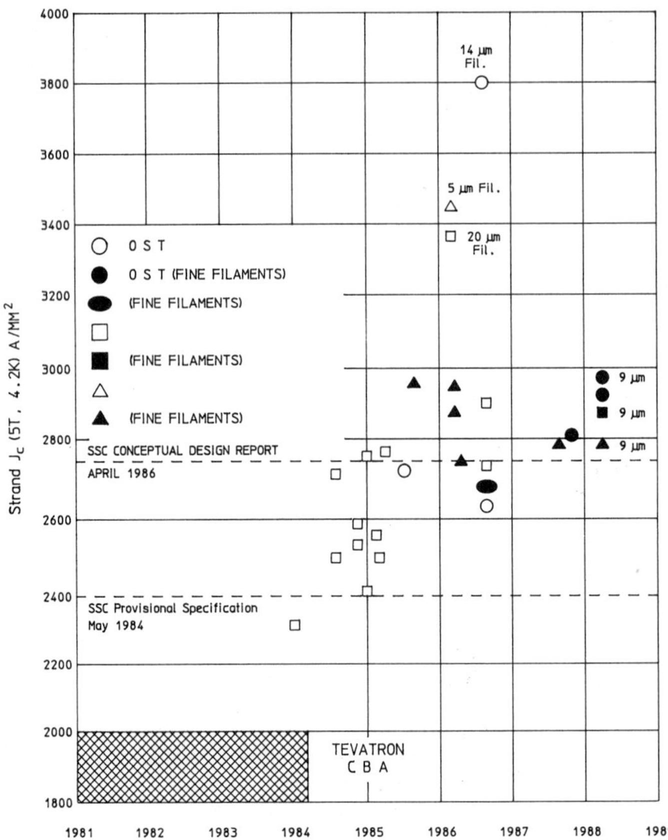

Fig. 1. Improvement of J_c over the Past Decade

early 1980. This composite was made of some of the leftover stock for the
TORE SUPRA composite. The NbTi filament was protected by a Nb barrier
from CuTi intermetallic formation and has NbTi/Nb filaments surrounded with
copper and CuNi to reduce the coupling losses. The outside of the basic
187 filament element was made of copper in order to remove the possible
heat generated into the helium in the final composite. There were 187 x
55 = 10,285 filaments in the final composite. Some of this composite was
drawn down to a fine wire in which the filament diameter was $1/2\mu$ and
studies were made to evaluate the ac properties of the composite. A short
length of this composite was made to 0.9 mm diameter wire in which the
effective filament diameter is 5μ and J_c - 2,300 A/mm^2 in a 5 Tesla background field was reported, which was quite a respectable value. The cross
section of this composite is shown in Fig. 2. Based upon the results of
the above work, SSC work for composite development of the fine filament
and high current density had been initiated.

SSC COMPOSITE

In the above section we have discussed a method to cope with CuTi
intermetallic formations on the surface of NbTi filaments in copper matrix,
which is one of the obstacles of the high performance fine filament composite. Another major aspect of the high current density fine filament is

Fig. 2.   1/2μ Filament NbTi Composite

in handling the mismatch of mechanical strengths between the hard NbTi
filament and soft copper matrix. As J_c of NbTi increases with the amount
of α -Ti precipitation, the NbTi filament would be mechanically harder
and the strength mismatch of filament to the soft copper matrix will be
more pronounced. This mismatch may cause the necking or sausaging of
filaments and may yield a somewhat deteriorated current density. One can
reduce this problem by minimizing the copper content between the filaments
but there is a limit to the reduction of copper content since the coupling
between the filaments can occur by the superconducting proximity effect.
With this background knowledge, we have made several billets for the
SSC program.

<u>Double Stacking</u>

     The first attempt we made was the double stacking approach, which was
similar to the TORE SUPRA composite. It appeared that the billet stacking
was rather simple compared to the single stacking of a large number of
filaments, although we were aware of the fact that the outer most filaments in each sub-element were going to be necked somewhat in the final
composite. The resultant composites yielded J_c ~2600 A/mm^2 in the inner
layer and ~2800 A/mm^2 in the outer layer at 5 T background field. The
specification was then 2400 A/mm^2. The spacing between the filaments was
chosen to be 1μ with a filament diameter of 5μ. It is interesting that
the current specification requires 1μ spacing with a 6μ filament diameter.
The resistive transition index value was rather low to be 25, which indicates that the filaments were sausaged. The cable made with these composites exhibited a current degradation of less than 5%.

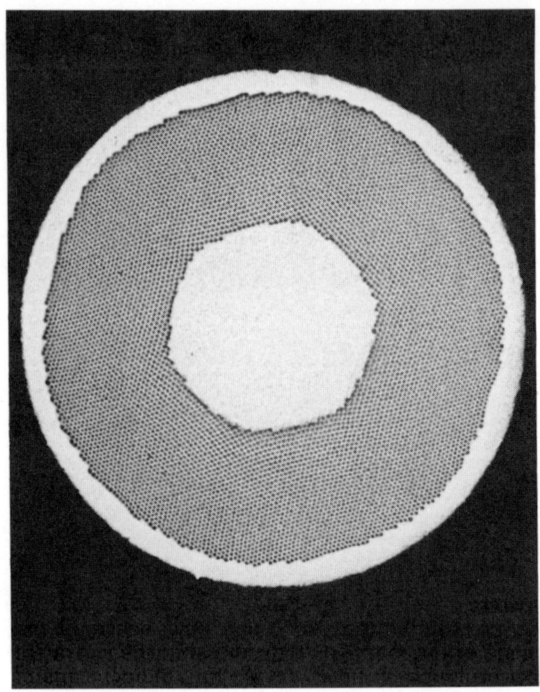

Fig. 3. SSC Inner Layer Composite

Single Stacking

As the specification of the SSC composite increased to 2750 A/mm^2 from 2400 A/mm^2 at 5 T, we changed our process to single stacking methods by using single core Cu/Nb/NbTi elements. We made two composites of different filament diameters, 9μ and 6μ, for the inner layer cable. Fig. 3 depicts one of the composites. Both composites met the specification of the SSC and the 9μ composite exceeded the 6μ composite in performance with a current density about 3000 A/mm^2.

Cabling

Cables for dipole magnet winding were made with these composites at the New England Electric Co. in Lisbon, New Hampshire. They exhibited a current degradation of less than 5%.

REMARKS

We have demonstrated that the superconducting composite can be fabricated for SSC. It is desirable to establish some performance margin for high volume manufacturing and progress is being made toward this goal.

SUPERCONDUCTING WIRE AND CABLE FOR THE SUPERCONDUCTING SUPERCOLLIDER

T.S. Kreilick, E. Gregory, D. Christopherson, G.P. Swenson and J. Wong

Supercon Inc., 830 Boston Turnpike, Shrewsbury, MA 01545

ABSTRACT

In the last few years the development of improved fine filamentary NbTi wires has been spurred on by the potential need for material to be used in the dipole and quadrupole magnets for the Superconducting SuperCollider (SSC). The result has been a substantial increase in the current carrying capacity of commercially available materials and a significant reduction in filament diameter. The manner in which these improvements have been achieved is outlined and the problems encountered discussed. The relative importance of factors such as homogeneity of the NbTi raw material, heat treatment and cold work cycles, reaction between the matrix and the filaments, filament spacing and uniformity of the filament arrav. on the critical current density obtainable, are discussed. Recent
s and cables for the SSC inner and outer

INTRODUCTION

The techniques employed by Supercon Inc. to fabricate wire and cable for the Superconducting SuperCollider are the culmination of a rigorous development effort. It is an ongoing program.[1] The objective of the NbTi strand development program at Supercon Inc. is to produce high quality fine filamentary materials reliably and economically. Reliability is defined in terms of reproducible current densities and adequate stability in a cable that meets all mechanical, dimensional and cleanliness requirements. The economic considerations are greatly reduced when thermo-mechanical processing of large quantities can be employed that yields strand and cable with long piece length. These issues must be addressed in production-size billets that produce quantities of material that are statistically representative of the conductor that will be required for the SSC. Such a knowledge base is being formulated at Supercon Inc. and it is the purpose of this paper to aid in the dissemination of this data.

Background

The first billet that Supercon Inc. fabricated for the SSC program was a 314 mm diameter single stack array of 4164 closely spaced Cu clad NbTi filaments.[2] Each 8.6 μm diameter filament was comprised of high homogeneity Nb 46.5 wt% Ti alloy surrounded by a niobium barrier to prevent the formation of brittle intermetallic compounds. The current that this composite conductor was capable of carrying was exceptionally high.[2,3]

The filament quality as measured by the slope of the resistive transition (n) was also very high with average values in the low 50's at 5T. The piece length, however, was quite poor. At a strand diameter of 0.81 mm, the longest lengths obtainable were approximately 1.5 km. Later analysis revealed an additional problem. Upon drawing the composite to smaller diameters to investigate the properties of finer filaments, the closely spaced filaments experienced proximity induced coupling at the SSC injection field of 0.3T.[4] The average filament spacing to filament diameter ratio (S/D) of this billet was 0.12.

The amount of information obtained from this first billet was substantial. Several successes were apparent, including the "single stack" technique, the use of a diffusion barrier and the small filament spacing to filament diameter ratio that enabled the fabrication of geometrically uniform filaments.[5] Further evidence for this last point was obtained in an experiment in which three billets were fabricated keeping all parameters the same except for the spacing between the filaments.[6,7] By reducing S/D, greater mechanical support is provided by one filament for its neighbors, thus enhancing filament quality and, therefore, current carrying capacity.

It was also apparent, from this first billet, that modifications of the design were necessary. Proximity coupling had to be reduced and several solutions were suggested. The easiest to employ, was simply to increase the spacing between the filaments. As a result, the SSC specification now sets a minimum spacing of 1 μm when the filament diameter is 6 μm (i.e. an S/D of 0.167). Alternatives to increasing the interfilamentary spacing are discussed elsewhere. These include resistive scattering by solid solution alloying of the copper with nickel[8,9] and flip-spin-scattering of the superelectrons by solute ions of manganese with localized magnetic moments.[9-15]

A second problem, with more serious economic implications, was the short piece length of the strand. More pieces require more processing time in the wire drawing and respooling of the strand, as well as, the cabling of that strand.

This question has been addressed in a process which began with a re-evaluation of Supercon's standard procedure for billet design, assembly and processing.[5,12] It is this introspection that has yielded the results that are presented here.

PROCEDURE

<u>Wire Fabrication</u>

The multifilamentary NbTi wire and cable described below are Supercon Inc.'s first seven pre-production billets designed for use in the outer windings of SSC dipoles and the first three pre-production billets designed for use in the inner windings.

All of the composite materials discussed incorporate homogeneous Nb 46.5 wt% Ti (±1.5 wt% Ti) alloy from the same vender, oxygen-free copper (CDA Alloy 101), and niobium diffusion barriers. Preparation of the NbTi alloy, the copper, and the niobium diffusion barrier includes inspection, machining and careful cleaning of each of the components. The monofilaments are assembled and then hot extruded to ensure a metallurgical bond between all components. After extrusion the monofilament material is drawn to either 3.3 mm (flat-to-flat) hexagonal rod to be used for outer billet restacks or to 2.75 mm (flat-to-flat) hexagonal rod to be used for inner billet restacks.

Monofilamentary materials are designed to yield a average filament spacing to filament diameter ratio in the finished conductor. The mathematical relationship is not straightforward as variables such as the processing history of the alloy, the type and the thickness of the diffusion barrier employed, the extrusion and wire drawing parameters, and other factors influence the outcome. The results of close approximations must be evaluated empirically until the variables listed above and others, not yet identified, are fully understood. Investigations are underway to better understand several of the process variables.[16,17]

Hexagonal monofilamentary rod is straightened and cut to the appropriate length for restacking in 314 mm diameter copper cans. The rods are chemically cleaned prior to assembly and great care is taken to maintain as perfect an array as possible during the assembly operation. Table 1 lists several design parameters for the ten multifilamentary billets. Changes in the number of filaments were attempts to yield more accurate copper to superconductor ratios. In the case of billet #2301-1, the result was quite close with values of 1.8:1 ± 0.01.

The composite billets are sealed by electron beam welding prior to isostatic compaction, and then hot extruded. Once extruded the rods are drawn using standard rod reduction and wire drawing techniques. Care is taken in die design and die reduction schedules. Figures 1 and 2 show billets 2071-3 and 2127-1, respectively, at an intermediate size (21.8 mm) during the reduction process.

Table 1
Design Parameters of SSC Outer and Inner Billets

Supercon Billet #	Billet Type	# of Filaments	Nominal Filament Diameter	Average Spacing/ Diameter	Billet Diameter
2071-1	outer	4164	6 μm	0.19	314 mm
2071-2	outer	4164	6 μm	0.19	314 mm
2071-3	outer	4164	6 μm	0.19	314 mm
2128-1	outer	4164	6 μm	0.19	314 mm
2128-2	outer	4164	6 μm	0.19	314 mm
2128-3	outer	4164	6 μm	0.19	314 mm
2301-1	outer	4134	6 μm	0.17	314 mm
2127-1	inner	7248	6 μm	0.19	314 mm
2127-2	inner	6852	6 μm	0.19	314 mm
2300-1	inner	6756	6 μm	0.17	314 mm

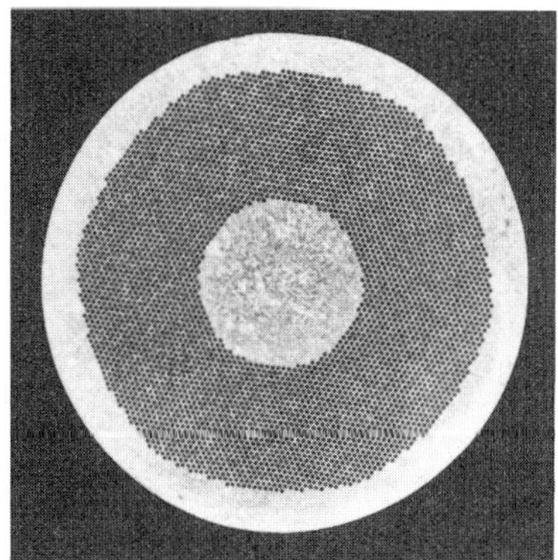

Figure 1   Cross-section of Billet #2071-3 at 21.8 mm diameter. 4164 NbTi filaments in a copper matrix.

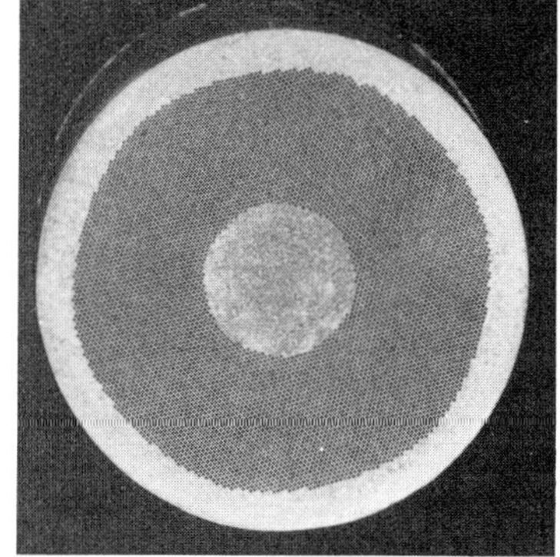

Figure 2   Cross-section of Billet # 2127-1 at 21.8 mm diameter. 7248 NbTi filaments in a copper matrix.

The seven billets of outer conductor listed in Table 1 yielded a total of 720 km (1874 kg) of strand at the final diameter of 0.65 mm. The average piece length was 6.37 km. Figure 3 shows a histogram for these seven billets. Seven pieces had lengths greater than 24 km. All strand specifications, as outlined in SSC-MAG-M-401, have been satisfied including the piece length requirements. The specification calls for no more than 10% of the strand to be less than 3.048 km in length. The actual percentage of material (from the seven outer billets) that was shorter than 3.048 km was 9.73%.

The piece length distribution for Billets 2127-1 and 2127-2 (518 kg) are shown in Figure 4. The two billets produced 130 km of strand at 0.81 mm diameter. The average piece length was 4.07 km. The piece length distribution for the inner strand is not as good as that of the outer strand. It is, however, much improved over earlier composites. 307 kg of billet 2300-1 are still in process. A heat-treatment study is underway to further optimize the current carrying capacity of this conductor. To date, short sample tests have revealed an Ic of 362 A at 7T.

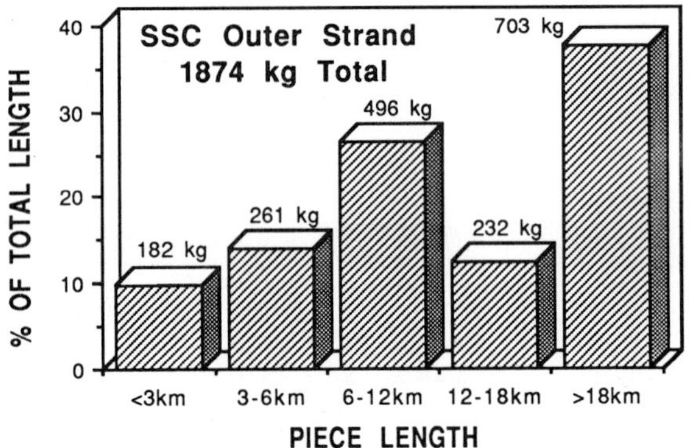

Figure 3   Piece length distribution of outer strand from billets 2071-1,2,3, 2128-1,2,3 and 2301-1

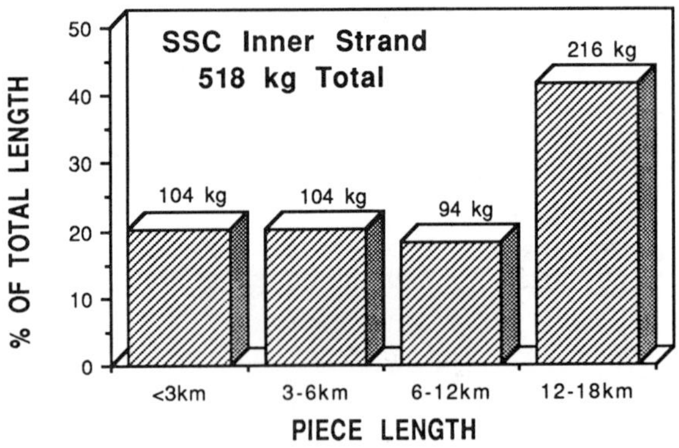

Figure 4   Piece length distribution of inner strand from billets 2127-1 and 2127-2

238

<u>Cable Fabrication</u>

The wire for both outer and inner windings was cabled into 30 and 23 strand keystoned Rutherford-type cables[18], respectively, as per Material Specification No. SSC-MAG-M-402. Table 2 lists pertinent data for the 30 strand outer cable. These first seven cables incorporate strand from billets 2071-1,2,3 and 2128-1,2,3. Strand from billet 2301-1 has not yet been cabled. Strand maps were generated for each cable identifying the location of all cold-welds. The strands were also sorted by the measured Ic with the average, in all cases, comfortably above the specified strand Ic of 285 A at 5.6T.

The first two outer cables were fabricated on an older cabling machine prior to the delivery of a machine designed specifically for the SSC program.[19,20] Note the higher degradation and the higher value for the RRR of the first two cables compared to the later cables. The change in RRR is accounted for in a different post-anneal of the copper prior to cabling. It is not clear if the greater degradation is due to this change, to the cabling machine itself or to the fact that this particular strand had an extra 30% twist. The degradation listed is the percent change in current carrying capacity of the cable as opposed to the average strand Ic x 30 strands. The margin refers to the degree to which the cable exceeded the outer cable Ic specification of 7860 A at 5.6T. Cable #SSC22-00007 incorporated strand that had not received a final anneal of the copper.

Table 3 shows data for the inner cables. Approximately half of the available strand has been cabled, to date, and only the first cable (#SSC12-00001) has been tested. The available margin is much smaller for inner cable (4.15%) than it is for outer cable (8.07%) when one examines the degradation allowed for in the material specifications. To state the obvious, the higher the current carrying capacity in the strand the higher the margin.

Figures 5 and 6 show cross-sections of 23 strand inner and 30 strand outer cables, respectively. The satellite photos show magnified views of several strands. Of particular interest are the strands at the thin edge of the cables. It is apparent that the filament annuli are not severely deformed during the cabling operation, as has been the case in the past. By altering the geometry of the strand and arranging the filaments closer to the center of the composite more copper is available to absorb the shock of deformation.

Table 2
<u>SSC outer cable data (30 strand)</u>

Cable No.	Machine	Cable Length	Cable Ic (5.6T)		Degradation	Margin	RRR
SC22-001	Kraft	842 m	8203 A		8.3%	4.4%	72
SC22-002	Kraft	866 m	8165 A		8.7%	3.9%	72
SSC22-00003	Dour	124 m	8621 A	(F)	4.6%	9.7%	57
SSC22-00004	Dour	3864 m	8338 A	(B)	7.7%	6.1%	58
SSC22-00005	Dour	4070 m	8631 A	(F)	4.3%	9.8%	61
			8584 A	(B)	4.9%	9.2%	60
SSC22-00006	Dour	3372 m	8495 A	(F)	5.2%	8.1%	59
			8481 A	(B)	5.4%	7.9%	59
SSC22-00007	Dour	1405 m	8138 A	(B)	4.2%	3.4%	39

Note: F = front, B = back

Figure 5   Cross-section of SSC inner cable #SSC12-00001.

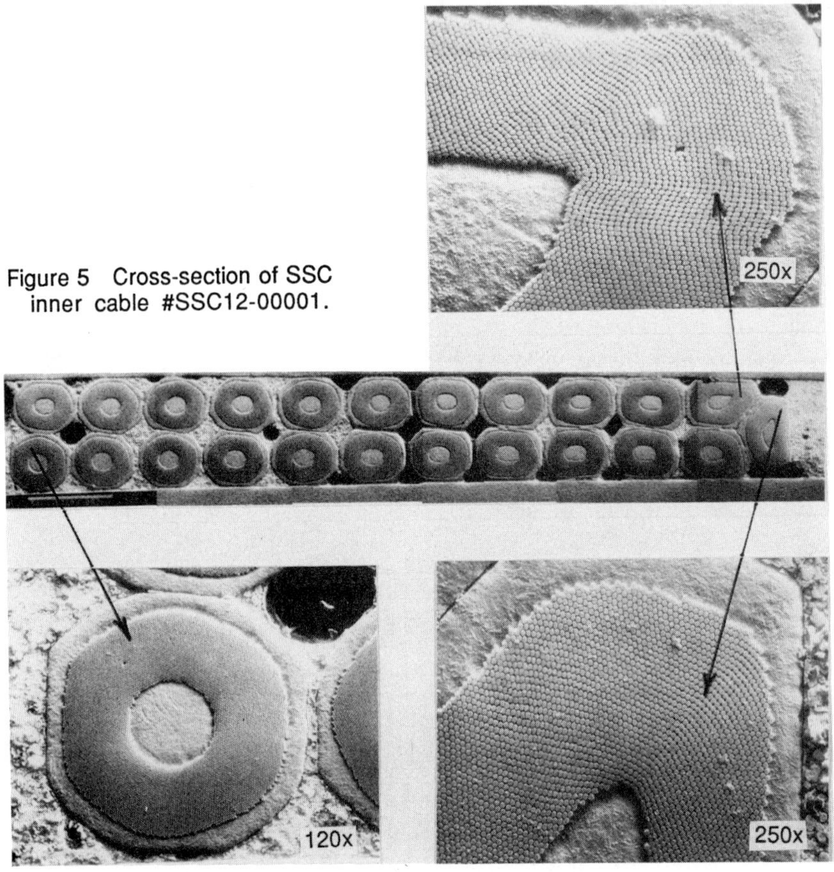

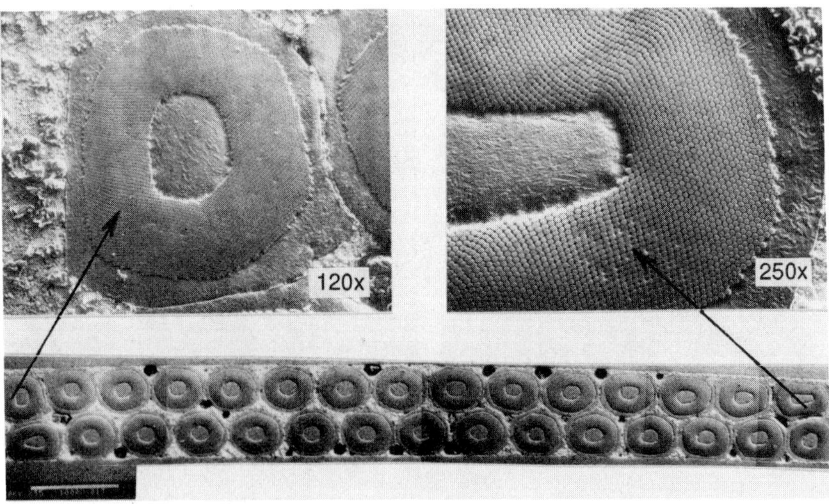

Figure 6   Cross-section of SSC outer cable #SC22-002.

SUMMARY

The work presented demonstrates that multifilamentary NbTi wire and cable, meeting the current SSC materials specifications can be produced in production quantities by Supercon Inc. Further improvements in piece length and current carrying capacity are possible with additional work.

Table 3
SSC inner cable data (23 strand)

Cable No.	Cable Machine	Cable Length	Ic (7T)	Degradation	Margin	RRR
SSC12-00001	Dour	1634 m	7476 A	1.0%	3.3%	35
SSC12-00002	Dour	1098 m	not yet tested	-	-	-

ACKNOWLEDGEMENTS

The authors would like to acknowledge the continued support of their colleagues at Supercon Inc., in particular M. DeFilippo for billet fabrication. Supercon Inc. would also like to acknowledge the financial assistance of the U.S. Department of Energy, through the Lawrence Berkeley Laboratory, and the Small Business Innovative Research program. In addition, the testing expertise of W. Sampson and his group at Brookhaven National Laboratory is gratefully acknowledged.

REFERENCES

1. E. Gregory, T.S. Kreilick and J. Wong, "Innovations in the Design of Multifilamentary NbTi Superconducting Composites for the SuperCollider and Other Applications", Paper III-D-9, IISSC, New Orleans, LA, February 9-10, 1989.

2. T.S. Kreilick, E. Gregory and J. Wong, "Fine Filamentary NbTi Superconducting Wires", Adv. Cryo. Eng., Vol. 32, pp. 739-745, eds. A.F. Clark and R.P. Reed, Plenum Press, 1986.

3. E. Gregory, T.S. Kreilick and J. Wong, "Fine Filamentary Materials for Accelerator Dipoles and Quadrupoles", in Proc. ICFA Workshop, P. Dahl ed., Brookhaven National Laboratory Report #BNL 52006, pp. 85-88, 1986.

4. A.K. Ghosh, W.B. Sampson, E. Gregory and T.S. Kreilick, "Anomalous Low Field Magnetization in Fine Filament NbTi Conductors", IEEE Trans., MAG-23, No. 2, pp. 1724-1727, 1987.

5. T.S. Kreilick, E. Gregory and J. Wong, "Geometric Considerations in the Design and Fabrication of Multifilamentary Superconducting Composites", IEEE Trans., MAG-23, No. 2, pp.1344-1346, 1987.

6. E. Gregory, T.S. Kreilick, A.K. Ghosh and W.B. Sampson, "Importance of Spacing in the Development of High Current Densities in Multifilamentary Superconductors", Cryogenics, 24, 7, pp. 178-182, 1987.

7. T. S. Kreilick and E. Gregory, "Further Improvements in Current Density by the Reduction of Filament Spacing in Multifilamentary NbTi Superconductors", Cryogenics, 27, pp. 402-403, 1987.

8. T.S. Kreilick, E. Gregory and J. Wong, "The Design and Fabrication of Multifilamentary NbTi Composites Utilizing Various Matrix Materials", J. Less Common Metals, 139, pp.45-51, 1988.

9. A.K. Ghosh, W.B. Sampson, E. Gregory, T.S. Kreilick and J. Wong, "The Effect of Magnetic Impurities and Barriers on the Magnetization and Critical Current of Fine Filament NbTi Composites", IEEE Trans., MAG-24, No. 2, pp. 1145-1148, 1988.

10. E.W. Collings, "Stabilizer Design Considerations in Fine-Filament Cu/NbTi Composites", Adv. Cryo. Eng., Vol. 34, pp. 867-878, eds. A.F. Clark and R.P. Reed, Plenum Press, 1988.

11. T.S. Kreilick, E. Gregory, J. Wong, R.M. Scanlan, A.K. Ghosh, W.B. Sampson and E.W. Collings, "Reduction of Coupling in Fine Filament Cu/NbTi Composites by the Addition of Manganese to the Matrix", Adv. Cryo. Eng., Vol. 34, pp. 895-900, eds. A.F. Clark and R.P. Reed, Plenum Press, 1988.

12. T.S. Kreilick, E. Gregory and J. Wong, "Influence of Filament Spacing and Matrix Material on the Attainment of High Quality, Uncoupled NbTi Fine Filaments", IEEE Trans., MAG-24, No. 2, pp. 1033-1036, 1988.

13. T.S. Kreilick, E. Gregory, P. Valaris and J. Wong, "The Mechanical and Electrical Effects of Adding Manganese to the Copper Matrix of Multifilamentary Niobium-Titanium Superconducting Composites", Proc. ICEC 12, Southampton, U.K., eds. R.G. Scurlock and C.A. Bailey, Butterworths, pp. 857-863, 1988.

14. E. Gregory, T.S. Kreilick, J. Wong, E.W. Collings, K.R. Marken, R.M. Scanlan and C.E. Taylor, "A Conductor with Uncoupled 2.5 µm Diameter Filaments, Designed for the Outer Cable of SSC Dipole Magnets", to be published in IEEE Trans., MAG-25, 1989.

15. R.B. Goldfarb, D.L. Ried, T.S. Kreilick and E. Gregory, "Magnetic Evaluation of Cu-Mn Matrix Material for Fine-Filament Nb-Ti Superconductors", to be published in IEEE Trans., MAG-25, 1989.

16. P. Valaris, T.S. Kreilick, E. Gregory and J. Wong, "Refinements in Billet Design for SSC Strand", to be published in IEEE Trans., MAG-25, 1989.

17. P. Valaris, T.S. Kreilick, E. Gregory and E.W. Collings, "The Effects of Processing on the Filament Array in Multifilament SSC Strand", Paper III-F-12, IISSC, New Orleans, LA, February 9-10, 1989.

18. R.M. Scanlan, J. Royet and R. Hannaford, "Fabrication of Rutherford-Type Superconducting Cables for Construction of Dipole Magnets", Cryogenic Materials '88, eds. R.P. Reed, Z.S. Xing and E.W. Collings, Proc. ICMC Conf., Shenyang, China, 1, pp.391-399, 1988.

19. J. Grisel, J.M. Royet, R.M. Scanlan and R. Armer, "A Unique Cabling Machine Designed to Produce Rutherford-Type Superconducting Cable for the SSC Project", to be published in IEEE Trans., MAG-25, 1989.

20. J.M. Royet, R. Armer, R. Hannaford and R.M. Scanlan, "An Industrial Cabling Machine for the SSC", Paper III-D-8, IISSC, New Orleans, LA, February 9-10, 1989.

DEVELOPMENT OF SSC CABLE IN FURUKAWA

Masaru Ikeda

Superconducting Products Department
The Furukawa Electric Co., Ltd.
2-4-3, Okano, Nishi-ku, Yokohama, 220 Japan

ABSTRACT

Furukawa Electric has been developing Rutherford cable to be used for
the Superconducting Super Collider (SSC) Project in the United States. The
fabricated cable of approximately 15 km in length could meet all of the SSC
specifications, and wound into SSC model dipole coils, which were successfully tested. In connection with development of the SSC cable, critical
current degradations due to twisting, annealing and cabling were studied.
It has been found that the cable annealed at an adequate condition has a
higher RRR value and the cable having a large keystone angle of up to three
degrees can be fabricated, without significant degradation.

INTRODUCTION

High performance superconducting cable is required for the SSC dipole
magnets. The minimum critical current density is as high as 2,750 A/mm^2 at
5 T in fine NbTi filaments (6 microns in diameter), which is much higher
than critical current densities which were achieved so far for practical
superconducting wires. The tolerances in dimensions are also severe in
comparison with any other Rutherford cables.

Furukawa Electric began development of the SSC cable in 1984, based on
much experience with Rutherford cable and backed with the great merit of its
vertically integrated production from raw material melting to insulation.
Approximately 15 km long SSC cable was fabricated and fully inspected in
accordance with the SSC specifications. The cable was wound into SSC model
dipole coils and tested by US laboratories and Japanese industries.

This paper describes the fabrication and test results of the SSC cable
and, in addition, relevant experiments on the development of the SSC cable.

FABRICATION EXPERIENCE OF RUTHERFORD CABLE

Our first experience with the Rutherford cable was with samples we
fabricated and submitted to LBL for the ESCAR Project. Thereafter, many kinds
of Rutherford cable have been fabricated. The representative information from
fabrication experience with Rutherford cables for accelerators is listed in

Table 1. QCS cable for the quadrupole magnets of the TRISTAN Project at KEK and RHIC cable for the BNL's Project are very similar to the SSC cable in filament diameter, strand diameter, number of strands and copper-to-non-copper ratio. The Rutherford cables are also used for other applications such as generators, fusion reactors and so on. At present, 25 km long cable to be used in the electo-magnetic thruster for ship propulsion is being produced. Products described here have been delivered to customers in a form of cable, not strand.

Table 1. Fabrication Experience of Rutherford Cables.

Year	User	Project	Strand dia. (mm)	Cu ratio	Fil. dia. (µm)	No. of strands
'75	LBL	ESCAR	0.65	1.4	10	15
'77	FNAL	ED/S	0.69	1.8	9	23
'78	Saclay	Synchrotron	0.69	1.8	8	23
'79-'80	KEK	π1 beam line	0.69	1.8	8	25
'79-'81	KEK	TRISTAN	0.71	1.8	9	27
'81-'84	DESY	HERA	0.83	1.8	10	24
'84-'88	KEK	QCS	0.68	1.1-1.6	4-9	27
'86	BNL	RHIC	0.65	1.8	5	30

Critical current densities of the cables listed in Table 1 are shown in Fig. 1. Most of the data on the critical current densities has been obtained with cabled strands, bacause the critical currents have been guaranteed and, therefore, measured for the cables. In the past, degradation of critical current due to cabling was about 5 %. In the figure, Jc of the strands before cabling is assumed to have been 5 % higher than that of the cabled strands.

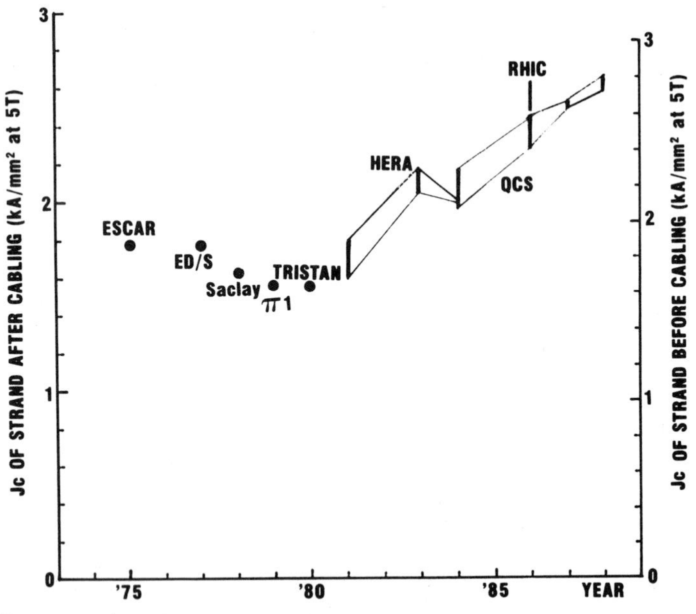

Fig. 1. Critical Current Density of Rutherford Cables.

Up to 1981, the level of the critical current density remained in the range of 1,600 to 1,800 A/mm^2 at 5 T. Thereafter, through HERA, QCS and RHIC cables, the critical current density increased to 2,800 A/mm^2 at 5 T. The QCS and the RHIC cables are very similar to the SSC cable in filament diameter and copper-to-superconductor ratio. The SSC cable which must have high critical current density has been developed parallel to the QCS and RHIC cables.

FABRICATION OF SSC CABLE

Parameters and Design

Table 2 shows parameters on the SSC cable fabricated, except small scale developmental lengths. Two kinds each of the inner and the outer cables were fabricated. Cable names are given as follows; "inner" and "outer" correspond to the inner and the outer cables, respectively, and "old" and "new" are correspond to the old specifications, which existed about two years ago, and to the recent specifications, respectively. The differences between the two specifications are mainly in filament diameter and spacing, copper-to-non-copper ratio for the inner cable, and twist pitch. Approximately 15 km long cable was fabricated in collaboration with KEK.

Table 2. Parameters on SSC Cables.

Cable name	"old-inner"	"new-inner"	"old-outer"	"new-outer"
Strand				
diameter (mm)	0.808	0.808	0.648	0.648
copper-to-non-copper ratio	1.3	1.5	1.8	1.8
number of filaments	12,835	7,250	6,630	4,160
filament diameter (μm)	4.8	6.0	4.8	6.0
filament spacing* (μm)	0.6	1.5	0.6	1.5
twist pitch (mm)/direction**	25/Z	12.5/Z	25/Z	12.5/S
Cable				
number of strands	23	23	30	30
cabling pitch (mm)/direction**	79/S	79/S	74/S	74/Z
narrow edge thickness (mm)	1.326	1.326	1.062	1.062
wide edge thickness (mm)	1.588	1.588	1.267	1.267
width (mm)	9.296	9.296	9.728	9.728
fabricated length (km)	5.4	1.0	6.5	1.2

The filament spacing* is given in the billet design.
The direction** shows the twisting and the cabling directions, where "Z" and "S" are clockwise and unclockwise, respectively.

In designing the extrusion billets, a double stacking method was employed for the "old" cables. In the "new" specification, the filament diameter and the copper-to-non-copper ratio for the inner cable were changed. Because the number of filaments decreased in accordance with this specification change, a single stacking method was employed for the "new" cables. It enabled filament spacings to have good margins over the specification and fabrication costs to be reduced.

Directions of twisting, cabling and taping of insulator for the "old" cables were subjected to the "old" specification, i.e., the inner and the outer cables had the same directions. In the "new" cables, a trial has been made so that the inner and the outer cables have opposite directions in order to reduce the twisting nature of the coil assembly. Of course, it ispossible that two halves of each coil have opposite directions from each other.

Fabrication Process

Furukawa Electric performs all processes, from raw material melting to insulation. Among such processes, in-process inspections are very important. For example, wire diameter measurements with laser micrometers, Eddy current tests for flaw detection and copper-to-non-copper ratio checking by electrical resistance measurement have been programmed.

Test Results on Cable

Final inspections are made in accordance with the specifications. Test results on critical currents and residual resistance ratios (RRR) are shown in Table 3. The critical currents of the cable was calculated from critical currents of several strands extracted from the cable. The obtained critical currents exceed specification requirements and almost exceed requirements corresponding to the critical current density of 2,750 A/mm^2 at 5 T. The RRR values are also sufficiently above the target values. Differences in the RRR values of each cable are understood depending on the existence of lump copper. All the other test results showed that the fabricated cable could meet specification requirements.

Table 3. Specifications and Test Results on SSC Cables.

Cable name	Test item	Specification	Target	Test result
"old-inner"	Ic(A) at 5 T	11,070*	12,700**	12,600-13,840
	Ic(A) at 8 T	4,380	5,120	4,970-5,320
	RRR	---	70	81-96
"old-outer"	Ic(A) at 5 T	7,610*	8,700**	9,280-9,820
	Ic(A) at 8 T	3,020	3,530	3,520-3,760
	RRR	---	70	97-118
"new-inner"	Ic(A) at 7 T	7,167**	---	7,214-7,253
	RRR	---	66	82-87
"new-outer"	Ic(A) at 5.6 T	7,860**	---	8,230-8,330
	RRR	---	70	85-99

Ic* and Ic** correspond to the critical current densities of 2,400 and 2,750 A/mm^2 at 5 T, respectively.

Dipole Magnet

By using these cables, SSC model dipole magnets were fabricated and tested by US laboratories and Japanese industries. One of the magnet having a length of one meter was wound from the "old" cables and tested by Toshiba Corporation. This magnet exhibited the best quench performance, where the first quench occurred at 7,000 A in coil current and 6.74 T in central field, and the second at 7,070 A and 6.82 T. In addition, full length SSC R&D dipole magnets were wound and successfully tested by US laboratories. These test results showed that the fabricated cable is suitable for the SSC dipole magnets.

RELATED EXPERIMENTS

Twisting

     The strands used for the "old" cable, which had been twisted with a
specified pitch of one inch, were further twisted in order to study critical
current degradation due to twisting. Test results on the critical current
degradation are shown in Fig. 2. The critical current degradations of the
further twisted strands are given as critical current change rates for the
strands twisted with a pitch of one inch.

     Degradations are observed, due to geometrical factors of filament elong-
ation. The outer strand is more sensitive to twisting by two times than the
inner strand. It may be understood that the outer strand is inferior in its
filament integrity than the inner strand.

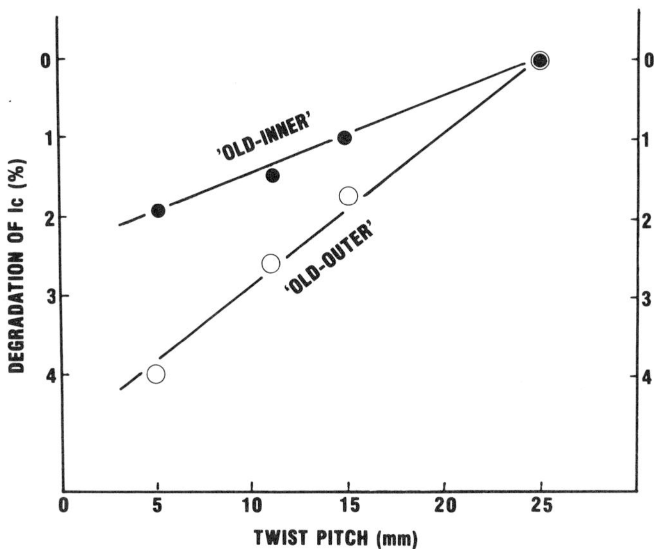

Fig. 2. Critical Current Degradation due to the Further
        Twisting of the "old" Strands used for the "old"
        Cables.

Annealing

     The RRR value lowered during cabling is easily recovered by annealing.
However, there is a possibility to degrade its critical current. Annealing
effects of the RRR value and the critical current were investigated. Strands
extracted from the "new-inner" cable were heat-treated at several temperatures for one hour. Test results are shown in Fig. 3. Without degrading the
critical current, it is possible to increase the RRR value to more than 100
through heat treatment of 250 to 300 °C for one hour. It is expected to
improve the stability of the cable, although the higher RRR value has little
effect on the stability of the coil because of the magneto-resistance effect.
It must be noted that final annealing also improves the twisting nature of
the cable.

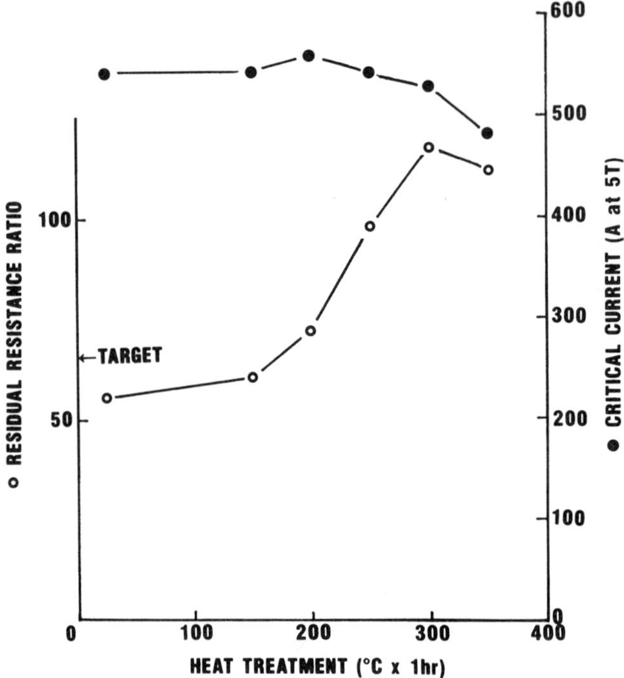

Fig. 3. Annealing Effects of the Inner Strands
extracted from the "new" Cable.

Cabling

Degradations of critical currents due to cabling were also investigated
by using the same strands as the inner ones used for the "old" and the "new"
cables, in collaboration with KEK. The trial-fabricated cables have the same
number of strands and the same cable width as the SSC inner cables, but have
the different keystone angle and the packing factor.

Critical currents on several strands extracted from the fabricated
cables were measured. In Fig. 4, the relationship is shown between the
critical current degradation and the packing factor with a parameter of the
keystone angle. The critical current degradations of the cabled strands are
given as critical current change rates for the same strands which were used
for cabling but not cabled. Packing factor (PF) is given as follows,
considering that the strand lay is inclined to the cable axis and assuming
that the cross sectional area of the strand remains unchanged during cabling;

$$PF = \pi d^2 n / 2 w (t_1 + t_2) \cos \theta \times 100 \ (\%)$$

$$\cos \theta = \sqrt{ 1 - (n d / p)^2 }$$

where, d is the strand diameter, n is the number of strands, t_1, t_2 and w
are the narrow edge thickness, the wide edge thickness and the width of the
cable, respectively, and p is the cabling pitch.

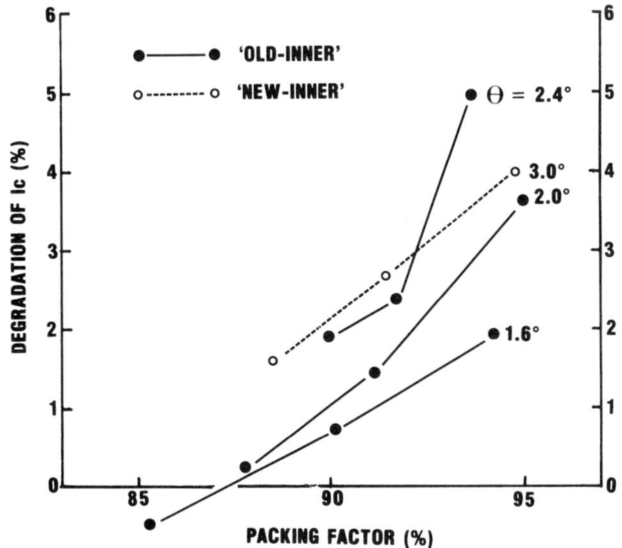

Fig. 4. Critical Current Degradation due to Cabling
of the Inner Strands used for the "old" and
the "new" Cables with Changing the Keystone
Angle (θ) and the Packing Factor.

Degradations of the critical currents increase as the packing factor
and the keystone angle increase. It is easily understood from the fact that
heavily deformed strands with a high packing factor and a large keystone
angle have low critical currents. From such a point of view, the "old"
strands are understood to be sensitive to deformation, because it is fabricated by a double stacking method which causes sausaging of filaments near
the copper sheath in the strand cross section. The most heavily deformed
cable is three degrees in keystone angle and 94.8 % in packing factor. This
cable exhibits only 4 % degradation of the critical current. It will be shown
in this Symposium by Dr. Ishibashi that such a large keystone angled cable
has a great advantage for ultra high energy accelerator dipole magnets.

CONCLUSIONS AND REMARKS

Based on our experience with Rutherford cable, SSC cable of approximately 15 km in length was fabricated and met all of the SSC specifications.
The fabricated cable was wound into the SSC model coils and successfully
tested. They have exhibited the best quench performance.

The related experiments will make it possible to improve performance
of the cable and the coil. (a) The RRR values can be increased through
annealing with negligible degradation of the critical current. (b) Large
keystone angled cable can be fabricated up to three degrees without
significant degradation of the critical current. (c) The inner and the outer
cables which have opposite directions of twisting, cabling and taping, can
reduce the twisting nature of the coil assembly.

High performance cable is required for the SSC dipole magnets. Therefore, specifications should be settled as soon as possible to establish the

fabrication process and the quality assurance program and, in addition, to
reduce the cost.

ACKNOWLEDGEMENTS

The author would like to acknowledge the continual encouragement of
Dr. Nishikawa, Director General of KEK. He is also grateful for the fruitful
discussion with Dr. Scanlan and Dr. Taylor of LBL, and Dr.Greene and Dr.
Sampson of BNL, and for the valuable comments from Prof.Hirabayashi and
Assistant Prof. Shintomi of KEK.

ELEMENTS OF A SPECIFICATION FOR SUPERCONDUCTING CABLE AND

WHY THEY ARE IMPORTANT FOR MAGNET CONSTRUCTION.

A. F. Greene and R. M. Scanlan

Brookhaven National Laboratory[*], Accelerator Development
Department, Upton, New York 11973 and Lawrence Berkeley
Laboratory[*], Superconducting Magnet Group, Berkeley
California 94720

ABSTRACT

The purpose of this paper is to point out several features of the
specification for SSC superconducting cable[1] and its insulation[2] that
are important for fabrication of dipole magnet coils. Among these are the
dimensions of the cable and insulation and their relevance for obtaining
coils with appropriate overall dimensions. Other important cable
properties are related to the twist direction of wire used to fabricate
it[3] and the opposite twist (or lay) direction of the cable. For some
coils it is easier to work with cable of a particular lay direction. In
conjunction with the ease of coil winding comes the requirement in the
specification for superconducting cable which restricts the cable surface
condition. The ease of winding coils is governed by the ability to bend
and twist the cable at the coil ends without having wires come out of
place, possibly later leading to insulation damage and a turn-to-turn
short.

CABLE AND INSULATION DIMENSIONS

The dimensions of the cable and insulation and their fabrication
tolerances are important elements for the construction of SSC dipole coils
with satisfactory overall construction tolerances. The bare cable
dimensions are described in Fig. 1 and the wrapped cable parameters are
given in Fig. 2. The relevance of these dimensions and tolerances to the
overall coil dimensions will be described below.

The variations of cable and insulation thickness are related to the
variation of overall coil azimuthal size and are, therefore, among the most
tightly controlled dimensions in the magnet. Azimuthal size is defined in
Fig. 3.   The azimuthal size is measured after the coils have been molded
and the pre-preg fiberglass insulation has been cured. A special fixture
and tooling are required to compare the coil size with a standard while
applying 10.0 kpsi or 8.0 kpsi stress to the surface of the inner or outer

---

[*] Work performed under the auspices of the U.S. Department of Energy.

coils respectively. These pressures are chosen because they are the
nominal desired stress values for the coils when they are later collared
during magnet assembly. In order to meet the requirements of magnet field
quality and to maintain an adequate assembly prestress window for the
coils, the azimuthal sizes of all coils should be within ± 0.002 in.

This coil size variation is difficult to achieve with thickness
variations of the coil components. Therefore, a method has been developed
to adjust the coil size to compensate for the variations of component
dimensions by molding the pre-preg fiberglass tape to different
thicknesses. The maximum adjustment required in the fiberglass tape is

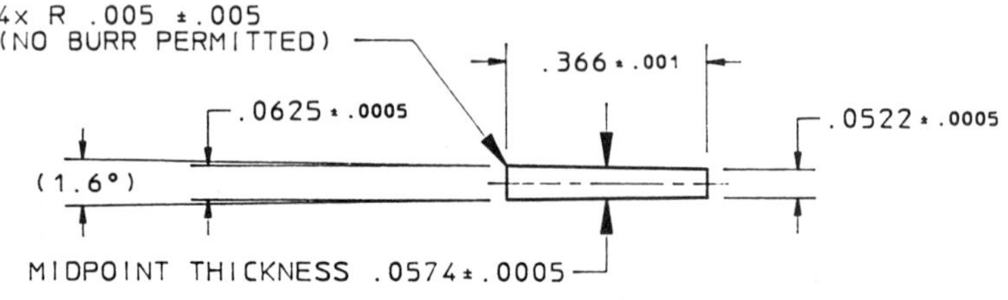

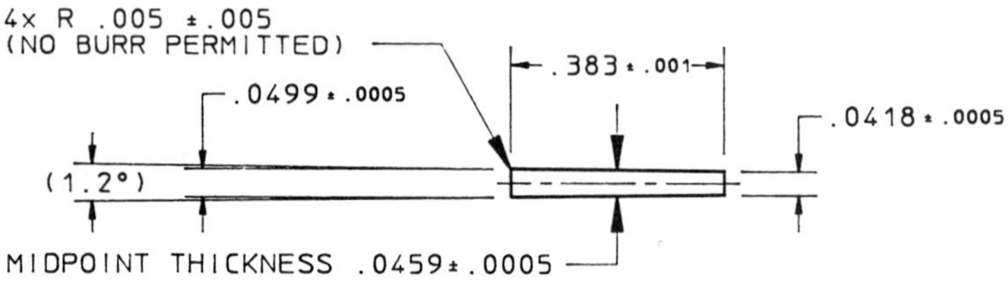

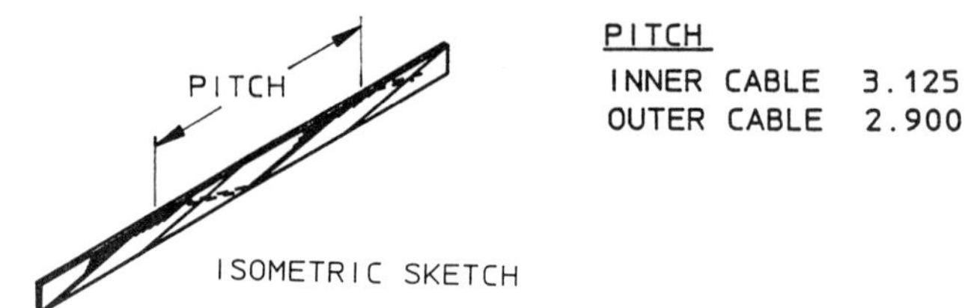

Fig. 1.   Dimensional and Other Parameters of the
           SSC Inner and Outer Layer Superconductor Cable.

approximately 0.0003 in. per layer out of 0.0045 in. uncured thickness. In
spite of this adjustment method, it is important to limit the variations of
the component dimensions as shown in Table I. The copper wedges used in
the coil fabrication are insulated in the same manner as the cable. The
accepted specification tolerance for the Kapton insulation is ± 10% per
layer. However, in practice the variation of Kapton thickness is ± 5% per
layer. From Table I it is clear that the variation of bare cable thickness
can contribute significantly to the variation of overall coil size and must
be tightly controlled.

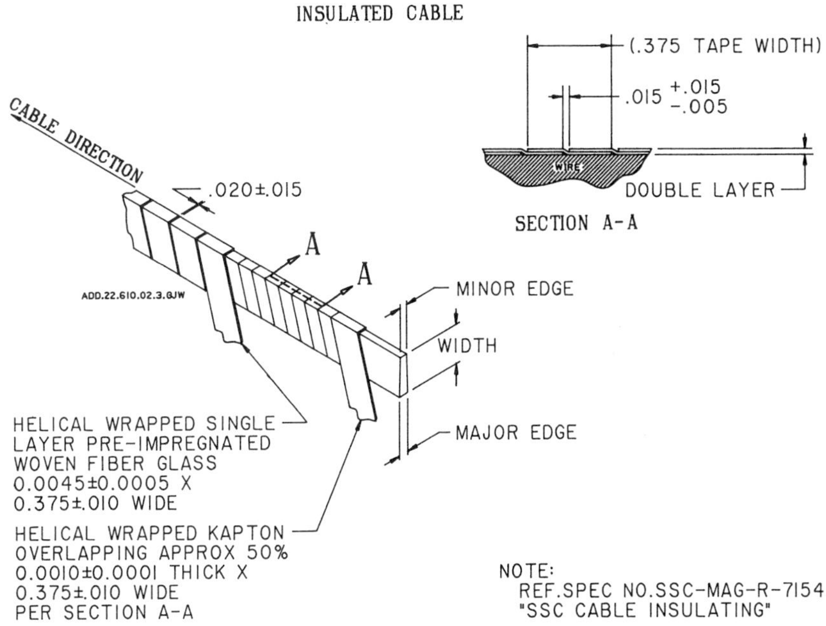

Fig. 2.   Wrapped Cable Parameters for the SSC Superconductor Cable.

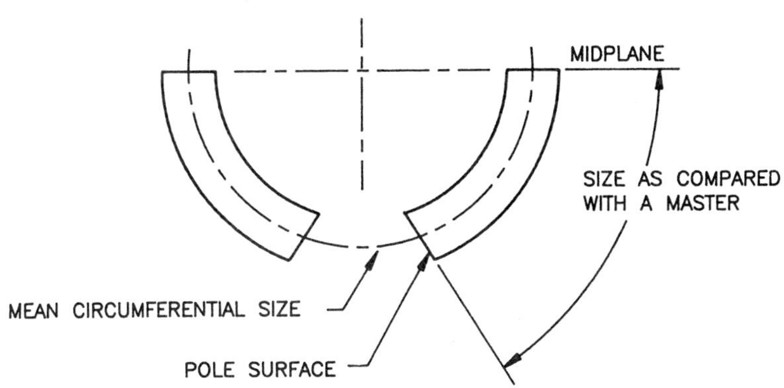

Fig. 3.   SSC Dipole Cross Section Definition of Coil Azimuthal Size.

The maximum variation of the cable keystone angle is ±0.1 degree. It is
necessary to control this dimension to maintain uniform coil prestress
across the width of the cable when the magnet is collared. Non-uniform
prestress may produce "creep" (or flow) of the Kapton insulation. This may
eventually lead to an electrical short between adjacent turns if the
"creep" is excessive.

Table I.   Estimated Variations of SSC Coil Component
            Thicknesses.

	Inner Coil (16 Turns, 3 Wedges) [Dimension-inches]	Outer Coil (20 Turns, 1 Wedge) [Dimension-inches]
Bare Cable (±0.00025 in./turn)	±0.004	±0.005
Wedges (±0.001 in./wedge)	±0.003	±0.001
Kapton (±0.0001 in. or 10% per layer specified/ ±0.0005 in. or 5% per layer actual)	±0.0076/±0.0038	±0.0084/±0.0042
Pre-Preg Fiberglass Tape (molded)	-	-
Totals:	±0.0146/0.0108	±0.0144/0.0102
Requirement:	±0.002	±0.002

It is also important to control the cable width within the specified
tolerance of ±0.001 in. This dimension controls the coil radial thickness.
Any variation beyond this tolerance may cause damage to the insulation
during coil molding or may cause excess friction between the insulated
cable and the molding fixture cavity during coil curing and sizing. Such
friction may produce an uneven azimuthal distribution of cable turns in the
coil and unacceptable variation of the magnetic field harmonics.

During manufacture of SSC cable it is normal practice to utilize a Cable
Measuring Machine to monitor the variations in the bare cable dimensions.
With a continuous measurement it is also possible to adjust, for example,
the rolled cable thickness and to maintain the required tolerance. In

Figs. 4-6 are shown the cable thickness, keystone angle and width measured
for a typical SSC cable using the Cable Measuring Machine; the estimated
measuring accuracies are ±0.0001 in., ± 0.01 degree, and ± 0.0001 in.,
respectively for the three dimensions.[4]

Upon receipt of the cable for use in fabrication of coils it is normal
practice to perform an incoming inspection of the cable dimensions.[5]
These measurements are done using different equipment shown in Figs. 7-9.
For these tests short sections of cable are cut from the end of each cable
reel and are subjected to conditions similar to those in the continuous
measurements by the Cable Measuring Machine. These measurements are simple
to perform and are useful as an independent check of those taken during
cable fabrication. Of course, since cable samples must be cut from the
reel, checks can only be made at the end of a cable length.

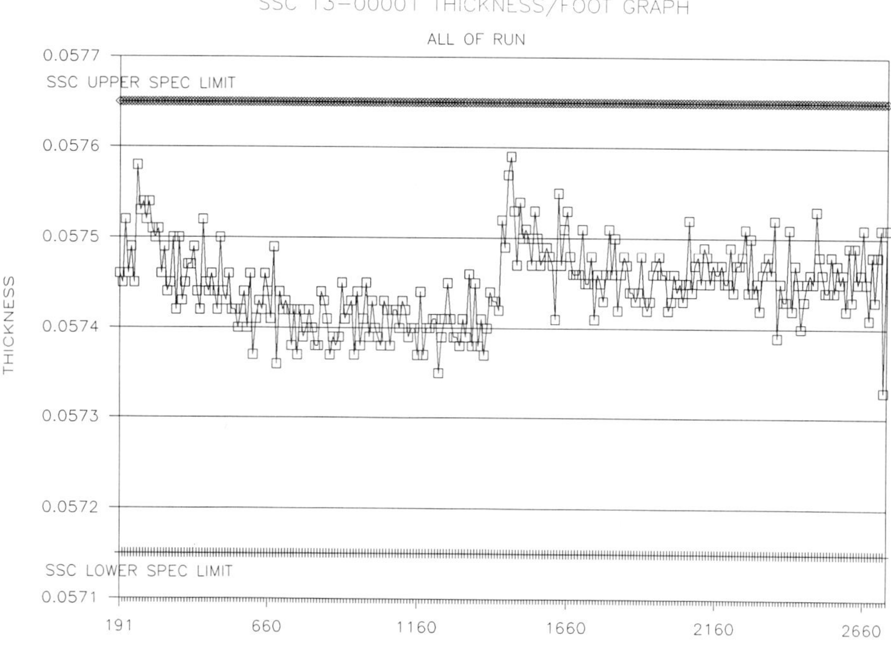

Fig. 4.   Measurements of an SSC inner cable thickness using the
           Cable Measuring Machine. Mid-thickness measurements
           are made continuously along the fabricated cable
           length.

The ten-stack fixture shown in Fig. 7 uses ten pieces of cable, each
approximately 4 in. long and assembled into the fixture with alternating
keystone direction. After calibrating the fixture with a standard block,
5 kpsi stress is applied to the coil stack. This stress is equivalent to
that applied with the Cable Measuring Machine.

The cable keystone angle is measured using the device shown in Fig. 8.
After calibration of the fixture with a standard, the cable is placed into
a slot machined with appropriate dimensions. Tension of 40 lb. and stress
of 2 kpsi are applied. The keystone angle is measured using sensors
mounted to the fixture.

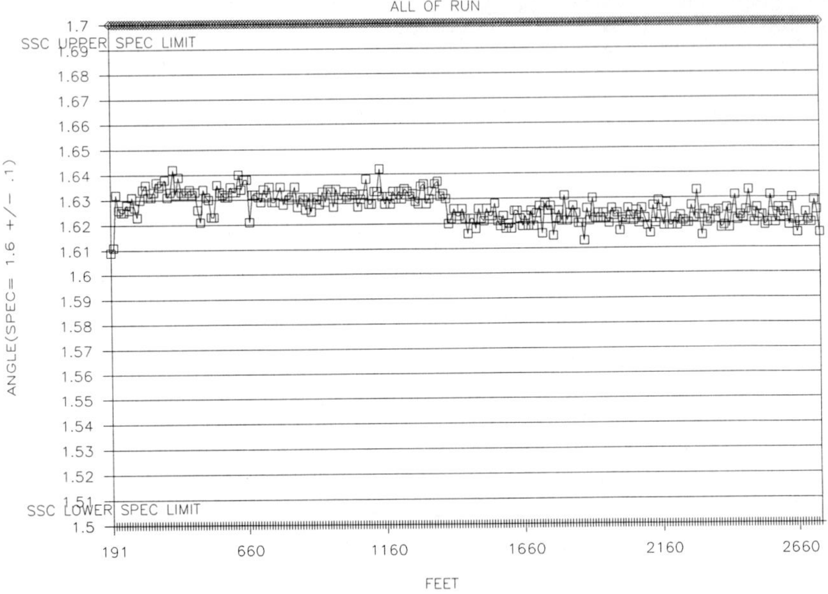

Fig. 5.   Measurements of an SSC inner cable keystone angle using the Cable Measuring Machine.

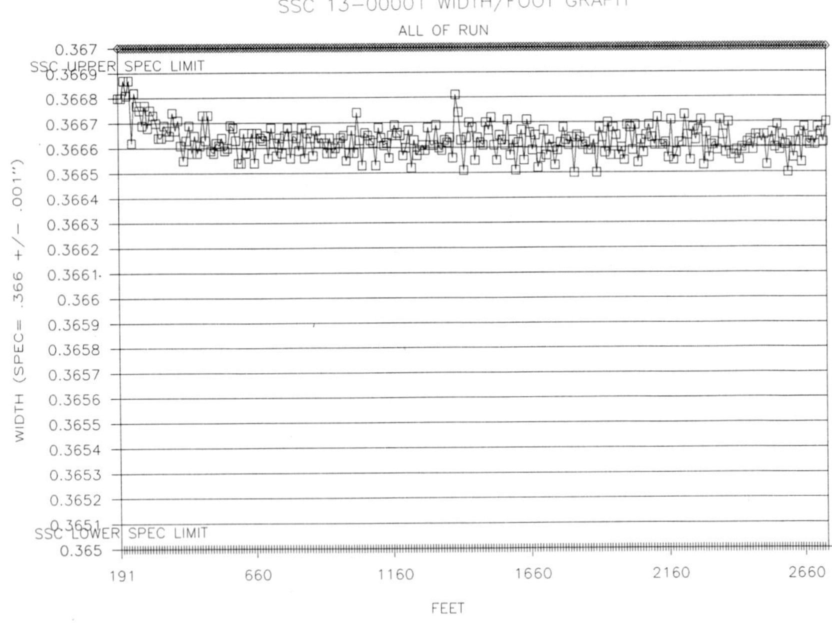

Fig. 6.   Measurements of an SSC inner cable width using the Cable Measuring Machine.

Fig. 7.   Ten-stack measuring fixture for incoming inspection
of SSC cable mid-thickness.

The cable width is checked using a device which allows a 3 ft. long
piece of cable to hang while suspending a 40 lb. weight; this is shown in
Fig. 9. The cable width is measured with a micrometer while it is under
tension. At the same time the residual twist of the cable is measured,
since the suspended 40 lb. weight is allowed to rotate. Measurement of an
excessive residual twist angle may indicate a problem with the cable. This
will be discussed further below.

CABLE MECHANICAL PROPERTIES

The direction of cable lay (or twist of the wires in the cable) is
relevant to the ease of coil winding. All wire used to fabricate SSC cable
has a "right twist", where the filaments in the twisted wire follow the
same direction as a right-hand screw thread. The wire twist direction has
been chosen to be the same for both SSC inner and outer wires to reduce the
opportunity for twisting errors. All SSC cable has a "left lay", where the
wires in the cable follow the same direction as a left-hand screw thread.
In order to fabricate a cable with good mechanical characteristics, it is
important to have the directions of wire twist and cable lay be opposite.
Those desirable mechanical characteristics include producing a cable with
minimal residual twist and having the surface of a cable be smooth with
wires tightly registered adjacent to one another. Residual twist of cable
will cause coils to be twisted, which is difficult to overcome during
magnet assembly. As mentioned above, residual twist in a cable is checked
during routine incoming inspection by suspending a 40 lb. weight from a
3 ft. length of cable. Experience has shown under these conditions that an
acceptable cable will have a maximum twist of 90° in the direction of cable
lay and 0° in the opposite direction.

There is a preferred direction for winding coils with a particular cable
lay. SSC inner coils are wound in one direction and outer coils are wound
opposite. This enables a simple splice between the inner and outer coils
which is completed during coil assembly.

Fig. 8.   Fixture for incoming inspection of SSC cable keystone angle.

Fig. 9.   Fixture for incoming inspection of SSC cable
width and residual twist.

The SSC outer coils are wound in the direction favorable to the left lay of the cable. This choice was made at the beginning of the R&D program because of concern in having wires in the outer cable come out of place while winding coil ends. SSC outer cable is more delicate because it is fabricated from more wires of smaller diameter (outer cable: 30 wires with 0.0255 in. diameter vs. inner cable: 23 wires with 0.0318 in. diameter). Therefore, because the inner coils are wound in the unfavorable direction, it is necessary to check the coil ends closely for wires which may come out of registration in the cable particularly while winding the first few turns. If a wire comes out of registration it produces a high-spot in the coil which will puncture the insulation when it is subjected to high prestress during magnet assembly. Such damage to the coil insulation will always lead to an electrical turn-to-turn short in the coil.

In Fig. 10 is shown a diagram of the initial stages of winding an SSC inner coil and the region where a wire crossover is most likely to occur. An explanation will be given of the reasons for a favorable or unfavorable winding direction for cable having a particular lay direction. Because of the geometry of a $\cos\theta$-type coil end, the cable lay is always tightened on one side of a coil end and loosened on the opposite side. It is more likely to produce a wire crossover while loosening the cable and also twisting it to conform with the coil end. However, since a choice must be made, it is best to loosen the cable on the side while starting to wind the coil end. Then it is possible to rely on the cable winding tension of 35 lb. to help register the wires against the side of the center post. On the opposite side of the coil end, where winding tension is of no assistance, it is best to tighten the cable lay.

As shown in Fig. 10, the location in the coil end where the cable is loosened is where wire crossovers are most likely to occur. Care must be taken by the coil winding technicians in the fabrication of the inner coils in these regions so there is no excessive twisting of the cable. Computerized control of the cable carriage and coil winding mandrel rotation will greatly reduce the opportunities for errors.

To fabricators of superconductor cable there is a requirement on cable surface condition which is included in the cable specification. The requirement states that, "... the cable wide face shall be uniform to within 25% of a single wire diameter...". If this requirement is not met, there is a serious risk of wire crossovers in the cable, especially when winding the coil ends. Experience during SSC R&D has shown that a possible cause of this condition, normally referred to as "popped wires", is the use of wires in a cable which are not all from the same billet geometry or have not been processed using the same procedures. Other possible causes are incorrect wire tension during cabling, wire or cable dimensions outside the required tolerance limits, or incorrect wire twist direction. In order to check for this condition of "popped wires", use is made of lump detectors to detect high spots in the cable of greater than 0.005 in. These detectors are installed in the equipment used for cable insulating and on the coil winding machines. See Fig. 11. Because the popped wires may eventually be hidden by the cable insulation, it is important that the technicians involved in that operation be aware of this potential problem.

CONCLUSION

Superconducting cable fabrication and coil manufacturing for SSC follow routine procedures which have been developed during the R&D program. However, some lessons have been learned of methods to prepare the conductor and of associated quality control steps that should be followed to assure that the cable and coils will be satisfactory. Adherence to these steps should enable a successful outcome of the construction project.

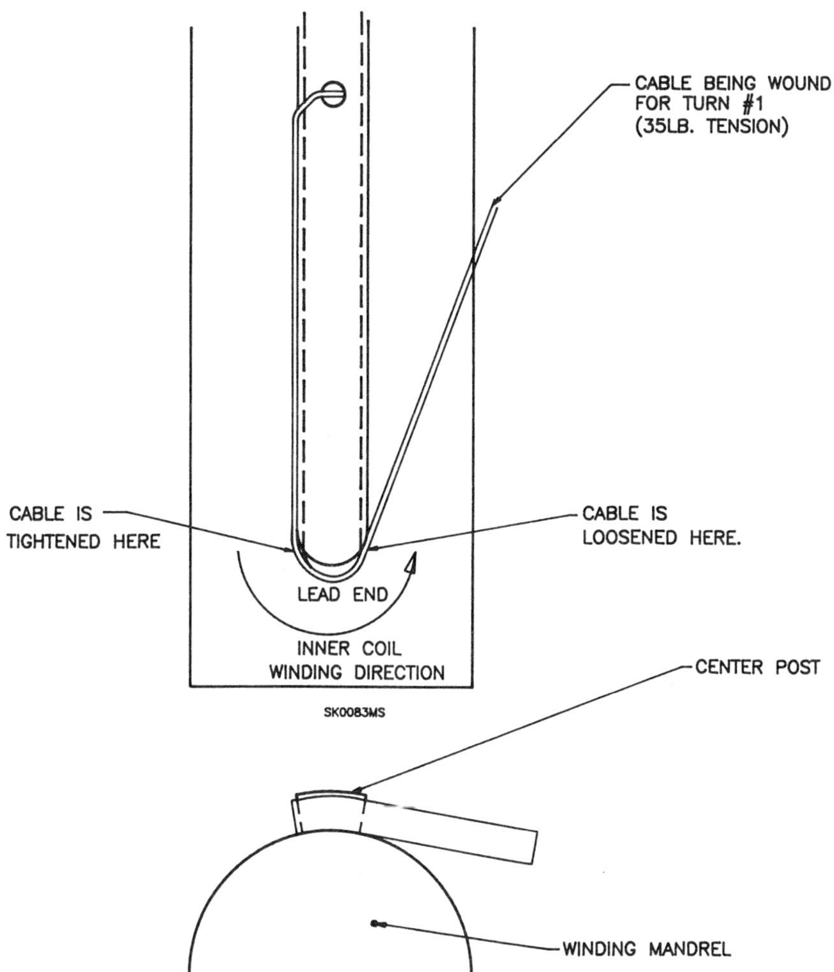

Fig. 10.   Cable characteristics while winding SSC inner coils.

Fig. 11.   Lump detector used in the insulating process and
during coil winding to look for irregularities in
the surface of SSC cable.

REFERENCES

[1]   SSC Specification SSC-MAG-M-4142, NbTi Superconductor Cable for
SSC Dipole Magnets.

[2]   SSC Specification SSC-MAG-R-7154, SSC Dipole Cable Insulating.

[3]   SSC Specification SSC-MAG-M-4141, NbTi Superconductor Wire for SSC
Dipole Magnets.

[4]   J.A. Carson, et.al., "A Device for Precision Dimensional
Measurement of Superconducting Cable", Proceedings of the ICFA
Workshop on Superconducting Magnets and Cryogenics, Brookhaven
National Laboratory, May 12-16, 1986, 162-65 (1986), P.F. Dahl,
Editor, BNL 52006.

[5]   SSC Specification SSC-MAG-R-7142, SSC Dipole Cable Test Methods.

QUALITY CONTROL TESTING OF CABLES FOR ACCELERATOR MAGNETS

M. Garber and W.B. Sampson

Brookhaven National Laboratory*, Accelerator
Development Department, Upton, New York 11973

ABSTRACT

A large number of cables have been tested for the CBA, HERA, and SSC
Projects. The short sample test procedures and apparatus are reviewed. A
simple rule for estimating cable performance from measurements on strands
taken from the cable is described. By using this "rule of thumb" cable
vendors can make reliable estimates without having to invest in the
substantial equipment required for full cable testing. Extracted strand
tests should be checked periodically by full scale tests on complete
cables.

INTRODUCTION

Short sample testing of superconducting cables used in accelerator
magnets is required in order to ensure that the individual strands have
not been unduly degraded during cabling. Occasionally this can be quite
large; an example of this is given below. Another factor is the large
increase in critical current density which has been achieved in recent
years[1], which has made wires with perfectly reproducible properties
difficult to manufacture. Mixing and matching of strands is used to
produce constancy in cable properties, but this must be verified by
short sample testing.

The first large scale program of short sample cable testing was
undertaken for the erstwhile ISABELLE-CBA Project about ten years ago.[2]
Test methods have been improved continuously since then. The development of cables carrying currents of order 10 kA at 6T, 4.2K, more than
double the current of a decade ago, has led to the need for special
mounting procedures and careful self field corrections to the measured
results. An earlier paper reviewed the important features of the BNL
cable test methods.[3]

Further details have been given in several conference papers. As
indicated in the following sections, a large investment in equipment and
refrigeration is required to test high current superconducting cables.

---

*Work performed under the auspices of the U.S. Department of Energy.

Thus, it is not feasible for each vendor to make these tests. However, a
simple correlation exists between the current of strands taken from a cable
and the full cable current. By utilizing this "rule of thumb", the vendor
can obtain prompt, in-house feedback between the manufacturing process and
short sample measurements. Examples are given below which show that a
relatively small number of strands need be sampled in order to get a
reasonable estimate of cable performance.

CABLE TEST METHOD

The V-I curve of a recently tested SSC cable is shown in Fig. 1. The
sample is mounted in a dipole magnet with the field oriented perpendicular
to the flat surface of the cable. The sample test length, i.e., that part
which is in the uniform field region, is 700 mm long. Each half of the
sample can be measured separately if, for example, it is desired to locate

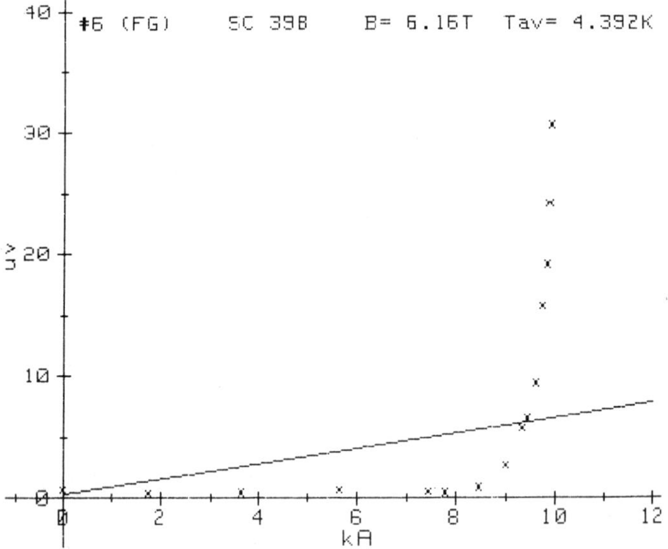

Fig. 1.   V-I curve for an SSC inner type cable.   Test length = 70.0 cm.
See Ref. 13 for physical description of the cable.

suspected flaws. The sample is long enough to provide adequate voltage
compared with the DC noise resolution, which is about 0.2 microvolts. This
sensitivity is achieved by means of a point by point determination of the
V-I curve as follows. The current is incremented in steps. At each value
the sample voltage is integrated for a time equal to 100 periods of line
voltage, thus filtering power supply noise. The magnet is operated in the
persistent mode so essentially no noise is contributed by the dipole field.

264

The straight line in Fig. 1 is drawn with a slope of 7 x 10^{-9} ohms. This corresponds to constant resistivity of 10^{-12} ohm cm. The intersection of this line with the V-I curve of the sample is the critical current, I_c. In practice the quantity $\rho = VA/I\ell$ is plotted vs I on a log-log plot. This curve is usually well approximated by a straight line, and I_c can be read off with good accuracy from the point whose ordinate is 10^{-12} ohm cm. The slope of this line, n, measures the steepness of the superconducting-normal transition, which depends on the uniformity of the NbTi microfilaments. Low n-values generally lead to low I_c's other things being equal.

I_c is often referred to as the "short sample" current in discussions of the performance of magnets made of the given cable. This is because it is a fact of experience that the magnet quench current is usually within a few percent of I_c (either more or less). The correlation is not exact, however, because of such things as variation of I_c along a cable, heating at cable joints in the magnet, variations in heat flow to the cable, etc. In any case, the quench current, I_q, in a short sample test is usually significantly greater than I_c, typically by about 10%. In general, the greater the Cu/SC ratio the greater the ratio of I_q to I_c.

In addition to the field, the temperature of the measurement must be reported. The sample and dipole magnet are placed in a helium bath. High current gas cooled leads necessarily require a bath pressure greater than standard atmospheric. As a result the temperature of measurement is about 0.2K above the normal boiling point of helium. It is desirable to refer measured values to a common reference temperature. This is done by means of the so-called Lubell formula.[7] A reference temperature of 4.22K is frequently chosen, although this is not necessary. The HERA project, for example, specifies 4.6K as reference temperature for short sample data. In order to distinguish between measured critical current values and those referred to some reference temperature we use the following notation: I_t is the critical current as measured at the helium bath temperature, T; I_c is the critical current at the reference temperature.

CABLE CLAMP ASSEMBLY

In order to observe V-I curves like that of Fig. 1, it is necessary to prevent any cable motion which would be generated by the large Lorentz forces which occur at high currents. Figure 2 shows the method by which this is done. A bifilar sample is used (two such are shown in the figure). The thin and thick edges of keystoned cables are placed as shown so that the top and bottom surfaces are parallel and the applied stress is uniform. The net Lorentz force on the assembly is zero. However, the perpendicular field produces forces on the members of the bifilar pair which try to make them slide apart in a shear like fashion. (The perpendicular field is oriented relative to the current direction so that the sum of the applied and self fields is a maximum at the thinner edge of the cable. This produces a tendency to shear apart rather than together. Cf reference 4.) In order to prevent this motion, a large compressive stress is required. The value required varies inversely with Cu/SC ratio. However, in order to standardize the procedure, a value of 10 kpsi is used for all cables. This is adequate for most of the cables commonly tested. Further details on the stress dependent behavior are discussed in reference 6.

Although the separation between cables in the bifilar pair is small, the couple created by the shear forces is enough to produce significant twisting of the assembly unless otherwise prevented. To do this the ends of the assembly are locked in position relative to the magnet bore tube by a key-slot arrangement.

CABLE TEST FACILITY

The sample holder discussed above is inserted in a dipole magnet
which is 1.2m long and has a 75 mm bore. This is an early R&D dipole
which is described in the literature.[8] It has been used continuously for
the last 12 years. The maximum central field is a little over 6T.

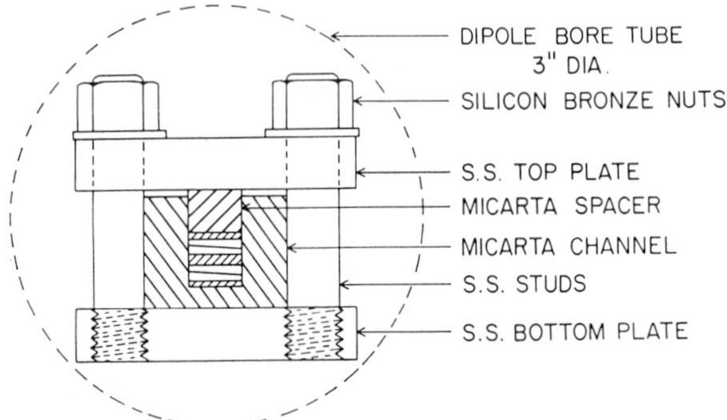

Fig. 2.   Mechanical Assembly. Two bifilar pairs of keystoned samples are
          assembled, as shown. The members of each pair are separated by
          Kapton, 0.010 in. thick. Instrumentation is located in G-10
          strips placed in contact with each sample. The thicknesses of
          the G-10 strips are 0.030, 0.100, and 0.030 in., respectively.
          The stainless steel studs are spaced 1.5 in. apart along the 48
          in. length of the fixture. The silicon bronze nuts are tightened
          to 200 in.-lbs. torque resulting in a mean pressure on the
          samples of approximately 10 kpsi.

The liquid helium dewar is 2.8m long and has an inside diameter of
600 mm. Helium is supplied from a 10,000 ℓ storage dewar, which is
connected to a 100 ℓ/hr liquefier. Gas recovery, storage, purification,
and compression are handled by a central BNL plant.

Thus, the cable test facility consists of four separate areas:

- the gas handling plant
- refrigerator and storage room
- test dewars and power supplies - see Fig. 3
- electronic measurement and controls lab.

It would not be feasible to duplicate this investment in plant and
operating cost at each cable vendor's facility. An alternative method of
assessing cable performance, which is described below, can more
economically serve to provide in-house short sample information to the
superconductor manufacturer. Before doing so we describe two examples of
quality control results, in the next two sections.

266

Figure 4 is an example of the customary quality control report for a cable, in this case the one discussed earlier in connection with Fig. 1. Measurements are made at several fields. The first part of the report lists the peak field, the helium bath temperature, the critical current I_t, the quench current I_q, the quality parameter, n, and the critical current calculated for the reference temperature (4.22K in this case).

The next part lists the results of resistance measurements at room temperature and 10K. These are conveniently measured in the apparatus described. A complete description of the procedure is given elsewhere.[5] The room temperature resistance, R295, is an important element of the cable specification as it is determined by the Cu/SC ratio. Conversely, the Cu/SC ratio may be calculated from the measured resistances. This calculated value and the residual resistance ratio, RRR, are listed.

Fig. 3. Dewars and power supplies. 1-test dewar. 2-He fill and recovery lines. 3-sample probe. 4-15 kA power supply. 5-4 kA dipole power supply. 6-magnetization test dewar. 7-assembly area.

The last part of Fig. 1 gives critical currents determined by fitting a straight line to the measured I_c-B points. This is a good representation of the I_c-B dependence for fields between 5 and 7T, for multifilamentary NbTi wire of the type used (46.5 wt % Ti). In making the fit to the cable data the slope is chosen to be the same as for the wire; wire data is usually more accurate since it is averaged over several wires. Special cable tests in which many points are measured and very great care is taken to reduce field and temperature errors confirm that wire and cable slopes agree. For routine production cable testing it is more efficient to measure points at a few fields and utilize the wire slope value. The resulting accuracy is about ± 150A.

Critical current densities are also listed. These are based on the specified or stated strand diameter and on the Cu/SC value determined from the resistance measurements.

```
Run 2066                 SC 398                10/25/88


Measurements

  B        T        It       Iq       n           Ic

 5.38     4.407    11547    12005     27         12371
 5.79     4.406    10166    10826     25         10944
 6.16     4.392     9356     9984     24         10064

R(295)=   25.5  uohms/cm        RRR= 69
R(10) =    .37      "           C/S= 1.42

Calculated results for T= 4.22

  B        Ic       Jc

 5.0     13284     2730
 5.6     11614     2387
 6.0     10501     2158
 7.0      7718     1586
```

Fig. 4.   Short sample test report (same cable as in Fig. 1)

AN EXAMPLE OF CABLE REJECTION

QC test results for several projects have been published.[2,9] From time to time cables are found to be below specification, necessitating diagnosis and correction of the underlying problem. An interesting example of one such case occurred during the production of cables for the HERA project. A detailed review of the complete run including test results for some 300 samples is given in reference 9. Early in the run the sequence shown in Fig. 5 was observed: a downward trend began at No. 28 and culminated in a clearly pathologic cable at No. 34. Fortunately the cabler was shut down at this point for other reasons. Thus, there was time for the manufacturer to investigate the cause, which turned out to be an improperly positioned mandrel. Gradual slippage of the mandrel led to increasing filament breakage and the observed steady decrease in I_c. After correcting this problem, no further difficulties arose during the remainder of the production run.[9]

ESTIMATION OF I_c FROM EXTRACTED STRAND DATA

Recently it was reported that a good estimate of the critical current of a cable is obtained by summing the critical currents of the strands extracted from the cable.[4] In doing this, the measured strand currents are used with no account taken of wire self field effects (strand measurements are described in Ref. 10). This "rule of thumb" has been tested for cables in which the current, the Cu/SC ratio, and the residual resistance ratio have different values. The results are shown in Table 1. The first four columns list the cable number and the measured values of Cu/SC, RRR, and I_c. Column 5 lists the degradation, $D = 100 \times (1 - I_c/I_w)$, where I_w is the sum of the critical currents of the wires used to make the cable. In most of the cases, I_w is approximated by measuring between 8 and 12 wires. The

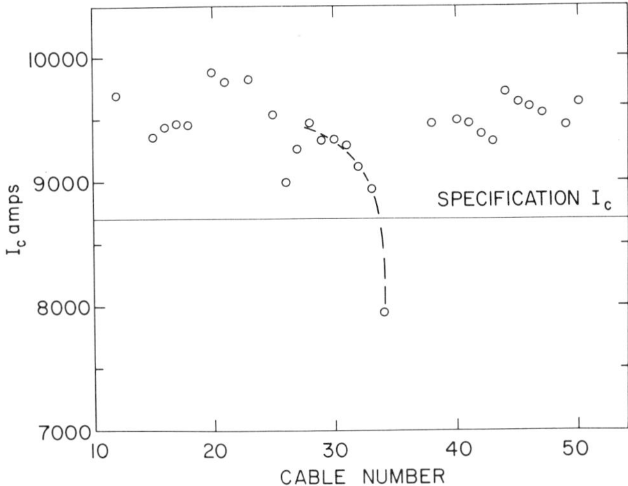

Fig. 5.   Critical currents of a sequence of cables during a production run for HERA (cf. reference 9).

average of these wire values is multiplied by the total number of cable strands to get I_w. In one case, cable 378, all 30 strands were measured because the value of Cu/SC varied considerably from strand to strand. In those cases where D values are not listed, pre-cabling wire samples were not provided. Two samples have zero or negative degradations. These come about partly from sampling error in determining I_w. In addition, neglect of self field effects in wires reduces I_w by a few percent.

The estimated critical current of the cable based on the extracted strand currents, I_{cx}, is similarly determined from a sample of strands. In most cases a relatively small number of strands was tested. Column 6 lists the percentage difference between I_{cx} and I_c, and column 7 lists the number of extracted strands, N_x. It appears from Table 1 that with Nx ~ 4 a reasonable estimate, good to about $\pm$ 2%, is obtained. Part of the difference between I_c and I_{cx} is due to the accuracy of the cable measurement, which is between 1 and 2% also.

The procedure by which I_{cx} is determined is the same as that by which I_w is determined. By estimating the degradation from wire measurements, i.e., from I_c and I_{cx}, geometrical and other experimental conditions are nearly identical; this is an advantage over the method of comparing wire and cable results. The values of D obtained by comparing I_{cx} and I_c are approximately equal to the differences between the values listed in columns 5 and 6 of Table 1.

A quantitative calculation explaining the rule of thumb would be hard to formulate. For one thing, the magnetic field distributions are quite different for the two measurement configurations, cable vs strand. For another, the voltage distribution along an extracted strand, varying as it does between highly deformed, kinked regions and regions which are only slightly deformed, is rather different from that along the cable.[11] It is fortuitous, therefore, that the rule works as well as it does.

Table 1.   Comparison of Cable and Extracted Strand Data (B=5T, T=4.2K).

Sample	Cu/SC	RRR	I_c(amps)	D(%)	$100[I_{cx}/I_c-1]$	N_x
RHIC Types[12]						
106	2.26	66	6660	13.1	2.7	2
107	2.26	66	6880	10.2	1.6	2
SSC Inner Types[13]						
351	1.50	34	11232	5.3	-0.6	8
351A	1.50	130	11621	-	-1.0	16
383	1.60	84	12828	4.0	0.8	3
395*	1.80	70	11130	0.0	-0.3	4
398	1.42	70	13284	-2.4	-1.5	4
399	1.54	82	13693	-	-2.9	4
6135	1.20	105	14160	-	-0.4	4
SSC Outer Types[13]						
369	1.70	68	9600	6.9	1.9	3
378	1.70	74	9160	7.8	1.2	2
380*	1.71	65	9164	5.0	1.5	4
385**	1.59	66	8955	9.3	0.4	30
6136	1.73	110	9660	-	1.1	4
HERA Types[9]						
L383	1.70	77	11973	-	0.0	4
1553	1.80	75	11397	-	-1.0	22

*These cables were made from a single continuous strand.
**This cable was made from strands in which Cu/SC varied from 1.24 to 1.71.
All strands, both before and after cabling, were measured.

SUMMARY

A large investment in equipment and refrigeration is needed for cable
critical current tests. This makes it impractical for wire and cable
manufacturers to undertake such tests themselves. Fortunately, a good
estimate of cable critical current can be obtained from extracted strand
measurements. A relatively small sample, ~4 strands, is adequate for this
purpose. Thus, it is possible to monitor cable production with modest
expenditures. Sample cables could be tested at a national facility such as
that at BNL in order to ensure the validity of the extracted strand
results.

ACKNOWLEDGEMENTS

We thank the cryogenic support group directd by D.P. Brown for their
vital support. Many helpful suggestions have been made in the course of
this work by K. Robins and A.K. Ghosh. We thank R.K. Maix and A.F. Greene
for discussions concerning the HERA cables.

REFERENCES

1. P.J. Lee and D.C. Larbalestier, "Determination of the Flux Pinning
    Force of a-Ti Ribbons in Nb46.5 Wt% Ti Produced by Heat Treatments
    of Varying Temperature, Duration and Frequency", J. Mater. Sci.,
    Vol. 23, pp. 3951-3957, 1988.

2. M. Garber, M. Suenaga, W.B. Sampson, and R.L. Sabatini, "Effect of
    Cu_4Ti Compound Formation on the Characteristics of NbTi Accelerator
    Magnet Wire", IEEE Trans. Nuc. Sci., Vol. NS-32, pp. 3681-3683,
    October 1985.

3. W.B. Sampson, "Procedures for Measuring the Electrical Properties of
    Superconductors for Accelerator Magnets", Proceedings of the ICFA
    Workshop on Superconducting Magnets and Cryogenics, Brookhaven
    National Laboratory, Publication No. 52006, pp. 153-156, 1986.

4. M. Garber, A.K. Ghosh, and W.B. Sampson, "The Effect of Self Field
    on the Critical Current Determination of Multifilamentary Supercon-
    ductors", IEEE Trans. Magn., Vol. MAG-25, pp. 1940-1944, March 1989.

5. W.B. Sampson, M. Garber, and A.K. Ghosh, "Normal State Resistance and
    Low Temperature Magnetoresistance of Superconducting Cables for
    Accelerator Magnets", IEEE Trans. Magn., Vol. MAG-25, pp. 2097-
    2100, March 1989.

6. A.K. Ghosh, M. Garber, K.E. Robins, and W.B. Sampson, "Training in
    Test Samples of Superconducting Cables for Accelerator Magnets", IEEE
    Trans. Magn., Vol. MAG-25, pp. 1831-1834, March 1989.

7. M.S. Lubell, "Empirical Scaling Formulas for Critical Current and
    Field for Commercial NbTi", IEEE Trans. Magn., Vol. MAG-19, pp. 754-
    757, May 1983.

8. A.D. McInturff, W.B. Sampson, K.E. Robins, P.F. Dahl, and R. Damm,
    "Prototype and Proposed "ISABELLE" Dipoles", IEEE Trans. Nuc. Sci.,
    Vol. NS-24, pp. 1242-1244, June 1977.

9. R.K. Maix, D. Salathe, S.L. Wipf, and M. Garber, "Manufacture and
    Testing of 465 km Superconducting Cable for the HERA Dipole Magnets",
    IEEE Trans. Magn., Vol. MAG-25, pp. 1656-1659, March 1989.

10. M. Garber and W.B. Sampson, "Test Methods for Short Sample Critical Current Determination of Twisted Multifilamentary Superconducting Wire", SSC Technical Note SSC-MD-89, Brookhaven National Laboratory, January 1985 (unpublished).

11. L.R. Goodrich, E.S. Pittman, J.K. Ekin, and R.M. Scanlan, "Studies of NbTi Strands Extracted From Coreless Rutherford Cables", IEEE Trans. Magn., Vol. MAG-23, pp. 1642-1645, March 1987.

12. RHIC Specification RHIC-MAG-M-4142, NbTi Superconductor Cable for RHIC Dipole and Quadrupole Magnets.

13. SSC Specification SSC-MAG-M-4142, NbTi Superconductor Cable for SSC Dipole Magnets.

AN INDUSTRIAL CABLING MACHINE FOR THE SSC*

John Royet, Rollin Armer, Roy Hannaford, and Ron Scanlan

Lawrence Berkeley Laboratory
1 Cyclotron Road
Berkeley, CA 94720

ABSTRACT

The SSC project will need the manufacturing of some 25,000 kilometers of keystoned flat cable. The technical specifications of the various cables to be produced are the result of five years of research and development work at LBL.

An experimental cable machine was built and run in the laboratory; many improvements were implemented and tested. Semi-industrial production of the various cables was performed, and the resulting cables were used and tested in the one-meter model magnets and 17.5 meter dipole prototypes.

From these experiments an industrial cabler specification was generated and used for an international RFQ. The winner of the contract is Dour Metal, a Belgium company that built the first industrial prototype which is now in a production line at New England Electric Wire Company.

In this paper we describe the main characteristics of the machine and give the first industrial production results of superconducting keystoned cable for the SSC project.

INTRODUCTION

We have already reported at the 1988 Applied Superconductivity Conference, San Francisco, CA, August 21-25, 1988 that Dour Metal, A Belgium company was the successful bidder for the industrial cabler prototype to be used in the SSC development program.

The machine was tested and accepted in Dour Metal plant on June 28, 1988.

This cabling machine was sent to the United States and installed in New England Electric Wire plant in Lisbon, New Hampshire. (See picture XBC 8810-10074)

*This work is supported by the Office of Energy Research, Office of High Energy and Nuclear Physics, High Energy Physics Division, Dept. of Energy under Contract No. DE-AC03-76SF00098.

This location was chosen due to the geographic situation of several superconducting wire manufacturers in the Northeast of the U.S. After installation in this new location, the cabler was tested again in order to detect possible adverse effects of the transportation, then harnessed to the production of the s.c. cables for the SSC development program, relaxing the LBL research machine from its production burden.

Fig. 1   XBC 8810-10074

CHARACTERISTICS REMINDER

The main characteristics of this cabling line are:

- high speed capability: up to 10 meters/minute
- up to 30 strands cable definition
- 5 planetary ratios in each direction
- equal path length for each strand
- constant wire tension
- broken strand detector
- quick emergency stop
- thermal stability and control
- accurate and stable mandrel position
- adjustable Turkshead position
- electronic monitoring and recording of the main cable parameters
- accurate monitoring of the caterpillar tension
- constant take up traction at variable speed with the associated dancer
- compatibility with the in line measuring machine developed at Fermi National Accelerator Laboratory

Some of the characteristics differ slightly from the LBL experimental machine:

The Dour Metal machine:

- is 2 times faster
- has 30 strands maximum possibility according to SSC specification (the LBL is 36 strands)
- has a planetary ratios geared permanently at ±1, ±2, ±2.8, ±4, and ±5.6. The LBL machine has optional ratios according to a wide gear reserve.
- The cable tension monitoring is excellent and continuous. (Air cushions and strain gauges).
- The motorized Turkshead was required in the specifications and will be the object of a detailed experimentation which was not possible on the LBL machine.
- The caterpillar is a very powerful one capable of one ton pulling force.
- The Dour Metal cabling line is equipped with a transverse take up with bobbin lateral motion and is associated with a dancer which keep the cable tension constant and regulates the cable respooling according to production rate of the cabler in any occurance, including emergency stop.
- The measuring machine has to be altered to match the high production rate of the Dour Cabler.

RESULTS AFTER 12 PRODUCTION WEEKS

During those 12 weeks the Dour cabler production was some 17 km of 23 and 30 strands cable to be used in the long magnets needed in the SSC development program.

This production length represents about 20% of the total amount of cable made on the LBL machine during the 3 years of development operation.

<u>Problems</u>

We have very few accidents to report:
- one broken strand and two "cross-over". However, as our research commitment was no cross-over, we have to improve the technique to avoid such defects:
    - Perfect wire respooling under constant tension.
    - Safety secondary brake device as close as possible to the Turkshead to keep a certain wire tension in case of broken wire up stream. This secondary brake is under development at LBL.
    - The amount of degradation of the Superconducting Wire was similar to the degradation observed on LBL machine: 1 - 5% depending on the type of wire.

<u>Motorized Turkshead</u>

At the occasion of those production runs, we have had the opportunity to harness the motor of the Turkshead (through the adjustable magnetic torque clutch).

The response of the line was immediate and accurate as expected: It was possible to reduce the cable tension between Turkshead and caterpillar by a factor 4. There was no measurable effect on the dimensional characteristics of the cable, however, the cable surface aspect seems to be slightly weavy with a reduced tension. This is not surprising because the cable tension after the Turkshead is a straightening factor of the strands composing the cable as the deformation process between the Turkshead rollers is not uniform due to several factors, the main of them being the keystone shaping, if the tension in this section is reduced the chances of local wire segment length equalization is reduced accordingly.

The main advantage of a power Turkshead is to reduce the collapsing tendency of a flat wide cable under too high a tension. The mechanical stability of the SSC cables 23 and 30 strands is good enough and the safety factor large enough and I do not think that such cable require the use of a power Turkshead. More experimental runs will be performed to document these preliminary results.

CONCLUSION

The design, construction, and installation of the cabling line in an industrial environment is a success for the technology transfer program which is one of the goal of the U.S. National Laboratories.

REFERENCES

1. J. M. Royet, Magnet Cable Manufacturing Technical Note, LBL-20129, July 1985.
2. J. M. Royet and R. M. Scanlan, "Manufacture of Keystoned Flat Superconducting Cables for Use in SSC Dipoles", proceedings of Applied Superconductivity Conference, Baltimore, MD, September 29-October 3, 1986; LBL-21295, September 1986.
3. R. M. Scanlan, J. M. Royet, and R. Hannaford, "Fabrication of Rutherford-Type Superconducting Cables for Construction of Dipole Magnets", ICMC Conference, Shenyang PR China, June 7-10, 1988, LBL-24321, May 1988.
4. J. Grisel, J. M. Royet, R. M. Scanlan, and R. Armer, "A Unique Cabling Machine Designed to Produce Rutherford-Type Superconducting Cable for the SSC Project", 1988 Applied Superconductivity Conference, San Francisco, CA, August 21-25, 1988, LBL-25777.

INNOVATIONS IN THE DESIGN OF MULTIFILAMENTARY NbTi SUPERCONDUCTING

COMPOSITES FOR THE SUPERCOLLIDER AND OTHER APPLICATIONS

E. Gregory, T.S. Kreilick and J. Wong

Supercon Inc.
830 Boston Turnpike
Shrewsbury, MA 01545

ABSTRACT

In recent work it has been shown that strand and cable, meeting the SSC specifications can be made on a production scale.

In addition, work has been carried out to develop finer filamentary materials with larger strand diameters and higher J_c's, capable of operating at higher fields. Some of the problems encountered and the solutions developed are discussed in an effort to illustrate to the accelerator magnet designers the properties of conductors that may be available in the near future.

INTRODUCTION

It has been shown that wire and cable meeting the present SSC specifications can now be made commercially in prototype production quantities [1]. It remains to improve the economics of production without sacrificing quality, to enlarge the "property window" and to ensure improved product reliability.

The decision to raise the strand current density (J_c) specification from 2400 A/mm^2 to 2750 A/mm^2 at 5T was made three years ago, at the time of issuance of the SSC Conceptual Design Report (March 1986) [2]. At that time it had already been shown, from the results of the program starting in late 1984, that J_c's in excess of this value were possible in filaments less than 5 µm in diameter [3]. These rapid improvements were attained by applying relatively simple concepts to the design of multifilamentary billets [4]. As is common in many development programs however, the principal problems encountered in converting these early successes into a commercially viable manufacturing process for producing even finer filamentary material were not those that had been widely anticipated, but ones which had been largely overlooked. In the period 1986-1988, these problems were also solved [4], and in the paper we will discuss some of them, their solutions and the possibilities that these open up for accelerator magnets in general.

THE PROBLEMS

Once material with a uniform array and a small spacing between the individual filaments had been produced [3], it appeared relatively simple to draw it down to a smaller size in order to determine how the properties (specifically J_c) would vary with filament size.

In 1986, research on this question showed that in filaments approximately 3 μm in diameter a J_c of 3450 A/mm^2 and an "n" of 42 could be obtained at 5T, while a J_c of 1350 A/mm^2 and an "n" of 28 could be obtained at 8T [5]. Although these results showed great potential, they were achieved in material which had serious practical limitations. Some of these were as follows:

1. It had been subjected to many multiple heat treatments.

2. The overall wire diameter was only 0.34 mm.

3. The average piece length was undetermined as only a small quantity had been produced.

4. When magnetization experiments were carried out, it was found that at injection field (0.3T) the wire behaved like one large superconductor despite the fact that it contained 4164 separate filaments [6].

In the last three years these problems have been addressed and some of the solutions reported upon [7, 8, 9]. We summarize them here in the hope that those concerned with the design of advanced accelerator magnets, will realize the opportunities that these solutions offer as well as the possible limitations that they impose.

<u>Filament Coupling</u>

It has been shown that for the best J_c values the spacing between the filaments (S) must be small [10, 11] relative to the filament diameter (D) but the most desirable value of S for any given diameter has not as yet been determined. In the 3 μm diameter filamentary material reported above, the value of S/D was 0.13, i.e. a spacing of ~400 nm. At the present time, this is assumed to be a maximum S/D to aim for, if a high J_c, 2.5 μm diameter filamentary material is to be developed. However, as mentioned above, when S is ~400 nm all the filaments are coupled by proximity-effect coupling [6]. When the material between the filaments is high purity copper, this coupling occurs (at fields of only a few gauss) until S > 1.5 μm. This critical value of spacing is, however, only 500 nm at the present injection field of the SSC (0.3T).

In order to meet SSC specifications with reasonable margins, an S/D ≤ 0.17 has to be maintained. In order to be sure that, with pure copper between the filaments, no coupling occurs in practical materials, it has been decided to limit the closest approach of the filaments to 1 μm. This limits the minimum filament size for the SSC conductor to 6 μm, i.e. D = 1/0.17. In 1986, based on these facts, it was realized that if conductors with properties improved over those of the present SSC specifications were to be developed, the coupling had to be reduced. This was done, without significantly reducing the stability of the conductor or the cable made from it, by replacing the OFHC copper between the filaments with Cu 0.5 wt% Mn [7]. The manganese atoms in solid solution maintain a magnetic moment that spin-flip-scatters the superelectrons, effectively decoupling the filaments, even at the lowest fields, until the spacing between them falls below 280 nm [8, 12].

If the only problem had been proximity effect coupling it would then have been possible to make uncoupled 2.5 μm diameter filaments very closely spaced and thus presumably capable of carrying a high J_c. The other problems are described below but the solution of the coupling problem means that one of the largest barriers to the production of fine filament material has been removed.

<u>The Overall Wire Diameter</u>

There are several aspects to this problem-

In the 0.34 mm diameter wire mentioned above, only 4164 filaments were required to give 3 μm diameter filaments.

It was pointed out early in the development of the SSC conductors [13] that for 0.81 and 0.65 mm diameter wires, ~40,000 and 23,000 filaments are required, respectively. In previous work, "double stacking" techniques had been used to achieve these high numbers, but the advantages of "single stacking" have been repeatedly pointed out, [14] making the problem one of stacking as many as 40,000 individual rods in as uniform an array as possible. The packing problem was addressed and solved on a non-automated basis as shown by the prototype billet containing 23,000 filaments in a 305 mm diameter can [8] (Figure 1). Techniques for automated packing exist and equipment could be purchased should a large demand for such material arise.

Figure 1

A more serious problem arises when an attempt is made to achieve the very highest J_c's in 0.81 mm diameter wire in particular: the availability of strain space. It has been repeatedly pointed out [4, 9, 15, 16] that, in commercial materials using annealed rods and the conventional heat treatment-cold work cycles, [17] the maximum allowable number of heat treatments is limited. When a heat treatment temperature of 420 °C is used, together with long times, it is possible to achieve around 3000 A/mm^2 at 5T with these heat treatments in multifilamentary wire but the 3450 A/mm^2 at 5T in the fine wire reported above, was achieved with seven heat treatments. If more strain space were available, it is tempting to believe that these high J_c's may be achieved in SSC diameter wire. A possible way of making this larger amount of strain space available has been reported in several recent publications [4, 9, 16]. The problem is that the extrusion step is usually carried out at an elevated temperature and this eliminates or reduces the effects of cold work and/or heat treatment carried out prior to the extrusion. Attempts were made in the past [13] to use hydrostatic extrusion as a way of extruding at lower temperatures, but problems of "sausaging" and lack of bonding obscured the results [18]. Since these problems are now better understood, it is possible to take advantage of low temperature processing so as not to lose the effects of the cold work and heat treatments obtained <u>prior</u> to extrusion.

Figure 2 and 3 show the results of some preliminary experiments carried out on material given heat treatments both before (HTB) and after (HTA) extrusion compared with material heat treated only after extrusion. When these comparisons were made using conventional extrusion at 600 °C, the benefits of additional heat treatments prior to extrusion were minimal (Figure 2). When extrusion was carried out hydrostatically at 200 °C, however, the advantage of the additional heat treatments carried out prior to extrusion were much more apparent (Figure 3).

Since the billets involved here were small (175 mm in diameter) and the conditions far from optimized, the J_c values were not outstanding. The results show, however, that a better utilization of strain space can increase J_c from 2650 A/mm^2 at 5T to 2900 A/mm^2 at 5T even in the 0.81 mm diameter wire.

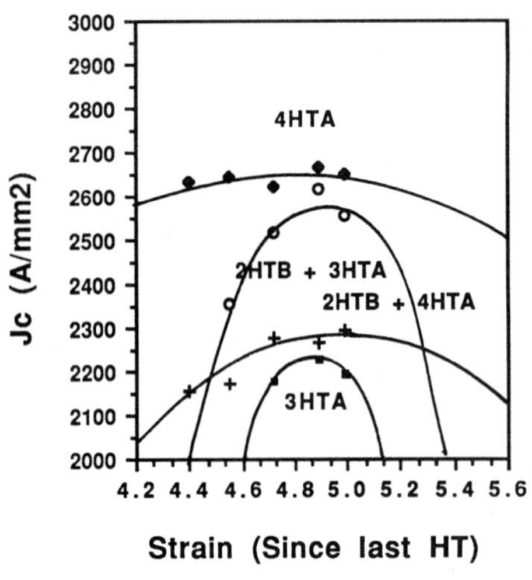

Figure 2. J_c vs. strain: conventional extrusion (600 °C).

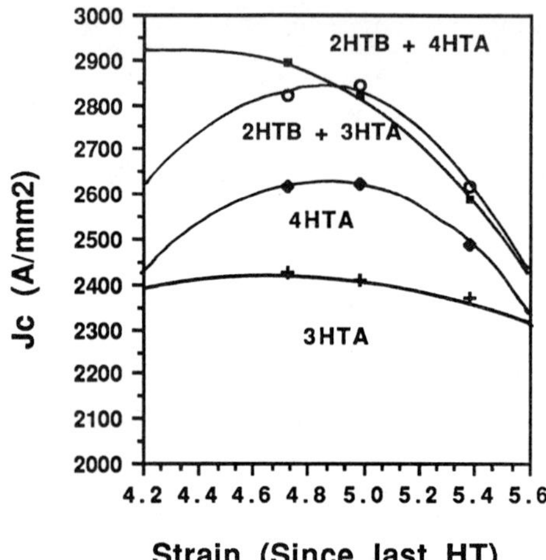

Figure 3. J_c vs. strain: hydrostatic extrusion (200 °C).

Piece Length

The piece length specification for Fermi type material was ~0.75 km. The tentative specification for the SSC is that only 10% of the material can be < 3 km. In addition, the wire must undergo stringent bend and spring back tests.

Problems have been encountered in the past in meeting all the specifications simultaneously and it is only in recent months that we have been able to report success with production quantities of SSC type wires [1].

It is important to realize that while the inner SSC strand contains only 7000 filaments, meeting the piece length specification in material of this type does not ensure that finer filament, higher J_c material, when made, will have the piece lengths necessary for the economical production of high quality cable. There are some indications that as the number of filaments increase, the ductility problem becomes more acute, and until finer filamentary material meeting the same specifications has been repeatedly produced in long piece lengths, it is unwise to speculate that this problem has been solved. It has been reported in the past [19] that as the spacing between the filaments is decreased and the J_c increased, there has been a tendency for the wire ductility to decrease and for the piece lengths achieved to be relatively short.

It remains to be determined whether ductility problems are due to a lack of bonding between the many interfaces, impurities, inclusions, or a change in the mechanical behavior of the composite when the spacing falls below a certain level. Another possibility is that the overall composite ductility is decreased as precipitation in the NbTi increases the strength [20].

Higher Field Magnets

These offer an even greater challenge to the conductor manufacturer than do the dipoles and quadrupoles in the main ring of the SSC. This machine would presumably be able to use higher field magnets, at least in the beam handling area, were they to become available.

One proposed accelerator that is considering higher field magnets for the main ring is the Large Hadron Collider (LHC) at CERN. The tentative conductor specification for the LHC is 1200 A/mm^2 at 11T. Perhaps more importantly, this has to be achieved in a larger strand than in the SSC, i.e., one which is 1.29 mm in diameter.

While the present requirement is for 15 μm filaments, smaller diameter filaments would obviously be desirable, particularly under injection conditions. If this requirement develops, the combination of fine filaments, a large diameter conductor, and high J_c's will pose a particular challenge.

While Nb$_3$Sn can be made with the required J_c at 4.2K and 11T, along with desirable flat J_c vs. field characteristics, coupling and bridging problems sometimes exist when such high J_c's are targeted [21]. The principal drawback to Nb$_3$Sn, however, is its inherent brittleness: Once reaction has taken place to form the compound, the greatest care must be exercised to ensure that no conductor damage occurs during magnet assembly and operation. Magnet reliability may therefore become a problem no matter how good the Nb$_3$Sn is or how it was made. However, if the ductile NbTi or ternary NbTiTa alloys can be operated in the region of 1.8-2.0K, they also can produce the necessary J_c's at 11T, albeit with a much steeper J_c-field characteristic than that of Nb$_3$Sn. Superfluid cryogenic technology has improved to such an extent in recent years that the operation of large devices in this 1.8-2.0K region now appears to be a practical reality [22, 23]. For these reasons, Supercon, Inc. has attempted to address some aspects of this 10T conductor problem in a recent paper [24].

14 filaments of 15 wt% Ta 44 wt% Ti Nb alloy were placed in each of a series of 50 mm diameter billets. These were extruded to 12.5 mm diameter rods. The finishing sizes of the wires were 0.25 mm to 0.15 mm, with the filaments 34 - 20 μm in diameter. All of the materials were given four heat treatments after extrusion and the properties are shown in Figure 4. LH represents the standard time and temperature normally used by Supercon, Inc. [24]. SH represents heat treatments at the same temperature but for half the time.

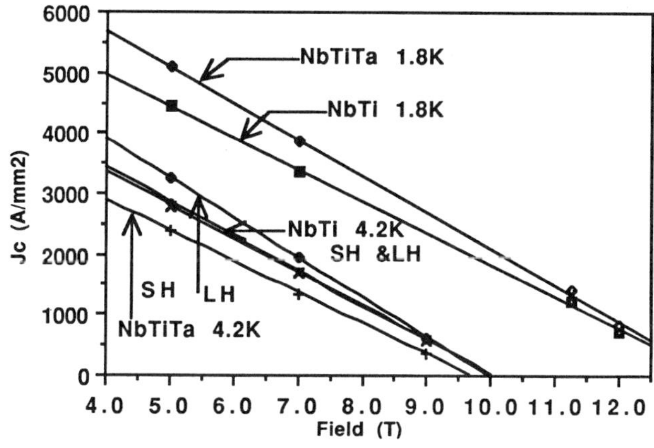

Figure 4

Since only a few data points are available, each representing the peaks of a series of strain vs. J_c curves, straight lines have been used to join the points. No reliable predictions of the Hc$_2$ can be made but the following tentative conclusions can be drawn provided that they are confirmed in further experiments

1. NbTiTa and NbTi both show markedly superior properties at 1.8K compared with those shown at 4.2K.

2. If the standard Supercon, Inc. heat treatment (LH) is used, NbTiTa is superior to NbTi at all fields at 1.8K, and below 9.5T at 4.2K.

3. If a shorter time heat treatment (SH) is used, the reduction of J_c properties for NbTi is far less than that for NbTiTa suggesting the precipitation in the ternary alloy is more sluggish than in the 46.5 wt% Ti binary alloy.

Values of 5084 A/mm^2 at 5T, 3841 A/mm^2 at 7T, 1410 A/mm^2 at 11.25T and 846 A/mm^2 at 12T were obtained in the fine wires of ternary material made from small billets. It is possible that by using larger billets containing NbTiTa, optimized heat treatments, and/or the low temperature extrusion discussed above, these properties, or superior ones, may be attainable in the diameter of wire required for the LHC. Recently, the Wisconsin Group has also reported on some recent work comparing NbTiTa with NbTi [25].

SUMMARY

Work carried out in recent years at Supercon, Inc. that addresses problems in the areas of fine filament coupling, J_c improvement in large diameter conductors and piece length is described. Also, relatively recent work on the binary NbTi alloy and a ternary NbTiTa alloy and their properties at 1.8 and 4.2K is briefly discussed. The paper points out that many techniques, some as yet not widely used, exist and can probably be applied to solve the problems of conductors for the next generation of accelerators.

ACKNOWLEDGEMENTS

The authors would like to acknowledge the support of their colleagues at Supercon, Inc. who performed the work reported in this paper. Supercon, Inc. would also like to acknowledge the financial assistance of the U.S. department of Energy (DOE), through the Lawrence Berkeley Laboratory, and the Small Business Innovative Research (SBIR) Program of DOE. Our thanks are also due to the Brookhaven National Laboratory, the University of Wisconsin and the Francis Bitter National Magnet Laboratory of M.I.T. all of whom assisted in obtaining much of the data reported.

REFERENCES

1. T.S. Kreilick, E. Gregory, D. Christopherson, G. Swenson and J. Wong, "Superconducting Wire and Cable for the Superconducting SuperCollider", Paper III-D-4, IISSC, New Orleans, LA, February 9-10, 1989.

2. R.M. Scanlan, J. Royet and R. Hannaford, "Fabrication of Rutherford-Type Superconducting Cables for Construction of Dipole Magnets", Cryogenic Materials'88, eds. R.P. Reed, Z.S. Xing and E.W. Collings, Proc. ICMC Conf., Shenyang, China, 1, 391-399, 1988.

3. T.S. Kreilick, E. Gregory and J. Wong, "Fine Filamentary NbTi Superconducting Wires", Adv. Cryo. Eng. Vol. 32, 739-745, eds. A.F. Clark and R.P. Reed, Plenum Press, 1986.

4. E. Gregory, "Conventional Superconducting Wire Technology", invited paper for publication in a special issue of the Proc. IEEE in superconductor applications scheduled to appear in the Fall of 1989, guest ed. T. Van Duzer.

5. E. Gregory, T.S. Kreilick and J. Wong, "Fine Filamentary Materials for Accelerator Dipoles and Quadrupoles", in Proc. ICFA Workshop, P. Dahl ed., Brookhaven National Laboratory Report BNL 52006 pp. 85-88, 1986.

6. A.K. Ghosh, W.B. Sampson, E. Gregory and T.S. Kreilick, "Anomalous Low Field Magnetization in Fine Filament NbTi Conductors", IEEE Trans., MAG-23, 2, 1724-1727, 1987.

7. T.S. Kreilick, E. Gregory, J. Wong, R.M. Scanlan, A.K. Ghosh, W.B. Sampson and E.W. Collings, "Reduction of Coupling in Fine Filament Cu NbTi Composites by the Addition of Manganese to the Matrix", <u>Adv. Cryo. Eng.</u>, A.F. Clark and R.P. Reed eds., Plenum Press, New York, 1988, vol. 34, pp. 895-900.

8. E. Gregory, T.S. Kreilick, J. Wong, E.W. Collings, K.R. Marken, Jr., R.M. Scanlan and C.E. Taylor, "A Conductor with Uncoupled 2.5 μm Diameter Filaments, designed for the Outer Cable of SSC Dipole Magnets", presented at the Applied Superconductivity Conference, San Fransisco, CA, Paper M4-4, August 21-25, 1988.

9. T.S. Kreilick, F.S. von Goeler, E. Gregory and J. Wong, "Enhanced Current Carrying Capacity Multifilamentary Superconductors Through the Use of Cold Worked NbTi alloy", presented at the Applied Superconductivity Conference, Paper MA-5, August 21-25, 1988.

10. E. Gregory, T.S. Kreilick, A.K. Ghosh and W.B. Sampson, "Importance of Spacing in the Development of High Current Densities in Multifilamentary Superconductors", <u>Cryogenics</u>, 24, 7, pp. 178-182, 1987.

11. T.S. Kreilick and E. Gregory, "Further Improvements in Current Density by the Reduction of Filament Spacing in Multifilamentary NbTi Superconductors", <u>Cryogenics</u>, 27, pp. 402-403, 1987.

12. A.K. Ghosh, W.B. Sampson, E. Gregory, T.S. Kreilick and J. Wong, "The Effect of Magnetic Impurities and Barriers on the Magnetization and Critical Current of Fine Filament NbTi Composites", IEEE Trans., MAG-24, Vol. 2, pp. 1145-1148, 1988.

13. R.M. Scanlan, J. Royet and C.E. Taylor, "Superconducting Materials for the SSC", <u>Adv. Cryo. Eng.</u>, Vol. 32, pp. 697-706, A.F. Clark and R.P. Reed, eds., Plenum Press, New York, 1986.

14. T.S. Kreilick, E. Gregory and J. Wong, "Geometric Considerations in the Design and Fabrication of Multifilamentary Superconducting Composite", IEEE Trans., MAG-23, Vol. 2, pp. 1344-1346, 1987.

15. M.I. Buckett and D.C. Larbalestier, "Precipitation at Low Strains in Nb 46.5 wt% Ti", IEEE Trans., MAG-23, Vol. 2, pp. 1638-1641.

16. E. Gregory, "Recent Advances in Commercial Multifilamentary NbTi Wires in the United States", Cryogenic Materials '88, eds. R.P. Reed, Z.S. Xing and E.W. Collings, Proc. ICMC Conf., Shenyang, China 1, pp. 361-371, 1988.

17. D.C. Larbalestier, A.W. West, W. Starch, W. Warnes, P. Lee, W.R. McDonald, P. O'Larey, K. Hemachalam, B. Zeitlin, R. Scanlan and C. Taylor, "High Critical Current Densities and Industrial Scale Composites Made From High Homogeneity Nb 46.5 wt% Ti", IEEE Trans., MAG-21, Vol. 2, pp. 269-272, 1985.

18. R.M. Scanlan, J. Royet and R. Hannaford, "Evaluation of Various Fabrication Techniques for the Fabrication of Fine Filament NbTi Superconductors", IEEE Trans., MAG-23, Vol. 2, pp. 1719-1723, 1987.

19. E. Gregory, "Recent Development in Multifilamentary NbTi Superconductors", Cryogenics 27, 6, 290-297, 1987.

20. S.O. Hong, E. Adam, E. Gregory, W. Maranick and F. Roemer, "Mechanical and Electrical Properties of High Current Density NbTi Conductors", IEEE Trans., MAG-19, Vol. 3, 758, 1983.

21. H.C. Kanithi, L.R. Motovidlo, G.M. Ozeryansky, D.W. Hazelton and B.A. Zeitlin, "Low Loss and High Current Nb3Sn Conductors Made by the Internal Tin Method", presented at the Applied Superconductivity Conference, San Francisco CA, Paper MJ-3, August, 1988.

22. R. Perin, "Progress on the Superconducting Magnets for the Large Hadron Collider", IEEE Trans., 24, 2, 734-740 (1988).

23. D. Ciazynski, J.L. Duchateau and B. Turek, "Conceptual Design of the Superconducting Cost System Cooled by Forced Flow He II for the Central Column of NET", IEEE Trans., 24, 2, 1440-1443(1988).

24. E. Gregory, T.S. Kreilick, F.S. von Goeler and J. Wong, "Preliminary Results on Properties of Ductile Superconducting Alloys for Operation to 10 Tesla and Above", ICEC 12, Southampton, U.K., eds. R.G. Scurlock and C.A. Bailey, Butterworth, Guilford, Surrey U.K., pp. 874-877, 1988.

25. D.C. Larbalestier, A.D. McInturff, P.J. Lee, J.C. McKinnell, R. Remsbottom, W. Stirch, P.M. O'Larey, W.K. McDonald, H.C. Kanithi and B.A. Zeitlin, "Development of High Current Density in NbTiTa Alloys", presented at the Applied Superconductivity Conference, San Francisco CA, Paper MA-1, August, 1988.

Cryogenics for Large Systems

Chairman:

R. Longsworth
APD Cryogenics

SSC REFRIGERATION SYSTEM DESIGN STUDIES

M. S. McAshan

SSC Central Design Group[*]
Lawrence Berkeley Laboratory
Berkeley, California 94720

Introduction

The SSC is an application of superconductivity on a scale vastly larger than any before. Its two collider storage rings are formed of some 10,000 superconducting magnets of many types, all of which must be maintained below the chosen operating temperature of 4.35 K under a wide variety of operating conditions. Thus an extensive cryogenic system is required that not only meets the magnets' operating requirements, but also provides for transient conditions such as cooldown, magnet quench, and periods for magnet maintenance. In addition, this system must function compatibly with scientific use of the SSC and run on a schedule consistent with the high beam availability that is necessary to the success of the SSC as a whole.

The ancestor of the SSC is, of course, the Tevatron,[1] and this heredity is expressed in the cryogenic system. The Tevatron has produced a body of successful superconducting magnet operating experience with beam and beam-loss heating. The Tevatron magnets are cooled by immersion in subcooled helium, the so-called single-phase flow, which is cooled in turn by heat exchange with boiling helium. This is the fundamental magnet cooling scheme adopted for the SSC.

The collider ring of the SSC is some fourteen times the circumference of the Tevatron and consists of two rings. Such a large-scale system, if it is to be constructed and brought into operation in a reasonable length of time, requires a parallel plan for tunnel construction and for magnet installation and commissioning. Thus a highly centralized cryogenic system is not appropriate. Instead, a system of units capable of independent operation, but interconnected for redundancy, more nearly matches the requirements. The choice of the number of units is a tradeoff between the economy of scale in the refrigeration plants and the costs of transporting the heat load. The present design of the SSC collider has ten cryogenic units or sectors, each just over 8 km in length, with an eleventh in the high energy booster ring of the injector.

---

[*] Operated by the Universities Research Association, Inc., for the U. S. Department of Energy.

Table I
Summary of SSC Cryogenic System Parameters

1.	Operating temperature, coils	4.35 K maximum
2.	Operating temperature, refrigerator	4.10 K nominal 4.00 K minimum

3. Total estimated refrigeration requirement (both rings and HEB) without redundancy and other allowances

4.15 K refrigeration		
	Synchrotron radiation	17.7 kW
	Other beam-related loss	4.1
	Static heat leak	10.6
	Other	4.4
Total		39.2 kW
4.15 K Liquefaction		144 g/s
20 K Refrigeration		50.2 kW
84 K Refrigeration		409 kW

4.	Total magnet mass	5.5×10^7 kg	
5.	Total liquid helium inventory	2.5×10^6 liters	
6.	Number of refrigerators	11	
7.	Number of air separation plants	2	

		Nominal	Full
8.	Refrigerator Capacity		
	4.15 K Refrigeration	4500 W	5620 W
	4.15 K Liquefaction	18 g/s	22.5 g/s
	20 K Refrigeration	6400 W	9600 W

9.	Air separation plant capacity	200 tons/day 2100 g/s

	Nominal	Installed
10. Electric power summary		
Refrigerator		
Compressor (4160 V, 90% pf)	3.0 MW	4.0 MW
Auxiliaries (440 V, 70% pf)	0.5 MW	0.5 MW
Air separation plant		
compressor (4160 V, 90% pf)	5.6 MW	
auxiliaries (440 V, 70% pf)	1.0 MW	
Site power		
11 refrigeration plants	38.5 MW	49.5 MW
2 air separation plants	6.6 MW	13.2 MW
Total	45.1 MW	62.7 MW

11. Storage capacity summary

Helium

Liquid helium	$11 \times 60,000$ gallons
Gas	$110 \times 60,000$ SCF (16 atm)
(liquid - 89% of inventory; gas - 11 % of inventory)	
Liquid nitrogen	$8 \times 20,000$ and $4 \times 55,000$ gal

(6 days' supply at 2000 g/s)

A system concept for the Super Collider cryogenics, together with an analysis of many of its operating conditions and requirements, is contained in the Conceptual Design Report (CDR) for the project, published in 1986.[2] Since then, systems and component modeling studies and development work on the dipole magnets and their cryostats have amplified the work of the CDR. At present a general review of the conceptual design of the collider is under way in preparation for the imminent detailed design. Here, I review the basic requirements for the cryogenic system of the SSC and describe some of its fundamental features, with the understanding that much is left to be accomplished in working out the complete design.

Layout of the SSC

The general layout of the SSC is illustrated in a number of places in these proceedings. The somewhat elongated shape of the collider tunnel is about 86 km in circumference and is made up of two kinds of regions: two arc regions and two much shorter interaction region (IR) clusters. The arcs are made up of a uniform magnetic lattice, the purpose of which is beam transport. The IR clusters contain the injection and abort utility regions and the interaction regions themselves, where the beams are made to collide. The two rings of magnets, installed in the tunnel, are in separate cryogenic envelopes and are positioned one above the other with a center-to-center distance of 70 cm. All of the cryogenic fluids flow in the magnet cryostat cross section; no parallel transfer lines are employed.

As mentioned above, the collider is divided into ten cryogenic sectors. Thus approximately every 8 km along the tunnel there is a service area, and at each one there is a helium refrigeration plant providing cooling to an 8-km length of both rings. The plants all have approximately the same capacity, and all have a complete set of auxiliaries and are capable of independent operation. The cryogenic systems of the sectors are, however, interconnected at the sector boundaries. It is possible, therefore, to pass refrigeration from sector to sector. This feature forms a basis for system redundancy.

It is also possible to send subcooled liquid nitrogen around the ring in the nitrogen shield of the cryostat. The concept for the nitrogen system includes air separation plants at the IR service areas and circulating pumps and subcoolers at each arc service area. Storage for liquid nitrogen is also located at each service area, and truck delivery to the stations is possible during times of system failure. The cold nitrogen gas from the subcooler is available for use in the refrigeration process and is exhausted to the atmosphere at room temperature.

The important system consideration here is the independence of the nitrogen supply from the helium refrigeration system, and final design decisions must include such considerations such as electric power rate structure, the local availability of liquid nitrogen, and the interest of vendors.

Basic Cooling Scheme

As with the Tevatron cryogenic system, the superconducting magnet coils are cooled by a single-phase helium flow, which is produced by circulation pumps at the refrigeration station and passes through the magnets in series. However, in the SSC system this stream is recooled periodically against saturated helium in heat exchangers separate from the magnets, rather than by means of the continuous heat exchange jacket that is built into the Tevatron magnets. The stream passes through a series string of magnets that is one-half sector, about 4 km in length, and is recooled 18 times. At the sector boundary this circulation is returned to the refrigeration station in a pipe running parallel to the magnet string and contained within the same cryogenic envelope. This returning stream supplies liquid to the recooler heat exchangers. The saturated gas from

Table II
Collider Cryogenic Heat Loads
(Redundancy and other allowances not included)

	Arc Sector	IR Sector	HEB Sector	Machine Total (kW)
4.15 K (watts)				
Synchrotron radiation	1980	1010	none	17.7
Other beam-related loss	300	700	300	4.1
Static heat leak	876	1411	730	10.6
Other	400	400	400	4.4
Ramping load				
Peak	(600)	(400)	(7000)	
Average	nil	nil	2400	2.4
Total	3556	3521	3830	39.2
4.15 K liquefaction (g/s)				
6.6 kA leads	6.32	13.35	5.07	82.3
1 kA leads		1.11		2.2
0.1 kA leads	3.00	2.40	4.50	33.3
Other	2.28	1.84	4.00	25.9
Total	11.6	18.7	13.6	143.8
20 K (Static)	4800	4920	2000	50.2
84 K				
Static Heat Leak	36300	37200	16000	381
Other	2500	2500	3000	28
Total	38800	39700	19000	409

evaporation in the recoolers is returned to the refrigerator in another pipe within the magnet cryostat. In each sector there are four magnet strings, two in each ring, connected to the refrigeration station.

There are two shields in the ring cryostat, one operating at 84 K, cooled by a circulation of subcooled nitrogen, and a second at 20 K, cooled by helium gas. There is only a single pipe in the cryostats of each ring for the nitrogen and for the 20 K helium. The coolant flows out from the refrigeration station in one ring and returns in the other. The 20 K shield pipe also acts as a service line for the magnet system. Valves connect this line with the single-phase flow path at 114-m intervals. These valves can be opened on magnet quench to vent the heated helium and to speed magnet recooling.

These various circulations can be unbalanced to produce flow across the boundaries between sectors. In this way liquid helium and nitrogen can be transferred from sector to sector, and refrigeration loads can be shifted from one sector to the next.

Cryogenic System Parameters

The three functions of the cryogenic system are refrigeration, convective heat transport, and cryogen inventory management. The basic parameters of the system are heat load and pressure drop budgets for each of the system's cooling loops. Several kinds of heat load and heat transport capacity estimates can be distinguished. The first tabulates the estimated heat loads of of the system components. The second estimates the extra capacity needed to provide for the operation of a sector under realistic and non-ideal conditions. The third contains information about extra capacity needed for inter-sector redundancy. It should also be noted that some unassigned capacity is needed for tradeoff and changes in the budget during the course of component design.

The following four tables list the fundamental parameters of the cryogenic system, together with heat-load budgets of several kinds and a pressure-drop budget for an arc sector.

Refrigerator Capacity Requirement

An outline of the refrigerator-load budget for an arc sector is shown in Table III. The most important point is that, of the approximately 900-watt load at 4.15 K, 575 watts is beam-related loss. This load varies with the ring operating conditions.

The unallocated load category allows for changes in the system during the course of design. Tradeoffs will be made when it is more economical to use more refrigeration rather than a more complicated component design. Allowance must also be made for changes in collider specifications in areas that affect the cryogenics, such as a change from distributed to lumped correctors, or an increase in the nominal ring current. It is a matter of judgment as to how much allowance should be included in the initial capacity of the refrigeration plant and how much should be left as part of a future upgrade.

The allowance for system performance is intended to cover system heat load due to failure of the system to meet its specifications, either because of manufacturing or installation error or because of component malfunction.

Reference has been made to the use of intersector redundancy in the SSC cryogenic system. Different kinds of loads can be shifted from sector to sector with varying ease. Liquid helium can easily be sent around the ring, and the supply is buffered by storage. The factor of 1.25 that has been applied represents a factor applied to the liquefaction capacity of the ring cryogenic system as a whole. The total load at 20 K in any sector can be shifted to the two adjacent sectors. Thus the factor of 1.5 applied in this case provides a 100 percent redundant 20 K system. On the other hand, only part of the 4 K

Table III
Outline of Heat Loads and Refrigeration Requirements
for an Arc Cryogenic Sector

	LHe (g/s)	4.15 K	20 K	84 K (watts)
Dipole				
Infrared		0.05	2.16	17.7
Supports		0.12	0.82	7.2
Connection and Instrumentation		0.15	0.32	2.1
(Synchrotron Radiation)		(2.34)		
Total static		0.32	3.30	27.0
4-km String (in one ring)				
216 Dipoles		69.1	713	5832
36 Quadrupoles		6.9	71	583
36 Beam Position Monitors		7.2	22	31
36 Spools				
Radiation and Support		12.8	128	1030
Piping and Valving		6.5	64	308
Vacuum Breaks		1.2	50	390
Isolation and End Boxes		26.0	96	400
6.6 kA Leads (4)	1.58	31.7	2	6
0.1 kA Leads (125)	0.75	21.3		
Safety Leads (72)	0.37	1.3		
Splices		29.0		
Subtotal	2.70	213.0	1146	8580
Synchrotron Radiation		505.0		
Pump Work		100.0		
Beam Microwave Loss		50.0		
Beam Gas Loss		25.0		
String Subtotal	2.70	893	1146	8580
Heat Loads for Individual Sector (includes both rings)				
4 Strings	10.8	3572	4584	34320
Storage and Distribution	0.8	20	253	2284
Purifier Operation				2500
Unallocated	1.0	678	830	6440
Quench and Cooldown Recovery	3.9			
Performance Allowance		1.5	230733	4456
Total Requirement	**18.0**	**4500**	**6400**	**50000**

Capacity of sector refrigerator including requirements for intrasector redundancy.

	LHe (g/s)	4.15 K	20 K	84 K (watts)
Redundancy Factor	1.25	1.25	1.50	
Adjusted Capacity Arc Refrigerator	22.50	5,620	9,600	
	(g/s)		(watts)	

refrigeration load can be shifted from sector to sector, so this kind of redundancy is of limited usefulness. The factor of 1.25 applied here reflects this limitation. It is necessary, therefore, to provide additional redundancy by duplication of equipment within the sector for the 4 K refrigeration.

Process Control in the SSC System

The concept of redundancy outlined above requires complex plant and system controls. Indeed, it is in the area of process control that the SSC requires new ground to be broken. The SSC cryogenic system as a whole—with its many interconnected refrigeration plants operating at several temperature levels and many operating modes, the dynamic nature of the loads, and the requirements for high availability and flexibility of operation—presents a process-control problem of unprecedented difficulty and complexity. In such situations, the problem can either be attacked head-on or mitigated to varying degrees. Complex control problems can be attacked through the use of sophisticated computer techniques. They can be mitigated by more conservative and costly hardware design. Computer capabilities are at present rapidly increasing, and their cost is rapidly declining. Thus a tradeoff develops between cost and control complexity. This situation is encountered throughout the design of the SSC. For example, a choice must be made between a more costly dipole with better magnetic field quality and a more complex and elaborate system of computer-controlled correction elements.

One of the most powerful tools available to assist in making control-system design decisions is system simulation. A system that can be convincingly simulated can be understood, controlled, operated, and improved in a systematic way. When the limit of the ability to simulate is reached, it is probably time to take a step away from the more complex control strategy. This criterion is a conservative one, but such a conservative stance is appropriate for the SSC.

Thus a strong effort in dynamic system simulation is needed to guide the design of the SSC cryogenic system. Work of this kind has begun, and software for dynamic simulation has been written. A description of this work and a report of results can be found elsewhere in these proceedings. There is much more to do as the increased resources of the new SSC Laboratory become available. Clearly a program of simulation must continue to be developed through the system and component design of the machine: the construction stage will require support of quality-control activities, the commissioning phase will require help in developing the operating procedures, and even the operating phase of the SSC will find simulation useful in providing the basis for a program of machine improvements.

Operating Schedule of the SSC Cryogenic System

The proposed operating schedule for the Super Collider, as nearly as it is now defined, calls for running 6000 hours per year, with an overall availability goal of 80 percent. Availability is defined to be the ratio of the actual useful time for an activity of the machine to the time scheduled for that activity. System availability studies[3] set a goal for the cryogenic system alone of 98 percent, which is consistent with the experience at the central helium liquefier of the Tevatron.[4]

The CDR describes an example operating schedule for the SSC. In this example the 6000 hours is broken down into two-week base operating cycles made up of ten days of physics operation, two days of machine studies, and two days of so-called maintenance and development (M&D) time. In this base cycle the availability of the cryogenic system for the ten-day operating period must be 98 percent. For the two days of studies, typically only one ring will be used, or the machine may run at reduced current or energy. The availability goal is still 98 percent, but the operating conditions are not the most severe.

Table IV
SSC Cryogenic System Loop Flow Characteristics

Nominal or Average Operating Conditions		80 K Shield	20 K Shield	Magnet String	Liquid Return	Gas Return
Pressure	(atm)	5	2.5	4	3	0.90
Temperature	(K)	84	20	4.25	4.425	4.3
Volume	(cc/g)	1.286	163.4	7.255	7.609	73.07
Flow Rate	(g/s)	750	100	100	85.6	26.8
Nominal Dimensions						
Line Diameter	(in)	3	2.5	4×1.25	1.77	3.334*
System Length	(km)	8.249	8.249	4.125	4.125	4.125
Pressure Drop Budget						
(ΔP in atm)						
Dipoles		0.801	0.598	0.122	0.146	0.0070
Quadrupoles		0.038	0.030	0.007	0.007	0.0008
Spools		0.082	0.080	0.010	0.015	0.0020
Recooler				0.054		
U - Tubes		0.076	0.059	0.009	0.007	0.0026
Unallocated		0.103	0.083	0.048	0.025	0.0026
Total		**1.100**	**0.850**	**0.250**	**0.200**	**0.015**

*Reducing (eccentric) to 2.709 in id in the interconnect region controlled correction elements.

During the two M&D days the collider will not run. At least some of this period, therefore, is available for refrigeration plant maintenance, such as deriming. It should be kept in mind that when no beam is in the rings the heat load is greatly reduced, so during this time some fraction of the refrigeration machinery can be taken off-line if necessary.

The annual cycle consists of 21 of these base cycles, yielding 210 days of physics operation and 42 days of studies per year. There will also be one long annual shutdown period of a month or more, when the magnet system will be allowed to warm up to nitrogen temperature. During this time the helium inventory will be stored, most of the helium refrigeration equipment will be shut down, and only the liquid nitrogen system will be kept operating.

It is to be understood that the actual operating cycle adopted for the SSC will be adjusted to the demands of the experimental program. In particular, it is probable that the base cycles will often be longer and fewer in number. However, this example is representative of the kinds of operating time and their proportions.

System Operating Modes

A number of system operating conditions can be identified that must be provided for in the design of the collider cryogenic system and refrigeration plants. Most of these are discussed in the CDR, but some are new or deserve new emphasis. These are briefly reviewed below:

Steady or Quasi-Steady Condition:

- Accumulation of Liquid Inventory: When the ring is not at helium temperature and inventory is in storage, there is a need to liquify boil-off helium. During start-up it will be necessary to be able to liquify helium delivered as gas or flashed from liquid delivery.

- Cooldown of Magnet System from 300 K to 80 K: This condition applies to the cooldown of entire strings. (The cooldown of sections is included in the category of magnet repair operations.) Clearly, every string will be cooled from 300 K at least once. There will be several cooldowns of strings in the ring each year.

- Cooldown of Magnet System from 80 K to 4.2 K: The process of cooling a string to 80 K is assisted by liquid nitrogen. Cooling below 80 K in a reasonable amount of time requires that turboexpander capacity producing 50 kW with input conditions at 80 K be available in each sector plant.

- Liquid Helium Inventory Handling: The inventory of each magnet string, about 40,000 liters of liquid, must be moved to or from storage in a time consistent with other operations, and the displaced gas must be dealt with. Warming the magnet system of the ring to 20 K and recooling it may be necessary to control the beam line vacuum.

- Magnet Conditioning: It is suggested that the strings of magnets in the SSC could be brought into operation by conditioning rather than training. This conditioning involves operating the string of magnets at a reduced temperature and cycling up the current to a point above the nominal operating current of the ring. The conditioning temperature that has been studied for this is 1.9 K, but it is hoped that 2.5 K will be satisfactory. A goal for the refrigeration plant, therefore, is to be able—with no additional machinery—to operate at 2.5 K under the heat load of one string. The beam-related losses need not be included in this, so the value from Table III is approximately 200 watts.

Table V
Fundamental Operating points for an Arc Refrigerator
(See Figure I)

Point	Description	T (K)	P (atm)	H (J/g)	Flow (g/s)
1	Liquid Helium Return	4.280	2.80	10.41	194.7
2	Liquid Helium Supply	4.150	3.50	10.20	408
3	Cold Gas Return	4.114	0.90	30.16	200
4	4 K Refrigeration Supply	4.500	3.50	11.56	242
5	4 K Refrigeration Return	4.114	0.90	30.16	242
6	20 K Shield Return	22.7	2.1	131.61	200
7	20 K Shield Supply	17.0	3.0	100.60	200
8	LN Shield Return	90.0	4.6	55.59	1900
9	LN Shield Supply	79.0	6.0	32.67	1678
10	Liquid Helium Make-up	4.40	1.17	10.66	13.3
11	Warm Gas Return	300.0	1.6	1573	13.3
12	Nitrogen Gas	78.5	1.14	228.4	222

- Nominal Operating Conditions (100 percent): The heat loads under normal operation are given in Table III.

- Operation Under Reduced Load (as low as 25 percent): As indicated in Table III, the heat leak is only about 25 percent of the nominal capacity of the 4.35 K refrigerator. Machine operation with reduced beam current or energy will have reduced refrigeration requirements.

- Failure Mode Operation (100 to 125 percent): Under conditions of equipment failure or magnet repair operation, load will be shifted to adjacent sectors or, in some cases, capacity will be made up by means of liquid injection.

Transient Operating Conditions

- Ring Filling Sequence: In normal physics operation, once or twice a day the rings must be refilled with protons. This is done by dumping the remaining beam, ramping the ring current down to the value for injection (1/20 the operating value), stacking each ring with new bunches of protons (which requires about 2 hours), and then accelerating back to 20 TeV (in 1000 seconds). During this operation, in which the heat load on the 4.35 K parts of the ring varies by a factor of four, temperature and temperature–current history control of the ring magnets is very important.

- Quench Recovery: A magnet quench results in dumping of the helium inventory of a half-cell into the 20 K shield piping and the warming of these magnets to an average of about 30 K. The refrigeration plant must be able to return the ring as quickly as is practical to the condition for filling. It may be necessary to adjust the temperature–current history of all of the magnets in the ring to accomplish this.

Refrigeration Plant Operating Requirements

Figure I shows a sector plant block diagram. Pictured are the four strings of magnets at the bottom and the refrigeration plant with one or more compressor trains at the top. Shown also are blocks for the storage and the circulation of the cryogens. Highlighted among the flows interconnecting the blocks are a set of illustrative process points, listed in Table V.

The first four points are attached to the 4 K cooling loops of the magnet system. Liquid helium is supplied (point 2) at the rate of 408 g/s and divided equally between the four strings. This flow passes down the 4-km string, is recooled in 18 places, and, at the end of the strings, passes back toward the refrigeration plant in the liquid return line (point 1). Feed streams of 50 g/s for each string for the recoolers are withdrawn from the liquid return flow and return to the refrigeration plant in the gas return line (point 3). Refrigeration is supplied by the flows (points 4 and 5) from the refrigeration plant. The helium circulation block obviously must contain a pump for circulation, a pump to return liquid from storage, and a subcooler. The subcooler serves to remove the pump work and also to separate the refrigerator operating conditions from the conditions of the helium flowing through the magnets. This is advantageous in providing flexibility for control and turn down of the refrigerator.

The large inventory of helium in the magnet string makes the time constants of this part of the system long. The thermal time constant mentioned in the CDR is about 2000 seconds. This is probably good from the point of view of control and of providing time for action in case of machine failure. The pressure time constant of the string is

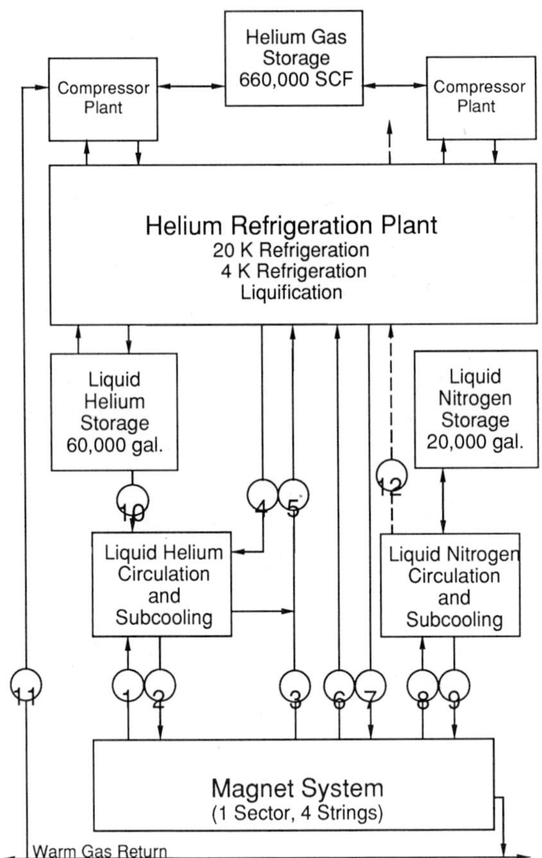

Figure I
Sector Refrigeration Plant Block Diagram

also long, which is not so convenient. The compressibility of helium is large: The adiabatic compressibility is 1.5 percent per atm. The inventory of a string of magnets is about 4×10^6 g and so, at the flow rate of 100 g/s, the pressure changes at a rate of 600 s/atm if the return flow is cut off. Note that in this condition work is being done on the string at a rate of 260 watts, a significant fraction of the nominal refrigerator load.

This control situation is the result of the large inventory of the magnet system compared with the flows associated with the refrigeration plant. There is an important difference between the SSC cryogenic system and other large systems that have been built, such as the Tevatron and the HERA ring, in which this ratio is smaller and the time constants shorter.

Points 6 and 7 are attached to the 20 K shield loop. The thermal time constant of this loop is also long, and its exact temperature is not particularly critical. Thus it is possible to turn off the 20 K refrigeration for a matter of hours with no important consequences for the operation of the magnet system. However, in order to take advantage of this flexibility, it is necessary to have some independence of the operating conditions of the turbine providing the 20 K refrigeration and the operating temperature of the shield. This is also needed for quench recovery. A large amount of cold fluid is vented into the shield system during quench; later, warm fluid passes into the shield during magnet recooling. The shield provides a way of storing this fluid and dealing with it over relatively long times. The 20 K refrigeration plant and its controls must allow this storage to be used.

The liquid nitrogen circulation in the ring connects with the sector station circulation and storage system at points 8, 9, and 12. The system here is operating at 42.5 kW.

Points 10 and 11 describe the sector liquefication load. Fluid is drawn from the single-phase flow, passes out through the many leads along the string, and returns to the refrigeration plant by way of the warm gas line that runs around the ring. As can be seen from Table III, only about half of this flows in the high-current leads and so is adjusted with ring current. Therefore the liquefication load of the sector is relatively constant and independent of ring operating conditions.

Major Issues in the Design of a Refrigeration Plant for the SSC

Keeping in mind the long list of operating conditions and the general problems presented by the collider, it is possible to identify several major issues that must be specifically addressed in any plant design for the SSC:

- High system availability: The cryogenic system must be doing what it is required to do when it needs to be done. Thus, to achieve high availability, not only reliability, but controlability is needed. The refrigeration plants for the SSC, unlike most process systems, do not always run at steady state. A variety of transient conditions must be routinely dealt with.

- Capacity adjustment: The 4.35 K load varies from about 0.25 to 1.25 of nominal capacity. Part of this can be absorbed by ballasting, but for economical operation of the collider, the refrigeration plant must easily and efficiently adjust to the load conditions. It is important to recognize that, in general, the liquefication load, the 4.35 K load, and the 20 K loads will need independent adjustment.

- Single-phase flow capacity and inventory handling capability: The single phase flow of 400 g/s in each sector must be adjustable independent of the refrigerator operating conditions. This flexibility is needed for a number of reasons: The large inventory of helium must be able to be moved in and out of storage conveniently. In addition, quench recovery, and possibly the need for high temperature stability at injection, make the ability to control the flow necessary.

- Quench tolerance: The refrigeration plant and system must tolerate the quench of single and multiple half-cells, and it must support rapid quench recovery.

- Contamination tolerance: There are 16 km of magnet coils and laminations connected to each plant. It is unreasonable to expect to clean up such a system to a high degree of purity. In addition, the extended and complex nature of the operations makes contamination sensitivity very undesirable.

- Expandability and flexibility: The collider is a scientific instrument, and the outcome of its operation cannot be predicted. Some very important phenomenon may be discovered at 15 TeV and, while it is being studied, a great deal of operating time will be without the synchrotron radiation heat load. On the other hand, this phenomenon may be found at 20 TeV, but at a very low rate. In this case some increase in the refrigeration capacity, in order to achieve higher ring currents, will be required.

Conclusion

In the preceding sections an overview of the SSC cryogenic system has been given, and a discussion of some of the many issues involved in specifying the system controls and refrigeration plant has been attempted. Significant progress in understanding these issues has been made since the publication of the CDR, primarily by investigating the system with static and dynamic process simulations. This work has led far enough so that system control strategies are being investigated[4] and a very preliminary set of refrigeration plant criteria have been derived.

Considerably more study and modeling of the system's behavior is needed before the job is finished. Under way at the moment is further study of inventory handling scenarios. The questions here are what controls and what capabilities are needed for the efficient moving of liquid into and out of storage. It is also important to understand exactly how inventory is to be moved around the ring.

Of the greatest importance for the design of the system as a whole is to get a clear understanding of temperature stability requirements. This can only be done by dynamic modeling of the magnet strings. The temperature as a function of time in each half-cell of a string during the ring filling operation must be calculated. At the moment it does not seem that dynamic refrigerator modeling is necessary for the SSC, but this may change.

As has already been said, the only practical way to understand the cryogenic system behavior and to extract from a complex set of interactions a clear and consistent set of requirements is through quantitative study using specific models. In this case, the final form of the models is the clearest expression of the nature of the system. This program is under way at the SSC Laboratory.

REFERENCES

1. H. T. Edwards, "The Tevatron Energy Doubler: A Superconducting Accelerator." *Ann. Rev. Nucl. Sci.* **35** (1985): 605-659

2. J. D. Jackson, ed. *Conceptual Design of the Superconducting Super Collider*, SSC report SSC - SR - 2020, 1986. (Available upon request.)

3. Ibid., Section 3.4.3, pp 87–90.

4. E. Shrauner, "Availability Study for SSC Cryogenic System: Model IV and IV, 2." SSC Report SSC-N-60, 1985.

DYNAMIC MODELING AND SIMULATION OF THE SUPERCONDUCTING
SUPER COLLIDER CRYOGENIC HELIUM SYSTEM

D. G. Hartzog, V. G. Fox, P. M. Mathias, and D. Nahmias
Air Products and Chemicals, Inc.
Allentown, Pennsylvania

M. McAshan and R. Carcagno
SSC Central Design Group
Lawrence Berkeley Laboratory
Berkeley, California

ABSTRACT

To study the operation of the Superconducting Super Collider (SSC)
cryogenic system during transient operating conditions, we have developed
and programmed in FORTRAN, a time-dependent, nonlinear, homogeneous,
lumped-parameter simulation model of the SSC cryogenic system. This
dynamic simulator has a modular structure so that process flowsheet
modifications can be easily accommodated with minimal recoding. It uses
the LSODES integration package to advance the solution in time. For
helium properties it uses Air Products' implementation of the standard
thermodynamic model developed by the NBS. Two additional simplified
helium thermodynamic models developed by Air Products are available as
options to reduce computation time. To facilitate the interpretation of
output, we have linked the simulator to the Speakeasy conversational
language.

We present a flowsheet of the process simulated, and the material
and energy balances used in the engineering models. We then show
simulation results for three transient operating scenarios: startup of
the refrigeration system from standby to full load; the loss of 4K
refrigeration caused by the tripping of one of two parallel compressors
in a sector; and a full-field quench of a single magnet half-cell. We
discuss the response of the fluid within the cryogenic circuits during
these scenarios.

INTRODUCTION

The Superconducting Super Collider (SSC) will contain about 10,000
superconducting magnets which must be maintained below the chosen
operating temperature of 4.35K over a wide variety of operating
conditions. This requirement will be accomplished by means of a
cryogenic system having eleven helium refrigeration plants, ten spaced
equally around the main ring and an eleventh to cool the final booster
ring. Each plant will be capable of independent operation but will be
interconnected for redundancy through cryogenic components within the
accelerator rings.

While individual refrigeration plant capacities are well within
current industry practice, process control of the SSC cryogenic system
represents a significant challenge in process control. It must not only
reliably produce and deliver the cooling required for normal operation of
the SSC, but must also handle transient operating conditions such as
cooldown, beam acceleration, and magnet quench. To meet this challenge,
the SSC Central Design group together with Air Products and Chemicals,
Inc. has undertaken two computer simulation projects of the SSC cryogenic
system. In the first, static process simulation was used to investigate
refrigeration plant performance for a variety of steady-state and quasi
steady-state operating modes (1). This work was followed by a program of
dynamic modeling and simulation to study system behavior during transient
operating modes, which is the subject of this paper.

This paper is organized into three sections. First, we describe SSC
cryogenic system. Then we discuss the development of a FORTRAN
time-dependent, nonlinear, lumped-parameter simulation model of the
cryogenic system. To allow running the dynamic simulator on a DEC VAX
8700 machine in a reasonable time frame we had to limit the scope of the
simulator to a small fraction of the total cryogenic system and had to
simplify the thermodynamic and individual equipment models used,
consistent with obtaining reasonably accurate results for the transient
operating conditions simulated. Finally, we present results of computer
simulations of three key transient operating scenarios: 1) loading a new
beam into the accelerator rings, 2) partial loss of refrigeration
capacity due to a helium compressor failure in one refrigeration plant,
and 3) full-field quench of a single half-cell.

PROCESS DESCRIPTION

The SSC cryogenic system consists of eleven large helium
refrigeration plants, plus numerous cryogenic components within the
proton accelerator rings (2). Ten helium refrigeration plants are
equally spaced around the 53-mile circumference ring, each servicing an
approximately 8 km ring sector. The eleventh refrigeration plant
services the high-energy booster ring. Figure 1 is a process flow
diagram of a helium refrigeration plant. Each refrigeration plant
consists of three parallel units: a helium liquefier to reliquefy warm
helium gas from the superconducting magnet leads, a 4K helium
refrigerator to cool the magnets, and a 20K helium refrigerator to cool a
20K vapor-cooled shield within the multilayer insulation of the magnet
cryostats. The three units are integrated with respect to helium liquid
and gaseous storage, liquid helium pumping, and, to a limited extent,
with respect to heat exchange and cold gas expansion.

Figure 2 illustrates a portion of a sector serviced by a helium
refrigeration plant. Single-phase supercritical helium at 4.15K and 4
atmospheres from the 4K refrigerator is pumped through four 4-km strings
of magnets, one upstream and one downstream of the refrigeration plant
for each of the two accelerator rings. The helium flows in series
through the magnets, which are grouped in half-cells consisting of six
dipole magnets and one quadrupole magnet. Each magnet is contained
within its own cryostat. After each set of two half-cells, the helium is
recooled in a recooler heat exchanger against boiling helium in order to
maintain the magnets below 4.35K. Each magnet string consists of 18 sets
of two half-cells. At the end of the string, the helium is returned
through a circuit running through each of the magnet cryostats. Along
the 4K liquid return circuit, small flows are withdrawn and expanded into
the recoolers. The low pressure saturated gas from the recoolers returns
through a 4K vapor return circuit, which also runs through the magnet
cryostats and then back to the 4K refrigerator. Each cryostat contains

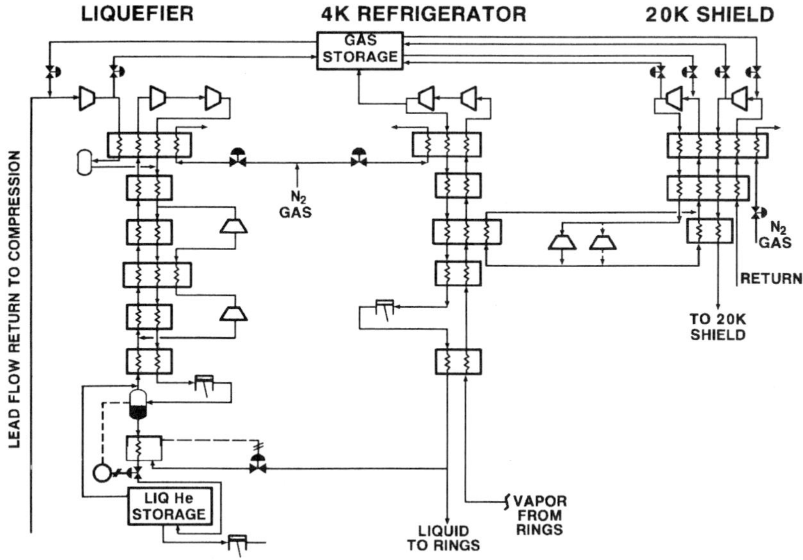

Fig. 1 Helium Refrigeration Plant Flow Diagram

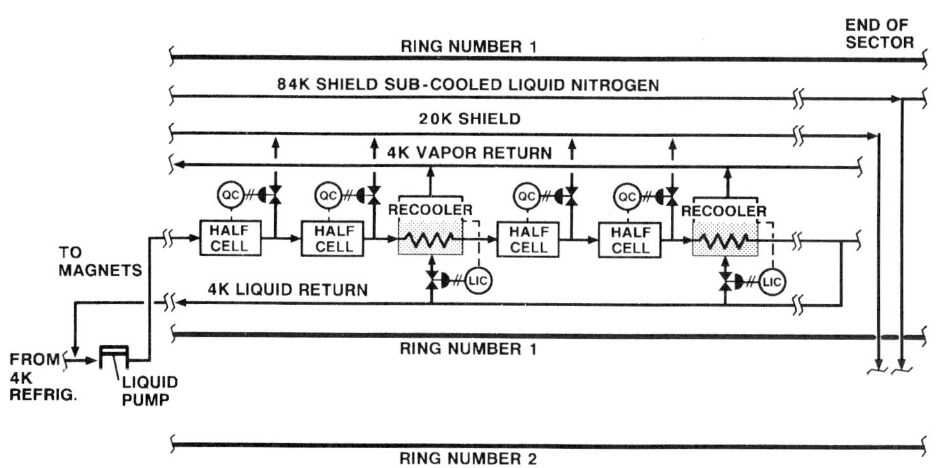

Fig. 2 Simplified Flow Diagram of Portion of an Accelerator Ring

vapor-cooled shields at 20K (helium) and 84K (nitrogen) with multilayer
insulation. The helium gas which cools the 20K shield circuit passes
from the 20K refrigerator into one ring, then to the second ring, and
then back to the 20K refrigerator.

At the end of each half-cell is a quench relief valve which connects
the 4K supercritical helium circuit with the 20K shield circuit. During
a quench of a magnet half-cell, excess pressure is let down into the 20K
shield circuit.

Each sector refrigeration plant operates independently of its neighboring plants. The cryogenic circuits of each sector are, however, interconnected at the sector boundaries by means of piping and valving with the corresponding circuits in the adjacent sectors. The refrigeration plants are sized at 150 percent of design so that, in the event of loss of refrigeration capacity in one sector due to equipment failure, redundancy is provided by shifting surplus helium refrigerant from one sector to another through the ring cryogenic circuits.

DEVELOPMENT OF THE DYNAMIC SIMULATOR

Hardware and Software Selection

The dynamic simulator was developed for both interactive and batch operation on a DEC VAX 8700 minicomputer, a computer conveniently available to the SSC Central Design Group.

Four basic approaches to dynamic simulation software were evaluated:

1. A process simulator package, specially oriented towards dynamic simulation, e.g., SPEEDUP

2. A real-time, interactive flowsheet oriented package, e.g., GEPURS

3. Augmented programming languages, e.g., ACSL

4. FORTRAN

Table 1 lists the advantages and disadvantages of each approach. In summary, both the real-time interactive approach, such as offered by GEPURS (General Purpose Real-Time Simulator) (3), and the augmented programming language, such as ACSL (Advanced Continuous Simulation Language) (4), had severe efficiency disadvantages compared with FORTRAN for this simulation requirement. Of greater promise was use of a process simulation package, of which SPEEDUP (5) is the only commercially viable example. This equation-based approach, while appearing modular to the user, integrates all equations simultaneously. Such a package has the advantages of built-in numerical analysis, quick model preparation if the unit modules are available, and a good user interface. However, at the time this project was conceived and undertaken, SPEEDUP had not yet demonstrated reliable capability, especially with complex thermodynamic calculations as is the case with helium. Therefore, FORTRAN was selected.

All calculations were done in double precision because of the sensitivity of the process to small changes in temperature in the thermodynamic region of interest. To keep development time down, we used a fixed-flowsheet approach but allowed for relatively easy modification of the code by using a parallel- modular structure. By parallel we mean that all the state equations are solved simultaneously. By modular we mean that the code which represents each piece of equipment is assembled into one subroutine. The system of ordinary differential equations which comprise the model are "stiff" so we used the backward difference algorithm in the 'LSODES' routine developed at Lawrence Livermore Lab (6). The Jacobian was evaluated numerically. Most of the I/O (input/output) is via files to keep the tedious process of data entry to a minimum. For post-processing of results, especially for graphics, we have created output files which can be read by the commercially available Speakeasy package (7).

Table 1
Software Development Approaches

<u>Process Engineering Packages - SPEEDUP</u>

- Advantages - specially oriented towards process simulation
  - good user interface
  - fairly quick model preparation
  - numerical analysis built-in
  - can build libraries of models easily

- Disadvantages - lack of demonstrated capability
  - problem too large (2000 variable maximum)

<u>Real-Time, Interactive Flowsheet Oriented Packages - GEPURS</u>

- Advantages - flexible, very easy to use
  - great user interface (menus, graphics, templates)
  - interact with model via several terminals
  - fast implementation if using exiting models
  - build model libraries in FORTRAN off-line
  - can be used as training simulator for operators

- Disadvantages - requires dedicated computer
  - must provide numerical algorithms
  - sequential solution of equations unstable for
      stiff system

<u>Augmented Programming Languages - ACSL</u>

- Advantages - flexible, FORTRAN-like language

- Disadvantages - time consuming model preparation
  - crude or no handling of algebraic equations
  - inefficient for large problems ($>$100 variables)

<u>Programming Languages - FORTRAN</u>

- Advantages - flexible
  - efficient
  - portable code

- Disadvantages - time consuming model preparation, usually from
      scratch

<u>Scope Limitations and Lumped-Parameter Assumptions</u>

Since we were implementing the code on a VAX 8700 minicomputer, not
a supercomputer, we had to keep the execution time within reason, on the
order of an hour or two. Thus we limited the size of the problem to two
adjacent sectors. This scope is consistent, however, with the transient
operating conditions to be simulated. We were able to ignore the 84K
liquid nitrogen- cooled shield and the helium liquefier since they are
coupled only very weakly with the main helium refrigeration circuit. We
also assumed that the helium is always in thermodynamic equilibrium, a
reasonable assumption for the relatively long time constants inherent in
the SSC cryogenic system.

A basic assumption, crucial to limiting the problem to manageable
proportions by eliminating the need to employ partial differential

equations, is that the "lumped-parameter" approach is adequate. A lump
or "node" represents the capacitance or volume of an individual process
vessel or group of vessels plus interconnecting piping. It can also
represent just sections of piping. Within a lump or node spatial
variations of density, pressure, and temperature are assumed to be
negligible. Therefore, homogeneous lumps can be modeled with first-order
nonlinear ordinary differential equations. The lumped parameter approach
is commonly used for process simulation studies, and appears to be
adequate for the SSC dynamic simulations undertaken for this project.

The specific lumping assumptions made are illustrated in Figure 3
and listed on Table 2. Each magnet half-cell was considered one node.
Therefore the dynamic simulator does not go down to the level of
individual magnets. We modeled only the helium fluid behavior in the
half cells; the heat loads resulting from synchrotron radiation, magnet
quench, and heat leak from ambient must be input to the program. The
boiling helium in each recooler is considered to be a node. However, the
helium flowing through the heat exchanger in the recooler is not modeled
as a node since holdup is quite small. Each 4K vapor return header is
modeled as 18 nodes corresponding to the number of recoolers in a
string. The vapor return headers are interconnected by a junction node
so that flow may be passed directly among them. Likewise a junction node
is used to distribute the 4K supercritical helium to each string, and a
suction node is used to collect the 4K liquid helium feed to the pump.
The 4K liquid return header in each string is treated as a series of 18
nodes, corresponding to the number of recoolers in the string.
Similarly, the 20K shield in each string is treated as a series of 18
nodes corresponding to the 18 pairs of half-cells.

All of the heat exchangers in the 4K and 20K refrigerators are
ignored. Their capacities (volumes) are lumped into respective
compressor suction and discharge nodes, which are assumed to be
isothermal.

<u>Other Modeling Details and Assumptions</u>

Liquid from storage is always assumed to be available to meet the
demand of the liquefier transfer pump, whose speed is manipulated by a

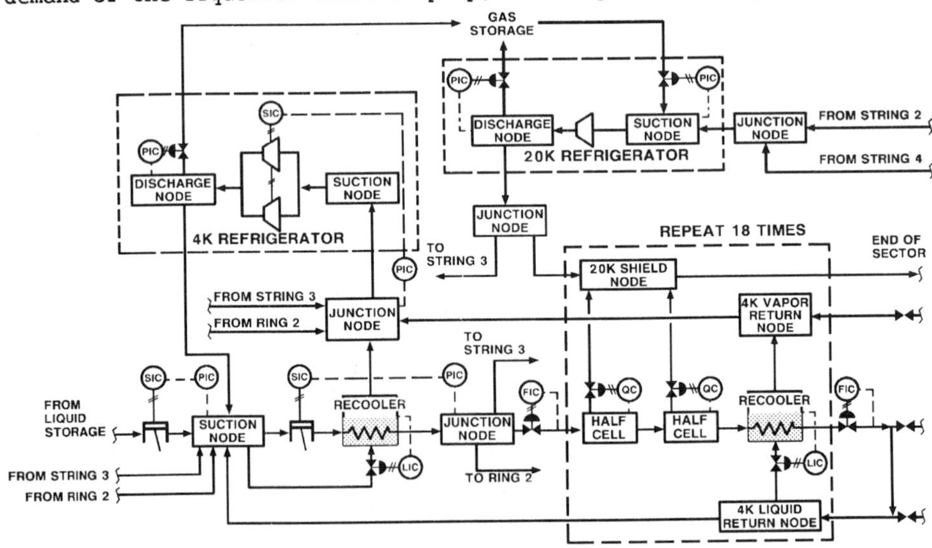

Fig. 3 SSC Helium Network Simulation Model Process Flow and
Instrumentation Diagrams for One Quarter Sector

Table 2
SSC Equipment Lumping Assumptions

Equipment Item	Assumptions
• Magnets	One Lump per Half-Cell Decoupled Electrical and Fluid Phenomena
• Vapor Return	One Lump per Two Half-Cell/Recooler
• Liquid Return	One Lump per Two Half-Cell/Recooler
• 20K Shield	One Lump per Two Half-Cell/Recooler
• Recoolers	One Lump on Shell-Side Instantaneous on Tube-Side
• Piping Junction	One Lump
• Refrigerators	Compressor Suction Side – One Lump Compressor Discharge Side – One Lump Isothermal

pressure controller on the suction node of the primary coolant pump. The liquefier transfer pump is allowed to run in reverse in order to remove excess liquid as dictated by the pressure controller. Any excess flow through the 4K refrigerator compressors may be shunted to gas storage via a pressure controller on the discharge. The inlet junction pressure of the 4K refrigerator compressor is controlled by manipulating the speed of the compressor. Interconnections between gas storage and the suction/discharge of the 20K refrigerator compressor are maintained by pressure controllers.

Rigorous mass and energy balances were written for all nodes. The mass balance is

$$d\rho/dt = \left(\sum_{I=1}^{N} w_I - \sum_{J=N+1}^{M} W_J \right)/V$$

Assuming that no work is done on the surroundings and that there are negligible changes in kinetic and potential energies, the energy balance is:

$$dU/dt = \left(\sum_{I=1}^{N} W_I(H_I - U) - \sum_{J=N+1}^{M} W_J(H - U) + Q \right)/\rho V$$

Flow between nodes was assumed to be proportional to the square root of the product of the pressure difference between the upstream and downstream nodes and the density of the upstream node. Valve trims were assumed to be linear.

The total heat input to all nodes, except the recoolers, consists of a fixed static heat leak plus, for the magnets only, a duty related to the beam synchrotron radiation. The radiation duty is either a constant or a specified function of time depending upon the operating scenario simulated. Using this approach is quite reasonable and keeps the model more tractable than it would be if the magnetic field interactions were incorporated in a rigorous sense.

The heat duty in the recoolers is assumed to be proportional to the log-mean temperature difference,

$$Q = U{\circ}A(\Delta T)_{1m}$$

307

Assuming that the thermal resistance of the heat exchanger wall is negligible, the overall heat transfer coefficient is given by

$$U_o = 1/(1/h_s + 1/h_t)$$

The tubeside coefficient is assumed to be proportional to the 0.8 power of the flow rate while the shellside coefficient is assumed to be proportional to the log-mean temperature difference to the 1.5 power (8).

Since the tubeside holdup is negligible, the tubeside outlet temperature was computed iteratively by Newton's method from the steady-state energy balance,

$$Q = U_o A (\Delta T)_{lm} = W_t C_p (T_{ti} - T_{to})$$

The heat transfer area and the liquid level in the recooler are assumed to be piece-wise linear functions of the shellside liquid inventory.

All pumps and compressors are reciprocating machines and are modeled as one stage, ideal positive displacement devices,

$$W = \rho S_N F_{max}$$

The enthalpy of the fluid exiting a pump is computed using a fixed overall adiabatic efficiency. Based on the simplified model of the refrigerators, the enthalpy of the fluid exiting a compressor is not required. The speed of a machine is assumed to lag its controller output signal in the first order sense,

$$dS_N/dt = (Y - X)/\tau$$

All controllers are proportional-integral. The only significant lag in a control loop is that of the valve positioning, which is assumed to lag the controller output signal in the first order sense.

Helium Thermodynamic System

A thermophysical-property software system has been developed to provide thermodynamic properties of helium for the dynamic simulation. The state of the fluid may be specified in a variety of ways and any of three models may be chosen for the prediction of helium thermodynamic properties.

The state specifications of the fluid that we provide cover most of the ways in which engineering calculations are usually done; for example, pressure-enthalpy, temperature-pressure, temperature-vapor fraction, pressure-vapor fraction and pressure-entropy. In addition, we have developed a density-internal energy specification which is especially useful for dynamic simulation since density and internal energy are the natural independent variables of dynamic simulation. Further, since this calculation is done on the order of one hundred thousand times for each scenario simulated, particular attention was developed to computational efficiency.

The accepted international standard for the thermodynamic properties of helium is the model developed by R. D. McCarty of NBS (9). While McCarty's model is extremely accurate, it may be too computationally intensive for dynamic simulation. Thus, two simplified but reasonably

accurate models have been developed and incorporated into the
thermodynamic system.

McCarty's model is actually a set of models, each optimized for a
particular region, with region-averaging to eliminate discontinuities at
the boundaries between regions (9), as shown in Figure 4. The first
simplified model uses one of the "local models" (for the subcritical
vapor and saturated liquid) for the entire region. This simplification
allows considerable savings in computer time with only a small sacrifice
in accuracy.

The second simplified model is based upon the Martin equation of
state (10), which, in turn, is a variation of the simple van der Waals
equation. According to the Martin equation of state, the pressure of a
pure fluid is given as a function of temperature and molar volume as
follows:

$$P = RT/(v-b) - a/(v+c)^2$$

In order to obtain an accurate description of helium, we have
allowed the parameters "a" and "b" to be temperature dependent and have
used the volume-translation idea proposed by Peneloux and Rauzy (11).

Figures 5 and 6 present comparisons between the McCarty NBS model
and the modified Martin equation for the pressure-enthalpy and
density-pressure diagrams of helium. We have found the simple model to
be an adequate representation of helium for dynamic simulation.

The density-internal energy flash becomes easy if the model is based
upon an equation of state. (This is the case for the simplified models,
but the region-averaging causes some difficulties for the McCarty NBS
model.) Here the flash is simply done, since density is an independent
variable of equations of state, by iterating on pressure at specified
density until the specified internal energy is obtained. However, the
solution must be checked to ensure that it is not in the unstable
(two-phase) region. For two-phase calculations, iteration is done to

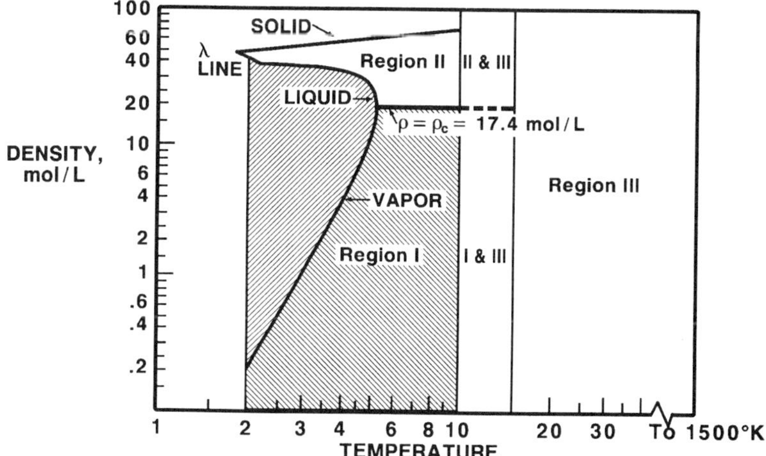

Fig. 4   Schematic ρ-T Diagram for Helium Showing Regions Represented
by Different Equations

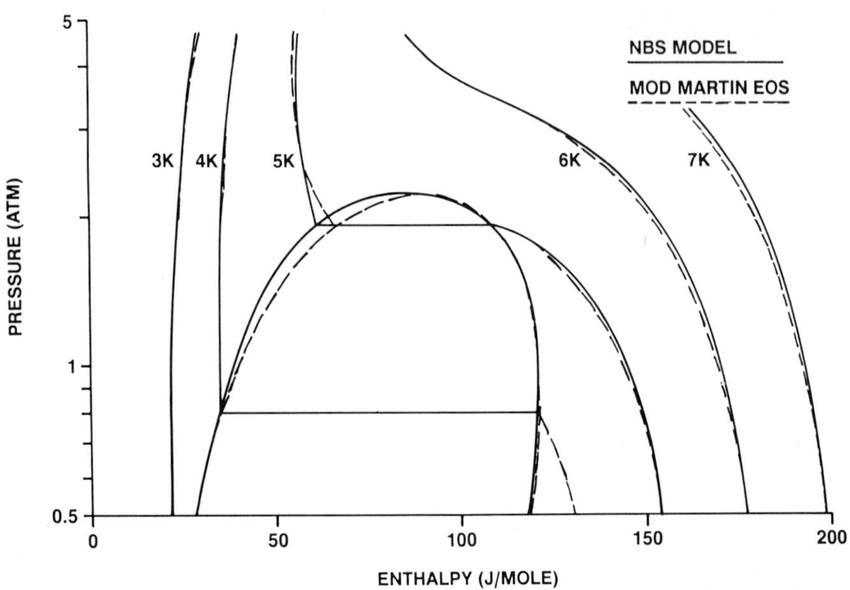

Fig. 5 Pressure-Enthalpy Diagram of Helium: Comparison Between NBS and
Modified Martin EOS

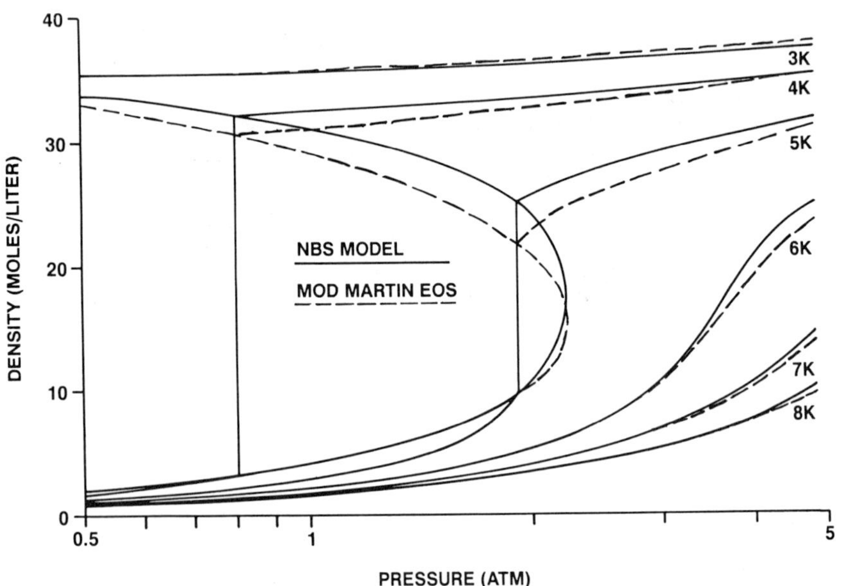

Fig. 6 Density-Pressure Diagram of Helium: Comparison Between NBS
Model and Modified Martin EOS

obtain the temperature that correctly gives the specified internal energy
and density.

The above is a brief summary of the thermodynamic system. More
details are available in a report submitted to the SSC Central Design
Group (12).

The operating scenarios simulated provide an analysis of the performance and accuracy of the various helium models. It is appropriate to provide key results of a timing study here. On the average, for a single-phase density-internal energy specification, the computer time on a VAX 8700 for the McCarty NBS model, the simplified NBS model and the modified Martin equation is 7.0, 0.66, and 0.24 milliseconds, respectively.

OPERATING SCENARIOS SIMULATED

Three transient operating scenarios were simulated:

1. Ring filling sequence - increasing the capacity of the 4K refrigerator from a standby to a normal operating mode as new proton beam is loaded into the accelerator rings.

2. Loss of 4K refrigeration capacity in one sector due to a partial compressor loss.

3. Full-field quench of a magnet half-cell.

In the sections that follow we describe each scenario, including the specific system transient response issue being addressed; we discuss how the computer simulation was performed; and we present simulation results.

Ring Filling Sequence

The SSC ring filling sequence consists of the steps of: 1) dumping the old proton beam, 2) adding new beam at 1 TEV (trillion electron volts), and 3) accelerating the new beam to 20 TEV. Ring filling is expected to occur once or twice a day (13). The transient response issue here is: will the cryogenic system smoothly track the resultant load changes and return to its initial steady state conditions?

To simulate this scenario, the refrigeration system initially was at normal full-load steady-state conditions. After 60 seconds at these conditions, the beam synchroron radiation load per half-cell was decreased linearly from 17.18 to 0 watts over the next 60 seconds. The total energy load per half-cell was maintained at the static heat leak of eight watts (for about four hours). By this time, the system was essentially at standby conditions, except for the vapor return nodes which have the largest time constants. After four hours, the synchrotron radiation per half-cell was increased using the following functionality:

$$Q = 17.18 \times 10^{-12} t^4, \qquad 0 < t < 1000$$

where, t is time in seconds beginning at 14,400. Since time was not allowed to exceed 1000, the beam synchrotron radiation per half-cell was back at its normal value of 17.18 watts at a process elapsed time of 4.3 hours. The simulation was concluded at a process elapsed time of 7 hours. About 55.2 minutes VAX 8700 CPU time was consumed.

The symmetry of this scenario allows the system dynamics to be found by investigating one string of one sector. The action of the 20K refrigerator and the 20K shield were ignored. Liquid returning to the inlet junction of the primary He pump, P2, from strings not calculated was set by symmetry. Symmetry was also used to set the flow of He from the 4K feed area exit node and the vapor flowing to the inlet junction of the 4K refrigerator.

In Figure 7, the total heat load per half-cell and the temperature
of Half-Cell 6B are plotted as a function of time. The temperature of
the half-cell approaches a new steady-state condition about three hours
after the old beam was dumped, and its temperature is back to its initial
value about two hours after acceleration of new beam began. This
temperature response is typical of all the half-cells. Therefore, we
have shown that the cryogenic system does closely follow load changes due
to the ring filling sequence.

There is a problem, however, with the two flow controllers on the
high-pressure 4K liquid helium circuit flowing through the magnets, one
controller being on the feed end and the other on the exit end of each
magnet string. In Figure 8, the positions of the two flow control valves
are plotted as a function of time. Neither valve returns to its initial
position; the feed valve ends up wide open while the exit valve is about
15 percent open, even though the helium flow is at its setpoint. This
situation arises because the use of two flow controllers allows too many
degrees of freedom. Flow is controlled by varying the resistance in the
line. The resistance required to achieve the desired flow rate can be

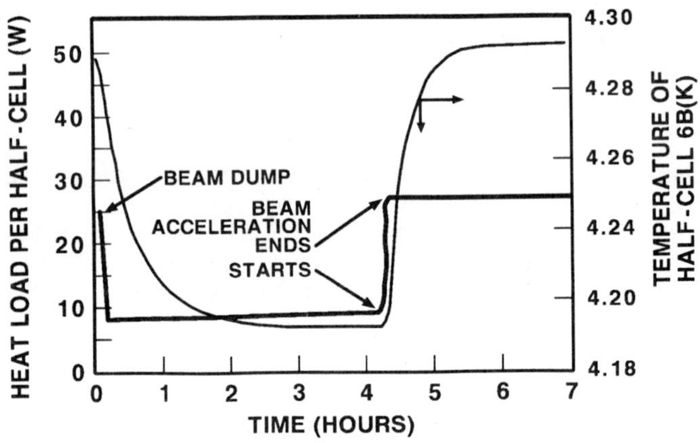

Fig. 7 Beam Filling – half cell heat load and temperature

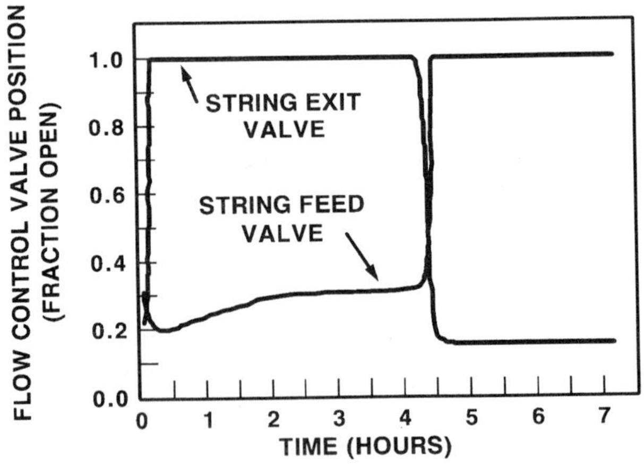

Fig. 8 Beam Filling – control valve position

obtained by either controller or by some combination of the two. This
problem can be remedied by replacing the exit flow controller with a
pressure controller.

Loss of Capacity in the 4K Refrigerator Compressor

In this scenario, half the compression capacity of the 4K
refrigerator of one sector is lost, and the 4K refrigerator compressor of
the adjacent sector is increased to its maximum capacity of 150% of
normal design. The transient response issues here are whether the
adjacent sector will make up for the lost refrigeration through the
connecting piping in the rings, and whether the resultant rise in magnet
temperature in the sector that has lost refrigeration capacity could
precipitate a magnet quench.

The scope of the simulation involves two connecting strings in two
adjacent sectors, and the 4K refrigerators in each sector. The 20K
shield and 20K refrigerator were ignored since their impact on this
scenario is minimal. Constant pressure boundary conditions were assumed
for the strings that were not simulated. This assumption introduces a
small error in the overall material balance, but allows the problem to be
simulated faster. The simulation was executed for three hours of elapsed
time, and consumed about 58 minutes of VAX 8700 CPU time.

Figure 9 is a simplified flow schematic of the adjacent sector
strings. At time zero, one of the two parallel compressors in the 4K
refrigerator of Sector A is tripped by setting the signal from its speed
controller to zero. Pressure then increases in the 4K vapor return line
to the Sector A 4K refrigerator, resulting in return gas flow from Sector
A to Sector B. This causes the Sector B 4K compressor to speed up to
handle the increased flow, which in turn increases the production of 4K
liquid in Sector B.

Figure 10 shows the impact of the Sector A compressor trip on some
liquid flows. The reduced compressor capacity quickly results in reduced
flow of 4K liquid to Sector A. The reduced flow causes the pressure to
drop, which in turn causes a pressure controller to activate the pump
which supplies liquid from storage. Contrary to what is desired,
essentially no liquid flows from Sector B to Sector A. Rather, the
excess liquid produced in Sector B is pumped to Sector B storage. Thus,
the proposed control scheme is not adequate.

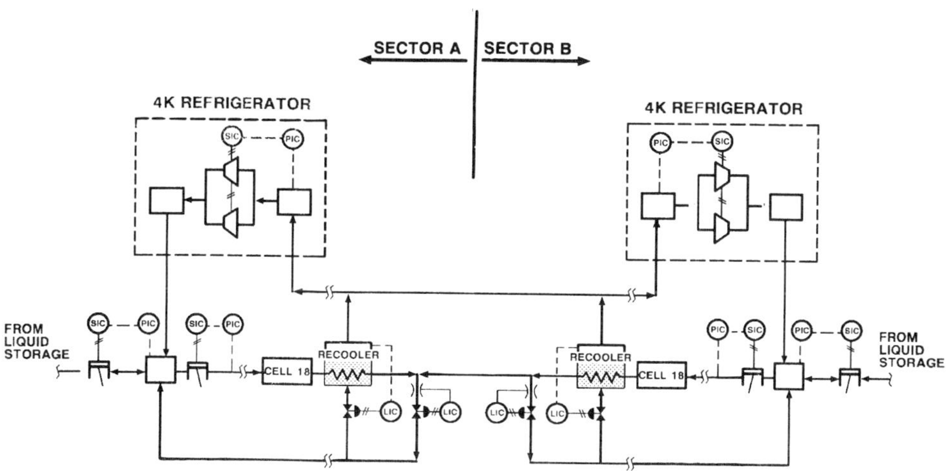

Fig. 9 Compressor Trip – simplified flow schematic

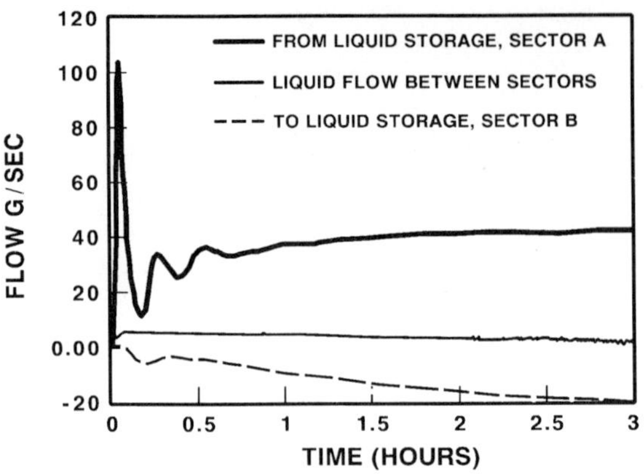

Fig. 10   Compressor Trip — liquid flow between sectors

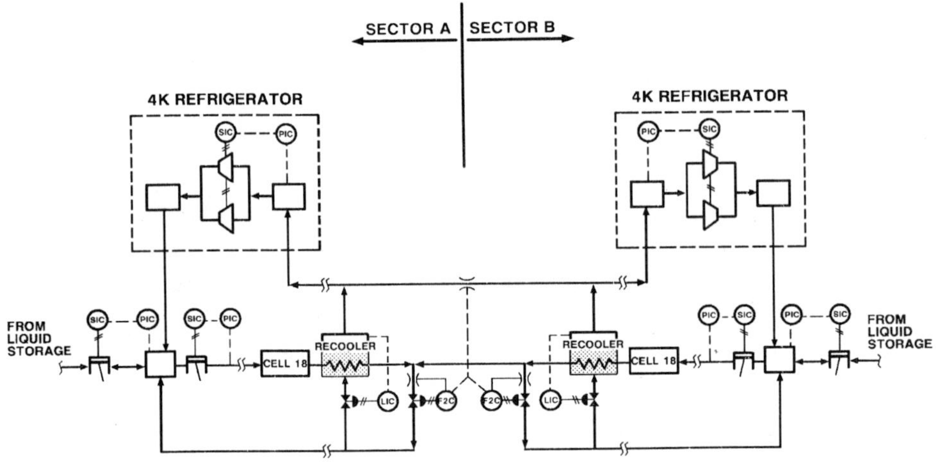

Fig. 11   Compressor Trip — Modified control schematic

In order to help move excess liquid from Sector B to Sector A, we
modified the control scheme to measure 4K vapor flow between Sectors A
and B. When vapor flow increases from Sector A to Sector B, indicating a
loss of Sector A compression capacity in one sector, liquid return is
throttled back on the Sector B 4K liquid return circuit to help direct
excess Sector B liquid to Sector A. This modified control scheme is
shown on Figure 11.

Figure 12 shows the results of the control scheme modification. Now
when 4K vapor flow increases from Sector A to Sector B, Sector B ships
liquid to Sector A, which is shown as a negative flow.

The second issue, whether magnet temperature rise in Sector A could
result in a magnet quench, is addressed on Figure 13 which shows Cell 18
temperature as a function of time for Sectors A and B. Cell 18
temperatures are the highest cell temperatures in each string. Cell 18
temperatures level off well below the selected maximum operating
temperature of 4.35K. Thus, there should be little risk of a magnet
quench.

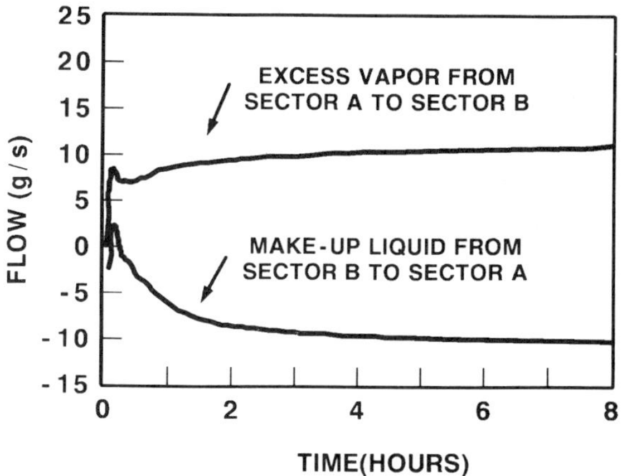

Fig. 12 Compressor Trip – Liquid and vapor flows between sectors

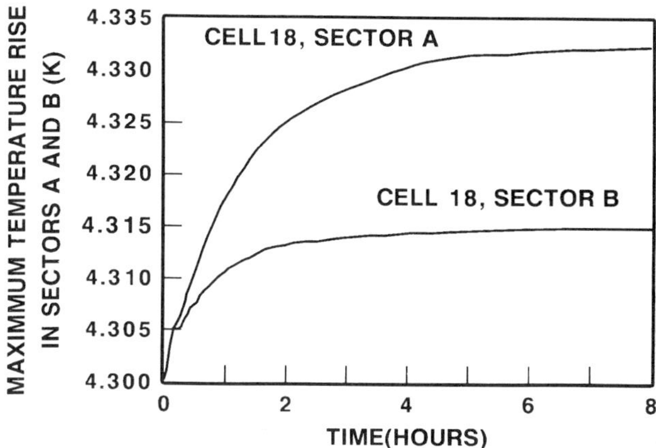

Fig. 13 Compressor Trip – magnet temperatures

Full-Field Quench of a Magnet Half-Cell

Magnet quenches can occur in a variety of ways including beam loss, vacuum failure, and conductor motion (2). In the event of a magnet quench, the remaining magnets in the half-cell are induced to quench by means of embedded heaters. Thus the half-cell is the minimum unit in the accelerator to undergo a quench. Upon a half-cell quench, the current to the other magnets of the ring is turned off within 20 seconds. The heat produced by a half-cell quench can vary, with the full-field quench producing the maximum. Multiple half-cell full-field quenches are also a possibility.

The event simulated here is a full-field quench of a single half-cell. The transient response issues associated with quenching are: will the cryogenic system, including magnet temperatures recover in an orderly way over a reasonable time, and can the quench propagate to adjacent half-cells? Since, as described above, we used a very simplified lumped-parameter approach to model the magnets, only the first issue can be addressed by the dynamic simulator -- and even these results must be regarded as preliminary. The issue of quench propagation requires a more sophisticated magnet model, one which addresses the interaction between the iron yoke and the helium within a magnet.

The scope of this simulation includes the 4K and the 20K refrigerators, the 4K feed area, one string of one sector, and the elements of the 20K shield which complete the shield circuit. There is no symmetry in this scenario. For convenience, and to prevent unnecessary expenditure of CPU time, constant pressure boundary conditions were used for sources and sinks not simulated. The simulation was run on a VAX 8650 and required about 120 minutes of CPU time.

Starting from normal operating conditions, a time varying power input given by

$$Q = 4,128 \, t \, e^{-t/40}$$

was imposed on Half-Cell 9A. This duty is plotted against time in Figure 14. The total energy delivered is the integral and has a value of 6.6 megajoules. The peak power and time constant are in agreement with magnet modeling work by General Dynamics (14).

In addition to the high heat load imposed on Half-Cell 9A, the opening and closing of the half-cell's relief valve into the 20K circuit is an important input. Figure 15 shows relief valve position as a function of time. The signal to open the relief valve as initiated at time zero. The valve position lags the signal with about a six second time constant. At 180 seconds, the signal to close the relief valve was initiated. The valve is essentially closed at about 220 seconds.

Figure 16 shows the flows between quenched Half-Cell 9A and its two neighbors, Half-Cell 9B immediately downstream and Half-Cell 8B immediately upstream. Initially the flow between 8B and 9A reverses and the flow leaving the quenched half-cell increases substantially as its helium inventory is depleted. At about 75 seconds, the flow between 8B and 9A becomes positive once again. At about 130 seconds the flow between 9A and 9B reverses to replace the depleted inventory in 9A, and remains reversed until the relief valve begins to close at 180 seconds. The flows return to their initial rate within an hour.

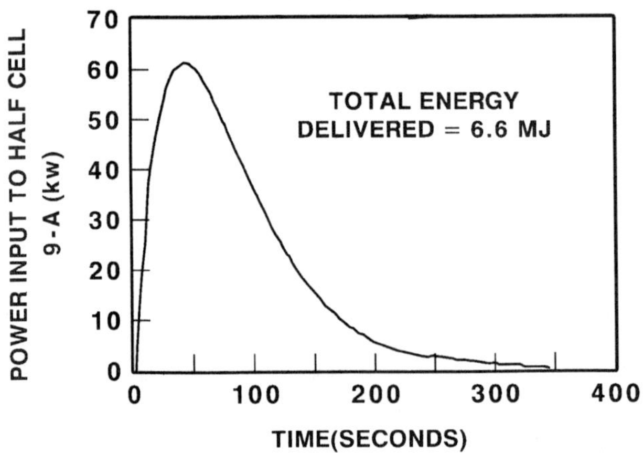

Fig. 14 Quench - heat duty vs time

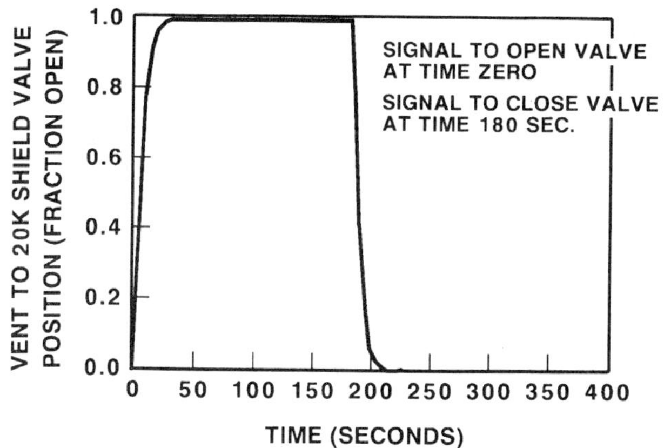

Fig. 15   Quench – relief valve position

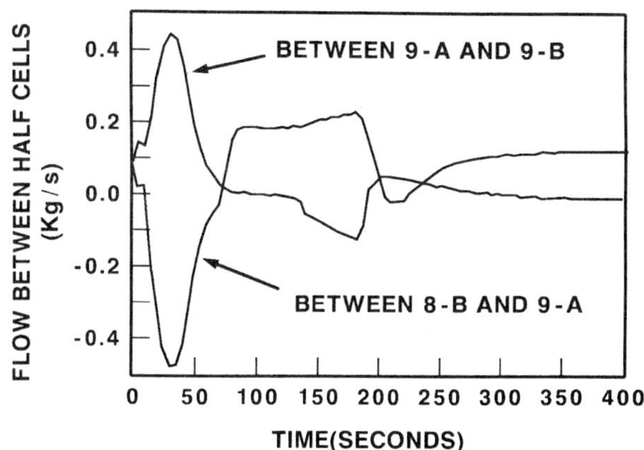

Fig. 16   Quench – flows between 9A and 9B and 8B

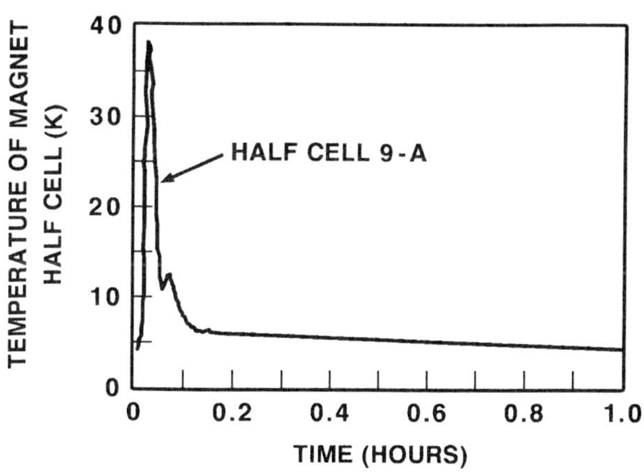

Fig. 17   Quench – 9A temperature

The temperature of Half-Cell 9A as a function of time is shown on Figure 17. The temperature jumps quickly to almost 40K and then drops to about 10K just as quickly. It takes about a half hour for the temperature to drop below 5K. Figure 18 shows that the upstream Half-Cells 8A and 8B experience much less severe temperature peaks and recover after about 1.5 hours. Figure 19 shows some half-cell temperatures downstream of the quenched half-cell. As expected, the magnitude of the temperature spikes decrease as you move downstream from 9A, and the spikes are displaced in time.

The overall conclusion is that the system can recover from a full-field quench of one half-cell in a stable manner over a reasonable time period. However, to more rigorously simulate quench phenomena, especially quench propagation between half-cells, further refinements in the equipment models within the dynamic simulator are needed.

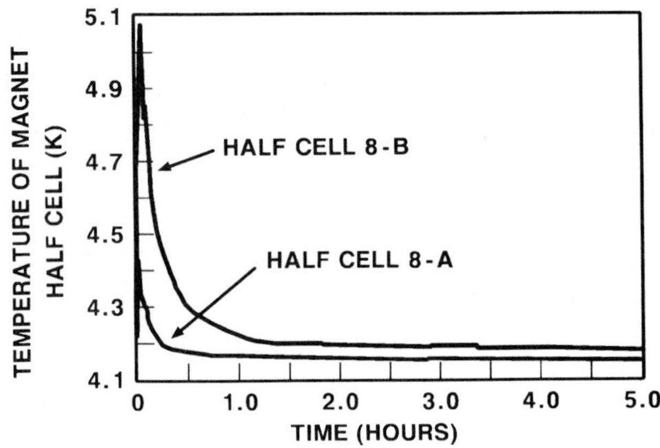

Fig. 18 Quench – temperature upstream half-cells

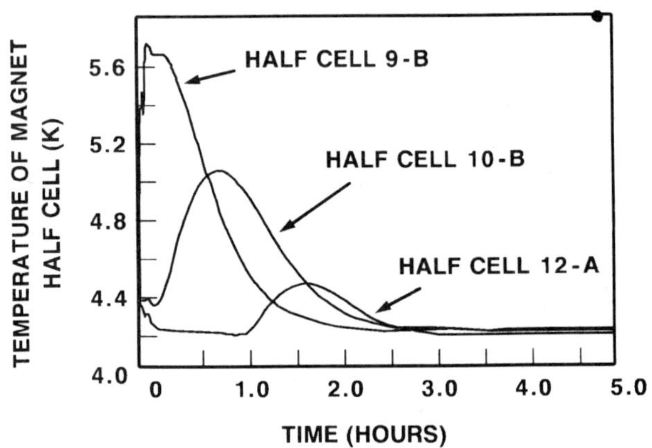

Fig. 19 Quench – temperature downstream cells

CONCLUSIONS

We successfully developed a simplified, robust dynamic simulator of the SSC cryogenic helium system. Simulation results for the key transient operating scenarios indicate that the proposed cryogenic system control scheme is reasonably adequate but with some minor modifications. While the quench scenario especially needs more work, simulations so far suggest that the system can recover from a full-field quench of one half-cell.

Additional work already planned or underway includes developing more rigorous models of the magnets and refrigeration plant heat exchangers, expanding simulation scope by extending boundary conditions, porting the simulator to a more user-friendly workstation environment, and evaluating implementation of the dynamic simulator into an improved version of the SPEEDUP commercial process simulator. An upgraded simulator would then be used to conduct more in-depth analyses of anticipated cyrogenic system operating conditions, and to refine the control scheme.

ACKNOWLEDGMENT

The authors would like to thank Peter Limon of the SSC Central Design Group for his many valuable inputs to both the management and detailed execution of this project.

NOMENCLATURE

A	=	heat transfer area, m^2
C_p	=	constant pressure specific heat, Joule/kg-K
F_{max}	=	prime mover maximum volumetric throughput, m^3/sec
H	=	specific enthalpy, Joule/kg
h	=	heat transfer coefficient, Watt/$(m^2$-K)
M	=	sum of all streams entering and exiting a node
N	=	number of inlet streams to a node
P	=	absolute pressure
Q	=	heat duty, watts
R	=	universal gas constant
S_n	=	normalized speed of prime mover
T	=	absolute temperature, K
t	=	time, seconds
U	=	specific internal energy, Joule/kg
U_o	=	overall heat transfer coefficient, Watt/$(m^2$-K)
V	=	node volume, m^3
v	=	molar volume
W	=	flow rate, kg/sec
X	=	valve fractional lift
Y	=	controller signal, fraction
ρ	=	node overall density, kg/m^3
τ	=	time constant, sec

Subscripts

I	=	inlet stream counter
i	=	inlet stream
J	=	outlet stream counter
lm	=	log-mean
o	=	outlet stream
s	=	recooler shellside
t	=	recooler tubeside

REFERENCES

1. "Dynamic Models and Simulator Development," contract report submitted to the SSC Central Design Group, March 1988.
2. "Conceptual Design of the Superconducting Super Collider," SSC-SR-2020, March 1986, SSC Central Design Group, c/o LBL, Berkeley, Calif.
3. Shinohara, T., "A Block Structured Approach to Dynamic Simulation," presented at First International Conference on Foundations of Computer-Aided Process Operations (FOCAPO-87), July 1987, Park City, Utah.
4. "ACSL-Advanced Continuous Simulation Language User Reference Manual," Mitchell and Gauthier Associates, Inc., Concord, Mass. (1975).
5. Pantelides, C. C., "SPEEDUP-Recent Advances in Process Simulation," Computers in Chemical Engineering, Vol. 12, No. 7 (1988).
6. Hindmarsh, A. C., "ODEPACK, A Systematized Collection of ODE Solvers," Scientific Computing, R. S. Stepleman et.at. (eds.), North-Holland, Amsterdam, 1983, pp 55-64.
7. Speakeasy IV Reference Manual, Speakeasy Computing Corporation, Chicago, Ill., copyright 1984.
8. Wu, K. C., "Recoolers for the SSC Reference Design D, " SSC Technical Note No. 25, March 13, 1985, High Energy Facilities, Brookhaven National Laboratory, Upton, N.Y.
9. McCarty, R. D., "Thermodynamic Properties of Helium-4 from 2 to 1500 K at Pressures to 10^8 Pa," J. Phys. Chem Ref. Data, 2(41), 923-958 (1973).
10. "Cubic Equations of State - which?", Inc. Eng. Chem. Fundam., 18(2), 81-97(1979).
11. "A Consistent Correction for Redlich-Kwong Volumes", Fluid Phase Equilibria, 8, 7-23(1982).
12. "Helium Thermodynamic System", preliminary report submitted to the Supercollider Central Design Group, March 1988.
13. McAshan, M. S., "Refrigeration Plants for the SSC," SSC-19, May 1987.
14. General Dynamics "Thermal Parametric Studies of the Production Dipole Magnets for the Superconducting Super Collider," CDG PIN No. 87-P-085, April 1988.

TEVATRON OPERATIONAL EXPERIENCES

B.L. Norris and J.C. Theilacker

Fermi National Accelerator Laboratory*, P.O. Box 500
Batavia, Illinois 60510

ABSTRACT

Fermilabs superconducting accelerator, the Tevatron has been operational for nearly six years. The history of its operation is presented. Several long shutdowns for superconducting dipole repairs are discussed. The dominant factor influencing the repair was conductor motion which fatigued the cable in the magnet ends. Borescoping and x-raying techniques were used to determine which magnet ends required repair.

Detailed downtime logs were kept for each of the running periods. A discussion of the sources of downtime and a comparison for different operating modes is presented.

INTRODUCTION

Fermilab is a National Laboratory which provides the tools necessary to support and perform high energy physics experiments. The tools are in the form of proton accelerators. In the early 1980's a superconducting accelerator, the Tevatron, was built as an upgrade to the original 500 GeV accelerator. The Tevatron was designed at twice the energy (1TeV), yet offered a reduction in operating costs over the 500 GeV conventional magnet accelerator. Magnet development was the principal concern of this project from 1972 to 1979. [1] Then, from its near-final design in the late 1970's the Tevatron required considerable research and development in the areas of large scale superconducting magnet production, large scale helium refrigeration and transport, vacuum technology, quench protection and controls systems. To date, it is the highest energy proton accelerator in the world.

The Tevatron is a 2 km diameter synchrotron consisting of nearly 1300 cryogenic components as shown in Table 1. Its primary components are the dipoles, quadrupoles and spool pieces [1]. The 6.3m bending dipoles utilize a NbTi alloy superconductor to achieve 4.4 Tesla. Spool pieces house a variety of components. These include; correction magnets, quench stopper, vacuum barrier, relief valves, and thermometry. Every other spool piece contains safety leads which bypasses current around a quenched cell of magnets.

Refrigeration for the Tevatron is supplied by a hybrid system consisting of a Central Helium Liquefier (CHL) connected to 24 satellite refrigerators by a 7 km LHe, LN_2 transfer line [2]. This system provides redundancy by relying more heavily on one system should a problem develop in the other. Also, large inventories of liquid helium stored at the CHL dewar system are available for fast magnet quench recovery or cooldown following magnet repair.

The quench protection scheme for the Tevatron is an active system; it requires prompt detection of a quench and active components to remove the current from the quenching magnets [3]. The microprocessor based controls for the quench protection, refrigeration, vacuum, and correction element systems communicate to a main control room through a high speed link [4]. This allows for centralized operations, alarming, and data logging of all systems. The main control room is manned around the clock, 365 days a year.

---

* Operated by Universities Research Association, Inc. under contract with the U.S. Department of Energy.

Since its initial commissioning in 1983, nearly six years of Tevatron operational experience has been realized. An outline of the operational history of the Tevatron is presented. Modes of operations as well as major shutdown projects are described. Over the past six years, a detailed operational downtime log has been kept. Trends of downtime are considered, both as a function of time as well as mode of operation.

OPERATION HISTORY

Figure 1 shows the operational history of Tevatron over the past six years with our projection to the end of calendar year 1989. Details of the experiences for the years 1983-1986 have already been covered by Martin [5] and will not be repeated here. Details will be confined to 1987-1989.

The first colliding beam physics run opened 1987. Counterrotating 900 GeV proton and antiprotons were collided for a center of mass energy of 1.8 TeV. The run lasted fourteen weeks and was interrupted once for a Tevatron dipole change. Broken conductor strands, due to motion, reduced the current carrying capability of the dipole, resulting in the magnet change.

Following the collider run there was a two week shutdown necessary to switchover to fixed target physics. During this time the colliding detector was moved from the collision hall to the assembly hall for maintenance and upgrade. Extensive maintenance of the accelerator systems also took place during this time. This included the replacement of five Tevatron components; three for suspected vacuum leaks, one for a low quench threshold, and one for power lead upgrade. Extensive maintenance to the cryogenic system was also accomplished including reciprocating expansion engine overhauls and system repurification at 80K. During the system repurification, we experienced a sitewide power outage. Fortunately, this had little impact on this mode of refrigeration system operation.

June of 1987 marked the return of 800 GeV fixed target physics. Although the run would last for 35 weeks, it proved to be interrupted by seven shutdowns for magnet changes and two sitewide power outages. Despite the interruptions, beam was delivered to the experiments at a higher rate than the 1985 run due to efforts to improve the beam intensity.

In March 1988 a three month shutdown began. Four major categories of work was scheduled for this shutdown, including: Upgrade of the main ring overpass at D0, shielding of the main ring beam pipe from the B0 collider detector, repair of Tevatron dipoles in a quarter of the accelerator, and general system maintenance.

The rash of dipole problems during the fixed target operation prompted the dipole repair program. Prior to the shutdown, methods for inspecting the dipole ends were developed. Two methods, borescoping and x-ray were chosen for determining the extent of repairs that would be necessary. X-raying could be done while the magnets

were warm or cold. Borescoping could only be done on warm magnets. Access for borescoping at one end was through a relief port, while the other end required disconnecting the liquid helium between magnets.

Four problem areas were being inspected at each end of a dipole.

Leads

> Leads were inspected for ties to prevent lead flexing during ramping. The preferred tie is Kevlar string, however some magnets used nylon ties. During repair, several magnets were found to have broken strands in the cable, the worst proved to have 12 of the 23 strands broken. Strand breakage may have been aggravated by low serial number magnets not having the G-10 conductor holddown block tumbled, to eliminate sharp edges.

Bore tube insulation

> The bore tube is insulated from the inside of the coil by spiral wrapped Kapton tape. At cryogenic temperatures the tape looses its ability to stick. This allows the tape to unravel at the upstream end of the magnet if it is not tied.

Coil Clearance

Inadequate clearance between the end of the coil and the single-phase terminating plate can result in contact during ramping. Repeated force on the end plate has resulted in a cracked weld on several dipoles, leaking helium to the vacuum space. Clearances were particularly a problem when round head screws were used instead of flat head on a G-10 lead holddown block. In several cases, screws were found to be backing out of the block. Clearances greater than 1.5 mm are adequate to avoid contact.

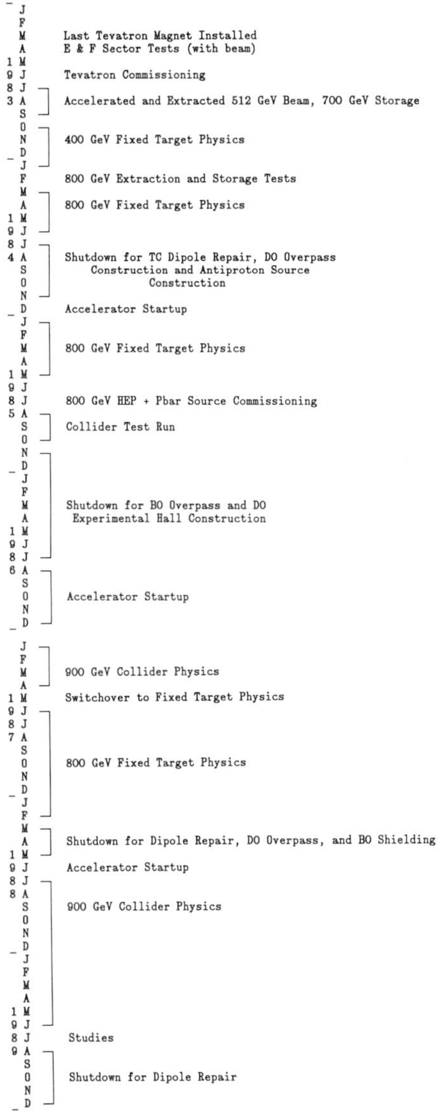

FIG. 1. Tevatron Operational History

Bolted Conductor Clamp

As the leads leave dipole, they are held by an L-shaped G-10 block. In low serial number dipoles the block halves were bolted together. Over time and thermal cycles, several bolted connections have become loose or separated altogether. This was later changed to a riveted connection. Magnets with bolted blocks were repaired.

Table I. Tevatron Components

Dipoles	777
Quadrupoles	224
Spool Pieces	201
Feedcans	24
Turnaround boxes	30
Bypasses	21
Other	17
	1294

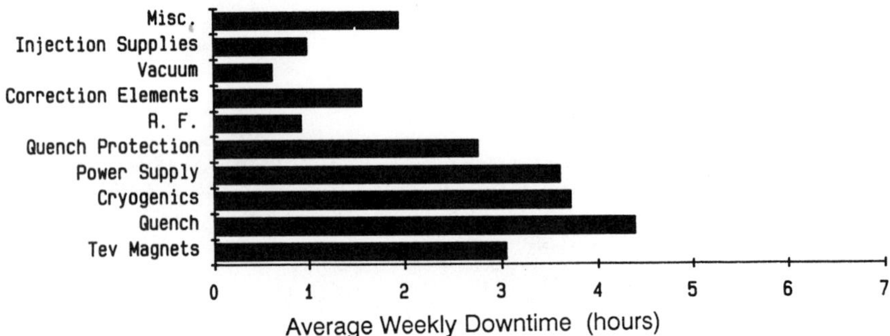

FIG. 2. Downtime for 1985 Fixed Target Run

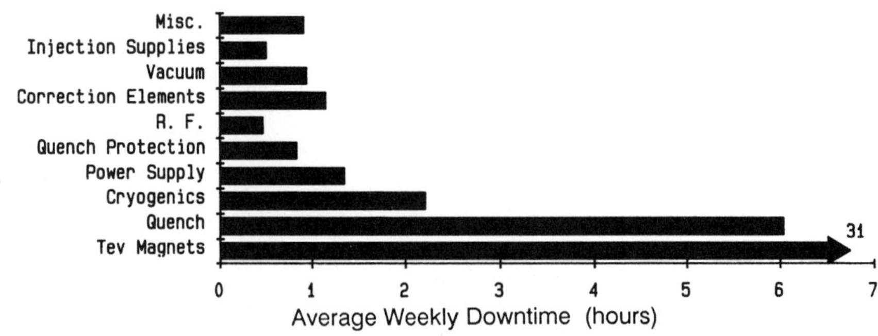

FIG. 3. Downtime for 1987 Fixed Target Run

During the inspection and repair, a black greasy material was found in several dipoles near the refrigerator feed point. It was found to have a high lithium content, such as the grease used on the refrigerator expansion engines. No grease has been found at the bottom of the expander cylinder. It is suspected that grease migrates down the cylinder to the point where it freezes. The reciprocating motion then grinds up the grease. It then travels with the helium stream where it drops out at the first sudden expansion and the first dipole.

Inspection of magnets began in areas of the ring with low serial number magnets (A-sector). Although areas of the ring with the lowest serial numbers tended to require more repairs, serial numbers alone were not definitive. In all, six sections were fully inspected and repaired (A1, 2, 3, 4, B1, E1) and one was partially repaired (E3). Table 2 shows the results of the dipole repair as reported by Hanna [6].

In June 1988 the system started for a second 900 GeV collider physics run. It is expected that we will continue this run until July 1989. At that time we will warmup the entire system to room temperature in order to inspect and repair the remaining two-thirds of the Tevatron dipoles.

Tevatron Downtime

Prior to discussing Tevatron downtime, it is important to understand several fundamental differences in the fixed target and collider operational modes. During fixed target physics, protons are injected, ramped to full field, "spilled" to the experimental areas, the magnets are ramped down, and the cycle repeats. The cycling of the magnets means that the refrigeration systems must satisfy both the static heat load of the magnets as well as AC losses within the collared coils (predominantly hysterisis). Injecting and extracting proton on each cycle (57 seconds) also increases the odds for stray beam to cause a quench.

During collider physics, magnets are ramped to full field and remain there for many hours. This significantly reduces the refrigeration load by "eliminating" AC losses but increases the liquefier load necessary for vapor cooled power leads. Liquefier loads tend to not be seen by the satellite refrigerators, only by the central liquefier. Since protons and antiprotons are injected or removed infrequently, beam induced quenches tend to be less frequent.

Downtime in the following charts have been converted to weekly average for direct comparisons of four different runs: 1985, 800 GeV fixed target, 1987, 900 GeV collider, 1987, 800 GeV fixed target and 1988, 900 GeV collider run (through January 31, 1989). Detailed downtime logs have been kept for these runs. However, it should be noted that the method used can result in hours being counted more than once. For instance, if a quench is caused by an injection kicker misfiring downtime will be logged under injection supplies and under quench.

Figures 2, 3, 4, and 5 show the weekly average downtime for ten Tevatron subsystems for the 1985 fixed target, 1987 fixed target, 1987 collider (through January 31, 1989) and 1988 collider runs, respectively. Comparing the two fixed target run (Fig. 2 and 3) shows an order of magnitude increase in downtime due to the recent TeV magnet problems. Fatiguing cable strands and welds as mentioned in the previous section were the cause.

The more recent fixed target run resulted in fewer magnet quenches, yet resulted in more downtime as shown in Figure 6. Longer quench recovery times in the '87 run were primarily due to two factors. First, more full field beam induced quenches (as opposed to injection quenches) occurred due to higher beam intensities and fast beam spill to the experimental areas. Secondly, long system recoveries in the A1 magnet strings resulted from an intermittent ground fault. The problem cleared several hours after the quench making it difficult to isolate. During magnet replacements elsewhere in the ring, components in the affected electrical cell of A1 were charged out. After two replacement episodes involving seven components, a dipole was removed which proved to have lead insulation damage.

In comparing the recent fixed target run with the collider run (Figures 3, 4, 5, and 6), two major differences in downtime are evident. First, the continual cycling of the magnets in fixed target mode emphasized the fatigue problems previously mentioned. This resulted in a factor of thirty higher downtime due to magnet replacement. Second, Figure 6 shows 40% more downtime due to magnet quenching in fixed target mode. This is due to the continual likelyhood of stray beam transport of beam in and out of the Tevatron increasing the inducing a quench. Figure 6 also shows similar recovery times for the '87 runs, since both runs were dominated (80%) by full field quenches.

Table II. Tevatron Dipole Repair

| | Dipole Type | | | |
| | TB Dipole | | TC Dipole | |
	Up	Down	Up	Down
Ends Considered	104	104	96	96
Repair Reason				
Leads not tied	85	22	0	28
Broken strands	7	0	0	0
L-block loose	13	8	2	11
G-10 block loose	22	25	24	12
G-10 block clearance	21	29	21	25
Beam tube Kapton	48	0	0	0
Inspection Type				
X-rayed	104	103	94	95
Bore scoped	104	84	8	78

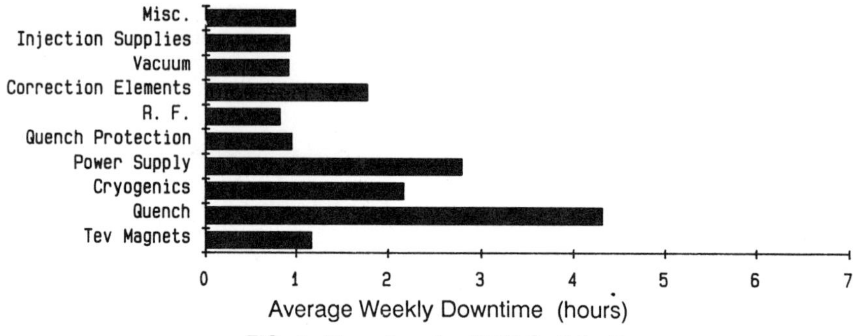

FIG. 4. Downtime for 1987 Collider Run

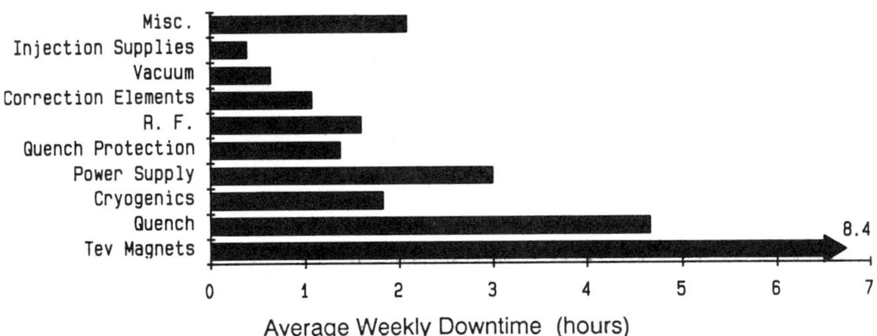

FIG. 5. Downtime for 1988 Collider Run (though January 31, 1989)

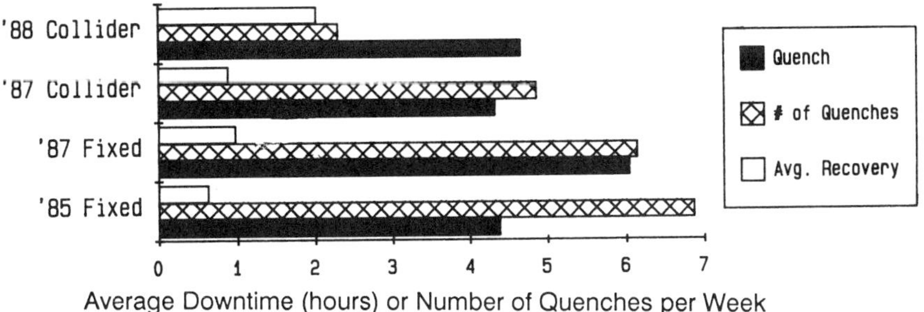

FIG. 6. Magnet Quench Downtime vs Operating Mode

During the 1988 collider run, improvements in beam transport and proton-antiproton store durations resulted in a factor of two fewer quenches than in the previous collider run. However, quench recovery times were a factor of two higher. This was due to a large number of multiple house quenches. There had been an intermittent problem with beam kicker magnets prefirings which forces the particle beams to hit magnets in various locations around the ring. The intensity of the beam is such tat quenches in four different magnet strings was not uncommon.

Figure 5 shows a high amount of downtime for Tevatron magnets during the current collider run. This represents the time necessary to replace dipoles on two separate occasions as well time required to access the Tevatron tunnel to clean the correction element power leads. Over the years, a black conductive substance has formed on the leads, causing occasional leakage current to ground. The leads have been cleaned and coated with a nonconductive sealer.

Improvements in downtime for the cryogenic, power supplies, and quench protection systems were realized between the '85 and '87 fixed target runs due to upgrade and hardening of components. Cryogenic systems were most influenced by three improvements. They include: The addition of a liquid helium pump at CHL to supply inventory from the dewar system to the satellite refrigerators during peak ring usage or CHL outages, improvements in satellite expander mean time between failures, and the addition of controls software which helped monitor and debug the system.

Each satellite refrigerator has one reciprocating expander operating during normal modes of operation. To date, over a million expander hours have been experienced at Fermilab. Mean time between failures have increased to over 6000 hours. Durations between overhauls of over a year (8760 hours) have been realized, with the longest registering in at over 13,000 hours.

SUMMARY

The Tevatron has just completed its sixth year of operation. This paper in conjunction with Martin [5] discussed the operational experiences throughout the six years. Magnet problems have been addressed during two long shutdown periods. Approximately one third of the dipoles have been repaired to date; mostly in areas with early production dipoles and areas prone to injection and extraction beam induced quenches.

The warm iron magnets of the Tevatron allow for a more rapid warmup and cooldown for magnet charges during operation. During a magnet change, one twenty fourth of the ring must be warmed to room temperature. To date there have been 110 such warmups, including two full ring magnet warmups. Fast cooldown following a magnet change is accomplished directly with a liquid helium cooldown wave. There have been no known problems directly associated with using this abrupt cooldown wave.

Problems associated with the Tevatron magnets have predominantly been in magnet ends. Typically, magnet end design takes a back seat to the coil design, as it should. However, it's importance is accentuated by the fact that it is the point at which "field" electrical and piping connections and thermal contraction takes place. The incorporation of magnet tests and life cycle tests can help find such problems earlier in a systems design or production.

ACKNOWLEDGEMENTS

The author would like to acknowledge the dedication that was required by all Fermilab employees in order to achieve five years of operation of the Tevatron. Special thanks to J. Crawford for supplying the downtime statistics which were presented.

REFERENCES

1.   Fermi National Accelerator Laboratory Superconducting Accelerator Design Report; (1979)

2.   C.H. Rode, "Tevatron Cryogenic System" 12th International Conference of High Energy Accelerators", (1983)

3.   P.S. Martin, "Design and Operation of the Quench Protection System for the Fermilab Tevatron", Fermilab Report, TM-1398, (1986)

4.   D. Bogert, et al., "The Tevatron Controls System", IEEE Transaction on Nuclear Science, Vol. NS-28, No. 3, p. 2204, (1981)

5.   P.S. Martin "Operational Experience with Superconducting Synchrotron Magnets", Proceedings IEEE Particle Accelerator Conference, Washington, D.C., p. 1379 (1987)

6.   B. Hanna, et al., "Non-Destructive Testing of Superconducting Dipoles in the Tevatron", to be submitted to the IEEE Particle Accelerator Conference, Chicago (1989)

CENTRIFUGAL PUMPS FOR THE SUPERCONDUCTING SUPER COLLIDER

W. Swift, J. A. McCormick, and W. D. Stacy

Creare Inc.
Etna Road
P.O. Box 71
Hanover, NH 03755

ABSTRACT

The current refrigeration system design for the SSC consists of multiple helium refrigerators and liquefiers, each servicing sectors of the SSC which are approximately 8.25 km in length. Within each cryogenic refrigeration sector, there are various requirements for the distribution of liquid and gaseous helium. The machines which are to provide this distributive function must be robust and highly reliable because of the expense associated with system downtime. This paper presents the preliminary design of a centrifugal pump for the SSC refrigeration system.

Operational and performance specifications are defined for the pumps based on SSC requirements. Trades between bearing support systems are reviewed and a design configuration is recommended. Performance models for the pumps are presented together with a discussion of how these models may be used in system studies.

INTRODUCTION

This paper presents the results of a preliminary design effort to establish the feasibility of using centrifugal pumps to circulate helium in the SSC refrigeration systems. The approach taken was to broadly define the SSC helium circulation requirements, assess centrifugal pump technology relative to these requirements, design a generic centrifugal pump capable of meeting the range of requirements, then develop a model of pump behavior which could be used for system studies and evaluation.

SSC Requirements

The cryogenic refrigeration and helium distribution systems for the SSC are not completely defined at this time. Three references [1], [2], and [3] served as the principal source of guidance in establishing requirements for the circulators. Additional information was obtained through discussions with the SSC Central Design Group. The results are reviewed in this section.

The SSC site consists of an elongated ring about 85.7 km in circumference which is divided into two arc regions and two interaction

Table 1. Circulation Requirements within
a Sector Cryogenic System

Condition	Circulator[1]	T_{in}—K	P_{ex}/P_{in}	Flow Rate g/s
Norm. Make Up	P1	4.25	3.5/2.0	12
Max.				16
Quench Recovery	P1	4.25	3.5/2.0	> 100
Circulation[2]	P2	4.25	4.1/3.3	200
(Low load)				60
(Maximum)				250
Circulation[3]	P2	4.25	2.4/1.8	167.2
(Low Load)				50.2
(Maximum)				209

Notes:

[1] P1 designates a make-up circulator, P2 provides
circulation through the ring sector

[2] Magnets at 4 atm.

[3] Magnets at 2 atm.

Table 2. System Circulator Requirements

CONDITION	CIRCULATOR	T_{in} (K)	P_{in} (atm)	ΔP (kPa)	m g/s	ΔH_{ad} J/g	Q cc/s
Normal	Make–up	4.25	2.0	152	12	1.17	93
Maximum	Make–up	4.25	2.0	152	16	1.17	123
Quench Rec.	Make–up	4.25	2.0	152	100	1.17	772
Nominal	Circulation	4.25	3.3	81	200	0.60	1481
Low Load	Circulation	4.25	3.3	81	60	0.60	444
Max Load	Circulation	4.25	3.3	81	250	0.60	1852
Nominal	Circulation	4.25	1.8	61	167	0.47	1300
Low Load	Circulation	4.25	1.8	61	50	0.47	390
Max Load	Circulation	4.25	1.8	61	209	0.47	1626

Table 3. Key Design Point Parameters for the Circulator

m	=	200 g/s	Flow Rate = 1540 cc/s
P_{in}	=	3.3 atm	Speed = 24,300 rpm
P_{ex}	=	4.1 atm	Diameter = 2.54 cm
ΔH	=	0.62 J/g	Shaft Power = 166 W
n_s	=	0.8	Overall Efficiency = 0.69
d_s	=	3.25	Heat Leak = 15 W

region (IR) clusters. At approximately 8 km intervals, there are service areas, each with a helium plant providing refrigeration to an 8 km length of each of two magnet rings. There are a total of eleven plants, one for each service area and one for the high energy booster ring of the injector complex. Each plant is equipped with helium liquid and gas storage, liquid nitrogen storage, cold box, compressor plant and a set of auxiliaries. It has been assumed that the compressor plants, cryogen and gas storage facilities are at the surface, while cold boxes, expanders and related equipment are underground at the tunnel level. It has been further assumed that there is little difference (significant to pump design) between the plants serving the IR sections and those serving the arc sectors.

Each refrigeration plant is centrally located in the sector and feeds two strings, one for each ring, in either direction. Each of the sector helium refrigeration plants is capable of independent operation at the rated heat loads of the ring. However, the cryogenic systems for the sectors are interconnected at their boundaries. It is possible to pass refrigeration from one sector to an adjacent one which provides for system redundancy.

<u>System Operating Modes</u>

Below is a brief list of the steady and transient operating modes of operation which are identified in reference [2].

Steady or Quasi Steady Modes

    Production and Accumulation of Liquid Helium Inventory
    Cooldown of the Magnet System from 300K to 80K
    Cooldown of the Magnet System from 80K to 4.2K
    Liquid Helium Inventory Handling
    Magnet Conditioning at 1.9K or 2.5K
    Nominal Design Operating Condition
    Operation at Reduced Load (as low as 25%)
    Failure Mode Operation

Transient Operating Conditions

    Ring Filling Sequence (Current Ramps)
    Quench Recovery

There are large variations in fluid flow, pressure, temperature and refrigeration requirements for the operating modes identified above. Estimates [3] have been made for circulation requirements under several operating conditions. These are listed in Table 1. The operating conditions listed in Table 1 have been converted to corresponding volumetric flow rates and head or enthalpy rise for use in establishing centrifugal circulator characteristics. These values are listed in Table 2.

CIRCULATOR PERFORMANCE

In order to generate a preliminary design of the SSC circulators, one must first determine the relationship between pump geometry, speed and the anticipated operating conditions. Since there is a broad range of operating conditions, and because several machines could be selected to pump in parallel, the range of design options is also fairly broad. The approach used was to first convert the range of operating conditions to a practical range of pump sizes and speeds, then generate representative performance characteristics for this range of machines. Finally, the design range is narrowed by considerations of mechanical performance, drive and shaft support alternatives.

Hydraulic Design

The performance of rotational fluid machines can be represented in reduced coordinates on maps which allow for comparison of performance

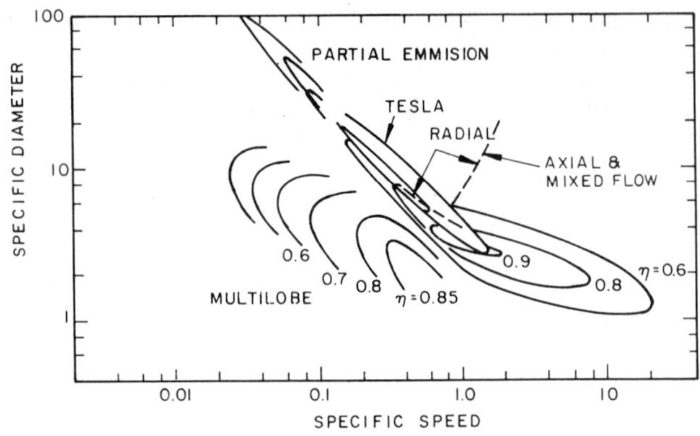

Figure 1. Best Efficiency Contours for Turbomachines Shown as a Function of Specific Speed and Specific Diameter. Adapted from [4]

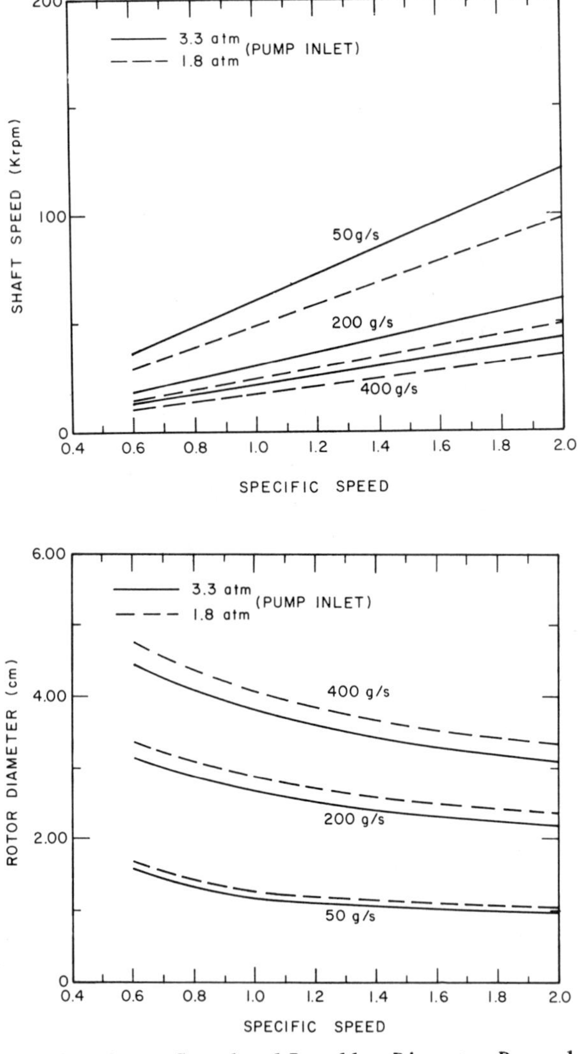

Figure 2. Circulator Speed and Impeller Diameter Depend on the Selection of Design Point Flow Rate and the Value of Specific Speed

options between different types of machines . The most common parameters used for this representation are specific speed, specific diameter and efficiency. Any one of several turbomachinery reference texts provide a good discussion of both the analytical and empirical basis for such representations. Basically, specific speed n_s and specific diameter d_s are parameters associated with geometric and performance relationships for fluid machines. A given machine will have a best efficiency at some combination of speed, flow and head. This best efficiency represents an optional match between fluid dynamic conditions, size, and rotational speed. One such map is shown in Figure 1 which is adapted from [4].

Specific speed is a dimensionless parameter which relates volumetric flow rate Q, adiabatic enthalpy rise (or drop) H_{ad}, and rotor angular velocity ω of a machine at its design or best efficiency point.

$$n_s = \omega \ (Q)^{0.5} / (H_{ad})^{0.75}$$

Specific diameter is also a dimensionless parameter relating rotor diameter to adiabatic enthalpy change and flow rate.

$$d_s = D \ (H_{ad})^{0.25} / (Q)^{0.5}$$

Therefore, to characterize a machine in terms of n_s and d_s one uses the diameter, flow, head and shaft angular velocity for a machine at best efficiency to calculate the n_s and d_s for the machine.

The contours in Figure 1 show lines of constant efficiency for various types of rotating machines. The figure shows that there are certain combinations of n_s and d_s which tend to result in machines with higher or lower efficiencies. The domain for centrifugal machines is within $n_s = 0.2 - 2$. Based on this map, which represents combined empirical data and analytical results, one can obtain varying efficiencies at design point for a given set of volumetric flow and adiabatic enthalpy rise conditions, depending on the speed and rotor diameter selected for the machine. The map was used as a starting point for determining shaft speed and rotor diameter.

The maximum attainable efficiencies for centrifugal or mixed flow machines occur over a range of specific speeds from $n_s = 0.6$ to $n_s = 2$. The values of specific diameter corresponding to these extremes are 4 and 2.7, respectively. These values, together with bounding values of flow rate and head or enthalpy from Table 2 were combined to estimate upper and lower limits of machine diameter and rotational speed. The results are shown in Figure 2.

Figure 2a shows the design speed of a circulator for inlet pressures of 1.8 and 3.3 atm for three flow rates (50 g/s, 200 g/s and 400 g/s) depending on the specific speed of the machine. The important conclusion to be drawn is that shaft speeds will be relatively high – values from about 10,000 rpm to > 100,000 rpm are predicted. These speeds are at the upper end of the practical range of use of conventional oil lubricated bearings. Gas or magnetic bearings offer attractive alternates because they reduce the possibility of contaminating the helium and they are ideally suited for rotational speeds of this magnitude.

Figure 2b shows the circulator impeller diameters which correspond to the speeds in Figure 2a. It is clear from this figure that even at the highest flow rate – 400 g/s, the machine will be relatively small. The largest predicted impeller diameters are about 4 cm (1.6"). Designs for a 50 g/s flow rate would require an impeller diameter of about 1 cm (0.4"). This is almost a lower limit from the standpoint of practical fabrication methods.

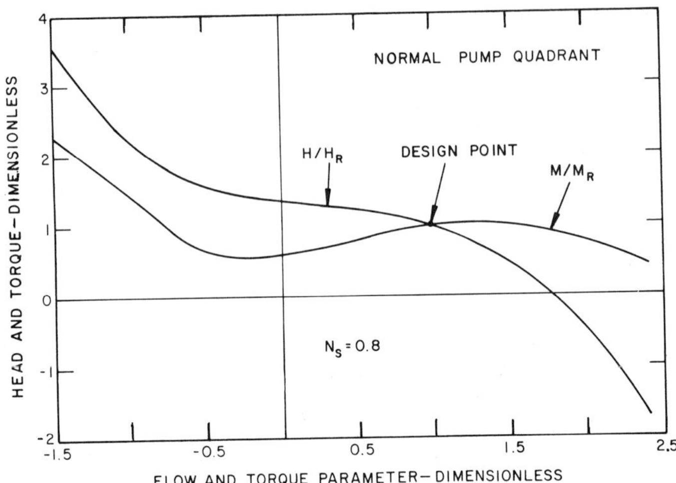

Figure 3. Predicted Performance of the Circulator

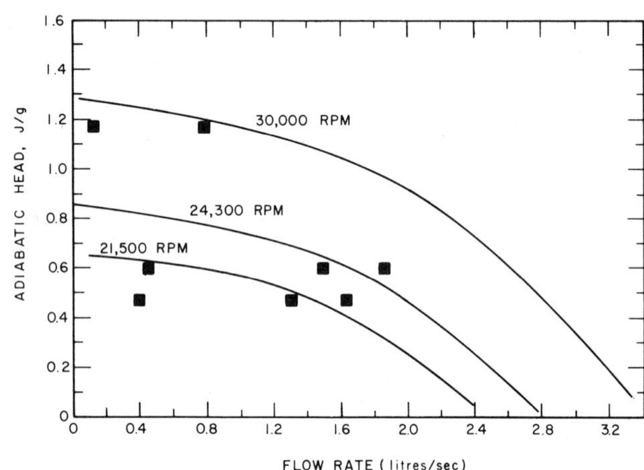

Figure 4.   Calculated Heat Rise and Overall Efficiency vs. Volumetric Flow Rate for the Helium Criculator at Three Shaft Speeds. (The Solid Symbols Correspond to Pressures and Flows in Table 2)

The preliminary design for the machine was selected at $n_s = 0.8$ and $d_s = 3.25$. If the machine is designed for these values at an operating condition of 200 g/s and 3.3 atm. inlet pressure, the resulting design speed and rotor diameter are 24,300 rpm and 2.54 cm, respectively. The corresponding ideal pumping power is 125 W.

These values are reasonable in terms of defining the machine for preliminary design purposes. Setting the design point at a lower specific speed has the implication of reduced efficiency; higher specific speeds would result in smaller rotors with some three dimensionality in the impeller design, and higher rotational speeds. The hydraulic efficiency at design conditions, predicted from Figure 1 is about 0.9. It is probably realistic to use 0.75 as a design target to account for additional losses due to leakage. This design point information, together with some empirical data is sufficient to estimate performance curves suitable for modelling. Table 3 summarizes the key design point parameters for the circulator based on the circulator P2 function. The value of overall efficiency in the Table is a combination of the estimated hydraulic efficiency 0.75 and a penalty for heat leak.

Predicted Performance

Analytical models for centrifugal pumps are not sufficiently well developed to accurately predict machine behavior at operating conditions far from design point. However, the empirical data of Donsky [5] has been used frequently to characterize the global performance of centrifugal pumps of several designs – generally for the purpose of evaluating transients in hydraulic systems. Figure 3 is adapted from the data in [5]. Donsky presented pump characteristics for machines of three specific speeds, 4.93, 2.78 and 0.66. Important features of these sets of characteristics have been used to interpolate to a specific speed of 0.8 for the purpose of estimating the shapes of the performance curves of the helium circulator. While this approach intxoduces some uncertainty in the results, the uncertainties are probably of the order of 5% in the shape of the curves – still good enough to be useful for system modelling.

Normalized head rise h and normalized shaft torque m are shown for a range of normalized flow rates q. Each curve represents the pump performance at rated speed N_r. These curves, together with conventional hydraulic scaling relations and assumptions about design values can be very useful in exploring options in both machine design point selection and system operating behavior. Some of these uses are briefly illustrated in the following paragraphs.

The hydraulic scaling relationships are derived from the following dimensionless variables for incompressible flow in turbomachines:

$$H/N^2D^2 \qquad\qquad Q/ND^3 \qquad\qquad M/\rho N^2D^5 .$$

The other useful variables are the normalized parameters shown in the figure:

$$h = H/H_r , \qquad\qquad q = Q/Q_r , \qquad\qquad m = M/M_r ,$$

and the efficiency:

$$\eta = \rho\, Q\, H\, /\, \omega\, M$$

where:

H is the head rise or enthalpy rise through the machine,
Q is volumetric flow rate,
N is rotational speed,

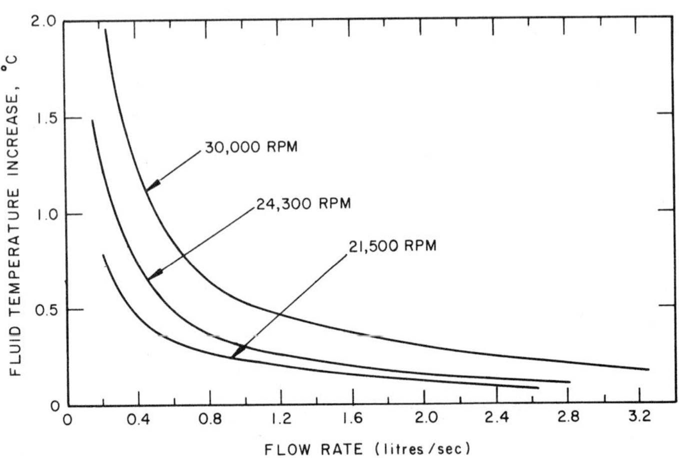

Figure 5. Net Circulator Efficiency Depends on the Hydraulic Efficiency and the Ratio of Heat Leak to Ideal Pumping Power

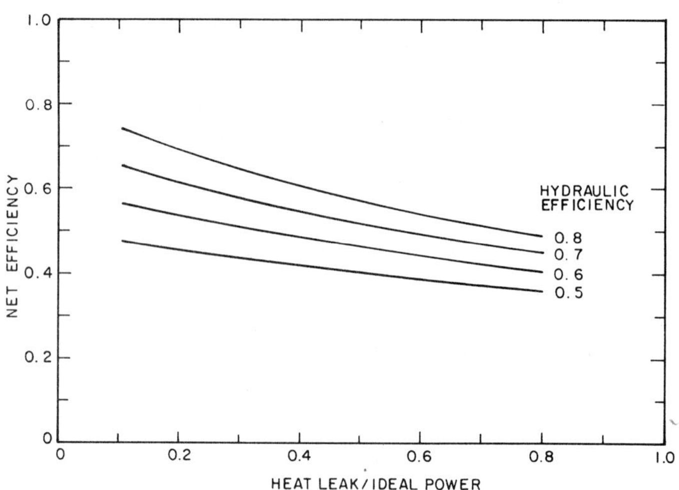

Figure 6. Overall Fluid Temperature Rise Estimated From The Circulator Efficiency Curves

D is impeller diameter,
M is shaft torque,
ρ is fluid density
ω is shaft angular velocity

In order to evaluate the behavior of a particular circulator design under some postulated operating scenarios, one need only to scale the performance of the machine from the generic curves.

Several dimensional curves have been constructed to illustrate how the pump will behave in terms of head rise and efficiency if the speed is varied to meet off design operating conditions. Figure 4 shows pump performance in the normal operating quadrant at three speeds. The speeds were chosen to meet the operating points listed earlier at a pump inlet pressure of 3.3 atm.

Figure 4a shows how the pump performance characteristic relating head rise to flow rate varies with operating speed. At any constant speed, the circulator will operate along one of the solid lines. The exact operating point will depend on the intersection of the flow/head characteristic with the system pressure loss characteristic. The solid symbols in the figure show that many of the preliminary operating conditions identified in Table 2 can be met by variations in shaft speed. The curves are generated by using the scaling relations applied to the curve in Figure 3.

Figure 4b is a similar representation showing pump efficiency varying with flow rate and speed. Again, the solid symbols relate to operating conditions in Table 2. It is clear from the figure that if the machine is designed for a best efficiency in the region of 1.2 to 1.6 l/s, operating efficiencies at the lower flows will be compromised somewhat. It is possible to shift the design in a way that the best efficiency point for the machine would be somewhat higher at the low flows, with a minor penalty at the high flows. Detailed system operating studies need to be performed before fine tuning the design point.

The efficiencies shown in this figure are the overall machine efficiencies, which include the hydrodynamic losses and the heat leak from the warm end of the machine. The motor, bearings and housing for the circulator are designed to operate at ambient temperature. Convective helium flow between the warm and cryogenic ends of the machine is controlled by a pressure balancing scheme across the shaft sleeve. Conduction is controlled by selection of materials and dimensions in a thermal transition section between the warm and cryogenic ends. For preliminary design purposes, the heat leak has been estimated at 15 W.

The net efficiency from the machine at any operating condition is given by the following relation:

$$\eta_o = (1/\eta_h + q/(m \, \Delta H))^{-1}$$

where:

η_o is the net efficiency,

q   is the heat leak,
η_h is the hydraulic efficiency,

m   is the mass flow rate, and
ΔH is the adiabatic head or enthalpy rise.

Figure 5 shows how the ratio of heat leak to ideal pumping power ($q/m\Delta H$) affects the overall machine efficiency for varying hydraulic efficiencies. Figure 6 illustrates further the effects of the combined pumping power and machine inefficiency on the temperature rise of the fluid as it passes through the circulator, for the same speeds and flow rates shown in Figure 4b.

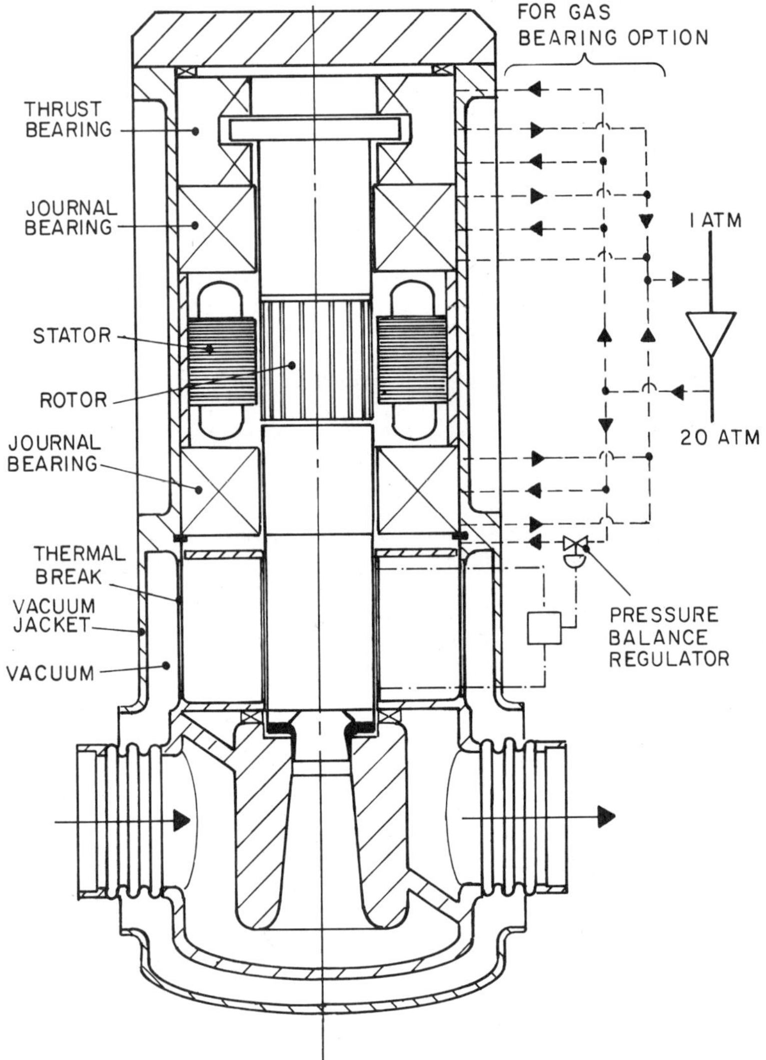

Figure 7. Cross Section Assembly Schematic of the
SSC Helium Circulator

The study in estimating performance characteristics and in establishing the impeller size, type and rotational speed has shown that a broad range of operating conditions can be achieved with good efficiency at sizes and speeds which are well within existing technology. The following section discusses the overall machine design and its bearing support system.

MECHANICAL DESIGN

There are two main features to the mechanical design which were evaluated – the drive means and bearing support system. A brushless, three phase induction motor was chosen over a turbine drive because of the relative flexibility and simplicity in the control of shaft speed. A pressurized gas bearing system is recommended for shaft support over an electromagnetic bearing system, principally because of the relative expense and complexity of the electromagnetic bearing system. This section describes the overall machine design and discusses the important elements of the drive and the bearings.

Figure 7 is a schematic cross section drawing of the circulator. The machine can be divided into three main sections: the warm end, which contains the motor and bearings, the cold end containing the impeller and diffuser and the thermal isolation section which provides structural alignment with reduced thermal conduction and convection. The unit is currently envisioned as an assembly which can be welded to the inlet and exit tubes and welded to vacuum jackets around these tubes. The cap at the upper end of the machine would be removable, allowing for access to internal components.

The warm end contains a three phase, adjustable speed, brushless induction motor. The motor rotor and stator are custom fabricated adapted from a conventional design used in oil bearing, high speed machining spindles. The motor rotor is fitted to a special shaft stub, whose lower end is a thin wall annular cylinder which is evacuated to reduce heat leak down the shaft. The warm end also contains the journal and thrust bearings.

An alternate means for supplying shaft power to drive the circulator was evaluated. A helium turbine, attached to the warm end of the shaft in place of the motor, has some advantages in terms of simplicity in fabrication. Furthermore, electrical motor controls would not be required. However, it is likely that the overall efficiency of the turbine drive would be substantially lower than the induction motor and controller. Furthermore, the complexity of pressure supply and speed regulation of the turbine may also be excessive in comparison with the adjustable frequency character of the electric motor. For the turbine drive, pressures at both the turbine inlet and exit would require monitoring and control, with valves to regulate circulator speed and torque. Special accumulators would be required to control turbine operation during losses of system pressure.

The major decision with respect to the mechanical design of the circulator was the selection of a suitable bearing system. For reliable long–term operation of the liquid helium circulator at high rotational speed, a robust non–contacting bearing system is required. Non–contacting bearings can be either gas lubricated or electromagnetic. The bearing system, shown schematically in Figure 7, consists of two radial bearings and a double–acting thrust bearing. On an individual level these bearings must be designed to support loads imposed by gravity, shaft unbalance, magnetic pull from the motor and pressure forces imposed by the system, while avoiding contact between rotating and stationary members. The load support mechanism for each bearing must be capable of damping oscillations associated with the mass of the shaft and the stiffness of the bearing. Drag forces imposed by the bearings on the rotating shaft must be acceptably small from the standpoint of circulator inefficiency. In the case of gas bearings, the design must suppress the tendency for half speed whirl at high shaft speeds.

Three bearing systems were evaluated:

1. Self—acting tilting—pad gas journal bearings and passive magnetic thrust bearing.
2. Active electromagnetic journal and thrust bearings.
3. Pressurized gas journal and thrust bearings.

The first arrangement is one used in helium circulators developed for Brookhaven and Fermi National Laboratories. For the SSC circulator, greater thrust load capacity is needed. For the tilting—pad gas bearings, good load capacity and whirl suppression require extremely small clearances between the shaft and the pads. As a result, they cannot tolerate the trace impurities and contaminants that are likely to develop under long—term operation in the SSC.

Both active electromagnetic and pressurized gas bearings can achieve good load capacity and high stiffness at relatively large clearances. Both systems have the further advantage that, unlike self acting bearings, bearing load carrying capacity is independent of rotational speed. With the larger clearances typical of these systems, contaminants can be accommodated with much less risk of seizure, making these bearing types good candidates for the SSC circulators. With pressurized bearings, filters can be easily fitted to the bearing supply, virtually eliminating the possibility of contaminants affecting bearing operation.

In the pressurized gas bearing recommended for this machine, helium at high pressure, taken from the compressor discharge in the refrigerator, enters the bearing clearance through a flow restrictor and discharges into a cavity adjacent to the bearing. Unit load capacity can be maximized through proper design of the restrictor. Load capacities typically range from 1/4 to 1/2 of the difference between supply pressure (upstream of restrictor) and discharge cavity pressure corresponding to a maximum unit load capacity of about 950 kPa (140 lb/in^{-2}).

CONCLUSIONS

The principal design features of a centrifugal circulator for the SSC have been presented. The machine is driven by a brushless, induction motor at speeds adjustable to > 30,000 rpm in order to accommodate a variety of flows and pressure requirements. The hydrodynamic design of the flow passages and the motor design are readily adaptable from existing technology. The shaft support system consists of a pressurized gas bearing system integrating a double acting thrust bearing and two journal bearings. The bearings and related pressure balancing circuit are critical to achieving the degree of robustness and reliability required for the SSC.

REFERENCES

1. SSC Central Design Group; *Conceptual Design of the Superconducting Super Collider*; SSC–SR–2020, March 1986.
2. McAshan, M.C.; *Refrigeration Plants for the SSC*; SSC Central Design Group, SSC–129, May 1987.
3. VanderArend, P.; *Refrigeration Plant Requirements and a Preliminary Plant Design*; SSC Cryogenics Workshop, June 1987.
4. Balje, O.E.; <u>Turbomachines. A Guide to Design, Selection and Theory</u>; New York, NY: John Wiley & Sons, 1981.
5. Donsky, B.; *Complete Pump Characteristics and the Effects of Specific Speeds on Hydraulic Transients*; <u>J. Basic Engrg., Trans. ASME</u>, Dec. 1961, p. 685.
6. Stacy, W.D.; *Performance Test Results for a Cold Helium Compressor*; <u>Cryogenic Properties, Processes and Applications 1986</u>, V92(251), A.J. Kidnay and M.J. Hiza, eds., AIChE, 1986, pp. 45–51.

LABARGE LIQUID HELIUM PLANT

R. R. Olson, Jr. and R. F. Pahade

Union Carbide Industrial Gases Inc.
Linde Division
Tonawanda, New York

ABSTRACT

New applications for helium have resulted in a significant growth
in helium demand. This growth has contributed to development of larger,
more efficient liquid helium production plants. The liquid helium plant
at LaBarge is sized for 4600 liters per hour of liquid helium production
(dewar mass gain basis) using two purification/liquefaction plant
trains. A liquid nitrogen forecooled refrigerator having a power consumption similar to the liquefier cycle would provide about 10,000
watts of refrigeration at liquid helium temperature. Operating experience with liquid helium production facilities has demonstrated reliable
commercial operation.

INTRODUCTION

New applications for helium, and growth of existing applications,
have resulted in a significant growth in worldwide helium consumption.
Use in applications as diverse as party balloons, welding gas mixtures,
magnetic resonance imaging and research and developement contribute to
this growth. To meet this demand, additional helium production facilities are needed. Cryogenic applications, plus the economies of liquid
product distribution dictate that these plants be large, efficient
liquid helium plants.

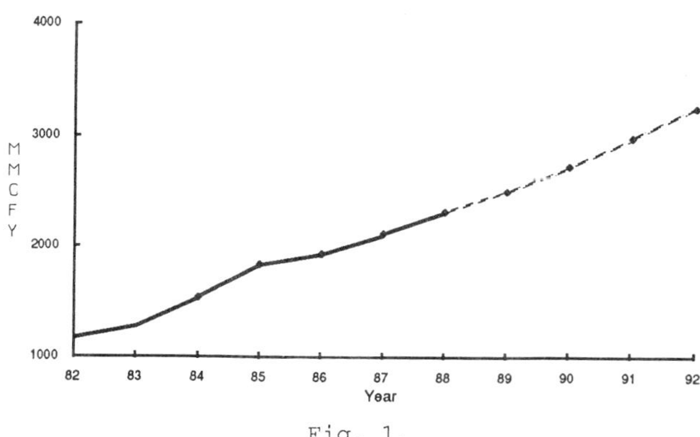

Fig. 1.

The dual-train liquid helium plant at LaBarge was designed for a total of 4600 liters per hour of liquid helium production (dewar mass gain basis). A liquid nitrogen forecooled helium refrigerator having a power consumption similar to the LaBarge liquefier cycle would provide about 10,000 watts of refrigeration at liquid helium temperature. The LaBarge plant is the largest liquid helium production facility in the world.

FACILITY DESCRIPTION

The LaBarge Helium production facility receives crude helium from two nitrogen rejection units (NRU's) associated with a dual-train natural gas processing facility. The crude stream, containing nominally 50% helium, is extracted from the NRU at a temperature near 100 K. A vacuum-insulated pipeline transmits the crude helium stream from the NRU to the helium production plant.

The helium production plant is designed to purify the helium contained in the crude stream, liquefy the purified helium, and store the liquefied product for subsequent distribution. Purification is accomplished in three steps, partial condensation purity upgrading, catalytic hydrogen removal, and final purification by pressure swing adsorption (PSA).

<u>Purity Upgrading</u>

Initial processing of the crude helium is conducted in a helium purity upgrader cold box. The upgrader cools the crude vapor to 80 K, until partial condensation of the stream improves the vapor helium purity to approximately 90%. The upgraded vapor is then warmed to ambient temperature levels prior to further processing. A small fraction of the refrigeration contained in the cold crude helium feed sustains the process. The bulk of the refrigeration is used to forecool a portion of the helium recycle stream for the helium liquefier, eliminating the need for liquid nitrogen forecooling or an additional turboexpander stage.

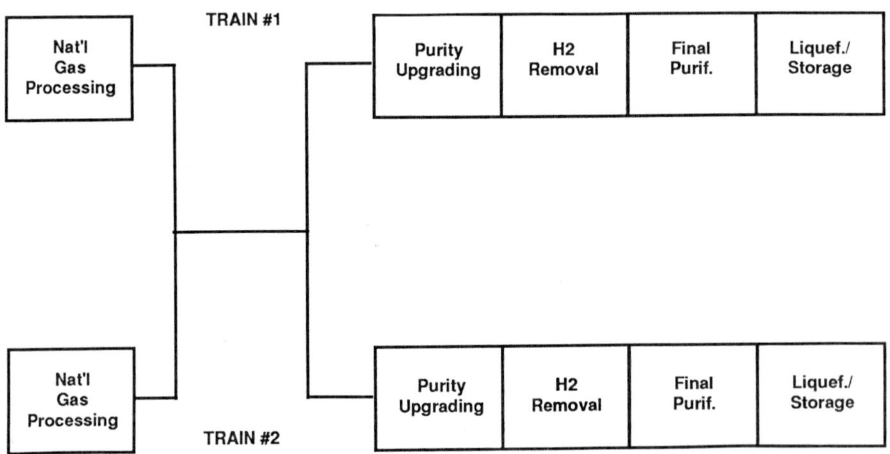

Fig. 2. LaBarge Helium Production Facility

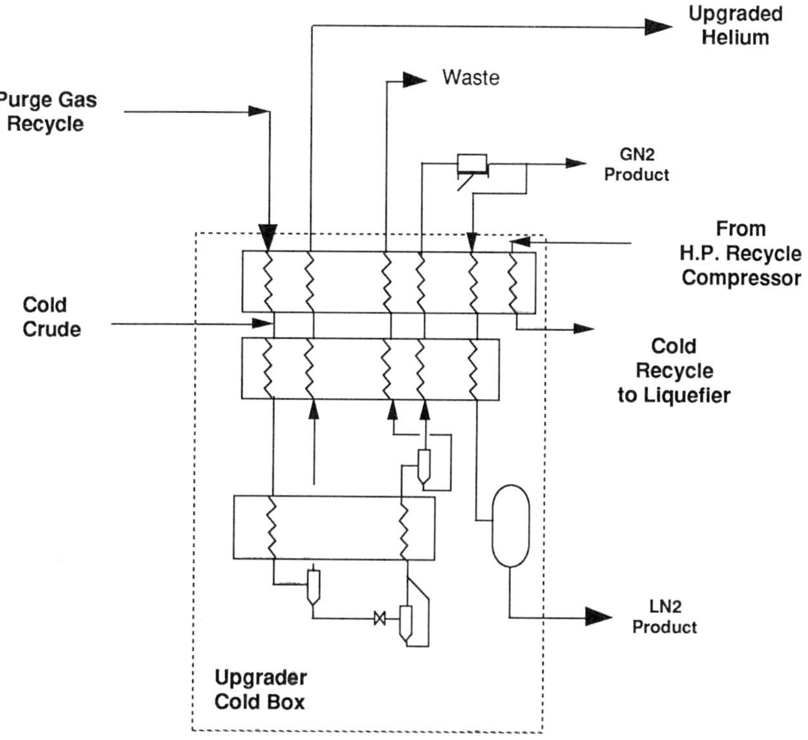

Fig. 3. LaBarge Purity Upgrader

The upgrader cold box, designed and built by Union Carbide Industrial Gases Inc., consists of a perlite-insulated casing containing multi-stream brazed aluminum heat exchangers, piping, valves, and separator vessels. Major functions of the upgrader cold box are to increase helium purity, reject contaminants recycled from the PSA unit, and recover refrigeration from the crude helium stream. The patented integrated upgrader process transfers the refrigeration to the helium liquefier[1]. In addition, nominal quantities of gaseous and liquid nitrogen are produced for in-plant use.

Hydrogen Removal

Upgraded helium is further purified by catalytic oxidation to remove small amounts of hydrogen contained in the crude helium stream. A small amount of air, containing oxygen slightly in excess of the stoichiometric amount required, is injected into the upgraded helium stream. This stream is then passed over a palladium catalyst, causing the hydrogen to be oxidized to water. The combusted helium stream is cooled to condense and removed some of the moisture. The remaining vapor-phase moisture is removed in the PSA unit.

Final Purification

Following hydrogen removal, the stream undergoes final purification in a pressure swing adsorption (PSA) unit, which removes all impurities except neon to a level less than 2 ppm. The pure helium gas is now ready to be liquefied.

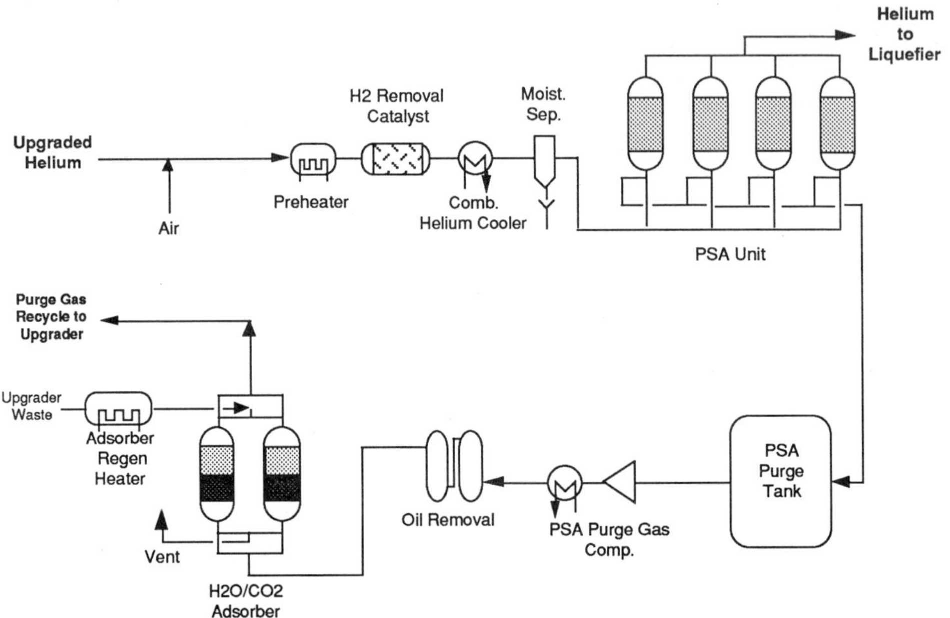

Fig. 4. LaBarge Purification System

The LaBarge pressure swing adsorption unit is an aumatically-controlled four-bed unit designed by Union Carbide Industrial Gases Inc. The PSA unit removes impurities by selective adsorption on an adsorbent bed. The adsorbent is regenerated by pressure letdown and purge. Multiple beds are used to provide continuous product flow. One bed is on-stream, while the other three beds are in various steps of the regeneration cycle. Pure helium from the depressurizing and on-stream beds are used to purge contaminants from the adsorbent and re-inventory the regenerated vessels.

Since the purge gas from the PSA vessels contains a significant amount of helium, the purge gas is recycled back to the purity upgrader cold box. The purge gas contains moisture and CO_2 which must be removed prior to the upgrader cold box processing. A dual bed thermally regenerated adsorber system provides this function. Recycling the PSA purge gas back to the upgrader cold box improves helium purification process recovery to values in excess of 98%.

<u>Liquefaction</u>

Helium gas from the PSA purifier is fed to the helium liquefier, designed and built by Sulzer Brothers Limited, where the stream is cooled down in brazed aluminum heat exchangers until it reaches the temperature level near liquid nitrogen saturation. At this point, any remaining trace air gas contaminants are adsorbed on an activated carbon adsorption trap. The gas is further cooled against warming recycled helium until the gas is near 20 K. Again, activated carbon adsorption traps remove neon and any residual hydrogen that may be present. The feed stream is then blended with the helium recycle stream prior to further heat exchange, expanded through the last refrigeration stage turboexpander and Joule-Thompson valve into the storage dewar.

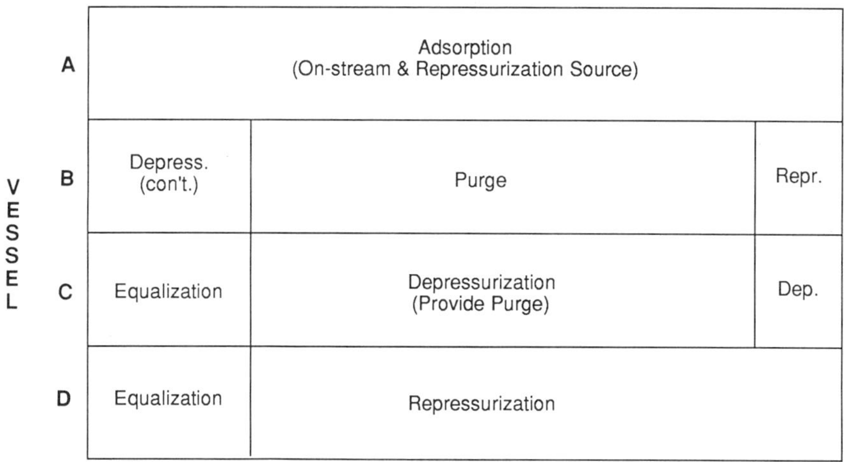

Fig. 5. Helium PSA

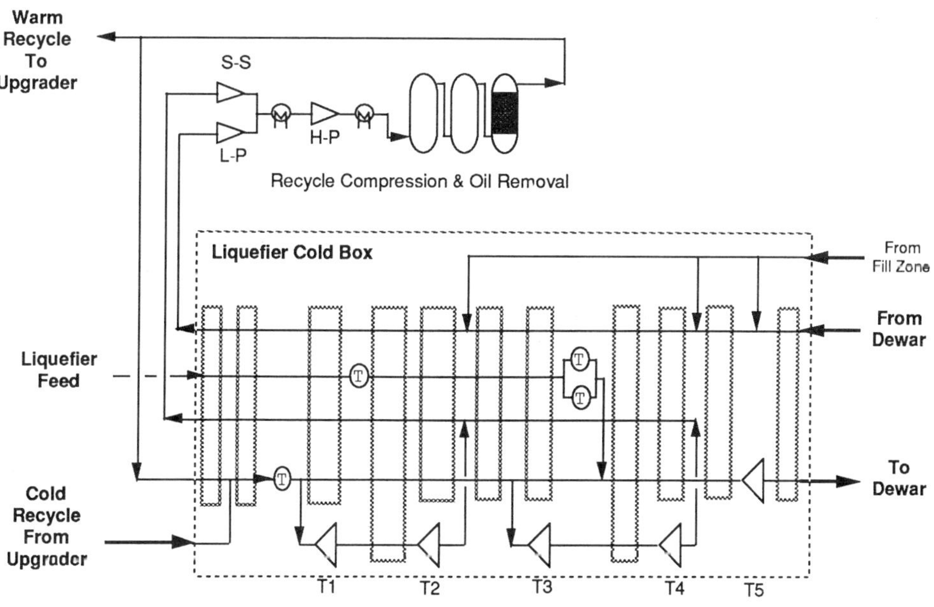

Fig. 6. LaBarge Liquefier System

<u>Liquid Storage</u>

Upon passing through the J-T valve, the cold helium fluid enters
the subcritical zone, and becomes a two-phase mixture. The two-phase
mixture is separated in the helium storage dewar. Liquid helium is
accumulated in the dewar, and the vapor portion of the stream, plus
displaced vapor from the dewar, is recycled to the cold end of the
liquefier.

The helium storage dewar, designed and fabricated by Union Carbide
Industrial Gases Inc., consists of a superinsulated, liquid nitrogen-
cooled thermally shielded horizontal vessel encased in a vacuum
jacket. The superinsulation is made up of highly reflective aluminum
foil and aluminized mylar, alternated with spacing material having a
low thermal conductivity. Thermal shielding is provided by a copper
jacket imbedded in the superinsulation, which is cooled by liquid
nitrogen.

These 60,000-gallon (227,000-liter) helium storage dewars are the
largest-size dewars in the world.

LIQUEFIER SYSTEM

<u>Cycle Description</u>

The liquefier supplies refrigeration at six levels to liquefy the
feed helium stream, which results in high thermodynamic liquefaction
efficiency. The first stage, approximately equivalent of liquid nitrogen forecooling, is recovered refrigeration from the cold crude helium
delivered to the plant from the NRU. Part of the discharge flow from
the high pressure helium recycle compression system is directed to the
upgrader cold box where it is cooled down against the warming crude
helium. The cooled helium recycle stream is then directed to the helium
liquefier heat exchanger system through vacuum-insulated piping, where
it is blended with the remainder of the high pressure recycle flow near
the 100 K temperature level.

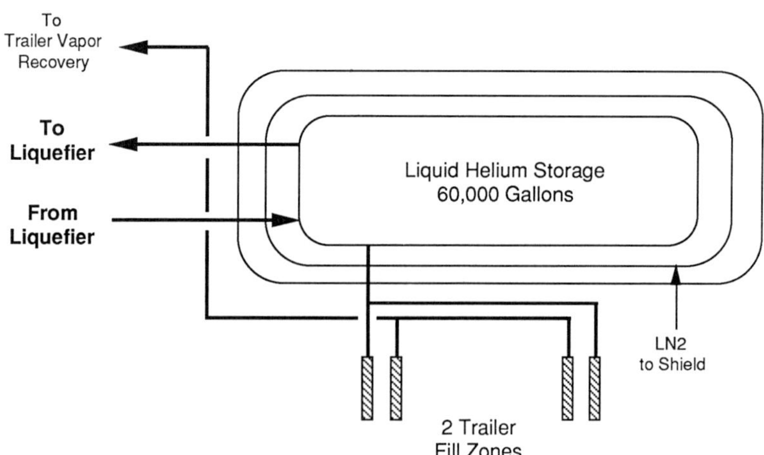

Fig. 7. LaBarge Helium Storage Dewar

The next two stages of refrigeration are supplied by two oil-brake
loaded turboexpanders T1 and T2. These turboexpanders are in series,
with heat exchange in between. Exhaust temperatures are approximately
75 K and 42 K, respectively. The exhaust flow of these turbines enters
a moderate pressure recycle side stream.

Blower-loaded gas-bearing turboexpanders provide the remainder of
the refrigeration. The next two units, T3 and T4, are arranged in series similar to T1 and T2, with heat exchange in between, and exhausting to the recycle side stream.

The last turboexpander, T5, exhausts at a pressure above the critical pressure. The exhaust stream undergoes further cooling against
saturated vapors returning from the storage dewar, until it is cooled
to under 6 K. This cooled supercritical fluid is then expanded to near
atmospheric pressure through the J-T valve, where a two-phase mixture
is formed. The liquid accumulates in the storage dewar, and the vapor
is returned to the liquefier low-pressure recycle stream.

Compression

Recycled helium leaves the liquefier at two pressure levels. These
two streams are fed to the suctions of two oil-flooded screw compressors. The low pressure stream is near atmospheric pressure, and is compressed in a Howden compressor driven by a 450 HP motor. The medium-
pressure side stream is compressed in a Howden compressor driven by a
600 HP motor. Both compressor casings, lube oil system, and bulk oil
removal system are assembled on a common skid. The equipment skid design and fabrication is by Reco. Discharge flows from both compressors
are combined and fed to the suction of the high-pressure recycle compressor.

The high-pressure recycle compressor increases the recycle stream
pressure to about 300 psia. The high pressure oil-flooded screw compressor is a Howden compressor driven by a 2000 HP motor. The high-
pressure compressor casing, lube oil system, and bulk oil removal system are assembled on an equipment skid, similar to the low-pressure and
side-stream compressors.

Oil Removal

Final traces of oil are removed from the recycle stream by passing
the gas through two stages of oil coalescers. These units are designed
to remove oil mist to a level of 0.01 ppm. While these coalescers are
effective in removing sub-micron aerosols, they do not remove any vapor-phase contamination. Oil vapor contamination is minimized by using
synthetic lubricants having very low vapor pressure, by pretreating the
oil to remove moisture and any light-end contamination, and by passing
the coalescer effluent through an activated carbon adsorption bed.

Cold Box

The helium liquefier, designed and fabricated by Sulzer Brothers
Limited, consists of a top-suspended vacuum/superinsulated vessel, containing brazed aluminum heat exchangers, two oil-lubricated and three
gas-bearing turboexpanders, piping, valves and adsorption traps. The
vacuum shell can be lowered for liquefier maintenance and repair. The
shell is fabricated in sections which can be fitted with castors, allowing the shell sections to be rolled off to the side of the pit to
make room for the next section. This minimizes the depth of the pit

required to accomodate the shell, and makes use of the space below the
adjacent liquefier cold box for shell section storage.

Turboexpanders

Liquefier process refrigeration is supplied by five Sulzer turboexpanders, T1 through T5. No reciprocating expansion engines are used.
Turbines T1 and T2 are oil-brake loaded expanders. T3 through T5 are
blower-loaded gas-bearing expanders. T1 through T5 performance parameters are listed below:

	Output Speed	
Turbine	KW	RPS
T1	18.5	2100
T2	16.0	2100
T3	13.0	2440
T4	6.8	1385
T5	1.9	2065

Operations

The liquefier has three different temperature-level feed points for
low pressure displacement gas returning from a liquid transport trailer
as it is being filled. Selection of the feed point is determined based
on the "condition" (temperature) of the trailer to be filled. Having
multiple feed points allows recovery of the refrigeration contained in
the vapor with minimum upset of the liquefier process. Each helium
liquefier has been designed to operate efficiently at 50% turndown, allowing the facility to operate at 25% load if required.

Since the natural gas processing plant was being started up concurrently, the helium plant startup was expedited by commissioning the
plant equipment in full recycle using a synthesized crude helium
stream. This allowed process tuning and debugging of minor problems
typically found during plant startup. When a stable supply of crude
helium was available, the helium plant was ready to process the feed-
stock. The startup progressed very well. Changes were made to the expander lube oil system, but no other field modifications were required.

On-stream factors for helium plants are typically in the 95 percent
range. These types of facilities are very sensitive to power interruptions, due to trip-out of compression equipment switchgear. To maintain
high on-stream factors, preventative and/or corrective maintenance
shutdown(s) of one-two weeks per year are necessary.

REFERENCES

1. R. R. Olson, U.S. Patent No. 4,666,481, "Process for Producing
    Liquid Helium", May 19, 1987.

Poster Session

Chairman:

M. A. Green
Lawrence Berkeley Laboratory

PASSIVE SUPERCONDUCTOR A VIABLE METHOD OF CONTROLLING
MAGNETIZATION MULTIPOLES IN THE SSC DIPOLE

Michael A. Green

M/S 90-2148
Lawrence Berkeley Laboratory
Berkeley, CA 94720

ABSTRACT

At injection, the magnetization of the superconductor produces the dominant field error in the SSC dipole magnets. The field generated by magnetization currents in the superconductor is rich in higher symmetric multipoles (normal sextupole, normal decapole, and so on). Pieces of passive superconductor properly located within the bore of the dipole magnet can cancel the higher multipoles generated by the SSC dipole coils. The multipoles generated by the passive superconductor (predominantly sextupole and decapole) are controlled by the angular and radial location of the superconductor, the volume of superconductor, and the size of the superconducting filaments within the passive conductor. This paper will present the tolerances on each of these factors. The paper will show that multipole correction using passive superconductor is in general immune to the effects of temperature and magnetization decay due to flux creep, provided that dipole superconductor and the passive correction superconductor are properly specified. When combined with a lumped correction system, the passive superconductor can be a viable alternative to continuous correction coils within the SSC dipoles.

BACKGROUND

The effect of superconductor magnetization on the quality of the magnetic field in a superconducting dipole was observed almost 20 years ago.[1] It has been observed that superconductor magnetization will produce higher normal multipoles such as sextupole, decapole, 14-pole and so on, even though the magnet was designed to produce none of these multipoles. From the beginning, when LBL and others started fabricating accelerator types of dipoles, it was recognized that the higher multipoles generated by the superconductor magnetization would have an adverse affect on the performance of accelerators at injection, if the injection field is low enough.

On the Fermilab doubler saver, the higher multipoles generated by magnetization were not a problem until the machine was used as a storage ring.[2] When the Tevatron is used as a fixed target accelerator, the circulating beam current is low enough that the effects of magnetization could be corrected using lumped correction elements. Colliding beam storage rings are strongly affected by low level sextupole and decapole at injection. This effect grows worse as the beam size is reduced and more protons are packed into the beam.[3] The SSC and HERA have proposed to correct out the magnetization sextupole and decapole using continuous correction elements down the bore of each dipole. This solution is quite expensive.

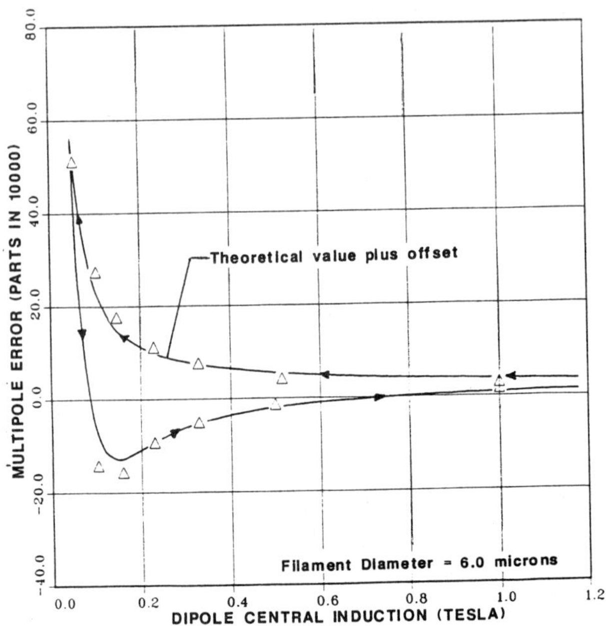

Figure 1.   A COMPARISON OF MEASURED ERROR AND THE THEORETICAL ERROR PLUS   THE
THE OFFSET AS A FUNCTION OF CENTRAL INDUCTION ON LBL DIPOLE D-15A-5
NORMAL SEXUPOLE TERM

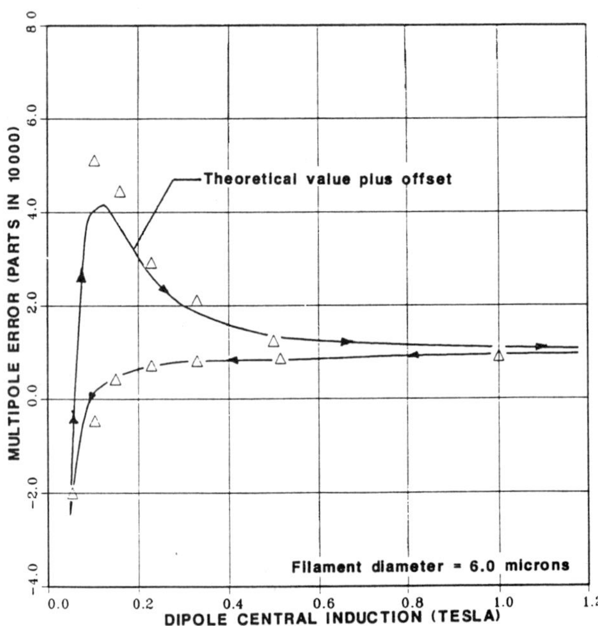

Figure 2.   A COMPARISON OF MEASURED ERROR AND THE THRORETICAL ERROR PLUS THE
OFFSET AS A FUNCTION OF CENTRAL INDUCTION ON LBL DIPOLE D-15A-5
NORMAL DECAPOLE TERM

352

Beam dynamics studies by the SSC-CDG suggest that if the magnetization sextupole can be reduced to 2 units (1 unit equals 1 part in 10,000) at an injection induction of 0.33T, the effects of magnetic field error on the stored proton beam can be controlled using lumped correction elements in each half cell of the machine lattice.[4] Another correction scheme proposed by Neuffer[5] would allow for the correction of sextupole up 5 or 6 units using lumped elements every three dipole magnets. If the magnetization sextupole can be reduced to one or two units in each dipole, the expensive continuous correction elements can be replaced by the lumped correction elements which are already needed to control the tune of the SSC.

BASIC THEORY

The field generated by circulating currents in a single filament of superconductor can be represented by the classical hydrodynamic doublet equation. In complex form this equation takes the following form:

$$H''^*(Z) = \frac{\Gamma e^{i\alpha}}{2\pi i (Z-Z_c)^2} \qquad\qquad 1$$

where H''^* is the complex conjugate of the field $H''(Z)$ at a point Z generated by a current doublet with strength Γ and doublet angle α at a location Z_c. Γ is nothing but the product of the circulating current and the average distance between the circulating currents. (For a fully penetrated round beam model conductor, the distance is 0.423 times the filament diameter.) Γ is proportional to superconductor magnetization and α is the angle of the flux line which generated the circulating current minus $\pi/2$. Both Γ and α are functions of the previous flux history of the superconducting filament.

Equation 1 can be expanded into a Taylor series about the origin. Since superconducting dipoles and quadrupoles of interest are symmetrical (they are symmetrical about the axes which are at $\theta = 0$, $\theta = \pi/2T$, π/T, and so on to 2π, where T is the fundamental multipole of the magnet in question (T = 1 is a dipole magnet and T = 2 is a quadrupole magnet)), the power series takes the following form:

$$H''^*(Z) = \sum_{N=1}^{\infty} a_n'' \, Z^{N-1} \qquad\qquad 2$$

where

$$a_n'' = -\frac{2T\Gamma}{\pi i} N \cos((N + 1)\, \theta_c - \alpha)\, r_c^{-(N+1)} \qquad\qquad 3a$$

when $N = T(2P + 1)$, $p = 0, 1, 2, \ldots$ and

$$a_n'' = 0 \qquad\qquad 3b$$

when $N \neq T(2P + 1)$, $p = 0, 1, 2$. We define θ_c as the filament angle from the x axis, r_c is radius from the origin to the filament and N is the series multipole (n = 1 is dipole, N = 2 is quadrupole, and so on). T, Γ, and α are previously defined. There is a similar power series for the image doublet in the iron, but it is not very important for this discussion.

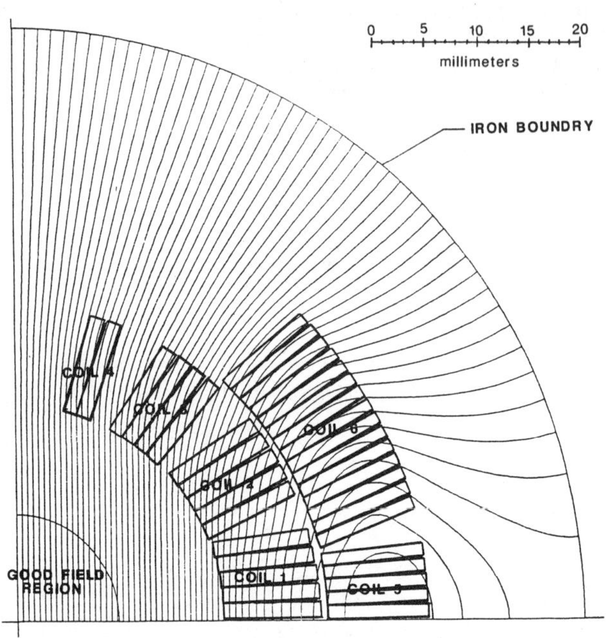

Figure 3. SSC DIPOLE CROSS-SECTION LBL NC-9: Showing Magnetic Flux LInes

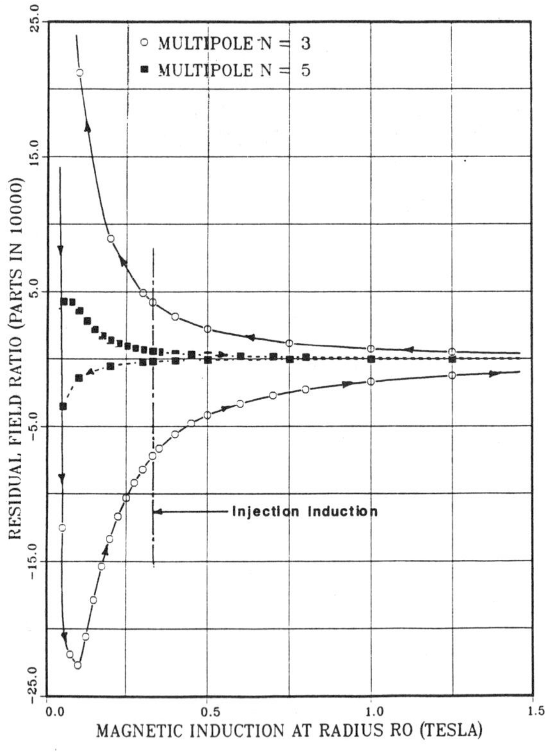

Figure 4. THE RATIO OF MAGNETIZATION SEXTUPOLE AND DECAPOLE TO THE TRANSPORT
CURRENT DIPOLE AS A FUNCTION OF DIPOLE CENTRAL INDUCTION WITHOUT
PASSIVE SUPERCONDUCTOR

It was said earlier that the doublet strength factor Γ is proportional to the superconductor magnetization. This relationship is as follows:

$$G = \frac{\pi D_f^2 M}{4} \qquad\qquad 4$$

where D_f is the superconducting filament diameter, M is the magnetization (Am^{-1}) and Γ is the ld product for the doublet (Am).

The magnetization of the superconductor contains four basic terms: 1) The bulk magnetization of the superconductor is proportional to filament diameter and J_c. 2) There is a magnetization due to surface effects such as h_{c1} and the vortex current.[7] This term is independent of J_c and filament diameter. 3) Coupling due to eddy currents between superconducting filaments and cable strands manifests itself as a flux change rate dependent term.[8] This term can be controlled by the twist pitch of the multifilamentary conductor and the transposition pitch of the cable. 4) There is magnetization due to tunneling between superconducting filaments which are in close proximity.[9,10]

For a superconductor with fully penetrated filaments which are spaced far enough apart to avoid proximity effects, the magnetization M will take the following form:

$$M = M_{f1} + M_{f2} \qquad\qquad 5$$

where

$$M_{f1} \approx \frac{2}{3\pi} D_f J_c [1 - \delta] \qquad\qquad 5a$$

and where

$$M_{f2} \approx H_{c1} - \frac{\ell n[(H - H_{c1}/2)\phi]}{\ell n[(H_{c2} - H_{c1}/2)\phi]} H_{c1} \qquad\qquad 5b$$

D_f is the filament diameter, J_c is the critical current density of the superconductor in the filament at a field H; H_{c1} is the lower critical field; H_{c2} is the upper critical field; δ is the fraction of the conductor current carrying capacity carrying transport current (δ cannot be larger than 1); and $\phi = \lambda/(2.07 \times 10^{-15})$ with λ the superconductor penetration depth (λ is about 2500 angstroms for Nb-Ti).

All of the magnetization terms except the h_{c1} and vortex effects will decay with time. Flux creep decays of long time constants have been observed in both the bulk magnetization and the proximity coupling terms.[11,12,13] The decay time constants for the filament twist pitch and cable transposition pitch dependent magnetization are generally short. Since these terms are small, this type of decay is not of concern in a magnet. The decay associated with flux creep is of concern because as much as half of the bulk magnetization can decay during injection into the SSC. This decay has a log time dependence.

The effects of magnetization on the magnetic field in an SSC dipole magnet was modeled using the LBL SCMAGØ4 computer program. This program shows good agreement with measurements of magnetization in dipole magnets.[14] Figures 1 and 2 show a comparison of the measured sextupole and decapole with normal sextupole and decapole calculated by the SCMAGØ4 code. The SCMAGØ4 has been used to calculate the effects of nonsymmetric magnetization and currents.[15] The asymmetries appear as skew terms and even normal terms in the magnetic field expansion.

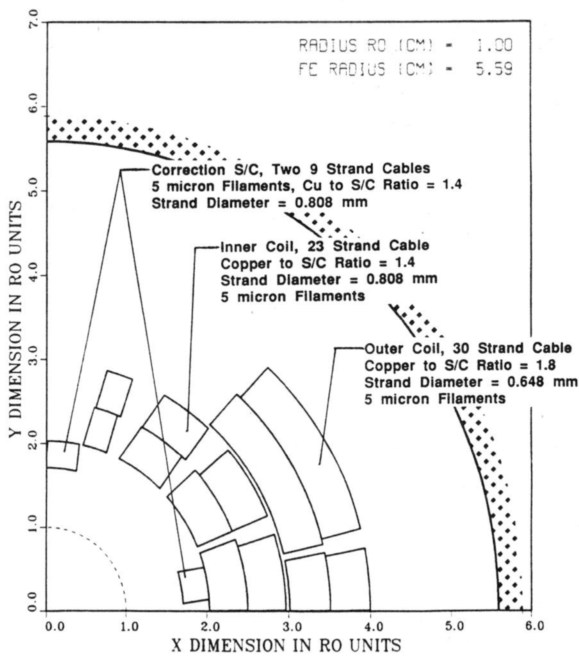

Figure 5. THE LBL NC-9 MODEL SSC DIPOLE COMPUTER MODEL CROSS-SECTION
PASSIVE SUPERCONDUCTOR CORRECTION CASE A

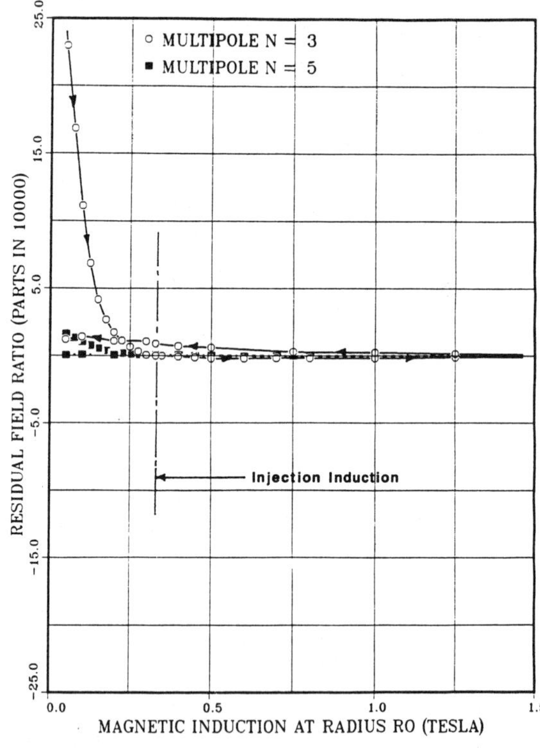

Figure 6. THE RATIO OF MAGNETIZATION SEXTUPOLE AND DECAPOLE TO THE TRANSPORT
CURRENT DIPOLE AS A FUNCTION OF DIPOLE CENTRAL INDUCTION WITH
PASSIVE SUPERCONDUCTOR CASE A

ELIMINATION OF MAGNETIZATION SEXTUPOLE AND DECAPOLE WITH PASSIVE ELEMENTS

The concept of using passive elements to eliminate the sextupole and the decapole in a dipole magnet is not new. The use of passive superconductor was first suggested by H. E. Fisk of Fermilab.[16] Ferromagnetic passive correction and correction using oriented permanent magnet materials has also been studied.[17] The oriented permanent magnet materials are expensive and difficult to manufacture so that the magnetization points in the correct direction.

Ferromagnetic correction using Mu metal (which has a saturation induction of about 0.65T) and soft iron (which has a saturation induction of 2.0T) is simple and not much metal is required to correct out the sextupole and decapole. The disadvantages of this approach are: 1) There is an offset in the sextupole and decapole at magnetic inductions above 0.5T. This offset is at its worst at a central induction of about 2T. 2) Ferromagnetic correction does not respond to changes in magnetization due to changes in temperature. 3) The decay of superconductor magnetization is not compensated for by ferromagnetic correction.

Compensation using passive superconductor works over a wide range of dipole central inductions whether the field is rising or falling. Compensation of the magnetization induced sextupole and decapole using a passive superconductor continues even when the temperature changes. It is probable that passive superconductor will compensate for the flux creep decay of the sextupole and decapole produced by magnetization.

Correction of dipole NC-9 with passive superconductor

Figure 3 shows the LBL NC-9 dipole cross-section with flux lines and the ratio of the magnetization sextupole and decapole to the central dipole induction for an uncorrected NC-9 dipole with superconductor which has 5 micron filaments. In Figure 4 one can see the positive magnetization sextupole as the central induction is reduced from 6.6T. When the central induction is reduced to zero and then increased, the positive sextupole decreases and becomes negative. When the central induction increases from zero to 0.1T (after coming down from 6.6T to zero), the magnitude of the negative sextupole decreases. At an injection induction of 0.33T, the sextupole ratio at a radius of 10 mm is -7.126 units (one unit is 1 part in 10,000). The decapole ratio at injection is about +1 unit.

Passive superconductor must create a positive sextupole of +7.16 units at injection. (The sextupole generated by the magnet coil should be corrected out by the sextupole created by passive superconductor.) A negative decapole of about one unit must be produced by the passive superconductor. To see how the passive superconductor works, look at Equation 3a. The passive superconductor, which is mounted inside the coil (in order to minimize the amount of passive superconductor needed to correct the field), sees nearly a perfect dipole field such that $\alpha = 0$ or $\alpha = \pi$. when $\alpha = 0$ Equation 3a takes the following form:

$$a_n'' = -\frac{2\Pi\Gamma}{\pi i} N \cos((N + 1)\, \theta_C)\, r_C^{-(N+1)} \qquad\qquad 6$$

This equation says that the sextupole term ($N = 3$) will vary as $4\theta_C$ and Γ. (The decapole term will vary as $6\theta_C$ and Γ.) In order to achieve an elimination of the magnetization field over a wide range of fields in the magnet, one must manipulate Γ by choosing the correction superconductor filament diameter, and the correct angle θ_C. Other higher multipoles can be manipulated by θ_C as well.

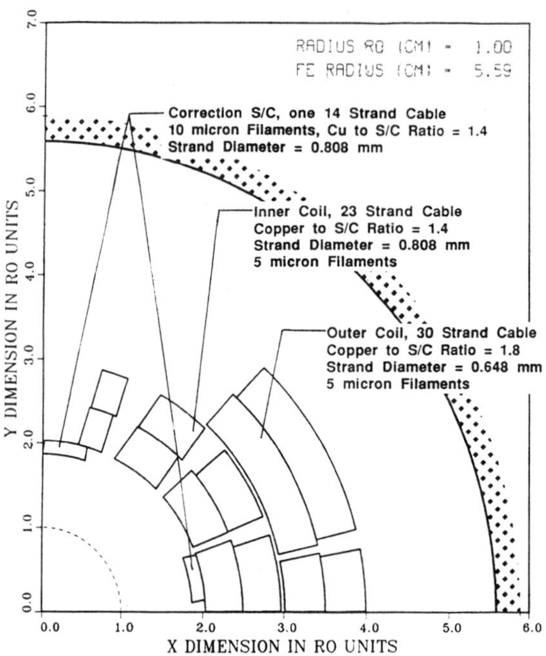

Figure 7. THE LBL NC-9 MODEL SSC DIPOLE COMPUTER MODEL CROSS-SECTION
PASSIVE SUPERCONDUCTOR CORRECTION CASE B

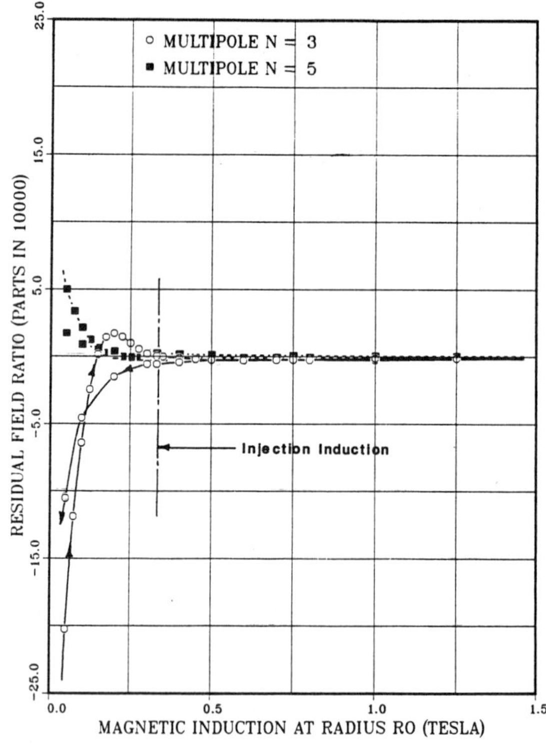

Figure 8. THE RATIO OF MAGNETIZATION SEXTUPOLE AND DECAPOLE TO THE TRANSPORT
CURRENT DIPOLE AS A FUNCTION OF DIPOLE CENTRAL INDUCTION WITH
PASSIVE SUPERCONDUCTOR CASE B

<u>Two correction methods</u>

Two correction schemes studied at LBL are presented here. In order to get the desired positive sextupole in the NC-9, dipole pieces of passive superconductor should be placed symmetrically about θ_c = 0, 90, 180, and 270 degrees (see Equation 5). If one wants to create a negative decapole as well as a positive sextupole, the passive superconductor at the midplane θ_c = 0 and θ_c = 180 must be split symmetrically with space between the conductor.

The first correction scheme, Case A, uses sixteen 9-strand cables with 9 strands made from inner cable strands (with a copper-to-superconductor ratio of 1.4 and 5 micron filaments). Figure 5 shows a quarter of the model NC-9 dipole with the correction, the Case A, passive superconductor. Figure 6 shows the ratio of magnetization sextupole and decapole to central dipole induction versus the central induction of the magnet. Table 1 shows the magnetization sextupole ratio for the dipole without passive correction cases A and B.

The second correction scheme, Case B, uses eight 14-strand cables with the 14 strands made from inner coil superconductor with 10 micron filaments and a copper-to-superconductor ratio of 1.4. Figure 7 shows a quarter of the model NC-9 dipole coil cross-section with the Case B passive correction superconductors. Figure 8 shows the ratio of magnetization sextupole and decapole to central dipole induction as a function of central induction. In both Case A and Case B the magnetization sextupole has been reduced by over two orders of magnitude at injection. In both cases, the sextupole and decapole are well within bounds so that the SSC accelerator can be corrected using lumped elements every half cell.

<u>The effects of temperature, conductor placement errors, magnetization mismatch, proximity coupling and magnetization decay</u>

The superconductor used to correct the magnetization multipoles in the dipole react to temperature changes in much the same way as superconductor in the magnet coils. The value of dJ_c/dT divided by J_c is about 0.208 k^{-1} for all of the superconductor in the magnet. In the range of temperatures expected in the SSC (from 4.2K to 4.5K), the change in sextupole passive correction with temperature is less than 0.02 units. (See Table 2 for a comparison of temperature effect on the quality of corrected field.) Even at 1.8K, satisfactory passive correction can be obtained.

A symmetrical one-degree placement error changes the correction superconductor magnetization sextupole by about 0.2 units (at a radius of 10 mm). (A one-degree error corresponds to a placement error of 0.32 mm.) If a one-degree error occurs on one superconductor block, a skew quadrupole of about 0.07 units (at a radius of 10 mm) is produced. The normal sextupole produced by a single conductor motion is of the order of 0.03 units. The allowable error in the placement of the passive superconductor is about 5 degrees (about 1.6 mm).

The magnet superconductor is expected to have a critical current density at 5 T and 4.2 K of 2750 Amm^{-2} $\pm$ 3 percent. The critical current density at 0.3 T is about 5.5 times larger than the critical current density at 5.0T.[18] Measurements on samples of many kinds of niobium titanium suggest that the variation of this ratio is about $\pm$10 to 15 percent. If the superconductor is carefully selected so that the passive superconductor has the same metallurgical structure as the magnet conductor, this variation is much lower. The critical current density of the magnet superconductor and the passive superconductor should be specified to have the same value at two different inductions (say 2 T and 5 T). The diameter of the superconducting filaments is expected to vary from magnet to magnet by less than 3 percent. It is reasonable to expect a magnet-to-magnet variation of the magnetization field multipole components to be about 5 or 6 percent. In a magnet system with passive superconductor the sextupole variation from magnet to magnet can be expected to be about 0.5 units.

Table 1

A COMPARISON OF SEXTUPOLE
RATIOS AT LOW FIELDS
WITH AND WITHOUT PASSIVE
SUPERCONDUCTOR CORRECTION

Central Induction# (tesla)	Sextupole to Dipole Ratio (units)*		
	without Correction	Correction Case A	Correction Case B
0.100	-22.69	11.14	-6.39
0.150	-17.85	4.17	0.17
0.200	-13.34	1.76	1.75
0.250	-10.30	0.67	1.00
0.300	-8.17	0.06	0.22
0.330	-7.16	0.01	0.04
0.350	-6.62	0.01	0.00
0.400	-5.57	-0.06	-0.10
0.500	-4.15	-0.19	-0.27
0.600	-3.30	-0.18	-0.27
0.800	-2.26	-0.15	-0.24
1.000	-1.68	-0.17	-0.25
1.250	-1.23	-0.10	-0.17
1.500	-0.94	-0.07	-0.13

\# Induction at the Dipole Center due to the Transport Current
* 1 unit = 1 part in 10000 taken at a 10 mm radius

Table 2

A COMPARISON OF SEXTUPOLE
RATIOS AT LOW FIELDS
WITH PASSIVE SUPERCONDUCTOR
AS A FUNCTION OF TEMPERATURE

Central Induction# (tesla)	Sextupole to Dipole Ratio (units)*		
	Correction 1.8 K	Correction 4.3 K	Correction 4.5 K
0.100	16.45	11.14	10.48
0.150	8.83	4.17	3.81
0.200	3.88	1.76	1.61
0.250	1.70	0.67	0.59
0.300	0.62	0.06	0.03
0.330	0.38	0.01	0.00
0.350	0.31	0.01	-0.01
0.400	0.12	-0.06	-0.07
0.500	-0.12	-0.19	-0.20
0.600	-0.12	-0.18	-0.18
0.800	-0.14	-0.15	-0.15
1.000	-0.22	-0.17	-0.16
1.250	-0.15	-0.10	-0.10
1.500	-0.12	-0.07	-0.07

\# Induction at the Dipole Center due to the Transport Current
* 1 unit = 1 part in 10000 taken at a 10 mm radius

Proximity coupling in either the magnet superconductor or the passive superconductor must be avoided. Proximity coupling can be eliminated by spacing the filament at least 1.2 microns apart in a copper matrix. If the matrix copper is poisoned with manganese, the filament spacing can be reduced. Recent experiments at the Brookhaven National Laboratory suggest that magnetization due to proximity coupling decays faster than does magnetization in the superconducting filaments by themselves.[19]

Recent experiments at LBL suggest that the magnetization decay rate is proportional to J_c and filament diameter. This suggests that the magnetization generated by the magnet superconductor and the passive superconductor will decay together. It is expected that the magnetization sextupole generated by the correction superconductor will continue to cancel the magnetization sextupole generated by the magnet coil superconductor as time progresses. This hypothesis has not been tested by an experiment.

CONCLUDING COMMENTS

Passive superconductor inside the coils of the SSC dipole can potentially greatly reduce the sextupole and decapole due to magnetization of the coil superconductor. This correction will occur over a wide range of magnetic inductions from 0.2 T to full field. (The 14-pole component can also be controlled by a more complex arrangement of the superconductor.) The passive superconductor should extend over the full straight section length of the superconductor dipole magnet. (There is no need to bring the passive superconductor over the dipole magnet ends.) The passive superconductor will increase the superconductor requirements of the SSC dipole by 4 to 5 percent depending on the case. (Less superconductor is required for passive correction than would be required to build powered continuous dipole correction coils.) The case of passive superconductor to correct the sextupole and decapole will permit one to eliminate the continuous powered correction coils down the bore of each dipole magnet.

ACKNOWLEDGEMENTS

The author thanks J. M. Peterson of the SSC-CDG for his many comments. The author also acknowledges the many conversations with W. S. Gilbert of the Lawrence Berkeley Laboratory and A. K. Ghosh of Brookhaven National Laboratory.

This work was supported by the Director, Office of Energy Research, Office of High Energy and Nuclear Physics, High Energy Physics Division, U.S. Department of Energy under Contract No. DE-AC03-76SF00098.

REFERENCES

1.   M. A. Green, IEEE Transactions on Nuclear Science NS-18 (3), p. 664, June 1971.
2.   H. C. Brown, et al, IEEE Transactions on Magnetics 21 (2), p. 979, March 1985.
3.   H. E. Fisk, et al, "Magnetic Errors in the SSC", Report of the SSC Central Design Group SSC-7, April 1985.
4.   A. Chao and M. Tigner, "Requirements for Dipole Field Uniformity and Beam Tube Correction Windings", SSC-N-183, May 1986.
5.   D. Neuffer, "A Novel Method for Correcting the SSC Multipole Problem", SSC-172, April 1988.
6.   M. A. Green, "Residual Fields in Superconducting Magnets", published in the Proceedings of the MT-4 Conference at Brookhaven National Laboratory, , p. 339, 1972.

7. W. J. Carr, IEEE Transactions on Magnetics, <u>MAG-21</u> (2), p. 335.

8. M. N. Wilson, <u>Superconducting Magnets</u>, Clarendon Oxford Press, Oxford, UK, pp. 174-181, 1983.

9. E. W. Collings, "Stabilizer Design Considerations in Ultrafine Filamentary Cu/Ti-Nb Composites", Proceedings of the 6th Workshop on Niobium-Titanium Superconductors, University of Wisconsin, Madison, WI, 1986.

10. M. A. Green, Advances in Cryogenic Engineering <u>33</u>, Plenum Press, New York, 1987.

11. D. A. Herrup et al., "Time Variations of Fields in Superconducting Magnets and their Effects on Accelerators", IEEE Transactions on Magnetics, <u>MAG-25</u>, No. 2, 1989.

12. M. R. Beasley et al., Physical Review 181, pp. 682-700, May 1969.

13. W. S. Gilbert et al., "Magnetic Field Decay in Model SSC Dipoles", IEEE Transactions on Magnetics, <u>MAG-25</u>, No. 2, 1989.

14. M. A. Green, "Field Generated Within the SSC Magnets Due to Persistent Currents in the Superconductor", Proceedings of the Ann Arbor Workshop on SSC Issues, LBL-17249, December 1983.

15. M. A. Green, "Aberrations Due to Asymmetries in Current, Filaments and Critical Current", LBL Engineering Note M6670, August 1987.

16. H.E. Fisk and A. D. McInturff, private communication on the use of a passive superconductor to correct the residual field higher multipoles.

17. M. A. Green, IEEE Transactions on Magnetics, <u>Mag-23</u>, No. 2, p. 506, 1987.

18. M. A. Green, "Calculating the J_c, B, T Surface for Niobium Titanium Using a Reduced State Model", IEEE Transactions on Magnetics, <u>MAG-25</u>, No. 2, 1989.

19. A. K. Ghosh, National Laboratory, private communication on the decay of proximity coupled currents, January 1989.

20. W. S. Gilbert, Lawrence Berkeley Laboratory, private communication on LBL measurements of magnetization field decay, January 1989.

A NEW METHOD TO CALCULATE CONDUCTOR MAGNETIZATION IN ACCELERATOR DIPOLES[*]

D. ter Avest and L.J.M. van de Klundert

University of Twente
Applied Superconductivity Centre
Dept. of Applied Physics
P.O. Box 217, 7500 AE ENSCHEDE, The Netherlands

ABSTRACT

During the ramping of superconducting dipole magnets as used in particle accelerators as the proposed SSC or CERN-LHC, magnetization in the conductor will manifest itself, perturbing the field homogeneity of the dipole. A numerical method was developed that covers the calculation of magnetic field inside the conductors, keystoned cables, current redistribution due to the variation of magnetic field with time as well as the multipole distribution in the dipole. For the calculation of current redistribution in a cable a keystoned cable geometry has been assumed, although the method applies to any geometry. The model incorporates saturation in the cable. In the case of an unsaturated cable the current redistribution can be described completely analytically. Also the model allows the calculation of the interstrand coupling loss as a function of the time-varying magnetic field. Finally, field remanency due to persistent currents in the filaments is described.

The model was converted into a computer code and applied to the design of a CERN-LHC 10T dipole magnet. A relation will be presented that describes the dependence of the multipole distribution on the time varying field as well as on a few material properties. Also the effect of persistent currents on the multipole distribution will be demonstrated.

INTRODUCTION

The field quality of dipole magnets is directly affected by the conductor parameters. For instance, the soldering of the conductor for better mechanical properties will diminish the interstrand resistance and therefore larger magnetization currents will occur during the ramping of the magnet, which leads to a poorer field quality. In most dipole magnets the conductor parameters will have to be a compromise

---

[*] These investigations in the programme of the Foundation for Fundamental Research on Matter FOM are supported, in part, by the Netherlands Technology Foundation STW, Utrecht, The Netherlands.

between ease of conductor production or fabrication of the magnet and allowable field errors. For instance, the Nb_3Sn superconductor produced by the so-called powder method [1], to be used for the Dutch contribution to the LHC project at CERN [2], has a very high current density [3]. This will enhance persistent current effects in the magnet and a limit will be put to the maximum allowable filament diameter.
Following the above discussion it appears to be important to examine quantitatively the effect of conductor parameters on the field quality. This paper describes the magnetization on cable level as well as on filament level and the magnitude of the field errors produced by these magnetization effects.

CABLE MAGNETIZATION

The effect of a time varying field on the current distribution in a cable during ramping was calculated assuming the cable is a thin layer with effective dimensions (Fig. 1) consisting of a continuum of multifilamentary superconductor. This reduces the problem to a set of two-dimensional equations. The twist angle ψ is defined in Fig. 2. For convenience, a local cartesian coordinate system (n,t,z) is used in the calculations as shown in Fig. 1. The n,t,z system consists of the normal (n) and tangential (t) direction to the surface of the current sheet, and the longitudinal direction of the cable (z).

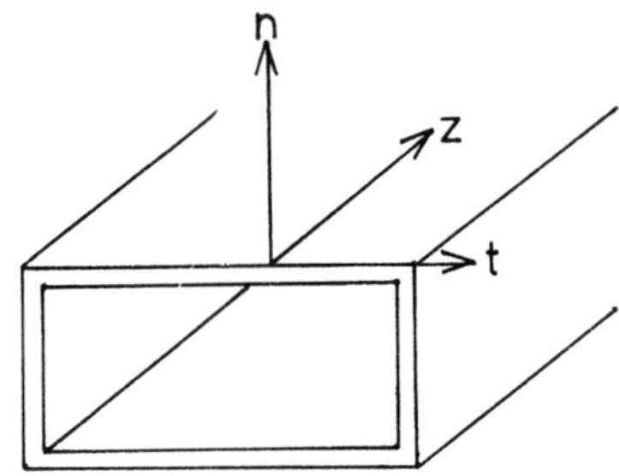

Fig. 1. Schematic representation of the cable as a thin current layer. The cartesian coordinate system (n,t,z) is shown.

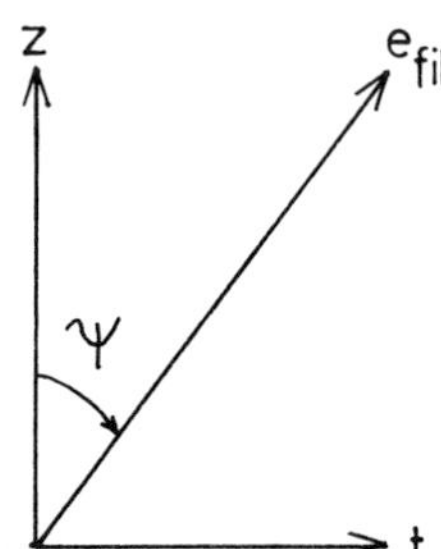

Fig. 2. Definition of the twist angle ψ, e_{fil} is the direction of the filament.

The surface current density $\bar{j}$ in the thin layer can be written:

$$\bar{j} = \bar{j}_{sc} + \bar{j}_{nn} , \qquad (1)$$

where $\bar{j}_{sc}$ and $\bar{j}_{nn}$ denote the superconducting and the normal fraction of the surface current densities.
The superconducting fraction expressed in the (n,t,z) system is

$$\bar{j}_{sc} = j_s \begin{pmatrix} 0 \\ \sin \psi \\ \cos \psi \end{pmatrix} , \qquad (2)$$

with $|j_s/j_c(|B_\perp|)| = 1$ saturated region,

$\quad\ |j_s/j_c(|B_\perp|)| < 1$ unsaturated region.

The normal surface current is described as

$$\overline{J}_{nn} = \overline{\overline{\sigma}} \; \overline{E} \tag{3}$$

in which $\overline{\overline{\sigma}}$ represents the electrical conductivity tensor. When assuming a conductivity $\sigma_{\parallel}$ parallel to the filaments and a conductivity $\sigma_{\perp}$ perpendicular to the filaments, it follows in good approximation that:

$$\sigma_{\parallel} = (1-\lambda)\sigma_0 \tag{4}$$

$$\sigma_{\perp} = (1-\lambda)/(1+\lambda).\sigma_0 \; , \tag{5}$$

with σ_0 the conductivity of the matrix material and λ the fraction of superconductor in the conductor. We can transform $\overline{\overline{\sigma}}$ into the (n,t,z) coordinates:

$$\overline{\overline{\sigma}} = \begin{bmatrix} \sigma_n & 0 & 0 \\ 0 & \sigma_{tt} & \sigma_{tz} \\ 0 & \sigma_{zt} & \sigma_{zz} \end{bmatrix} , \tag{6}$$

where the components of $\overline{\overline{\sigma}}$ can be written as follows using eq. 4 and 5:

$$\sigma_n = \sigma_0(1-\lambda)/(1+\lambda) \tag{7}$$

$$\sigma_{tt} = \sigma_0(1-\lambda)(1+\lambda \, \sin^2\psi)/(1+\lambda) \tag{8}$$

$$\sigma_{tz} = \sigma_{zt} = \sigma_0\lambda(1-\lambda) \, \sin \psi \, \cos \psi \, /(1+\lambda) \tag{9}$$

$$\sigma_{zz} = \sigma_0 \, (1-\lambda)(1+\lambda \, \cos \psi)/(1+\lambda). \tag{10}$$

Combination of the eq. 1-3 and 6 gives:

$$j_n = 0 \; , \tag{11}$$

$$j_t = j_s \sin \psi + \sigma_{tt} \, E_t + \sigma_{tz} \, E_z \; , \tag{12}$$

$$j_z = j_s \cos \psi + \sigma_{zt} \, E_t + \sigma_{zz} \, E_z \; . \tag{13}$$

For the calculation it is necessary to know $\overline{B}$ and $\dot{B}_n$ in a sufficient number of points on the circumference of the cable. The Maxwell equations applying to this problem are:

$$(\nabla \times \overline{E})_n = - \dot{B}_n \; , \tag{14}$$

$$(\nabla . \; \overline{j}) = 0 \; , \tag{15}$$

$$(\nabla . \; \overline{B}) = 0 \; . \tag{16}$$

In the unsaturated region there exists no electric field parallel to the filaments, i.e. $E_{\parallel} = 0$. This gives:

$$E_t = - \cot g \; \psi \; E_z. \qquad \text{(unsaturated region)} \tag{17}$$

Equation (13) gives:

$$\partial_t E_z = - \dot{B}_n(t) \; . \tag{18}$$

Of course, the function $\dot{B}_n(t)$ must fulfil equation (16). Now, demanding that

$$\oint E_z \, dt = 0 \, , \tag{19}$$

gives the solution E_z^0 for a cable without saturated regions. In the case that saturated regions occur a constant E_0 must be added to E_z:

$$E_z = E_z^0 + E_0 \tag{20}$$

In order to find E_t eq. (17) is used in the unsaturated region and eq (15) in the saturated region ($j_s = j_c$):

$$j_c \sin \psi + \sigma_{tt} E_t + \sigma_{tz} E_z = \text{constant},$$

or

$$E_t = - \cotg \psi \, [E_z^A + E_0] + (E_z^A - E_z^0) \, \sigma_{tz}/\sigma_{tt}$$
$$+ (j_c^A - j_c) \sin \psi/\sigma_{tt}, \tag{21}$$

where A denotes a point at the boundary of a saturated and an unsaturated region. Eq. (15) can now be transformed into:

$$\oint E_t \cdot dt = 0. \tag{22}$$

Substituting (17) and (21) into (22) it results:

$$E_0 = \frac{\int_s [(\cotg \psi - \sigma_{tz}/\sigma_{tt}) \, (E_z - E_z^A) + \sin \psi \, (j_c^A - j_c)] dt}{T \cotg \psi} \tag{23}$$

where T is the circumference of the cable, and s the length of the saturated region. Remark that $j_c(B_\perp)$ and E_z^0 are functions of t. In the unsaturated region eq. (15) can be solved to find:

$$j_s = j_c^A - \frac{(E_z^0 - E_z^A)}{\sin \psi} (\sigma_{tz} - \sigma_{tt} \cotg \psi) \, . \tag{24}$$

Now using (23) $E_z(t)$ can be solved as a function of the position of the boundary between saturated and unsaturated regions. Subsequently $E_t(t)$ can be calculated in both the saturated and unsaturated region. Together with equation (24), $E_z(t)$ and $E_t(t)$ are used to calculate

$$\Delta j_z = j_z(t) - j_z(t, \dot{B}_n = 0) \, ,$$

which represents the current redistribution. Also the transport current can be extracted:

$$I_T = \oint j_z \, dt =$$

$$\sigma_{zz} E_0 T + \int_u j_c^A \cos \psi \, dt - \tag{25}$$

$$\int_u (E_z - E_z^A) (\sigma_{tz} - \sigma_{tt} \cotg \psi) \cotg \psi \, dt -$$

$$\int_u E_z \sigma_{zt} \cotg \psi \, dt + \int_s j_c \cos \psi \, dt +$$

$$\frac{\sigma_{zt}}{\sigma_{tt}} \sin \psi \int_s (j_c^A - j_c) \, dt +$$

$$\sigma_{zt} \int_s E_z^A \left(\frac{\sigma_{tz}}{\sigma_{tt}} - \cotg \psi\right) dt - \frac{\sigma_{zt}^2}{\sigma_{tt}} \int_s E_z \, dt \, ,$$

where s and u are the lengths of the saturated and unsaturated regions respectively.

In practice, one can introduce the transport current I_T to find the position of the boundary and subsequently Δj_z. A computer code was written that calculates $j_z(t)$ for an arbitrary number of $\dot{B}_n$-points along the circumference with variable spacing between the points. As long as a sufficient number of $\dot{B}_n$-points is known the calculation can be done for an arbitrary geometry of the cable. For the dipole geometry, as can be seen in Fig. 3, a map of field values inside the conductors was generated using Biot-Savart. This method gives sufficiently reliable results as could be checked with finite element programs [4].

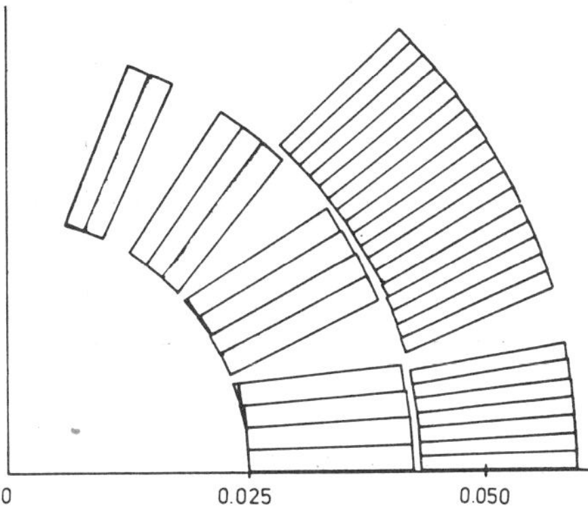

Fig. 3. Cross section of one quadrant of the dipole magnet straight section, showing the conductors.

The magnetization will be most profound in the case there exists no saturation in the thin layer. For this case the multipole change ΔB_m as a function of the input parameters $\dot{B}_n$, σ_0, ψ and λ was calculated. The default multipole components ($\dot{B}_n = 0$) are listed in Table 1. Fig. 4,5 show the results for the sextupole (B_3) and decapole (B_5) as a function of $\dot{B}_n$ and σ_0 at a reference radius in the beampipe (radius 0.025 m.) of 0.02 m.

The calculation of the coupling loss is rather straightforward. The loss per cable P_{cable} amounts:

$$P_{cable} = \int \bar{J} \cdot \bar{E} \; dt = \int \bar{J}_t \cdot \bar{E}_t \; dt + \int \bar{J}_z \cdot \bar{E}_z \; dt \; .$$

FILAMENT MAGNETIZATION IN SOLID AND HOLLOW FILAMENTS

The magnetization in the filaments due to the Meissner effect will be described here. With respect to the special case of hollow Nb_3Sn filaments in the cable as is produced by the so-called powder method [1], the doublet theory will be applied to hollow filament cables. However, the theory will also allow the choice of solid filaments. A more extensive comparison between hollow and massive filaments can be found in literature [5]. Consider a filament of radius R_f with an inner region (radius r_0) of normal conducting material. Assume that the boundary separating the parts with opposite superconducting currents is half of an ellipse cutting the x-axis at x_0, see Fig. 6. The filaments are assumed to point in the z-direction as twist will have a relatively

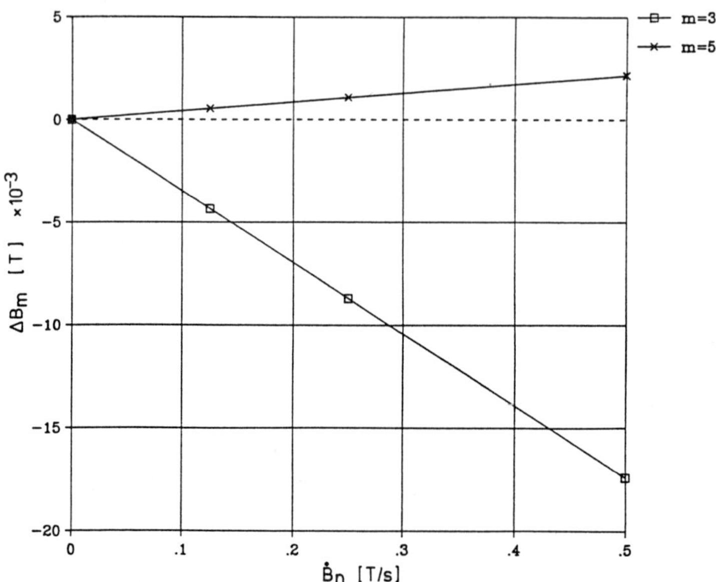

Fig. 4. Multipole change ΔB_m as a function of the dipole ramp rate $\dot{B}_n$ at a reference radius in the beam pipe of 0.02 m.

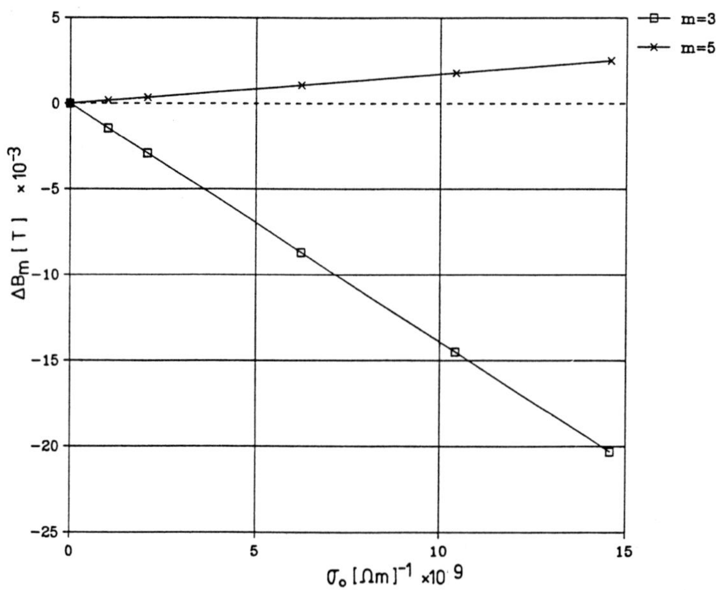

Fig. 5. Multipole change ΔB_m as a function of σ_0, the electrical conductivity, at a reference radius in the beam pipe of 0.02 m.

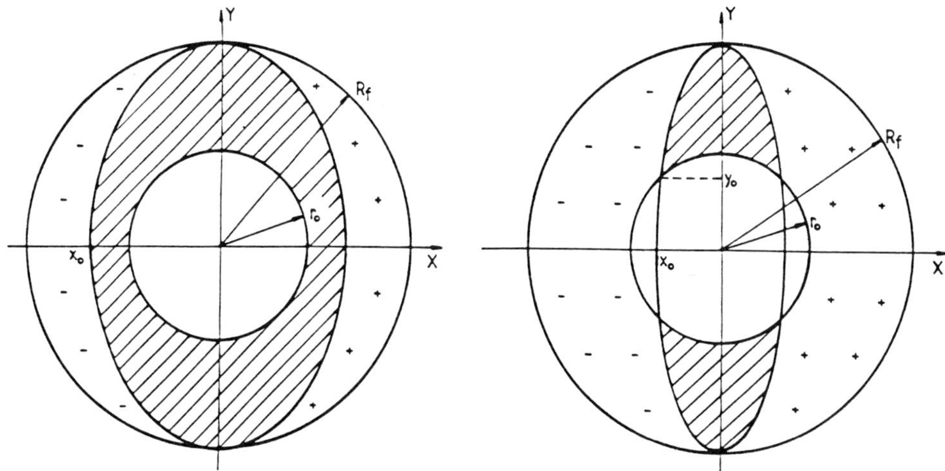

Fig. 6. Current distribution in the hollow filament showing the magnetization currents (++, --) and the transport current (shaded area). $E_z = 0$ along the left hand side of the ellipse ($x_0 < 0$) in case a positive transport current flows in the z-direction. $|x_0| > r_0$ large transport current, $|x_0| < r_0$ small transport current.

small effect on the filament magnetization. Due to a varying external field $\dot{B}_n$ an electric field in the z-direction will appear:

$$E_z = \dot{B}_n(x - x_0\sqrt{R_f^2 - y^2}/R_f) \ . \tag{26}$$

The superconducting transport current density amounts to:

$$j_z = j_c(B)f(x_0) \ , \tag{27}$$

with

$$f(x_0) = [2\frac{x_0}{R_f}\ \text{sign}(x_0)\ \arccos(y_0/R_f) - 2\left[\frac{r_0}{R_f}\right]^2 \arccos(y_0/r_0)]/\pi$$

$$y_0 = R_f\sqrt{\frac{r_0^2 - x_0^2}{R_f^2 - x_0^2}} \qquad\qquad |x_0| < r_0$$

$$f(x_0) = \frac{x_0}{R_f}\ \text{sign}(x_0) - \left[\frac{r_0}{R_f}\right]^2 \qquad\qquad r_0 < |x_0| < R_f .$$

The magnetization can be described analogously:

$$M = -\frac{4}{3\pi}\ R_f j_c(B)g(x_0) \tag{28}$$

with

$$g(x_0) = 1 - \left[\frac{x_0}{R_f}\right]^2 - \left[\frac{r_0^2 - x_0^2}{R_f^2}\right]y_0/R_f \qquad\qquad |x_0| < r_0$$

$$g(x_0) = 1 - \left[\frac{x_0}{R_f}\right]^2 \qquad\qquad r_0 < |x_0| < R_f$$

Alternatively the magnetization can be written resulting from a doublet:

$$M = Ia/(\pi R_f^2) \tag{29}$$

where I is the doublet current and 'a' the distance between the two currents. The penetration field B_p is defined as the external field change for which the filament is fully penetrated, i.e. there is maximum screening in the centre of the wire and $\dot{B}_{centre} = \dot{B}_{extern}$. Although B_p can only be calculated numerically, the maximum centre field induced by the screening currents can be used as a good approximation. For a hollow filament without transport current B_p can be calculated analytically when the magnetic field B is considered to be constant over the filament and using the fact that the filament is just penetrated when the ellipse in Fig. 6 reaches the point $x = r_0$ ($x_0 = r_0$). It follows that:

$$B_p = \frac{2}{\pi} \mu_0 j_c(B)(R_f - r_0) \ . \tag{30}$$

The magnetic field B_s at the centre of the filament due to the screening currents is a function of $e = x_0/R_f$:

$$B_s = 2 \ \mu_0 j_c(B)R_f h(e)/\pi \ , \tag{31}$$

with:

$$h(e) = 1 - k(e) \ \arcsin(\sqrt{1-e^2}) \qquad\qquad r_0 < |x_0| < R_f$$

$$h(e) = 1 - y_0/R_f + k(e)\left[\mathrm{atan}(y_0 k(e)R_f) - \mathrm{atan}(k(e))\right] +$$

$$\frac{y_0}{R_f} \left[\frac{1}{2} \ \ln(1 + (\frac{y_0}{x_0})^2(1-e^2)) - \ln(\frac{r_0}{x_0})\right] \qquad |x_0| < r_0$$

$$k(e) = \sqrt{1-e^2}/e \ .$$

Except for the virgin curve of the superconductor ($B < B_p \simeq 50$ mT for the used conductors), the local current density in the superconductor will be non-zero. It is therefore assumed that everywhere in the cables $B \geq B_p$ which is not fulfilled for only a minor part of one cable in the outer layer at the median plane, where the local direction of B_y changes sign (independent of the value of the centre field of the dipole). A fully penetrated filament has screening currents in only one direction per half filament, a partly penetrated filament has screening currents in both directions in one half of the filament (e.g. due to an oscillating power supply or simply when ramping up and down).

A computer code was written that treats both cases of penetration for a filament with transport current. In the case of a fully penetrated filament the value of the transport current is used to calculate f and x_0, from which g can be calculated. In the case of a partly penetrated filament a second x_0' is introduced for the new field change. This results in a negative contribution to be added to the magnetization already calculated in the fully penetrated case. The value of h(e) is used to calculate the field change. If it is assumed that the magnetic field is constant over the surface of one strand, the average value of the magnetization can be calculated for each strand and therefore an effective doublet per strand is found. The doublet current per filament for a fully penetrated filament is:

$$I_d = (I_{max} - I_T)/ \ 2 \ , \tag{32}$$

where I_{max} is the transport current for a saturated filament and I_T the transport current. For a partly penetrated filament this value must be

corrected with the induced reversed screening current. In order to
calculate the effect of the magnetization on the multipoles, the values
of I, a and the field direction are used to determine in each strand the
effective doublet. To find the field dependent critical current density

Table 1. Multipole components
for the structure of Fig. 3
(non-optimized) in the case of
no magnetization, calculated
at a reference radius in the
beampipe of 0.02 m.

n	B_n (T)
1	0.500
3	-0.398e-4
5	-0.134e-5
7	0.475e-3
9	0.300e-3

Table 2. Conductor parameters for the inner and outer keystoned cables
(1,2 resp.) as used in the calculation.

	cable 1	cable 2
dimensions:		
height [mm]	17.0	17.0
width [mm]	2.18/2.64	1.44/1.81
strands per cable	26	40
filaments per cable	23040	6912
outer radius filament, R_f [μm]	10	15.5
inner radius filament, r_0 [μm]	6	10.5
fraction sc in strand, λ	0.5	0.5
twist angle filaments, ψ [rad]	0.1	0.1
electrical conductivity, σ_0 [Ωm^{-1}]	$6.25.10^9$	$6.25.10^9$
Kim parameter j_0 [A/m^2]	$1.0.10^{10}$	$1.0.10^{10}$
Kim parameter B_0 [T]	2.0	2.0
transport current for B_1 = 0.5 T [A]	730	730

at low field we used the so-called Kim relation:

$$j_c(B) = j_0 B_0/(B_0 + |B|) , \qquad (33)$$

as is commonly used for NbTi in the lower B regime, but it also holds
for Nb$_3$Sn as can be checked with measurements [6]. The values for j_0 and
B_0 that were used can be found in Table 2. Results are shown in Fig.
7,8.

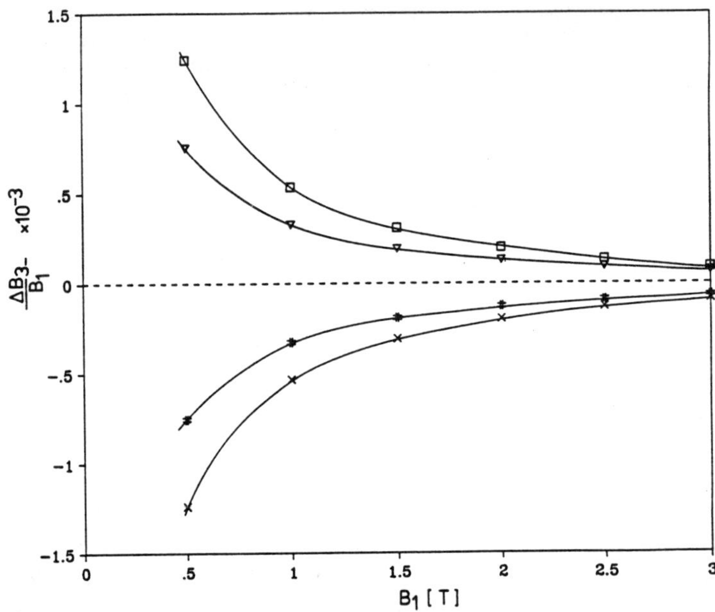

Fig. 7. The relative sextupole change $\Delta B_3/B_1$ as a function of the dipole field B_1 for a hollow filament (□ ramping up, X ramping down), and for a massive filament with the same area of superconductor (▽ ramping up, #ramping down)

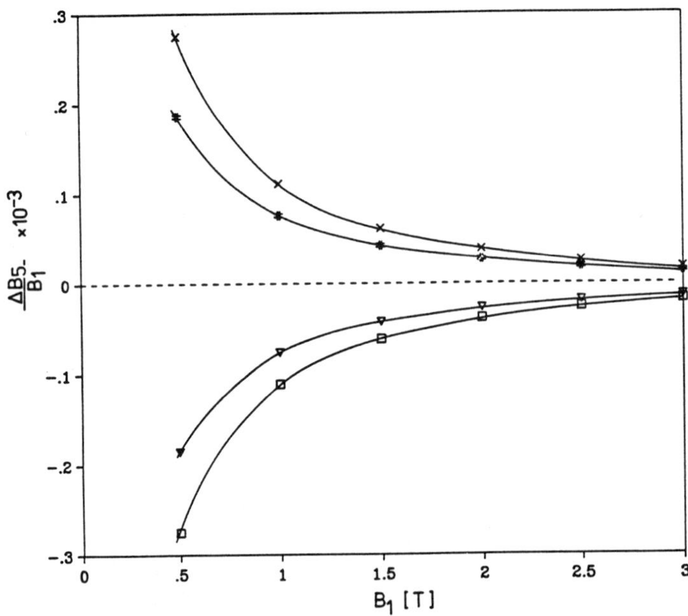

Fig. 8. The relative decapole change $\Delta B_5/B_1$ as a function of the dipole field B_1 for a hollow filament (□ ramping up, X ramping down), and for a massive filament with the same area of superconductor (▽ ramping up, #ramping down)

RESULTS

The results of the calculations of cable magnetization can be found in Fig. 4,5. They show that in the case of no saturation in the thin layer, the change in multipole components has a linear dependence on $\dot{B}_n$ and σ_0. In general it is found that:

$$\Delta B_m \propto (\sin^2\psi)^{-1} \left[\frac{1-\lambda}{1+\lambda}\right] \dot{B}_n \; \sigma_0 \; = F(\psi,\dot{B}_n,\sigma_0,\lambda)$$

Of course, as there is no saturation, ΔB_m will not depend on the dipole field level. The above relation makes it possible to determine the effect of a change of the value of one or more parameters directly by using Fig. 4 and 5. For instance, it was assumed that the matrix material of the conductor is copper. By the introduction of copper-nickel layers in the matrix, the value of σ_0 will decrease and the resulting multipole components can be determined using the above relation. Allowable values of the conductor parameters can be obtained as a function of the required field homogeneity.

The results of the calculations of filament magnetization are shown in Fig. 7,8. Input parameters for the calculation were taken according to Table 2. Fig. 7,8 show the dependence of the relative sextupole change $\Delta B_3/B_1$ and the relative decapole change $\Delta B_5/B_1$ on the central field $B_1 = B_y(0,0)$, for a hollow and a solid filament with the same area of superconductor. It is clear that in this case the field error is largest for the hollow filament which has larger dimensions and therefore a larger magnetic moment. The relative multipole change $\Delta B_m/B_1$ is more than one order of magnitude larger than the actual value B_m/B_1 for the plotted multipoles. Although the values of the multipoles are still small compared to the dipole field B_1, they vary strongly with the central field for $B_1 < 1T$. Comparing the results of Fig. 7,8 with the results of the cable magnetization (Fig. 4,5) shows that the effect of filament magnetization is much more profound in the relevant $\dot{B}_n$ range ($\dot{B}_n \leq 0.01$ T/s), provided the value of the matrix electrical conductivity is equal to or less than the value assumed in the calculation (copper, RRR=100). The interstrand coupling loss was calculated for the structure in Fig. 3 and amounts to 24 mW per unit of length of dipole when $\dot{B}_n = 0.01$ T/s.

CONCLUSIONS

A model was presented that makes it possible to calculate the effect of cable magnetization and filament magnetization on the dipole magnetic field quality for an arbitrary lay-out of conductors in the dipole. Relations were shown that allow easy checking of the multipole components for a new set of conductor parameters. For the used values of the conductor parameters and dipole ramp rate the filament magnetization has more effect on the field quality than the cable magnetization. The calculated effect of filament magnetization at $B_1 = 0.5$ T (injection field of the LHC accelerator) expressed in the change in multipoles amounts to about $0.6*10^{-3}$ T for the sextupole and $1.4*10^{-4}$ T for the decapole, independent of the dipole ramp rate. As the multipole changes are at least one order of magnitude larger than the actual multipole value, there seems only little to gain in the reduction of the filament diameter as the multipole change has a linear dependence on the filament diameter.

REFERENCES

1. C.A.M. van Beijnen et. al., Multifilament Nb_3Sn superconductors produced by the ECN technique, IEEE Transactions on Magn., MAG-15, no.1, jan 1979, pp. 87-90.
2. H.H.J. ten Kate et. al., On the development of a 1 meter 'twin aperture' 10T Nb_3Sn dipole magnet for the CERN LHC, paper at this conference, Proc. IISSC New Orleans, Feb. 8-10 1989.
3. E.M. Hornsveld et. al., Development of ECN-type niobium-tin wire towards smaller filament size, ICMC 1987 St.Charles USA, pp 493-498.
4. D. ter Avest, 3-D magnetic field and Lorentz force computations in an LHC dipole using TOSCA, to be published at the 11th. Magnet Technology Conference, Japan, August 28, 1989.
5. L.J.M. van de Klundert, Hollow filaments with a normal conducting inner region, internal report University of Twente
6. M.A.Green, Generation of the J_c, H_c, T_c surface for commercial superconductors using reduced-state parameters, LBL-24875/SSC-N-502 (internal report).

ON THE DEVELOPMENT OF A 1 METER "TWIN APERTURE"

10 T Nb$_3$Sn DIPOLE MODEL MAGNET FOR THE CERN LHC [*]

H.H.J. ten Kate, C. Daum[+], L.J.M. van de Klundert, H. Bode[@]
W. Hoogland[+], T.A. Roeterdink[#], F. van Overbeeke[&] and R. Perin[o]

University of Twente, Applied Superconductivity Centre,
POB 217, 7500 AE Enschede, (+) NIKHEF-H, Amsterdam;
(#) ECN, Petten; (&) HOLEC, Ridderkerk; (@) SMIT WIRE,
Nijmegen, all in the Netherlands; and (o) CERN, Geneva,
Switzerland

ABSTRACT

The possible realization of the new 16 TeV hadron collider LHC of
CERN presupposes the availability of the techniques to construct 10
tesla twin aperture dipole magnets either by using conventional NbTi
magnet technology at 1.8 K or by using a Nb$_3$Sn conductor at 4.2 K. At
present both alternatives are under development at CERN in collaboration
with European laboratories and industries. A new initiative to explore
the Nb$_3$Sn route was started recently in the Netherlands by the Applied
Superconductivity Centre at the University of Twente and NIKHEF, the
National Institute for Nuclear and High Energy Physics. A 4 years R&D
program was started to design and to manufacture a 1 meter Nb$_3$Sn dipole
model magnet in collaboration with other research institutes and
companies in the Netherlands. The magnet now under development will be
the first one meter Nb$_3$Sn LHC dipole model magnet with a twin aperture
designed for use at a nominal field of 10 T. The paper reports on
general aspects of the design of the magnet and the present state of the
developments.

INTRODUCTION

The possible realization of a new Large Hadron Collider (LHC)
providing proton-(anti)proton collisions up to an energy level of about
16 TeV is studied at CERN. This new machine should provide a much higher
level of collision energy as a next step towards new discoveries and
more understanding in particle physics.

An elegant and economical solution would be to combine the proposed
LHC with the present infrastructure of the LEP accelerator. Moreover,

---

[*] These investigations in the programme of the Foundation for
Fundamental Research on Matter FOM are supported, in part, by the
Netherlands Technology Foundation STW, Utrecht, the Netherlands.

this solution enables to perform proton-electron colliding experiments
in the case the LEP and LHC accelerators are combined.

The existing LEP tunnel has a circumference of 27 km and can house
the LHC when it is placed on top of the LEP machine. As a consequence of
this, the required level of magnetic field in the accelerator dipole
magnets for LHC has, for obvious reasons, to be about 10 T in order to
attain the 16 TeV centre-of-mass proton-proton collision energy.
Moreover, due to the restricted space in the tunnel above the LEP
machine, it is required to combine both proton channels into a single
into a single compact dipole magnet instead of the installing two
separate and thus independent accelerator rings. These so-called

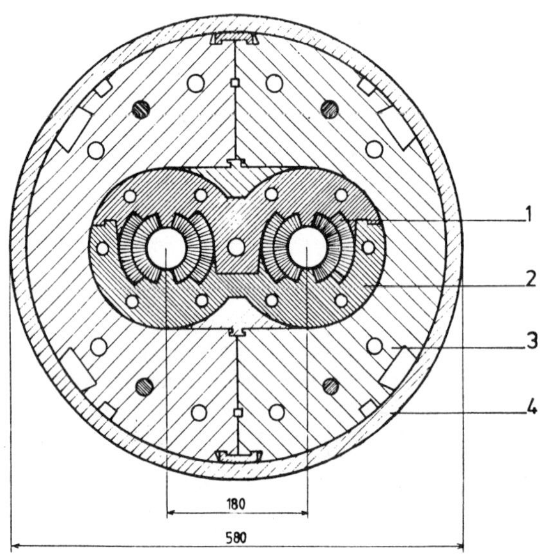

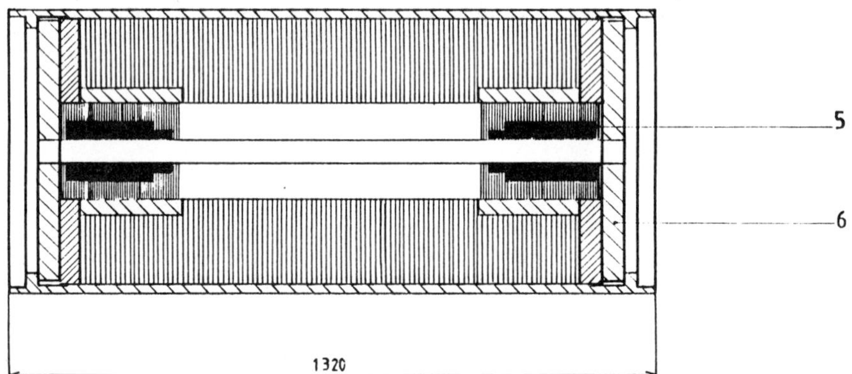

Fig. 1. View of the cross section and the longitudinal section of a
10 T twin aperture LHC dipole magnet showing the main parts:
(1) coils, (2) collars, (3) cold iron yoke, (4) the outer
shrinking cylinder enclosing the cold part of the the magnet,
(5) coil head sections, and (6) end-flange and end-plates.

"two-in-one" or "twin aperture" dipole magnets are now under development at CERN in collaboration with European laboratories and industries.

In figure 1, both the cross- and longitudinal sections of such a 1 meter twin aperture model dipole magnet are shown schematically. Such a dipole magnet consists of 2 beam channels each with their own set of coils. They have a common force-retaining collaring structure and a common magnetic yoke linking the magnetic fields in both channels. As a consequence both coil sets are being housed in a single cryostat. Figure 1 only provides a schematic view. Details of construction as well as the cryogenic parts and cryostat are not shown.

The obvious advantages of the twin aperture concept are the reduced space requirements and costs. On the other hand the principle magnetic coupling as caused by the relatively small distance between the beam channels and the common yoke, is a disadvantage.

In the proposed LHC ring there are about 1860 main dipole magnets each with a length of approximately 10 meter. For more information about the general aspects of LHC and the magnet technology the reader is referred to references [1-3].

NbTi OR Nb_3Sn FOR ATTAINING 10 T

The field level of 10 T in the dipole magnets can be attained in two ways. The first possibility is to use a NbTi/Cu composite superconductor for the coil windings. The magnet is then cooled with superfluid HeII at an operating temperature of 1.8 K. Due to the reduced temperature the critical field of the NbTi superconductor is enhanced with about 2 T to about 13.7 T . The obtained critical field is sufficient provided the required critical current density in NbTi of at least 1300 A/mm^2 at 11 T can be realized in a commercial large scale production of the superconductor.

The major disadvantages of the NbTi route of development are the complexity of the required large scale 2 K cooling system, the limited temperature margin, the extremely low enthalpy of the conductor and the construction materials in the magnets at this temperature of 2 K. In particular, the latter property means that a coil wound with NbTi conductors is extremely sensitive to energy releases caused by (particle) beam losses the coil structure itself. The relatively high sensitivity to heat releases that can introduce quenches in the magnet could reduce the efficiency of operation of the accelerator to an unpractical level when it operates near its maximum energy.

The second method is to use a Nb_3Sn superconductor at an operating temperature of about 4.5 K. Nb_3Sn has a much higher critical temperature of 18 K in stead of the 9 K of NbTi. Moreover, its critical field is, dependent on the exact composition, between 20 and 25 T at 4.2 K instead of the 13.7 T of NbTi at 1.8 K. Furthermore, the enthalpy at the operating temperature of 4.5 K is about a factor of 40 larger than NbTi at 1.8 K.

The obvious advantages are partly negated by the much more difficult production procedures such as the required application of the so-called "wind-and-react" process and vacuum impregnation of the coils. The extreme brittleness of Nb_3Sn sets a clear maximum of about 0.2-0.3 % to the allowable strain in a practical multifilamentary Nb_3Sn conductor. This means that the structural analyses of the magnet has to be very accurate. One has to make sure that after the heat treatment of the coils, in order to obtain the Nb$_3$Sn in the conductor, during all the production steps, during cooling-down and the further operation procedures, this strain level will never be exceeded.

LHC MAGNET DEVELOPMENT PROGRAMME

Both the NbTi and Nb$_3$Sn routes are developed at CERN [2] though most of the efforts to develop LHC magnets have been concentrated on the NbTi/1.8 K version because in that case the magnet technology is more or less straightforward and based on the experiences with, for example, the TEVATRON and HERA magnets. Moreover, in contrast to using a Nb$_3$Sn conductor, there are several possible manufacturers which could supply a NbTi conductor and in industry there is much more experience with winding NbTi than with winding Nb$_3$Sn conductors.

A 1 meter single aperture model magnet using a NbTi conductor was constructed by ANSALDO and tested at CERN in 1988 up to a field of about 9.3 T at an operating temperature between 1.6 and 2 K [4].

Four 1 meter 10 T twin aperture model magnets are ordered now by CERN in the European industry. One of these NbTi magnets will be made in the Netherlands by HOLEC [5].

The third activity at present, is the construction of a 9 m long twin aperture NbTi magnet by ABB using the existing HERA tooling, coil geometry and conductors. The purpose of this is to gain experience with the twin aperture concept in combination with a full size magnet with a length of 9 meter. The aperture size in this magnet is 75 mm and the expected maximum field is about 7.5 T.

However, the main disadvantage of this route remains the present limits of the critical field and current density in the commercially available NbTi conductors. When using NbTi it will be very difficult to attain a clear and stable 10 T operation of the complete accelerator with its series connection of about 1860 main dipole magnets. Especially the sections of the accelerator which are exposed to an increased level of radiation are critical parts.

In parallel to this the Nb$_3$Sn route is developed within two projects. CERN and the ELIN company in Austria collaborate since October 1986 in a joint development of a 1 meter, also single aperture, 10 T Nb$_3$Sn dipole model magnet [6]. At this moment the coils for this magnet are being wound and tested. Recently the first half coil produced by ELIN was tested at CERN in a so-called mirror configuration in which a magnetic field of 10.2 T was achieved.

The second project is the one presented here for the first time. In a collaboration between the University of Twente, the National Institute for Nuclear and High Energy Physics NIKHEF and CERN, a twin aperture 10 T Nb$_3$Sn model magnet will be designed, constructed and tested in the Netherlands. For this project a conductor development programme has been set up with the Netherlands Energy Research Foundation ECN for their powder method Nb$_3$Sn conductor. It is envisaged to use this material in the magnet. With this Nb$_3$Sn material a superior average critical current density can be realized in comparison with other Nb$_3$Sn superconductors. A further collaboration exists with the industrial partners SMIT WIRE for cable insulation and HOLEC for assistance with winding the Nb$_3$Sn coils.

DESCRIPTION OF THE DIPOLE MAGNET

The criteria which have led to the present "2-in-1" geometry as it is shown schematically in figure 1, are the nominal magnetic field of 10 T on the particle beam, the small aperture size and the required intra-beam distance to meet the restrictions set by the available space in the LEP tunnel. The resulting design shows a 2-shell coil system with graded current density, surrounded by a structure of laminated collar pieces and a vertically split and laminated cold iron yoke. This package is enclosed by an outer shrinking cylinder that handles the major part of the forces.

The main parameters of the dipole magnets as determined and optimized to meet the mentioned requirements, are listed in Table 1.

Table 1. Main parameters of the 9.5 meter long LHC dipole magnets.

nominal magnetic field on the beam =	10.0	T
maximum field in layer 1 =	10.2	T
maximum field in layer 2 =	8.7	T
full magnetic length =	9.5	m
coil inner diameter =	50	mm
coil outer diameter =	122	mm
distance between beams =	180	mm
overall mass =	18	tons
cold mass =	15	tons
mass of superconductor =	600	kg
stored energy at 10 T =	684	kJ/m
operating current =	16	kA
max. Lorentz force in x-direction =	4.6	MN/m
max. Lorentz force in y-direction =	-1.2	MN/m
max. Lorentz force on coil head =	0.17	MN

The various model magnets now under construction have a 1:1 scale magnet cross section. They have a length of about 1 meter which means that within the model coil programme the length of the straight central part of the magnets has been reduced in order to keep down the costs of development.

The magnetic field is generated by two (almost) identical coil systems which are positioned around the two beam channels. The coil systems have a two-shell geometry and consist of 4 sub-coils connected in series. In both shells a very specific distribution of conductors is required in order to obtain a good field quality.

To approximate the ideal cosine distribution of current around the beam pipe while optimizing the maximum possible nominal field with the given critical current density of the superconductor, the conductors are grouped in blocks that are separated by wedges. An example of a possible conductor distribution that meets the LHC requirements is shown in Figure 2. Here two different conductors are applied, one for each coil. Both conductors carry the same current and they have the same width of about 17 mm but a different height.

The coil support structure has to provide first, a sufficient rigidity and a sufficiently high dimensional accuracy to guarantee the field quality during the entire operating cycle, and second the superconducting coils have always to be under compression to avoid coil or conductor displacements that can initiate a quench in a coil which is

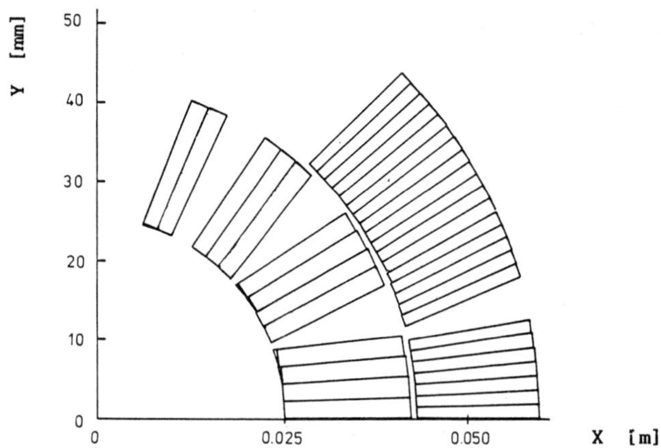

Fig. 2.   Example of a conductor distribution in the coils applying a two shell geometry. The picture shows one quadrant of a coil in an LHC dipole magnet.

fatal to accelerator operation. In contrast to the case of the usual 5 and 6 T dipole magnets, the required pre-stress on the coils for 10 T operation is so high that it is not practical nor optimal to use just the collars to provide pre-stress. Therefore, clamping of the coils in the LHC dipole magnets is provided by the combined action of collars, iron yoke and the outer cylinder in the so-called "mechanically hybrid" support structure. The interested reader is referred to reference [7] for a detailed explanation of this complex clamping arrangement. Here we confine ourselves to remark that about 30 % of the horizontal force F of 4.6 MN per meter at 10 T is taken by the collars. The remaining and thus major part is taken by the outer cylinder by which the iron yoke is the transfer medium. The stress level in the windings is about 60 MPa maximum which is a tolerable value.

THE Nb_3Sn CONDUCTOR

The practical realization of the LHC dipole magnets with the two shell geometry presupposes the availability of large-aspect ratio Rutherford type of cables made of Nb_3Sn strands. In order to reach the 10 T nominal field the keystoned cables for the inner and outer shell should have a current density of at least 1300 A/mm^2 at 11 T and a copper part for stabilization not less than 50 %. The preliminary parameters of the two cables as required for the inner and outer coils are given in Table 2.

At this moment three practical methods for Nb_3Sn wire production exist. The three types of conductors are respectively called bronze, internal tin and powder method conductors.

The bronze route conductors have the advantage of being available on an industrial scale but, inherent to the presence of bronze and Ta in the wire section, the wire is less performing in terms of the average current density. The LHC specification for the average current density to attain 10 T can only be reached by reducing the amount of copper in

Table 2. Parameters of two Nb$_3$Sn cables for the inner and outer coils
          respectively.

		inner coil cable	outer coil cable
strand diameter	[mm]	0.90	1.35
percentage copper	[%]	50-60	50-60
number of strands		24	36
strand twist length	[mm]	30	30
RRR value	[-]	>100	>100
cable dimensions	[mm^2]	2.18/2.67 × 16.4	1.45/1.77 × 16.4
cable pitch	[mm]	<120	<120
cable length (1m magnet)	[m]	4 × 25	4 × 40
nominal current	[kA]	≈16 @ 10.2 T	≈16 @ 8.7 T
critical current	[kA]	>17 @ 11 T	>18 @ 8.5 T

the wire section to about 35 % instead of the specified 50-60 % [2]. In
spite of this disadvantage, the bronze type Nb$_3$Sn wires and cables are
available at present and therefore the first Nb$_3$Sn single aperture model
magnet is being wound with this material [6]. As it was mentioned
before, a first successful result has been obtained with a maximum field
test of a half coil made by ELIN.

Using the so-called internal diffusion method it is probably
possible to just meet the 10 T specification with 50 % copper in the
wire section, but at this state of developments no practical length of
conductor is available. In any case, with these materials, it is at
present not likely to go substantially beyond the critical current
specifications in order to further increase the magnetic field or the
safety margin.

The third method is the so-called ECN powder process [8]. This is
characterized by the use of hollow Nb tubes with a filling of NbSn$_2$
intermetallic powder. The niobium tubes are surrounded by copper
hexagons and then stacked. The assembled billet is drawn to the final
diameter and twisted without any intermediate heat treatment. The
niobium tube acts as diffusion barrier like the Ta shell in the bronze
conductors in order to avoid a contamination of the copper with tin
during the final heat treatment for diffusion, necessary to obtain the
superconducting layer of Nb$_3$Sn. In this way a pure copper matrix is
achieved without using bronze and Ta that unfortunately both do not
contribute to superconductor stabilization.

As a consequence, the average current density in the ECN type
powder method wire is, with the same amount of stabilizing copper, much
higher than with the bronze process. The diameter of the Nb$_3$Sn ring
formed after heat treatment is about 67 % of the diameter of the initial
Nb tube. A disadvantage of this type of wire production is that, at the
present state of developments, it is hardly possible to make wires with
sub-micron filaments.

During the past years wires with 36 and 192 filaments were made in
a production quantity of about 500 kg mainly for application in ECN
related projects as the SULTAN coils and NET prototype conductors. These
conductors have relatively thick filaments with diameters between 30 and
80 μm when the wire diameter is around 1 mm.

The next step in the development process was to reduce the filament
size in order to make applications possible in which low ac loss or
a low filament magnetization are essential. Double stack ECN type of

wires are now under development. Wires with 324, 684 and 1332 filaments
have already been manufactured on a laboratory scale [8]. A further
increase of the number of filaments is in progress. A brief collection
of characteristic wire parameters is given in Table 3.

Table 3. Comparison of parameters of characteristic ECN type Nb_3Sn wires
having a wire diameter of 0.85 mm [8].

number of filaments	[-]	36	192	324	684	1332
percentage of copper	[%]	50	52	59	59	55
reaction time @ 675°C	[hrs]	75	45	45	32	16
Nb tube diameter	[μm]	72	45	29	21	15
Nb_3Sn fil. diameter	[μm]	48	31	20	14	10

Note that the necessary time for heat treatment reduces with the
filament diameter. This is an important benefit in comparison to bronze
and internal tin conductors where the time of heat treatment is long and
hardly dependant on the filament size. With an industrial production of
many magnets in parallel, time consuming heat treatments of 75 to 100
hours of each coil are tiresome and expensive.

The wires are made in a production unit at ECN Petten in which the
whole range of billet assembly, wire drawing and cabling is covered. As
an example of the internal lay-out the cross section of the 1332
filaments wire is shown in figure 3.

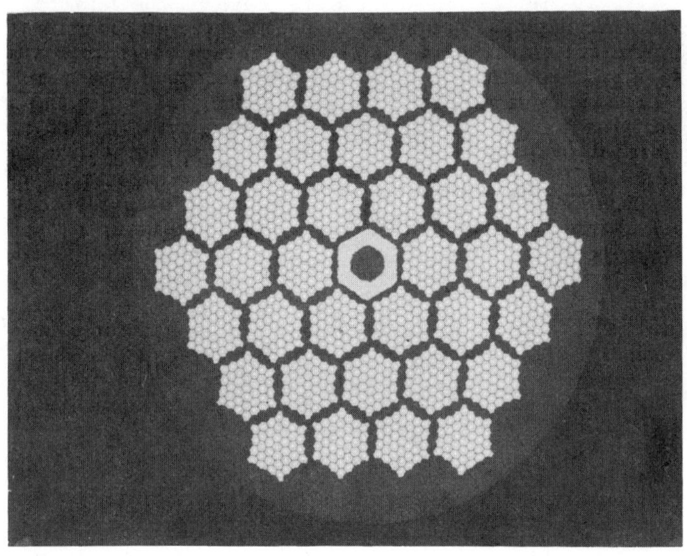

Fig. 3. Cross-section of an ECN-type Nb_3Sn wire, bare diameter
0.85 mm, 1332 filaments in pure copper, filament diameter
10 μm, 55 % copper.

The non copper critical current densities obtained in the wires
are, dependent on the amount of powder in the wire, in the range 1500 to
1800 A/mm^2 at 11 T which means an overall current density of 700 to
800 A/mm^2 with a copper part between 50 and 60 %.

Using the 192 filaments wire a prototype conductor according to the
LHC cable sizes for the outer coils was manufactured and tested recently
[9]. A cross section of the cable is shown in figure 4. The maximum

current of the cable was investigated in our laboratory in the field range 7 to 13 T using a superconducting rectifier [16-18] for the current supply up to 50 kA. The cable showed an excellent performance. The quench current of the cable at the proposed working point of 8.7 T is 30 kA which is about 100 % over the nominal operating current [9]. This result clearly demonstrates the potentials of the Nb$_3$Sn powder method conductor.

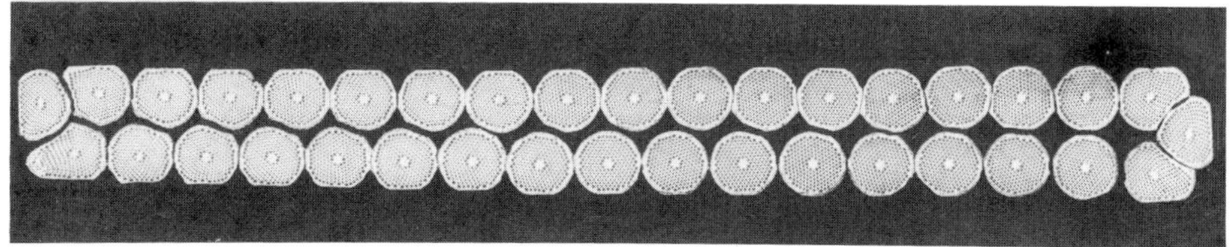

Fig. 4.   Photograph of the cross-section of a prototype keystoned LHC cable with ECN-type Nb$_3$Sn wires. The cable has 36 strands of 0.90 mm diameter and a size 1.45/1.77 * 16.40 mm^2. The number of filaments is 192 per strand. They are embedded in a pure copper matrix that is about 55 % of the wire section.

In the framework of the LHC project, a measuring programme is in progress in our laboratory to determine the current carrying capacity of several pieces of LHC-type of Nb$_3$Sn cables in the field range 7 to 15 T. Results of these investigations will be presented elsewhere [10].

DEVELOPMENT PROGRAMME FOR THE 1 METER Nb$_3$Sn MODEL MAGNET

The research and development programme for the Nb Sn LHC model magnet started in April 1988 and was approved officially by the main financier the Netherlands Technology Foundation STW in September 88. The programme includes an integral design study of the LHC dipole magnet components in order to generate the best possible solution for the Nb Sn magnet within the boundaries set by the general LHC magnet specifications. In parallel to the specific design and construction work to realize a one meter magnet within a period of 2 years, also more general aspects not directly related to magnet production will be studied as filament and cable magnetization and ac loss during magnet ramping.

Important subjects and components which are to be developed are:

*   the superconducting wires and after that the cable. The filament size in the ECN type conductors has to be reduced to below 15 μm in order to reduce filament magnetization. A regular production of strand material with about 2000 filaments has to be developed.

*   calculation of stationary and dynamic filament and cable magnetization effects. The ECN type of superconductors have hollow filaments which means an increased level of magnetization in comparison with solid filaments. First results can be found in [14] and [15].

*   three dimensional calculation of magnetic fields and forces. Two methods are applied. An approximation has been made using the method of image currents with saturated iron [12]. A more exact result is obtained using the FEM programme TOSCA in which the contribution of the yoke with all fields can be calculated [13]. In order to illustrate

this a representation of a half coil in TOSCA is shown in figure 5.

*   three dimensional computation of coil and conductor deformation in order to make sure that never the maximum reversible stress level of Nb Sn is exceeded. These calculations are done with the FEM programs ANSYS and SUPERTAB.

*   method of coil winding, giving the accurate dimensions, heat treatment, vacuum impregnation, coil assembly, collaring procedures and magnet assembly. All these items have to be reconsidered carefully in order to prevent damage to the Nb Sn in the conductor.

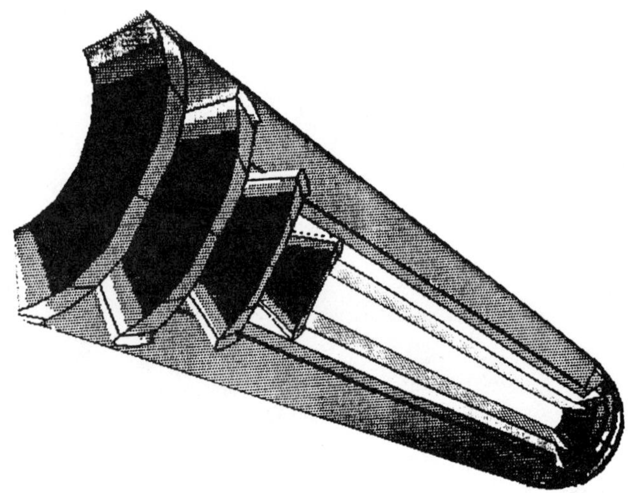

Fig. 5.   Impression of the 3-D geometry of a half coil in a 1 meter model LHC dipole magnet. Picture generated by TOSCA, a FEM programme to calculate magnetic field and lorentz forces.

*   examination (effects of heat treatments and thermal contraction) of materials for cable and coil insulation, longitudinal wedges, end spacers etc. When using Nb Sn it is extremely important to check the thermal behaviour, all the materials which are in contact with the conductor and coils in order to prevent damage due to for example a difference in local and integral thermal contraction in the temperature range 4 to 1000 K.

*   strain, voltage and temperature sensors, sensing coils etc., and actuators as heaters to be included in the magnet in order to diagnose the magnet during test and to provide quench protection.

*   magnet test station, quench protection system, magnet powering,

test procedures, etc. A pre-test of the magnet in the Applied Superconductivity Centre at the university foreseen using a superconducting rectifier [16-18] for powering the magnet.

The practical work for this project has been started in August 88. A new winding machine and collaring press have been manufactured and installed [15]. Design and laboratory tests of the magnet components are in progress. The obvious critical activity is the production of the superconducting cable. Coil winding should start in November 88. A first cryogenic test is planned for Autumn 1990.

CONCLUSIONS

In a collaboration between the University of Twente, NIKHEF, ECN, HOLEC, SMIT WIRE and CERN a Nb_3Sn model dipole coil is being designed and constructed. It will be the first 10 T Nb_3Sn twin aperture dipole model magnet and the second Nb_3Sn magnet that is made in the framework of the CERN LHC development programme.

It is intended to use the so-called ECN type of Nb_3Sn wire in order to obtain the highest possible field in the magnet due to the superior average current density of this material in comparison to NbTi at 1.8 K or in comparison to other commercial Nb_3Sn materials.

In principle the current carrying capacity of the powder method ECN-type of conductors is sufficient to attain 11 T operation applying the present 2-D geometry, provided the mechanical structure and the conductor can handle the enhanced forces. This principle possibility is the most striking advantage of using Nb_3Sn instead of NbTi.

The magnet development programme has started in April 1988 and will run for 4 years. A first test of the magnet is scheduled for autumn 1990.

ACKNOWLEDGEMENTS

The authors wish to acknowledge the Netherlands Technology Foundation STW for its financial support of the project. Without this it would not be possible to carry out the development programme as it is presented here. Furthermore, we thank the colleagues Dick ter Avest, Bennie ten Haken, Sander Wessel, Rens Dubbeldam, Hans Geerinck and Wim van Emden, who have joint the project team in the course of 1988. Their work and motivation are essential to gain success.

REFERENCES

1.   G. Brianti, K. Hübner, "The Large Hadron Collider in the LEP-tunnel", CERN report 87-5, 1987.
2.   R. Perin, Progress on the superconducting magnets for the Large Hadron Collider, IEEE Transaction on Magnetics, MAG-24, 734 (1988), CERN LHC-Note 63.
3.   D. Leroy, R. Perin, G. de Rijk, W. Thomi, Design of a high-field twin aperture superconducting dipole model, IEEE Transaction on Magnetics, MAG-24, 1373 (1988), CERN LHC-Note 62.
4.   R. Perin, D. Leroy, G. Spigo, The first, industry made, model magnet for the CERN Large Hadron Collider, Proc. ASC San Francisco 21-25th August 1988, IEEE Transactions on Magnetics, MAG-25 (1989).

5. F. van Overbeeke, R.L. Dubbeldam, H.H.J. ten Kate, Industrial fabrication of a 10 Tesla twin aperture dipole magnet for LHC at HOLEC, to be published at the 11th Magnet Technology Conference, Japan, August 28, 1989.

6. S. Wenger, F. Zerobin, A. Asner, Towards a 1 m long high field Nb_3Sn dipole magnet of the ELIN-CERN collaboration for the LHC-project, Proc. Appl. Superc. Conf., San Francisco, August 21-25, 1988, IEEE Transaction on Magnetics, MAG-25 (1989).

7. D. Hagedorn, D. Leroy, R. Perin, Towards the development of high field superconducting magnets for hadron collider in the LEP tunnel, Proc. MT-9 Zurich 9-13 Sept. 1985, CERN LHC-Note 31.

8. E.M. Hornsveld, J.D. Elen, L.A.M. van Beijnenand, P. Hoogendam, Development of ECN-type Niobium-tin wire towards smaller filament size, Proc. ICMC , June 14-18, 1987, Advances in Cryogenic Engineering, vol. 34, 493 (1987).

9. E.M. Hornsveld, W.M.P. Franken, J.A Roeterdink, J.A. Eikelboom, S. Wessel, B. ten Haken, H.H.J. ten Kate, Fabrication and testing of a niobium-tin cable designed for the outer coil of an LHC dipole, to be published at the 11th Magnet Technology Conference, Japan, August 28, 1989.

10. H.H.J. ten Kate, B. ten Haken, S. Wessel, T. Roeterdink, E. Hornsveld, Critical current measurement of prototype cables for the CERN LHC up to 50 kA and 13 tesla using a single step superconducting rectifier, to be presented at the 11th Magnet Technology Conference, Japan, August 28, 1989.

11. D. ter Avest, C. Daum and H.H.J. ten Kate, Magnetic field and lorentz forces in an LHC-dipole magnet. 3-D analysis using the fem programme TOSCA, to be published at the 11th Magnet Technology Conference, Japan, August 28, 1989.

12. C. Daum, D. ter Avest, Three-dimensional computation of magnetic fields and lorentz forces of an LHC dipole magnet using the method of image currents, to be published at the 11th Magnet Technology Conference, Japan, August 28, 1989.

13. D. ter Avest, L.J.M. van de Klundert, A new method to calculate conductor magnetization in accelerator dipoles, paper at this conference, Proc. IISSC New Orleans, Feb. 8-10, 1989.

14. D. ter Avest, L.J.M. van de Klundert, Filament and cable magnetization in superconducting accelerator dipoles, to be published at the 11^e Magnet Technology Conference, Japan, August 28, 1989.

15. H.H.J. ten Kate, C. Daum, L.J.M. van de Klundert, W. Hoogland, T.A. Roeterdink, F. van Overbeeke, R. Perin, Development of a 10 T Nb_3Sn "twin aperture" model dipole magnet for the CERN LHC, to be published at the 11th Magnet Technology Conference, Japan, August 28, 1989.

16. H.H.J. ten Kate, Superconducting Rectifiers, Thesis University of Twente, April 14, 1984.

17. H.H.J. ten Kate, J. Knoben, H.A. Steffens, L.J.M. van de Klundert, A 25 kA, 0.5 kW thermally switched superconducting rectifier, Proc. MT-8 Grenoble 1983, Journal de Physic C1, vol. 45, 471 (1984).

18. H.H.J. ten Kate, W. Uyttewaal, B ten Haken, L.J.M. van de Klundert, The Twente high-current conductor test facility, first results on critical current and propagation in two cables, Proc. CEC/ICMC conferences 1987, Advances on Cryogenic Engineering, vol. 33, 1988, Plenum Press, N.Y..

ANALYTICAL SOLUTIONS TO SSC COIL END DESIGN

R. C. Bossert, J. S. Brandt, J. A. Carson,
H. J. Fulton, and G. C. Lee

Fermi National Accelerator Laboratory
Technical Support Section
Batavia, Illinois 60510

J. M. Cook

Argonne National Laboratory
Advanced Photon Source
Argonne, Illinois 60439

ABSTRACT

As part of the SSC magnet effort, Fermilab will build and test a
series of one meter model SSC magnets. The coils in these magnets will
be constructed with several different end configurations. These end
designs must satisfy both mechanical and magnetic criteria. Only the
mechanical problem will be addressed. Solutions will attempt to
minimize stresses and provide internal support for the cable. Different
end designs will be compared in an attempt to determine which is most
appropriate for the SSC dipole. The mathematics required to create each
end configuration will be described. The computer aided design,
programming and machine technology needed to make the parts will be
reviewed.

ANALYTICAL SOLUTIONS TO SSC COIL END DESIGN

When winding a superconducting $\cos\theta$ magnet coil, special attention
must be paid to the end configuration. Parts must be made to confine
the conductors to a consistent shape. This shape must be defined and
described to both the parts manufacturers and those analyzing the
magnetic field.

Past Experience With Superconducting Magnet Ends

Over one thousand Tevatron dipoles have been produced at Fermilab.
The path which the conductors take as they are wound around the ends on
these magnets is defined as the intersection of a sphere and a cylinder
(see Figure 1). It can be generated by simply drawing a circle with a
compass on the surface of a cylinder. At the center of each turn, the
cable is held vertical, that is, perpendicular to the beam path in the
yz plane (see Figure 1). This shape has the advantage of being easy to
define and inspect. The disadvantage of the "vertical" end is that it
results in high internal stresses in the cable. The amount of stress

induced in the cable is inversely proportional to the bore diameter.
These stresses, although high, were within acceptable limits on the
Tevatron magnets. The Tevatron bore diameter is three inches.

    The SSC magnets have a much smaller bore diameter than the Tevatron (4
cm). As a result, stresses become unacceptably high. Many problems
result from these high stresses. They are listed in Table 1.

Table 1. Problems in Winding COSΘ Magnet Ends

A.   Turn-to-Turn Shorts.

      Cables which are forced into an unnatural path will cause breakdown of
      the kapton insulation between turns.

B.   Degradation of Strands.

      High stresses cause the strands to stretch, and in the worst cases,
      break. Strands will also come "out of lay" causing the cable to take a
      shape other than its intended keystone shape.

C.   Difficult to wind.

D.   Difficult to maintain magnet-to-magnet consistency in conductor
      placement.

E.   Tendency of cables to move when magnet is powered, causing quenches.

F.   Tendency of turns to move into the bore after curing.

      The cable tries to take a position which is less stressful than that
      into which it was wound. This usually results in the inner coil turns
      moving into the bore area (see Figure 2).

Solutions

    It is necessary to create an end configuration which produces less
internal stress in the cable. Internal stress is proportional to the amount
of strain in the cable. "Strain" is defined as the amount of deformation per
unit length to which the cable is subject as it is wound around the end. The
symbol for strain is $\Delta L/L$, where ΔL is the deformation and L is the total
perimeter of the end. There are two different approaches to reducing strain.

    1.   Decrease the difference between the inside and outside perimeters of
          the respective turns (decreasing ΔL).

    2.   Increase the total perimeter of the end by making the end flared
          (increasing L).

    The following geometries will attempt to decrease strain by the first
approach.

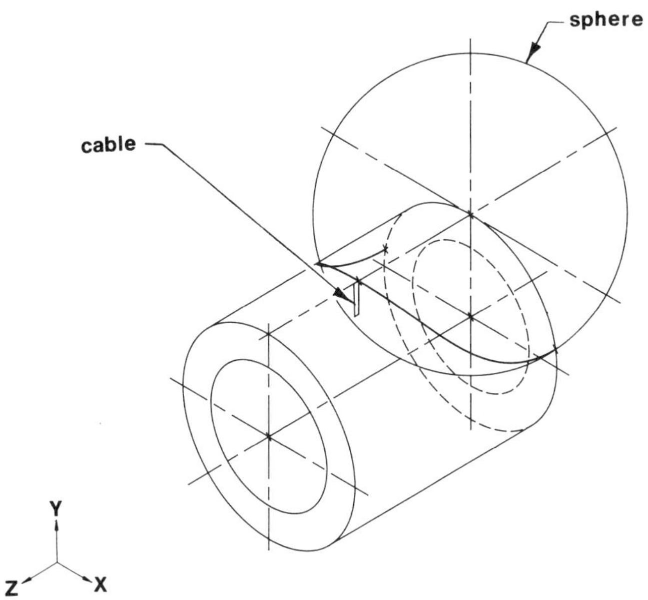

Fig. 1.   Intersection of sphere and cylinder.

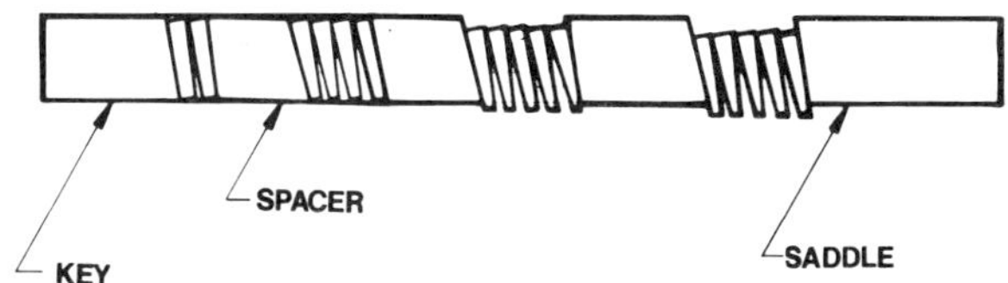

Fig. 2.   Turns moving into bore.

Creating Geometries

A.   Method of Defining Base Curve

The base curve is defined as the path which an individual conductor takes as it is wound around the end of a coil. It is represented by a one dimensional line in three dimensional space.

1.   Ellipse on Cylinder Method.

We begin by laying out an ellipse in the flat conditon (see Figure 3). b1 is equal to the arc length of the outer surface of the coil cross section at that point. a1 is chosen for magnetic reasons. The ellipse is then wrapped on the cylindrical surface.

The shape of the ellipse can be adjusted to conform to empirical evidence by changing the "order" or the exponent in the ellipse equation. A "superellipse" with an exponent of three,

$$\frac{x^3}{a^3} + \frac{y^3}{b^3} = 1$$

for example, has a "squarer" shape than an ellipse with an exponent of two

$$\frac{x^2}{a^2} + \frac{y^2}{b^2} = 1$$

(see Figure 4). It is important, due to the incomplete nature of our models, to be able to adjust the shape of the base curve to conform to empirical evidence. Some of the shapes, such as the ellipses on cylinders, are generated purely for geometric reasons. Even the best models are seriously flawed. All current models, for example, assume the cable is an infinitely thin, homogeneous strip, and not the composite of shapes and materials which really make up a cable.

The ellipse on cylinder method of defining a base curve has several advantages. It is mathematically simple and easy to define. It readily lends itself to manipulation by changing the order. Its disadvantages are that it does not take into consideration any actual stresses in the cable.

2. Elastica

A planar elastica is defined as a plane curve into which the central line of a thin elastic rod can be bent by means of forces and couples applied to its ends only. The curve then satisfies the condition that it have the minimum possible strain energy, subject to certain end constraints. In our case we have a cylindrical surface instead of a flat plane, complicating the situation slightly. The curvature of the elastica has a new component perpendicular to the constraining surface. The strain energy corresponding to this component, which was zero when the constraining surface was a flat plane, must now be added to the classical strain energy corresponding to the component

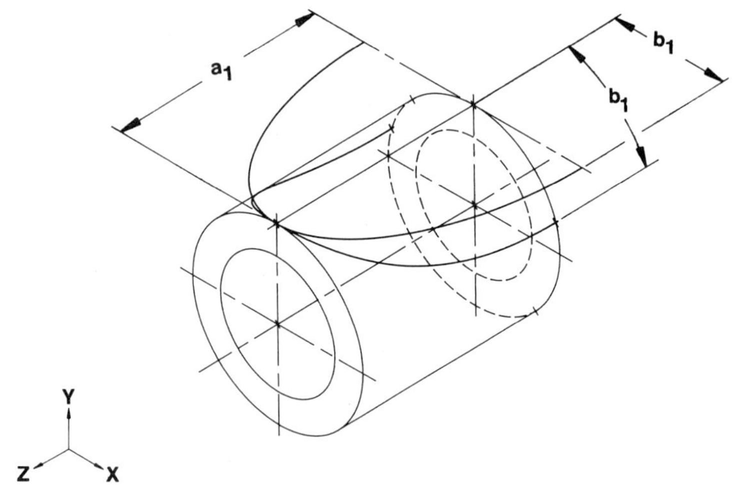

Fig. 3. Ellipse to be wrapped on cylinder.

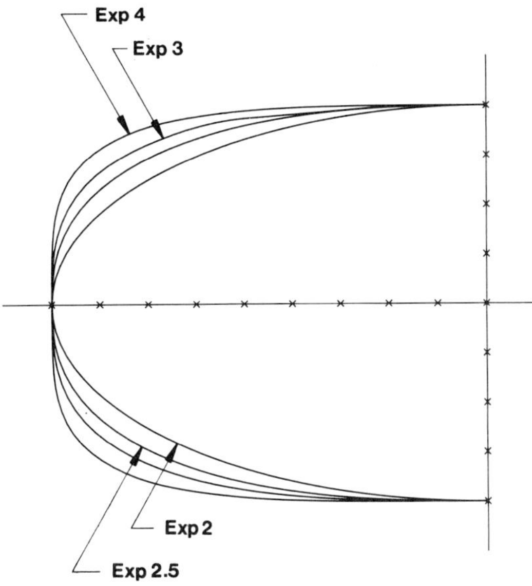

Fig. 4. Superellipses.

of curvature parallel to the constraining surface. The base curve is
therefore defined as a smooth line segment with minimum total strain
energy. It is confined to a cylindrical surface and must satisfy
certain initial and final conditions. These conditions are:

a. The line must begin at a point on the surface of the cylinder and
    be pointing in the direction of the positive z axis (see Figure
    5).

b. The line must end at a point on the top center of the cylinder
    (at a value of x=0) and be pointing in the direction of the
    negative x axis (see Figure 5).

The elastica has the advantage of being based on elasticity theory
rather than pure geometry. It may create less stress in the cable
than the ellipse on cylinder method. It has the disadvantage of
being less amenable to mathematical manipulation. Manipulation of
this type, based on evidence obtained from inspecting previously
wound coils, may be necessary given the incomplete model.

B. Base Curve Position

The base curve may be placed on either the inside or the outside radius
of the layer (see Figure 6). In either case a "gap" appears on the other
radius. This happens because the cables are laying at an angle and no longer
require as much radial clearance as they would if they were stacked
vertically. Fermilab's SSC coils use the outside radius for the base curve.
This allows us to use the gap on the inside radius to provide internal
support for the conductor. A "shelf" is attached to the spacer, filling the
empty space and keeping the turns from moving into the bore.

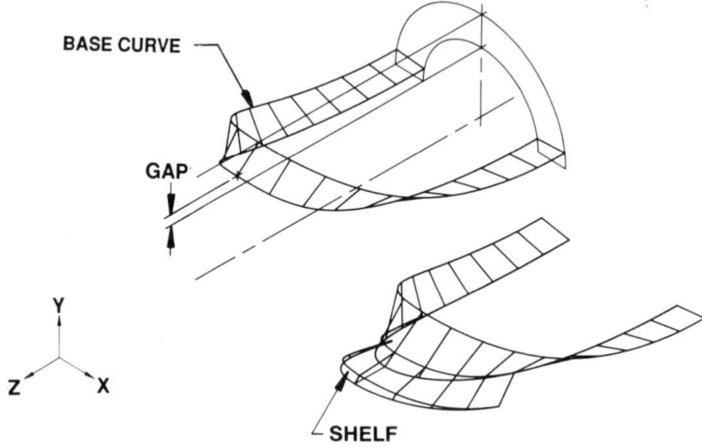

Fig. 5. Elastica

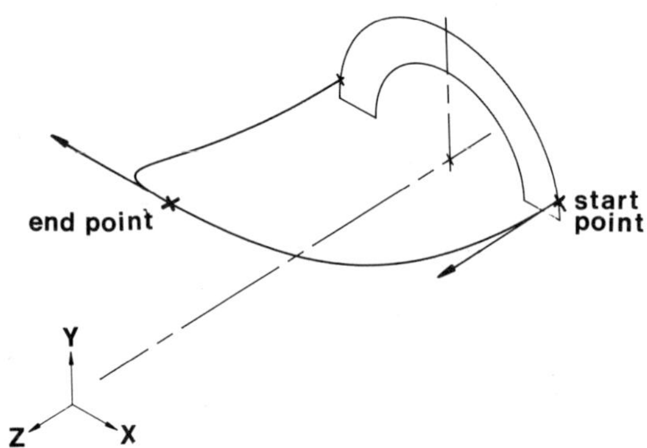

Fig. 6. Base curve positioned on outside of layer.

C. Method of Defining Surfaces

    1. Connecting Lines Between Equally Spaced Points.

        a. If Base Curve Is Ellipse On Cylinder.

We begin with a base curve (see Figure 7). a_1 is the longitudinal radius of the ellipse. b_1 is the azimuthal radius of the ellipse. The perimeter of this ellipse is calculated (L_1. We can now solve for another ellipse. We know L_2 because it must be equal in length to L_1. b_2 of our second ellipse is equal to the arc length of the inner surface of the coil cross section. We solve for a_2. The two ellipses are then wrapped around their respective cylinders (see Figure 8). Eleven equally spaced points are drawn on each ellipse half. The points are then connected making ten "facets". The eleven lines are then cut off at a distance of one cable width from the top surface. A curve is drawn connecting the new points. This gives us a new perimeter, L_3. L_3 is shorter than L_1 or L_2 and lies between L_1 and L_2. Then by an iterative process we repeat this procedure increasing a_2 in small steps until $L_3 = L_1$ giving us an average $\Delta L/L$ of zero.

392

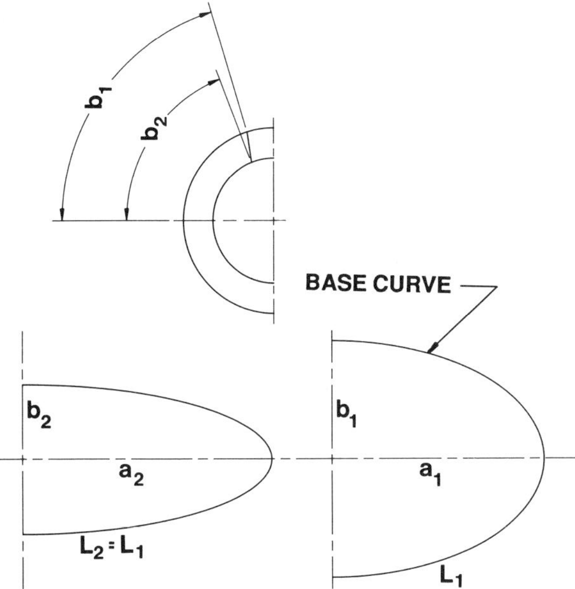

Fig. 7. Ellipse on cylinder as base curve.

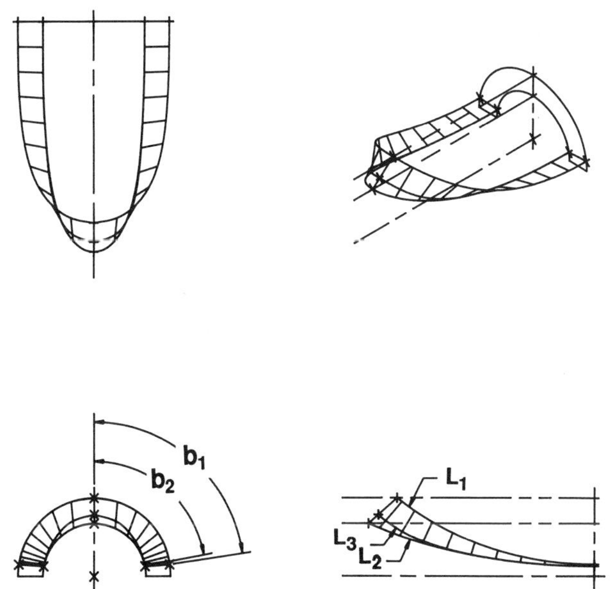

Fig. 8. Ellipses wrapped around their respected cylinders.

b.   If Base Curve Is Elastica.

We create two elasticas, one representing the inside edge of
the turn and one representing the outside edge of the turn.
An equal number of equally spaced points are placed on each
curve. Straight lines are then drawn between the
corresponding points. The resulting surface is then
"trimmed" to the thickness of the cable.

This method is geometrically simple. It does, however, have
undesirable qualities. The surface which is created is not
"developable", that is, one which can be laid onto a flat
surface without distortion. We are making an attempt to
create an end configuration which results in a developable
surface.

2.   Creating a Developable Surface.

A base curve is wrapped on a cylinder (curve L_1 on Figure 9). A
set of equally spaced points is placed on the curve (P_1, P_2,
...Pn). Vectors are drawn from the points (V_1, V_2, ...Vn) in
such a way that they sweep out a certain surface called by
differential geometers the rectifying developable of the curve.
It is by definition perpendicular to the direction of curvature
of the curve at every point. The cable is modeled by a strip in
this surface along the curve.

Two simplifications in our mathematical model allow us to
approximate in this way a two dimensional elastic strip by a one-
dimensional elastica. First and most drastic, the torsional
rigidity of the strip (its resistance to torques with axes
parallel to the curve) is taken to be zero. Second its flexural
rigidity about an axis perpendicular to the strip is taken to be
infinite. The flexural rigidity of the strip about its other
axis is then equal to the rigidity of the elastica. In the
Fortran program implementing this model the first two assumptions
are relaxed slightly: the torsional rigidity is small but non-
zero and the first flexural rigidity is large but finite.

After the surface is created in this manner, it is trimmed at the
appropriate cable width (L_3).

It is not certain that the developable surface creates a less
stressful condition for the cable than connecting the straight
lines. Analysis is still needed to determine whether the
developable surface is truly more desirable.

D. Method of Stacking Cables

Independent of the method of defining the shape of the individual
surface for a turn, there are two methods of stacking a cable within
an end. These are the "individually determined" and "grouped"
methods (see Figure 10).

1.   Individually Determined.
In this method each turn has its surface calculated individually.
This results in the turns not laying directly upon each other.
As turns get farther from the pole, the geometry requires that
they be stacked at an increasingly larger angle (see Figure 10).
This creates small spaces between each turn. These spaces
typically become filled with epoxy.

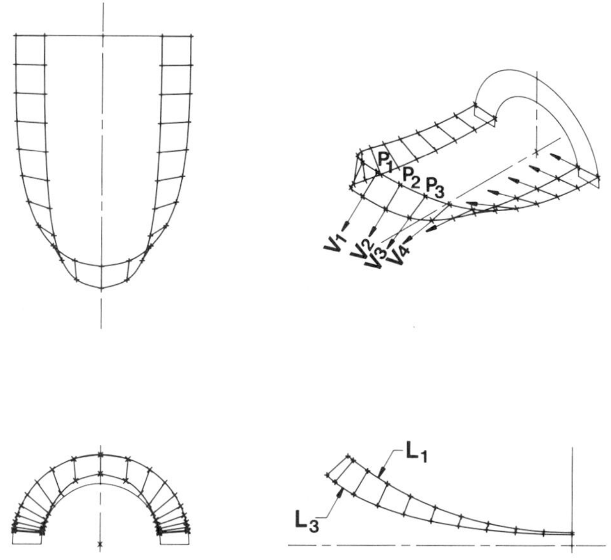

Fig. 9.   Developable Surface.

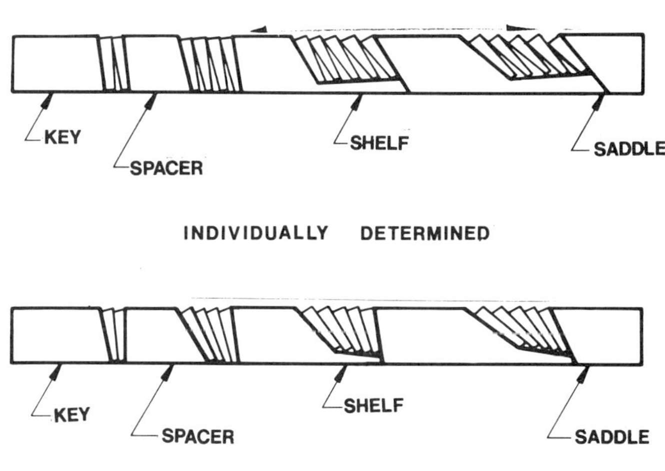

Fig. 10.   Cable stacking methods.

2.   Grouped.

In the grouped method only the center of each current block is a
calculated surface. The turns within each block are layed
directly on each other with no spaces between them. As the turns
get farther away from the center of the current block, their
internal stresses become progressively higher.

The advantage to the individually determined method is that the
stress in each turn is minimized. The advantage to the grouped
method is that the spaces betwen turns are eliminated, further
restricting cable movement.

Creating the Drawings

Creating these end geometries would be impossible without modern
computer aided drafting and design. Software exists for such functions
as "laying an ellipse on a cylinder", "creating a spline in three
dimensions", and "placing equally spaced points on a curve". Solid
modeling programs allow the designer to create these various parts,
visually inspect and analyse them, then send the data to a numerically
controlled machine.

Much effort has been put into automating the design of these parts.
Having a clear mathematical description of the curves allows us to write
programs which create the curves based on an initial input. The designer
must answer several prompts concerning the two dimensional coil cross
section and the longitudinal current block positions on the ends. Curves
are created by the program. The output consists of a data file of
coordinate points from which three dimensional surfaces are defined and
tool paths automatically created for the numerically controlled machine.

Such a program has been completed for an end design based on straight
lines connecting ellipses. An upcoming program will deal with elasticas
and developable surfaces.

Manufacturing

All parts are presently being produced by numerically controlled
machines. Raw material is G-10 tubing. Both Fermilab's machine shops
and outside vendors are making parts.

The most efficient way of machining these parts is by using a five
axis machine with a cylindrically shaped cutting tool. All surfaces
could be machined in a single pass by this method. Due to the
unavailability of five-axis machines, we have been presently producing
them with three-axis machines. We must use a spherically shaped tool and
cut the surfaces in several passes. This process requires more
complicated programming and longer machining time than would be required
by the five-axis machine.

It is clear that machining parts is unacceptable for quantities as
large as would be required for the SSC. Other fabrication techniques
must be pursued for the production magnets.

Fermilab is attempting to make molded parts. Several materials are
under consideration. Some success has been achieved using phenolic with
50% glass fiber. The material is hydraulically transferred into a mold
cavity and cured at 350 degrees farenheit for 40 seconds. Molded parts
have not yet been used in a magnet.

<u>Conclusion</u>

There are many possible solutions to the problem of winding the end of an SSC coil. The Fermilab short model program is choosing several and attempting to compare them in the most experimental way possible. One meter coils will be built with various combinations of "ellipse vs. elastica", "straight line vs. developable" and "grouped vs. individually determined" configurations. These magnets will be tested cold. The coils will also be potted and sectioned to closely examine the placement of conductors. Analysis of the results will determine which type of end will be used in the Fermilab SSC long models.

REFERENCES

Cook, J. M., 1989, "Program Bend (a program creating surfaces using
        elasticas and developable surfaces)", Argonne National
        Laboratory.
Lee, G. C., 1989, "Autoend Program (a program creating surfaces by
        straight lines connecting ellipses)", Fermi National Accelerator
        Laboratory.

STRAIGHT ENDS FOR SUPERCONDUCTING DIPOLE MAGNET USING

"CONSTANT PERIMETER" GEOMETRY*

John Royet

Lawrence Berkeley Laboratory
1 Cyclotron Road
Berkeley, CA 94720

ABSTRACT

The ends of the SSC Dipole magnets are a very critical aspect of the superconducting cable windings needed for this large project. The internal coils, where the radius at the pole is as small as 3/10 of an inch for the first turn, are difficult to form with the very stiff cable, and a high tension is needed.

The curing operation on the coils is performed in a heated forming press which applies an important additional stress on the superconducting wire and insulation. A new design of this sensitive region of the magnets was performed at LBL, and several prototypes were built and tested. In this paper the construction method used to solve some of the most critical problems is exposed along with a description of the experimental work in progress.

INTRODUCTION

The ends of superconducting dipole or quadrupole magnets, mostly those where the bore diameter is less than 5 or 6 times the cable width are difficult to wind due to the small bending radius of the cable around the pole. However, the construction of such magnets with straight ends remains attractive as compared with magnets with flared ends, especially due to hardware simplification and cost reduction.

The main difficulty is still the small bending radius of the superconducting cable the hard way to fit a small bore without additional superconductor degradation and possible turn to turn shorts as well as of the high cable stress and the resulting thickness growth.

One way to ease high stress in cables is to invoke the use of "constant perimeter" during winding.

CONSTANT PERIMETER

By definition, we require both edges of the cable (the inner and outer) to assume the same overall length during winding. As a result, the cable will not bend the hard way reducing swelling and high stress points.

---

*This work was supported by the Director, Office of Energy Research, Office of High Energy and Nuclear Physics, High Energy Physics Division, U.S. Department of Energy, under Contract No. DE-AC03-76SF00098.

In producing a constant perimeter path model I would take a thin flat ribbon of elastic material, steel as an example, form it as a 180° loop and try to fit one edge on a cylinder surface as a cable should be around the bore and mark the ribbon perimeter. It is clearly seen that the inner edge of the ribbon on the cylinder is placed axially farther out on the cylinder than the outer edge. As the length of those two edges remains the same, (the ribbon cannot store any strain) the ribbon has to titlt as it comes around. The maximum tilting angle is observed at the loop summit or pole. The amount of tilting is reduced for those turns which emerge close to the midplane. See Fig. 1.

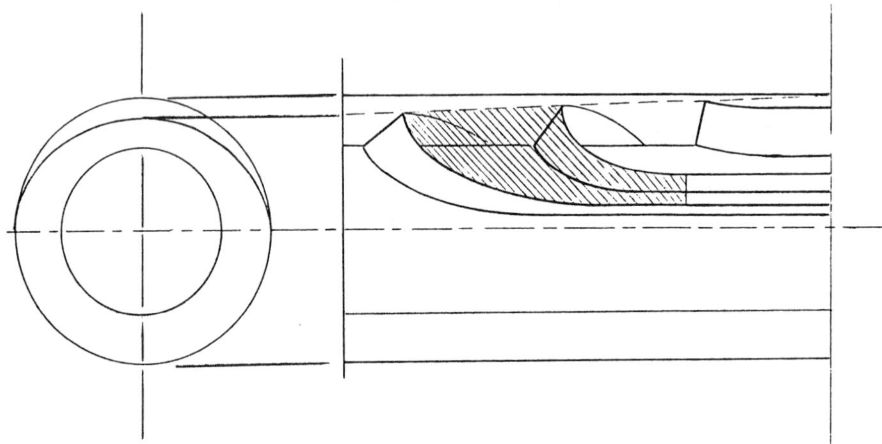

FIG. 1   CIRCULAR INNER CYLINDER

The constant perimeter geometry for the cable at the ends of the dipoles has to be accommodated in the space included between two cylinders: the smallest being the inner diameter of each layer and the largest the outside diameter of that layer. If one of these cylinders is circular the other one is neccessarily of an odd shape due to the reduction of the apparent cable radial dimension in the tilted areas. Due to the electrical insulation required on the inner surface of the dipole magnet, the best geometry seems to be an elliptical inner shape see Fig. 2.

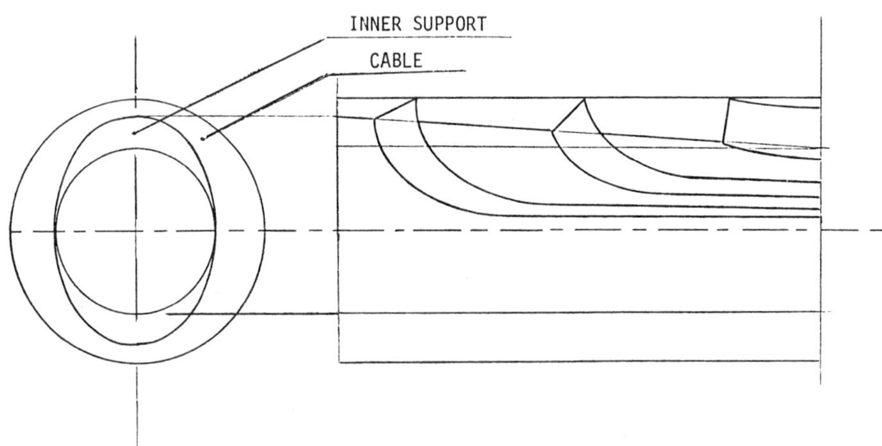

FIG. 2   CIRCULAR OUTER CYLINDER WITH CONOIDAL
INNER SUPPORT

DIPOLE WINDING

The superconducting cable is wound the usual way along the straight section and then is forced on conical supports at each end. The cable bundle pattern at the magnet end is a prolongation of the straight section. The copper wedges, however, cannot be used due to the changing shape needed to generate the magnetic field characteristics needed and are replaced with machined spacers.

END SPACERS

The end spacers have several important functions and characteristics: They should:
- confine the superconducting cable at the required position in order to obtain the magnetic field uniformity needed.
- contribute to and enchance the dielectric strength of the end windings.
- exhibit high flexibility during winding without risk of breaking.
- have reasonable resistance to radiation.

MATERIALS

Several different materials were tried for end spacers, all of them of the plastic nature. The use of plastic spacers was required due to the fact that hey had to be fitted in the windings prior to curing. At this stage, the magnet shape and dimensions are slightly different than during the final collaring assembly.

Silicon rubber, epoxy, and fiberglass reinforced epoxy have been used, none of them with full success. A recent experiment performed with LEXAN[1] mold injected parts resulted in the closest possible dimensions and shapes corresponding to the assembled coil. The model parts used to build the molds were hand carved in epoxy resin. Another experiment was performed with machined parts from a LEXAN[1] block using a numerically controlled milling machine programmed with the dimensions of the hand carved models. The resulting end spacers were very similar in shape, but not in cost.

Thoee end spacers were strong enough to be deformed elastically and inserted in place during the coil winding operation performed at a lower cable tension than with conventional straight end design.

During the press curing which is performed at a temperature close to the LEXAN[1] softening point the end spacers flow and adapt to the superconducting cable bundles. When the coil is cooled, the LEXAN[1] solidifies and fits tightly, filling the entire space between cable blocks and even strands.

This in turn reduces any possible superconducting wire motion due to high Lorentz forces when the magnet is energized.

One model magnet that was built using these type ends was tested and operated satisfactorily.

NEXT DEVELOPMENT

The junction of the conical end support with the straight part of the magnet, should be as smooth as possible then, the cone support thickness at this point should be almost zero. In order to achieve an external pressure resistant assembly we need some kind of internal support: stainless steel tubing was used in the original model magnet. A combination of a stainless steel bore liner with the conical support seems to be one of the next path to explore.

---

[1]LEXAN is a trade name of a General Electric product.

REFERENCES

Meuser, R. B., "End Effect in Superconducting Beam Transport Magnet", IEEE Trans. Nucl. Sci., NS-18, pp. 667-669, (1971).
Carson, J., Fermi National Accelerator Lab Sept. 1985: S.S.C. Dipole Magnet 1 Meter, 4 cm Bore. Constant Perimeter Saddle Design.
Hassenzahl, W., Caspi, S., Gilbert, W. S., Helm, M., and Laslett, L. J., "Field Quality of the End Sections of S.S.C. Dipoles", 1986 Applied Superconductivity Conference, Baltimore, MD, Sept. 29 - Oct. 3, 1986. LBL Report LBL-22208, SSC-MAG-106, SSC-N-250.

AN ALTERNATE END DESIGN FOR SSC DIPOLES*

Craig Peters, Shlomo Caspi, and Clyde Taylor

Lawrence Berkeley Laboratory
1 Cyclotron Road
Berkeley, CA 94720

ABSTRACT

Experience in the SSC dipole program has shown that fabrication of cylindrical coil ends is difficult. Cable stiffness requires large forces to maintain the proper position of the conductors in the end during winding. After winding, the coil ends remain distorted and significant motion of the end conductors is required to force the coil end into the molding cavity. Local mechanical stresses are high during this process and extra pieces of insulation are required to prevent turn-to-turn shorts from developing during the winding and molding steps. Prior to assembly the coil end is compressed in a mold cavity and injected with a filler material to correct surface irregularities and fill voids in the end.

LBL has developed an alternate design which permits the conductors to be wound over the ends using minimal force and technician coersion. The conductors are placed on a conical surface where the largest diameter over the outer layer conductors is 10 cm. No coil end spaces or insulation pieces between turns are required. The conductor geometry was analytically optimized to meet SSC multipole requirements for the ends. The first 1-m dipole utilizing this end geometry has been constructed and successfully tested. Design and construction data are presented. Also, model test results, including training and multipole measurements of the end, are given.

INTRODUCTION

The forming of straight dipole coil ends in SSC prototype models has historically proven difficult. Large forces are required to position the conductor during coil winding and during subsequent insertion of the wound coil into the molding cavity. This is due to the resistance of the superconducting cable to being formed to the small radii required in the ends of small aperture straight ended magnets. The inherent stiffness of the cable is increased by the glass tape wrap used on the cable over the Kapton insulating wrap. The cable insulation is scuffed and traumatized during winding and molding resulting in occasional turn to turn shorts in the ends of the formed coils that sometimes appear or later during coil assembly. At BNL and LBL the occurrence of these shorts has been reduced but not eliminated by installing extra insulation and fiberglass strips between turns in the ends during coil winding.

Another problem with straight ends is the tendency for the molded coil ends to lose their shape following removal from the molding fixture due to high residual stresses in the

*This work was supported by the Director, Office of Energy Research, Office of High Energy and Nuclear Physics, High Energy Physics Division, U.S. Department of Energy, under Contract No. DE-AC03-76SF00098.

wire. As the shaped end tries to flatten out, the insulation is further damaged. Also assembly of the mis-shapened end is made more difficult.

Two fixes to the problems of straight ends have been tried. At BNL the coil ends are filled or potted as the final step in the molding process. This eliminates voids and helps the end to maintain its shape; the filling material allows the conductors in the end to be well clamped and supported after assembly. However, the basic difficulty in winding and molding still remains - that is turn to turn shorts can only be prevented by operator skill and the use of insulating strips and separators between turns.

A second fix has been recently tried at LBL and at FNAL. This is the constant perimeter approach which attempts to reduce cable stresses by tilting the cable in the end so that the inner and outer edge perimeter lengths are the same. The tilting of the conductor in the pole region shortens the cable radial height and allows introduction of a thin tapered cone to both support the winding during assembly and give it a precise cylindrical outside diameter. Introduction of this approach has only slightly improved the windability of straight coil ends. This is partly because the existing straight end uses a semi-constant perimeter geometry. Also because of the small bending radius, there are still significant stresses in the cable even with the constant perimeter geometry. At least three carefully shaped end spacers are required in each end of the inner coils to place the turns close to their constant perimeter shape. It is likely that separators and/or separate insulating strips between turns will be continued to be required for electrical reliability.

At LBL we have designed and built a dipole model with a new end geometry in which the problems of the straight end are for the most part eliminated. The magnetic and mechanical designs have been done consistently to produce an end which is easy to wind and assemble, requires relatively few individual pieces, and whose magnetic multipoles stay within the SSC specifications.

FLARED END DESIGN

The alternate design proposed at LBL is referred to as the flared end design. The primary goal of this design was to produce a winding that was first of all more electrically rugged and reliable than the present designs and at the same time meet the SSC end multipole requirements. The secondary goal was to produce a more manufacturable end, requiring less technician time and skill to wind and mold.

Since the problems of electrical reliability and winding difficulty for straight SSC dipole ends result primarily from high forces required to bend the conductor around the small radii inherent in small bore dipole magnets, the approach was to increase the bending radii of the end conductors as much as possible.

The natural way to increase conductor radii is to wind the end conductors on a conical surface. The geometry chosen is a simple conical surface in which the cone axis is tilted slightly with respect to the dipole axis as shown in Figure 1. (We observed that a coil wound on a non-tilted cone surface drooped toward the coil axis when the coil was removed from its supporting cone.) After experimenting with several shapes, a 19 deg. cone with its axis tilted downward at 5 deg., shown in Figure 1, was chosen as the basic geometry for LBL's flared end. The inner and outer surface of each conductor layer each have this geometry, that is, a 19 deg. cone with its axis tilted downward at 5 deg. The four conical surfaces of each half of the dipole have a common axis and are concentric cones. The radial thickness of the space separating the inner coil outer surface and the outer coil inner surface was chosen to be 1.25 mm, enough to provide adequate space to accommodate small variations in overall length that might occur in manufacturing. This separator is explained more in the Coil Assembly section below.

Tooling

The tilted cone geometry results in straightforward procedures for designing tooling. Dimensions were computed using a simple geometry program and verified using 2-D CAD layouts. Machining of the winding cones and molding cavities for the end were done using conventional milling machines and lathes in most cases. Several pieces were machined

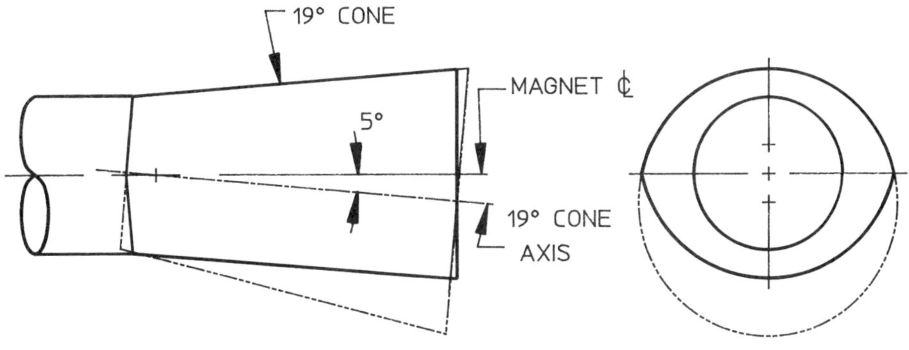

Figure 1   Flared End Geometry

using 3-axis CNC machines because of the simplicity of programming the shapes involved. A flared end coil molding cavity is shown in Figure 2.

<u>Magnetic Optimization</u>

With the conical surfaces now chosen the optimum conductor placement was selected that would produce magnetic multipoles within the SSC end multipole requirements. Two constraints were imposed for mechanical reasons; rule one: the conductor was permitted to tilt a maximum of 10 deg. from an orientation normal or perpendicular to the conical surface. The purpose here was to maintain adequate thickness of each coil end to eliminate complications that would arise from radial thickness deficits. Rule two was to have no separate coil blocks in the ends, so that the ends have no spacers and the conductors pass over the pole as one block. This was to keep the design simple with a minimum number of parts to fabricate and install during coil winding.

Programming of the end geometry is described in Ref. 1. A "wire frame" for the purpose of optimizing the magnetic field geometry was used where each turn is represented by one wire as shown in Figure 3. The computation assumes the current is concentrated at the wire. Ramp and return sections of the end are defined as in Figure 4. The computations assume the cable cross-section dimensions in the end remain the same as in the straight section, i.e. no distortion of the cable cross-section. The "teardrop" determines the position of the first conductor turn. The two parameters that could be varied in the computations to adjust multipoles were the length and width of the teardrop. Rule two above necessitates that the straight section copper wedges be terminated at the start of the end and feathering wedges be used in the ramp section of the end to gather the four straight section conductor blocks into one block for the return trip over the pole. Rule one above necessitates a wedge at the midplane of the ramp section in order to keep the conductors in this section standing nearly normal to the conical surfaces.

<u>Final End Shapes</u>

For our first model, the feathering wedges, teardrop, and end shoe were machined from sections of G-10 tube stock (in production, these parts would be injection molded). The basic conical surfaces were machined conventionally on a lathe. The contoured faces that contact the cable were machined using an N.C. controlled three axis milling machine programmed using data from the "geometry" program described above. The final shapes of these end parts was determined assuming the conductor thickness was the same in the end as it was in he straight section.

<u>Coil Winding & Molding</u>

Machined aluminum winding cones having the conical external surface tilted at the prescribed 5 deg., with a cylindrical bore are attached to a cylindrical winding mandrel shown in Figure 5. A steel winding pole attached to the mandrel terminates at the leading edge of the winding cone. The machined G-10 teardrop, described above, is accurately fixed on the winding cone using two 0.8 cm diameter pins. The teardrop inner tip is

405

Figure 2 Flared End Molding Cavity

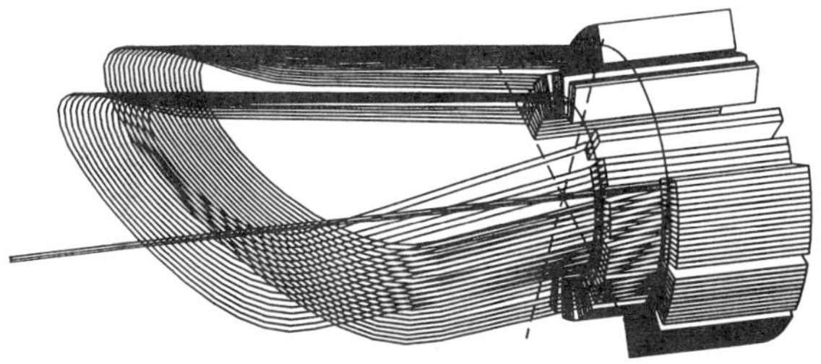

Figure 3 Wire Frame End Geometry

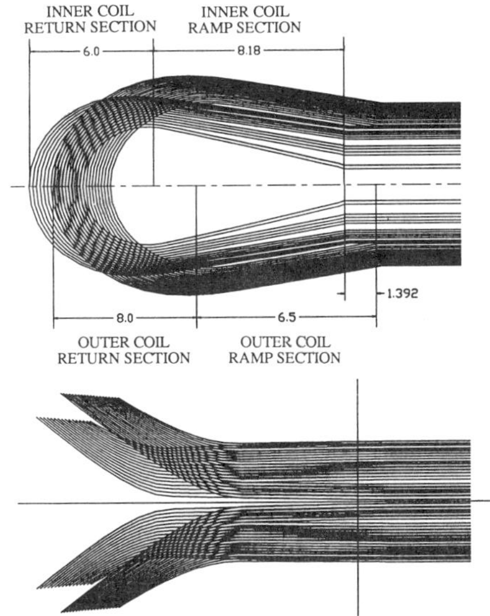

Figure 4 End Ramp & Return Sections

supported in the axial direction against the end of the steel winding pole. During winding, the inner coil lead is routed radially inward through the teardrop and winding cone into the end of the cylindrical winding mandrel, Figure 6.

Cable tension is set at 18 kg during coil winding. As each turn is wound onto the mandrel, a spring loaded clamp, is used to draw the conductor up against the teardrop at the end of the straight section; this creates a "reverse" bend in the cable. It is not necessary that the conductor be held tightly against the teardrop during winding. The effect of leaving "sag" in the conductor at this point is to cause in the area the innermost turns of conductor to tighten firmly against the end of the teardrop as the coil is compressed in the molding fixture. This is a desirable effect in constrast with the tendency in straight ended coils for the conductor to loosen at the end as the coil is compressed in the molding fixture. During winding, a minimal amount of mechanical tapping is used to aid in firmly positioning the conductors at the end. After winding and before insertion into the molding fixture, the conductor conforms to the conical winding surface with a maximum gap under the conductor of only about 2 mm. (for the inner coil). Figure 5 shows a molded inner coil viewed from the end. The ability of the coil end hold its as-wound shape is apparent.

As mentioned above the final shapes of the end parts were determined assuming the conductor thickness would be the same in the end as it is in the straight section. The cable in the ramp section of the end compressed to the same thickness as the cable straight section conductor during the molding step. However, no axial load is applied to the return portion

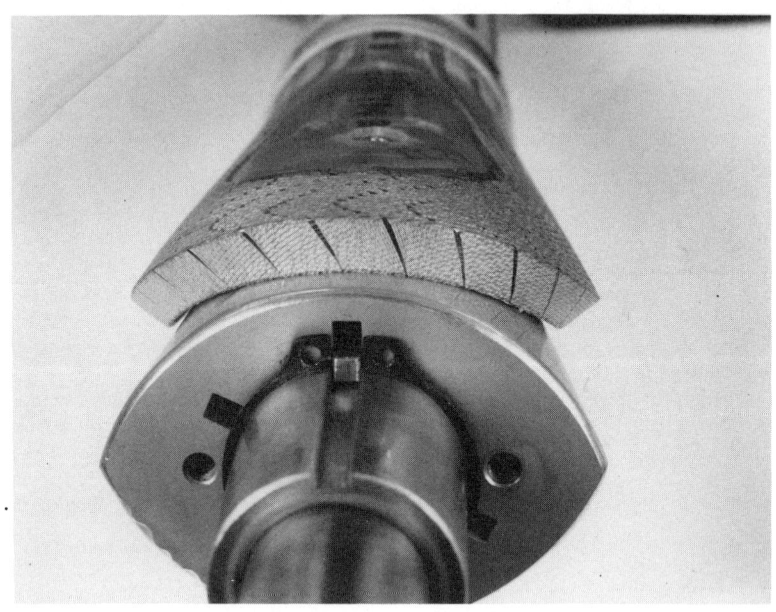

Figure 5 Coil End Winding Cone

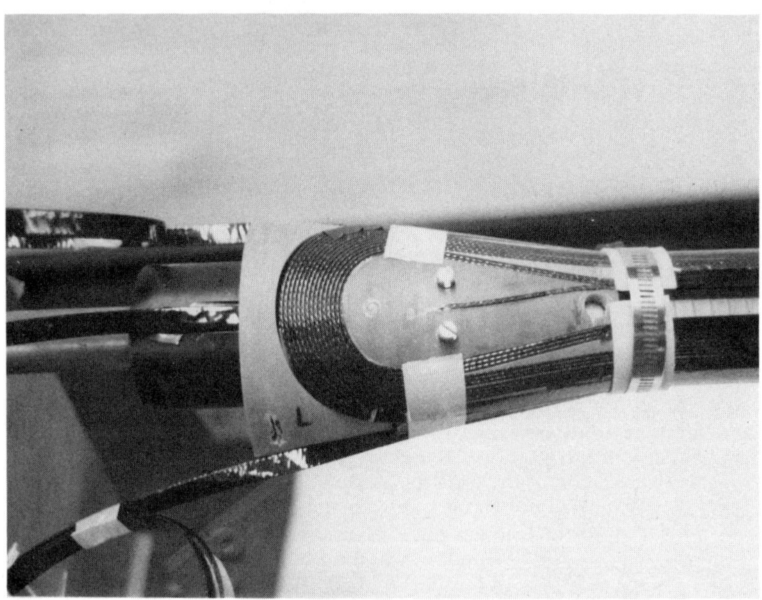

Figure 6 Coil Pole Lead Routing During Molding

of the end during molding. As a result, the conductor thickness at the end of the return section increases slightly. The end of the molded inner coils were measured and the average conductor thickness at this point was found be larger than the nominal by 17% (about 0.28 mm total for 16 turns compared to about 0.45 mm for straight ended coils, ref. 1).

<u>Coil Assembly</u>

Figure 7 shows an exploded view of the flared end coil assembly. The two aluminum end cones are positioned and aligned by a fixture at each end of the assembly table. The G-10 teardrops are precisely located on the cones by s.s. assembly pins, Figure 8. At the lead end the coil lead from the inner turn is routed radially outward through the G-10 teardrop as shown in Figure 8. The end shoe contour was trimmed to fit the final coil end shape. After mounting the coil, teardrops, and end shoes, the coil end is clamped and wrapped with a thin (0.25 mm) flat braid of Kevlar to hold it in position; this step might be omitted for a production magnet, but is used here as a convenient assembly aid. The entire inner layer including ends is wrapped with teflon sheet to minimize friction between layers. A 1 mm-thick molded end separator is placed over the inner coil end held in and position by the assembly pins in the end cone. The purpose of this separator is to allow adjustment of the relative axial position of the inner and outer coil ends. This is not necessary in 1-m models but in full length 17-m models the coils may have slightly varying lengths after being removed from the molding tooling. If this occurs, it may be accommodated at assembly by using a thinner or thicker separator. If necessary, a 1 cm relative length change of the inner coil compared to the outer coil in either direction, could be conveniently accommodated by this method.

The outer coil is mounted similarly to the inner coil with alignment pins to position the coil end. Both inner and outer coil leads egress from the outer conical surface of the outer coil teardrop, Figure 9. Pole extension blocks, shown in Figures 7 & 10, are positioned at the inner end of the coil teardrops. These blocks have the same inner profile as the collars and provide a short "straight" section to lead into the flared end. The G-10 insulator, Figures 7 & 10, has a short cylindrical section which transitions to the conical section. The inner surface of this insulator is precisely contoured using a N.C. machining to provide proper positioning and clamping of the coil end as it transitions from the straight to flared section. A second function of this G-10 extension block is to allow the Kapton ground plane insulation to extend about 1.5 cm beyond the end of the last collar pack. Insulation between inner and outer coils and between outer coil and ground is provided by the coil separator and end insulator, respectively.

The coil ends in the first model were clamped radially using a bolted S.S. clamp, Figure 11. The azimuthal coil pressure applied to the end at assembly was approximately the same as for the straight collared coil section. Subsequent flared end dipole models will utilize aluminum collars, Figure 10, to clamp the coil ends, rather than the bolted clamps used here. After assembly of yoke and shell, the coil ends are loaded axially to 10,000 lbs. by pushing on the coil end shoes with a 2.5 cm thick s.s. end plate. End plate loads were monitored during assembly, cooldown, and operation of the magnet.

HARMONIC ANALYSIS

The concept of harmonic analysis for an "end" is based on a reduction of the 3 dimensional field components to 2D problem. This is done by integrating the field components with respect to the axial magnet dimension which eliminates the integrated z directed field and reduces the problem to an harmonic analysis of integrated 2 dimensional field. Such a detailed analysis was documented in Appendix A (Ref. 1) and the results given as 2 sets of field harmonics one for the straight section and one for the end both having the same fundamental (e.g. the same central dipole field). In this way construction errors resulting in "end harmonics" are totally detached from field errors in the straight section and corrections can therefore be made directly to the "end". Magnetic measurements can also be easily converted into "end harmonics" through a simple transformation. The "end" computations were done on a geometry with no iron and were aimed at providing a geometry with low integrated harmonics. The predicted and measured harmonics are tabulated in Table A, B, and C for the individual layers-inner, outer and both together respectively.

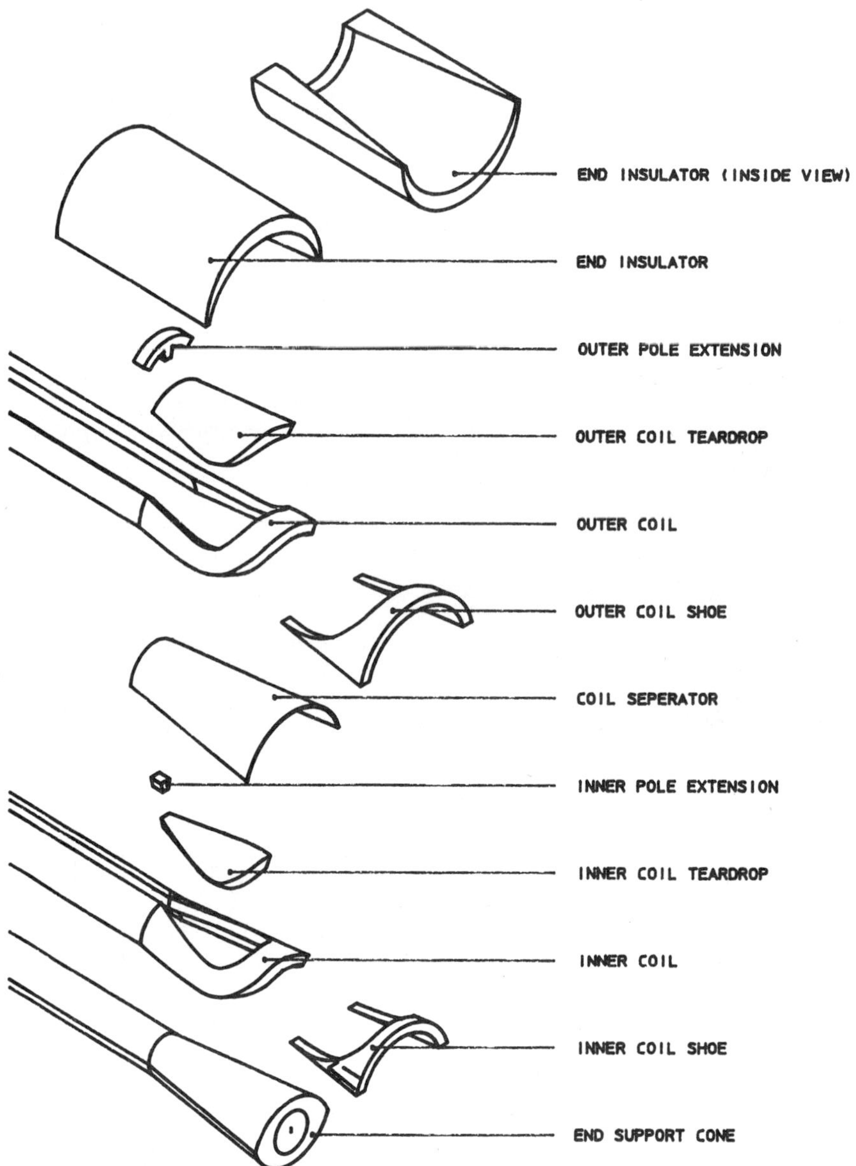

Figure 7 Exploded View of Flared End Coil

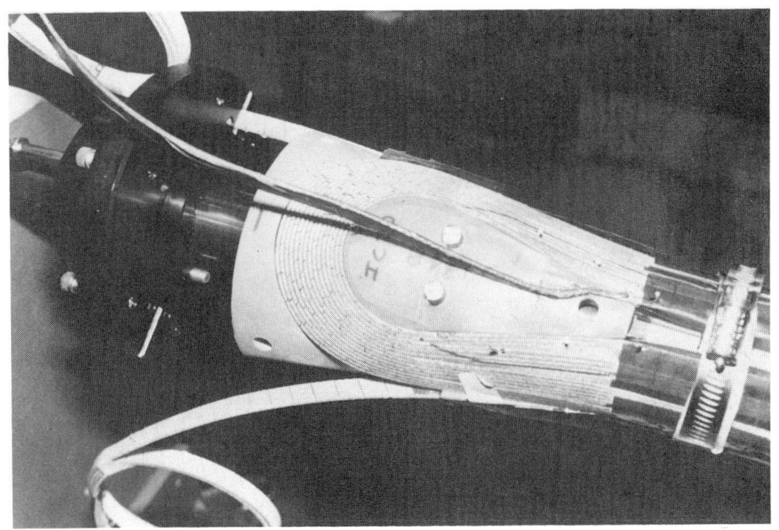

Figure 8 Inner Pole Lead Exiting Coil End

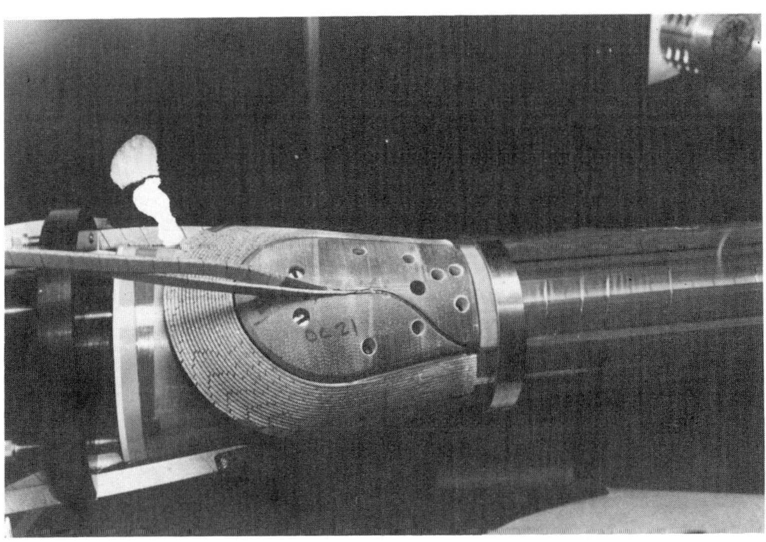

Figure 9 Pole Leads Coming Out of Coil End

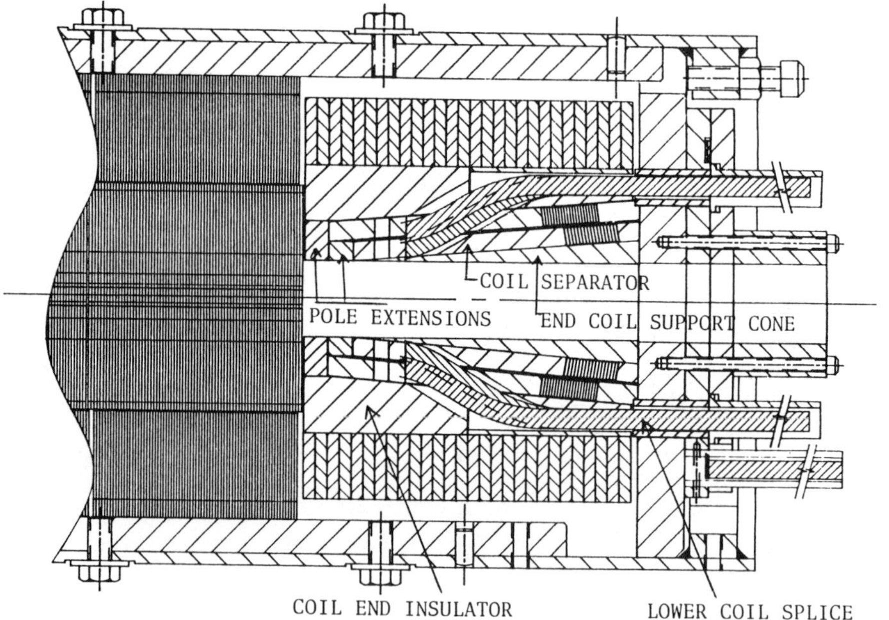

Figure 10 section View of Flared End Magnet

Figure 11 Coil End Clamped Used on First Flared End Model

Table A

Harmonic	Predicted units for a 12.28 cm	Measured No Iron (a 12.03 cm end)
sextupole	-89.93	-80.4
decapole	-17.44	-21.2

Comparison between computed and measured harmonics for the inner layer only (no iron)

Table B

Harmonic	Predicted units for a 10.41 cm	Measured No Iron (a 10.26 cm end)
sextupole	-101.09	-92.15
decapole	-9.13	-7.67

Comparison between computed and measured harmonics for the outer layer only (no iron)

Table C

Harmonic	Predicted units for a 11.326 cm	Measured No Iron (a 11.06 cm end)	Measured with iron (a 9.70 cm end)
sextupole	0.001179	-1.068	8.0
decapole	-13.5327	-14.94	-12.35

Comparison between computed and measured harmonics for the two layer dipole.

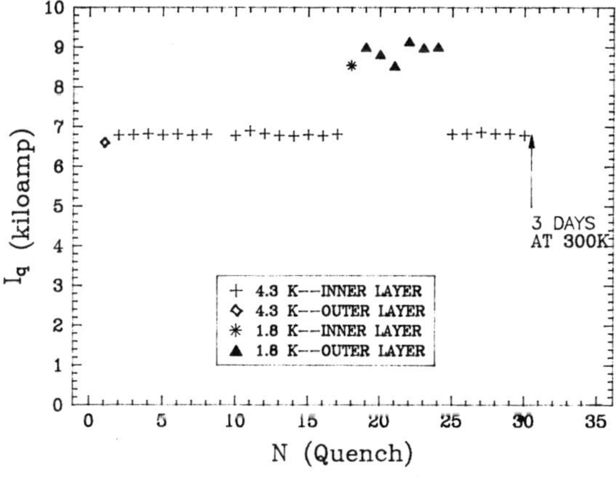

Figure 12 First Flared End Model Training Results

(Note that when iron was included, it did not cover the end, but rather terminate near the end of the straight section). Training behavior at 4.3K is excellent and does not appear to be related to coil ends. The magnetic field uniformity of the tapered end is satisfactory for SSC operation.

MAGNET TRAINING

The initial cryogenic testing took place between February 8th and 17th, 1989. Memory tests of training were carried out, at 4.3K, on March 1st. Figure 12 shows the training history.

At 4.3K, the first quench was at 6581A in the straight section of the pole turn of the outer layer. The second and all subsequent quenches were in the straight section of the pole turn of the inner layer, at 68 ± 20A. This is the same plateau current reached in magnet D15A-3, which has identical superconducting cable in the inner layer. The magnetic field uniformity of the tapered end is satisfactory for SSC operation.

At 1.8K, the first quench was at 8544A in the inner layer, somewhere between turn 4 and turn 16 (midplane), from the pole. The next six quenches were between 8545A and 9150A, but in the outer layer return end, in the pole turn.

After the 10 day thermal cycle, the first retraining quench took place in the return end of the outer layer, but only 1% below the previous plateau valve. The next four quenches were in the straight section of the pole turn of the inner layer at (6804 ± 4) A.

CONCLUSION

The results of the effort develop an alternate coil end design for the SSC Dipole Magnets has been encouraging. The flared end design has proven relatively simple to wind and mold.

The larger winding radii have reduced mechanical stresses and insulation trauma and has produced an electrically rugged coil end.

The number of individual spacers and insulation pieces reuired to wind and assemble flared coil ends in fewer than the present straight end coils.

The use of CAD for this design enabled accurate computation of geometries for the coils and spacers for prediction of magnetic harmonics and eventual fabrication of the spacers reuired for the coil ends.

The prediction and measurement of coil end magnetic harmonics agreed well and fall within the SSC specification for the ends.

REFERENCES

1. S. Caspi and M. Helm, "A "Flared End" Design for the SSC Dipole", June, 17, 1988, SSC-MAG-205, LBID-1417.

COIL MEASUREMENT DATA ACQUISITION AND

CURING PRESS CONTROL SYSTEM FOR SSC DIPOLE MAGNET COILS

Carl E. Dickey

Technical Support, Superconducting Magnet Production
Fermi National Accelerator Laboratory
Batavia, IL

ABSTRACT

A coil matching program, similar in theory to the methods used to match
Tevatron coils, is being developed at Fermilab. Modulus of elasticity and
absolute coil size will be determined at 18-inch intervals along the coils
while in the coil curing press immediately following the curing process. A
data acquisition system is under construction to automatically acquire and
manage the large quantities of data that result. Data files will be
transferred to Fermilab's VAX Cluster for long term storage and actual coil
matching. The data acquisition system will also provide the control
algorithm for the curing press hydraulic system. A description of the SSC
Curing Press Data Acquisition and Controls System will be reported.

INTRODUCTION

Members of Fermilab's Technical Support Section are now involved in the
design and construction of a combined development laboratory and production
facility for the SSC 17 Meter Dipole Magnet. The facility, located in
Fermilab's Industrial Center Building, will include a coil winder, a curing
press, a collaring press, and a yoke and skinning press. The design of
production tooling for the SSC Dipole is largely being based on experience
gained during magnet development and production for Fermilab's
superconducting synchrotron, the Tevatron. The facility will feature
automatic control and data acquisition for virtually all processes. From a
control and data acquisition standpoint, the curing press process has proved
to be the most complex and problematic. This paper will present and discuss
the data acquisition and control solution which we have designed to operate
the SSC Curing Press.

The Curing Process

The Superconducting Super Collider will require the production of 8,000
dipole magnets. (See Figures 1 and 2.) Each dipole will require two inner
and two outer coils. Therefore, the SSC will require the production of
approximately 32,000 superconducting coils, each 16.6 meters in length.

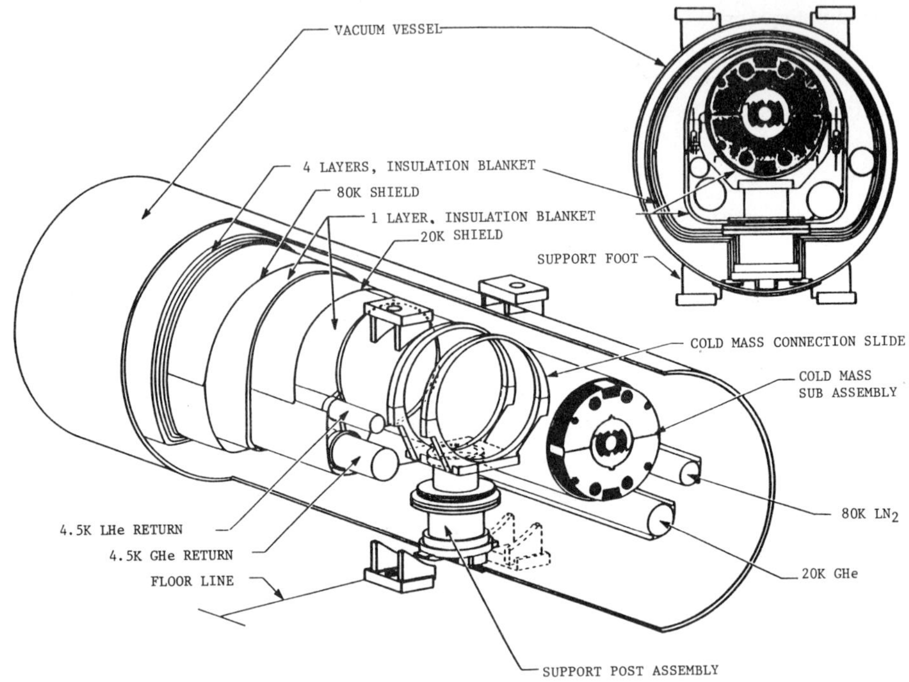

Fig. 1   SSC DIPOLE

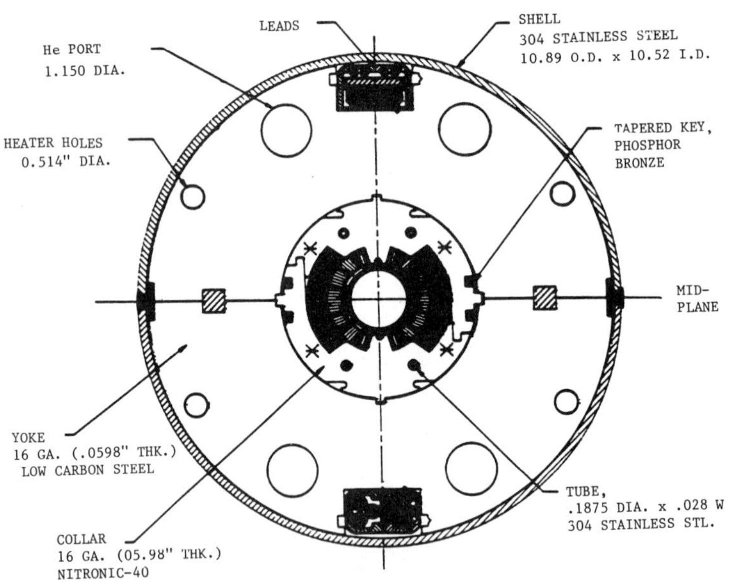

Fig. 2   SSC DIPOLE CROSS SECTION

Inner and outer dipole coils will be produced in a similar manner.
Basically, the coil will be wound on a mandrel from a stock of
superconducting cable. (See Figure 3.) The superconducting cable will have
been previously insulated and wrapped with epoxy impregnated, thermal setting
tape. After the winding process has been completed, the coil is ready to be
cured.

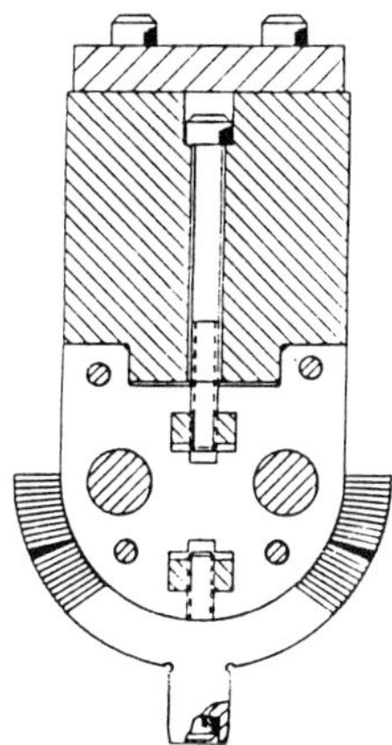

Fig. 3   COIL IS WOUND ON A MANDREL

During the curing process, the coil is confined at high pressure in a
precision mold. The curing process bonds the superconducting cable wraps
together and forms the coil into a unified structure. To cure the coil, the
coil and mandrel will be assembled into the curing tooling. (See Figures 4
and 5.) The curing tooling assembly, which contains the coil, is placed in
the curing press. (See Figure 6.) The press is brought to bear on the
curing tooling which loads the coil. The tooling is heated to approximately
250 degrees farenheit causing the epoxy to polymerize. After a sufficient
period of time has passed to insure that the epoxy has been thoroughly cured,
the tooling is cooled. At this point, a superconducting coil has been cured
and in theory is ready for the collaring process. In practice, the situation
is not so simple.

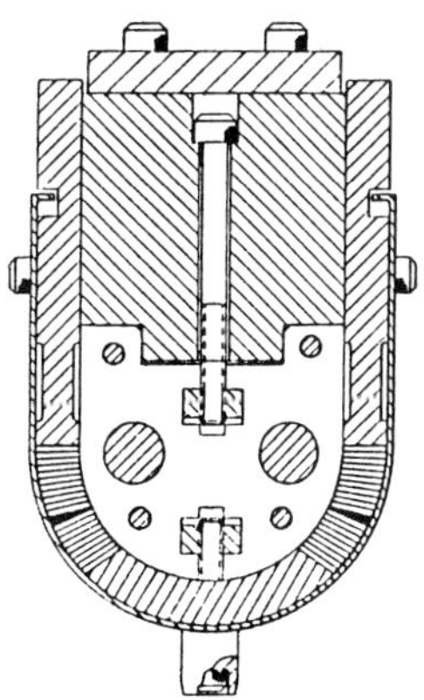

Fig. 4   COIL ASSEMBLY WITH SIZING BARS AND RETAINER

417

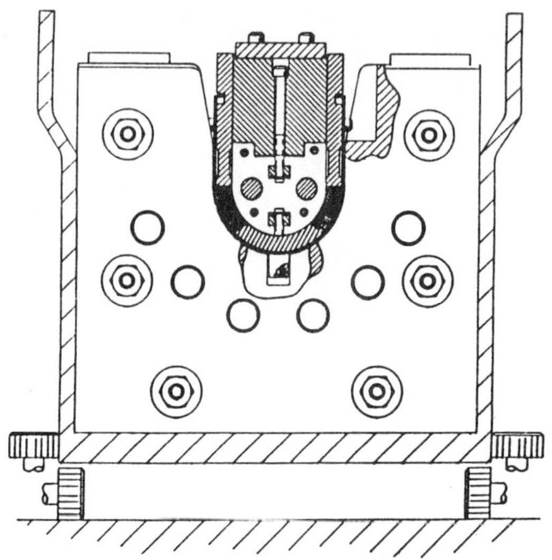

Fig. 5   COIL ASSEMBLED INTO TOOLING

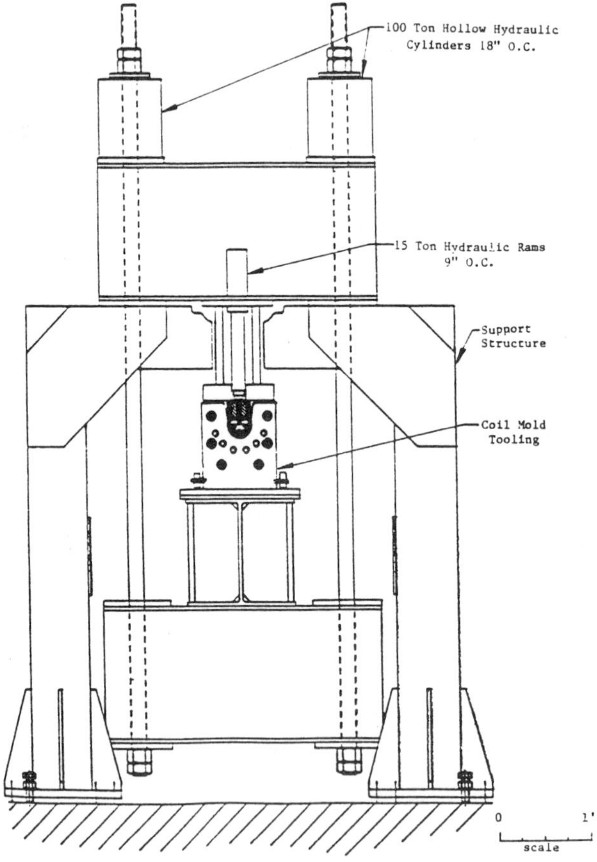

Fig. 6   THE SSC CURING PRESS

<u>Parting Plane Offset and Coil Matching</u>

The SSC dipole magnet will be constructed by collaring two sets of inner and outer coils together. (See Figures 7, 8 and 9.) The collaring process forces the coils together and maintains sufficient load upon the coils to help insure a stable geometry under full field excitation. Unless the two sets of coils are matched in terms of absolute size and modulus of elasticity, upon collaring, the planes of contact between the coils will part from the horizontal dividing plane of the free state. (See Figure 10.) As this "parting plane" increases, so does the harmonic energy of the magnetic field. Low harmonic energy is a figure of merit for a dipole magnet.

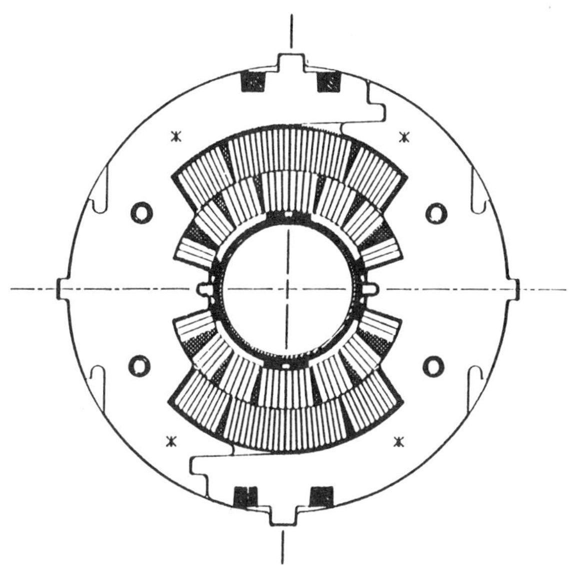

Fig. 7   COLLARED COIL ASSEMBLY

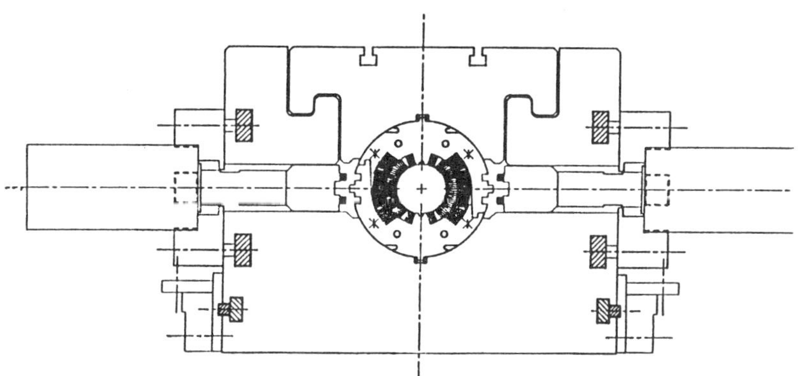

Fig. 8   COLLARING TOOLING

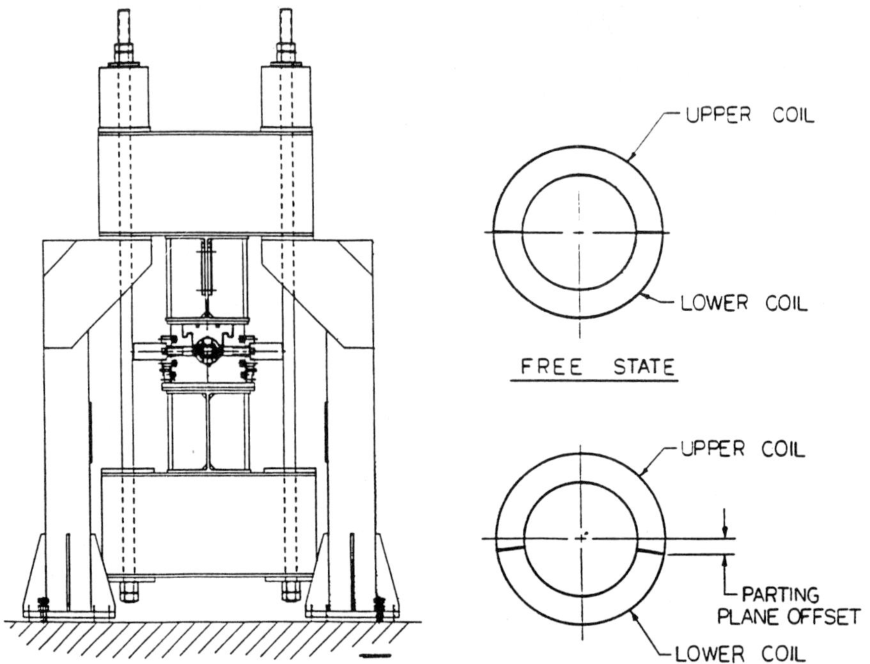

Fig. 9   COLLARING PRESS   Fig. 10   PARTING PLANE
                                                   ILLUSTRATION

In the past there have been two successful methods used to minimize the parting plane offset. These techniques involve "matching" coils to form collaring sets. The first method involves generous amounts of "artistic ability". Basically, the absolute size and modulus of elasticity for two coils are measured. Then, using the measured data and previously acquired curing press experience, one of the coils can be recured to match the other. This technique is especially difficult since the recuring process changes both the size and modulus of the coil simultaneously. Considering the generous amounts of time and experience needed for the "artistic method", it is difficult to imagine successfully matching 32,000 coils in this manner.

The second method calls for a number of coils to be produced as uniformly as possible. Absolute size and modulus of elasticity data will be compiled for each coil. Coils will be matched into collaring sets. This method might be called the "statistical method" since it is expected that many matchable coils will be produced in a large population of sufficiently narrow distribution. As previously discussed, the SSC Dipole Project will require that a large number of coils be manufactured. In addition, Tevatron coil production experience indicates that coils can be produced in a sufficiently narrow distribution of absolute size and elastic moduli to insure that "statistical matching" will not be difficult.

CURING PRESS DATA ACQUISITION AND CONTROL

Statistical matching requires that large quantities of coil data be acquired, stored and compared. In fact, it is expected that each coil will be characterized by more than 18,000 pieces of information. Obviously, coil information should be automatically acquired, processed and stored. Traditionally, the acquisition of coil data has taken place after the coil has been cured and removed from the curing assembly. However, there are advantages to acquiring coil data immediately after the curing process while

420

the cooled coil remains assembled into the press. First, if data acquisition takes place within the curing press, a step in the production process will be eliminated. Not only will time be saved, but the coil will not have to be handled as often. Secondly, to acquire coil data using the curing press itself will require that the curing press be fully instrumented and automatically controlled. If the curing press instrumentation and control system could be designed to work at 250 degrees farenheit, we not only have the ability to automatically acquire data from a cooled cured coil, but for the first time we also may observe and control the curing process itself. We may be able to better control the coil's size and elastic modulus. It is possible that a "smart" curing press will enable the band of coil distribution to be narrowed and thus increase the frequency of matchable coils. Therefore, we believe that performing data acquisition within the curing press will lead to more efficient production of better dipoles.

Modulus of elasticity is the relationship between applied stress and resultant strain. To obtain modulus of elasticity information with the curing press, the press instrumentation and control system must be able to apply known stresses to the coil and determine the associated resultant strains. A "load cell" is a device, usually strain gauge bridge based, which can be used to readback applied stress. Figure 11 illustrates the load cell which has been developed to provide stress readback for the curing press. This load cell is basically a standard Sensotec Model 53 which has been modified to provide temperature compensated output up to 275 degrees farenheit, easy mounting in the curing press and 15,000 lbs. of load sensing in a small package which possesses a 50% overload margin. The Sensotec Model 53H is a strain gauge transducer using a 350 ohm full bridge electrical configuration.

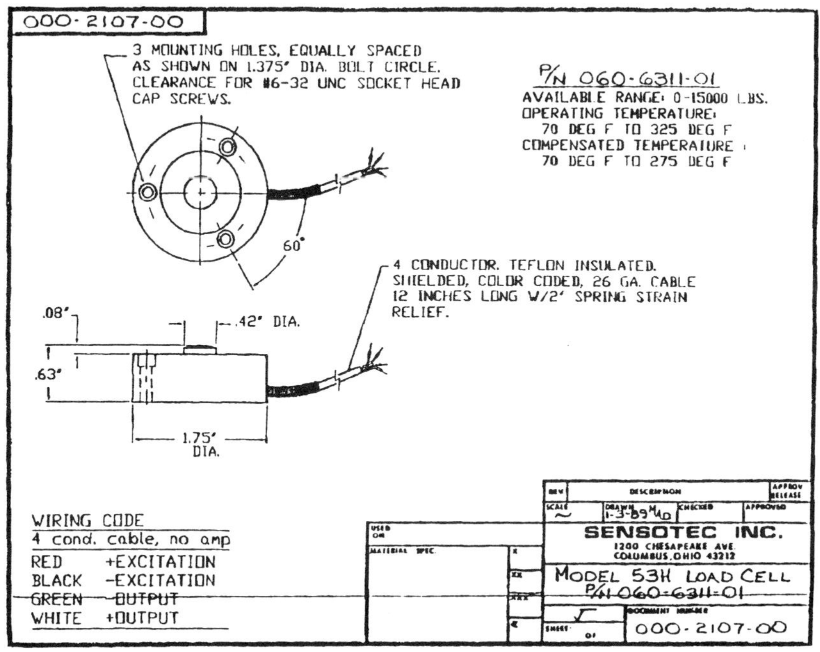

Fig. 11   SENSOTEC MODEL 53H LOAD CELL

Strain readback will be accomplished using a device known as a "DC gage head". A DC gage head is basically an AC linear variable differential transformer (LVDT) which has been built into a hermetically sealed package and provided with a spring return indicator shaft. In addition, the DC gage head includes a hybrid circuit which provides AC excitation for the internal LVDT and signal conditioning of the LVDT secondary signals to provide a linear DC output. The hybrid eliminates much of the external circuit complexity typically associated with LVDT's. Figure 12 shows the Schaevitz Model GCD-121-250 DC gage head which has been selected for use in the curing press. The gage head system will also be used to determine the absolute size of the coils. The absolute coil size will be established by taking gage head readings at full press loading while the upper press block is in contact with the specific tooling. As the load is caused to diminish, the upper press and the specific tooling seperate. The difference between the gage head readings at full load and at zero load will determine the absolute size of the coil.

GC/GP Series

SPECIFICATIONS BY MODEL

AC-Operated Units (2.5 kHz)

Model Number	GCA-121-050 GPA-121-050	GCA-121-125 GPA-121-125	GCA-121-250 GPA-121-250	GCA-121-500 GPA-121-500	GCA-121-1000 GPA-121-1000
Gaging Range	±0.050" (±1.27 mm)	±0.125" (±3.17 mm)	±0.250" (±6.35 mm)	±0.500" (±12.7 mm)	±1.000" (±25.4 mm)
Phase Shift (Degrees)	+6	+5	+5	+2	+1
Sensitivity (mV/0.001"/V Input)	5.0	2.4	1.7	1.1	0.9
Impedance (Ohms)					
Primary	430	1710	800	900	900
Secondary	950	1820	940	1150	2100
Pretravel (Nominal)	0.26" (6.6 mm)	0.30" (7.6 mm)	0.06" (1.5 mm)	0.18" (4.5 mm)	0.01" (0.25 mm)
Minimum Overtravel	0.15" (3.81 mm)	0.15" (3.8 mm)	0.15" (3.8 mm)	0.90" (22.8 mm)	0.10" (2.5 mm)
Spring Load Over Gaging Range	3.5 to 5.8 oz. (99 to 164g)	3.5 to 5.8 oz. (99 to 164g)	3.5 to 5.8 oz. (99 to 164g)	3.2 to 8.0 oz. (91 to 227g)	3.2 to 8.0 oz. (91 to 227g)
Dimensions					
A (±0.01"/0.25 mm)	1.70" (43.1 mm)	2.55" (64.7 mm)	3.41" (86.6 mm)	5.09" (129 mm)	7.35" (187 mm)
B (±0.03"/0.76 mm)	4.33" (109 mm)	5.14" (130 mm)	6.10" (154 mm)	10.75" (273 mm)	13.01" (330 mm)
C (±0.02"/0.50 mm)	3.27" (83 mm)	4.12" (104.6 mm)	4.99" (126 mm)	8.27" (210 mm)	10.53" (267 mm)
Weight	2.2 oz. (64g)	2.9 oz. (85g)	3.17 oz. (90g)	5.0 oz. (142g)	7.5 oz. (213g)

DC-Operated Units

Model Number	GCD-121-050 GPD-121-050	GCD-121-125 GPD-121-125	GCD-121-250 GPD-121-250	GCD-121-500 GPD-121-500	GCD-121-1000 GPD-121-1000
Gaging Range	±0.050" (±1.27 mm)	±0.125" (±3.17 mm)	±0.250" (±6.35 mm)	±0.500" (±12.7 mm)	±1.000" (±25.4 mm)
Sensitivity (mV/0.001"/V input)	200	80	40	20	10
Pretravel (Nominal)	0.30" (7.62 mm)	0.35" (8.8 mm)	0.18" (4.5 mm)	0.20" (5.08 mm)	0.10" (2.5 mm)
Minimum Overtravel	0.39" (9.4 mm)	0.14" (3.5 mm)	0.03" (0.76 mm)	1.00" (25.4 mm)	0.10" (2.5 mm)
Spring Load Over Gaging Range	3.5 to 5.8 oz. (99 to 164g)	3.5 to 5.8 oz. (99 to 164g)	3.5 to 5.8 oz. (99 to 164g)	3.2 to 8.0 oz. (91 to 227g)	3.2 to 8.0 oz. (91 to 227g)
Dimensions					
A (±0.01"/0.35 mm)	2.46" (62.5 mm)	3.30" (83.8 mm)	4.17" (105.9 mm)	5.86" (148.8 mm)	8.11" (205 mm)
B (±0.03"/0.76 mm)	5.08" (129 mm)	5.90" (149.8 mm)	6.77" (171.9 mm)	11.53" (292.86 mm)	13.76" (349 mm)
C (±0.02"/0.50 mm)	4.02" (102 mm)	4.87" (123.6 mm)	5.74" (145.7 mm)	9.05" (229.8 mm)	11.29" (286 mm)
Weight	2.5 oz. (71g)	3.2 oz. (93g)	3.5 oz. (100g)	5.5 oz. (156g)	8.0 oz. (227g)

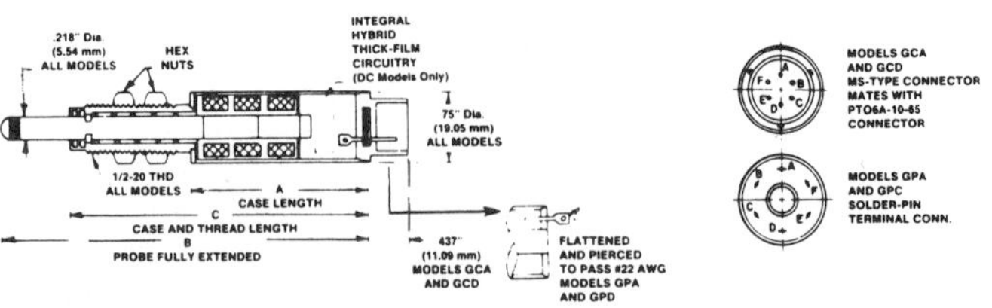

Fig. 12 SCHAEVITZ MODEL GCD-121-250 DC GAGE HEAD

Figure 13 illustrates the relative positions of the curing press gage heads and load cells. Gage heads will be located outside the tooling. A convective boundary layer flow exists along the outer tooling surfaces. This vertical flow will keep the gage heads cool. The load cells will be located in the 1.5 inch thick steel upper press block assembly and positioned to contact the sizing bars which directly load the coil. (See Figure 4.) The sizing bars will be decoupled to isolate the load cell from sizing bar side effects. (See Figure 14.) A total of 150 gage heads and 150 load cells will be distributed at 18 inch intervals along both sides of the full coil length.

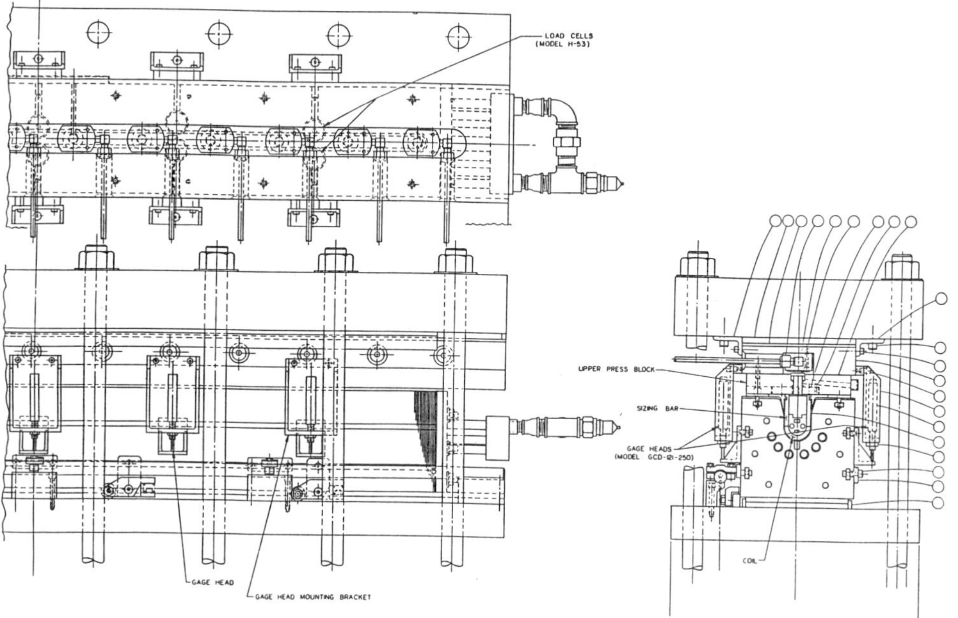

Fig. 13 GAGE HEAD AND LOAD CELL LOCATIONS

The gage head system has been designed to permit automatic calibration. Gage heads will be assembled onto the press in positions which are "close" to the perfect positions. Full open and full closed position readbacks will be obtained and the computer will establish the calibration. The load cells will be calibrated using a fixture shown in Figure 15. A small hydraulic cylinder will be mounted on a base and connected to a Lebow high accuracy load cell. The cylinder will be energized and caused to load individual Sensotec load cells, installed into the upper press block, through the Lebow load cell. The outputs of both load cells will be compared by the computer and calibration will be determined. The Lebow load cell will serve as the calibration standard.

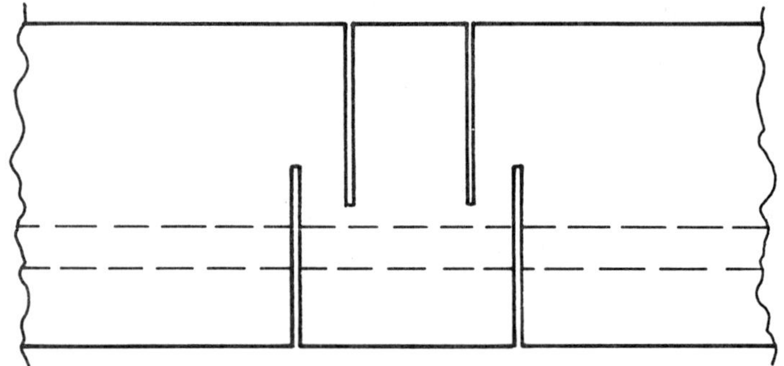

Fig. 14   DECOUPLED SIZING BAR

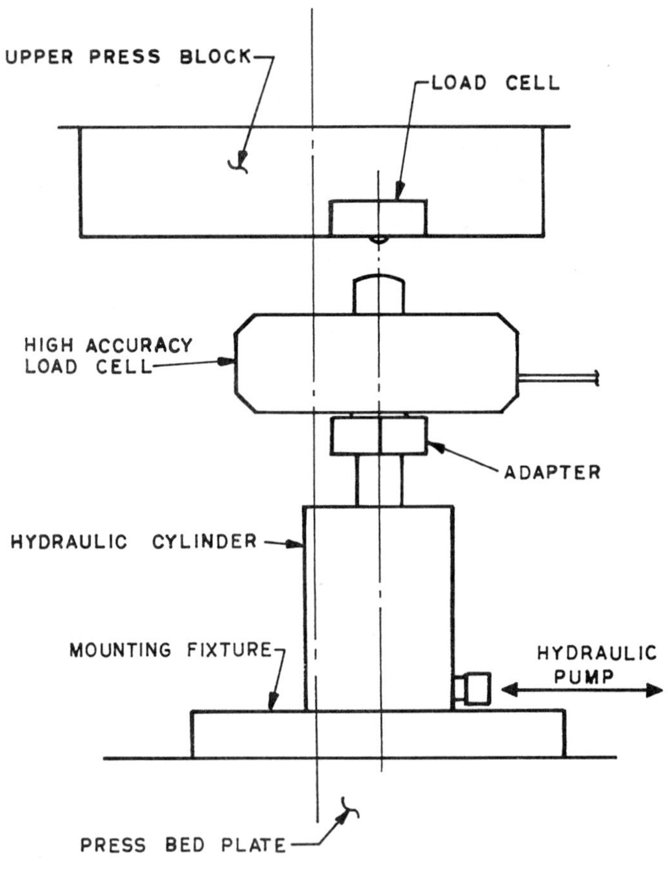

Fig. 15 LOAD CELL CALIBRATION FIXTURE

It is important to note that the instrumentation layout illustrated in Figure 13 includes no instrumentation in the specific tooling. The instrumentation will be assembled into the curing press and calibrated. Specific tooling can be moved in and out of the curing press without affecting the instrumentation. Removing the instrumentation from the production process should decrease assembly time and increase instrumentation accuracy and reliability.

The curing press will include over 300 instrumentation transducers. Stress and strain measurements will be obtained for ten increasing and ten decreasing stresses per measurement cycle. Three measurement cycles will be performed per coil. Therefore, more than 18,000 measurements will be required to characterize each coil.

After careful consideration, the Hewlett Packard 3852A Data Acquisition and Control System was selected to control the press, perform the necessary measurements, process the data and route the information for storage in Fermilab's VAX system. The HP3852A supports modules to measure and condition the load cell strain gauges, measure gage heads, measure coil resistance during the curing process, measure and compensate thermocouples for temperature readback and control, and provide digital I/O, DAC output, and stepper motor control.

Figure 16 shows the overall data acquisition and control system for the SSC Curing Press. Basically, strain gauge, voltage, and thermocouple multiplexers reside in the HP3852A Data Acquisition and Control Mainframe or the HP3853A Extender Chassis. These multiplexers feed into HP44701A 5 1/2 digit integrating voltmeters. The HP3852A's send information over IEEE 488 (HP-IB) to HP37204A extenders. The extenders permit data to be sent up to 1200 meters to the HP360CH control computer. The computer, tape drive, disc drives, printer and plotter will be housed in a remote location which is isolated from the industrial environment. The HP360CH uses the MC68030 microprocessor running at 25MHz. Control software will be developed using HP BASIC UX. A local area network (LAN) is being established to permit communication over ETHERNET via a THINLINE. This system will permit communication with the curing press operating system from remote terminals.

The majority of controls problems that have been encountered concern control of the hydraulic system of the press. The control system must permit proper valve sequencing and hydraulic pressure regulation up to 10,000 psi. Figure 17 is a schematic layout of the curing press hydraulics. Figure 18 shows the typical control arrangement for the various hydraulic valve solenoids. Basically, common collector digital signals from a HP44723A High Speed Digital I/O Module will control solid state relays, which in turn place 115 VAC power on selected valve solenoids. (See Figure 16.)

In the past, hydraulic pressure has been regulated manually using the Enerpac V-152 Pressure Control Valve. (See Figure 19.) However, for efficient data acquisition, hydraulic pressure must be automatically controlled. A number of valve/actuator systems were investigated. None of the systems could provide pressure control at 10,000 psi. Studies were performed that indicate that the Enerpac V-152 valve displays highly repeatable pressure regulation with respect to the rotational position of the valve spring loading screw. Consequently, this points to the possibility of pressure regulation by controlling the valve's position. Figure 20 is a drawing of the valve actuator which has been designed to position the pressure control valve. A high power stepping motor supplies torque to the pressure control valve. Due to the design of the valve, a spline coupling is needed to take up the linear motion of the valve's loading screw. A limit switch carriage has been incorporated into the design to electrically

SSC CURING PRESS
DATA ACQUISITION AND CONTROL SYSTEM
ETHERNET
LOCAL AREA NETWORK (LAN)
LOCAL AREA NETWORK (LAN)
300 MB DISK DRIVE EXPANDABLE 900 MB
HP 9000 MOD. 360 CH 68030 -25 MHZ UNIX WITH BASIC UX
1/4 INCH 133 MB TAPE DRIVE
HP 9000 MODEL 318 M WORK STATION
HP 480 PRINTER OR HP 7550 A PLOTTER
HPIB
HP 37204 A EXTENDER
SERIAL ≦1200m
HP 37204 A EXTENDER
HP 480 PRINTER OR HP 7550 A PLOTTER
CURING PRESS HYDRAULIC SOLENOIDS
HPIB
1 HP 3852 A & 4 HP 3853 A
4 HP 44701 A 5 1/2 DIGIT INTEGRATING VOLTMETER
HP 44723 A HIGH SPEED DIGITAL I/O
1 MB EXTENDED MEMORY
24 HP 44718 A STRAIN GAUGE MUX.
6 HP 44705 A RELAY MUX.
1 HP 44708 A THERMOCUPLE RELAY MUX..
HP 44714 A STEPPER MOTOR CONTROLLER
HP 44727 B DAC (12 BIT)
150 LOAD CELLS
150 GAGE HEADS.
6 TYPE J THERMO- COUPLE
CURING PRESS HEATING SYSTEM
LIMIT SWITCH
ENCODER
SLO-SYN 6180 PT TRANSLATOR DRIVE
SLO-SYN M172-FF401 STEPPER MOTOR

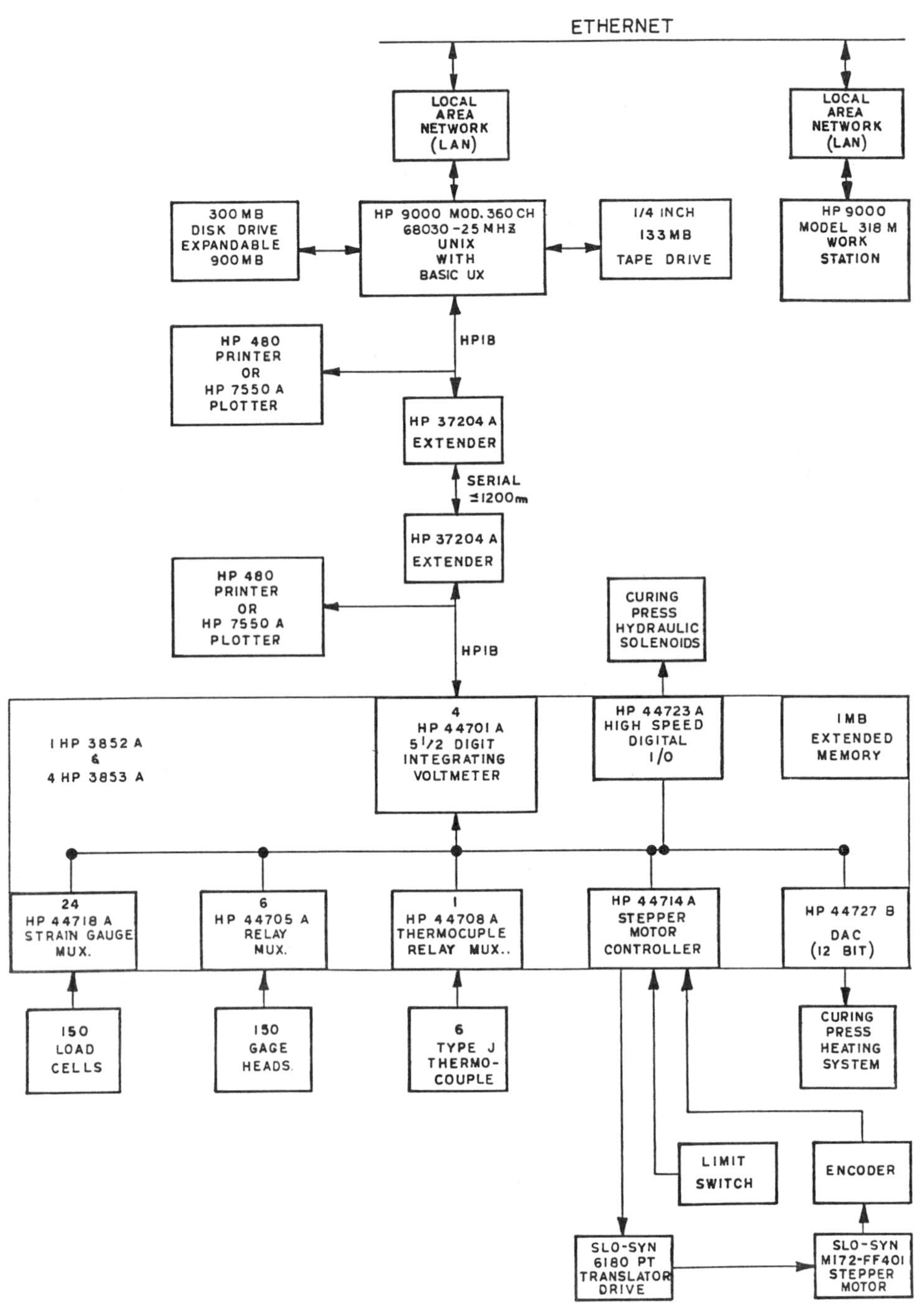

Fig. 16

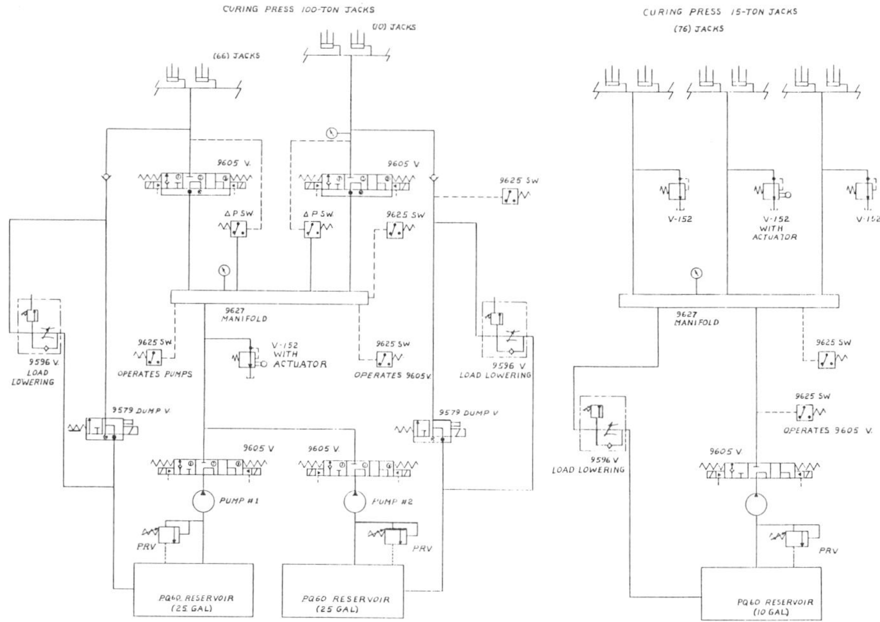

Fig. 17   SSC CURING PRESS HYDRAULICS

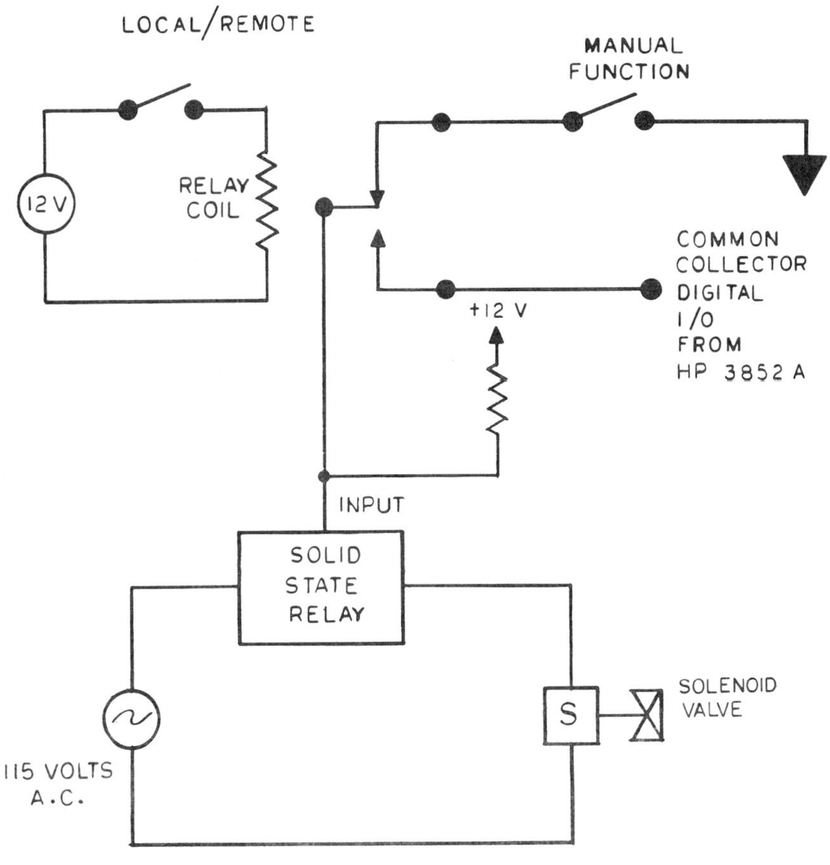

Fig. 18   TYPICAL SOLENOID CONTROL

VALVE OPERATION

System pressure is applied to the piston Ⓒ The hydraulic pressure on piston Ⓒ will act against the two inch compression spring Ⓑ and lift the cone off the seat so that the oil flows thru orifice Ⓐ to tank. The oil passes thru the orifice until the oil pressure is equal to the compressive force from the bias spring. At that time the cone will move upon the seat, and the process repeats this cycle over again. Just enough oil from the system is allowed to pass thru the orifice to tank which will maintain the system pressure as set by the relief valve.

VALVE FUNCTION

V-152 RELIEF VALVE serves the purpose of limiting the pressure developed by the pump in a hydraulic circuit, thus limiting the load that may be developed or imposed upon other components. To increase setting, turn the handle clockwise.

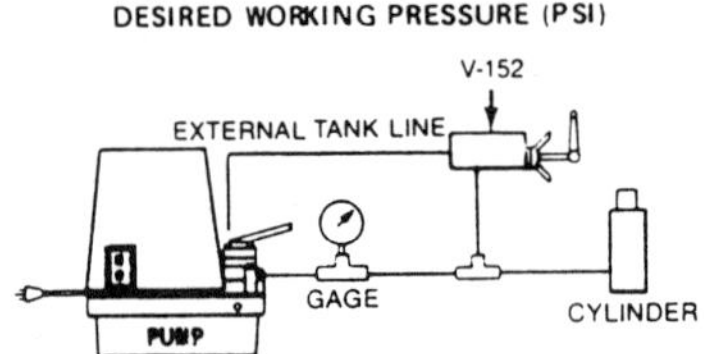

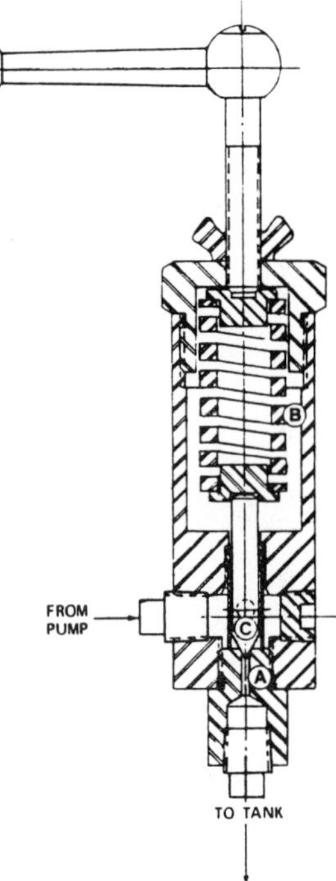

Fig. 19   ENERPAC V-152 VALVE

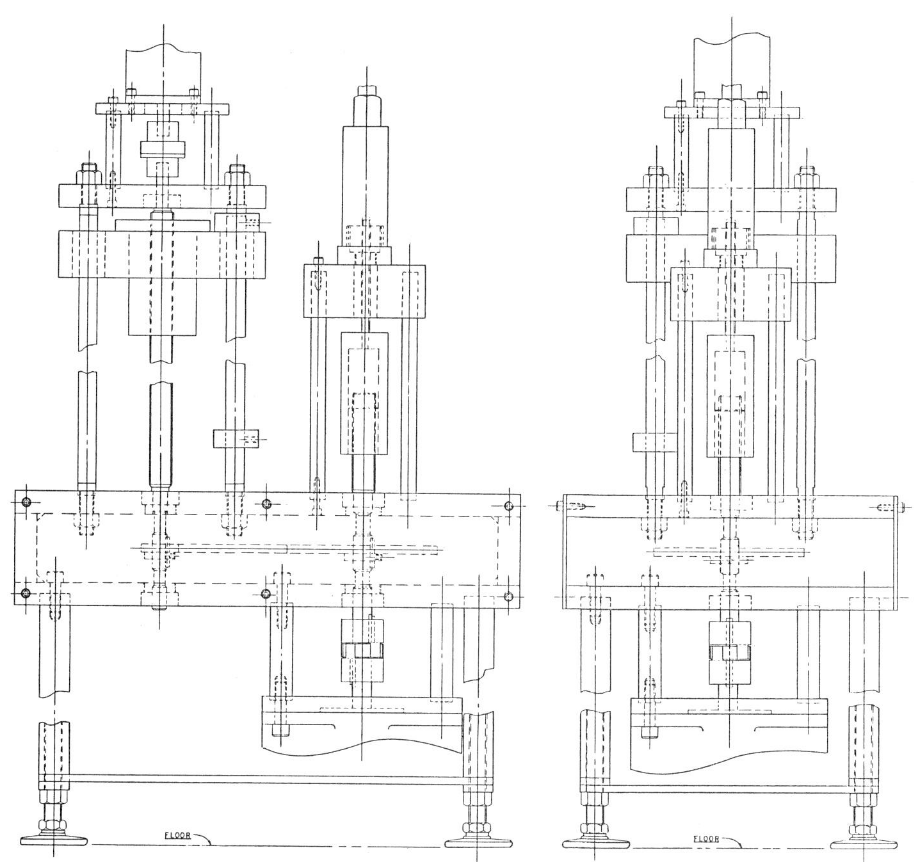

Fig. 20 HYDRAULIC SYSTEM ACTUATOR

indicate the valve's full open limit and full closed limit. Ball bearings, recirculating ball screws and splines have been used throughout the design to insure low friction. An optical encoder is used to provide position feedback. The system will provide 4800 discrete regulation points between the full open and full closed valve positions. The actuator will be controlled using a HP44714A Stepper Motor Control Module, which will be located in the HP3852A Mainframe. The HP Stepper Motor Control Module will feed into a Superior Electric 6180-PT Translator Drive. The Translator Drive decodes the pulsed trapazoidal motion profile output of the HP44714A, amplifies the signals and directs the power pulses to the SLO-SYN M172-FF401 stepper motor. Figure 16 shows the actuator control system.

CONCLUSION

Construction of the SSC Dipole Factory at Fermilab is well underway. Computer control and data acquisition are being designed into many of the production processes. We are enthusiastically looking forward to the development of the SSC Dipole.

DESIGN PRINCIPLES FOR PROTOTYPE AND PRODUCTION

MAGNETIC MEASUREMENTS OF SUPERCONDUCTING MAGNETS

Bruce C. Brown

Fermi National Accelerator Laboratory
P. O. Box 500
Batavia, Illinois 60510

ABSTRACT

The magnetic field strength and shape for SSC superconducting
magnets will determine critical properties of the accelerator systems.
This paper will enumerate the relations between magnetic field properties
and magnet material selection and assembly techniques. Magnitudes of
various field errors will be explored along with operating parameters
which can affect them. Magnetic field quality requirements will be
compared to available measuring techniques and the relation between
magnetic field measurements and other quality control efforts will be
discussed. This will provide a framework for designing a complete magnet
measurement plan for the SSC project.

INTRODUCTION

The quality control measurement[1] of superconducting accelerator
magnets[2] has several interesting aspects, of which some are unique to the
accelerator problem and others are special to the superconducting magnet
problem. This paper will attempt to define a strategy for viewing this
problem and supply an overview of its engineering and management aspects.
Specific results and analyses will be included to make the discussion as
concrete as possible and to provide a reference for further work.

Superconducting magnets such as the dipoles[3,4] designed for the SSC[5]
have been classified as ''coil dominated'' magnets.[6] The critical
magnetic field properties are largely set by the current flowing through
the coils and the coil geometry. Iron yokes which surround the coil
create enhanced fields which can be described with an image current model
at fields below iron saturation. A more detailed analysis is required at
high fields to describe the iron effects after the onset of saturation.
In addition to the dominant transport current fields, hysteretic fields
due to magnetic materials (both ferromagnetic (yokes) and ferrimagnetic
(superconducting cable)) are very important at low fields. Both eddy
current and flux creep effects cause time dependences of magnet fields.
To plan an effective magnetic testing program, an overview of these
effects is necessary. Since various properties determine different field
errors, techniques must be devised to separate these effects and relate
measured properties to fabrication techniques and materials properties.

Table 1.   Uses of Magnet Measurements

*Production Quality Control

Use	Typical Measurement
Assembly Tooling	Measurement of Field Angle
Acceptance Testing	Maximum Current (Strength) Strength vs Current Field Uniformity
Optimize Installation	Field Uniformity Field Strength vs Current

*Prototyping and Special Magnet Design Tests

*Special Tests for Accelerator Systems

In Table 1 are listed various uses for magnet measurements of accelerator magnets. Let us examine these purposes for measurements and some of the measurements themselves. Then we will begin examination of the design considerations for hardware and software to be used in magnet measurements.

As part of the process of magnet fabrication, some of the assembly steps are more easily carried out using ''magnetic tooling'' (magnetic measurements performed to allow critical alignments or adjustments).[7] Most such steps will be carried out with factory based measurement equipment. However, in the fabrication of the Fermilab Tevatron dipoles, in addition to the field angle determination during assembly, adjustments of the relative positions of the iron and coil during final testing allowed a significant improvement in field uniformity. Field angle measurements of dipole and quadrupole magnets provided alignment data for tunnel installation which substantially reduced the magnet assembly precision requirements.

Acceptance testing of the produced magnets will determine their suitability for use in the accelerator. Typically, measurements of the properties required (performance tests) are the preferred technique for acceptance testing. However, it may be possible in the case of superconducting magnets to provide cheaper or more sensitive tests in other ways. Total project cost may be reduced by performing specially designed tests to portions of the magnet during assembly. In addition, some tests may be desired which can reveal production line malfunction which will be corrected for ''good engineering practice'' since determining the true effects on the accelerator (material fatigue, long term insulation failure...) may be more expensive than correcting the manufacturing problem.[8]   The result of acceptance testing will be accepted magnets, rejected magnets, and perhaps some magnets which will be repaired.

The 20 TeV synchrotron proposed for the SSC project will perform as a single tool. Thus, the product of the magnet builders is not a set of 10,000 magnets but instead is ONE machine. This distinction was important for the existing Fermilab Tevatron since the properties of the

produced magnets would have produced an inferior accelerator if assembled
with randomly chosen magnets from the assembly line. Instead, in a
process referred to as ''shuffling'' but which could better have been
called ''stacking the deck'', properties of sets of magnets were matched
to provide improved accelerator properties.[9] Magnetic measurements were
performed on each magnet to make this possible. More stringent magnet
fabrication requirements or additional magnetic correction schemes would
have been the (more expensive) alternatives. The usefulness of this
requirement will be determined by the relationship between the
accelerator design and the achievable tolerances on materials for and
assembly of the magnets.

In addition to these production requirements, there will continue to
be a series of measurements required which must be implemented as special
tests. These prototype measurements will need to be carried out both to
resolve any magnet design questions which arise but also to provide
information about the magnets for the design of related systems such as
power supply systems or quench protection systems. These will not only
require different and additional resources, but may also need to be
integrated into production measurements, either because of requirements
to achieve sufficient statistics, because the measurement is closely
related to existing measurements or simply to achieve the same high
quality of data collection and management achieved by the production
measurement system. This requirement for special tests along with the
assurance that the initial set of production measurements will require
small adjustments in order to achieve all project goals, demands a
flexible measurement system implementation. Modern tools allow a well
defined and well monitored production measurement system which is also
flexible.

In order to accomplish the various tasks in Table 1 we will need a
variety of magnet measuring tools and systems. Each system will require
its own special considerations concerning

Implementation
Record Maintenance and Reporting
Measurement Quality Assurance

The importance of attempting to create a common set of tools for
handling this diverse set of measurement requirements is driven by the
need to minimize development costs and operating complexity. In
addition, the complete understanding of a given set of measurement tools
is challenging enough to suggest that the number of tools developed be
minimized. After solving the problems for the 8000 dipoles, there will
remain important measurements for main quadrupoles, correction packages,
interaction region quadrupoles and the other special magnets required for
the SSC main synchrotron. Measurements will also be required for the
magnets in the injector accelerators. A well considered system rooted in
well considered tools is critical to the success as well as minimizing
the cost of this endeavor.

In this paper we will examine some of the measurements which have
been developed or planned, explore the effects which control the desired
properties but also place requirements on the techniques for measurement
control, discuss the relation between magnetic measurement quality
control tests and other available QC possibilities, and finally discuss
various aspects of measurement design and data management design to allow
this effort to be effectively managed and to integrate successfully with
the overall requirements of the SSC accelerator project.

Table 2. Measurements Applied to Tevatron Superconducting Magnets

<u>Measurements under Cryogenic Conditions</u>

Measurements of Maximum Current Capability and Quench Protection

    Ramp Quench   Linear Ramp to Quench
    Cyclic Quench Multiple Ramps to Each Target Current
    Heater   Integrity of Quench Heater Circuits
                   Full Energy Quench of Unprotected Magnet

High Current Magnetic Measurements

    Harmonic Measurements of Field Quality
    NMR Measurements of Dipole Uniformity
    Integrated Field Strength and Field Angle Measurement

Miscellaneous Electrical and Low Current Tests

    Hi Pot (electrical insulation test)
    Field Angle with AC System
    Field Angle from Monitor System (Yoke Coil)

<u>Measurements at Room Temperature</u>

Electrical Tests

    Coil Inductance and Resistance
    Hi Pot (electrical insulation test)

Magnetic Tests

    Field angle with AC system
    Field angle with Monitor System (Yoke Coil)
    Field Uniformity (AC system)
    Field Strength (AC system)

Mechanical Tests

    Cryostat Position Measurements
    Pipe Offsets
    Yoke Shape

SOME TYPICAL MEASUREMENT TECHNIQUES

Let us briefly examine some of the measurements which are performed on superconducting accelerator magnets.[10] In Table 2 we have outlined the quality control tests done on the Fermilab Tevatron Magnets. Some were implemented to assure that the magnets met minimum requirements for the accelerator, some to check the operation of critical devices or to adjust some variable assembly feature and some recorded properties of assembled magnets to allow their optimal use in the accelerator.

In order to evaluate the maximum strength (maximum current) for a superconducting accelerator magnet, one may excite the coils with a prescribed excitation pattern and record the current at which the magnet is not able to sustain the superconducting state. When a portion of the coil ''quenches'' to the resistive state, the whole magnet will eventually become resistive. The measurement procedure must record the

current and voltage development of this quench event in order to study
the quench propagation, heat development and voltages created. These are
typically measurements made to accuracies of 0.1% at best. Time
resolutions of milliseconds are useful. A critical aspect of these
measurements is the careful record of the cryogenic state of the magnet
and the excitation history.

Measurements to determine the shape of the magnetic field have been
carried out using rotating coils at fixed excitation currents. The
results are expressed using a standard harmonic expansion of the magnetic
field. By making use of sense coil configurations which were insensitive
to the dominant (dipole or quadrupole) field, precise measurements (10^{-4}
to 10^{-6} of the dominant) can be carried out with low precision
electronics (12 bit ADC). Important limitations due to power supply
constraints and probe motion imperfections can be avoided in this way
also. The magnetic field shape will be sensitive to the magnetic history
and cryogenic conditions of the magnet so these will be recorded by the
measurement system. Measurements through all parts of the hysteresis
cycle may be needed to assure proper materials selection and magnet
assembly. The results of these measurements must be compared with the
system requirements for the accelerator. Alternative techniques are
under consideration including different rotating coil techniques, NMR
arrays and arrays of Hall Probes are being examined but the
considerations of this paper apply equally to all of the above.

The strength of the magnetic field can be recorded in various ways.
The required information is the integrated strength along the centerline
of the magnet. The dipole strength is $\int B\,dl$ whereas the quadrupole or
sextupole strength is the integral of the relevant derivative of the
field. Dipole strength has been successfully recorded using NMR probes
for the body field in conjunction with Hall Effect Probes in the end
fields where the field uniformity is not sufficient to observe the
resonance for NMR. Stretched wire probes and rotating coils have been
used for a variety of measurement systems to measure the integrated
strength of both dipole, quadrupole and higher order correction systems.
Required accuracies are typically 10^{-4} or better for dipoles with
somewhat less accuracy required for quadrupoles. Again, the magnetic
history will affect these measurements and must be carefully controlled
to provide suitable results.

In addition to these ''generic'' measurement requirements, each
magnet design will contain features which should be monitored and
controlled. Often the magnetic measurement system will provide the most
suitable technique for quality assurance. Such measurements will be
integrated into the production measurement flow.

A MEASURERS MODEL OF A SUPERCONDUCTING MAGNET

Design tools exist to allow accelerator scientists to interact with
magnet designers to provide a suitable design and manufacturing plan for
creating superconducting magnets. However, the construction of a
magnetic measurement system will profit from a model for the magnet which
is more heuristic but more amenable to understanding the various effects.
With such a model in hand, the results of measurements should be better
understood, allowing a measurement plan to be developed which will reveal
all interesting aspects of the measurements <u>and</u> the magnets. With such a
plan in hand, the magnet measurement scient<u>ist</u> in combination with the
manufacturing engineer can plan a complete quality assurance system
including mechanical measurements during fabrication and assembly (e.g.
sizes and stresses), electrical measurements such as inductance and
magnetic measurements of components and of the final assembly.

The nature of such a model is shown in Table 3. We can calculate the dominant effects with only the first two items since they dominate the field at all excitation levels. However, we notice that to determine them accurately, we will need to make measurements at intermediate fields so that other effects can be minimized. Our separation of the effects of the iron into three distinct contributions has not been convenient for magnet designer's codes. But the effects are very distinctive. The saturation effects turn on quite sharply in most conventional as well as superconducting magnets and can characterized quite distinctively at a limited range of (high) currents. Effects of mechanical deformation must be kept small in the design to avoid fatigue as well as to maintain suitable field quality. Separating such effects from ones due to iron saturation is difficult.

The fields created by the magnetization of both the iron and the superconductor[11] are most important at low fields. In fact, both are most easily studied by noting that they are nearly independent of field level (they create number of Gauss at a reference radius which changes less than a factor of two or three over the whole current range) When plotted as a fraction of the dominant field, they will, of course, fall quickly at high fields and are therefore of much less importance to the accelerator there. However, these effects depend critically on the magnetic history of the test magnet so this must be carefully recorded and controlled. In particular, for measurements at a fixed current, the current must approach the target current smoothly since small overshoot will product large changes in the hysteretic fields.

The effects of flux creep on the fields of accelerator magnets has only been appreciated recently.[12] This effect adds considerable difficulty to attempts to make _precision_ measurements of magnetic fields since the observed effects appear to depend sensitively on the detailed magnetic history. Although the effects are fairly small, they have been observed to create important changes in the operation of the Fermilab Tevatron.[13] Although Table 3 lists them as being important at low

Table 3. Sources of Magnetic Field

Effect	Calc Tool	Range	Hysteresis	Temp. Effects
Current in Coil	Amperes Law	All Fields	none	No
Infinite μ Iron	Image Currents	All Fields	none	Small
Iron Saturation ($\mu < \beta$)	(requires detailed calc)	Hi Fields	???	Small
Iron Remanent Field	Magnetic Charges	Low Fields	YES	Small
Persistent Currents in Superconductor	Doublet Theory	Low Fields	YES	Yes
Mechanical Deformation	Finite Element Code + Ampere	High Fields	NO ??	No
Flux Creep in Superconductor	Low Fields	YES	Small	

fields, their effects at high field will depend on the details of the SSC
accelerator design and the significance of such effects cannot be ruled
out a priori. It is perhaps suitable to suggest that this effect be
taken as representative of perhaps several important small effects which
<u>require</u> that the magnet measurement system be designed with a
comprehensive view to provide detailed measurements and flexible
implementation so that new effects can be studied and then monitored as
required.

By studying these effects in detail, on can obtain an understanding
suitable to allow the design of the measurement system details and to
prepare analysis systems which will provide the required monitoring of
results so that both the quality of the magnets and the quality of the
measurements can be assessed.

SOME CONSIDERATIONS OF MEASUREMENT SYSTEM HARDWARE

The measurement system will require a variety of tools. All will be
familiar to the accelerator builder but the special tasks associated with
the quality assurance aspect of the magnet measurement facility will
require special attention to these features so that suitable systems can
be chosen for this purpose. It is often advantageous to apply some of
the systems which will be utilized in the accelerator complex to the
magnet measurement problem in order to provide system testing experience
for them. However, this must not get in the way of the principle
purpose: quality magnet measurements. To illustrate this point, we will
contrast in Table 4 typical power supply and current readout requirements
for measurements and accelerators.

Table 4

POWER SUPPLY FEATURES

Feature	Magnet Meas. Req.	Accelerator Req.
Current Stability	0.0001 for 200 Sec.	.00002 for 20000 Sec.
Ripple	<0.3% Typical	<.005% Typical
Overshoot	<0.1 A Max.	0.1 A Reproducible
Ramp Rate	Variable (per meas.)	As specified
Setting Accuracy	<1% Typical	<0.005% Typical
Setting Precision	0.1 A Typical	0.01 A Typical
Readout Accuracy	<0.01%	<0.1% Typical
Control at Zero Current	Important	Irrelevant

A number of other hardware considerations will distinguish the high
volume, precision Quality Assurance magnet measurement system from other
familiar hardware systems. The significance of the measurement for the
life of an accelerator makes it useful to tie the measurement to absolute
references via calibration precedures. When industrial vendors are
involved it is recognized that it then becomes commercially as well as
technically useful to tie these calibrations to NBS standards. In order
to assure that the hardware used for the measurement is well identified
(so that its calibration is known), an electronic identification which
can be read by the data collection system is needed. Simple systems

suitable for existing measurement systems have been in use for some time
at Fermilab. Systems which are suitable for expansion to a large scale
effort need to be identified and implemented.

Magnetic field measurement techniques must be selected for which
problems of calibration, data acquisition, analysis and monitoring are
understood.   However, the same concerns must be shown for control and
readout systems for cryogenics, current, probe position and magnet
orientation (as examples). ''In place'' calibration is desired.
Redundant measurement allow the monitoring of stability and reliability
to be carried out within the nominal data flow. The optomization
criteria must take note that the cost of additional routine measurements
carried out by the computer is very low, additional measurements
performed routinely by the measurement personnel is still inexpensive but
calibration and especially diagnostic work carried out by the scientific
staff is very costly in both direct cost and opportunity cost.

DATA COLLECTION ENVIRONMENT

Implementation of the measurements required for the SSC project is
clearly a project which deserves the use of modern computer software and
hardware tools to make the implementation meet its goals and remain cost
effective. An appropriate place to start consideration of these tools is
the axiom which states that all computer applications need to be designed
so that the people who use them find the computer to be their assistant
rather than being made to feel that they are the computer's assistant.
In each phase of the design one needs to focus on the person who is to be
served by the process under design. The widest context which we will
consider here is the accelerator scientist for whom the measurement data
is obtained. The most restricted view is that of a technician who
calibrates measurement equipment or a materials management clerk who
maintains product and equipment inventory information. Table 5 lists
some of the personnel involved in this process. Their domains of view
are described with terms intended to be suggestive but not at this point
analytical.

Table 5. Domains of Interest of Measurement System Users

User	Domain of View
Accelerator Scientist	Measurement Analyzed Results (Production and Prototype) Inventory of Products
Measurement Scientist	Measurement Results Raw, Reduced and Analyzed Equipment Status
Measurement Manager	Measurement Status and Results Inventory of Products Equipment Status
Measurer	Equipment Status Local Product Inventory Measurement Results - Raw and Reduced
Data Analysis Clerk	Measurement Results - Reduced and Analyzed
Equipment Technician	Calibration System Status Equipment Status

The point of view adopted in the measurement system created for the AntiProton Source project[14] is that the principle design goal was to allow the Accelerator Scientist and the Measurement Scientist (engineers and physicists) to have full control of all information they required to evaluate the quality of the magnets and measurements by making it available through the computer system. This implied the need to provide databases to control the calibration information used in the measurements, configuration files to provide consistent information on equipment and settings, careful design of the results files to provide all the information which must flow through each step of the measurement process. It requires that the computer read all status information electronically to eliminate, where possible, errors due to operator data entry. (Experience at the Fermilab MTF indicates that error rates for entry of status information can at best approach 99% accuracy). This requires a degree of planning in both the hardware and software implementation so that the required ''near 100%'' status readout is possible.

The programs were written as single threads of code which directed the measurement process from beginning to end. It was required that the data be recorded in both Raw form and Reduced form (calibrations applied) by the initial data collection program. The requirement to record the raw data is widely utilized in any application where data volume permits. This will allow a reprocessing of the data with updated calibration information or revised programs if that should be required. Data analysis may proceed in one or more passes to provide a variety of required results. All data was organized to allow use of data management tools (in this case DATATRIEVE)[15] for general data viewing and monitoring. Data management software has been written using both traditional and data management languages. The only data recorded outside of this procedure was operator commentary which was recorded in an electronic logbook.

The measurement system utilized for measurement of prototype SSC 17 m dipole magnets[16] at Fermilab has employed some significantly different styles than those used for the production measurements described above. In particular, the very much more extensive instrumentation requirements of this phase of measurement has been provided for using multitasking software. This has provided the ability to make measurements of the (cryogenic) environment of the magnet under test with separate code from the process which was directing the magnet test sequence.

Based on this experience, some of the deficiencies of these systems have been observed. The degree of automation available in the AntiProton Source system was limited by available hardware automation. In anticipation of systems with further automation in the hardware, a plan has been outlined and partly implemented which meets some of the problems inherent in the previous systems and to allow for a higher degree of automation in the data collection and data management. The AntiProton Source System was originally used for storage ring magnets which were intended to be used at a single current for which a prescription of a pre-ramp (hysteresis ramp) was sufficient. However, as this system has been used for accelerator magnets, the schemes available for recording the results at different portions of the excitation cycle have required ad hoc changes to the procedures and no integrated method to record hysteresis or current control information was available. Recently, an incremental addition has been implemented to provide a single record structure in which measurement history (programs requested, files written, measurements approved), current history and the electronic

logbook can all be written to a single file which is still easily
accessed with the existing data viewing tool (DATATRIEVE). The intention
is to provide two classes of data output: status data which accumulates
in a chronological log and run data which are written as records of

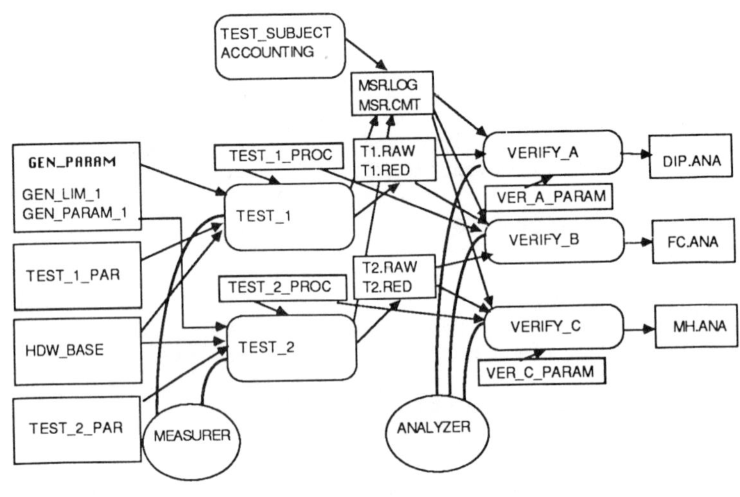

Figure 1. Plan for information flow in magnet measurement systems.

prescribed structure to a separate file for raw and reduced results for
each measurement type. It is believed that this structure MAY be
sufficient for a fairly automatic measurement and analysis procedure
based on command files. Figure 1 outlines the data flow of the
previously mentioned system and a system which has now been partly
implemented for a more automatic measurement system.

The criteria for deciding between further complexity of a given data recording structure and the complexity of adding additional structures requires some discussion. It is likely that the limitation of importance is the channel capacity and buffer capacity of the human perception process. A useful discussion can be found, for example, in Chapter 19 of James Martin's <u>Design of Man-Computer Dialogs</u>.[17] We learn there that for each level of abstraction, up to about seven items can be recalled and manipulated in our senses at one time. To extend the capacity to deal with complexity, we are advised to group materials in to manageable patterns which can then be manipulated further. This suggests that we are wise to select a limited number of files or databases and to provide modest levels of complexity in each.

Based on experience with existing measurement system and the programming effort required to produce them we can conclude that several classes of commercial software are likely to be interesting for constructing Quality Assurance Magnet Measurement Systems: Databases, Forms Management/Data Entry, Graphics, and perhaps statistical packages will be useful. Since the essence of automation is extensive software support of all aspects of the process, the volume of software will be substantial and will certainly justify investment is superior software development tools. It is ASSUMED that the hardware will be based on as many standardized items as possible. Commercial electronic hardware based on widely accepted IEEE or ANSI standards will certainly be the core for the data acquisition system. Mechanical hardware will require careful engineering since comparable national or international standards exist only at component, not system levels for mechanical hardware. The choice of standard (or at lease commercial) software is essential to the implementation of flexible, responsive and maintainable measurement software systems.

SUMMARY AND CONCLUSIONS

The tools required to implement measurement systems for SSC or other superconducting magnets are available. Although new measurement techniques may be considered as part of a new system design, the principles which are required for system planning can be enumerated now. The detailed plan for quality assurance of the component magnets of a superconducting ring must integrate quality assurance at all stages of the fabrication and acceptance of the materials and the magnets. The quality of the magnetic field will depend upon the materials and the assembly. The details of whether to spend more on testing of cable, insulation and laminations or on room temperature or cryogenic temperature testing of the magnets will depend upon the details of the magnet design. For example, the relative shape of the coil when warm or cold depends in detail on the system for coil support within the magnet. Support schemes have been proposed in which the coil is substantially deformed when warm in order to achieve sufficient support when cold. Such a design would suggest very different testing scenarios from those used for the Tevatron or HERA. Details of magnetization may be alternatively studied by detailed materials measurements or by measurements of the finished magnets. The principles which determine the sensitivity of various measurement options are outlined above.

The overall objective of the Quality Assurance project is to provide a reliable and well understood accelerator. To achieve this will certainly require magnetic measurements at many points. However, the requirement of a well understood accelerator which runs for long operating cycles with little down time can be acheived in different ways. With magnets whose manufacturing process <u>guarantees</u> magnets with precisions well beyond what is needed combined with some ''go/no-go'' type testing during or perhaps after the manufacturing process one can mimize the magnetic testing. Accelerator plus magnet designs are viable in which the manufacturing tolerances are much too tight to be achieved with large margins. For these designs, one will need more detailed testing.

The entire enterprise of assembling a large accelerator will require a data management structure which allows the Accelerator Scientist to have access to all the information he needs to monitor the progress of accelerator construction. A data flow plan for the measurement task has been outlined which is suitable for a measurement system with substantial automation. The measurement systems for SSC will grow along these lines in order to provide the quality measurement information required by the accelerator construction enterprise.

DISCLAIMERS AND ACKNOWLEDGEMENTS

This work is an exploration of the issues concerning quality assurance magnetic measurements. It represents accumulated experience of projects at Fermilab related to the SSC needs but not management policy of either Fermilab nor SSC management. I would like to express my thanks to my Fermilab colleagues with whom I have explored these ideas including John Carson, Gene Fisk, Moyses Kuchnir, Mike Lamm, Paul Mantsch, Peter Mazur, Rich Orr, John Peoples, Jim Strait, Frank Turkot, and especially Ray Hanft and Dave Harding. Special thanks are also due to the measurement and development staffs of the Fermilab Magnet Test Facility. Fermilab is operated by Universities Research Association, Inc., under contract with U.S. Department of Energy.

REFERENCES

1. General considerations of Magnet measurements are discussed in W.V. Hassenzahl, Ed., Magnet Measurement Workshop, SSC-SR-1025 SSC Central Design Group, Lawrence Berkeley Laboratory, One Cyclotron Road, Berkeley, CA 94720, Dec 1986; B.C. Brown, Proceedings of ICFA Workshop on Superconducting Magnets and Cryogenics, edited by P.F. Dahl, Brookhaven National Lab, May 1986, p 297; P. Mantsch, op. cit.
2. See for example, R. Palmer and A. V. Tollestrup, Superconducting Magnet Technology for Accelerators, in J. D. Jackson (ed.) Annual Review of Nuclear and Particle Science, Vol 34, 247-284
3. P. Dahl et al., IEEE Trans. Magn., MAG-23, 1215 (1987)
4. R. C. Niemann et al., IEEE Trans. Magn., MAG-23, 490 (1987)
5. Conceptual Design of the Superconducting Super Collider, SSC-SR-2020, SSC Central Design Group, Lawrence Berkeley Laboratory, One Cyclotron Road, Berkeley, CA 94720
6. K. Halbach, Nucl. Instr. & Meth. 74, 147 (1969); K. Halbach, Nucl. Instr. & Meth. 78, 185 (1970).
7. As an example see, M. Kuchnir and E. Schmidt, IEEE Trans. Magn., MAG-24, 950 (1988)
8. Problems with over-closed or under-closed collars on the Fermilab Tevatron Dipoles discovered by longitudinal NMR scans revealed collaring press assembly errors which were straightforward to correct. The effect of the field errors on accelerator properties was not significant.
9. L.P. Michelotti and S. Ohnuma, IEEE Trans. Nuc. Sci. NS-30, 2472 (1983)

10. Measurement systems for the Tevatron are described in many
    unpublished memos and B.C. Brown, et al., IEEE Trans. Nucl.
    Sci., NS-30, 3608 (1983); W.E. Cooper et al., IEEE Trans. Nucl.
    Sci., NS-30, 3602 (1983); A.D. McInturff et al., IEEE Trans.
    Nucl. Sci., NS-30, 3378 (1983); R.K. Barger et al., Adv. Cryo.
    Engr., 31,657 (1986)
11. B. C. Brown et al., IEEE Trans. Magn. MAG-21, 979 (1985)
12. R. W. Hanft et al.; in Proc of the 1988 Applied
    Superconductivity Conference, San Francisco, California,
    August 21-25, 1988;H. Brueck et al. Unpublished Memo, DESY,
    Hamburg Aug 1988; W. S. Gilbert et al. in Proc of the 1988
    Applied Superconductivity Conference, San Francisco, California,
    August 21-25, 1988
13. D.A. Herrup et al.; in Proc of the 1988 Applied
    Superconductivity Conference, San Francisco, California,
    August 21-25, 1988
14. B.C. Brown et al., IEEE Trans. Nucl. Sci., NS-32, 2050 (1985)
15. DATATRIEVE is a product of Digital Equipment Corp.
16. K. McGuire, et al., Adv. Cryo. Engr., 33, 1063 (1986)
17. James Martin, Design of Man-Computer Dialogs, Prentice-Hall,
    Inc. Englewood Cliffs, N.J., 1973

REVIEW ON THE INDUSTRIAL FABRICATION OF THE SUPERCONDUCTING HERA QUADRUPOLE COILS

D. Krischel*, J. Böer[+], H. Fechteler[+] and F. Sommer[+]

* Interatom GmbH
Accelerator and Magnet Technology
D-5060 Bergisch Gladbach (FRG)

[+] Siemens AG
UB Kraftwerk Union
D-4330 Mülheim, FRG

Abstract

The superconducting quadrupole coils for HERA had been developed and successfully tested as laboratory prototypes by CEN-Saclay under contract of DESY, Hamburg. We report on the fabrication under industrial conditions and on test results of collared preseries and series coils. Special emphasis is laid on the consequences resulting from slight but relevant changes in material properties for the series compared to the prototypes like mechanical tolerances, permeability, insulation thickness or sensitivity to temperature during curing.

Introduction

The HERA proton ring to be built at DESY, Hamburg needs more than 220 superconducting quadrupole magnets to focus the proton bunches. Interatom/Siemens in late 1986 got the contract for the fabrication of 122 quadrupoles, 31 of which have a different length compared to the standard of 1.861 m. Under contract of DESY the magnets had been designed and 4 prototypes had been built by CEN-Saclay, France. The fabrication know-how was to be transfered from CEN to industry within a preseries of 5 magnets.

Superconducting Cable

The soldered superconducting cable is of the Rutherford type with 23 NbTi/Cu-strands and $9.5 \pm 0.03 * 1.5 \pm 0.02$ mm (average) outer dimensions. The short sample critical current is above 6900 amps (5.5 T, 4.6 K). The cable had been fabricated by Vacuumschmelze, Hanau, FRG.

Fig. 1. Winding machine

Table 1. Overview on the HERA quadrupole fabrication

Fabrication Step	Material	Tools	Quality Assurance Measurements
I Winding of two* layer coil * first layer is cured second layer is wound on top	– Rutherford type s.c. cable with Kapton plus prepreg insulation – GFRP and copper spacers	– winding machine – mandrel	– winding tension – geometric dimensions
II Curing		– mould 1. layer – mould 2. layer – curing press – central heating system	– total length – Young's modulus – resistance, shorts – self-inductance – temperature – air moisture
III Assembling and Collaring a quadrupole out	– Kapton reinforcements – collar laminations from 1.4429 stainless steel (fine blanked)	– mounting table – collaring press	– resistance, shorts – high voltage test, 1 KV
IV Electrical connection between 4 coils and end collaring	– collar laminations without "nose" – keys – end caps	– turnable hold structure	– total length – resistance – high voltage test 5 KV to mass
V Warm field measurement		– field prove device (DESY supply)	– 26 values per quad. – data transfer to DESY and CEN – cold test at CEN preseries and selected series magnets

Fabrication

An overview on the quadrupole fabrication is given in table 1 including
quality assurance measurements.

The winding machine is designed with stationary storage spools and
rotation mandrel, Fig. 1. For the series fabrication the spool with the
cable for the second layer was mechanically fastened on top of the laminated mandrel to facilitate guiding of the cable. For the prototypes the
spool was hanging from above and had to be turned around by hand with
each winding.

The specified insulation originally consisted of two layers of 25 micron
kapton tape wound with 49 % overlapping and b-stage prepreg glassfiber
tape with 21 % epoxy resin content. The cable we received was within the
geometric tolerances, but compared to the cable used at CEN for the
prototypes the dimensions were at the upper limit.

After winding the very first layer it was found that the mould of the
curing press could not be completely closed. With all the plus tolerances
of 11 windings added up the layer did no more fit to the mould dimensions.
To overcome the problem several tests were made with one or two thinner
kapton tapes of different thickness in combination with different contents of epoxy resin in the prepreg tape. After intense discussions a two
tape insulation was thought to be more reliable. The curing temperature
was varied between 160° C and 180° C together with the duration of 1 to
5 hours at maximum temperature. As a result the thickness of the kapton
tapes was reduced to only two times 12.5 microns. 24 % of epoxy resin was
found to fulfill all demands on the stiffeness, helium transparency and
shape of the coils. The reduced thickness of the kapton tapes, however,

led to unexpected shorts after collaring within the first following
coils. Local reinforcements with kapton and a slightly adapted shape of
some coil head spacers were successfully introduced to exclude further
problems.

During curing the coils were kept at a temperature of 160° C for 2 hours.
Within a temperature - instead of time - controlled procedure the coils
are shaked three times to settle the windings mechanically within the
mould as long as the prepreg has not yet finally reacted.

A suitable mould release was found to minimize the excess of epoxy resin
on the mould.

Fig. 2. Curing press

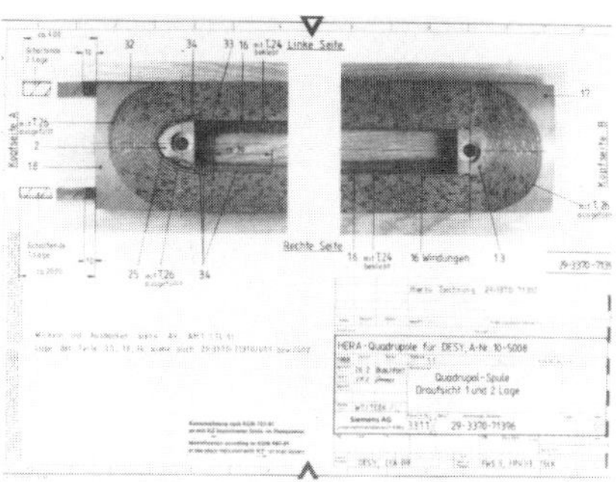

Fig. 3. Fabrication drawing

In addition to the descriptions received from CEN/DESY /1/ drawings were
produced in our shop where the specific component is shown in colour
photographs. In the drawings it is pointed to special materials and
handling. This procedure proved to be very helpful to assure correct
execution of each step during winding and assembling.

The magnetic permeability of the AISI316L austenitic collar laminations
to be used for the preseries quadrupoles was measured to $1.02 \leq \mu \leq 1.08$.
The specified value was ≤ 1.02. The material had been ordered by CEN with
the same specification as for the prototypes. The higher permeability,
however, could not be accepted by DESY with regard to field quality.
Besides one magnet with "$\mu = 1.08$" collar sheets the preseries fabrication continued with material DESY could supply from other resources.
For the series quadrupoles a new melt of steel DIN 1.4429 corresponding
to AISI316LN with small tolerances in the content of relevant elements
was ordered. Before fine blanking DESY arranged a cold test of sample
material to guarantee collar laminations with $\mu \leq 1.02$. After blanking
of collar laminations from a second new melt, however, about 15 % were
found to have severe mechanical defects like burrs, scratches or chips at
the surface. To prevent any defects on the superconductor cable those
laminations were sorted out from the others with proper surfaces. During
this period the series production was continued with winding coils in
stock. The material problems of course demanded for a lot of changes in
planning of personnel and logistics.

For the fabrication of shorter quadrupoles the winding mandrels were
reduced to the specified length.

Fig. 4. Collaring of a quadrupole

Test results

With the before described changes in material and fabrication all
preseries quadrupoles were successfully cold tested at CEN. Along with
series production another five quadrupoles have undergone cold test at
CEN before being mounted into their cryostat. Warm field measurements
were done with a device supplied by DESY. The field harmonics of all
series quadrupoles were within the specified tolerances. Test data have
been transmitted on line to DESY/CEN.

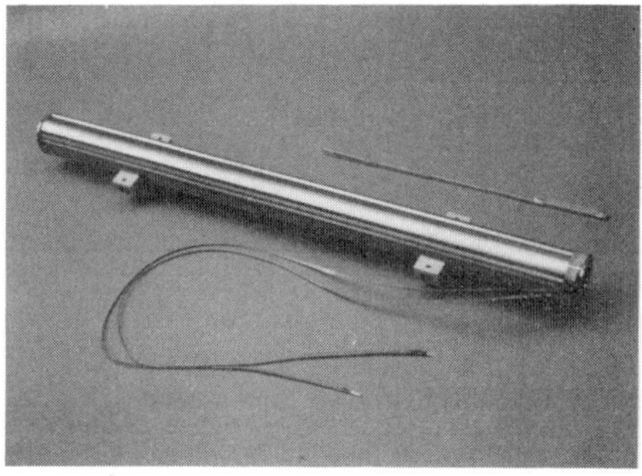

Fig. 5. Quardupole completed with collar laminations,
end caps and current leads

In January 1989 the magnet fabrication has been successfully finished.

We would like to mention that the development of new solutions and final
fabrication procedures have been jointly decided on in a very cooperative
manner by CEN, DESY and our group.

References

/1/ S. Wolff, J. Perot et al, Request for Proposal of DESY B2.917, speci-
    fication of the HERA quadurpole coils, DESY, Hamburg, June 1986

THE EFFECTS OF PROCESSING ON THE FILAMENT ARRAY IN MULTIFILAMENT SSC

STRAND*

P. Valaris,[*] T. S. Kreilick,[*] E. Gregory,[*] and E. W. Collings.[°]

[*]Supercon, Inc., 830 Boston Turnpike, Shrewsbury, MA 01545
[°]Battelle Memorial Institute, Columbus, Ohio 4320

ABSTRACT

Recent Supercon, Inc. investigations have been seeking to determine the extent to which processing affects the strand and its desired properties at the final wire size. The focus of this investigation is to determine the effects of processing in two main areas. The first being how the overall copper/copper-manganese to superconductor ratio changes due to processing. This was examined by the use of wet chemical analysis. The second being how the filament quality is affected by the selection of the spacing to diameter ratio (S/D) and the processing. These apparent effects are examined by the use of SEM-image analysis (SEM-IA) techniques. Such investigations enable the billet designer to compensate for expected processing effects. Wire from three billets is investigated and compared.

INTRODUCTION

Supercon, Inc. has continued to investigate image analysis as an analytical technique in the determination of strand quality from a processing point of view with much promise. In the endeavor to better understand processing effects and determine design characteristics, image analysis has permitted us look directly at strand components under consideration.[1] This being the primary benefit of image analysis over a "n value" survey which is a by-product of Jc testing.

As reported in previous papers[1-10], geometric uniformity becomes an increasingly important factor when small spacing between the filaments and fine filaments are desired. Of equal importance is meeting design specifications namely filament size and the overall copper to superconductor ratio at final wire size as well as current density specifications. Filament size, filament quality, interfilament spacing and a geometrically uniform array are interdependent in the sense that varying any one component can effect one or all the others. For this reason the undesirable effects of processing on the above have to be pin-pointed and the designer be made aware of them. In this effort image analysis has been determined to be an increasingly useful tool for assessing the matrix area of the strand. Where available, "n values" will be compared to the image analysis assessment of filament quality.

In the following paper two main areas are investigated, the first being variation in the copper to superconductor ratio and the second being the reduction of "sausaging" in fine filaments by reducing the S/D. In addition, image analysis as an investigative technique has been further developed and utilized in the following investigation.

---

*Work supported by DOE contract # DE-ACO1-87ER80437

<u>Method and Equipment Used to Perform Image Analysis</u>

Feature analysis, a package of the image analysis software, is the feature that is used to perform the analysis on the matrix array of a given strand. The three main parts of the analysis system are the electron microscope, the analysis system and the image analysis software. Once the backscatter electron image of the sample has been resolved as desired, then a binary image of this is acquired by the analysis system. This image is then used for feature analysis where the one fundamental parameter selected is "Feature area". From this a host of raw data can be obtained, namely:
-the individual filament area
-the matrix alloy area
-the ratio of the area of the number of filaments to that of the background (matrix alloy area) and finally a histogram of the filament area distribution. This data is then used to investigate and determine the quality of given strand.

<u>Equipment Used</u>

ISI Microscope
EDS - Kevex Super 800
Kevex Super 800 Analysis System with Kevex Feature Analysis Software-Binary Image is formed of 512 (H) x 256 (V) pixels

VARIATION IN THE COPPER TO SUPERCONDUCTOR RATIO

<u>Loss of Copper</u>

In order to determine the Cu/Sc ratio consistency of SSC strand, for any given billet, samples are taken at a wire diameter of 0.330 cm. (0.130 inches) from different parts of the strand yield. These are then drawn down to fine sizes as listed in Table 1b and the Cu/Sc ratio determined.This is done by wet chemical analysis. To broaden this investigation three billets of about the same Cu/Sc ratio but with differences in the other general properties were examined. Vital statistics for the three billets used in this investigation are listed in Table 1a. Values of the Cu/Sc ratio obtained are listed in Table 1b and are plotted against wire size in Figure 1. From this analysis we are able to conclude the following.

The overall Cu/Sc ratio varies with wire size. This variation becomes increasingly apparent at around 0.038 cm. (0.015 inches) and below where the decline becomes more pronounced. This apparent change suggests a loss of ~5 percent should be anticipated of the total amount of copper in a production size billet. Figure 1 shows the variation of Cu/Sc ratio with wire size of three different 23K samples; that is samples from different parts of the same billet. The 23K-Collings sample was independently investigated by E. W. Collings of the Battelle Memorial Institute. The remaining sample investigations were made at Supercon, Inc. By having determined the Cu/Sc ratio at the time of billet assembly we are able to extrapolate to the right of Figure 1, beyond the values that were actually investigated.

At final wire size the Cu/Sc ratio may vary by two to three percent depending on the front from which the sample is taken. At 0.0648 cm. (0.0255 inches), the Cu/Sc ratio variation between series samples 23K-a and 23K-b is approximately two percent. Also from a series of spot checks of production material the apparent spread at 0.0648 cm. (0.0255 inches) was determined to be 3.4 percent.

<u>Note</u>: In making the Cu/Sc ratio determinations, E. W. Collings of the Battelle Memorial Institute determined the density of the Nb 46.5 Wt. % Ti to be 6.096 gm cm^{-3}. At Supercon, Inc. the density used for Nb 46.5 Wt.%Ti was 6.04 gm cm^{-3}, the value specified by the manufacturer. The copper/copper-manganese density was calculated from specific density values depending on the manganese weight percent content. This difference in the Nb-Ti densities should be noted when assessing the 23K-Collings curve in Figure 1 and data in Table 1, a and b.

Table 1a. Vital Statistics for billets used in the investigations.

	6A	23K	6B
Wt % of Mn in Matrix Copper	0.11	0.50	0.66
Number of Filaments	1700	23,000	1700
Cu-CuMn/Sc at Billet Size	1.96 :1	1.87 : 1	1.85 : 1
Billet Size (cm.)	7.62 (3 inches)	30.48 (12 inches)	7.62 (3 inches)
Average S/D	0.135	0.195	0.135
Local Ratio at 0.0648 cm.	~0.2882:1	~0.4280:1	~0.2882:1
Processing Status	Cold Worked	Heat Treated	Cold Worked
Filament Size at 0.06477 Cm	~12 μm	~2.5 μm	~12 μm
Filament Size at 0.0216 Cm	~2.7 μm	~0.8 μm	~2.7 μm

Table 1b. Cu/Sc Ratio Data for Figure 1.

Wire Size (Cm)	COPPER TO SUPERCONDUCTOR RATIO					
	6A	23K Collings	23K-a	23K-b	6B-a	6B-b
0.302		1.941				
0.205					1.824 ±0.004	
0.193				1.903 ±0.003		
0.122	1.862 ±0.003					
0.117				1.903 ±0.003		
0.074				1.897 ±0.003		
0.065	1.850 ±0.005	1.912	1.867 ±0.003		1.817 ±0.006	
0.053		1.909			1.816 ±0.004	
0.044			1.858 ±0.004			
0.043				1.875 ±0.004		
0.038	1.835 ±0.004	1.897			1.810 ±0.006	
0.0305			1.840 ±0.008			
0.0274		1.885			1.803 ±0.006	1.779 ±0.003
0.0216	1.818 ±0.003					1.775 ±0.003
0.0193			1.834 ±0.008			
0.0163					1.794 ±0.007	1.760 ±0.003
0.0147				1.856 ±0.005		

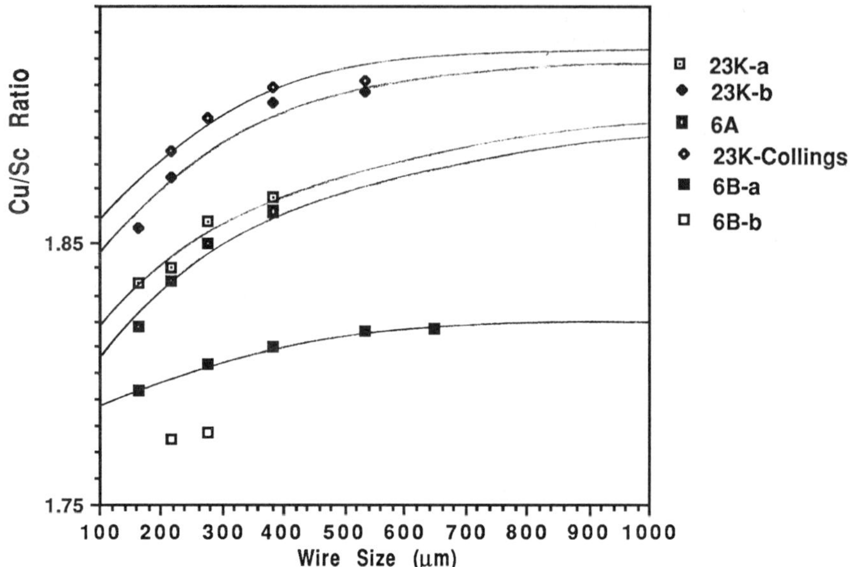

Figure 1. Showing variation of Cu/Sc ratio versus wire size.

REDUCING "SAUSAGING " IN FINE FILAMENTS BY REDUCING THE S/D

Ensuring Good Filament Quality

In the early part of 1988, Supercon, Inc. produced a full size billet with approximately twenty three thousand filaments. This billet is referred to as the 23K billet. Characteristic of this billet is the 2.5 micron filaments at 0.06477 cm. (0.0255 inches), a largely uniform array [2] and an average S/D of 0.195 among other specifications listed in Table 1a. Recent surveys of the filament array of this billet have shown that the filaments are to some extent "sausaged" [1,5,8 and 10] (see Figure 2d), hence the strand is unable to perform as well as expected. This apparent "sausaging" is depicted by the broadness in histogram Figure 2d, and affirmed by an "n value" of 19. The reason for the "sausaging" has been attributed to a relatively large S/D value of ~0.195, for the size of filament. The condition of the diffusion barrier at the final intermediate precipitation heat treatment sizes has also been called into question and is currently being investigated. This assertion is backed up by the survey of billet 6B, which has an average S/D of ~0.135 and much improved filament quality (see histogram in Figure 5b), although it should be noted that billet 6A and 6B are 7.62 cm. (3 inch) trial size billets.

Technique Used to Compile the Histogram

Cross-sectional mounts were made of strand from billet 23K and 6B at wire sizes which yield filaments ranging from ~14 microns to ~2.5 micron. These samples were then polished to enable an SEM backscatter image to be formed of the clarity shown in Figure 3a, and 4a and b. A binary image is the acquired, as in Figure 3b and feature analysis used to survey the matrix area. Generally between 15 and 30 filaments are surveyed in any one frame. This number of filaments has been found to be optimum as it offers a large number of pixels per filament giving the acquired data more resolution.   Areas surveyed are those close to the core, in the midst and along the outer edge of the matrix annulus. Raw data for each histogram is comprised of at least 180 filaments surveyed in the 23K samples and 60 filaments in the billets 6A and 6B. The number of filaments surveyed amounts to 1-3% of the total and is considered a fair representation of the entire filament array. The data was sorted into bins so that the histograms could be generated. The histogram thus being an assessment of filament quality. The series of histograms in Figure 2 and 5 trace filament deterioration verses wire size. A strain of ~1 separates the histograms. Figure 3a and 4a and b, showing contrast required for SEM-IA.

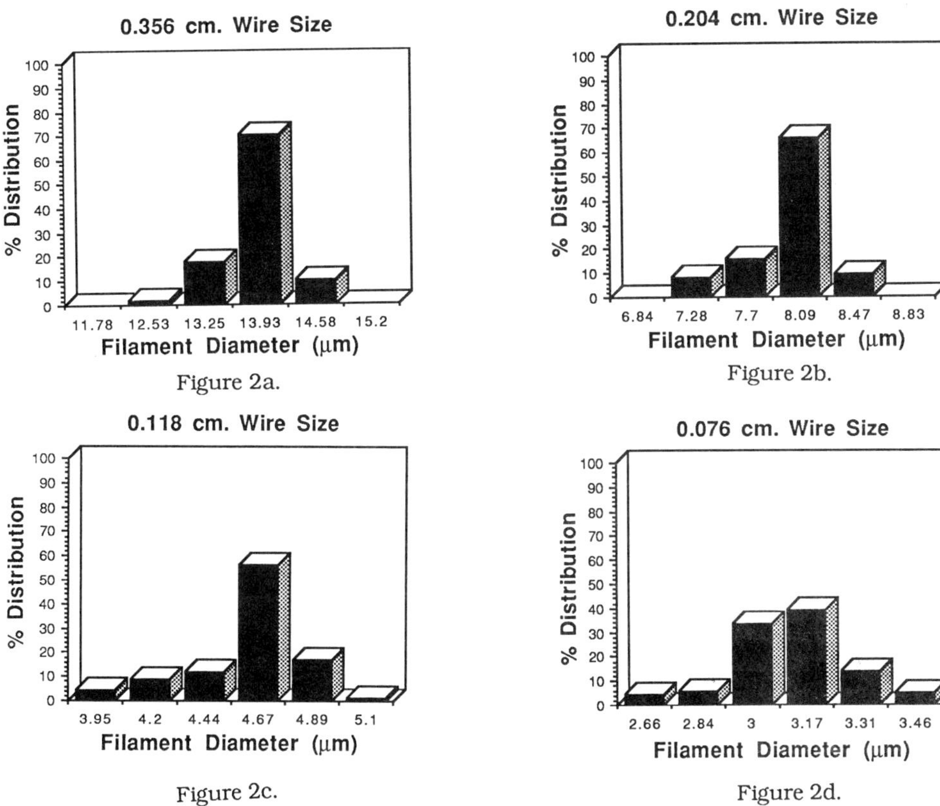

Figure 2a.

Figure 2b.

Figure 2c.

Figure 2d.

Figures 2, a-d. Showing Filament Degradation With Wire Size. Filament Area Distribution in Figure 2d Correspons to an "n value" of 19.

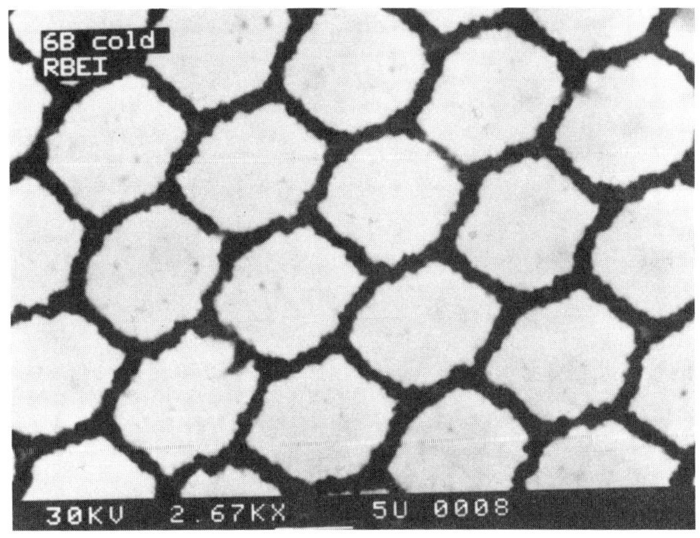

Figure 3a.

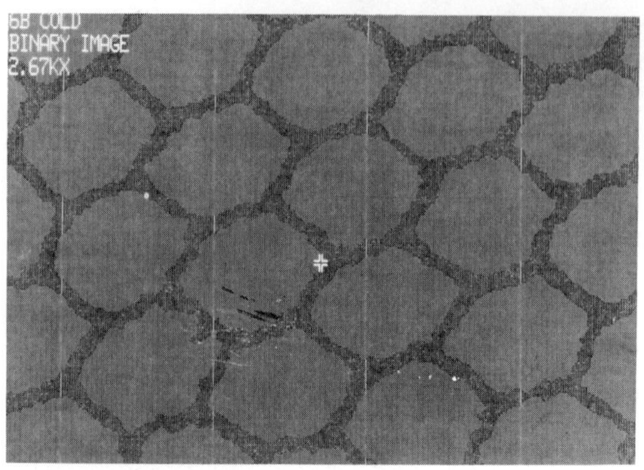

Figure 3b.

Figure 3. Showing the Clarity of a SEM Image and Its Respective Binary Image Scan.

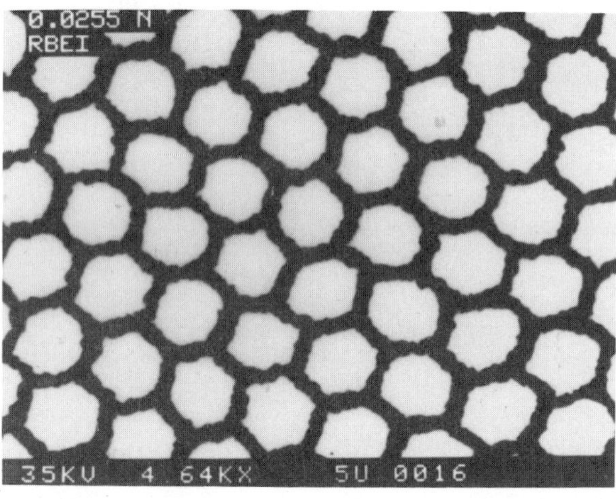

Figure 4a. Matrix Array Sample of the ~23K Material.

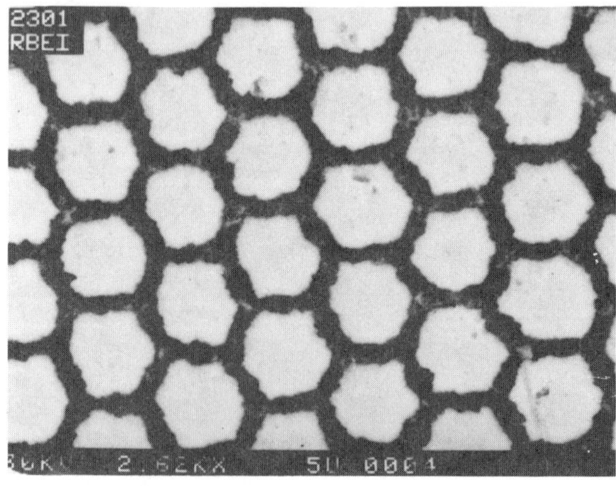

Figures 4, a-b. Showing the Matrix Array of 3 Evaluated Areas. Figure 4b. Matrix Array Sample of a ~4K 6 μm. outer Material.

<u>Forming the Histogram</u>

The center bin is chosen to contain filament areas that are within 5 percent of the average central filament area; that is each bin has a range of 10 percent. The filament area of each central value of the adjoining bins vary by steps of 10 percent from the average central value. These bins once more contain filaments with areas that vary by 5 percent either way from the central value of that bin. Thus all bins have a range of 10 percent.

<u>Assessing Data</u>

A gradual filament deterioration of the 23K material,which features an average S/D of ~0.195, is portrayed by histograms in Figure 2. Figure 2c in particular shows the filament quality at almost the specified final wire size. The actual final wire size being 0.0648 cm. (0.0255 inches) and yielding 2.5 micron filaments. The broadness of the distribution in Figure 2c shows the depressed quality of the filaments. The average central bar represents only 40 percent or less of the filaments in in the matrix array. The areas of the filaments in general are spread over a 70 percent range which in turn translates to filament sizes ranging from 2.66 microns to 3.46 microns, obviously not optimum. In contrast to the above, samples surveyed from billet 6B which feature an average S/D of ~0.135, show that filament quality has markedly improved at ~2.5 microns and smaller. Almost 80 percent of the filaments are in the average central bar, with the range of the filament area spread being only 30 percent. See Figure 5 , a and b.

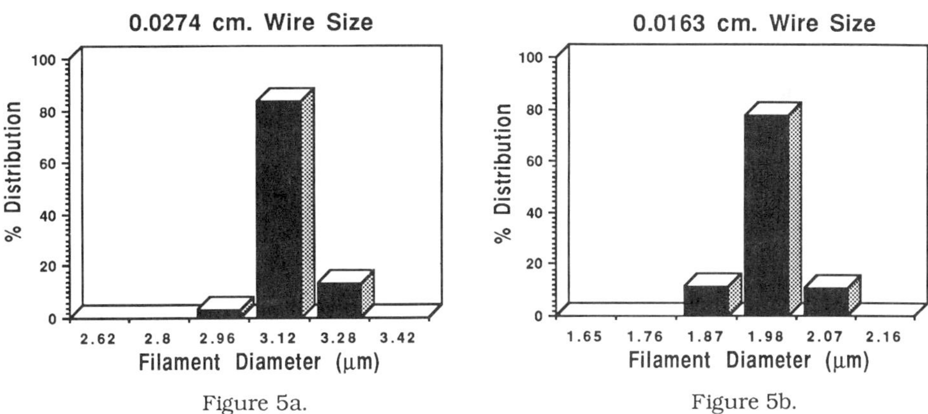

Figure 5a.   Figure 5b.

Figure 5, a and b . Showing the improved quality of Billet 6B which has an S/D of ~0.135.

<u>Acknowledgements</u>

The authors are grateful to the Supercon, Inc. staff for their technical assistance, in particular A. Samgkhum, for sample preparation. Supercon, Inc. would also like to acknowledge the financial assistance of the U.S. department of Energy (DOE), through the Lawrence Berkeley Laboratory, and the Small Business Innovative Research (SBIR) Program of DOE. Our thanks are also due to the Brookhaven National Laboratory, the University of Wisconsin and the Francis Bitter National Magnet Laboratory of M.I.T. all of whom assisted in obtaining much of the data reported.

SUMMARY

Important points that the designer needs to consider are:
- Copper losses of ~5 percent are to be expected in the processing of a production size billet.
-The Cu/Sc ratio of material from any given billet may vary by up to 3.4 percent over the piece length of the conductor.

- Reducing S/D tends to significantly improve 2.5 μm. diameter filament quality. Image analysis used on carefully prepared samples is a useful way of assessing the components of the matrix array.

The matrix array component which requires more investigation, especially where fine filaments are concerned is the diffusion barrier-this investigation will be the subject of a later paper.

REFERENCES

1. P. Valaris, E. Gregory, T. S. Kreilick and J. Wong, "Refinements In Billet Design For SSC Strand", Presented at the <u>Applied Superconducting Conference</u>, San Francisco, California. August 21-25, 1988.

2. T. S. Kreilick, E. Gregory, J. Wong, R. M. Scanlan, A. K. Ghosh, W. B. Sampson and E. W. Collings, "Reduction of Coupling in Fine Filament Cu-NbTi Composites by the Addition of Manganese to the Matrix," <u>Advances in Cryogenic Engineering</u>, A. F. Clark and R. P. Reed, eds., Plenum, New York, 34, 895. (1988).

3. E. Gregory, " Recent Developments in Multifilamentary NbTi Superconductors, " <u>Cryogenics</u>, 27,6, 290, (1987).

4. E. Gregory, "Conductor development for the Superconductor Super Collider (SSC)," <u>Advances in Cryogenic Engineering</u>, A. F. Clark and R. P. Reed, eds., Plenum, New York (1988), 34, 867.

5. E. Gregory, T. S. Kreilick, A. K. Ghosh, W. B. Sampson, "Importance of Spacing in the Development of High Current Densities in Multifilamentary Superconductors," <u>Cryogenics</u>, 27, 4, 178, (1987).

6. T. S. Kreilick, E. Gregory, "Further Improvements in Current Density by Reduction of Filament Spacing in Multifilamentary NbTi Superconductors," <u>Cryogenics</u>, 27,7, 401 (1987).

7. T. S. Kreilick, E. Gregory, J. Wong, "Influence of Filament Spacing and Matrix Material on the Attainment of High Quality Uncoupled NbTi Fine Filaments," <u>IEEE Trans</u>. 24,2, 1033, (1988).

8. T. S. Kreilick, E. Gregory, J. Wong, "Geometric Considerations in the Design and Fabrication of Multifilamentary Superconducting Composites," <u>IEEE Trans</u>. Mag-23, 2, 1344, (1987).

9. T. S. Kreilick, E. Gregory, J. Wong, "The Design and Fabrication of Multifilamentary NbTi Composites Utilizing Various Matrix Materials," <u>Journal of the Less Common Metals</u>, 139, 45, (1988).

10. T. S. Kreilick, E. Gregory, J. Wong, "Fine Filamentary NbTi Superconducting Wires," <u>Advances in Cryogenic Engineering</u>, A. F. Clark and R. P. Reed, eds., Plenum, New York, 32, 739 (1986)

SUPERCONDUCTIVITY AND THE MAGNETIC ELECTRON BOND

Peter Szurek

Sentek

Novi, Michigan

ABSTRACT

The concept of the magnetic electron bond as the
fundamental characteristic of superconductivity was first
introduced during a presentation at the 1988 Winter Annual
Meeting of the American Society of Mechanical Engineers.
Postulates describing the role of the electron and the magnetic
bond were suggested to explain in a consistent manner known
observations. What may becoming clear is that a boundary set of
conditions may exist above and below the transition temperature
at which a material superconducts. Prior to recent history,
scientists have concentrated on postulating, experimenting, and
learning about the set of conditions that exist above the
transition temperature, which has set the standard for todays
quantum theory. Above the transition temperature we have
learned about the interrelationships that exist between the
electron, a small magnetic and negatively charged body, and the
nucleus, a large positively charged body. By grouping common
general characteristics due to the interaction between the outer
shell electrons and the nucleus of different elements, three
bond types have been established, covalent, ionic, and metallic.
We may now be in the process of determining those conditions
that lie below the transition temperature, a realm where charge
effects may no longer dominate magnetic effects. This may
involve updating the quantum theory to reflect those conditions
that exist above and below the transition temperature. The
following discussion reviews, updates, and attempts to answer
some preliminary questions regarding postulates that may define
some of the conditions that lie below the transition temperature. As an introduction, figure 1 depicts what may occur to
loosely held outer shell electrons below the transition
temperature due to increased inner electron shielding.

QUANTUM MECHANICS

Among the properties that we have learned about the
electron when above the transition temperature is that the
electron in an atom has a set of four quantum numbers, a level

(n), sublevel (l), magnetic orbital (ml), and magnetic spin
number (m2). The preceding statement is a major premise of
modern quantum theory. A few points to remember concerning the
quantum numbers are as follows (1): as (n) increases, the
energy of the electron increases and as (n) increases, the
distance of the electron cloud to the nucleus increases; the
general geometric contour of the electron cloud varies with (l)
and as (l) increases, the energy of the electron increases
slightly; the orientation of the electron cloud with respect to
a given direction is specified by (ml) when a strong magnetic
field is imposed and has very little effect on the energy level
of the electron; and last the spin association of an electron
about its own axis with respect to the orbital spin axis is
specified by (m2) and within an orbital in an atom, two
electrons have opposite spin.

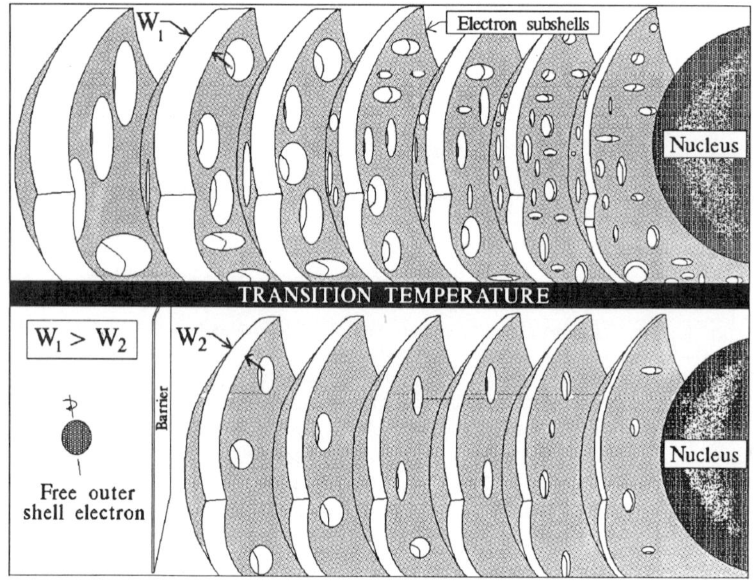

Fig 1. Representation of shielding above and below transition temperature

Of the four quantum numbers, the level and sublevel are
principally charged related, resulting in different energy
levels of the electron and geometry within the atom due to
electrostatic attraction between the positively charged nucleus
and negatively charged electron. The orbital and spin are
principally magnetically related, resulting in orientational
characteristics rather than variations in energy levels and
geometry.

Two of the fundamental properties of an electron then are
its charge, negative, and that it is magnetic(2). Based on
these two fundamental characteristics, the electron may form
several types of bonds; a bond being defined as the means or
mechanism by which atoms or groups of atoms are combined in
molecules(3). I would like to further suggest a definition of
bonding as the degree of positive charge association of the
outer shell electrons with the nucleus of atoms. The type of
bond being a measure of the shielding of the outer shell
electrons from the nucleus by the inner shell electrons.

458

ELECTRON SHIELDING AND BONDING

The following discussion with regards to bonding considers
what happens to elements that do not possess the charge and
magnetic balance of certain elements as indicated by their
quantum number such as Helium, Neon, Argon, Krypton, Xenon, and
Palladium.

Atoms of different elements have nuclei of varying size and
positive charge potential, and have a different number of
electrons orbiting the nucleus. The characteristics and stability of elements varies widely depending on the combination of
these factors. For instance, hydrogen has a stable nucleus
being relatively small, yet is highly reactive with other
elements due to its largely unshielded electron shell. Uranium
on the other hand has many electrons to provide shielding of the
nucleus, but has such a large number of protons and overall
nucleus that the nucleus becomes unstable.

Covalent Bond

Hydrogen with only one electron and little shielding of the
nucleus is highly reactive and may combine with several elements
to form a strong stable covalent bond; elements with a higher
positive charge potential form stronger bonds than elements with
lower charge potential. For example, the covalent bond strength
between two hydrogen atoms is 104 Kcal/mole while the bond
strength between a hydrogen and fluorine atom is 135 Kcal/mole
(1). The fluorine atom with nine protons versus the hydrogens
one is able to more strongly attract the shared electron than
the single proton of hydrogen resulting in a higher bond
strength. Fluorine, although not as reactive as hydrogen, is
still relatively reactive having only nine electrons orbiting
and thus shielding the nucleus. As the size of the nucleus
increases and the number of electrons orbiting and surrounding
the nucleus increases the ability to form a covalent bond, which
depends upon the sharing of electrons, decreases.

Along this same line of reasoning, elements with atoms that
share multiple electrons exhibit stronger bond strengths than
ones that share a single electron. For example, the triple bond
strength of nitrogen is 225 Kcal/mole, the double bond strength
of oxygen is 118 Kcal/mole, and the single bond strength of
fluorine is 37 Kcal/mole. As the discussion on the covalent
bond is completed, we may say that elements with the smallest
atoms have minimal inner electron shielding; the outer shell
electrons have maximum association with the positive nucleus.

Ionic Bond

As the size of the elements atom increases to the point
where the inner electron shield becomes significant, the ability
to form a covalent bond, or sharing of electrons, becomes lost.
Charge attraction between the elements of medium sized atoms
still exists, however is not strong enough for sharing to occur.
Instead, one atom becomes a donor of an electron(s) and one atom
a receiver of an electron(s).

When this electron transfer occurs, an ionic bond forms due
to the resulting charge imbalance. For example, referring to
figure 2, when sodium and fluorine react, the sodium atom

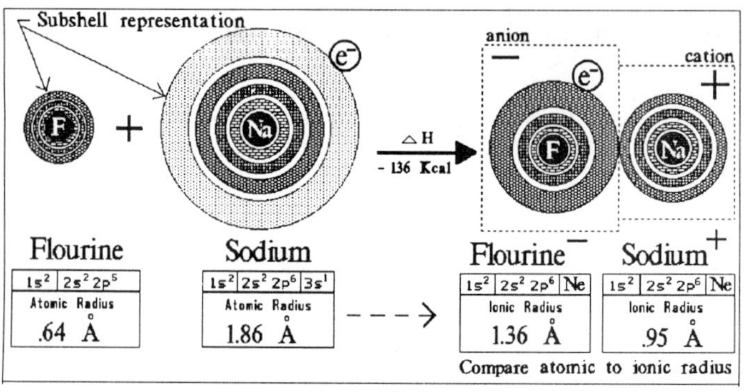

Fig 2. Ionic bond with electron transfer

donates an electron to fluorine. The original sodium atom had a charge balance of eleven protons and electrons; by donating an electron to fluorine it now has a charge imbalance of eleven protons and ten electrons. This charge imbalance creates electrostatic attraction and an ionic bond of 136 Kcal/mole as the sodium nucleus tries to reclaim the lost electron to maintain charge balance. Fluorine in receiving the electron has successfully completed its 2p sub level at a cost of increasing its atomic radius from .64 angstroms to an ionic radius of 1.36 angstroms. This is an 860% increase in volume. Likewise the sodium atom has reduced its atomic radius from 1.86 angstroms to an ionic radius of .95 angstroms, an 87% decrease in volume. In the ionic state, the fluorine and sodium atom both have ten electrons, yet the sodium nucleus has better inner electron shielding as evidenced by its smaller ionic radius. Since there is an actual transfer of an electron, the sodium atom if surrounded by fluorine atoms does not discern which fluorine atom has received this electron and thus a uniform crystal structure takes place. In essence, each neighboring donated electron supplies a percentage of the correcting charge imbalance. As expressed earlier by example of the bond strength difference between the covalent bonds of hydrogen-fluorine and hydrogen-hydrogen, a similar comparison may be made for ionic bonds. When potassium with 19 protons combines with chlorine the resulting bond is 104 Kcal/mole; this compares to 98 Kcal/mole when sodium with 11 protons combines with chlorine. Again the atom having a higher positive charge potential forms a stronger bond. As the discussion on the ionic bond is completed, we may say that elements with medium sized atoms have increased inner electron shielding; the outer shell electrons have less association with the positive nucleus.

Metallic Bond

The ability to form an ionic bond, or electron transfer, may depend on having a balanced need between two atoms. The size of the atom is one factor since it determines how many electrons are orbiting the nucleus; the size of the atom has to be big enough, provide enough shielding to preclude forming a covalent bond, or sharing of electrons. Another factor may be the change in volume that occurs when there is electron transfer; the donating atom may achieve improved shielding although it has one less electron since a reduction in volume occurs. The receiving atom has less electrons and a greater need of orbiting bodies, at the expense of an increase in

volume. The donating atom is referred to as the cation and the
receiving atom the anion.

As the size of the element increases whereby it has a
greater number of electrons, the anions need for additional
electrons may diminish; the atom has a sufficient number of
orbiting bodies that increasingly shield the nucleus. In
essence, when the number of electrons reaches a certain
quantity, the nucleus has a harder and harder time keeping track
of its outer most electrons when in the presence of similar
sized atoms, due to its substantial inner electron shielding.
At this point, the outer shell electrons may have little
association with the positively charged nucleus. These outer
shell electrons become mobile and under an electromotive force

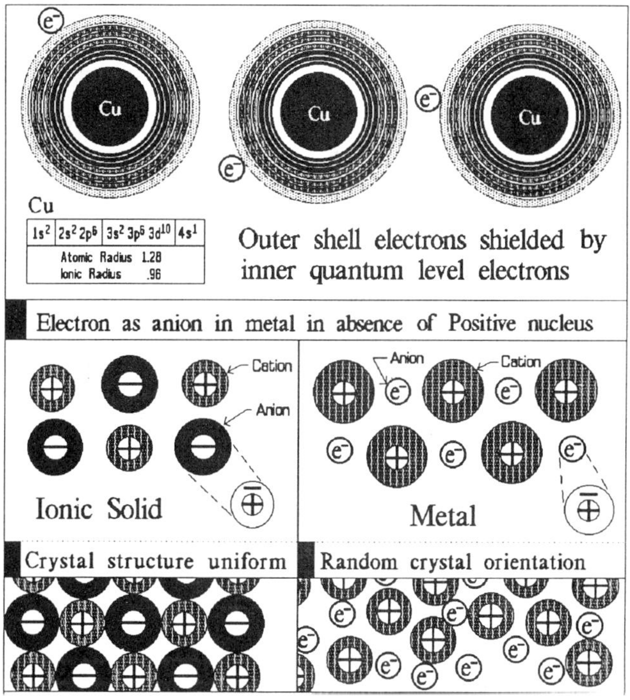

Fig 3. Metallic bond with electron exchange

no longer undergo a single transfer as for the ionic bond, but
multiple transfer, or electron exchange between the nuclei.
When this occurs, a metallic bond exists between the atoms(1).
Referring to figure 3, for the metallic bond, the nucleus and
inner shell electrons may be considered the cation and the
unpaired outer shell electron the anion. As in the ionic bond
there is a charge differential. For example, the nucleus of
copper contains 29 protons and 28 electrons, excluding the
unpaired 4s1 outer shell electron. The nucleus may be considered
the cation. In this model representation, the outer shell
electron contains one electron and zero protons and may be
considered the anion. An ionic solid forms a crystal structure
that is relatively uniform as both the cation and anion are
similar in size. However, for this model of the metallic solid,
the cation is significantly larger than the diminutive anion, or
electron. This size differential results in a random crystal
orientation. As the discussion on the metallic bond is
completed, we may say that elements with large atoms have
several levels of electrons and substantial inner shell

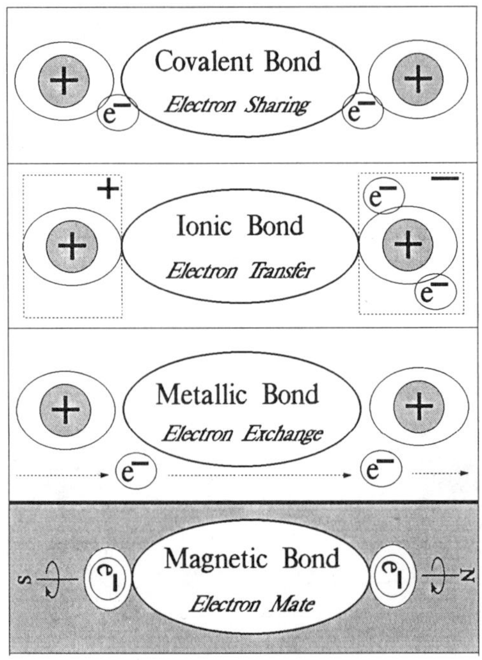

Fig 4. Bond Summary

shielding; the outer shell electrons have little association with the positively charged nucleus.

Bond Summary

For the three bond types, we may observe a trend of increased inner shell electron shielding with atom size and a decreasing association of the outer shell electrons with the positively charged nucleus. Referring to figure 4, as discussed, as we move from smaller to larger elements the nuclei's grip on the outer shell electron diminishes. This is evidenced by the different and similar characteristics between the covalent, ionic, and metallic bonds.

Prior to discussing aspects concerning the possibility of a fourth bond, a magnetic bond, please consider the trends that have been established among elemental size, shielding, and bond characteristics, and of observed phenomena of superconducting and non-superconducting materials. I have formulated a few questions below that I hope will generate questions from you, the reader, as this may lead to a better understanding of superconductivity.

A question then arises, what may happen when a nucleus loses total control of its outer shell electron(s)? What happens when we combine the structure of the ionic bond and the mobility of the metallic bond? What happens when we combine or incorporate elements of the largest atoms, whereby there is maximum inner electron shielding and minimal association of the outer shell electrons with the positively charged nucleus? If the positively charged nucleus loses control of the outer shell electrons, of the four quantum numbers, which major

characteristic will dominate, charge or magnetic? If it's
magnetic characteristics, what differences in electron behavior
may we expect from the electron when not bound or dominated by
charge effects? Four quantum numbers have been defined for an
atom to explain conditions that occur above the transition
temperature of a superconducting material. If a different set
of conditions occur below the transition temperature, what are
its conditions and how shall we define them?

INTRODUCTION TO DISCUSSION OF THE MAGNETIC BOND

     Upon presentation of the concept of the magnetic electron
bond at the ASME conference, a good question was raised which
will serve as an introduction to the discussion of the magnetic
electron bond. I suggested that above the transition temperature the electron behaves according to present day quantum
mechanics obeying the rules concerning the magnetic spin quantum
number whereby two electrons in a magnetic orbital have opposite
spins, +1/2 and -1/2; however, below the transition temperature
I suggested that the electron may pair with another electron,
and the pair may align themselves along a common axis, their
magnetic north-south axis, and spin in the same direction with a
net magnetic adding effect. The question posed was as follows:
How can electrons magnetically combine along the same axis,
spinning in the same direction, when this would violate a
fundamental law of quantum mechanics, the magnetic spin quantum
number?

     Since according to the Pauli exclusion principle there can
be only up to two electrons within a given orbital in an atom,
each with opposite spin, one may suggest that this is a
condition which is adhered to above the transition temperature,
when the electron is influenced by the nucleus of the atom; a
condition that occurs when the diminutive negatively charged
electron is in the presence of a large positively charged
nucleus. When below the transition temperature, a new set of
conditions may be adhered to, similar but different, since the
electron may no longer be influenced by the large positively
charged nucleus of the atom. Below the transition temperature,
the loosely held outer shell electrons may be acting more as an
independent magnetic body.

     Above the transition temperature, a condition of the
electron being within an orbital and orbiting about the nucleus,
the electron has opposite spin to the second electron within the
same orbital. This would be intuitively a good condition since
it would avoid electron collisions and spread the two electrons
out to cover a maximum amount of volume around the nucleus for
good shielding, both of which contribute to good stability. It
also may allow for each electron to be in their lowest possible
ground state as they orbit the nucleus, a fundamental premise of
quantum mechanics(1).

     Below the transition temperature, a condition may exist
where the loosely held outer shell electrons may not be within
any specific orbital and may not orbit about any specific
nucleus; these electrons may have the same spin orientation of
nearby electrons. They may have the same spin for the same
reason electrons in orbitals have the opposite spin, to allow
the electron to be in the lowest possible ground state and most

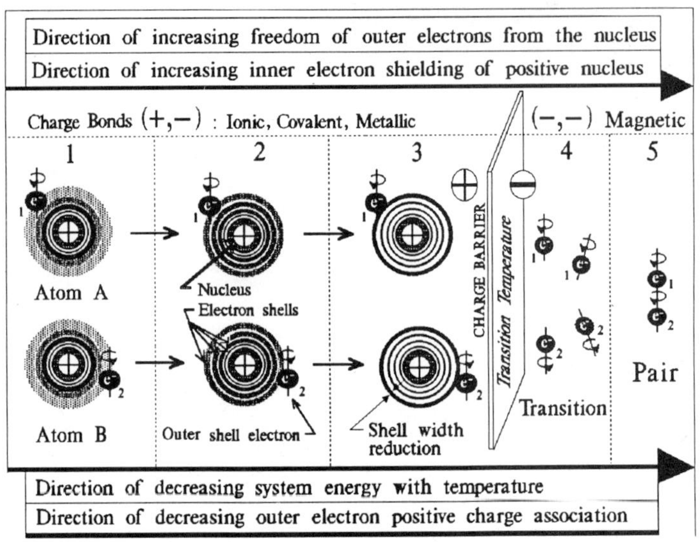

Fig 5. Relationship of positive and negative charges;
how it may effect magnetic bond formation

stable state. This statement would be consistent with the
fundamental premises of quantum mechanics.

Lets take a minute to refer back to the discussion on the
three bonds and the trend of greater electron "freedom" from the
nucleus for the outer shell electrons as we move from the
covalent to ionic to metallic bond. In each bond the outer
shell electron moves to a state of greater independence, from
being shared, to being transferred, and finally to being
exchanged by the nuclei. In the metallic case, for discussion
purposes I suggested looking at the outer electron as an anion
and the nucleus as a cation to explain electron mobility and
crystal structure; this may suggest the electron is already
experiencing some signs of freedom and independence. It is then
logical to suggest that this trend may continue and under the
right set of circumstances the outer shell electron may achieve
independence; the point at which independence occurs may be the
superconducting transition temperature. In addition, below this
transition temperature, the electron in the absence of charge
related properties may adhere to magnetic properties, forming a
magnetic bond. This is diagrammed in figure 5.

This may be one of the fundamental reasons for zero
resistance below the transition temperature; the outer shell
electron may be essentially sealed off by the inner shell
electrons of its original host nucleus and surrounding nuclei.
There is no longer any electrostatic attraction, or holding
force, between the electron and the nucleus to be overcome.
Materials with different elements within a lattice, having outer
shell electrons of different energy levels and inner shell
shielding, different atomic radii, and different inner atomic
spacing between atoms, may result in different transition
temperatures within the material. When the transition
temperature is experimentally observed, it may be for the
material as a whole, as one entity, and not for localized atomic

regions. Unless we have a perfectly organized uniform crystal
structure, within the material there may be local transition
zones correspondingly.

SHIELDING

     Before discussing aspects of the suggested magnetic bond I
would to express a comment or two regarding shielding. The
nucleus of the atom has a shield and that shield consists of
electrons. Likewise, the "nucleus" of an electron has a shield
and that shield consists of magnetic lines of force. An atom
with a greater number of electrons has in general greater
shielding; likewise an electron that expounds a greater number
of magnetic lines of force may have greater shielding. Two
atoms of the same kind may have a different degree of shielding
according to its state (bond), i.e. a sodium atoms atomic radius
is much larger than its ionic radius, and thus is less shielded
as the electrons must shield a greater amount of free volume.
Similarly two electrons may have a different degree of shielding
according to its state (bond), i.e. magnetically paired
electrons may combine to form higher magnetic fields, and as a
result may achieve greater shielding than individual pairs.

MAGNETIC ELECTRON BOND

     A suggested general statement concerning the magnetic
electron bond, if it exists, is as follows: Materials which
incorporate elements with the largest atoms may form structures
whereby the loosely held outer shell electrons may become
independent of the positive nuclei and magnetically mate.
Magnetic electron bonding may occur when the electrons positive
charge association has been minimized, when charge effects have
given way to magnetic effects. The transition temperature being
the point at which charge effects have given way to magnetic
effects.

     A suggested definition of characteristics of the magnetic
electron bond was previously presented in a paper presented at
the ASME Winter Annual Meeting. The papers reference number may
be found under REFERENCES at the end of this paper(4). The
remainder of the discussion here will be a review and update of
some of the suggested characteristics. This will be subdivided
into two sections: Characteristics of the electron and
characteristics of the magnetic electron bond.

<u>Characteristics of the electron</u>

     The following are suggested characteristics of the
electron.

     <u>Composition</u>. Referring to figure 6, the electron may
consist of a thin flexible outer shell, have plasma internally,
and an anti-matter:matter generator at the core.

     The flexible outer shell may allow for expansion and
contraction due to sources of heat energy addition and
subtraction to achieve long term stability; it may allow for
quick reaction from this external energy transfer and
flexibility to sustain collisions. The skins geometry may be

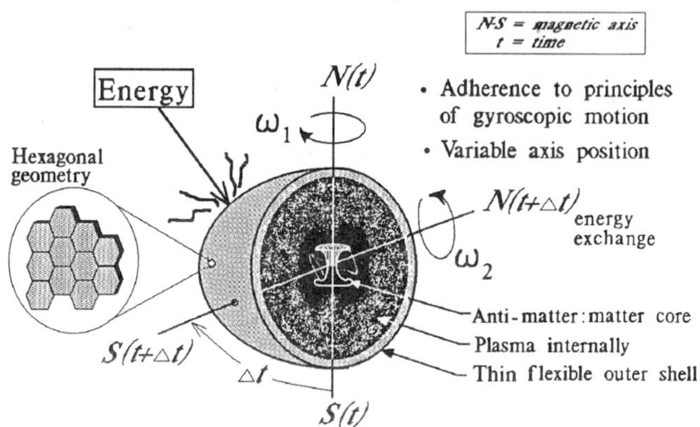

Fig 6.   Composition and reaction of electron
due to energy exchange

hexagonal in nature, allowing for strength yet flexibility. If
the legs contract or bend due to temperature changes the skin
around the hexagon may move inward or outward like a diaphragm.

Internal plasma may allow the electron to have an energy
transfer medium from the shell to the anti-matter:matter
generator at the core, and may also serve as a buffer between
the shell and core for stability. An influx of energy or energy
transfer from the plasma to the generator may drive the
frequency response higher and create more vibration by the
electron; an outflow of energy or energy transfer from the
generator to the plasma may drive the frequency response lower
and create less vibration by the electron. The plasma may be
the fundamental characteristic responsible for the electrons
high heat carrying capacity.

The center of the electrons core may contain an
anti-matter: matter generator; the anti-matter may be
responsible for the electron having a negative charge. Matter
not containing any appreciable anti-matter may be positively
charged. This condition of some types of matter having
anti-matter, the electron, and some types of matter not having
anti-matter, the proton, may result in the attractive behavior
of two bodies, being similar to the way in which hot air always
flows towards cold air.

Referring to figure 7, we may consider the electron to be
an electro-magnetic pump which has a cycle passing from an
anti-matter region or low side in pump terminology, through a
boundary zone which would be considered the valve, and a matter
region or high side. A result of the pumping action, or cyclic
response, may be the generation of magnetic lines of force. The
magnetic lines of force may be expounded from the matter region
and return into the anti-matter region, thus forming a magnetic
axis. As the pump motion and action may follow a non linear
path, this may result in torque and spin of the electron. The
shape of the motion may be similar to the bell of a musical
trumpet or the flower of a petunia. In this suggested
explanation, starting at point (A) of figure 7,

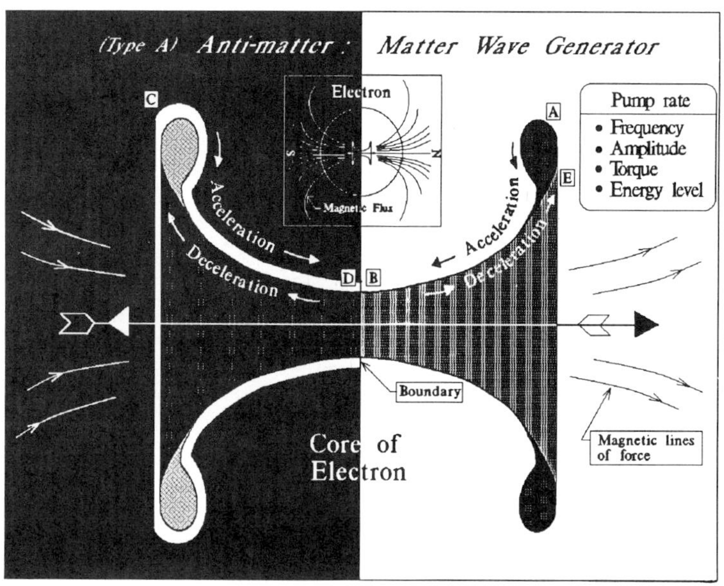

Fig 7. Electron as electro—magnetic pump

the wave may absorb some energy and accelerate down towards the
boundary zone, point (B), pass through the boundary and begin to
use up some of its energy, generating magnetic lines of force,
resulting in deceleration of the wave, point (C); the generated
magnetic lines of force may accelerate and drive the wave
through the transition zone, point (D). As the wave begins
absorbing energy, it begins to decelerate towards point (E)
possibly releasing plasma if the wave is growing and absorbing
plasma if the wave is receding. The wave reaches point (A)
where the cycle is repeated. The amplitude and frequency of the
wave and thus vibrational characteristics may be dependent on
the amount of energy exchange occurring between the shell, the
plasma, and the wave. This diagrammed in figure 8.

When an electron is in a particular energy level within an
atom, this may be another way of saying that the electrons
anti-matter:matter wave generator is operating within a certain
frequency band, within a certain amplitude range, within a
certain curve envelope, and confined to a certain axial length
of travel. The intent here is to be consistent with
Schrodingers wave equation(1). The higher the temperature the
more non uniform the wave reaction may become which may induce
increased levels of vibration. Further questions along this
line of thought are as follows: Does the wave have an
underlying frequency or subwave operating along the primary wave
that creates pumping action of the wave(figure 8)? Does the
subwave become more unstable or non uniform with increasing
temperature causing more random vibration? Are there different
types of anti-matter: matter generators, the electron being of
one type, type A, along the lines of blood where there is type
A, B, O, etc.?

Gyroscopic motion. Following the principles of gyroscopic
motion, the magnetic axis of the electron may vary its position
with respect to its surface at any given moment in time due

to external energy transfer(5). This would allow the electron
to most efficiently and quickly equalize the energy imbalance to
maintain stability.

     Spin velocity. Changes in plasma temperature which may
change the electrons size may result in changes in spin velocity
following the rules of moment of inertia of a mass that spins
about its own axis.

Characteristics of the magnetic electron bond

     Based on the postulates expressed for the electron, I will
next review and update discussion of the suggested magnetic
bond.

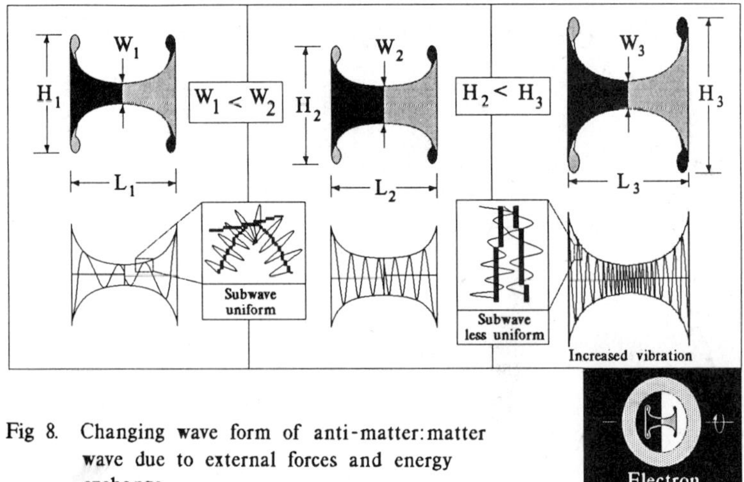

Fig 8. Changing wave form of anti-matter:matter
        wave due to external forces and energy
        exchange.

     Axis alignment. Referring to figure 9, two electrons that
have magnetically paired may align themselves along their
north-south magnetic axis, giving rise to one common axis. The
magnetic lines of force expounded by one electron would feed
into its paired electron. The mid section around the two
electrons may contain a magnetic ring similar to the Van Allen
belt around the Earth.

     Spin orientation and field strength. Magnetically paired
electrons may spin in the same direction with a net magnetic
adding effect (2). Pairs that spin with the same relative spin
velocity may exhibit a higher magnetic field than pairs of
differing velocity. This may result in a stronger magnetic bond
and may be analogous to the different covalent bond strengths
characterized by electron sharing. Synchronous pairs may
exhibit the strongest field and have magnetic lines of force
that are more continuous over time.

     Dipole link and chains. Pairs may link by dipole
attraction to form chains of higher magnetic fields analogous to
Van der Waals forces between molecules(1). The force of the
dipole link being much weaker than the magnetic bond. A chains
magnetic field strength may be higher than individual pairs due
to having more continuous magnetic lines of force over time and
closer inter-electronic spacing. A synchronous chain may
exhibit the strongest field possible for a given number of

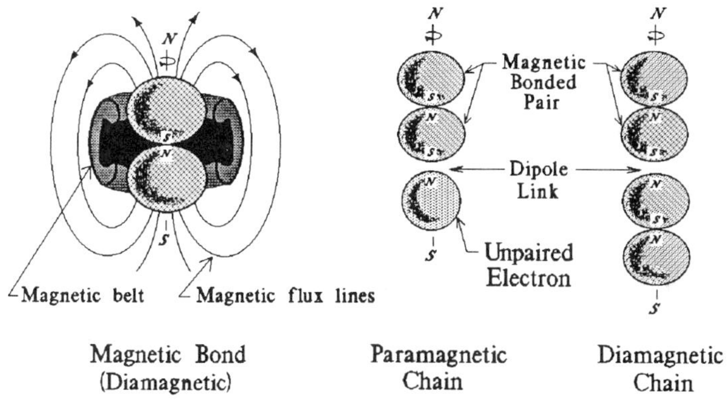

Fig 9. Diamagnetic electron bond and diamagnetic
and paramagnetic chain

electrons. If paired electrons chain via a dipole link, then a
linear structure would enhance the length of chains and field
strength. Following the principle of aligned fields, the
magnetic effects of these linear electron chains add up; the
longer the chain the larger the field may be(2).

If with lower temperature, the anti-matter:matter wave
reaction or flow becomes more uniform, whereby disrupting
vibrations have been reduced, electron pairing and chain
formation among the outer shell electrons may be more favorable.
In addition, these loosely held outer shell electrons may pair
with reduced temperature as the inner electron shield of atoms
stabilizes and intensifies. The shield may intensify and
stabilize with reduced temperature as the wave function of each
inner shell electron becomes more uniform, having lower
vibrational characteristics and energy, resulting in a reduction
of the average electron shell width and the atoms ionic radius.
These reductions would contribute to shield intensification.
Lets recall one of the key points of the angular momentum
quantum number stated earlier, with the same principal quantum
number, as the angular momentum quantum number increases, the
energy of the electron increases slightly. As we then remove
energy and reduce vibration of the electron, the amplitude of
the electron about a mean free path may be reduced.
Incorporating larger and larger elements into a material may
lessen the need for shield intensification through temperature
reduction as there is an increasing number of inner shell
electrons to provide shielding. Also, the larger elements may
create greater inter-atomic spacing for reduced charge
attraction(+,-).

A chain with all electrons paired may exhibit a diamagnetic
characteristic, being slightly repelled by a strong magnet, a
chain that has an electron unpaired on either end may be
paramagnetic, and slightly attracted to a strong magnet. This
is diagrammed in figure 9. A material may be diamagnetic or
paramagnetic depending on the percentage of each type of chain.
A chain may combine to form a ring if there is an unpaired
electron on each end (toroid). This may not be feasible in
ceramic materials, but may be possible in other materials, being
similar to the benzene ring formed by carbon atoms(6).

Under an electromotive force, the chains may move most
efficiently along linear channels where there are no
obstructions(7). Pathways that deviate from linear flow impart
inefficiencies, energy transfer that may overcome the weak
dipole links causing chains to separate, with an end result of
lower current density.

CONCLUSIONS

Recent discoveries of materials that superconduct at much
higher transition temperatures than previously recorded may lead
us to a new and better understanding of the conditions that lie
above and below the transition temperature at which a material
superconducts. This may result in an update of the present
quantum theory to incorporate ideas that are consistent,
enhance, and extend its fundamental suppositions. Among the
ideas expressed in this paper are that above the transition
temperature, the interrelationship of electrons with the nucleus
of atoms is dominated by charge effects giving rise to three
types of bonds, covalent, ionic, and metallic. Below the
transition temperature, the interrelationship may not exist
between the outer shell electrons and nuclei due to inner
electron shielding, giving rise to a magnetic bond among
electrons. The structure below the transition temperature may
include linear electron chains that may be highly dependent on
the physical make up of the electron. The electron may have at
its core an anti-matter: matter wave generator that determines
when and how electrons combine to form a magnetic bond and when
and how electrons combine to form covalent, ionic, and metallic
bonds. If a magnetic electron bond exists between the loosely
held outer shell electrons below a materials superconducting
transition temperature, we may study the periodic table of
elements in search of elements that when properly combined may
have characteristics that maximize stabilization along with
providing higher current densities at higher transition
temperatures.

REFERENCES

1. Masterton, W.L., and Slowinski, E.J., Chemical
Principles, 3rd ed., Saunders, Philadelphia, 1973, pp. 25-229.
2. Smith, R.J., Circuits, Devices, and Systems, 3rd ed.,
Wiley, New York, 1976, pp. 158-162, 549-565.
3. Webster, N., Webster's Unabridged Dictionary, Deluxe
2nd ed., Simon & Schuster, New York, 1983, pp. 206
4. Schoenung, S.M., Superconductivity Applications and
Developments, MD-Vol. 11 #G00440, The American Society of
Mechanical Engineers, New York, 1988, pp. 61-67.
5. Baumeister, T., Avallone, E.A., and Baumeister III, T.,
Standard Handbook for Mechanical Engineers, 8th ed.,
McGraw-Hill, New York, 1978, pp. 3-2-3-23, 6-150, 15-2-15-84.
6. Streitweiser Jr., A., and Heathcock, C.H., Introduction
to Organic Chemistry, 1st ed., Macmillan, New York, 1976, pp.
5-37, 169-175, 325-331, 593, 611-615.
7. Fitzgerald, K., "Superconductivity: Fact vs Fancy,"
IEEE Spectrum, May 1988, pp. 30-41.

R. C. Bossert, R. C. Niemann, J. A. Carson, W. L. Ramstein, M. P. Reynolds and N. H. Engler

Fermi National Accelerator Laboratory
P.O. Box 500
Batavia, IL 60510

ABSTRACT

Installation of superconducting accelerator dipole and quadrupole magnets and spool pieces in the SSC tunnel requires the interconnection of the cryostats. The connections are both of an electrical and mechanical nature. The details of the mechanical connections are presented. The connections include piping, thermal shields and insulation. There are seven piping systems to be connected. These systems must carry cryogenic fluids at various pressures or maintain vacuum and must be consistently leak tight. The interconnection region must be able to expand and contract as magnets change in length while cooling and warming. The heat leak characteristics of the interconnection region must be comparable to that of the body of the magnet. Rapid assembly and disassembly is required. The magnet cryostat development program is discussed. Results of quality control testing are reported. Results of making full scale interconnections under magnet test situations are reviewed.

SSC MAGNET MECHANICAL INTERCONNECTIONS

Introduction

Several full size SSC model magnets have been assembled at Fermilab.[1,2] The interconnection area is an integral part of these magnets. They have been connected to test stands and to each other in string tests.

Procedure

The interconnection is developed from the beam tube outward. The seven mechanical connections to be made are shown in Fig. 1.

Power leads and electical connections inside the cold mass are made first. The cryostat connections are then made. Order of connections with their estimated times to make in a tunnel situation are shown in Table 1:

Table 1.   Interconnection Procedure

Connection:

1.) Weld Beam Tube ------------	setup - 5 minutes weld - 2.5 minutes each total (2 welds) - 15 minutes
2.) Weld Cold Mass Bellows ------	setup - 5 minutes weld - 15 minutes total (2 welds) - 40 minutes
3.) Weld Shield and Return Pipes -	setup - 2.5 minutes each weld - 2.5 minutes each total (8 welds) - 40 minutes
4.) Mount, and Insulate Interconnection Shields ------	mounting - 5 minutes attaching - 10 minutes insulating - 5 minutes total (2 shields)-40 minutes
6.) Weld Vacuum Vessel --------	setup - 5 minutes weld time - 40 minutes total (2 welds) - 90 minutes

<u>Total Connection time</u>:   3 hours, 45 minutes

Disconnection:

1.) Break Vacuum Connection -------	setup time - 5 minutes cutting time - 30 minutes cleanup time - 5 minutes total (2 cuts) - 80 minutes
2.) Remove Shields and Insulation -----------------	remove shield - 10 minutes remove insulation - 5 min. total (2 shields) - 30 min.

3.) Break Shield and Return
    Pipes --------------------
	setup time - 5 minutes
	cutting time - 5 minutes
	cleanup time - 3 minutes
	total time (8 cuts) -
	         104 minutes

4.) Break Cold Mass Connection -----	setup time - 5 minutes cutting time - 15 minutes cleanup time - 5 minutes total (2 welds) - 50 minutes
5.) Break Beam Tube Connection ----	setup time - 5 minutes cutting time - 5 minute cleanup time - 5 minutes total (2 welds) - 30 minutes

<u>Total disconnection time</u>: 4 hours, 54 minutes

It is obvious that a substantial amount of time will be devoted to making interconnections. Efficiency in making and breaking these connections is therefore critical.

Piping System Connections

A. Beam Tube. The beam tube is the first connection to be made. It consists of a bellows assembly with a stainless steel flange on each end. This assembly consists of two bellows of different diameters with an isolating vacuum between them. This extra vacuum space isolates the internal beam tube area from the liquid helium space, insuring a seal even if one bellows should fail. The bellows assembly is inserted between two mating flanges on the beam tubes and welded with a computer controlled weld head. Two welds are made, one on each end of the bellows. These welds are made by consuming disk shaped washers which are placed between the flanges (shown in cross section in Fig. 2). The area internal to the weld is purged with an inert gas. Nitrogen is being used as the purge gas for the beam tube connection.

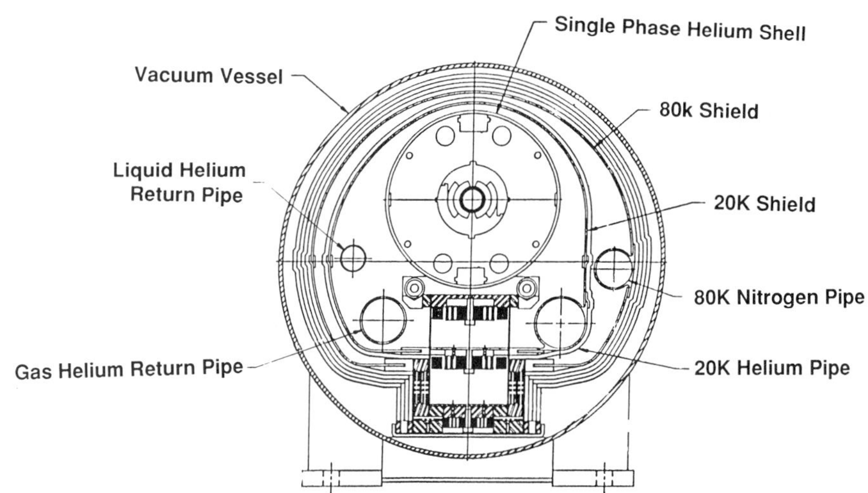

The Interconnection is developed from the beam tube outward. There are seven mechanical connections to be made.

Fig. 1. Seven cryostat connections

The weld system consists of an orbital water-cooled welding head with a 220 V power supply. The head is made to clamp onto a pipe as shown (see Fig. 3). With minor modifications to the collets it can be made to clamp to the beam tube flanges. The tungsten electrode is driven by a rotating gear. The weld cycle has been previously programmed into the unit so that the operator has only to press a button to run the appropriate program. A typical weld program includes times and currents for the initial tack welds, upward current ramp, weld cycle, and downward current ramp. The weld cycle includes time, current and motor speeds for both the high and low end of the "pulsed" arc. Up to 99 steps (called "levels"), are available. Pre and post purge times are also programmable. A typical weld program is shown in Table 2.

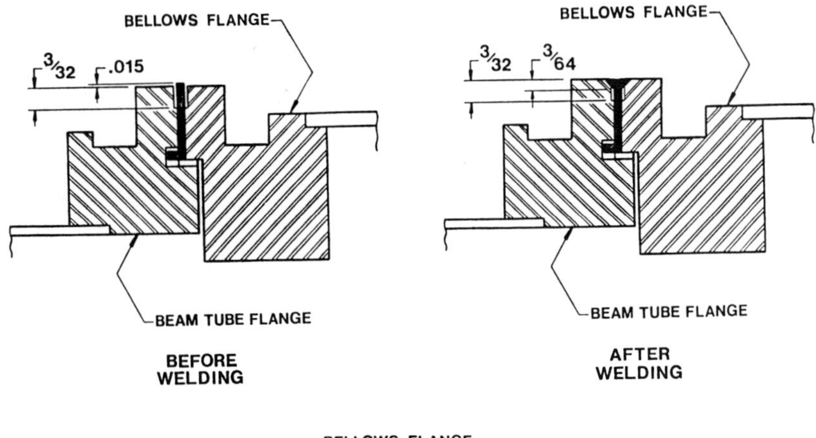

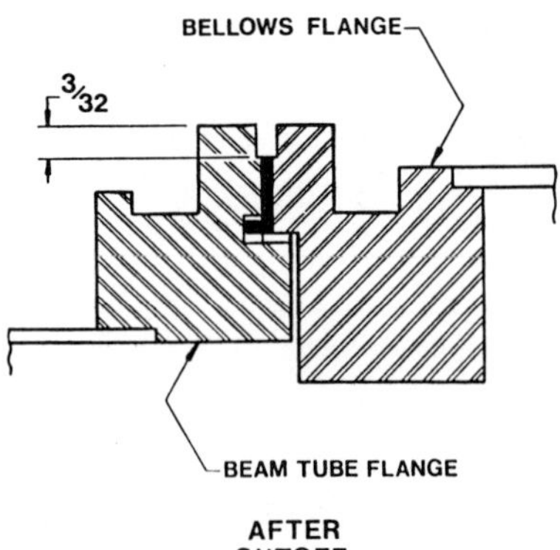

Fig. 2. Beam tube weld

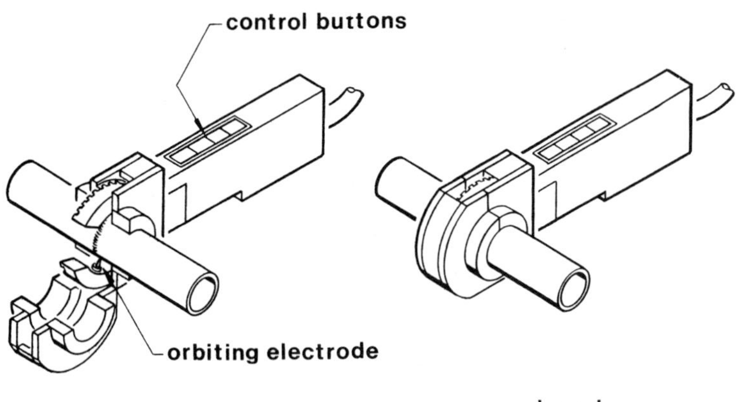

Fig. 3. Beam tube weld head

Table 2.

Weld Procedure #1625-10/0   1 5/8 Sleeve S.S. Fermi SSC
Pre-purge gas 35.00 sec
Strike Current 20.00 amp   Post-purge   20.00 sec

| LEVEL NO. SEC. | LEVEL TIME SEC. | CURRENT SLOPE SEC. | MOTOR SLOPE SEC. | TIME SEC. | --------HIGH-------- | | TIME SEC. | --------LOW-------- | |
					CURRENT AMP	MOTOR RPM		CURRENT AMP	MOTOR RPM
1	14.56	7.00	4.00	O.28	89.0	0.71	0.28	27.0	0.71
2	14.56	7.00	0.00	0.28	88.6	0.71	0.28	27.0	0.71
3	14.56	7.00	0.00	0.28	88.0	0.71	0.28	27.0	0.71
4	14.56	7.00	0.00	0.28	87.6	0.71	0.28	27.0	0.71
5	14.56	7.00	0.00	0.28	87.0	0.71	0.28	27.0	0.71
6	14.56	7.00	0.00	0.28	86.6	0.71	0.28	27.0	0.71
7	14.56	7.00	0.00	0.28	86.0	0.71	0.28	27.0	0.71
8	12.00	12.00	12.00	0.15	48.0	1.42	0.15	27.0	1.42
FINAL	6.00	6.00	6.00	0.12	8.0	2.60	0.12	4.00	2.60

The beam tube is removed by cutting out the section of the consumable washer which is fused to the flanges. This cut is made by an orbital pipe cutter as shown in Fig. 4.   The cutter is a two piece unit which clamps onto the pipe. It is specially designed to fit the low clearance application, as are the cutters for all the other cryostat connections. A cutting tool similar to that used on a lathe revolves around the pipe. The tool bit is attached to a ring gear which is driven by a variable speed motor. A cam feed system allows the operator to vary the tool feed from 0 to .006 inches per revolution.   Both pneumatic and hydraulic motors are available. Power requiremints for air are 35 cfm at 80-100 psi. The hydraulic requirements are 4-5 gpm at 1500 psi. Air is presently being used at the test facility. It has not yet been determined which medium will be used in the SSC tunnel.

Several beam tube bellows have been welded and cut off from test stands. Clearance problems for the weld head and pipe cutter have at times been encountered, and have made hand welding and grinding necessary.   We have found that alignment is critical to the welding unit. Alignment problems have not been encountered on the test stand, since the beam tube connection is used as the reference to align the magnet to the stand. Compressing the bellows to the proper length for insertion into the opening has also proved to be more difficult than was anticipated. Mechanical fixtures work, but are cumbersome and difficult to use.

B. Cold Mass Connection. The cold mass connection is then made. The bellows is stored on the return end of the cold mass. The technician must slide the bellows into position for welding. Bellows and sleeves are stainless steel. Again there are two welds to be made, one on each side of the cold mass bellows. These are fillet welds, with filler material to be added (see Fig. 5). The weld head for this connection is water cooled with a tungsten torch. The internal purge gas is helium. The entire weld head revolves around the pipe (see Fig. 6). It is driven by a chain which is mounted to a guide ring. The guide ring is firmly clamped to the pipe (cold mass). An automatic arc gap control keeps the tungsten electrode always at the proper distance from the weld. Filler material is fed at a controlled rate from a metal spool. All weld parameters are pre-programmed as in the beam tube weld.

The cold mass bellows is removed by cutting out the fillet welds and sliding the bellows back into the storage position. An orbital pipe cutter similar to the beam tube cutter is used to remove the welds. Whereas the beam tube cutter feeds in the radial direction (like a lathe cutoff tool), the cold mass cutter feeds in the longitudinal direction (like a turning operation). It is clamped to the end sleeve of the cold mass and runs concentric with the sleeve within .005 inches. This is achieved by use of a spring-loaded tracking module. This cut is made without lubrication, as are all the cutoff operations in the interconnection region. Air requirements for this cutter are 85 cfm at 80-100 psi. Hydraulic requirements are 8-10 gpm at 1500 psi.

Cold mass bellows have been welded and cut off successfully with the automatic equipment in an interconnection mockup at Fermilab. Clearance problems have resulted in an inability to use the longitudinally feeding pipe cutter for the cold mass on the test stands. This problem will be resolved in the next cryostat iteration.

C. Shield and Return Pipes. Two shield and two helium return pipes surround the cold mass (see Fig. 1). These pipes range from 1.90 to 3.5 inches in ouside diameter. They are all welded and cut off in the same manner. The bellows for these pipes are permanently attached to the lead end of the magnet. They are welded on the return side by using a "sleeve" assembly as is shown in Fig. 7. The corner of the sleeve is melted and fused to the pipe. The sleeve serves as a consumable

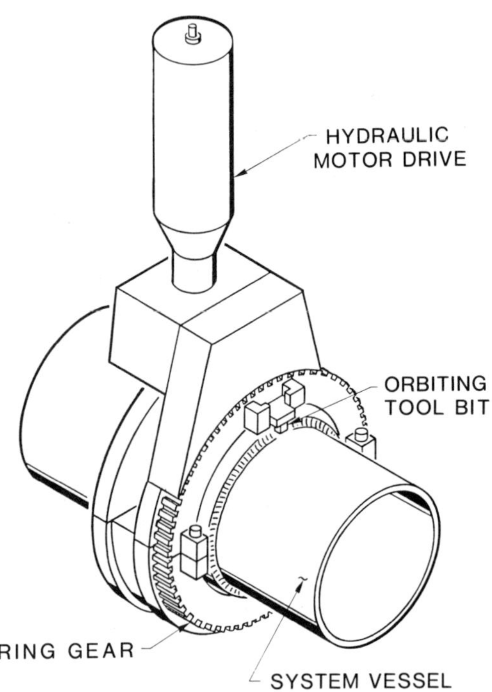

Fig. 4. Beam tube weld cutter

material for welding as well as a mechanism for aligning the pipes for the welding unit. It is stored on the bellows end flange. The technician slides the sleeve over the gap between the bellows and it's mating pipe on the return end of the magnet. A new collet must be inserted into the same weld head that was used for the beam tube. The weld head is then clamped in the proper position on the pipe. A weld program designed for the specific joint is called. Purge gas is helium for the 20K shield and helium return pipes and nitrogen for the liquid nitrogen line. After welding, the head is placed on the other end of the sleeve, and the operation is repeated. The welds are removed by longitudinal-feed orbital pipe cutters. These cutters are identical in operation to the one that is used for the cold mass, but similar in size to the one used for the beam tube. Air and hydraulic requirements are the same as for the beam tube cutter. Figure 8. shows the SSC Interconnection area with a small orbital pipe cutter mounted in position on a helium return pipe.

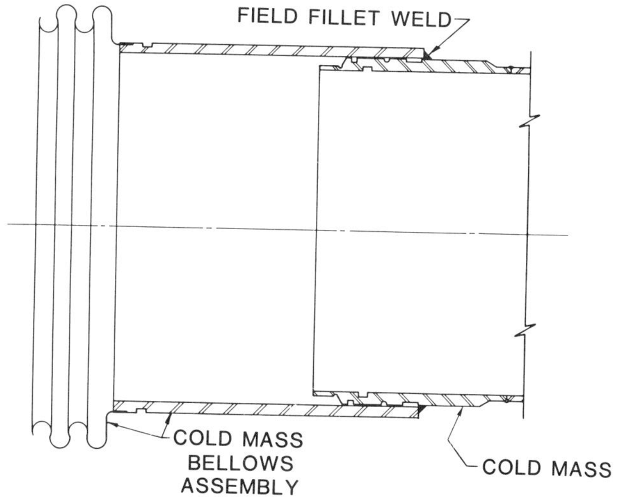

Fig. 5.   Cold Mass Weld.

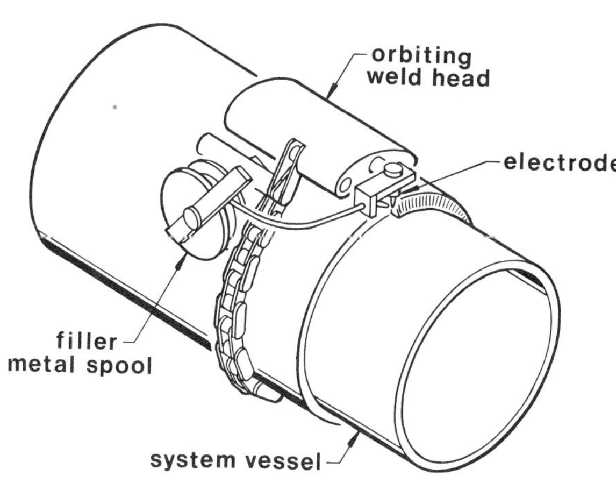

Fig. 6.   Cold Mass Welding Unit

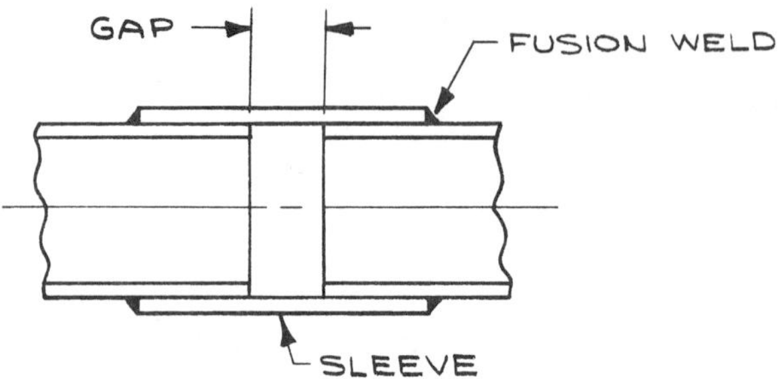

Fig. 7. Shield and return pipe weld

Many welds using the sleeve assembly have been made and cut apart at Fermilab, both on and off the test stand. This method has several desirable features. The sleeve serves to align the pipes for welding, eliminating the problem of weld programs shutting down due to pipe misalignment. Bellows compression is no longer necessary, because it does not need to be compressed into a specific slot as is the beam tube bellows. Longitudinal tolerances in pipe position are taken up by the gap between pipes, allowing a smaller bellows travel to be specified.

D. Vacuum Shell. The vacuum shell connection is shown in Fig. 9. The weld and cutoff prodeedure is similar to the cold mass connection, except that the shell material is carbon steel instead of stainless. The bellows is stored on the return end of the vacuum vessel. The technician slides it into position. The same weld head is used as the cold mass. The appropriate weld program is called, and the fillet weld is made. The weld is removed by a longitudinal-feed pipe cutter of the proper diameter. Air and hydraulic requirements are the same as that for the cold mass cutter.

Fig. 8. SSC interconnection area

No vacuum vessel connections have yet been made by this procedure. An o-ring connection is presently being used at the Fermilab cold magnet test facility.

Bellows

Bellows are required for all interconnection joints. The bellows are installed in their compressed state and expand to compensate for the contraction of the cryostat when cooled down. They also accomodate any differences in position, both axially and radially, of the magnet ends. They must function at the temperatures dictated by their particular system. All but the vacuum bellows must be non-magnetic. They must satisfy the SSC magnet requirement of 20 thermal cycles with a reasonable safety factor. Specifications concerning pressure, longitudinal travel, axial offsets, and physical dimensions must be met. Table 3 lists the specifications for the interconnection bellows presently being used.

Table 3.   Bellows Specifications.

System	Inside Diameter	Outside Diameter	Travel	Maximum Pressure	Maximum Axial Offset	Minimum Life Cycles
Beam Tube	1.600	3.125	3.0	300 psi ext. 1*10e11 torr int.	not specified	1000
LHe Ret.	1.50	1.875	2.125	300 psi int. 0 external	.100	1000
80K Shield	2.375	3.250	3.125	300 psi int. 0 external	.100	1000
He Gas Ret.	2.709	3.500	2.125	300 psi int. 0 external	.100	1000
20K Shield	3.000	3.875	3.125	300 psi int. 0 external	.100	1000
Cold Mass	11.80	12.80	2.125	300 psi int. 0 external	.040	1000
Vacuum Vessel	26.00	28.00	1.37	75 psi int. 15 psi ext.	.200	not specified

(all dimensions in inches)

Bellows are made of 316L stainless steel. This material offers a good combination of mechanical strength, weldability and corrosion resistance.

A quality assurance testing program for SSC bellows and welded joints has begun. Testing of welded joints has not yet taken place. One bellows of each type except the vacuum bellows has been tested. The bellows that were tested were made of 321 stainless steel. An earlier cryostat design used this material. The 316L should have improved extension-compression cycle life. A more comprehensive test program using the 316L bellows will be initiated. Twenty samples of each type will be tested. After this testing is complete, the test program will be reevaluated. Table 4 shows the procedure and results of the testing completed to date.

Table 4.   Bellows Testing.

Proceedure:

1.) Initial Leak Check
2.) 50 cycles at nitrogen temperature
3.) Vacuum Leak Check
4.) Hydrostatically pressurized to 20
        atmospheres internally
5.) Hydrostatically pressurized to 25
        atmospheres internally
6.) Vacuum Leak Check
7.) Repeat steps 2 through 6 until bellows
        fails or until 1000 cycles is reached.

Results:

Bellows	Cycles to Failure	Bellows Material
LHe Return	175	321 Stainless
He Gas Return	459	321 Stainless
20K Shield	No Failure at 600	321 Stainless
80K Shield	100	321 Stainless
Cold Mass	No Failure at 100	321 Stainless

Interconnection Shields

The 20K and 80K thermal shields must continue through the interconnection area. Thermal shield "bridges" are installed for this purpose. The shield bridges are mounted as shown in Fig. 10. The top half is placed over the space between the two magnet shields. It is supported by the magnet shields. The bottom half is then hinged into the top half as shown. Predrilled rivet holes are matched on the other side. Rivets then hold the halves together. The shield bridge is "thermally attached" to the magnet shield on one side and overlapped, but not attached on the other. The thermal attachment presently consists of a series of small fillet welds as shown. A crimped connection is under consideration for future models. The overlap allows the shields to move with respect to each other when they contract by 2.75 inches during cooldown.

The shield bridges are each made of .062 inch thick 1100-0 aluminum. This material is very soft and easy to form. It is also extremely high in thermal conductivity at 20 and 80 K. Since the shield bridges are thermally attached on one side only, and not attached to the 20K and 80K pipes, the temperature gradient through them is different than that on the magnet shields. It is longitudinal, and significantly longer than the azimuthal path for the shields inside the magnet. A material with higher conductivity than the 6061-T6 aluminum of the magnet shields is therefore necessary. Although not strong enough to use for the magnet shields, 1100-0 aluminum is structurally sufficient for the short bridge in the interconnection area.

Insulation

The shield bridges must be insulated. They are insulated in exactly the same manner as the shields inside the magnet.[3] Junctions between the magnet and the interconnection insulation blankets are made by aluminum tape. The blanket junction is made in a "stairstep" configuration as shown in Fig. 11. No provision for magnet shrinkage in the interconnection area is necessary. This has been accomodated for in the cryostat body.

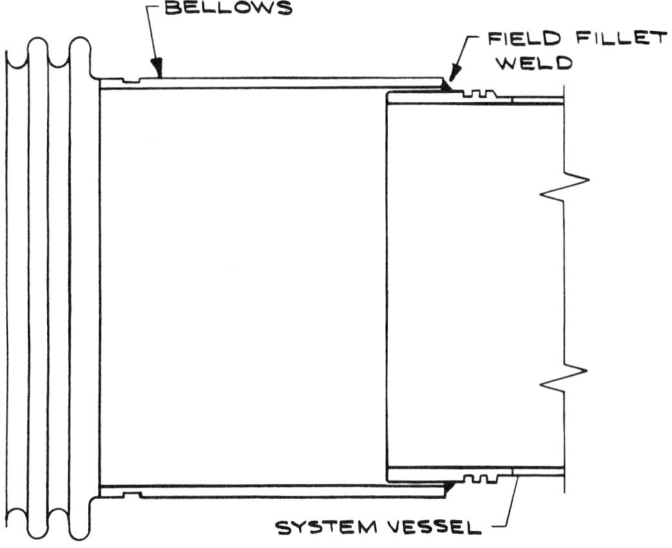

Fig. 9. Vacuum shell connection

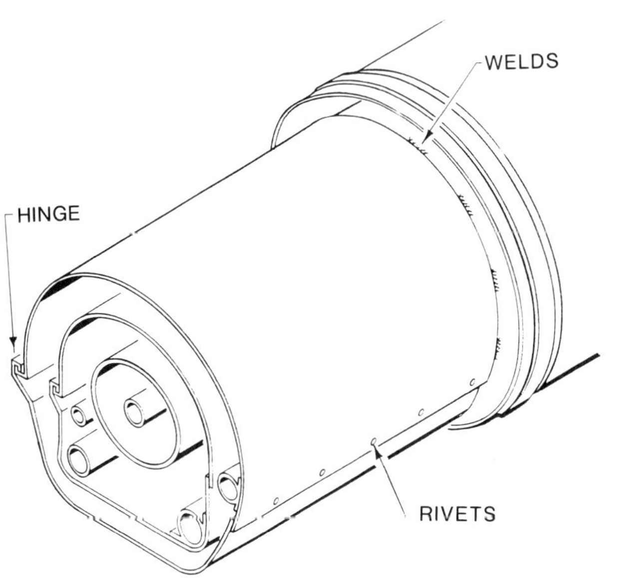

Fig, 10. Interconnection Shields

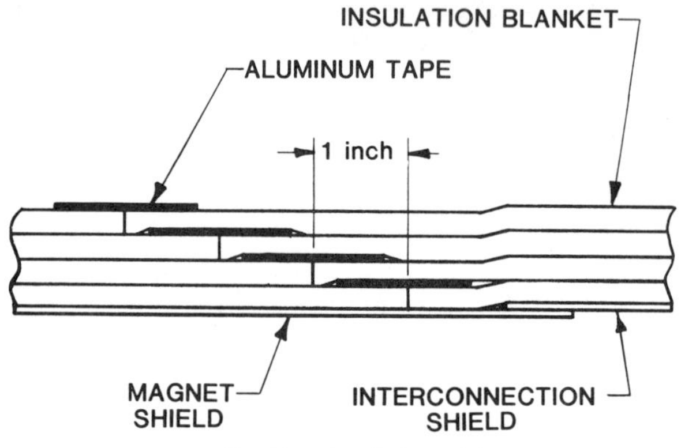

Fig. 11.   Insulation Joint

Interconnection Model

Fermilab constructed a full size interconnection model for the initial cryostat design. It has been used to practice connecting and disconnecting magnets. The model has several flaws: 1.) It is not located in a full size tunnel cross section. 2.) It consists only of one magnet, not two stacked vertically as is in the real tunnel. 3.) No electrical connections are included in the model. 4.) It is a static model, representing only the warm, installed position of the magnet.

The model is fine for practicing welding and cutoff. It does not, however, have the ability to replicate the movements within the interconnection due to thermal contraction. Design of a new model which corrects these deficiencies is underway.

CONCLUSION

The magnet interconnection area consists of many interrelated systems. The initial design and testing has been completed. Continuing development is necessary including verification of times involved in making interconnections, testing of components, and leak checking under tunnel conditions.

REFERENCES

1. Niemann, R.C., et al., "Superconducting Super Collider Second Generation Dipole Magnet Cryostat Design," Applied Superconductivity Conference, August 22-25, 1988, San Francisco, Ca.

2. Engler, N.H., et al., "SSC Dipole Magnet Model Construction Experinece," presented at the 1987 Cryogenic Engineering Conference, St. Charles, IL, June 14-18, 1987.

3. Gonczy, J.D., et al., "Multilayer Insulation (MLI) in the Superconducting Super Collider a Practical Engineering Approach to Physical Parameters Governing MLI Thermal Performance," presented at the International Superconducting Super Collider Symposium, New Orleans, LA, February 8-10, 1989.

MULTILAYER INSULATION (MLI) IN THE SUPERCONDUCTING SUPER
COLLIDER - A PRACTICAL ENGINEERING APPROACH TO PHYSICAL
PARAMETERS GOVERNING MLI THERMAL PERFORMANCE

J.D. Gonczy,   W.N. Boroski, and R.C. Niemann

Fermi National Accelerator Laboratory
P.O. Box 500
Batavia, Illinois 60510

ABSTRACT

Multilayer insulation (MLI) is employed in cryogenic devices to control
the heat load of those devices. The physics defining the thermal performance
of an MLI system is extremely complex due to the thermal dynamics of
numerous interdependent parameters which in themselves contribute differently
depending on whether boundary conditions are transient or steady-state. The
Multilayer Insulation system for the Superconducting Super Collider (SSC)
consists of full cryostat length assemblies of aluminized polyester film,
fabricated in the form of blankets, and installed as blankets to the 4.5K cold
mass, and the 20K and 80K thermal radiation shields. Approximately 40,000
blankets will be required in the 10,000 cryogenic devices comprising the SSC
accelerator. Each blanket will be nearly 56 feet long by 6 feet wide and will
consist of as many as 32 reflective and 31 spacer layers of material.
Discussed are MLI material choices, and the physical parameters which
contribute to the operational performance of MLI systems. Disclosed is a
method for fabricating MLI blankets by employing a large diameter winding
mandrel having a circumference sufficient for the required blanket length.
The blanket fabrication method assures consistency in mass produced MLI
blankets by providing positive control of the dimensional parameters which
contribute to the MLI blanket thermal performance. The fabrication method
can be used to mass produce prefabricated MLI blankets that by virtue of
the product have inherent features of dimensional stability, three-dimensional
uniformity, controlled layer density, layer-to-layer registration, interlayer
cleanliness, and interlayer material to accommodate thermal contraction
differences.

INTRODUCTION

The objective of the SSC MLI system is to limit cryostat heat leak
from thermal radiation and residual gas conduction to the low levels specified
by the SSC magnet systems design criteria.[1] The MLI system must maintain
these low heat leak levels during the 20 plus years lifetime of the SSC while
in an environment of nuclear irradiation and for extended periods of
operational upset conditions including thermal cycles and poor insulating
vacuum.

Crucial to the development of an MLI system for the SSC is a solid understanding of the characteristics which contribute to the overall performance of the MLI system. One must assess each parameter's influence on the performance level of the MLI, and under what operating conditions, if any, are the dominate parameters made less prevalent. Laboratory measurements continue to work toward resolving these questions by attempting to quantify the components of heat transfer in MLI. However, a definitive MLI system has yet to be designed, and there continues to be philosophical differences between partisans of specific MLI systems; each advocate having his particular system certified with laboratory results. While each physically different MLI system offers specialized performance characteristics, many other desirable traits are sacrificed. The final MLI solution may be found in a hybrid compromise of many MLI systems.

Equal in importance with laboratory measurements are concerns regarding the installation of the MLI system in the SSC cryostat. These concerns are generally recognized during cryostat operation as differences between the calculated heat leak to the cryostat, based on laboratory measurements, and the actual heat leak measured in the field. At the root of these differences is the relative inability to duplicate, in a practical manner at installation, the same MLI geometry and boundary conditions as was tested in the laboratory.

Historically, heat leak differences that varied by factors of 2 or 3 were regarded as MLI systems having good results. However, this celebrated view of heat leak in an aggregate of 10,000 SSC cryostats is not affordable, since the ultimate operating cost of the SSC depends principally on the ability to prevent heat leak to the cryogens. An essential requirement of the SSC MLI system is that the MLI geometry measured in the laboratory is, in fact, duplicated 40,000 times in the 10,000 SSC cryogenic devices.

Apart from the laboratory measurements, the challenges to developing a practical MLI system for the SSC which is both predictable and consistent in thermal performance are: 1) to design an MLI system that lends itself to mass fabrication and mass installation techniques, and which automates decision making during installation of the MLI blanket in the SSC cryostat; 2) to develop a fabrication method that uses ordinary production skills; and 3) to develop an apparatus to achieve the blanket design in a cost effective manner. These challenges have been accomplished and are described below.

SSC MAGNET CRYOSTAT OVERVIEW

<u>Major Cryostat Components</u> - The major components of the SSC cryostats are the cryogenic piping, the cold mass assembly which includes the superconducting magnets, the thermal radiation shields, the support system for suspending the cold mass assembly and thermal shields, the multi-layer insulation system, the vacuum vessel, and the interconnections between cryostats.[2] See Fig. 1.

In each SSC cryostat, the cold mass assembly housing the superconducting magnets is surrounded by several regions of progressively higher temperature. The region directly surrounding the cold mass assembly is the 4.5K region which is cooled by cryogenic piping containing liquid helium at 4.3K. An MLI blanket containing five reflective layers is spiral wrapped around the main body sections of the cold mass to a thickness of ten layers.

A second region known as the 20K region surrounds the 4.5K region. The 20K region is cooled by cryogenic piping containing gaseous helium at 20K. The 20K region is bounded by an aluminum thermal shield around which an MLI blanket assembly of thirty-two reflective layers is wrapped.

A third region known as the 80K region surrounds the 20K thermal shield. The 80K region is cooled by cryogenic piping containing liquid nitrogen at 77K. Bounding the 80K region is another thermal shield formed from aluminum sheet. Two separately installed MLI blanket assemblies each consisting of thirty-two reflective layers are wrapped around the 80K shield.

A vacuum containment vessel at room temperature (300K) surrounds the 80K shield.

PARAMETERS AFFECTING MLI HEAT TRANSFER

<u>Radiation Heat Leak and Thermal Boundaries</u> - Heat transfer by thermal radiation is from hot to cold, and varies directly as the fourth power temperature difference between the communicating surfaces. Thermal radiation also varies directly with a material property called surface emissivity. Shiny aluminized surfaces like those used in MLI reflect substantially more thermal radiation than they absorb; and for heat flowing through the MLI to become constant, the heat that is absorbed by each MLI layer must be radiated away or emitted. Since the reflective aluminized layer absorbs a relatively low level of thermal radiation, it consequently radiates thermal energy at a low level. The physical characteristics and surface condition of the material which combine to reflect thermal radiation, and therefore to emit heat at low levels, is described as a measure of the emissivity of the surface. As such, low emissivity surfaces are good inhibitors of thermal radiation. The surface emissivity of the MLI reflector is a quality-controlled property of the MLI and is described to the MLI vendor in a materials specification.

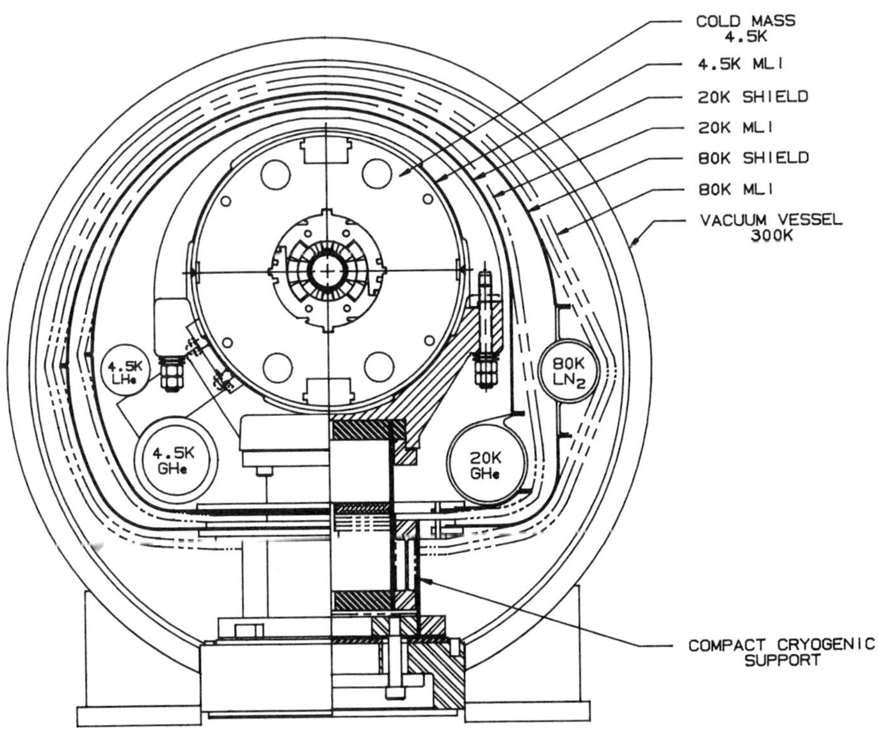

Fig. 1. SSC Cryostat Cross-section

A functional MLI system consists of many layers of low emissivity surfaces which reflect thermal energy by providing multiple adjacent layers of low emissivity surfaces which are relatively close in temperature. The layers also serve as a barrier to impede gas conduction. A low thermal conductive material is often used as a spacer to distance each reflective layer as the reflective layers are installed uniformly to encapsulate the cold surface. The heat flux through the MLI varies inversely with the number of reflective layers, and is a function of the layer density or number of reflective layers per unit thickness of MLI. MLI systems are more cost-effective when installed as multilayer blankets rather than as many individually installed layers.

<u>Insulating Vacuum and Residual Gas Conduction</u> - A portion of the heat that is transferred in a cryostat occurs by means of residual gas conduction. To limit the heat leak contribution by gas conduction to a negligible amount, the concentration of gas molecules is reduced by evacuation of the gases to an insulating vacuum level. At this insulating vacuum level, the residual gases because of their reduced number are more likely to collide with the surfaces of their confinement boundaries and not so much with each other. Heat transfer is conducted by individual gas molecules traveling between hot and cold confinements, and the amount of heat transferred by this molecular flow varies directly with the number of gas molecules present as measured by the residual gas pressure.

A paramount objective of the SSC design criteria is the ability to establish and maintain a stable insulating vacuum; and since the main impediment to a credible vacuum is the evacuation of interstitial blanket spaces, serious thought regarding evacuation was given to the blanket design. In spite of an apparent advantage to the evacuation process that a perforated MLI blanket might offer with its larger conductances for gas flow, it is noted that as the evacuated gases approach molecular flow, the perforations become ineffectual. It has been demonstrated that interstitial pressures in perforated MLI remain above 1×10^{-4} mmHg during room temperature evacuations.[3] The judgement at this time is for non-perforated reflective materials instead of the perforated for the SSC blanket design, although the design option to perforate part or all of future MLI systems for the SSC remains open. The considerations here are twofold. Firstly, the cost of perforating the MLI can be as much as one and a half times the cost of the aluminized material itself. And secondly, the SSC cryostats are certain to experience vacuum perturbations and perhaps situations with prolonged poor vacuum conditions due to leaks.

The initial disadvantage of a longer pump-down time for a non-perforated MLI has been discounted in favor of the long term advantages of the blanket in providing greater resistance to upset conditions. Granted, the initial pump-down time would be longer with the non-perforated blanket due to its poor evacuation conductances, but there is ample time to establish the pre-cooldown vacuum as the 53 miles of accelerator rings are put in place. By similar reasoning, the non-perforated blanket would show less sensitivity to transient vacuum conditions in that the diffusion of gases through a non-perforated blanket would take longer because of the poor conductance to gaseous flow in the reverse direction. An objective of the MLI blanket design is to foster a dampening effect to heat flow so that critical time required to recover from an upset condition is increased.

Table 1 lists for comparison some important material property considerations for MLI.

Table 1. Comparison of MLI Material Properties

	POLYESTER	POLYIMIDE	POLYAMIDE
TRADE NAME[†]	MYLAR, REEMAY, DACRON	KAPTON	NYLON, CEREX
MOISTURE ABSORPTION [7,8,9]	0.4% @ 50% Rh REEMAY SPUNBONDED 0.5% @ 98% Rh	1.3% @ 50% Rh ————	8.0% @ 50% Rh CEREX SPUNBONDED 3–5% @ 95% Rh
OUTGASSING [4]	DACRON NET 1.1 × 10^{-4} g/g DOUBLE ALUMINIZED MYLAR 2.6 × 10^{-3} g/g	———— DOUBLE ALUMINIZED KAPTON 3.1 × 10^{-3} g/g 3.9 × 10^{-3} g/g	NYLON NET 4.0 × 10^{-2} g/g ————
RADIATION DAMAGE [6,7]	5.7 × 10^{8} RAD (50% MAX. MECHANICAL)	5.0 × 10^{9} RAD (50% MAX. MECHANICAL)	7.0 × 10^{7} RAD (50% MAX. MECHANICAL)
IRRADIATION GAS [6,7]	3–5 ml/g @ 10^{9} RAD H_2(70%), CO_2(20%), CO(10%)	————	20–25 ml/g @ 10^{9} RAD H_2(52%), CO(20%), CO_2(12%), N_2(8%), O_2(3%)

[†] DUPONT DE NEMOURS & CO; REEMAY INC; JAMES RIVER CORP.

<u>Outgassing and Irradiation Gas Evolution</u> - In a vacuum, gases are boiled from material surfaces in what is called "normal outgassing" of those surfaces. Since all cryostat components including the MLI must be maintained in an insulating vacuum, cleanliness constraints must be asserted to keep the vacuum space substantially free of contaminants. Excessive boiling from high vapor pressure contaminants in the vacuum system will increase the time and energy necessary to establish the cryostat design vacuum. During evacuation at room temperature, the major gas constituent of normal outgassing is water vapor which comprises approximately 95% of the evolved gases.[4] Gas evolving from materials during nuclear irradiation (operation of the accelerator) is also a significant consideration which must be addressed when choosing materials to be used in the cryostat vacuum envelope.

<u>MLI Material Choices</u> - The plastic materials used in the SSC blanket design are comprised entirely of polyester plastics. See Table No. 2. The blanket incorporates 32-reflective layers of double aluminized polyethylene terepththalate (PET) film, each separated by a single thin spacer layer of spunbonded PET material. A thick layer of spunbonded PET material covers the blanket top and bottom, and positions polyester hook and loop fasteners at the blanket edges. The fasteners are secured to the cover layer by sewing. The multiple blanket layers and cover layers are sewn together as an assembly along both blanket edges. Polyester thread is used in all sewing processes.

<u>Dimensional Parameters Considered by the Blanket Design</u> - Each MLI blanket for the SSC will be fitted with openings through which the support system for the cold mass and shield assemblies will penetrate. These holes also serve as evacuation ports which access the interstitial spaces between the many blanket layers. Interlayer registration of the holes through the MLI blanket must be maintained during the logistics of fabrication, handling, and installation of the 56 feet long by 6 feet wide MLI blankets. Also critical to the thermal performance of the MLI system is a uniform MLI layer density. Uniform layer density using previous MLI fabrication techniques has been extremely difficult to maintain at installation. Proper alignment of the blanket layers in the length and width dimensions and control of MLI layer density is assured by means of sewing together the multiple layers of the blanket along each blanket edge for the entire length of the blanket.

Table 2. SSC MLI Blanket Materials

BLANKET COMPONENT	MATERIAL	DESCRIPTION	COMPANY
REFLECTOR	ALUMINUM via VACUUM DEPOSITION	ALUMINIZED METAL COATING; TWO SIDES 350 ANGSTROMS THICK, D.C. RESISTANCE OHMS/SQUARE = 0.9 OHMS TOLERANCE (+0.1)(−0.0) EMISSIVITY <0.03.	MULTIPLE VENDORS
REFLECTOR SUBSTRATE	POLYETHYLENE TEREPHTHALATE	FLAT FILM, 1 MIL THICK, NO PERFORATIONS	DUPONT & CO.
SPACER	POLYETHYLENE TEREPHTHALATE	SPUNBONDED POLYESTER, 4 MIL THICK, 0.5 OZ/YD^2	REEMAY, INC.
COVER LAYERS	POLYETHYLENE TEREPHTHALATE	SPUNBONDED POLYESTER, 9 MIL THICK, 1.35 OZ/YD^2	REEMAY, INC.
HOOK FASTENER	POLYESTER	HOOK #80, WHITE #012 WIDTHS: 1 INCH & 2 INCH	VELCRO USA, INC.
LOOP FASTENER	POLYESTER	LOOP #2000, WHITE #012 WIDTH: 1 INCH	VELCRO USA, INC.
THREAD	POLYESTER	V125 WHITE	BELDING CORTICELLI THREAD CO.

During the operation of the SSC, the multilayer insulation blankets are subjected to extreme temperature gradients along their thicknesses causing the layers nearest the cryogenic structure to experience dimensional contraction to a greater extent than the layers furthest from the cryogenic structure. Prior fabrication techniques have failed to account for the dimensional response of the blanket over the entire temperature range of the insulated cryogenic structure. The SSC blanket fabrication method accommodates MLI material thermal contraction in its length and width dimensions to temperature decreases as low as 4.5 Kelvin.

MLI BLANKET FABRICATION

Pictured in Fig. 2 is a large diameter winding apparatus used to fabricate the MLI blankets for the 4.5K cold mass, and the 20K and 80K shields. The apparatus consists of a rotatable mandrel with a fixed diameter of approximately 18 feet and an outer surface that is crowned with a convex cross-section. Accessory supply spools hold the blanket reflective and spacer materials. The function of the apparatus is to spiral wrap the appropriate number of MLI blanket layers around the mandrel. Since each wrap of MLI material increases the circumference of the mandrel, each successive layer of material is slightly greater in length than the preceding layer. Likewise in the transverse direction across the crowned surface, each successive layer is wound on an increasing circumference. The finished blanket is then bound together at its edges by rotating the blanket through a stationary sewing machine. A single cut is made directly across the width of the MLI, and the blanket is removed from the mandrel. The resulting MLI blanket has sufficient length and width for an SSC blanket assembly. In addition, there is locked between the parallel sewn seams: dimensional stability, three-dimensional uniformity, controlled layer density, interlayer registration, interlayer cleanliness, and extra material in successive layers in the length and width directions to aid thermal contraction, since the last layer wrapped on the mandrel is designated for installation as the first layer installed against the cold surface to be insulated. Fig. 3 shows an 80K inner MLI blanket for the SSC Dipole Magnet Design B Cryostat.

Fig. 2.   Large Diameter Winding Apparatus for MLI Blanket Production

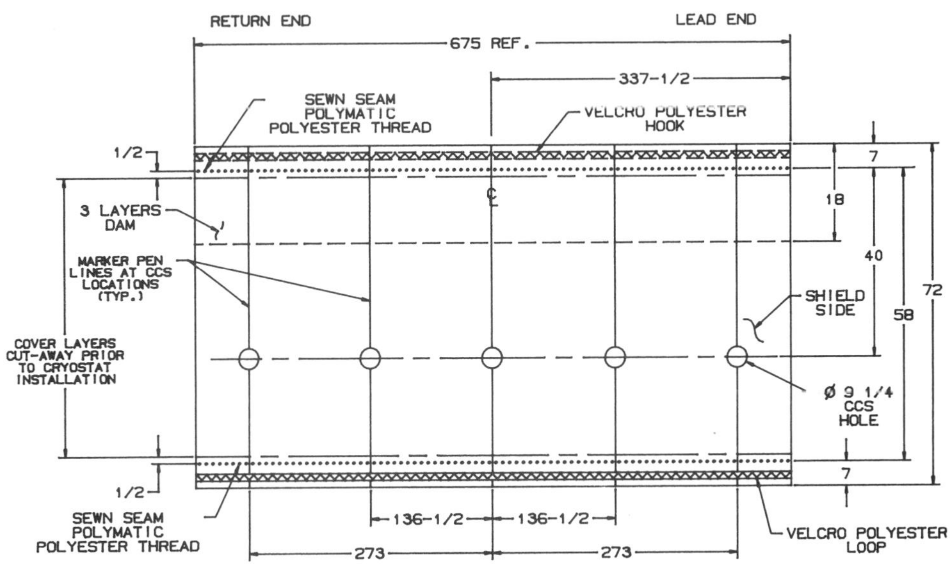

Fig. 3.   SSC Dipole Magnet 80K Inner Blanket

BLANKET INSTALLATION

The tightness of the blanket wrap during installation in the SSC cryostat is fixed by the blanket design. Parallel velcro strips sewn to the outer blanket layers serve to secure the blanket at installation. The velcro strips are separated by a distance determined empirically for each blanket (20K, 80K inner and 80K outer) by trial fitting blanket sections onto actual shields and adjusting the velcro separation until the desired blanket fit around the shield was obtained. Since both edges of the blanket are sewn, the MLI material locked between seams is caused to wrinkle or wave as the blanket is wrapped around the shield. This wrinkling occurs as a result of the regimented layers being unable to separately slide across each other as would be the case, for example, if only one edge of the blanket were sewn. Because of layer-to-layer registration, these wrinkles or waves usually have a uniform thickness along the wave that is approximately the same thickness as in the main body of the blanket. The MLI layer density is therefore little affected by the wrinkles. Furthermore, a favorable element of these wrinkles is that they provide added material for thermal contraction of the blankets.

At installation, the MLI blanket is wrapped around the shield such that edges of the blanket overlap. As the edges are drawn together, tension on the blanket from pulling the blanket against the shield is taken by the sewn seams and the cover layers. The material located in the greater blanket area between the sewn seams is isolated from the tension by the seams. Next to the shield surface, opposite edges of the lower cover layer are secured to each other by the full-width engagement of the velcro strips. The overlapped edges of the MLI layers are then joined along the entire blanket length using a stepped-butted-joint geometry. See Fig. 4. The upper cover layer edges are drawn together over the stepped joint and secured by the full-width engagement of the upper velcro strips. The completed blanket installation and stepped-butted-joint are doubly secured from opening by the two velcro pairs.

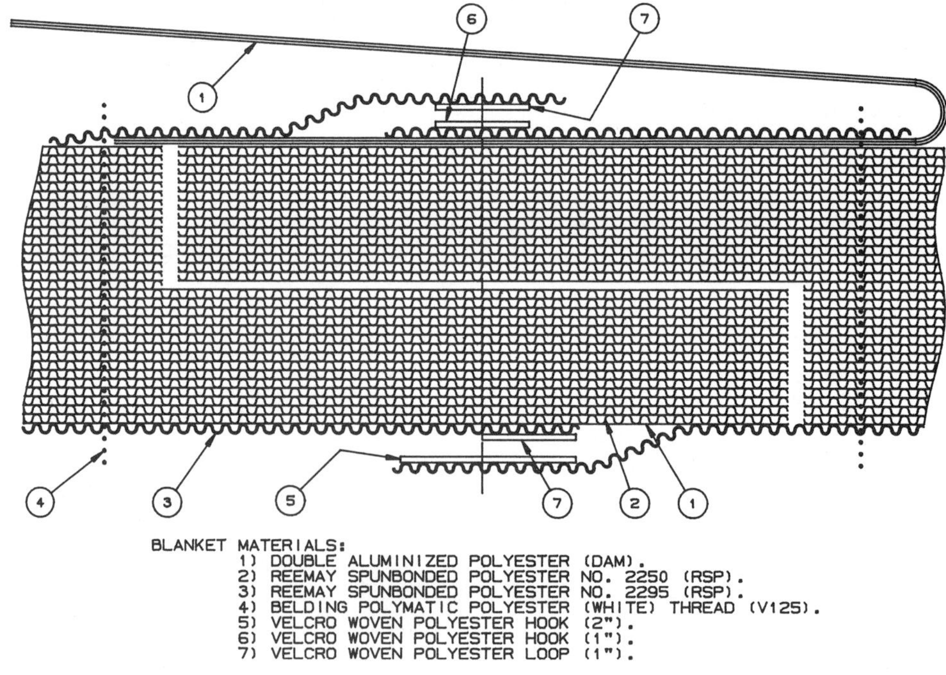

Fig. 4. MLI Stepped/Butted Seam

DISCUSSION

A unique feature of this blanket design and installation technique is
that it eliminates the need for MLI tape (aluminized adhesive tape) to secure
the blanket. This distinction is immediately realized as a gain to the
ultimate obtainable vacuum that the SSC might expect. The elimination of
MLI tape from the cryostat vacuum space would preclude: 1) adhesive
outgassing from the tape; 2) virtual leaks of trapped gases from between the
tape and the taped surface; and 3) increased deterioration of the MLI
materials due to reaction of the evolved gases with chemical radicals formed
in the MLI by irradiation during the lifetime of the SSC accelerator.

CONCLUSIONS

The apparatus and blanket fabrication method is a cost effective means
to mass produce dimensional uniform MLI blankets:

1) it reduces the number of production personnel necessary to fabricate
   a finished blanket;
2) it eliminates layer by layer handling of the MLI materials, thereby
   greatly reducing labor costs and insuring interlayer cleanliness;
3) it allows precise location and insertion of penetrations or holes
   in the MLI to be done during blanket fabrication while the MLI is
   on the winding mandrel.

Incorporated with the blanket design are fasteners which automate
decision making during installation of the MLI blankets into an SSC cryostat.
The blanket design guarantees that each MLI installation is identically
reproduced in a straight-forward manner. By virtue of the blanket design,
the apparatus, and the fabrication method, the MLI installation geometry in
each of the 10,000 cryostats will be dimensional uniform. By providing
positive control of the dimensional parameters which contribute to the MLI
blanket thermal performance, consistent and predictable operational
performance of the MLI system is replicated in the 10,000 SSC cryostats.

REFERENCES

1. SSC Central Design Group, Superconducting super collider magnet system
   requirements, SSC-MAG-D-101 (JULY 23, 1987).
2. R.C. Niemann, et. al., Model SSC dipole magnet cryostat assembly,
   presented at the IISSC, February 8-10, 1989, New Orleans, LA.
3. J.W. Price, Measuring the gas pressure within a high performance insulation
   blanket, in: "Advances in Cryogenic Engineering," Vol. 13, Plenum Press,
   New York (1967).
4. A.P.M. Glassford and C.K. Liu, Outgassing rate of multilayer insulation
   materials at ambient temperature, J. Vac. Sci. Technol., 17:696.
5. W. Burgess and Ph. Lebrun, Compared performance of Kapton and Mylar
   based superinsulation, in: "Proceedings of the Tenth International
   Cryogenic Engineering Conference," Helsinki, Finland (1984).
6. H. Burmeister, et. al., Test of multilayer insulations for use in the
   superconducting proton-ring of HERA, in: "Advances in Cryogenic
   Engineering," Vol. 33, Plenum Press, New York (1987).
7. M.H Van de Voorde and C. Restat, Selection guide to organic materials for
   nuclear engineering, CERN 72-7 (1972).
8. DuPont & Co., Industrial Films Div.; and, Textile Fibers Dept.
9. James River Corporation, Nonwoven Division.

STATUS OF SUSPENSION CONNECTION FOR SSC COIL ASSEMBLY

E.T. Larson, T.H. Nicol, R.C. Niemann, R.A. Zink

Fermi National Accelerator Laboratory
Batavia, Illinois 60510

ABSTRACT

Superconducting Super Collider dipole magnets require an integrated
suspension system to meet the structural and thermal requirements outlined in
the design criteria. Sliding suspension connections which retain the cold mass
assembly during static and dynamic loading, while allowing axial motion during
thermal contraction are an integral part of this magnet suspension system.
Variations from the original prototype design have been tested and their
performance compared. The results of these evaluations, and areas of future
investigation are described.

INTRODUCTION

Dipole magnets for the Superconducting Super Collider (SSC) require an
integrated suspension system which meets the demanding criteria of low heat
leak, high structural strength, alignment stability, high reliability, and low cost
associated with zero maintenance cryostat systems.[1]

One of the components in that suspension system is the slide connection
which must transfer loads between the cold mass and the compact cryogenic
supports (CCS).[2] These loads may come from shipping, quench or seismic
sources. There are five connections to the cold mass, four are of a sliding
design allowing rotational and axial degrees of freedom.[3] The center connection
is fixed, locking the angular and axial position of the cold mass. The five
connections share lateral and vertical loads, while the center connection transfers
any large axial loads to the HYBAS support system.[4] During cooldown the
slide connections accommodate the cold mass axial contraction, which is over
3/4" on the connection farthest from center. The slide connection provides a
minimal resistance to axial motion to help reduce any large imposed bending
moments on the post supports. The connections must also fit within the close
geometry of the cryostat without causing thermal contact shorts, and be able to
withstand the cryogenic and radiation environments associated with
superconducting accelerator magnets.

DESIGN CONSIDERATIONS

Design A Experience

The original Design A version of slide connection consisted of two parallel rods with sleeve bearings which supported a guiding tie bar. This bar could slide axially while supporting small blocks which held insert pins that penetrated the cold mass skin, see Fig. 1. The insert pins acted both as fiducial and structural support members, requiring a socket to interface with the magnet iron yoke and a seal weld on the outer shell of the cold mass.

Field experience with this system demonstrated the difficulty of positioning the fiducial pins accurately and repeatedly. Only ten pins 0.25 inches in diameter support the weight to the cold mass assembly, with each pin requiring a weld and leak check step, adding to the complexity of the cold mass. The parallel rods of the slide required accurate positioning to prevent binding during axial motion, and developed high wear rates due to high contact stress between

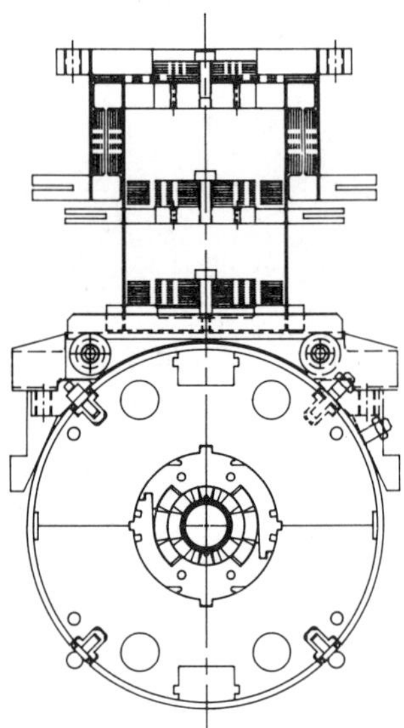

Fig. 1.   Design A Slide Connection

the rod and tie bar. A temporary solution was implemented when a bushing was inserted between the parts reducing friction and wear rates considerably. The bushing consisted of Teflon and lead captured in a sintered bronze matrix attached to a steel backing plate. The material, commercially known as DU, proved to be an excellent material for reducing friction and wear and could withstand the stringent environment of the cryostat. Tests were made on Design A assemblies at cryogenic temperatures to verify the performance of the DU bushings. The test results showed an increase in coefficient of friction under cryogenic conditions but the increase from 0.10 to 0.25 was acceptable, see Fig. 2.

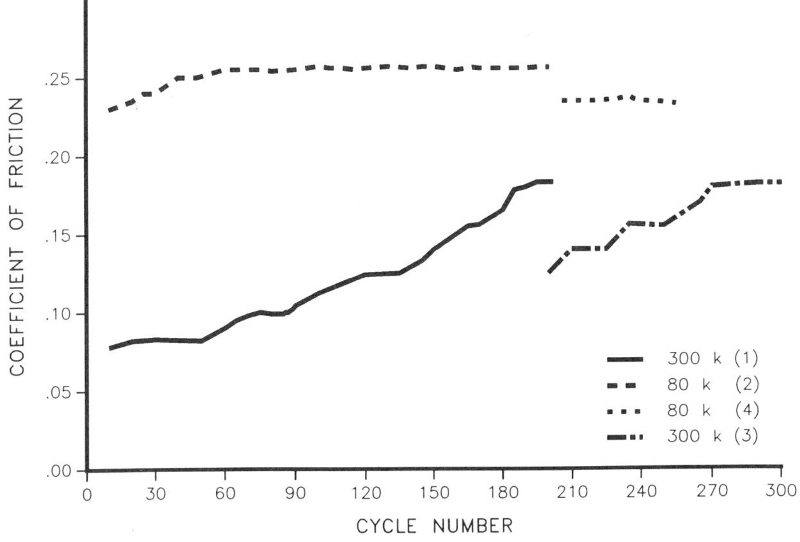

Fig. 2. Design A Friction vs. Equivalent Thermal Cycles

While the Design A version was successful in the magnets it was installed in; there were several areas of the design which could be improved upon. The following design areas were principle motivations for changing to the Design B connection.

A   Eliminate the need for multiple weld penetrations in the cold mass skin.

B   Improve the structural support to withstand dynamic loads.

C   Improve bearing performance and reduce wear rate.

D   Provide angular adjustment of the cold mass during assembly.

E   Separate the alignment and support functions.

Design B Experience

Prototype models of Design B slide connections were fabricated to test the concept of applying bearings directly to the cold mass skin. The prototypes consisted of a frame assembly that would support four DU bearing pads, see Fig. 3. The frames lower half was attached to the 4K ring on the top of the CCS by a split ring bolted at four locations. The upper half of the frame contains the cold mass during loading and is attached to the lower half by four half inch bolts. Each bolt has four Belleville washers stacked in series giving an effective spring rate of 6,500 lbs/in. The washers are compressed 0.050 inches during assembly, yielding an effective preload of 1,300 lbs. These washers act as tensioning springs which maintain uniform loading of the bearings and compensate for any radial differential thermal contraction effects between the cold mass and the slide connection.

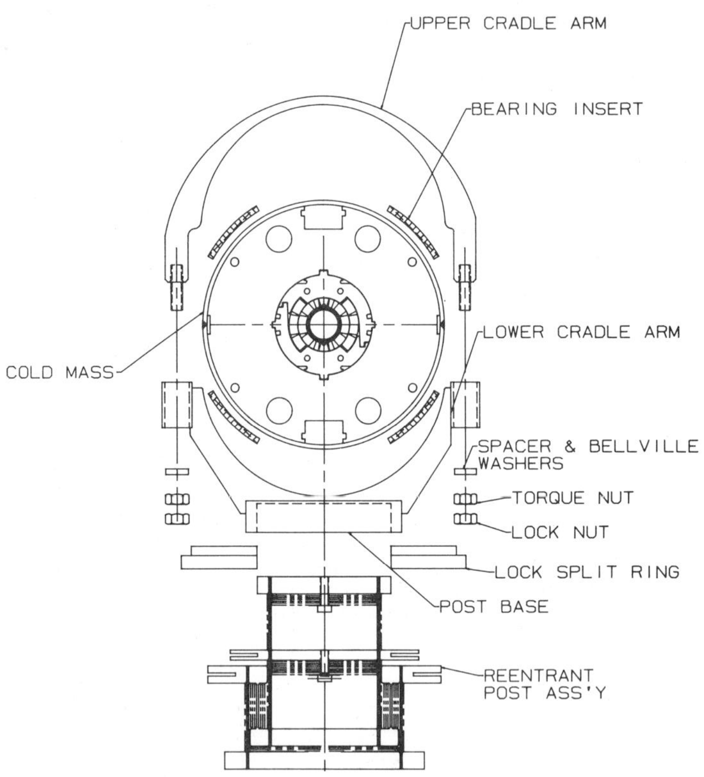

Fig. 3.   Prototype Design B Slide Connection

Since DU bushings had proven themselves in the Design A slide connections it was logical to use them in Design B but with a larger contact surface to reduce localized wear effects noted in Design A.   The bearings were offset 45 degrees from the vertical and horizontal planes.   The cold mass has distortions in these planes due to features in the iron yoke and welding techniques used during assembly.   These locations were selected to minimize any effect these distortions might have on alignment and bearing performance.

The DU material requires a nominal 20 micro inch contact surface finish for optimum performance.   Tests on short machined mandrels proved that unacceptable friction and wear would result from surfaces which did not meet the 20 micro inch criteria.   Concern over surface finish initiated a search for material to use in the cold mass skin.   Stainless steel sheet with a stock "2B" finish met the surface finish criteria and was available as a standard product.   Use of the Design B concept would require better surface finish on the cold mass outer skin than was presently being provided by the stock 3/16" plate already in use.

DEVELOPMENT AND TESTING PROGRAM

The prototype assemblies were tested at all the load levels specified in the design criteria and successfully passed these tests.[3]   These prototype Design B assemblies used very thin sections in the vertical plane due to the geometric constraints of the cryostat.   This resulted in distortions from welding during assembly.

As a result of these problems investigations were made into developing a cast version of the assembly in order to avoid any assembly welding, and to provide a complex stock geometry for machining that would minimize scrap. The material selected for the castings was CF 8, a stainless steel having composition and properties that closely match 304 series stainless. The cast versions were of similar geometric cross section as the prototypes, but the thin framework was eliminated in favor of a more uniform section that could be easily cast, see Fig. 4.

An opportunity to test the prototype slide connections came when magnet D0004 was made available for mechanical tests, after manufacture it was found to be defective. Sections of the cold mass skin were removed and replaced with 2B finish material to provide a suitable surface to test the prototype Design B slide assembly. The center connection which is normally fixed was replaced with a sliding connection, thus allowing the entire cold mass to be moved axially. Ambient tests were made to determine the performance of the slides on this full scale model. Axial motion which simulated successive thermal contraction cycles was provided by a hydraulic actuator. The force required to move the assembly, the deflection of the CCS and slide connections, and the amount of wear the DU bearings experienced were all noted during the testing. Figure 5 plots the force to move the assembly vs. the number of simulated thermal cycles. Deflection of the CCS's was uniform indicating equal distribution of axial loads, with negligible deflection of the slide connection itself relative to the CCS. There was visible transfer of the outer layer of DU material to the surface of the cold mass skin, considered normal during the break in of the bearing. This transfer establishes a free moving layer between surfaces that enhances performance.

Since the cryostat is over 58 feet in length, with a very long time constant to reach thermal equilibrium, it was impractical to do full length cryogenic tests of the slide connections at liquid helium temperatures. Therefore, tests were designed to demonstrate the feasibility of the slide connections performance under simulated cryostat conditions using a short sample cryostat which was fabricated specifically for the task. This would yield the necessary data for evaluation at a reasonable cost, allowing extrapolation to liquid helium conditions. These tests were designed to measure the changes in performance under vacuum at 1×10^{-6} torr, and at liquid nitrogen temperature, since the majority of material property changes and thermal contraction effects have occurred by 80K. A short mandrel that simulated the cold mass was cooled to 80K in a short dewar. External hydraulic actuators would allow axial motion to be induced at cold temperatures and thus simulate hundreds of thermal cycles in a short time frame. The tests showed that the effective coefficient of friction dropped slightly under vacuum, before cooling, and the coefficient of friction increased by a factor of 3 at 80K, see Fig. 6. There was a slight decrease in friction under vacuum, which could be attributed to a drier environment, but the overall effect is less than 10 percent. The cast versions provide an extremely stable foundation for the bearing pads, reducing the deflection of the upper half and subsequently reducing the sliding resistance by more than 30%. This is a significant improvement over the prototype Design B. The coefficient of friction remains stable over several hundred cycles, the actual number of cooldowns for magnets is estimated at 20 or less over the life of the accelerator. These test results are consistent with earlier data taken on DU bushings used in the Design A version. The results of these cold tests indicate an effective coefficient of friction of 0.30 at helium operating temperatures, which translates into lateral forces of 1300 lb. per support point when the cold mass is cooled to operating temperature. This lateral force is well within the capability of the CCS suspension system which is rated for 3,150 lb lateral loads at each support point.[4]

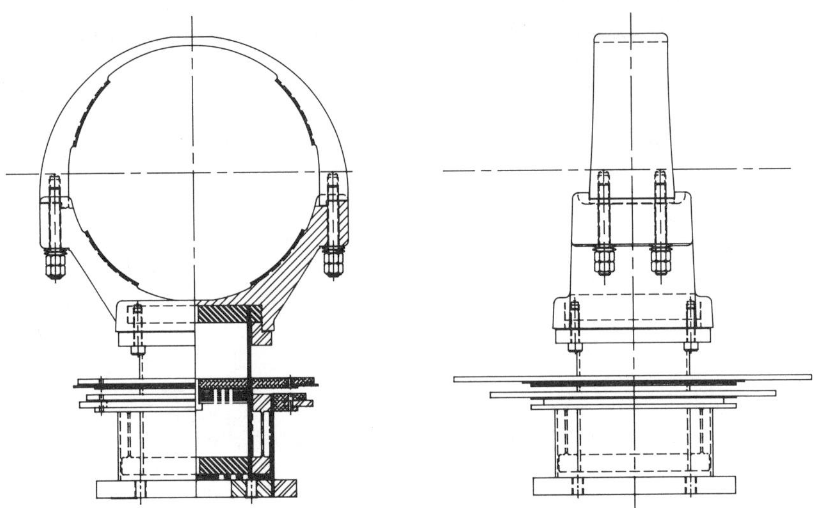

Fig. 4.   Design B Slide Connection Casting

Tests of the outgassing rate of the slide connection were made to establish the magnitude of gas influx into the vacuum system. A high vacuum chamber was used to measure the pressure rate of rise for cast versions of the slide connection. The assemblies were cleaned in solutions of freon and ethyl alcohol to remove surface contamination, but no other cleaning steps were taken. The 80 liter vessel experienced a pressure rate of rise of 5.5×10^{-9} Torr/sec with the connection at 300K resulting in an outgassing rate of 2.43×10^{-7} $W/m^{2.5}$ This is in the normal range for stainless steels which have no bakeout or other special surface treatments. Since the surface area of the connection is small in comparison to the vacuum vessel, and the outgassing rate at operating temperatures is negligible compared to room temperature conditions, the connection does not impose a significant load to the vacuum system.

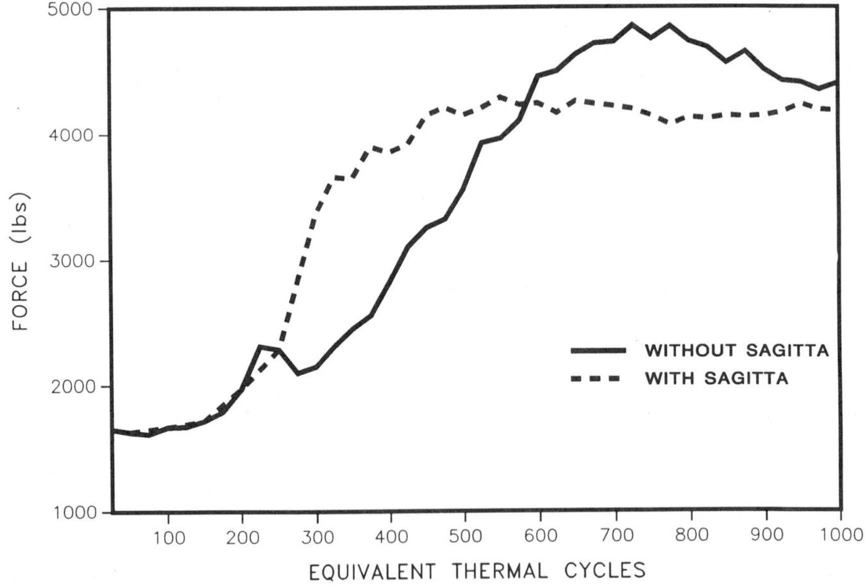

Fig. 5.   D0004 Full Length Slide Tests

Heat leak measurements of the slide connection and CCS were made to quantify any changes which the slide connection might contribute. The heatmeter showed slightly higher heat leak which could be attributed to residual gas conduction effects. The 5-8 mW addition is small compared to the 120 mW budgeted heat leak to 4.5K.[6]

ALIGNMENT AND SUPPORT

The Design A version of the slide connection combined the alignment and support functions by supporting off the fiducial insert pins. This created problems since the pins were not designed to carry high loads and the slide connection assembly tolerances were held to alignment fixture levels ±0.001".

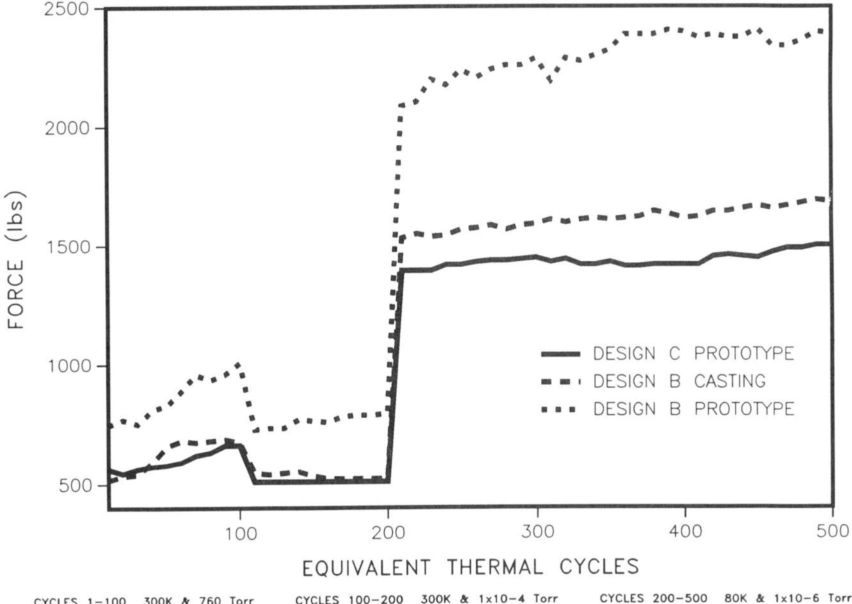

Fig. 6. Cold Slide Test Comparison

The Design B version allows one to separate the alignment and support functions. The fabrication tolerances are relaxed on the slide connection to ±0.010" and the fiducial inserts shifted axially 6" from the support points. The fiducials presently in use are the same as Design A but are now independent of the support system. While Design B fabrication tolerances are relaxed the support must still repeatably position the magnet center within five mils, relative position stability is now the objective versus absolute position accuracy. The new cold mass assembly tooling being fabricated at FNAL is not operational, so the insert pin method used in the original BNL tooling is still being used for alignment. When the new FNAL tooling is available there will be no penetrations in the cold mass skin and the external features of the cold mass will be the fiducial reference. This will allow a full length reference versus a five position reference for magnet alignment.[7]

PROGRAM HISTORY

To date there have been seven cryostats assembled using the Design A slide connection, one using the prototype Design B version, and four using the final Design B version. Table 1 lists the magnet assemblies which have used each version of the slide connection.

Table 1

Magnet Assemblies	Design Version
HLM II, D0001, D0002, D000X DD000Z, DD0010, DD0012	Design A
DD0014	Design B Prototype
DD0011, DD0013, DD0015, DD0016	Design B Casting

FUTURE DIRECTIONS

During assembly of Design B cryostats several problems have occurred with interferences between the upper half of the slide connection and the inner 20K shield. Stand-offs made of G-10 were installed to minimize thermal shorts at these interference points. A new upper half of the connection was prototyped which would provide a nominal 3/16" profile above the cold mass skin. The design uses rolled sheet stock welded to mounting blocks which can be directly interfaced with the existing lower half of the slide connection, see Fig. 7. The rolled plate has four holes which pick up tabs on the DU bearings. The tabs are formed by rolling back the DU itself, creating a captured geometry when installed between the cold mass and the upper half of the connection. The existing design uses a recessed pocket in the casting to capture the bearings, requiring two machining steps. The same method of capturing the bearings using tabs can be applied to the lower half. Tests of the new low profile upper half, in conjunction with the cast lower half, have shown even lower resistance to axial motion. The design is seen as a candidate for the Design C cryostat.

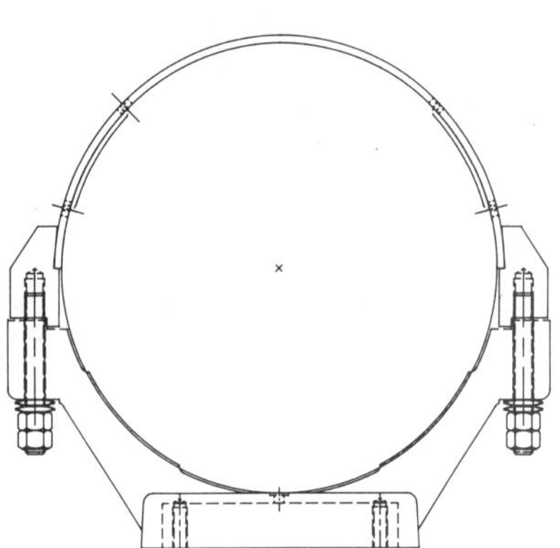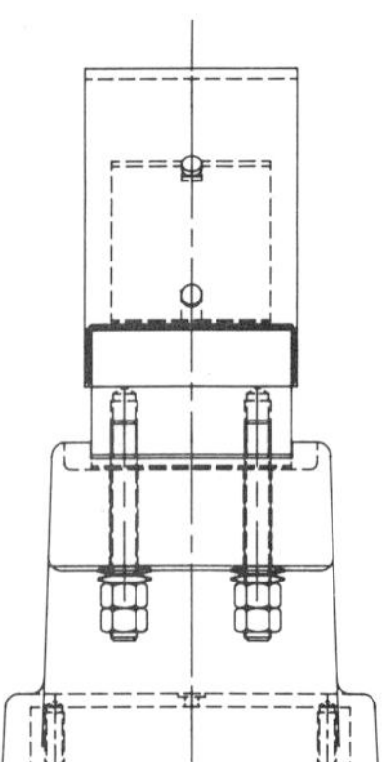

Fig. 7. Design C Prototype Slide Connection

ACKNOWLEDGEMENTS

The authors would like to express their appreciation to G. Czop, E. Hanson, M. Ruschman, C. Schoo, G. Turkot and J. Tweed for their contributions in the laboratory evaluations.

The work presented was performed at Fermi National Accelerator Laboratory which is operated by Universities Research Association Inc., under contract with the U.S. Department of Energy.

REFERENCES

1. SSC Central Design Group, Superconducting super collider magnet system requirements, "SSC-1000," October 1986.

2. R.C. Niemann, et al., Design, construction and performance of a post type cryogenic support, in: "Advances in Cryogenic Engineering," Vol. 31 Plenum Press, New York (1986), p. 73

3. E.T. Larson., et al., Improved design for a SSC coil assembly suspension connection, in: "Advances in Cryogenic Engineering," Vol. 33, Plenum Press, New York (1988) pp. 235-241.

4. T.H. Nicol., et al., SSC magnet cryostat suspension system design and performance, in: "Advances in Cryogenic Engineering," Vol. 33, Plenum Press, New York (1988) pp. 227-234.

5. E. Hanson.,"Outgassing tests on the SSC iron core saddle"- Fermilab 1988 (unpublished)

6. J.D. Gonczy., et al., Thermal performance measurements of a graphite tube compact cryogenic support for the SSC, presented at the ICEC12, Southampton, (1988).

7. J.A. Carson.,"SSC dipole magnet cryostat cold mass assembly construction alignment procedures"- Fermilab 1987 (unpublished)

STATUS OF TESTS OF DOUBLE-SIDED SOLID STATE MULTISTRIPE DETECTORS

P. L. Skubic, G. R. Kalbfleisch, and M. A. Lambrecht

Physics and Astronomy Department
University of Oklahoma, Norman, OK, 73019

C. D. Wilburn

Micron Semiconductor, Ltd., Lancing, Sussex
BN158UN, England

ABSTRACT

Two double-sided multistripe silicon detectors purchased from Micron Semiconductor, Ltd. have been bench tested. The devices are 40 mm x 40 mm x 300 μm thick, with 40 stripes on each side, in directions orthogonal to each other. Interstripe resistances of 250 $\pm$ 100 kΩ were measured on the non-diode (ohmic) side when the detectors were fully depleted. A Sr^{90} source and charge sensitive amplifiers were used to measure charge correlations between orthogonal stripes on opposite sides of the detectors as a function of bias voltage. At a full depletion voltage of 42 V, the difference in pulse heights from opposite-side stripes gave a Gaussian distribution consistent with the amplifier noise, indicating that the detectors work well as two dimensional position sensitive devices.

INTRODUCTION

High resolution solid state microstripe detectors have been used extensively in recent high energy physics fixed target experiments. Typical detectors having a 300 μm thick depletion depth and a pitch of 50 μm have been used to measure particle track positions with an accuracy of $\pm$9 μm.[1] The use of such detectors has recently been extended to colliding beam experiments. For such applications, there are advantages if the non-diode (ohmic) side of the detector is also segmented and both sides are read out (double-sided).[2] For example, if orthogonal segmentation is used on the two sides, information in two views can be obtained with one half the thickness, thus decreasing errors due to multiple scattering and at the same time reducing the amount of material which causes gamma conversions. Also if the ionization charge is digitized on both sides, ambiguities due to multiple tracks in the same detector element can often be resolved.[2,3] For colliding beam experiments at the SSC, double-sided solid state microstrip detectors would therefore be useful for vertex detectors which require very precise position measurements for track reconstruction. Double-sided detectors have recently been tested by several groups. Charge sharing has been shown to work well on the ohmic side of a double-sided detector with a pitch of 25 μm for readouts every 200 μm.[4] An ohmic side position resolution of 52 μm was obtained with a 500 μm pitch detector in beam tests with 250 GeV/c pions.[3]

MONTE CARLO CALCULATIONS

Monte Carlo simulations were done to investigate the application of microstripe detectors to colliding beam experiments at the SSC where tracks are incident on the detectors at large angles with respect to normal incidence. The simulation assumed a 200 μm thick detector with 25 or 50 μm pitch. Tracks passing through the detector at various angles relative to 90° were allowed to form clusters of hits with charge fluctuations consistent with a measured "Landau" distribution. The hits were then subjected to the following signal processing algorithms: "TRUE" for the original charges; "ADC" for charge measurements with an infinitely precise ADC readout device assuming a 2 standard deviation threshold cut; and "DISCRIMINATOR" for a readout employing discriminators only with a threshold cut of 4 standard deviations.

Fig. 1 shows the error (rms.) in calculated position as a function of angle relative to normal incidence for 50 μm pitch (a) and 25 μm pitch (b) detectors with every stripe read out. An interesting case is 25 μm pitch (DISCRIMINATOR) where the position resolution deteriorates rapidly for angles above 30° relative to the normal. For this case, where a 4 σ threshold cut is assumed, a 50 μm pitch detector is preferable for angles above 35°. This effect is due to the higher probability of hits to have charges below threshold as the angle increases.

Fig. 2 shows a similar calculation for a more realistic case where the ADC is assumed to have a charge resolution of 0.1 fC. The calculations were done for hit-defining thresholds of 0, 2, and 4 standard deviations above the charge resolution. For each hit above threshold, the adjacent stripes were also read out as would be the case for the CDF SVX readout chip.[5] We assume that every stripe is connected to the readout chip in Fig. 2(a), and that every other stripe is connected to the readout chip in Fig. 2(b) where charge sharing is used to calculate positions. The error is less than 20 μm for angles up to 80° for the 0 σ and 2 σ threshold cuts, although noise degrades the position resolution for the 0 σ case. For the 4 σ cut with every stripe read out, the resolution becomes rapidly worse above 40° due to loss of hits

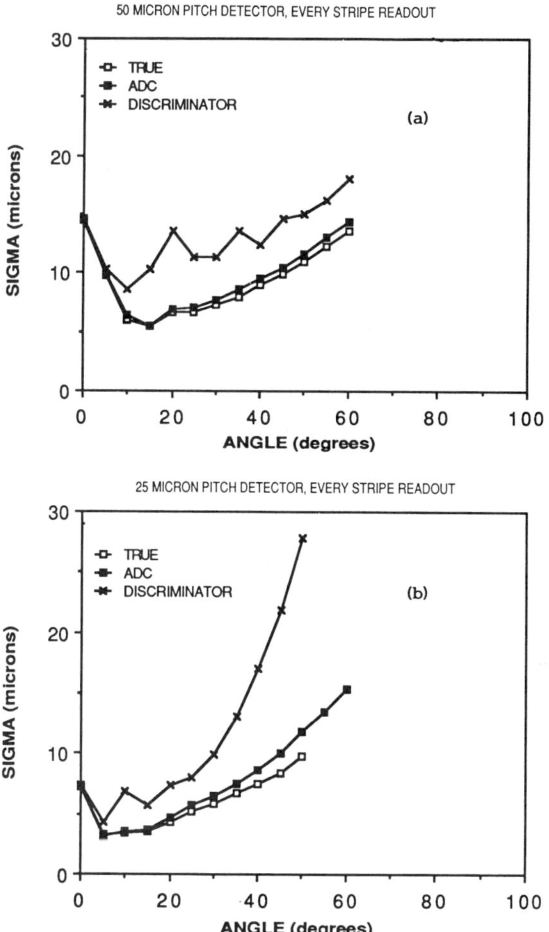

Fig. 1.   Error (rms.) in calculated position *vs*. angle relative to normal incidence for the Monte Carlo simulation described in text. Hit clusters were obtained from tracks passing through a 200 µm thick microstripe detector with 50 (a) or 25 (b) µm pitch stripes.

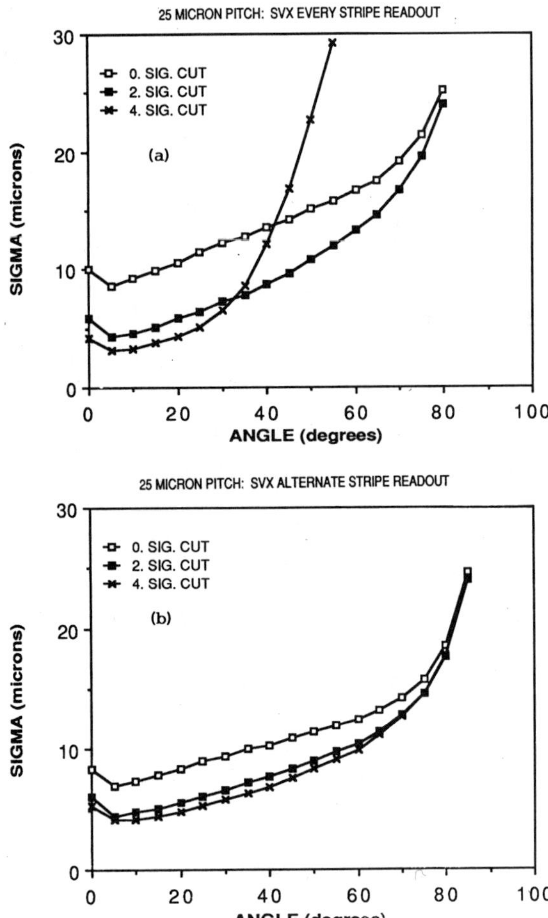

Fig. 2.   Error (rms.) in calculated position *vs.* angle relative to normal incidence for the Monte Carlo simulation described in text. Hit clusters were obtained from tracks passing through a 200 μm thick microstripe detector with 25 μm pitch with every (a) or alternate (b) stripes read out.

above threshold, while the 4 σ cut with alternate-stripe readout gives good resolution up to 75°.

EXPERIMENTAL RESULTS

Two double-sided multistripe silicon detectors purchased from Micron Semiconductor, Ltd. (design BB) have been tested by our group. The devices are 40 mm x 40 mm x 300 µm thick, and have 40 stripes on each side in orthogonal directions. The stripes have a "pitch" (i.e., center-to-center spacing) of 1 mm, and are 900 µm wide separated by a 100 µm gap.

Our initial results are presented in Ref. 2. The unbiased interstripe resistances on the ohmic side were found to be about 2.5 kΩ. The full depletion bias voltage was 42 V and the ohmic side interstripe resistances under overbiased conditions (50 V) were measured to be (250 ± 100) kΩ. The detector was then connected for dynamic signal tests with Sr^{90} betas which deposit on average 90 keV in dE/dx energy loss in 300 µm of silicon. All stripes were connected by 1 MΩ resistors to positive bias (ohmic side) or ground (junction side) and were AC coupled to the 80 amplifiers. The amplifier system consisted of LeCroy (LRS) 8 channel HQV810 hybrid charge sensitive pre-amplifiers followed by delay line shaping with dual ECL 10115 amplifier stages. The outputs were integrated and digitized by 80 channels of LRS 2285 ADC's and a scintillation counter was used to generate a 400 ns gate.

The rms. noise (sigma) of each channel was measured as a function of bias voltage. The junction side stripes had a constant noise of between 50 and 75 ADC counts for voltages from 25 V to 55 V. The ohmic side stripes get rapidly noisy below full depletion (~400 ADC counts at 25 V) and have a constant noise above full depletion (35 V to 55 V) in the range 110-150 ADC counts. The rise in the noise for the ohmic side is roughly consistent with a 250 kΩ interstripe resistance at full depletion and a 2.5 kΩ interstripe resistance without bias (a factor of 3.3 on the noise for each factor of 10 less resistance).

Fig. 3 shows the pulse height spectrum for betas from the Sr^{90} source for the junction (a) and ohmic (b) sides. The signal to noise ratio for the ohmic side is typically 10:1. We have investigated charge sharing for the case of alternate-stripe readout. Fig. 4(a) shows the pulse height distribution for trigger stripe "6" alone where stripes "5" and "7" are floating. The lower peak is due to charge lost to readout stripes "4" and "8". Fig. 4(b) shows the total pulse height ("4"+"6"+"8") read out on all three stripes. We see that the lower peak is no longer present which shows that the charge on the floating stripes is successfully collected by the readout stripes.

About 60% of the detectors' area was covered by the trigger counter and beta source. The charge correlations between the ohmic and junction sides were studied as a

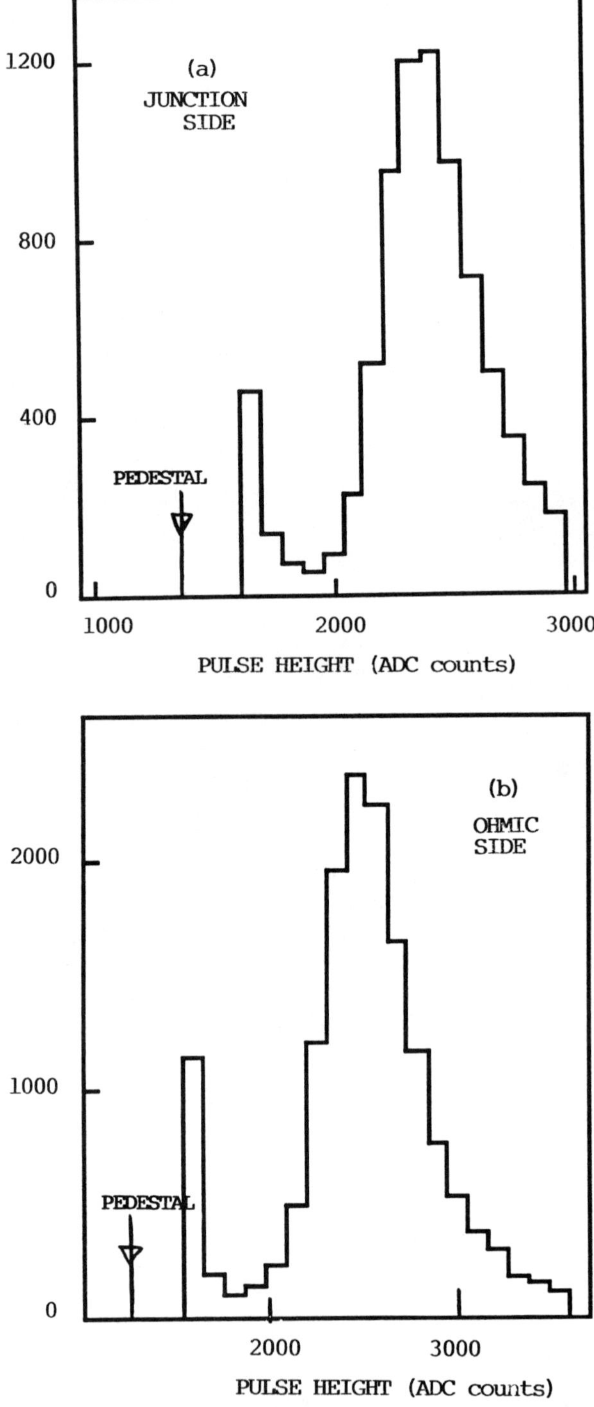

Fig. 3. Pulse height distribution in ADC counts for junction side (a) and ohmic side (b) stripes.

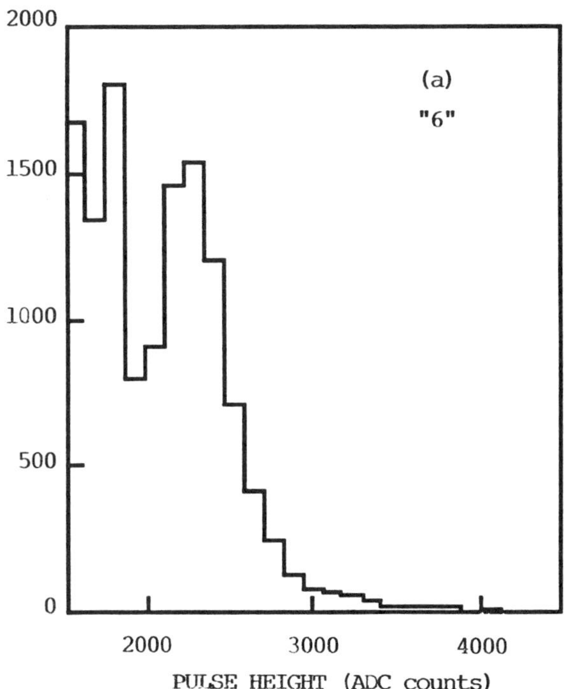

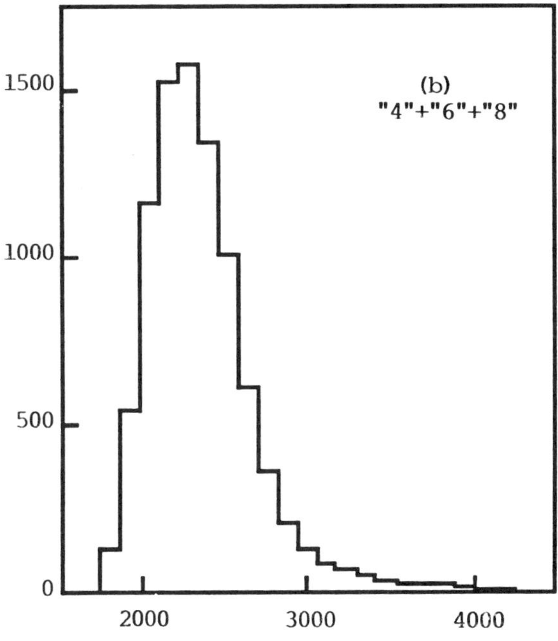

Fig. 4.   (a) Pulse height distribution for one ohmic side trigger stripe ("6) on a detector with every other stripe read out. (b) Pulse height distribution for the sum of three adjacent ohmic side readout stripes ("4"+"6"+"8") on a detector with every other stripe read out.

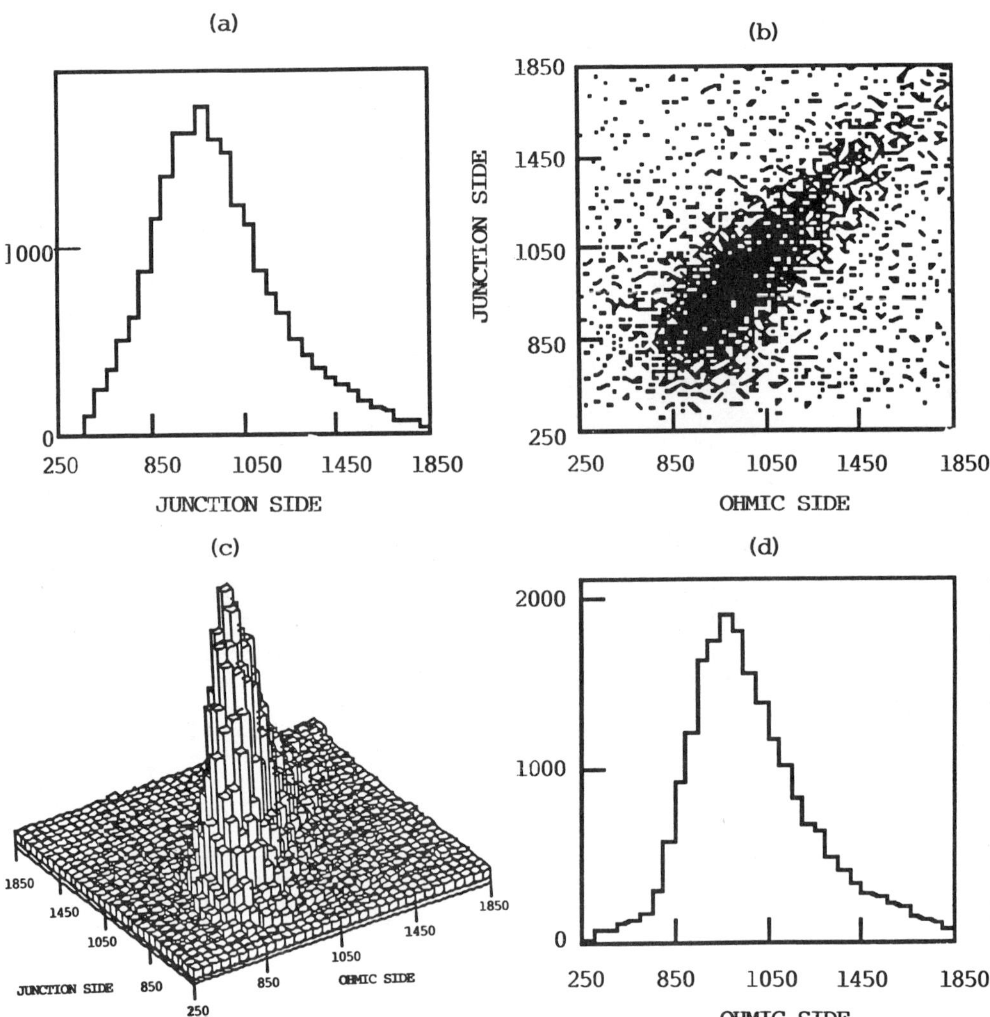

Fig. 5.   (b,c) Scattergram showing pulse height correlations between ohmic and junction sides of a double-sided multistripe detector. (a) Junction side projection of scattergram. (d) Ohmic side projection of scattergram.

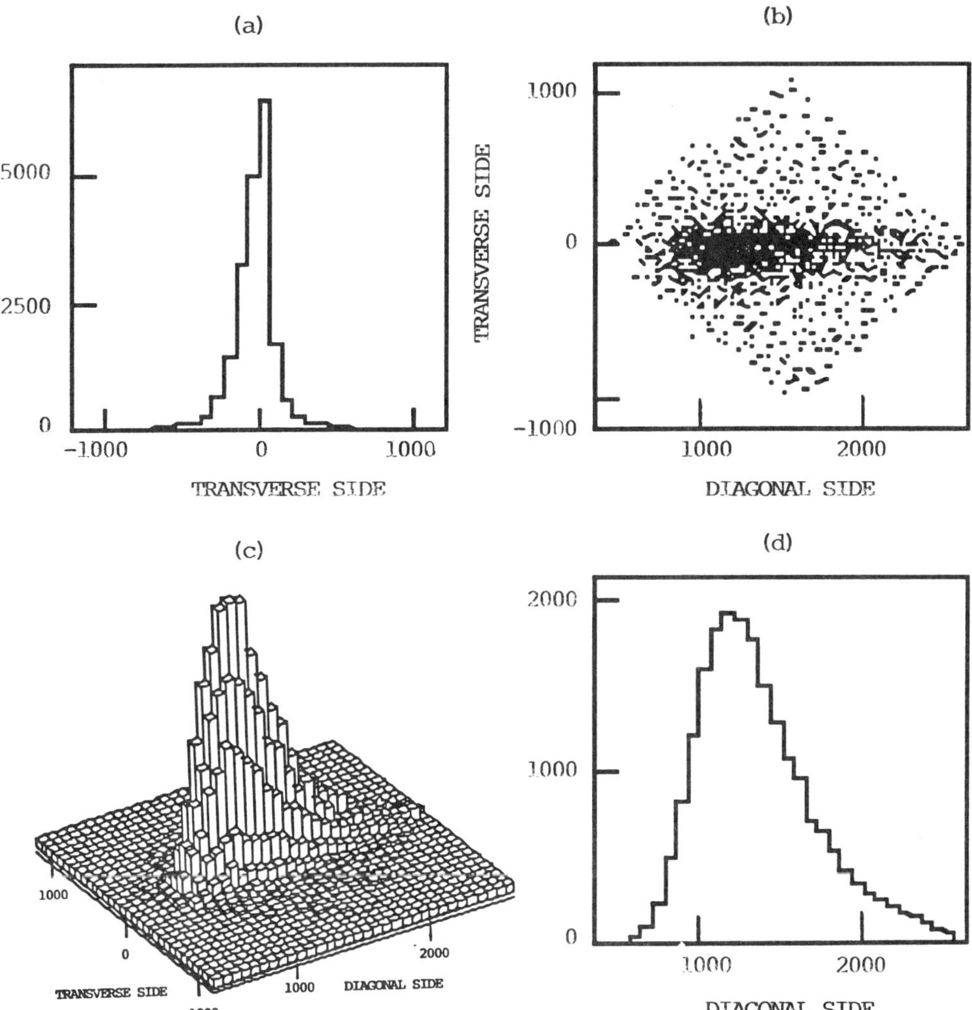

Fig. 6.   (b,c) The same scattergram as in Fig. 5 rotated where the abscissa is the diagonal distance from the orgin in Fig. 5 and the ordinate is the distance transverse to the diagonal. (a) Transverse side projection of scattergram.   (d)   Diagonal side projection of scattergram.

function of voltage. For example, the correlated signal pulse
heights for a bias voltage of 35 V are shown in the
scattergram of Fig. 5. This scattergram is shown rotated in
Fig. 6 where the abscissa is the diagonal distance from the
origin in Fig. 5 and the ordinate is the distance transverse
to the diagonal. The transverse distribution is narrow which
shows that the signals are well correlated. The transverse
distribution (with finer bins) gave a good fit to a Gaussian
shape with a standard deviation of 106 ADC counts and a χ^2 of
15 for 12 degrees of freedom. The widths of such
distributions were found to decrease with increasing bias
voltage and become approximately constant for voltages above
full depletion (above 35 V). The signal to transverse width
ratio above full depletion is ten to one and is dominated by
the ohmic side noise width below full depletion as expected,
thus showing that the two signals are correlated within the
noise performance of the amplifiers.

CONCLUSION

     Solid state microstripe detectors will be useful for
vertex tracking at the SSC. Monte Carlo simulations indicate
that good position resolution can be obtained for tracks
incident on microstripe detectors at angles up to 60° with
existing readout circuits. Double-sided detectors have been
made by several vendors and signal to noise ratios of from 6
to >10 have been obtained for ohmic side segments.[2-4] Tests on
two Micron Semiconductor, Ltd. multistripe detectors show that
they work well as position sensitive detectors and the charges
collected on the two sides have been measured to be equal
within the noise performance of the amplifiers.

REFERENCES

1.   N. Ushida, et al., "Hybrid Emulsion Spectrometer for the
     Detection of Heavy Flavor States", to be submitted to
     Nucl. Instr. and Meth., (April 1989).

2.   G. R. Kalbfleisch, P. L. Skubic, M. A. Lambrecht, C. D.
     Wilburn, IEEE Trans. Nucl. Sci., NS36, 272 (1988).

3.   G. Apollinari, et al., IEEE Trans. Nucl. Sci., NS36, 46
     (1988).

4.   G. Batignani, et al., IEEE Trans. Nucl. Sci., NS36, 40
     (1988).

5.   F. Bedeschi, et al., IEEE Trans. Nucl. Sci., NS36, 35
     (1988).

AMORPHOUS SILICON DEVICES FOR HIGH ENERGY PARTICLE DETECTION

J. Xi, R. E. Hollingsworth, and A. Madan
Glasstech Solar, Inc. (GSI)
Wheat Ridge, CO 80401

R. Y. Zhu
California Institute of Technology
Pasadena, CA 91101

ABSTRACT

Hydrogenated Amorphous Silicon (a-Si:H) p-i-n diodes with very thick intrinsic layers have been fabricated and tested to explore the feasibility of particle detection in an SSC environment. Diodes with a thickness up to 42 μm have been successfully fabricated on metal coated glass substrates. Direct single particle detection has been achieved for 122 keV gamma rays. A signal was obtained for 2.3 Mev beta rays which are minimum ionization particles. The devices were neutron irradiated with an integrated flux of $10^{14}/cm^2$, resulting in a minor sensitivity drop which was fully recoverable by thermal annealing.

INTRODUCTION

The Superconducting Super Collider (SSC) project requires the development of low cost particle detectors which can handle high event rates and can survive in a high radiation environment. Existing particle detectors fail to meet at least one of these stringent requirements. Crystalline silicon semiconductor detectors are being used in other high energy physics experiments [1] and have been proposed for use in a general purpose SSC environment. [2] However, existing crystalline silicon detectors are expensive and subject to radiation damage at the radiation levels anticipated in SSC detectors. [3] Hydrogenated amorphous silicon (a-Si:H) has the potential to meet the stringent criteria posed by the SSC project. This thin film semiconductor is produced by plasma enhanced chemical vapor deposition (PECVD) and is routinely used in large area applications such as inexpensive solar cells, thin film transistors to drive large area liquid crystal displays, imaging, electrophotography, etc. GSI has developed the manufacturing equipment [4] with the capability of inexpensively depositing a-Si:H on substrates as large as 1 m^2.

We report here the direct detection of 122 keV gamma rays using a-Si:H p-i-n diodes of 42 μm in thickness. These types of devices were irradiated with neutrons of flux $10^{14}/cm^2$ which resulted in a minor decrease in the sensitivity and was fully recovered with thermal annealing. With further development, detectors of this type should be able to detect single minimum ionizing particles. The hydrogen content of a-Si:H (about 12%) could also lead to an e/h value close to one and therefore good energy resolution.

EXPERIMENTAL

The devices studied were thin film amorphous silicon p-i-n diodes with intrinsic layer thicknesses between 2 and 42 μm. The devices were grown on glass substrates coated with Cr or transparent conductive tin oxide with the following structure: substrate/p-i-n or substrate/n-i-p. The devices were fabricated in GSI's commercially available state of the art ultrahigh vacuum multi-chamber plasma enhanced CVD deposition system. Individual deposition chambers were used for the p, i, and n layers to eliminate dopant contamination of the intrinsic layer. The p layer used was a boron doped a-SiC:H made from a gas mixture of silane (SiH_4), methane (CH_4) and diborane (B_2H_6). The n layer used was a phosphorus doped a-Si:H made from a gas mixture of silane and phosphine (PH_3). Both doped layers were approximately 400 Å thick. The intrinsic layer was made from pure silane for low deposition rate material (2 Å/sec) or from pure disilane (Si_2H_6) for higher deposition rates of 10-20 Å/sec. The deposition conditions used were those which generally lead to high solar cell efficiencies (11.2% for a SiH_4 device). [5] The device thickness was directly measured with a stylus profilometer. The thickness over the 100 cm^2 substrate typically varied by ±5%. The devices were completed by thermally evaporating semitransparent Ag dots through a mechanical mask.

The bias dependent collection efficiency of photo-generated carriers was examined to determine the voltage necessary for complete carrier collection. Devices with semitransparent contacts were illuminated with monochromatic light with wavelengths of 500, 650, and 750 nm, corresponding to approximate penetration depths of 0.1, 1.0 and 25 microns respectively. It is well known that in a reverse biased p-i-n diode, holes are swept toward the p layer and electrons toward the n layer; hence holes which are generated near the p layer and similarly, electrons which are generated near the n layer, are quickly and easily extracted since they have only a short distance to travel before they are collected. Therefore, the response due to the short wavelength light in a collection efficiency experiment will be dominated by the carrier which has to traverse the full thickness of the intrinsic layer. The long wavelength illumination approximates to the uniform generation of charged carriers by minimum ionizing particles. In this experiment the incident light was mechanically chopped at 18 Hz and the AC signal was measured with a lock-in amplifier as a function of the DC voltage bias; saturation of the AC signal as the DC bias is increased indicates complete collection of the photo-generated carriers. Since there are differences in the transmission of the contacts, then this experiment gives only a relative measure of the internal collection efficiency.

The particle response of the diodes was examined for 122 and 136 keV gamma rays from [57]Co and 2.3 MeV beta particles from a ^{90}Sr source which are nearly the minimum ionizing particles. An EG&G Ortec 142A charge sensitive preamplifier and an Ortec 673 spectroscopy amplifier were used to amplify and shape the signal. A Lecroy 3001 multichannel analyzer was used to record the spectra. The gamma measurements were performed with the a-Si:H diode providing a trigger for the electronics. The beta particle measurements were performed as a coincidence measurement with a crystalline silicon diode providing the trigger. The crystalline silicon diode was also used to calibrate the sensitivity of the electronics.

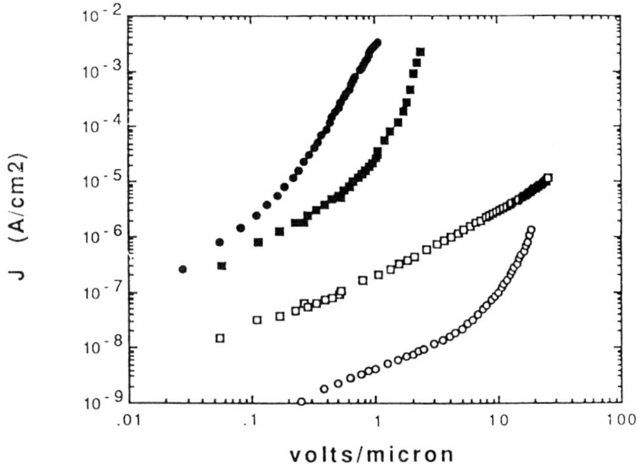

Figure 1. Forward and reverse current voltage characteristics of a 40 μm silane p-i-n (circles) and a 40 μm disilane n-i-p (squares). The solid symbols refer to the device under forward bias voltage.

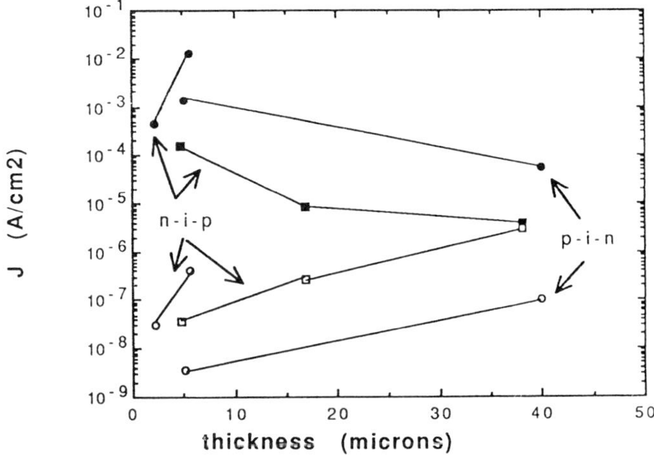

Figure 2. Forward and reverse current at a single applied field as a function of device thickness. The squares are for disilane n-i-p type devices. The circles are for silane devices. The solid symbols are for a forward bias of 0.4 V/μm; the open symbols are for a reverse bias of 10 V/μm.

RESULTS AND DISCUSSION

A. Current-Voltage and Capacitance - Voltage Characteristics

The current voltage characteristics of 40 μm diodes deposited at 2.2 and 21 Å/sec are shown in figure 1. The reverse current of the low deposition rate sample increases approximately linearly up to an applied field of almost 10 V/μm but superlinearly above that field with a breakdown occuring at 25 V/μm. The high deposition rate sample exhibited a larger reverse current. With an applied forward bias voltage the high deposition rate device exhibited a lower current density at a given applied field. This behavior is probably due to a larger density of localized states in the intrinsic layer of the high deposition rate device.

In general the reverse current increases and the forward current decreases for a given applied field as the intrinsic layer thickness is increased, as shown in figure 2. This could be due to changes in the intrinsic amorphous silicon film properties as the deposition thickness increases. Further, the high deposition rate devices generally exhibit worse I-V characteristics for all thicknesses.

The capacitance of the devices was found to be nearly independent of the applied voltage as is to be expected. Table 1 shows the capacitance dependence on i-layer thickness and the dielectric constant as determined from the capacitance measurements which was found to be 12.5 ± .5.

B.   Collection Efficiency Results

Figure 3 shows the collection efficiency of 5 μm thick devices fabricated using SiH_4 (low deposition rate) and Si_2H_6 (high deposition rate) gases with illumination at a wavelength of 500 nm through both the p and n-layers. For illumination through the p-layer, the response was found to be virtually flat with applied bias indicating a complete collection of the electrons for electric fields which was only slightly stronger than the built-in field of the device. For illumination through the n layer, the signal was found to increase strongly with bias voltage until it saturated near a field of 20 V/μm. The Si_2H_6 sample required a slightly higher field to reach saturation. Figure 4 shows the collection efficiency of these devices with uniformly absorbed light of 750 nm. In this case, complete collection occurs at a lower applied field of 7 V/μm, independent of the direction of illumination.

Complete carrier collection could no longer be achieved for the low deposition rate device of 40 μm thickness, as shown in figure 5. For this device, illumination was only possible through the n layer. Holes could not be fully collected, but since hole transport dominates for illumination through the n layer, it was not possible in this instance to determine whether complete electron collection occurred. It should be noted that the long wavelength response increased by a factor of 2 when the field was changed from 10 V/μm to 20 V/μm. Note however, that the largest signal to noise ratio for particle detection occurred at a field of 10 V/μm implying that only half of the generated carriers were collected in this situation.

Table 1   Typical Capacitance measured for p-i-n a-Si Detectors Made from SiH_4

Sample #	613-1-5	736-12-2
Area:	.12 cm^2	.12 cm^2
Thickness	5.7 μm	40 μm
Capacitance	230 pF	37 pF

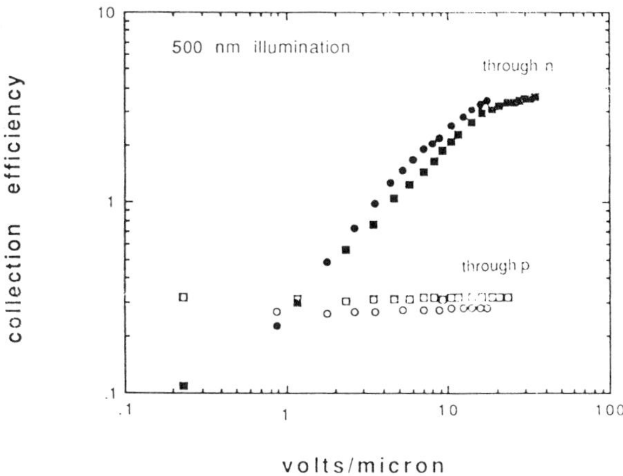

Figure 3. Collection efficiency, in arbitrary units, for a 5 μm thick devices fabricated from silane (circles) and disilane (squares) using strongly absorbed light with a wavelength of 500 nm.

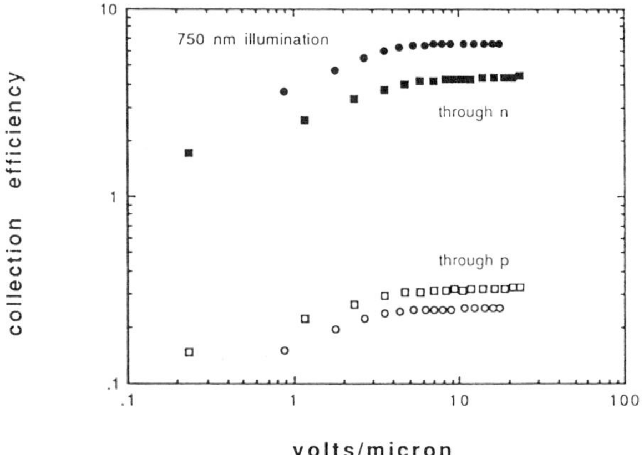

Figure 4. Collection efficiency using weakly absorbed light with a wavelength of 750 nm for the devices shown in figure 3.

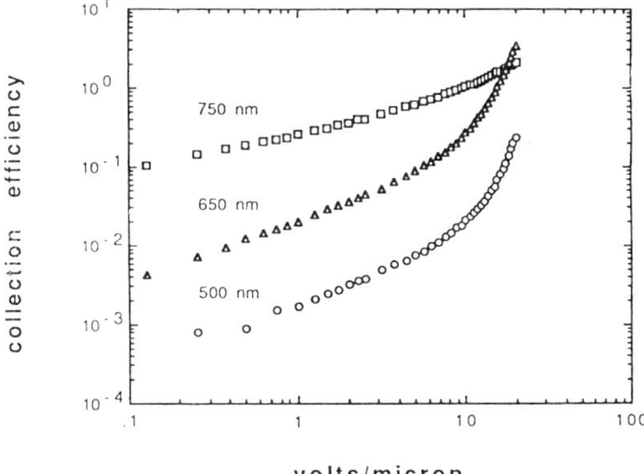

Figure 5. Collection efficiency, in arbitrary units, of a 40 μm thick device fabricated from silane. Illumination is through the n layer.

C.   Particle Detection

Figure 6 shows a typical gamma spectrum from the SiH_4 (low deposition rate) sample. The 122 keV gamma signal was clearly seen above the noise pedestal. It should be noted that the noise corresponds to approximately 17 keV and hence the energy resolution is insufficient to resolve the 136 keV gamma particle. Figure 7 shows the spectrum from the Si_2H_6 (high deposition rate) sample. Since these devices exhibit a high leakage current and correspondingly large noise, the remainder of this discussion concerns the low deposition rate sample.

Figure 8 shows the signal, noise, and signal/noise ratio for different amplifier shaping times. The monotonic increase of the noise with shaping time indicates that the dominant contribution to the noise is the leakage current. The best signal/noise ratio is 7.4 at 1 μsec. It should be noted that the collection of the generated carriers requires about 2-3 μsec since the limitation is the low mobility of holes; in these types of devices the mobility of electrons is 1.4 cm^2/Vsec [6] and hence the electrons should be extracted in about 30 nsec assuming a uniform electric field of 10 V/μm and a thickness of 40 μm.

Figure 9 shows the signal, noise and signal/noise ratio as a function of applied voltage for the same device examined above. The signal should increase with bias as long as the shaping time is shorter than the charge collection time since the carriers have a larger velocity at the higher voltages and are therefore swept out in a shorter time. An additional contribution to the signal is the likely larger collection efficiency at the higher voltages. The noise increased suddenly at 500 V, but the reason was not clear since there was no corresponding large increase of the leakage current. Other devices show a similar increase in the noise between 500 and 700 volts.

Coincidence measurements were also made to detect 2.3 MeV beta rays from a ^{90}Sr source. The signal was inferred from the shift in the peak of the noise pedestal and gave a signal to noise ratio of about 0.3 for a near minimum ionization particle. This corresponds to the collection of approximately 390 electrons with a shaping time of 2 μsec at an applied voltage of 400 V (10 V/μm). Assuming a collection efficiency of 0.5 implies a generation of 20 electrons/micron for a minimum ionizing particle. Samples with better sensitivity are needed to resolve the discrepancy between this value and the value of 50 electrons/μm reported by Perez-Mendez. [6]

D.   Radiation Resistance Testing

One of the low deposition rate 40 μm thick samples was neutron irradiated with a total neutron flux of 10^{12} fast n/cm^2 (> 100 keV), 10^{13} epithermal n/cm^2 (0.5 eV-100 keV) and 10^{14} thermal n/cm^2 (< 0.5 eV) and Table 2 summarizes the results. The leakage current increased by approximately a factor of 4 and the signal from 122 keV γ rays decreased about 20% after irradiation. Modest annealing at 130°C gave a complete recovery of the performance.

Further work is necessary to distinguish the damage effect by fast neutrons, which is believed to be the major damage source for the SSC environment. [3]

CONCLUSION

The direct detection of 122 keV gamma rays and the complete recovery
of the device by neutron damage by low temperature annealing are
encouraging since it suggests that intensive development of the amorphous
silicon as a high energy particle detector is warranted.   In order to
meet the requirements for an SSC calorimeter, two improvements are
needed.   The sensitivity needs to be improved to allow the detection of a
single minimum ionizing particle for calibration purposes and the hole
mobility needs to be improved to allow for fast response times.   Both of
these requirements appear to be acheivable.

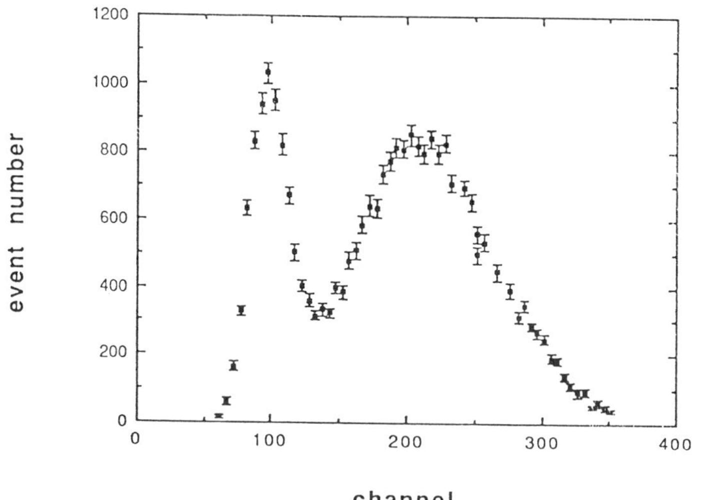

Figure 6. 122 keV gamma spectrum from a
40 μm silane p-i-n device with an applied
field of 400V and 1 μsec shaping time.

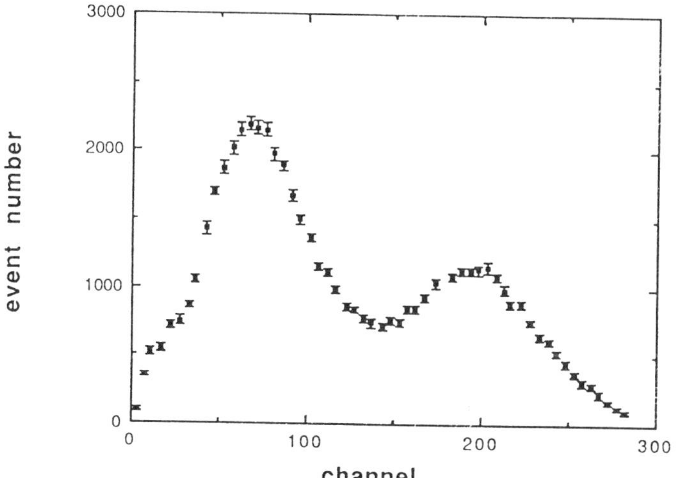

Figure 7. 122 keV gamma spectrum from a
42 μm disilane n-i-p device with an applied
field of 600 V and a 6 μsec shaping time.

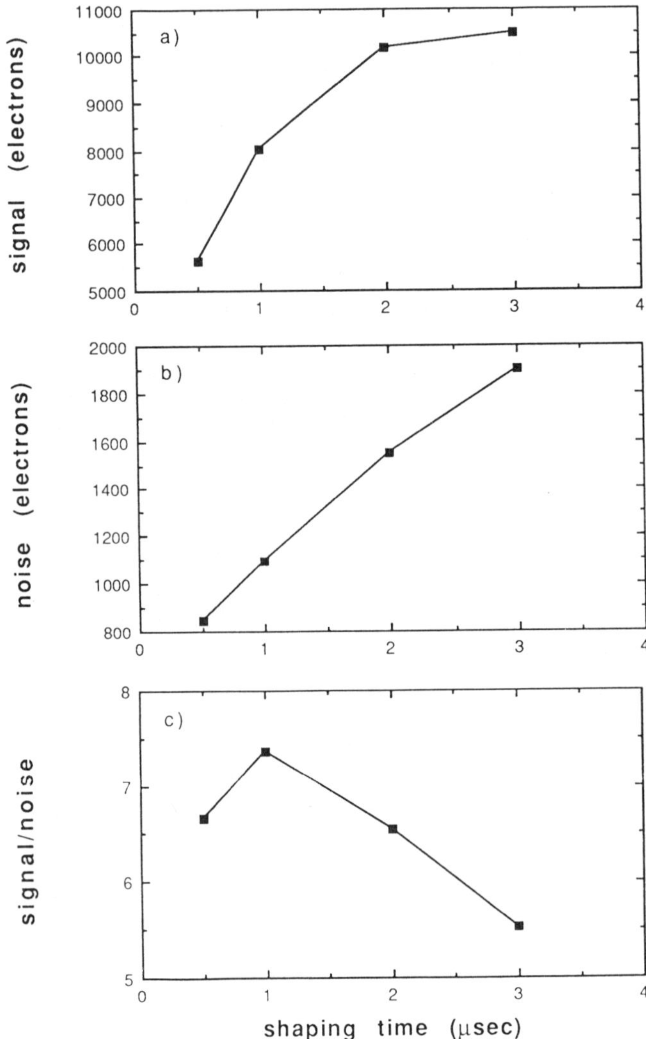

Figure 8. (a) Signal, (b) noise, (c) signal to noise ratio of a 40 µm silane p-i-n to 122 keV gamma rays as a function of shaping time with an applied voltage of 400 V.

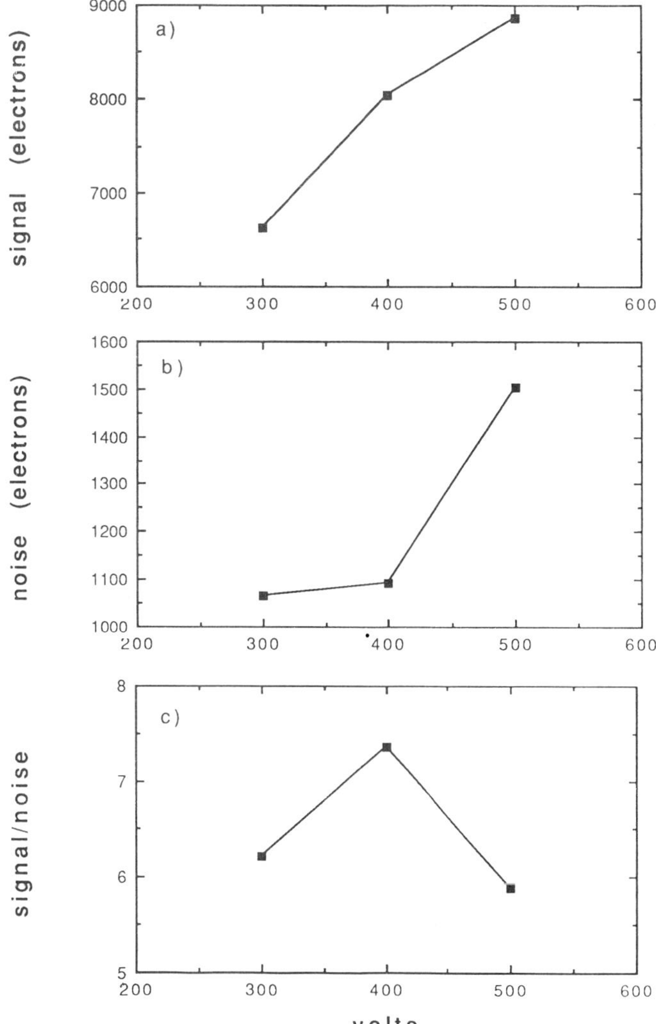

Figure 9. (a) Signal, (b) noise, (c) signal
to noise ratio of 40 μm silane p-i-n to 122
keV gamma rays as a function of voltage
using a shaping time of 1 μsec.

Table 2

Radiation Damage and Recovery by Thermal Annealing

	Before Irradiation	After Irradiation	After Thermal Annealing	
			1 hr. 130°C	15hr. 130°C
Reverse Leakage Current (A)	13×10^{-9}	50×10^{-9}	27×10^{-9}	13×10^{-9}
Signal Peak (Channel)	104	80	105	116
Noise Level (Channel)	18.8	17.3	15.8	18.8

As to the first point, the detection of a single minimum ionizing particle requires a thicker device which exhibit a complete collection of the generated carriers with little increase in the noise. Calculations show that it should be possible to fully deplete a 60 μm thick device if the i layer possessed a density of localized states of 5×10^{14} cm^{-3}, the value measured on a GSI produced sample of 12 μm thickness. [5] In this case relatively straight forward process optimization should yield full depletion with noise on the order of 1000 electrons. If we take the generation rate for minimum ionizing particles to be 50 e/μm [5] it then yields a signal to noise ratio of 3 for a minimum ionizing particle.

As to the second point discussed above, hole mobility needs to be also increased. Typical hole mobilities of intrinsic a-Si:H films are in the range of 0.005-0.01 cm^2/Vsec, which yield collection times on the order of microseconds. [7] Recent work involving a careful control of the hydrogen content of the films has led to hole mobility as high as 0.2 cm^2/Vsec. [8,9] It is expected that the hole mobility could be increased to 0.5 cm^2/Vsec and the material be capable of withstanding 20 V/μm in which case hole collection times on the order of 60 nsec for a 60 μm thick device can be expected.

The anticipated improvements in the sensitivity and the response time could make a-Si:H a technically attractive material for a SSC calorimeter. As mentioned earlier there is virtually no limitation to the size of the substrate that a-Si:H can be deposited upon; since this type of material is used is also an inexpensive solar cells, then it is possible that this could be an economic attractive material of choice for the application under discussion in this paper. Preliminary estimates suggest that a cost of less than $1/cm^2 could be possible for amorphous silicon diode detectors when produced in the quantities required for a SSC calorimeter. This low cost combined with the good technical characteristics could make it a prime candidate as a SSC calorimeter.

SUMMARY

A-Si:H thin film diodes with very thick i layers (42 μm) have been fabricated on glass substrates. A signal to noise ratio exceeding 7 has been achieved in 122 kev γ ray detection experiments. The device also

demonstrated good radiation resistance. By further improving the i layer
quality as well as optimising the device structure, it is anticipated
that minimum ionization particles can be detected. To meet the SSC
experiment performance requirements, it is necessary to increase the hole
mobility of the intrinsic material.

ACKNOWLEDGMENTS

We would like to thank C. Marshall and P. Bhat for making samples.
This work was supported by SBIR Program, Phase I, 1988, project No. DE-
AC05-88ER 80644, of the U. S. Department of Energy.

REFERENCES

[1] R. Beuttenmuller, H. W. Kramer, T. W. Ludlam, V. A. Polychronakas,
     V. Radeka, E.Chisi, C. W. Zabjan, F. Piuz, J. S. Russ, A. Tschulik,
     and M. I. Esten, Nucl. Instr. and Method, A252, 471 (1986).

[2] J. Kirkby et al. Proceedings of the Workshop on Experiments,
     Detectors, and Experimental Areas for the Supercollider, Berkeley
     (1987) p. 388.

[3] D. E. Groom, "Radiation Levels in SSC Detectors", Summer Study on
     High Energy Physics in the 1990's, Snowmass, CO, 1988.

[4] A. Madan, S. Muhl, and B. von Roedern, Materials Research Society
     Symp. Proc. 118, 587 (1988).

[5] P. K. Bhat, C. Marshall, J. Sandwisch, H. Chatham, R.E.I. Schropp,
     and A. Madan, presented at the 20th IEEE Photovoltaic Specialists
     Conference, Las Vegas, Nevada, Sept. 1988.

[6] V. Perez-Mendez, S. N. Kaplan, G. Cho, I. Fujieda, S. Qureshi,
     W. Ward, and R. A. Street, Nucl. Instr. and Method A273, 127 (1988).

[7] R. A. Street, J. Zesch, and M. J. Thompson, Appl. Phys. Lett. 43,
     672 (1983).

[8] N. Shibata, K. Zukuda, H. Ohtoshi, J. Hanna, S. Oda, and I. Shimizu,
     Materials Research Society Symp. Proc. 95, 225 (1987).

[9] J. Hanna, A. Kamo, M. Azuma, N. Shibata, H. Shirai, and I Shimizu,
     Materials Research Society Symp. Proc. 118, 79 (1988).

ADVANCED COMPOSITE STRUCTURES FOR SUPERCONDUCTING SUPER COLLIDER

Vicki Lynn Morris

Structural Composites Industries (SCI)
325 Enterprise Place
Pomona, California 91768

ABSTRACT

Advanced composite structures are currently being utilized in
numerous cryogenic systems. They offer low thermal conductivity, high
strength and stiffness, lightweight, low outgassing, integral fittings,   and
other advantages.   Several types of fibers and resins are available to meet
extreme structural and thermal requirements.   Typical properties for filament
wound composite laminates are provided. Design considerations for various
configurations, such as straps, tubes, and struts, are discussed.   Typical
applications are   dewars for spacecraft, coolers for   satellite instruments, and
superconducting magnets for medical imaging and high-energy physics.   The
cryostat developed by Fermi National Accelerator Laboratory for the SSC dipole
magnets has advanced composite support posts and anchor tie bars.

Figure 1   Filament Wound Carbon Composite Inner Tube and S-Glass Composite
Outer Tube of Cryostat Cold-Mass Support Post for SSC Dipole Magnet

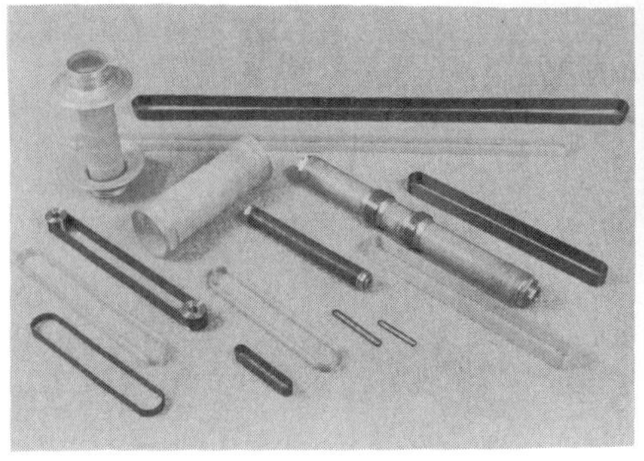

Figure 2   Typical Composite Structures for Cryogenic Applications

Figure 3   Filament Winding

INTRODUCTION

Filament wound composite structures have an important role to play in the rapidly expanding cryogenics world. They offer high mechanical strength and stiffness with low thermal conductivity for many aerospace and industrial applications.

Considerations in selecting the composite material and configuration include tension, compression, flexural, shear and torsion loads; fatigue, deflection and vibration requirements; creep; temperature gradient; installation geometry; and, of course, cost trade-offs.

MATERIALS

Fibers

Several types of fibers are candidates for use in low thermal conductance structural members. Fibers normally utilized are glass, carbon, Kevlar, and more recently, alumina. Glass and alumina offer the lowest thermal conductivity above 100 K so are often the materials of choice for applications involving ambient to cryogenic temperatures. However, below about 80 K carbon offers equal or lower thermal conductance. Thus, some cryogenic systems utilize glass supports between ambient and 300 K and carbon supports between 80 K and 4.5 K.

In general, high modulus fibers provide best fatigue strength while high strength fibers allow minimum cross-section for cryogenic supports. Some typical fiber properties are as follows:

Fiber	Tensile Strength (ksi)	Tensile Strength (msi)
E-Glass	500	10
S-Glass	665	12
Kevlar 49	525	18
Alumina	210	28
T-300 Carbon	450	33
IM-6 Carbon	650	40
HM-S Carbon	360	50
T-1000 Carbon	1016	43

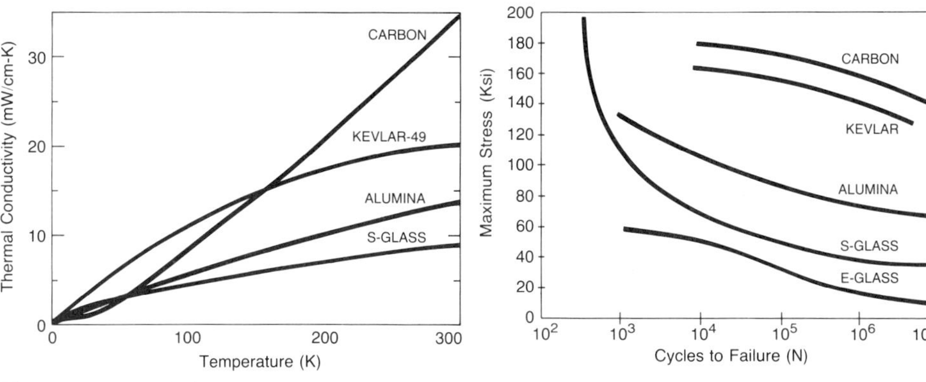

Figure 4   Relative Thermal Conductivities of Unidirectional Epoxy Composite Materials

Figure 5   Relative Fatigue Behavoir of Unidirectional Epoxy Composite Materials

Structures fabricated with any of these fibers have high strength-to-weight ratios compared to metal ones.

Fatigue strength, of course, depends on configuration as well as material properties.

Resins

Resins selected for cryogenic applications must have high strength, good interlaminar shear properties, and serviceability at temperatures ranging from cryogenic to about 350°F.   Proper resin choice results in non-brittle composite structures at low temperatures.

In addition, many cryogenic systems require low outgassing for vacuum compatibility.   The criteria normally used for selection are that a maximum of 1.0 percent mass loss and a maximum of 0.10 percent volatile condensible material are permitted per ASTM E-595 and NASA SP-R-0022A.   Test results for typical cryogenic resins are shown below.

Resin	SCI REZ 081	SCI REZ 100
Total Mass Loss (TML), %	0.13 - 0.24	0.17
Collected Volatile, Condensible Material (CVCM), %	0.00	0.01
Water Vapor Recovered (WVR), %	0.03 - 0.04	0.07

Composite Laminates

At SCI, advanced composite structures are manufactured by filament winding.   Hundreds of thousands of continuous fibers are precisely oriented in predetermined patterns to provide strength and stiffness in exactly the amount and direction needed.   The resin holds the filaments in place and facilitates load transfer from one fiber to the next.   Final resin cure is usually at 250 to 350°F.

Typical unidirectional laminate data for various fibers with SCI REZ 100 (66 volume percent filament) at room temperature are as follows:

	S-2 Glass	Kevlar-49	IM-6 Carbon	T-1000 Carbon
Strengths (ksi)				
Long., (tension), F_x	230	220	417	540
Long., (comp.), F_x'	140	44	275	350
Trans., (tension), F_y	9	4	7	7
Trans.,(comp.), F_y'	20	20	30	30
Shear, S_{xy}	12	7	14	14
Moduli (msi)				
Long., E_{xx}	8.10	12.63	26.43	27.47
Trans., E_{yy}	1.30	0.83	1.30	1.30
Shear, G_{xy}	0.80	0.31	0.75	0.75
Poisson Ratio	0.25	0.34	0.27	0.27
Ultimate Strains				
Long., (tension),e_x	0.028	0.018	0.015	0.019
Long., (comp.), e_x'	0.017	0.004	0.013	0.012
Trans., (tension), e_y	0.007	0.005	0.004	0.008
Trans., (comp.), e_y'	0.015	0.016	0.015	0.014
Shear, e_{xy}	0.012	0.028	0.023	0.023

Composite material strength increases about 30 percent in going from ambient to cryogenic temperature; modulus increases by about 10 percent.

The coefficients of thermal expansion for S-Glass laminates are 1.4×10^{-6} from 75 to 250°F and 1.6×10^{-6} from 75 to -320°F. Carbon laminates have much greater dimensional stability, displaying essentially no contraction or expansion over very wide temperature ranges.

Advanced composite structures have low vibration susceptibility because of their high structural dampening properties. They generally do not exhibit creeping tendencies.

<u>Age Life in Vacuum</u>

SCI recently evaluated S-Glass composite suspension supports that had been on several spaceflight missions and enclosed in evacuated dewars for over eight years. Inspection revealed insignificant dimensional changes, and testing showed no degradation of tensile or fatigue strengths. In fact, the average ultimate tensile strength was slightly higher than the average ultimate tensile strength of pre-flight lot sample test parts.

CONFIGURATION

Design

The philosophy used to define the configuration of a thermally optimized composite structure is to push its performance to the practical limit to meet the load and fatigue requirements with minimum cross-section and longest length within the confines of the cryogenic system geometry. Typical shapes consist of straps, tubes, struts, posts, rings and shells.

Straps

Support straps, for example, carry only tension loads. Thus, their sizing is dictated by the required tensile ultimate and fatigue strengths and the minimum allowable resonant frequency of the support system. Fatigue strength is generally the design driver.

SCI has fabricated straps in sizes ranging from 1 to 32 inch long, 0.04 to 1 inch wide and 0.012 to 0.173 inch thick with 0.14 to 1 inch pin diameter.

Ultimate failure is caused by combined tensile and bending stress at the straight leg to bend radius transition region.

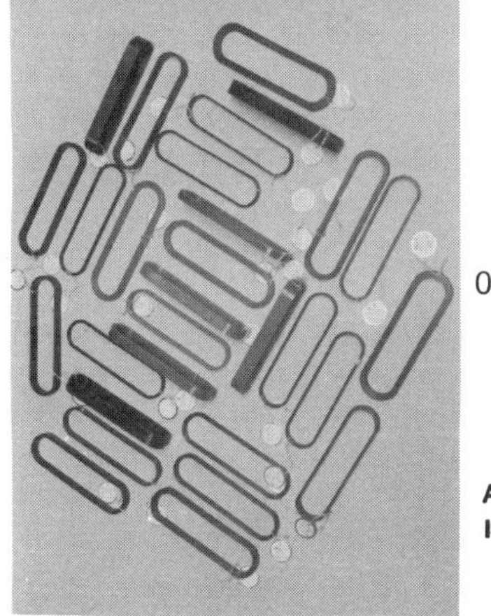

Figure 6   S-Glass and Carbon Support Straps

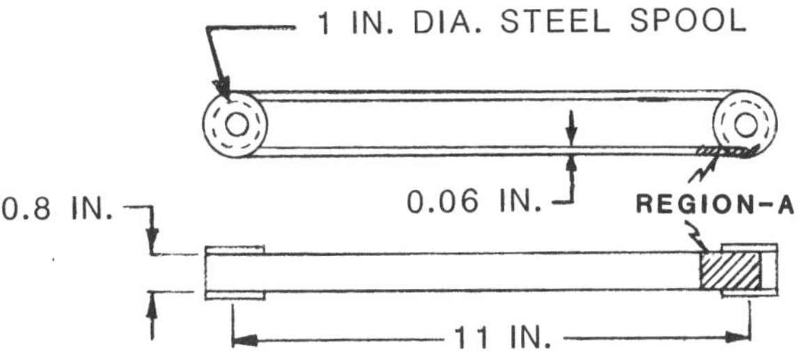

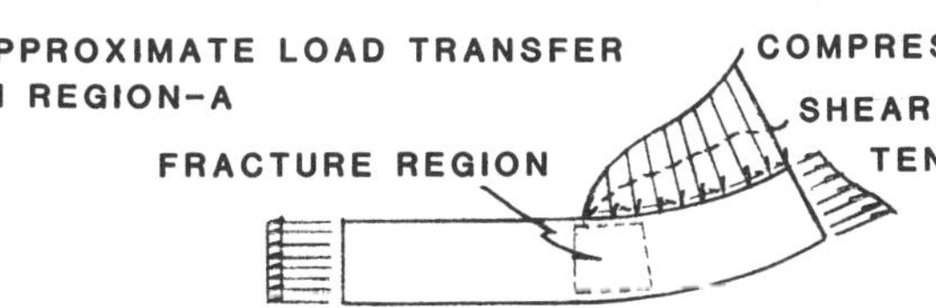

Figure 7   Typical Composite Support Strap

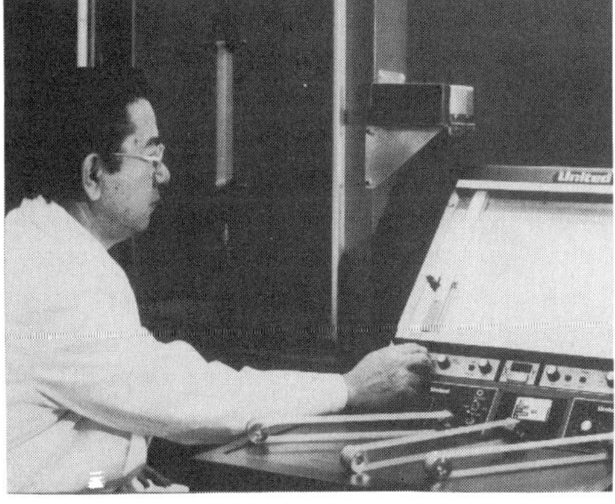

Figure 8 - Tensile Testing of Composite Support Strap

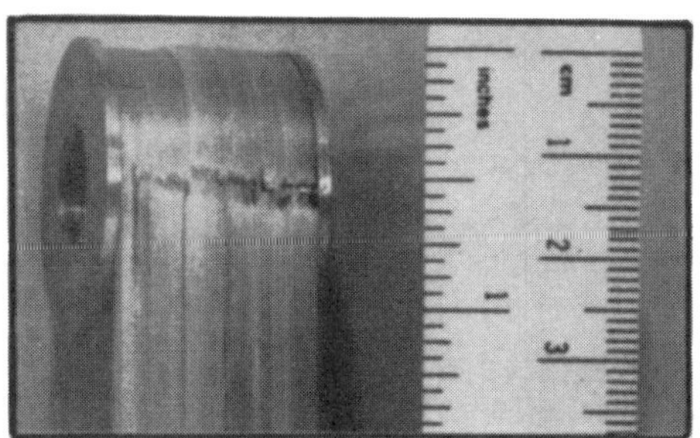

Figure 9   Typical Ultimate Failure of Composite Strap

The composite tensile stress at failure (static strength) is calculated by dividing the breaking load (lb) by the cross-sectional area (sq in) of the strap. The delivered strap static strength is a function of strap geometry--- specifically, the ratio of the load pin diameter to the thickness (D/T) ratio. Delivered strength is higher for "thinner" than for "thicker" composites. Thus, D/T can be utilize to predict expected static strength of straps.

Straps have a linear load-deflection relationship to their static strength. Because they are often preloaded in tension to prescribed levels for specific applications, precise knowledge of their spring constant (K) is sometimes essential; $K = P/\Delta L$, in which P is the tensile load and ΔL is the axial deflection between load pin centers.

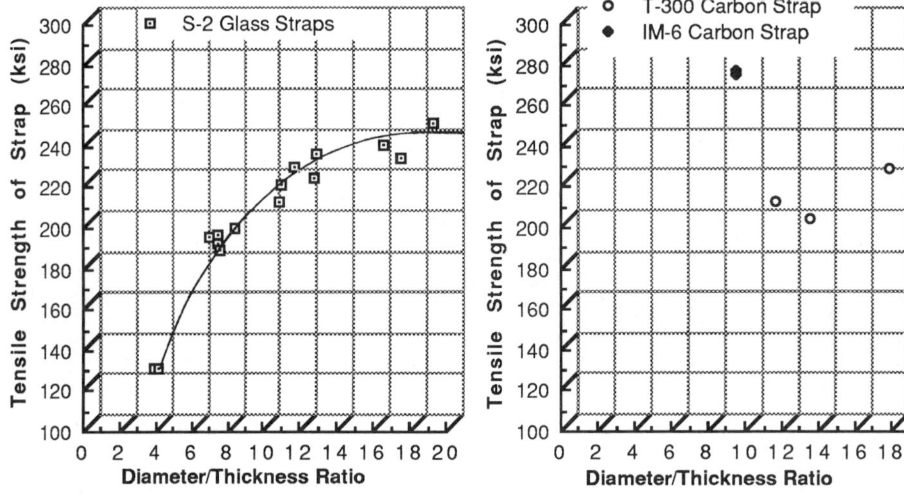

Figure 10   D/T Ratio vs. Tensile Strength for S-Glass Straps

Figure 11   D/T Ratio vs. Tensile Strength for Carbon Straps

For S-Glass straps at room temperature, a maximum cyclic stress/static strength ratio of 0.5 results in about 1000 fatigue cycles while a ratio of 0.2 results in over 1,000,000 fatigue cycles. Fatigue resistance is improved at cryogenic temperatures with the stress level required for failure at a specified number of cycles approaching twice the value required at room temperature.

Composites fabricated with higher modulus fibers, such as alumina and carbon, yield even higher fatigue strengths than S-Glass. One explanation is that there is less damage at the fiber-matrix interface because there is less interfacial strain than lower modulus composites for the same cyclic stress level.

Figure 12   Advanced Composite Tubes, Posts and Rings

<u>Struts, Tubes and Other Configurations</u>

Filament wound composite struts, tubes, posts, rings and shells can be designed to carry tension, compression, flexural and torsional loads. Structural efficiency is optimized by tailoring orientations and thicknesses of filaments and by utilizing high fiber content. Precision control of thermal expansion and dimensional stability are possible by proper choice of materials and wrap patterns.

Figure 13   Large Size Filament Wound Composite Tubes

Figure 14   Square Filament Wound Tubes

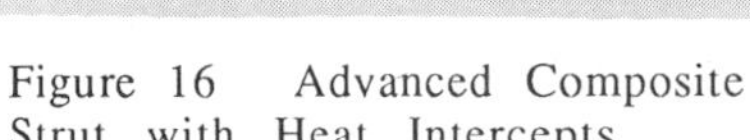

Figure 16   Advanced Composite Strut with Heat Intercepts

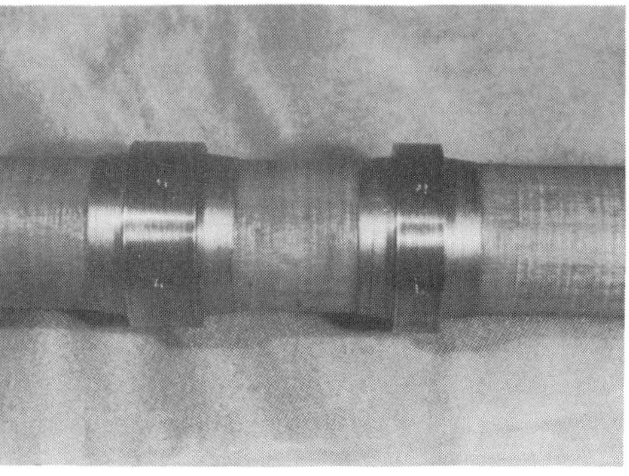

Figure 15 - Filament Wound Shells with Integral Fittings

Large size, one-piece or low-part-count structures in co-cured assemblies can be filament wound.

Integral metal or other fittings can be wound into composite structures, and heat intercepts are easily added for cryogenic applications.

APPLICATIONS

Some of the cryogenic systems currently utilizing advanced composite structures are: hydrogen, oxygen and helium dewars for spacecraft; coolers and refrigerators for satellite instruments; superconducting magnets for medical imaging and high-energy physics.

<u>Spacecraft and Satellites</u>

Aerospace dewars operating at cryogenic temperatures are a well established technology.

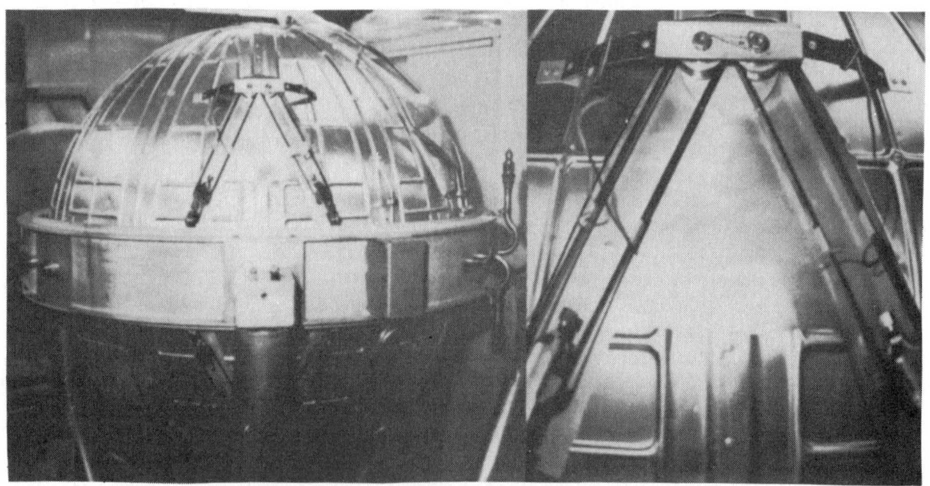

Figure 17   Filament Wound Straps on Space Shuttle PRSA Tanks

The Power Reactant Storage Assembly (PRSA) tanks of the Space Shuttle Orbiter store the oxygen and hydrogen used to generate electrical power and make-up oxygen for the crew compartment. Each tank is a dewar consisting of two concentric spheres. The inner sphere, or pressure vessel, contains the stored fluid at cryogenic temperatures and supercritical pressures. The outer sphere allows maintenance of a high vacuum between the two spheres for thermal insulation. Twelve filament wound support straps, supporting the pressure vessel from the outer shell, are located within the vacuum space between the two spheres. The straps are preloaded to as high as 2000 lb to prevent compression loading during Space Shuttle launch vibration and $\pm$ 5g constant acceleration loads. Total static supported weight is about 90 lb for the oxygen tank and 180 lb for the hydrogen tank.

The Infrared Astronomical Satellite (IRAS), launched in 1983, successfully completed its mission with a lifetime of 10 months. Its 24-inch telescope had to be cooled to reduce thermal background noise for the mapping of celestial infrared sources. Cryogenic temperature was achieved using super critical helium stored in a toroidal main cryogen tank that was supported with filament wound thermal isolation straps.

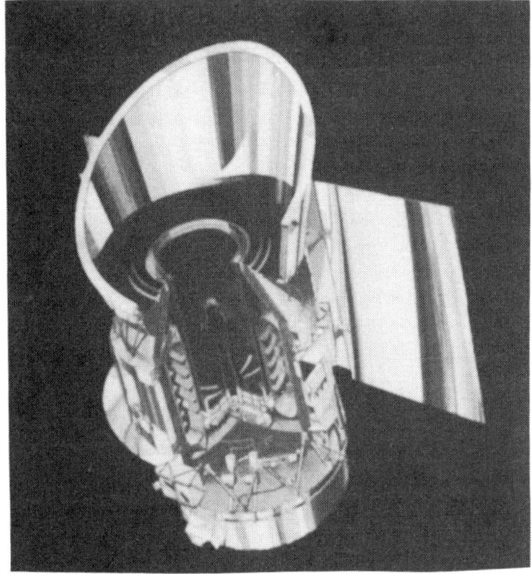

Figure 18   Composite Support Straps on IRAS

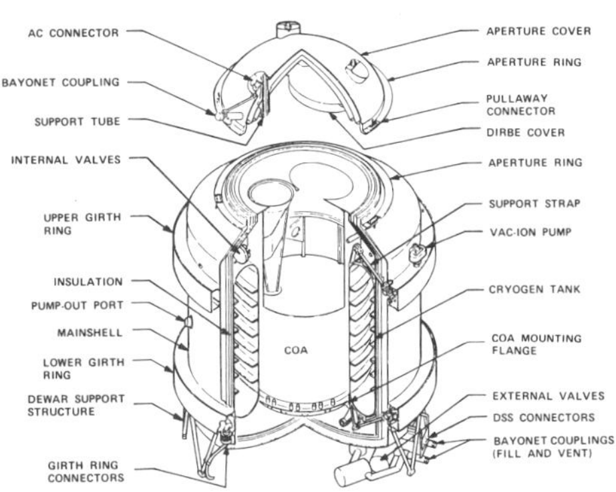

Figure 19   Filament Wound Straps on COBE

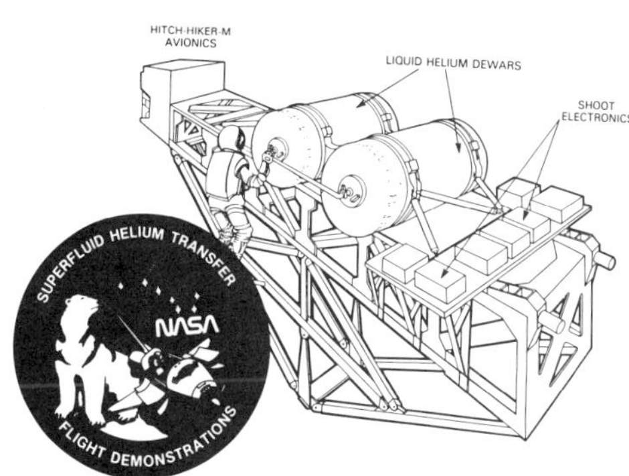

Figure 20   Filament Wound Straps for SHOOT

Figure 21   Composite Straps on OAMP

The Cosmic Background Explorer (COBE) dewar is expected to have a 14 month on-orbit lifetime.

Smaller size, advanced composite suspension bands are used for satellite instruments on the Focal Plane Assembly (OAMP), Geostationary Operational Environmental Satellite (GOES), Galileo Spacecraft, Thematic Mapper of Landsat D Spacecraft, High Energy Astronomy Observatory (HEAO), and other systems.

During the next ten years, several space missions will require superfluid helium replenishment at two or three year intervals to extend their lifetimes. Thus, the Superfluid Helium On Orbit Transfer (SHOOT) flight demonstrations will provide the technology and critical operations required for liquid helium transfer in space. The two supply dewars consist of identical vacuum vessels mounted side by side. Each of these vessels enclose a 55 gallon liquid helium tank that is suspended from the outer shell with six filament wound thermal isolation support straps.

<u>Superconducting Magnets</u>

The suspension system in a superconducting magnet performs two essential functions: (1) It resists internally and externally generated structural loads imposed on the cold mass assembly to ensure that the position of the assembly is stable over the operating life of the magnet; and (2) It serves to insulate the cold mass from heat conducted from the outside world.

The main magnet of a magnetic resonance imaging (MRI) system is often housed in a liquid helium tank that is supported by composite straps and struts for high cryostat mechanical and thermal performance. Some MRI systems are installed in mobile vans so must tolerate the mechanical loads and thermal exposures associated with road transport for distances exceeding 100,000 miles in many climates.

Stringent refrigeration requirements, large structural loads, and minimum radial space have lead to the development at Fermilab of a reentrant tube cryogenic support for the dipole magnets of the SSC. It consists of two concentric composite tubes, an inner carbon tube and an outer S-Glass tube. The connection between the two tubes is via a metallic tube located in the vacuum space between them. It connects the top of the glass tube to the bottom of the carbon tube and allows the inner tube to reenter the outer, increasing the conductive · heat path.

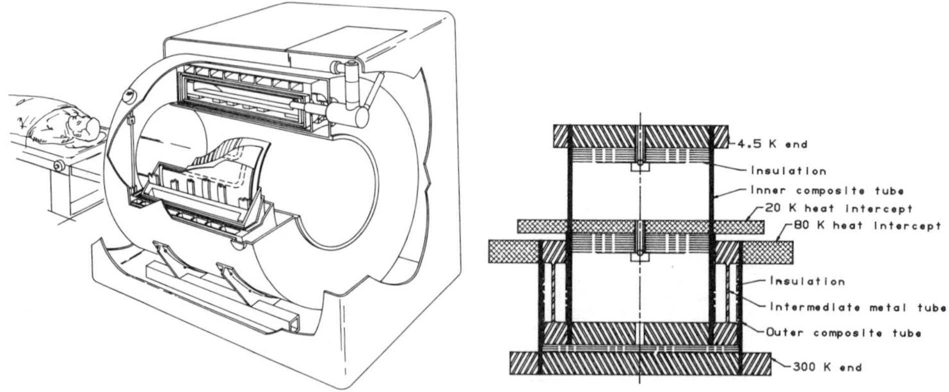

Figure 22   Superconducting MRI Magnet with Composite Supports

Figure 23   Cross-Section of an SSC Cryostat Support Post

The cryostat with carbon and S-Glass tubes meets SSC thermal and mechanical requirements. The heat leak into the 4.3 degree portion of the magnet has been measured at less than 0.2 watt over about 55 feet, even less than the design requirement of 0.3 watt.

Filament wound carbon composite was selected to meet the SSC cryostat anchor tie bar needs of 4.5 K operating temperature, over 18 msi elastic modulus, over 40 ksi allowable tensile strength, and dimensional stability.

CONCLUSIONS

Spacecraft and satellite requirements for advanced composite cryogenic structures are continuing to grow. Technology to improve the hold time of dewars is being developed, including new composite materials and configurations for support structures.. With more sensitive, cooled detectors, instruments in space will be able to collect even more advanced scientific data.

There are new applications for magnetic resonance; imaging and spectroscopy techniques are advancing rapidly. Compact, rugged, low-cost magnets with optimized suspension systems are needed to make possible broader applications in medicine, as well as industrial applications in process monitoring and non-destructive evaluation.

Low thermal conductivity, high mechanical performance, lightweight structures will be required for numerous applications in the energy, electronics, and transportation areas.

Advanced composite, filament wound structures can meet these diverse needs as well as SSC performance requirements.

REFERENCES

1. Hopkins, R.A. & Payne, D.A., Advances in Cryogenic Engineering Materials, Vol. 33, p. 925, 1988

2. Kriz, R.D. & Sparks, L.L., Advances in Cryogenic Engineering Materials, Vol. 34, p. 107, 1988

3. Morris, E.E., Composites for Extreme Environments, ASTM STP 768, p. 95, 1982

4. Nichol, T.H., Neimann, R.C., Gonczy, J.D., Advances in Cryogenic Engineering Materials, Vol. 33, p. 227, 1988

5. Rios, P.A. & Rhodenizer, R.L., Advances in Cryogenic Engineering Materials, Vol. 33, p. 9, 1988

DEVELOPMENT OF IMPROVED, RADIATION-RESISTANT PLASTIC AND LIQUID SCINTILLATORS FOR THE SSC

C. Zorn, M. Bowen, S. Majewski, J. Szaban, and R. Wojcik

Department of Physics, University of Florida
Gainesville, Florida 32611

C. Hurlbut, and W. Moser

Bicron Corporation, Newbury, Ohio 44065

ABSTRACT

Plastic scintillators available on the market suffer from radiation damage at doses much lower than the ones expected at the SSC. The two major effects are induced transmission losses and damage to the scintillating fluors. The dominant radiation-induced absorption effects in the blue part of the transmission spectrum can be remedied by the use of large Stokes shift fluors emitting in the green region above 500 nm. With these fluors, the newly produced scintillators based on polyvinyltoluene and polystyrene have greatly enhanced radiation resistance. In cases when even the improved plastic scintillators will fail, liquid scintillators can be used. When damaged, the active material can be replaced without dismounting detector modules. Contrary to expectations, our tests showed that standard liquid formulations are damaged at rather low radiation doses of 1-5 Mrad. Newly developed formulations are significantly more radiation resistant.

INTRODUCTION

Design studies of the proposed SSC accelerator facility have brought to the foreground the previously ignored problem of radiation damage to the detecting elements of a high energy physics spectrometer [1,2]. Annual doses of 10 Mrad are expected in the forward regions of an interaction point. Common elements in such spectrometers are plastic and liquid scintillators, particularly as elements of calorimeters. Such detectors will be especially prone to high radiation doses due to electromagnetic and hadronic showers. It is imperative that studies of current organic scintillators be carried out in order to ascertain the extent of the problem and alleviate it.

We have been performing an initial study of both plastic and liquid scintillators with the dual goals of understanding the phenomenology of radiation damage in such materials, and finding a solution to the problem. This report is a summary view of our current results. Some additional details not presented here concerning the experimental studies can be found elsewhere [3,4,5,6].

The plan of the rest of the paper is to present two sections giving an overview of the experimental methods used in the studies. This is followed by two sections concerning the observed phenomena, one on plastics and the other on liquids. Finally we close with some conclusions and mention some other avenues of investigation.

EXPERIMENTAL PROCEDURE

(a) Handling of Samples

Samples of the plastic and liquid scintillators were provided by Bicron Corporation. The plastics were machined into right circular cylinders 1.5 cm in diameter and 1 cm in thickness. All sides were highly polished. The standard reference scintillator was BC408, a premium plastic scintillator manufactured by Bicron Corporation, with an emission spectrum peaking at 425 nm. The experiments involved having the samples placed in sealable aluminum canisters through which a specific atmosphere could be circulated. The "standard" atmospheres were argon (an inert gas) and air (oxygenated atmosphere) although other possibilities were tried during the course of the studies (namely oxygen, humidified air, nitrogen, and helium).

The samples were irradiated by a 150 Ci ^{60}Co source at a rate of approximately 80 krad/hr up to dosages of 10 Mrad or so. After the final irradiation, the samples were allowed to recover, again in the same atmospheres as during irradiation. For some of the studies, the samples were allowed to recover at a higher ambient temperature (50°C) in order to accelerate recovery. In addition, some samples were heated in a small oven at 50°C during irradiation in order to study the effect upon the extent of radiation damage.

The liquid scintillator samples were poured into small open vials (2 cc) under controlled gas atmospheres of argon, nitrogen, oxygen, and air. The samples had been previously saturated in the appropriate gases by vigorously bubbling gas through the scintillator. The liquid-filled vials were sealed into aluminum canisters and placed into a 600 Ci ^{60}Co irradiator. As with the solid samples, controlled atmospheres were provided for each canister. The dose rate for the liquid samples was measured to be 33 krad/hr, and the typical dose was 5 Mrad.

(b) Experimental Measurements

Scintillation light output, fluorescence emission, and optical transmission were the standard quantities used to gauge the extent of radiation damage during irradiation and recovery. All measurements were normalized to the pre-irradiation data for the samples.

Scintillation pulse-height spectra were measured in a setup wherein either a ^{137}Cs or ^{54}Mn gamma source (1 μCi) was used to excite the sample. The scintillation light was detected by several types of phototubes including the standard bialkali types (EMI 9813QKB, Hamamatsu R1332Q), and the yellow-enhanced multialkali varieties (EMI 9816KBM, Hamamatsu R1333). The pulse height spectra were stored on a LeCroy 3001 qVt before being transferred onto an IBM PC-AT microcomputer via a CAMAC interface. In order to maximize the signal, the liquid samples were poured into quartz cuvettes wrapped in teflon tape and the bottom of the cuvette was coupled to the phototube faceplate via optical grease.

The fluorescence emission spectra were measured with a Hewlett-Packard 1046a Fluorescence Detector. Again the data was transferred and stored on the IBM PC-AT. For some fluorescence measurements with the plastic samples, an alternative detector was constructed which utilized a continuous x-ray source (X-Ray Technologies, Inc.) to excite the sample. The details of this device have been reported elsewhere [5]. The transmission spectra were measured with a Hewlett-Packard 8452a Diode Array Spectrophotometer. No correction was made for any reflectance at sample/air interfaces, and so the transmission data does not indicate the absolute transmittance. Again, the data was transferred and stored on a IBM PC-AT (via a GPIB interface).

EXPERIMENTAL OBSERVATIONS

(a) Plastic Scintillators

Before seeking alternative scintillators, it is useful to investigate methods of increasing the resistance level of a standard scintillator. Since radiation damage is due to chemical changes in the scintillator, it is reasonable to assume that the nature of the surrounding gas atmosphere may have a critical role to play in the chemical processes. Additionally, the ambient temperature will affect the rate of the reactions.

With this in mind, samples of the standard scintillator BC408 were irradiated to doses of up to 6 Mrad in different atmospheres at room temperature before being allowed to recover in the same atmospheres but at an elevated temperature of 50°C in order to increase the rate of recovery. Figure 1 displays the results of this irradiation. During irradiation, there are no significant differences in the extent of damage due to the different atmospheres. This changes radically during the recovery period. The sample in argon is seen to recover to a significantly higher level than the one left in oxygen.

In another experiment, the BC408 samples were irradiated at the elevated temperature of 50°C during irradiation as well as during recovery. As figure 4 shows, the scintillation loss is much greater for a sample irradiated in air as compared to a sample irradiated in argon. Again, the sample in argon recovers to a greater extent that the one in air, although both samples recovered to a far lesser extent than the corresponding samples irradiated at room temperature.

It is clear from the data that oxygen plays a strong role in radiation-induced chemistry. The difference in the results for room temperature and heated irradiations indicate the importance of sample size and dose rate. The sample size will affect the extent to which the ambient gases can diffuse into the samples and hence participate in any chemical reactions. Out of necessity, the samples are submitted to an accelerated testing procedure after which the recovery process is observed. In a real situation within a particle physics spectrometer, the scintillator will be irradiated at a much lower dose rate. Since the recovery process requires a period of about one month for a plateau to be achieved, it is imperative that radiation damage be assessed after recovery has been completed.

We note here one possible caveat. It is known from studies of the effects of radiation upon the mechanical properties of polymers that the presence of oxygen during low dose irradiation can produce much more damage than expected in an accelerated test [7]. This is presumably due to the fact that atmospheric oxygen cannot diffuse into the sample during the accelerated test at a rate enabling it to participate in the chemistry. As expected, accelerated tests are valid in an oxygen-free environment.

In a future experiment, we plan to use a vacuum chamber to slowly remove all dissolved oxygen from the scintillator since it is likely that the oxygen already present in the sample will play a dominate role during irradiation. If the presence of oxygen is found to be highly destructive to the optical qualities of scintillators at low dose rates, then it will be imperative to remove the oxygen in and around the scintillator. As a final note, we point out that Holm and Wick [8], in a study of the polystyrene based scintillator, SCSN-38 [9], found that the final scintillation output, after recovery, was a function of the total dose, and not the dose rate (which varied from 30 Gy/hr to 10 kGy/hr where 1 Gy = 100 rad). This is one demonstration of the possibility that accelerated radiation damage tests may be closely related to the damage seen in actual experimental situations.

The current data indicates that the fluorescent component of BC408 was probably not significantly damaged. Figure 2 displays a comparison of the scintillation loss to the integrated transmission loss in BC408 and undoped PVT. If there has been any damage to the fluors, then they have not added to the optical losses. In addition, as seen in figure 3, the transmission curves reach the same cutoff point (about 400 nm) even during irradiation. If a significant number of fluorescent molecules were damaged, then one would expect the cutoff point to move below 400 nm (i.e., the fluor is no longer present to absorb light below 400 nm). Another study

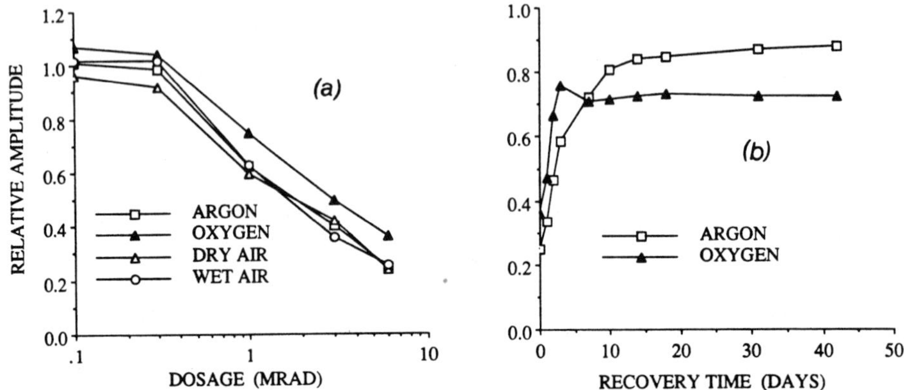

FIGURE 1. BC408 plastic scintillator irradiated in different atmospheres. The pulse height data are derived from a [54]Mn gamma source. The samples were heated at 50°C during recovery. Note the difference in recovery between those samples left in argon and oxygen atmospheres.

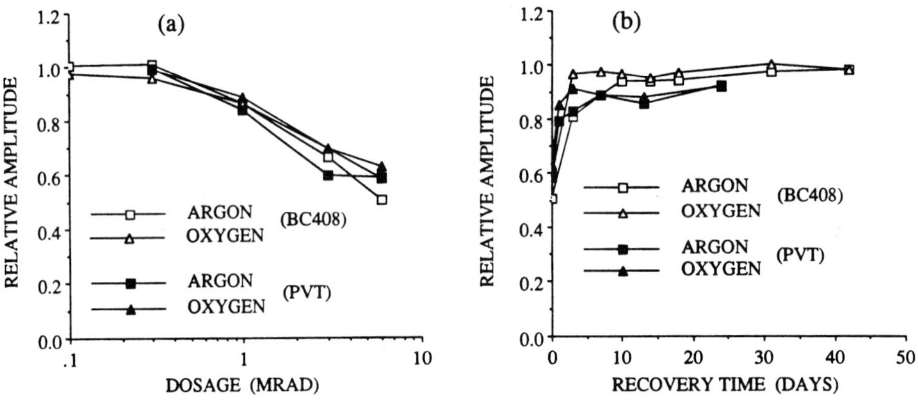

FIGURE 2. Comparison of changes in transmission (weighted by the fluorescent emission spectrum and the phototube response curve) for BC408 and undoped polyvinyltoluene (PVT).

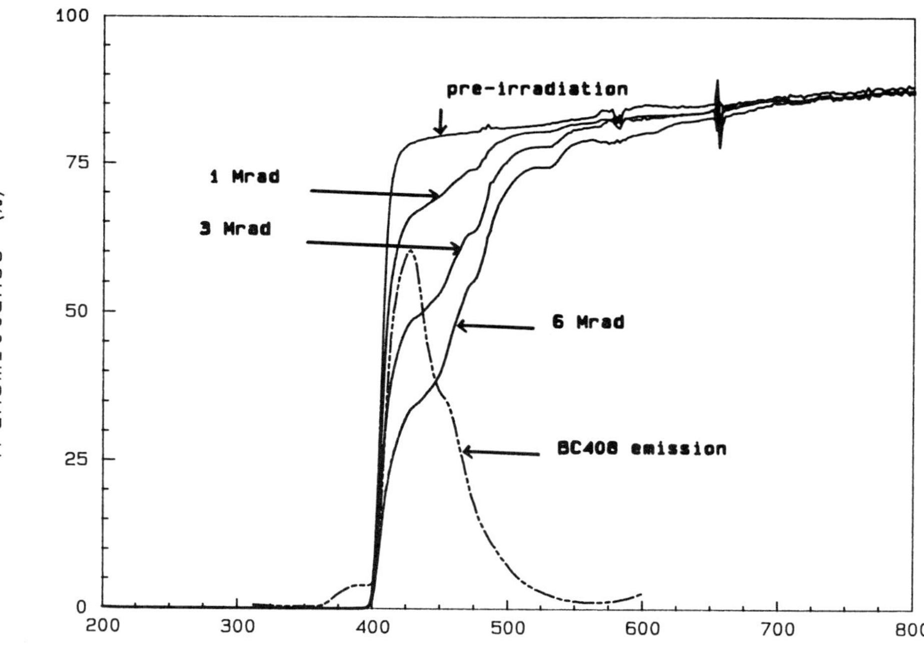

FIGURE 3. Transmission spectra for a BC408 sample (in argon) during irradiation. The transmission curves corresponds to pre-irradiation, 1 Mrad, 3 Mrad, and 6 Mrad. Superimposed is the measured fluorescence emission spectrum for BC408. Note that damage to the plastic base is seriously affecting the amount of light output of the damaged scintillator. This suggests that one should shift the emission spectrum above 500 nm.

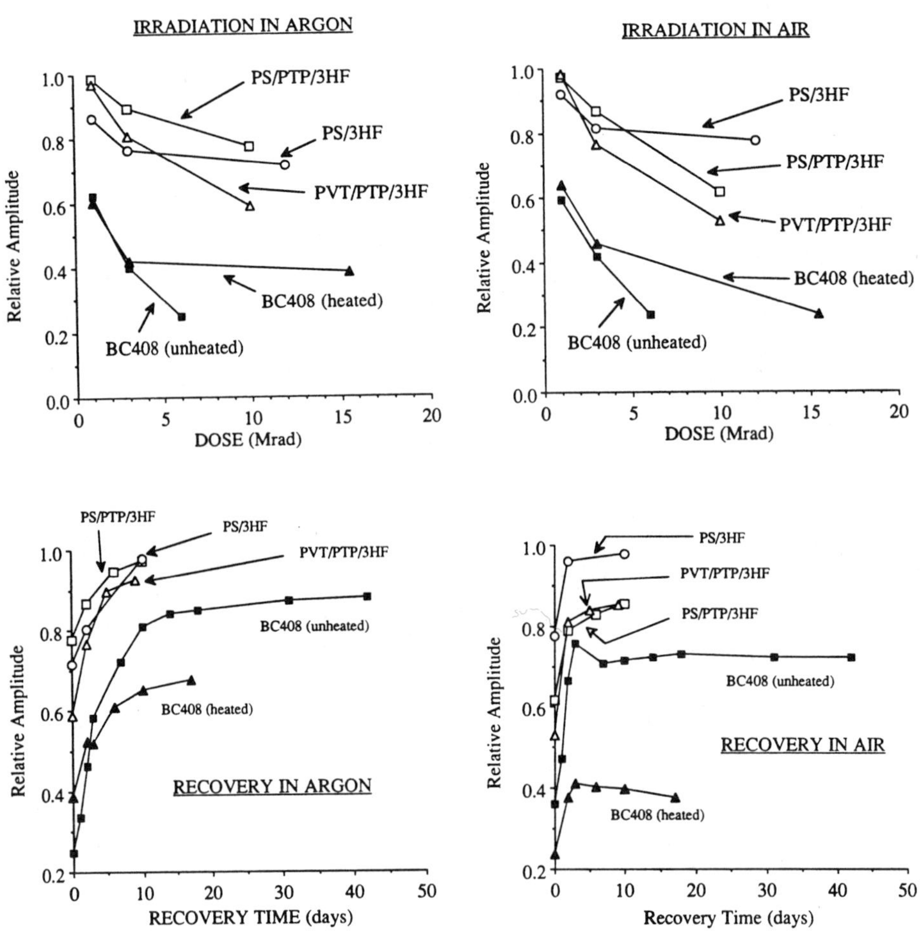

FIGURE 4. Scintillation test results for polystyrene (PS) and polyvinyltoluene (PVT) based scintillators doped with 3-hydroxyflavone (3HF). For comparison, the results for BC408 are presented as well.

[10] utilizing front face fluorescence seems to support this hypothesis of fluor stability.

Figure 3 displays the loss in transmission for BC408 during irradiation. Superimposed is the fluorescence emission of the scintillator. It is clear that assuming one uses radiation stable fluors, then a shift of the emission spectrum to above 500 nm should markedly increase the radiation resistance of the scintillator.

We have tested this idea by irradiating various polystyrene and PVT based scintillators doped with the large Stokes shift fluor 3-hydroxyflavone (3HF) [11,12]. The resultant emission peak is shifted to 530 nm. Figure 4 displays the irradiation and recovery data based on the scintillation tests. For comparison, the results with BC408 are presented on the same plots. We are investigating other possible fluors with large Stokes shifts that we hope to report on in a future paper.

(b) Liquids

Two standard liquid scintillators were irradiated to 5 Mrad in inert atmospheres. BC517L is a mineral oil based scintillator which displays better radiation resistance than BC505, a pseudocumene based scintillator. However, the latter forms a more efficient scintillator as the pulse height spectra in figure 5 shows. Despite the changes in scintillation output after irradiation, a comparison of the transmission spectra of both scintillators displayed no discernible differences (figure 6). Upon irradiation of the undoped solvents (figure 7), the explanation for the loss in scintillation efficiency became evident. First, figure 7 shows that the mineral oil is superior in transmission to pseudocumene after 5 Mrad. Secondly, it is evident that the loss in transmission affected the radiative transfer of the primary fluor's emission to the secondary waveshifter. This effect must be taking place in the case of the plastic scintillators as well. However the extent of the absorptive changes in the plastic drastically affects the secondary emission as well.

Unlike plastics, there is no recovery observable in the liquid scintillators. In fact, no recovery in the transmission spectra of the base materials was observed. Here again, as with plastics, the problem lies with the base material more so than with the fluors. As an illustration of this, two samples of liquid scintillator were irradiated, one pseudocumene (PC) doped with para-terphenyl (PTP), and the other PC/PTP with tetraphenylbutadiene (TPB) added as a waveshifter. Figure 8 shows the marked improvement in radiation resistance when the waveshifter is added.

As with the plastics, a comparable 3HF doped liquid scintillator was irradiated (PC/3HF). Figure 9 displays the pulse height spectrum. Within measurement error, no discernible damage is noted. Although the liquids display no recovery, it is clear that such a material can be replaced within detector modules by suitably intelligent design.

Another large Stokes shift fluor of interest is 1-phenyl-3-mesityl-2-pyrazoline (PMP) [13,14]. Figure 10 shows the results when irradiated up to 25 Mrad. This also illustrates the importance of shifting the primary emission above 400 nm before using a secondary waveshifter. An important problem with this fluor is its high sensitivity to the presence of oxygen during irradiation. Either the scintillator will have to deoxygenated, or preferably, an alternative stable fluor with a large Stokes shift needs to be synthesized.

SUMMARY, CONCLUSIONS AND FUTURE PLANS

Assuming that the fluors are radiation hard, it is clear that the loss in scintillation light output due to high doses of radiation is caused by major changes in the absorptive character of the scintillator solvent. Hence it is imperative that the emission spectrum of the scintillator be shifted to wavelengths above 500 nm. As a corollary, scintillators utilizing at least one waveshifter should be doped with a large Stokes shift primary whose fluorescence emission is preferably above 400 nm (e.g., PMP). By necessity, the fluors themselves must be stable against irradiation.

Our current plans involve several avenues of investigation. First, we have begun to irradiate samples in a 3 MeV electron beam accelerator at Florida State

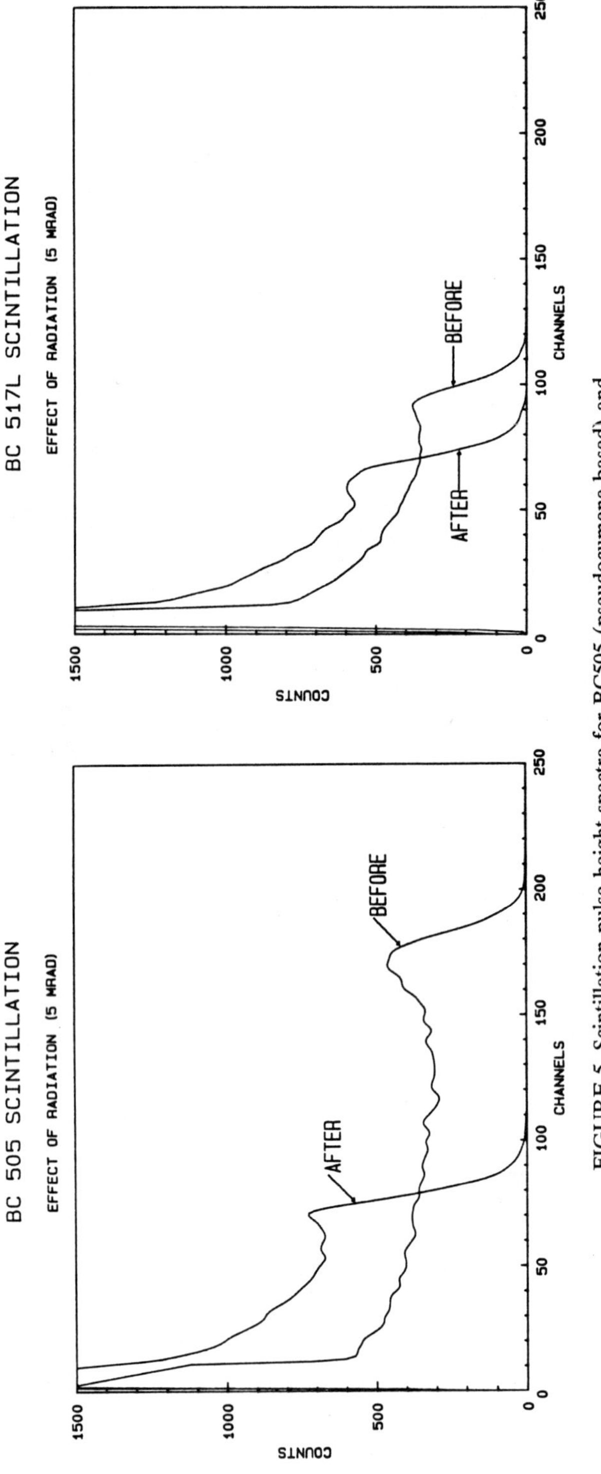

FIGURE 5. Scintillation pulse height spectra for BC505 (pseudocumene based) and BC517L (mineral oil based) scintillators before and after irradiation with 5 Mrad from a ^{60}Co source.

544

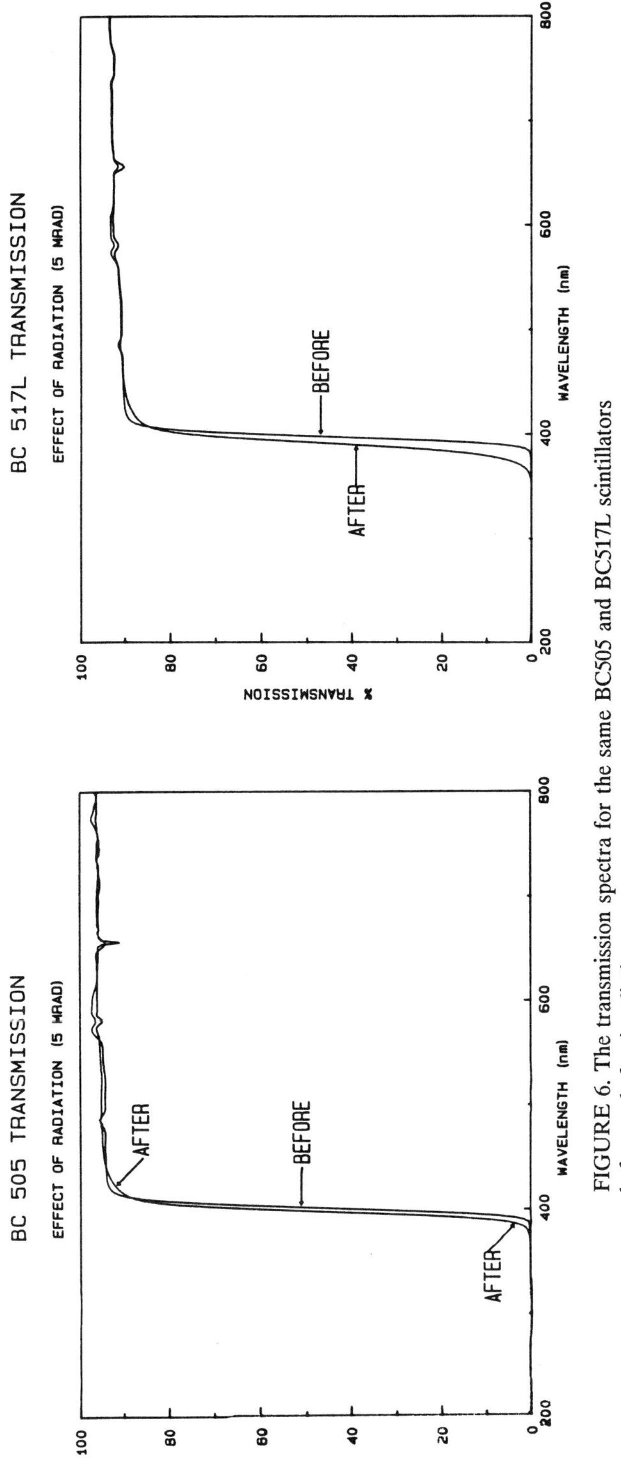

FIGURE 6. The transmission spectra for the same BC505 and BC517L scintillators before and after irradiation.

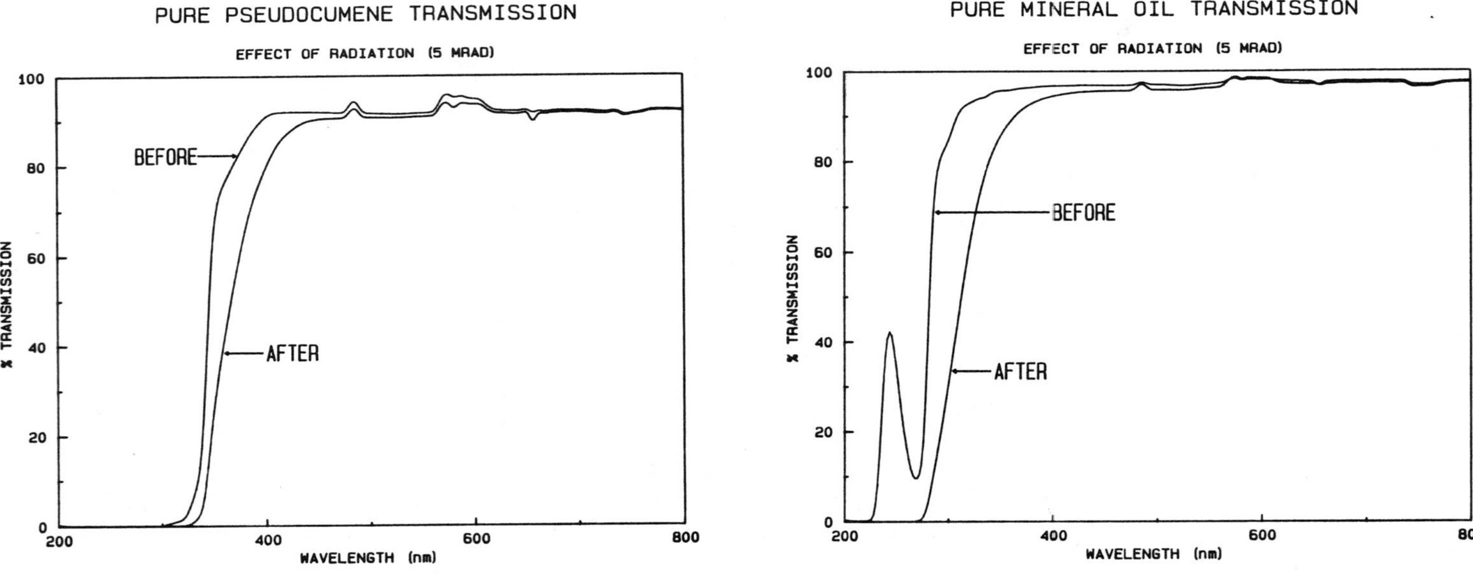

FIGURE 7. Transmission spectra for irradiated samples of pseudocumene and mineral oil. Note that the mineral oil has better transmission properies after irradiation than the pseudocumene has before irradiation. Hence the superior radiation resistance of BC517L. These results also suggest that the problem lies in the absorption of the primary radiative emission before it is absorbed by the secondary fluor.

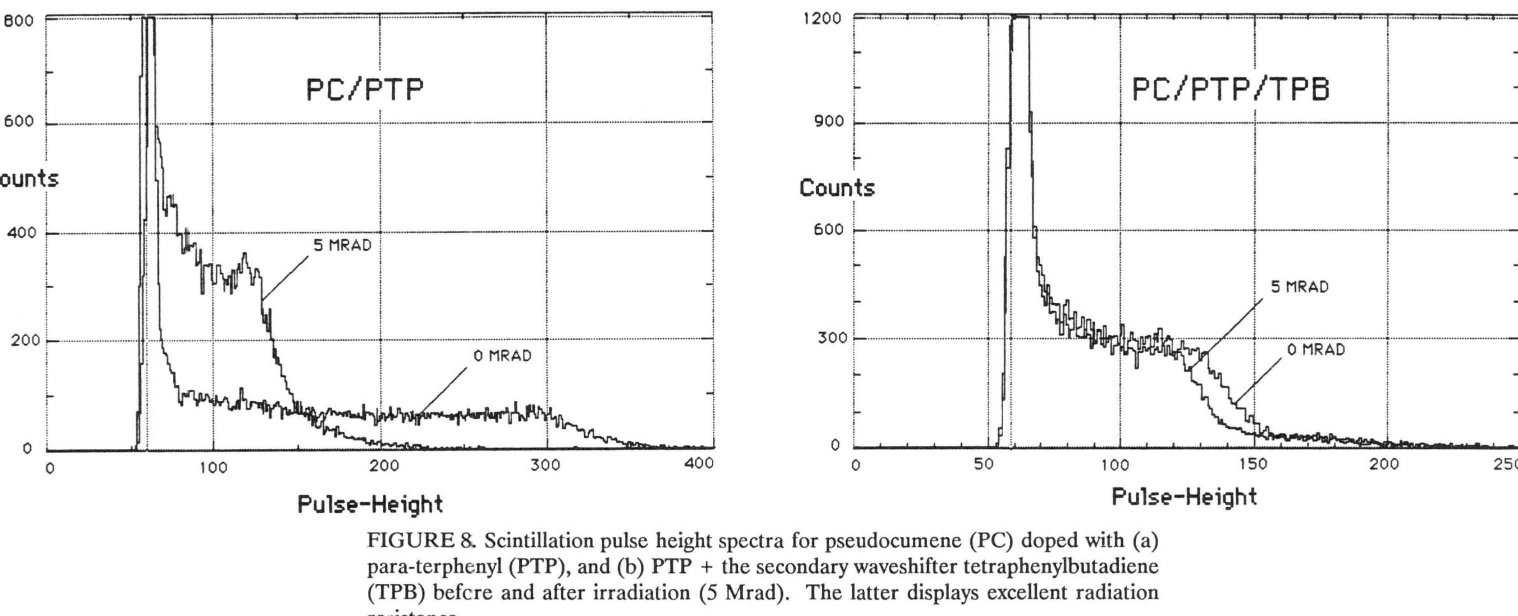

FIGURE 8. Scintillation pulse height spectra for pseudocumene (PC) doped with (a) para-terphenyl (PTP), and (b) PTP + the secondary waveshifter tetraphenylbutadiene (TPB) befcre and after irradiation (5 Mrad). The latter displays excellent radiation resistance.

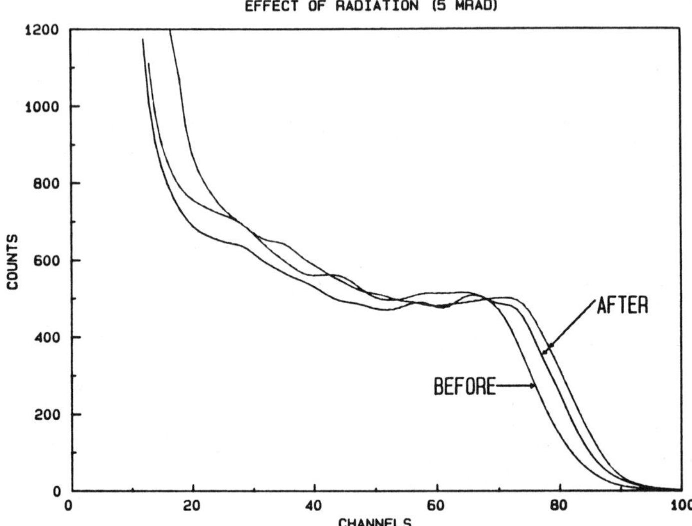

FIGURE 9. Using the idea of shifting to longer emission wavelengths, a pseudocumene sample doped with 3HF was irradiated to 5 Mrad. As the pulse height spectra shows, there is no discernible damage.

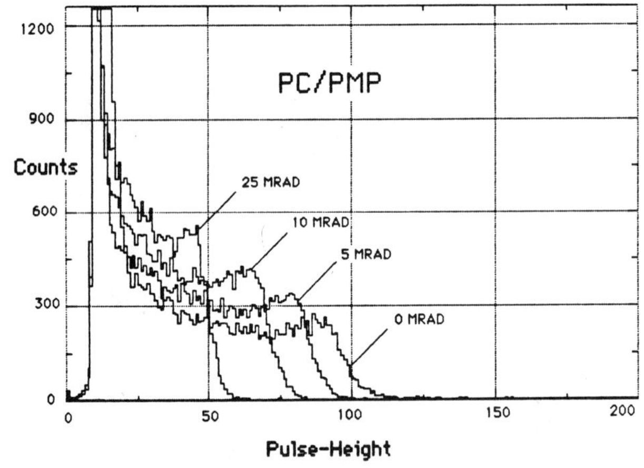

FIGURE 10. The fluor 1-phenyl-3-mesityl-2-pyrazoline (PMP) is of current interest as its large Stokes shift allows an emission at about 425 nm without the need of a secondary waveshifter. The graph shows the change in pulse height spectra as a PC/PMP liquid scintillator sample is irradiated to 25 Mrad. Note that as PMP is highly sensitive to oxidation, the sample was prepared and irradiated in an oxygen-free atmosphere.

University in Tallahassee, Florida. This facility allows for much higher dose rates than currently available with the ^{60}Co gamma source. In addition, this facility will allow us to irradiate larger samples (say 50 cm x 1 cm x 5 cm) of scintillator in order to ascertain the relative merits of several scintillators when used in more realistic sizes (our current work is with small 2 cc samples). Of current interest in experimental particle physics is the use of optical fibers as a component of calorimeters [15]. We are actively engaged in testing various fibers.

Although the 3HF doped scintillators displayed the significance of waveshifting as a method of enhancing radiation stability, it should be noted that 3HF is, by modern standards, a somewhat slow fluor with a decay lifetime of 6 ns [16], and has a low quantum efficiency as well [17]. Hence we are seeking other candidates with suitable properties of large Stokes shift, radiation stability, high solubility, and fast (1-2 ns) decay lifetimes.

All of our current studies have focussed on the optical changes in scintillators without any investigation of the detailed chemical causes. In the near future, we will begin a collaboration with R.L. Clough of Sandia National Laboratories with the intent of examining the chemical reactions responsible for the radiation damage phenomenology. It is hoped that with increased understanding will come methods of alleviating the damage incurred by high dose irradiations.

Finally, we are investigating the possibility of finding and using more radiation resistant base materials. Our current research on this matter has centered on one member of the polysiloxane family of polymers [18,19,20]. We are continuing this line of investigation by searching for other suitable materials for plastic and liquid based scintillators.

ACKNOWLEDGEMENTS

We thank Adam Cox of Bicron Corporation for the preparation of the many samples used in our studies, and Steve Garland for his assistance in the calibration of the fluorescence detector. Our gratitude goes out to Dr. R. Hanrahan of the Department of Chemistry at the University of Florida for his assistance in the use of the ^{60}Co irradiator. In addition, we thank Dr. M. Kasha of Florida State University for suggesting the use of 3HF. This study was supported at the University of Florida by U.S. Department of Energy contract DE-FG05-86ER40272.

REFERENCES

[1]   D.E. Groom, "Radiation Levels in SSC Detectors," to be published in the proceedings of the 1988 Snowmass Conference (1988).

[2]   Report of the Task Force on Radiation Levels in the SSC Interaction Regions, SSC Central Design Group SSC-SR-1033 (1988).

[3]   C. Zorn et al., *Nucl. Instr. Meth.* **A271** 701 (1988).

[4]   C. Zorn et al., *Nucl. Instr. Meth.* **A273** 108 (1988).

[5]   C. Zorn et al., "A Pilot Study of the Radiation Resistance of Selected Plastic Scintillators," to be published in Nuclear Instruments and Methods (1989).

[6]   S. Majewski et al., "A Pilot Study of Radiation Damage in Liquid Scintillators," submitted to Nuclear Instruments and Methods (1989).

[7]   R.L. Clough, "Radiation Resistant Polymers," Encyclopedia of Polymer Science and Engineering, John Wiley and Sons, New York (1988).

[8]   U. Holm and K. Wick, "Radiation Stability of Plastic Scintillators and Wavelength Shifters," to be published in the proceedings of the 1988 IEEE Nuclear Science Symposium.

[9]   produced by Kyowa Gas Company, Japan.

[10]   A. Bross et al., "Radiation Damage to Scintillators," contribution to the Workshop on Scintillating Fiber Detector Development for the SSC, Fermilab, Batavia, IL, Nov. 14-16, 1988.

[11]   D. McMorrow and M. Kasha, *J. Phys. Chem.* **88** 2235 (1984).

[12]   C.L. Renschler and L.A. Harrah, *Nucl. Instr. Meth.* **A235** 41 (1985).

[13] H. Gusten et al., *J. Phys. Chem.* **82** 459 (1978).

[14] S. Majewski et al., "Radiation Damage Tests of a PMP Scintillator," submitted to Nuclear Instruments and Methods (1989).

[15] J. Kirkby, CERN-EP/87-60 preprint (1987)

[16] J. Flournoy, private communciation.

[17] M. Kasha, private communication.

[18] M. Bowen et al., "A New Radiation-Hard Plastic Scintillator," to be published in Nuclear Instruments and Methods (1989).

[19] M. Bowen et al., "A New, Radiation-Resistant Plastic Scintillator," to be published in the proceedings of the 1988 IEEE Nuclear Science Symposium.

[20] M. Bowen et al., "Preliminary Results with a Polysiloxane-based, Radiation Resistant Plastic Scintillator," to be published in the proceedings of the Workshop on Scintillating Fiber Detectors, Fermilab, Batavia, IL, Nov. 14-16, 1988.

A CRYOGENIC TEST STAND FOR FULL LENGTH SSC MAGNETS WITH SUPERFLUID CAPABILITY

T. J. Peterson and P. O. Mazur

Fermi National Accelerator Laboratory[*]
Batavia, Illinois

ABSTRACT

The Fermilab Magnet Test Facility performs testing of the full scale SSC magnets on test stands capable of simulating the cryogenic environment of the SSC main ring. One of these test stands, Stand 5, also has the ability to operate the magnet under test at temperatures from 1.8K to 4.5K with either supercritical helium or subcooled liquid, providing at least 25 Watts of refrigeration. At least 50 g/s flow is available from 2.3K to 4.5K, whereas superfluid operation occurs with zero flow. Cooldown time from 4.5K to 1.8K is 1.5 hours. A maximum current capability of 10000 amps is provided, as is instrumentation to monitor and control the cryogenic conditions. This paper describes the cryogenic design of this test stand.

INTRODUCTION

Fermilab's Magnet Test Facility has been modified to include two test stations suitable for testing full length (17 meter) prototype SSC dipole magnets (Figure 1). The 1500 Watt helium refrigerator which provides liquid nitrogen and 4.5K helium to these test stands is described in detail most recently in "Operational History of Fermilab's 1500 W Refrigerator Used for Energy Saver Magnet Production Testing".[1] At this facility all of the more than 1000 superconducting dipoles and quadrupoles were cold tested prior to installation in the Tevatron. Of the original six test stations fed by the 1500 Watt refrigerator, three remain for testing new Tevatron components, and two special SSC magnet test stations have been built. These differ from the Tevatron magnet test

---

[*] Operated by Universities Research Association, Inc. under contract with the U.S. Department of Energy.

Fig. 1.   Photograph of the Magnet Test Facility at Fermilab
with SSC dipoles at the two SSC magnet test stands.

stations not only in having an interface which mates to the SSC magnets, but in providing flow to a low-temperature helium gas shield (absent in the Tevatron) and in having the capability to cool the magnet below 4 Kelvin.

The first of these SSC test stands is described in "Cryogenic Instrumentation of an SSC Magnet Test Stand". [2] It is capable of providing up to 30 grams/sec of 4 atm helium at temperatures as low as 3 Kelvin. A second test stand, capable of providing a larger range of helium flow rates and temperatures, is now operational. It can provide up to 50 grams/sec of 4 atm helium down to 2.3 Kelvin, and also has the capability of cooling full length SSC magnets with stagnant or forced-flow pressurized superfluid (helium II) down to 1.8 K. The latest report of SSC magnet test results from these test stands is "Tests of Full Scale SSC R&D Dipole Magnets". [3]

Other 1.8K magnet cooling facilities which have previously been described in the literature include a dual bath system holding about 90 liters of pressurized superfluid[4] which is in principle like the "Claudet bath".[5] A 1.8K test dewar, also of the dual bath design, is presently used for testing short LHC magnets at CERN. It contains about 530 liters of superfluid and has 9 Watts of cooling capacity at 1.8K.[6] The Francis Bitter Magnet Laboratory at MIT has a dual bath cryostat for cooling the superconducting portion of high field hybrid magnets to 1.8K.[7]

552

Fig. 2.   Photograph of Stand 5

This report describes some of the unique cryogenic features of Fermilab's 1.8K magnet test stand, (Figure 2) which, due to its location at the magnet test facility, is refered to as Stand 5. The following is a list of those features which we would like to highlight.

1.   Heat exchangers were designed to accomodate three operating modes: (a) subcool up to 100 g/s of normal helium (helium I), (b) provide a forced flow of superfluid at a lower flow rate, and (c) to remove heat returned to the subcooler by the high apparent thermal conductivity of stagnant superfluid.

2.   Special check valves are located at each large-diameter penetration (e.g., cooldown lines and quench reliefs) to the superfluid space which reduce the heat inleak to acceptable levels.

3.   10 KA current leads are mounted in a separate, demountable can for quick replacement.

4.   Pumping to obtain the low pressures for the saturated bath of superfluid in the subcooler is provided via room-temperature pumps: a roots blower backed up by a liquid ring pump.

SYSTEM DESCRIPTION

Figure 3 is a simplified flow schematic for Stand 5. For forced-flow magnet cooling the pressurized (2 to 4 atm) helium enters the test stand from the 1500 Watt refrigerator where it splits, part going into the shell side of exchanger 1 (HX1), a simple finned tube-in-shell exchanger which precools the supply via counterflow heat exchange with the return flow. After HX1 the flow splits again, part to the JT exchanger (JTX) where the flow which maintains the liquid level in the subcooler (HX2) via the liquid level valve is precooled by the pumped boil-off vapor, and part to the tubes in the bath in HX2 where it is cooled to near bath temperature. The other branch of flow above HX1 cools the base of the power leads (part goes up the leads for the usual counter-current cooling) and then goes to the 4.5K shield.

For cooling with stagnant superfluid the liquid return valve is closed, subcooler liquid level is maintained as in the forced flow cooling mode, power lead and shield cooling is also the same, and the pressurized stagnant superfluid is cooled via heat exchange with the saturated bath. HX2 (Figure 4) has two sections in series: first, a coil of finned copper tubing to cool forced flow of up to 100 g/sec of normal helium, and second, seven parallel vertical copper tubes containing the pressurized superfluid, with large inner diameter and surface areas to improve heat transfer through the superfluid from the magnet and reduce the delta-T due to the Kapitza resistance. These seven vertical tubes are 66 cm long, 2.5 cm diameter, and have 0.5 cm tall fins to enhance the surface area in the low-pressure bath. The total inner surface area is about 3000 square cm, and the outer surface area is about 14500 square cm. Pipes totaling 80 square cm of cross-section and about 60 cm in length connect these seven copper tubes to the magnet, permitting a large heat flux from the magnet in stagnant superfluid.

The region of piping which is designed to contain superfluid is separated from the 4.5K helium region by specially designed check valves (Figure 5) where a significant pressure drop is not acceptable, such as at the cooldown line and quench relief lines. We expect a temperature of 2.17K at the top of the plug in this check valve and process temperature (about 1.8K) below the plug. For this situation the total heat leak through the check valve would be about 0.5 Watts if the plug fit into its seat with an average gap of .004 inches. We expect a much better fit than that, in which case the heat leak of 0.1 Watt through the bleed hole in the center of the plug may dominate the heat leak. This bleed hole allows pressures above and below to equilibrate when mass is trapped above the check valve.

In other places a tube of length and diameter to minimize heat flux into the superfluid while allowing sufficient flow with acceptable pressure drops in forced-flow conditions is utilized. We chose to use a 1.5 meter length of 1.1 cm inner-diameter tubing, which, if one end were at T-lambda (2.17K) and the other at 1.8K would transport 1.2 Watts of heat (calculations based on reference 8), and with 50 g/s flow of 1.4 atmosphere liquid helium has a 1.0 psid pressure drop.

Connections for an optional circulating pump are shown in figure 1. These are special bayonet connections designed for superfluid as well as helium I, but no tests with a circulating pump have been performed yet.

The current leads are mounted to the main vacuum vessel in their own container (the vertically oriented can above the interface to the magnet in Figure 1) and connected via a soldered splice of the cable and flanged helium line and vacuum connections. This entire subassembly of two power leads can be easily replaced with a spare one after warmup of the feed can assembly. The power leads consist of a copper rod with a copper fin helically wound around it and brazed to it for good thermal contact. A tight-fitting G-10 tube contains the flow in the helical path. The pressure drop of a few psid for the flows required for high currents (up to 10000 amps) is acceptable due to the 2 to 4 atm pressure flow to the magnet. This results in turbulent flow and good convective heat transfer to the helium gas. The superconductor is contained in the 4.5K shield-flow pipe for a short distance to the ceramic feedthroughs (as indicated in figure 1) where it passes through to the superfluid space.

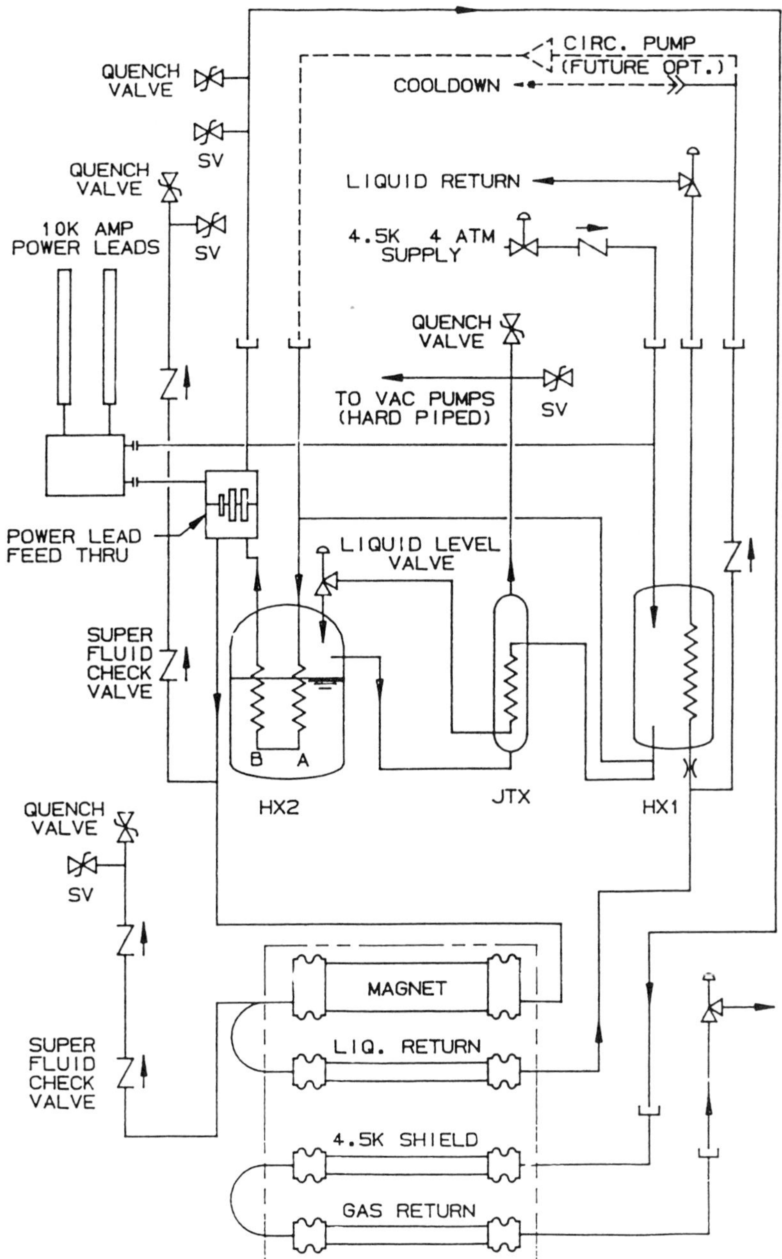

Fig. 3.   Flow schematic for Stand 5.

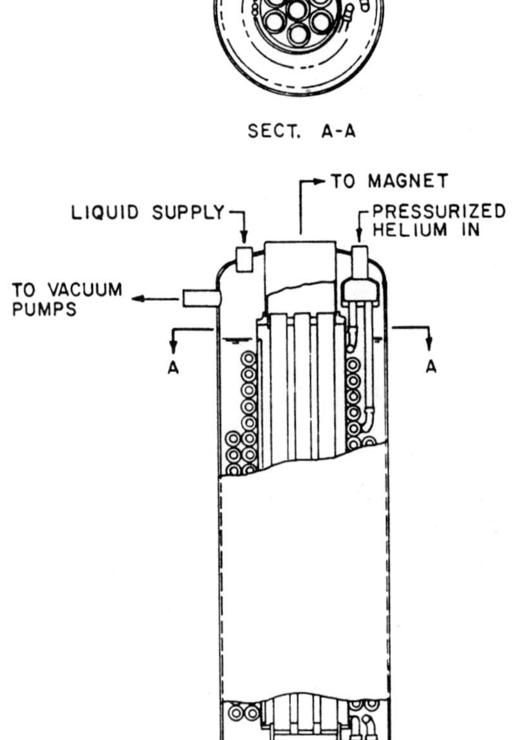

Fig. 4. The low temperature subcooler, HX2.

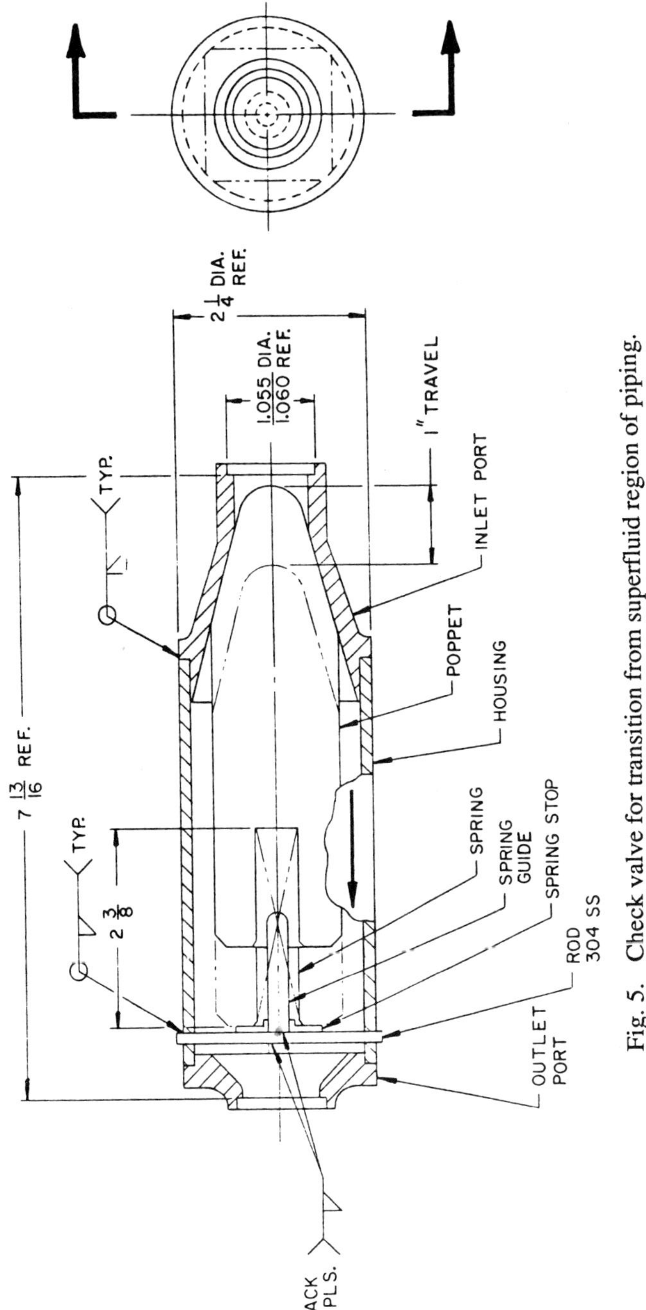

Fig. 5. Check valve for transition from superfluid region of piping.

Fig. 6. Photograph of vacuum pumping system.

The vacuum pumping system (Figure 6) consists of a vacuum jacketed transfer line carrying the 4K, low pressure gas from the outlet of JTX to a warmup heat exchanger (not shown in figure 1) and the room-temperature vacuum pumps. Warm, 20 atmosphere helium compressor flow passes through a series of spirally wound coils of finned tubing while the pumped flow passes over these tubes in the shell-side of the heat exchanger. Although the option exists to use the cold, high pressure flow, we simply pass it through a heater and send it back to the compressor interstage.

The two-stages of vacuum pumping consist of: first, an oil injected blower (40 horsepower moter, 1750 RPM, Kinney model KMBD3201) cooled with the same oil as is in our screw compressors, UCON LB165; and second, a compound liquid ring pump (60 horsepower, Kinney model KLRC-951) which also uses UCON LB165. Since this discharges into a line going to our screw compressor suction, only bulk oil knockout is provided. This system is rated for 2.85 g/s of helium flow at 12.3 torr pressure. At about 70 torr and above the blower is off and we pump with the liquid ring pump only.

CONCLUSION

This system has successfully cooled a 17 meter long SSC dipole magnet to 1.77K and powered it to over 7000 amps at that temperature. Cooldown from 4.5K to 1.8K took about 1.5 hours. Numerous tests of SSC dipoles in helium I from 2.3K to 4.5K with various flow rates up to 50 g/s have also been performed.

REFERENCES

1.	R .K. Barger, et. al., Operational History of Fermilab's 1500 W Refrigerator U sed for Energy Saver Magnet Production Testing,	in: "Advances in Cryogenic Engineering" Vol. 31, Plenum Press, New York (1986), p. 657.

2.	K. McGuire, et. al., Cryogenic Instrumentation of an SSC Magnet Test Stand, in: "Advances in Cryogenic Engineering" Vol. 33, Plenum Press, New York (1988), p. 1063.

3.	J. Strait, et. al., Tests of full Scale SSC R&D Dipole Magnets, to be published in the proceedings of the 1988 Applied Superconductivity Conference. Also catalogued as Fermilab	TM-1545 and SSC-N-538.

4.	R. P. Warren, et. al., A Pressurized Helium II-Cooled Magnet Test Facility, in: "Proceedings of the 8th International Cryogenic Engineering Conference", IPC Business Press, England	(1980), p. 373.

5.	G. Claudet, et. al., The Design and Operation of a Refrigerator System Using Superfluid Helium, in: "Proceedings of the Fifth	International Cryogenic Engineering Conference", IPC Business Press, England (1974), p. 265.

6.	F. Haug, et. al., Cryogenics of the 1.8K Test Station for 10 Tesla Superconducting Magnet Models, in: "Proceedings of the 12th International Cryogenic Engineering Conference", Butterworth,	Guildford, England (1988).

7.	M. J. Leupold and Y. Iwasa, Subcooled Superfluid Helium Cryostat for a Hybrid Magnet System, in: Cryogenics, Vol. 26	(November, 1986), p. 579.

8.	G. Bon Mardion, et. al., Practical Data on Steady State Heat Transport in Superfluid Helium at Atmospheric Pressure, in: Cryogenics, Vol. 19 (January, 1979), p. 45.

FERMILAB R&D TEST FACILITY FOR SSC MAGNETS

J. Strait, M. Bleadon, R. Hanft, M. Lamm, K. McGuire,
P. Mantsch, P. O. Mazur, D. Orris, and J. Pachnik

Fermi National Accelerator Laboratory
P. O. Box 500
Batavia, IL 60510

ABSTRACT

The test facility used for R&D testing of full scale development dipole magnets for the SSC is described. The Fermilab Magnet Test Facility, originally built for production testing of Tevatron magnets, has been substantially modified to allow testing also of SSC magnets. Two of the original six test stands have been rebuilt to accommodate testing of SSC magnets at pressures between 1.3 Atm and 4 Atm and at temperatures between 1.8 K and 4.8 K and the power system has been modified to allow operation to at least 8 kA. Recent magnets have been heavily instrumented with voltage taps to allow detailed study of quench location and propagation and with strain gage based stress, force and motion transducers. A data acquisition system has been built with a capacity to read from each SSC test stand up to 220 electrical quench signals, 32 dynamic pressure, temperature and mechanical transducer signals during quench and up to 200 high precision, low time resolution, pressure, temperature and mechanical transducer signals. The quench detection and protection systems is also described.

INTRODUCTION

Tests[1-6] of full scale R&D dipole magnets[7,8] for the Superconducting Super Collider (SSC)[9] have been carried out at the Fermilab Magnet Test Facility. This facility has previously been used for the production testing of superconducting magnets for the Tevatron. In this paper we will describe the cryogenic operation of the two SSC test stands, the high current DC power system, the quench protection system, the test stand and magnet instrumentation, and the data acquisition system. Because testing of SSC magnets has concentrated so far on quench performance and mechanical behavior, systems for magnetic field measurement have not been fully developed and will not be described here.

TEST STANDS

Two test stands, with somewhat different capabilities, have been constructed for the testing of SSC magnets, replacing two test stands previously used for testing Tevatron magnets. Each consists of two end

Table I

Test Stand Characteristics

Characteristic	Stand 4	Stand 5
Temperature range	3.2-4.8 K	1.8-4.8 K
Helium pressure	1.3-4.5 Atm	
Helium mass flow	30 g/s	50 g/s
Maximum magnet current	8 kA	9 kA
Voltage tap feed-through wires	200	200
Low voltage feed-through wires	330	330

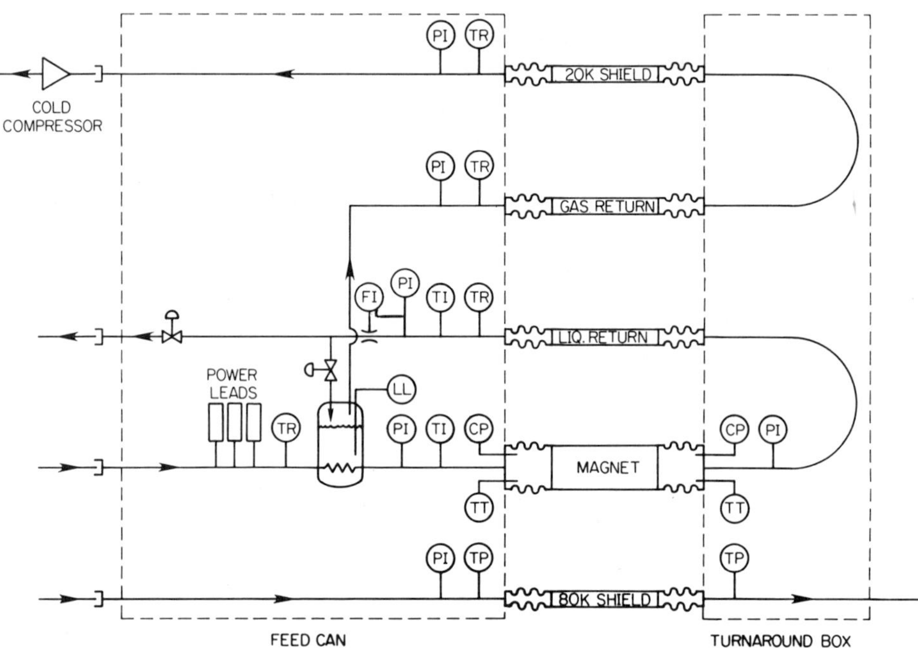

Fig. 1. Simplified flow schematic for test stand 4 showing locations of instrumentation. TR is a pair of carbon resistors. TT is a triplet of resistance thermometers: germanium, encasulated carbon and platinum. TP is a platinum resistor. TI is a vapor pressure thermometer. PI is a room temperature pressure transducer. CP is a cold pressure transducer. FI is a differential pressure transducer. LL is a liquid level probe.

boxes, whose ends simulate neighboring magnets in a magnet string, and a structure for supporting and aligning the magnet under test. The end boxes serve to route cryogens through the magnet, bring pressure tap capillary tubes to a room temperature gauge panel, and bring magnet power leads and instrumentation wires to the magnet under test. The characteristics of the two test stands are summarized in Table I.

Figure 1 is a flow schematic, including the locations of cryogenic instrumentation, of the first of two stands to be built (Stand 4). Helium at a temperature of approximately 4.6 K and a pressure between 1.3 Atm and 4 Atm is provided by a CTI-1500 refrigerator.[10] The helium enters the feed end box through the liquid supply line, where some of the flow is diverted to cool the power leads. The helium then passes through a boiling liquid subcooler, whose shell-side pressure sets the magnet test temperature. After passing through the magnet, the helium is routed by the return end box through the liquid return line of the magnet cryostat and back to the feed end box. Some of the flow passes through a J-T valve into the shell side of the subcooler, while the remainder returns to the refrigerator. Boil-off gas from the subcooler passes through the gas return line of the magnet and then the 20 K shield; in this mode of operation this shield actually operates at approximately the same temperature as the magnet. The helium passes through a cold compressor,[11] is warmed to room temperature and returned to the compressor intake of the refrigeration system. When the magnet is being operated at 4.4 K or above, the cold compressor is turned off. With the cold compressor on, the pressure on the shell side of the subcooler can be lowered to approximately 0.3 Atm, allowing the testing of magnets at temperatures as low as 3.2 K. The 80 K shield is cooled by liquid nitrogen which is vented to the atmosphere after a single pass through the magnet.

Temperatures are measured at the points indicated in Fig. 1 by resistance thermometers and vapor pressure thermometers. Pressures are measured by pressure transducers and mechanical gages at the room temperature ends of capillary tubes and by cold pressure transducers immersed directly in the helium in the two magnet interconnect regions. Helium mass flow through the magnet is measured with a venturi flow-meter in the liquid return line in the feed end box. The cryogenic instrumentation of this test stand has been described before.[12]

Figure 2 is a flow schematic for the more advanced test stand (stand 5). The cryogenic operation of this test stand is described fully in a companion paper[13] in these proceedings and will be described briefly here. Helium from the refrigerator passes first through a counter-flow heat exchanger (HX1), the other side of which contains the outlet flow from the magnet. The main flow then goes through the tube side of a boiling liquid subcooler (HX2) and then to the magnet. Helium exiting the magnet is routed back through the magnet liquid return line, passes through HX1, and is returned to the refrigerator. A fraction of the flow from the first heat exchanger is diverted to the shell side of HX2 to maintain the liquid level. Boil-off gas from the subcooler is warmed to room temperature and sent to a large vacuum pumping system. For 4.4 K testing, this pump is turned off; the pump is capable of lowering the pressure in the subcooler to 0.01 Atm, allowing testing of magnets to temperatures as low as 1.8 K. A fraction of the flow is diverted before the first heat exchanger to cool the power leads and is then sent through the 20 K shield, back through the gas return line and is returned to the refrigerator. This sets the 20 K shield at approximately 4.5 K. The 80 K shield is cooled by liquid nitrogen which is vented to the atmosphere at the outlet. Test stand instrumentation, indicated in Fig. 2, is similar to test stand 4.

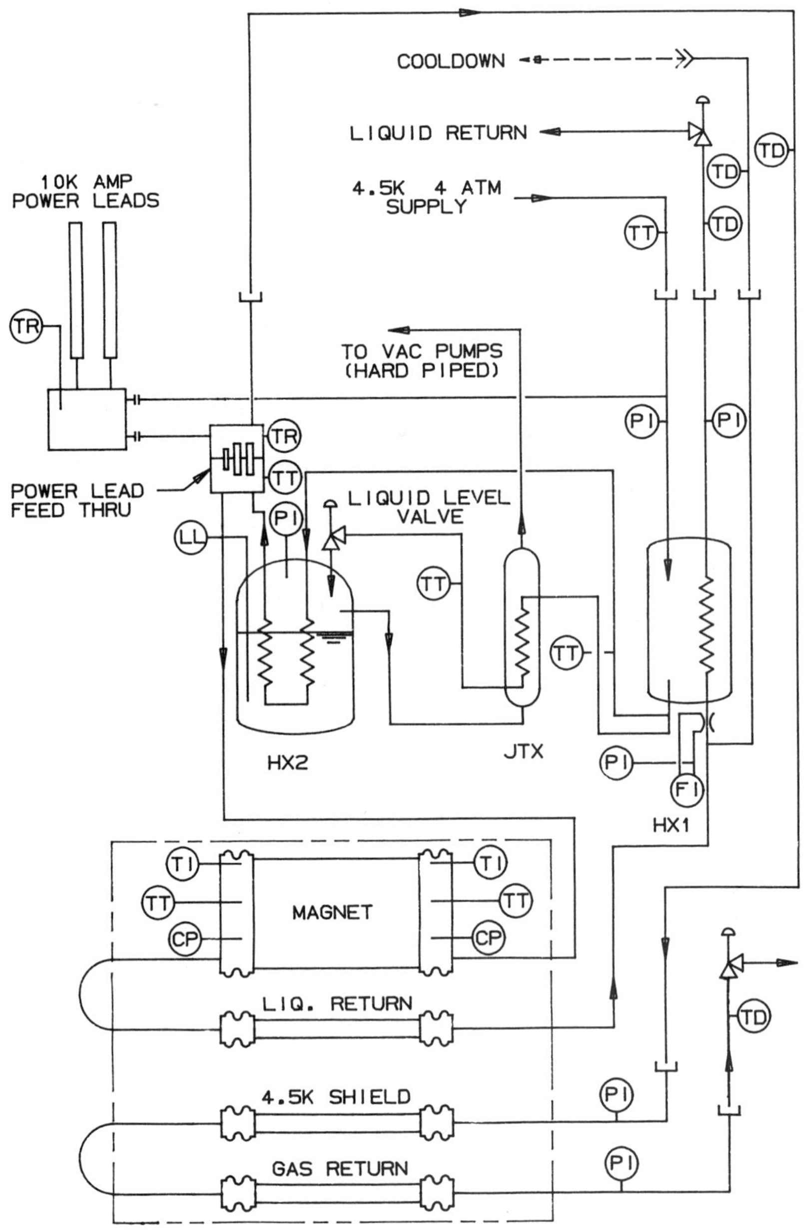

Fig. 2.   Simplified flow schematic for test stand 5 showing locations of instrumentation.   The notation is the same as in Fig. 1, with the addition that TD is a pair of resistance thermometers: germanium and encaulated carbon.

POWER SYSTEM

A simplified schematic diagram of the power system is shown in Fig. 3. The system is based on two Transrex 500 kW power supplies, each with its own filter, series SCR, and dump resistor circuits, which are put in parallel near the test stands. Current is directed to the test stand in use by mounting the appropriate copper plates in a "switch-yard." The LC filter consists of the 8 mH choke (a Fermilab main-ring magnet) and the parallel 22500 μF and 4500 μF capacitors. The diode in the filter protects the electrolytic capacitors from being back-biased during quenches, while the 3 Ω resistor conducts enough DC current to keep the diode on.

Magnet current is measured both by a shunt[14] and a zero-flux current transformer[15] (transductor). The time derivative of the current is measured using an air-core transformer ("dI/dt coil"). The primary is a 5-turn solenoid in the main bus and the secondary is a 1000 turn solenoid inside the primary, yielding a mutual inductance of 0.11 mH. The dI/dt signal is used for subtraction of the inductive component of the magnet voltage both for quench protection (see below) and for analysis of quench data. A soft ground is provided through a 1 Ω and a 25 Ω resistor. Voltage across the 1 Ω resistor is used to detect ground faults, while the voltages across both resistors are monitored by the quench data acquisition system.

Under normal conditions, the series SCR Q2 is gated on while Q1 is off. When a quench is detected (see next section), the power supply is turned off, Q2 is gated off and Q1 is gated on. The -400 V on the

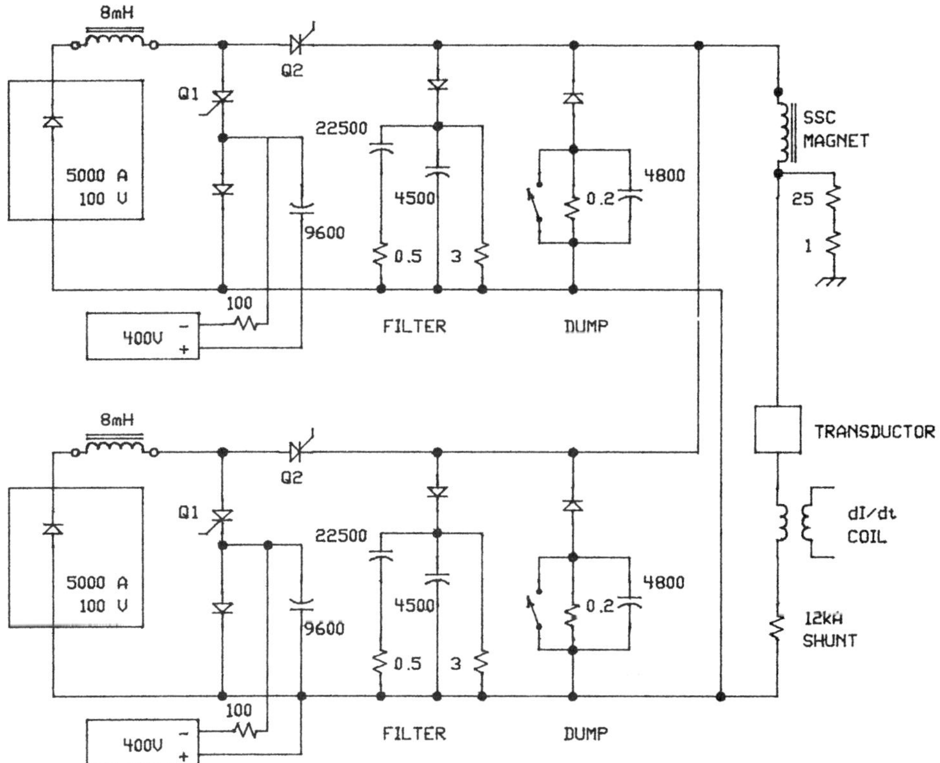

Fig. 3. Simplified power system diagram. All resistances are indicated in Ohms and all capacitances in μF.

9600 μF capacitor back biases Q2 to turn it off, while Q1 and the diode in series with it provide a discharge path for the energy stored in the 8 mH choke. The magnet discharges through the dump resistor. A switch allows the dump resistor to be shorted, causing the magnet to absorb its full stored energy. This is equivalent to using a single diode across the magnet leads for quench protection. Most tests of SSC magnets have been performed in this mode.

QUENCH PROTECTION SYSTEM

Magnet and test stand quench protection is provided by a set of analog comparators, which detect non-zero resistance in the magnet coil or power leads, excessive resistance in the gas-cooled copper leads, or non-zero ground current. The protection system requires the detection of a non-zero resistive voltage while being insensitive to the inductive component. A quench detection circuit module (QDC) accepts two signals, amplifies or attenuates each independently and sets a latch if the difference between the amplified signals lies outside an adjustable window centered on zero. The relative gains are adjusted so that the difference signal has no inductive component.

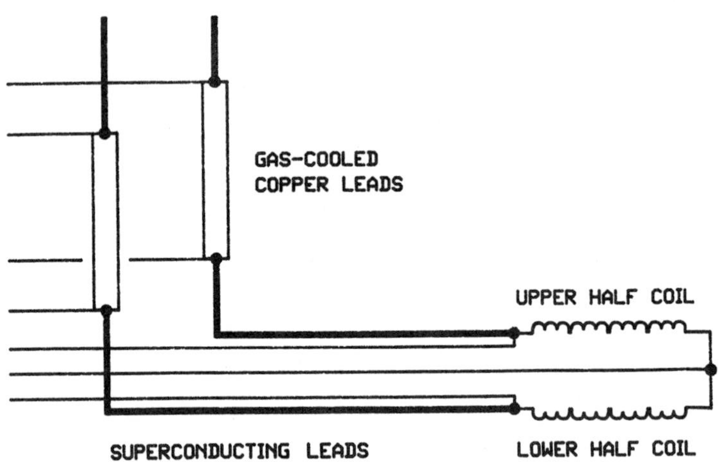

Fig. 4. Voltage taps used in the quench detection system.

Figure 4 shows the voltage taps on the magnet and power leads used in the quench protection system. Signals from the whole main coil, the two half coils, the two superconducting power leads, the two gas-cooled copper power leads, the 1 Ω resistor between the negative bus and ground (see Fig. 3), and the trim coil (not shown), are brought to the control room through isolation amplifiers. Signals from the two SSC test stands are routed through a relay multiplexer to a common set of quench detection circuits.

Table II(a) lists the quench detection circuits used, their inputs and thresholds and Table II(b) indicates the actions taken when each is triggered. The most sensitive QDC takes the difference between the upper and lower half coils. To protect against the possibility of a symmetric quench, a second QDC measures the voltage across the whole main coil and

Table II

a) Quench Detection Circuits

QDC	Inputs			Threshold	Group
1	Upper Half Coil	-	Lower Half Coil	0.25 V	A
2	Whole Main Coil	-	dI/dt Coil	4.0 V	A
3	Whole Main Coil			11 V	A
4	Superconducting Leads	-	dI/dt Coil	0.01 V	B
5	Copper Leads	-	Shunt	0.03 V	B
6	Trim Coil			0.5 V	C
7	1 Ohm Ground Resistor			0.10 V	C

b) Magnet Quench Protection

QDC Group	Action	Delay Range
A	Phase back power supply SCRs	0-2 sec
	Energize energy extraction system	0-2 sec
	Energize protection heaters	0-2 sec
	Turn off correction coils	0
B	Phase back power supply SCRs	0-2 sec
	Energize energy extraction system	0
	Energize protection heaters	0
	Turn off correction coils	0
C	Ramp main current to 0 at 200 A/sec	0
	Turn off correction coils	0

uses the signal from the dI/dt coil (see Fig. 3) to cancel the inductive
component. Because the frequency response of the superconducting magnet
(considered as an inductor) is different from that of the dI/dt coil, they
respond differently to the power supply voltage ripple, so the threshold
cannot be placed as low as on the first QDC. The third QDC monitors
the total magnet voltage and protects against excessive ramp rates that
might result if the power supply were to go out of regulation. Between the
bottom of each gas-cooled copper power lead and the magnet, the current is
carried by 2 (stand 4) or 3 (stand 5) Tevatron superconducting cables,
which do not have any additional copper stabilizer. Quench detection circuit
4 measures the sum of the voltages across the positive and negative
superconducting leads and subtracts the inductive component using the dI/dt
coil. The fifth QDC protects the gas-cooled leads from thermal run-away
that could result from inadequate helium flow. A signal proportional to the
current and is subtracted from the sum of the lead voltages, placing a

window about the nominal resistance of the leads. Quench detection circuit 6 monitors the voltage across the correction coil uncorrected for inductive effects. Ground current is detected by QDC 7 which is sensitive to voltage across the 1 Ω resistor in the system ground.

When any one of the quench detection circuits 1 through 5 (groups A and B in Table II) is triggered, several actions are taken to protect the magnet and power leads. Both the main and correction coil power supplies are turned off, the dump resistor is switched across the magnet leads, and protection heaters (strip heaters) in the magnet are energized to cause the quench to propagate throughout the coil. To allow studies of quench protection methods and of quench propagation, each of these actions may be delayed independently by up to 2 seconds for quenches detected by the first three QDCs (group A). Because lead quenches may take a substantial time to be detected, most of the delays are bypassed if QDC 4 or 5 (group B) is triggered. Quench detection circuits 6 and 7 (group C) cause the main coil current to ramp to zero at 200 A/sec and the trim coil power supply to be turned off. No delays are allowed. If the trim coil quenches, there is no need to quench the main coil. If a ground fault occurs without a quench, it is undesirable to quench the magnet or switch in a dump resistor, since either of these actions would generate considerable voltage to ground in the coil.

MAGNET INSTRUMENTATION

Recent SSC magnets[4-6] have been heavily instrumented with a variety of measurement devices which must be accommodated by the test facility. Strain gage based devices and linear potentiometers have been used to measure stress, force and motion in various magnet components, resistance thermometers have been used to measure the thermal performance of the cryostat[16] and several magnets have included up to 133 voltage taps for quench localization and propagation measurements. Coil azimuthal stresses at the collars have been measured with both direct compression and bending beam[17] load cells. Direct compression load cells[17] have been used to measure the axial force between the coil end and the magnet end plate, while bending beam motion transducers and linear potentiometers have been used to measure the deflection of the end plates, motion of the end of the coil relative to the cold mass skin, and overall length changes of the magnet with both temperature change and excitation. Strain gages have been applied directly to the cold mass skin and to many cryostat components[16] to measure stresses due to thermal effects and excitation. Depending on the application, results may be reported as strain, stress, force or displacement. The strain gages may be wired as full- or half-bridges with voltage or current excitation, or may be read as individual gages by a 4-wire resistance method.

DATA ACQUISITION SYSTEM

The data acquisition system can be divided into two main sub-systems: one for electrical quench signals and one for cryogenic[12] and mechanical signals. The two systems conform to similar philosophies: front-end electronics are located as close to the test stand as is practical, amplified signals are brought to digitization electronics and data are collected and transmitted via serial CAMAC to programs running on a MicroVax II.

Hardware

To reduce the load on the MicroVax and eliminate need for the MicroVax to perform real-time operations, intelligent readout electronics with local memory are used for digitization. A complete set of electronics exists

for each test stand;[18] multiplexing between the two test stands is done by
the readout programs selecting the appropriate CAMAC modules. Fast
digitization of signals during quenching is provided by 32-channel, 12-bit
analog-to-digital converters[19] (ADCs) with 1024 words of memory per
channel. These are operated in a "ring buffer" mode, with digitization
stopping after a pre-selected number of data samples have been taken
following a quench trigger. Currently nine such modules are used for each
test stand for electrical signals and one for cryogenic and mechanical signals.
The sampling clock frequency and number of post-trigger samples can be
controlled on-line by the magnet tester. Clock periods used typically range
from 0.2 to 2 msec while normally 128 or 256 pre-trigger samples are
stored. Cryogenic and mechanical data are measured with higher precision
but poorer time resolution by a digital voltmeter[20] (DVM) with signals
multiplexed through a relay scanner.[21] Once a scan is started, the DVM
and scanner trigger each other and data are stored in memory in the
voltmeter; the MicroVax need only initiate a scan and collect the data at
the end. Up to 1400 channels can be multiplexed into one DVM in this
manner. We currently have hardware to read 360 channels from the two
test stands.

The number of channels that can be read is ultimately limited only by
the number of feed-through wires. On each test stand there are
approximately 330 low voltage instrumentation wires and 200 voltage tap
wires available for magnet data taking (in addition to a smaller number for
monitoring signals within the end boxes). In addition, signals from
instrumentation in the insulating vacuum (for example, strain gages mounted
on the cold mass skin or thermometers mounted on cryostat components)
are typically brought out through the magnet cryostat wall, increasing the
number of signals possible.

To be able flexibly to accommodate the different configurations of
instrumentation on the R&D magnets, a modular system has been developed
for voltage tap, strain gage, thermometer, and linear potentiometer readout.
Voltages across segments of the magnet coil are measured by constructing
signals from pairs of voltage taps. The ADCs are protected from the
potentially large common- and differential-mode voltages by transformer-
coupled isolation amplifiers.[22] A single width NIM module includes two
isolation amplifiers and a differential amplifier, which gives the difference
between the two input signals. The inductive voltage from two coil
segments of equal inductance is eliminated in taking the difference, allowing
greater sensitivity to small resistive voltages. The gains of the individual
and difference amplifiers are independently selectable by a combination of
internal and external switches.

Double width NIM modules provide constant voltage or current
excitation for up to eight full- or half-bridge strain gages. Constant voltage
modules provide for remote sensing of the excitation voltage, while constant
current modules include a precision, high stability, resistor for measuring the
excitation current. The strain gage voltage is routed to front panel
connectors for digitization by the digital voltmeter. These modules also
include bridge completion resistors for half-bridges and optional amplifiers
with switch selectable gains of up to 500 to allow dynamic readout by
ADCs. The modules are configured for a specific test by internal terminals.

Other double width NIM modules provide constant 2.5 mA excitation
for up to 12 individual gages read by the 4-wire resistance method. Again
the current is measured by a precision resistor in each module. In this
readout method, thermal and magnetic field effects on the active strain gages
are corrected by subtracting the apparent strain measured on a compensating
gage, which is in the same environment as the active gage but is unstrained.

If the active and compensating gage are excited by the same current source, the effect of small uncertainties in the absolute calibration of the current measuring resistor will tend to cancel. To allow dynamic readout, the analog difference between the active and compensating gage voltages can be amplified. The configuration of the module is set by internal jumpers.

Similar modules[12] provide excitation and optional amplification for resistance thermometers, except the current is in the range 1-25 μA. Excitation and signal conditioning for pressure transducers is provided by commercial electronics.[23]

<u>Software</u>

Three cooperating programs[12] on the MicroVax are used to collect and store the data. Cryogenic and mechanical transducers are read through the DVM-scanner system and one 32-channel ADC by a program called CRYO_MONITOR, which responds to requests from the other two programs. CRYO_LOGGER makes requests for high resolution quasi-static data at fixed time intervals, typically every 10 minutes, logs the data to disk and displays the cryogenic data on a terminal screen for use by the refrigerator operators and magnet test personnel. Both raw voltages and values after conversion to physical units are recorded and each type of data is written in both binary and ASCII formats. Complete configuration and calibration data are written at the beginning of each file and whenever CRYO_LOGGER is restarted.

The magnet test and measurement program SSC controls the power supplies, measures the main and correction coil currents from shunts or transductors via digital voltmeters, collects, analyses and logs quench data from the ADCs, collects (via CRYO_MONITOR) and logs strain gage and other cryogenic data as a function of magnet current, and has structures to allow for magnetic field measurements when such systems become more completely developed. An ASCII "log" file records time-stamped information about the test, including changes in magnet current, the writing of data files, the occurrence of quenches and operator comments. Magnet and power system electrical signals from each quench are written as raw ADC counts in a binary file, which contains full configuration and calibration information to allow off-line programs to convert the data to physical units. The quench data files also contain information about the cryogenic state of the magnet before the quench, results of simple on-line analysis of the data (e.g. quench current, maximum voltages and $\int I^2 dt$) and operator comments. Mechanical and cryogenic data as a function of magnet current are written in ASCII files, which have a format similar to the ASCII files written by CRYO_LOGGER.

ACKNOWLEDGEMENTS

Fermilab is operated by the Universities Research Association under contract with the United States Department of Energy. Many people have contributed to the design, construction and maintenance of this facility. We wish to acknowledge the contributions of members of the staff of the Fermilab Technical Support Section, in particular, R. Barger, A. Bianchi, B. C. Brown, R. Dachniwskyj, E. Desavouret, K. Dillow, T. Fritz, J. Garvey, D. Hartness, S. Helis, F. Johnson, E. Justice, G. Kirschbaum, D. Krause, M. Kuchnir, D. Lewis, D. Massengill, R. Nehring, A. Rusy, E. Schmidt, H. Stahl, J. Tague, D. Validis, F. Wilson, and W. Zimmerman. Visitors from the SSC Central Design Group, Brookhaven National Laboratory and Lawrence Berkeley Laboratory have helped with the debugging of many systems while in the process of testing SSC magnets. We acknowledge in particular the contributions of A. Devred, J. Kaugerts and J. Tompkins of CDG, P. Wanderer of BNL and J. Zbasnik of LBL. The test stand end boxes were designed and built by the Fermilab Accelerator Division Cryogenic Systems group under the direction of T. J. Peterson and J. Theilacker. D. Howard of the Fermilab Accelerator Division Electrical Engineering Support Group designed the modules for 4-wire resistance strain gage measurements.

REFERENCES

1. J. Strait, et al., Full length prototype SSC dipole test results, IEEE Trans. Magn. MAG-23:1208 (1987).
2. J. Strait, et al., Tests of Prototype SSC Magnets, in: "Proc. of the 1987 IEEE Particle Accelerator Conference," E. R. Lindstrom and L. S. Taylor, eds., p. 1540 (1987).
3. J. Strait, et al., Tests of Prototype SSC Magnets, IEEE Trans Magn. 24:730 (1988).
4. J. Strait, et al., Tests of Full Scale SSC R&D Dipole Magnets, presented at the Applied Superconductivity Conference, San Francisco, CA, August 22-25, 1988.
5. J. C. Tompkins, et. al, Performance of Full-length SSC Model Dipoles, presented at the International Industrial Symposium on the Super Collider, New Orleans, LA, February 8-10, 1989.
6. A. Devred, et. al, Quench-start localization in Long SSC Dipoles Using Voltage Taps Technique, presented at the International Industrial Symposium on the Super Collider, New Orleans, LA, February 8-10, 1989.
7. P. Dahl, et al., Construction of cold mass assembly for full length dipoles for the SSC accelerator, IEEE Trans. Magn. MAG-23:1215 (1987).
8. R.C. Niemann, et al., Design, construction and test of a full scale SSC dipole magnet cryostat thermal model, IEEE Trans. Magn. MAG-23:490 (1987).
9. "Conceptual Design of the Superconducting Super Collider," SSC-SR-2020, SSC Central Design Group, Lawrence Berkeley Laboratory, Berkeley, CA (1986).
10. R. Barger, et al., Operating History of Fermilab's 1500 W Refrigerator Used for Energy Saver Magnet Production Testing, Adv. Cryo. Engr. 31 (1986).
11. CCI, Allentown, PA.
12. K. McGuire, et al., Cryogenic Instrumentation of an SSC Magnet Test Stand, Adv. Cryo. Engr. 33:1063 (1988).
13. T. J. Peterson and P. O. Mazur, A Cryogenic Test Stand for Full Length SSC Magnets with Superfluid Capability, presented at the International Industrial Symposium on the Super Collider, New Orleans, LA, February 8-10, 1989. (Also a Fermilab preprint, TM-1562.)
14. Empro Manufacturing Co., Inc., Indianapolis, IN.
15. Hazemeyer B.V., Hengelo, Holland.
16. T. H. Nicol, Structural Performance of the First SSC Design B Dipole Magnet, presented at the International Industrial Symposium on the Super Collider, New Orleans, LA, February 8-10, 1989.
17. A. D. Anerella, et. al, Measurement of Internal Forces in Superconducting Accelerator Magnets with Strain Gauge Transducers, presented at the Applied Superconductivity Conference, San Francisco, CA, August 22-25, 1988.
18. Six ADC modules for multiple voltage tap readout on highly instrumented magnets are shared between the two test stands. Nine eight-conductor cables must be changed at a patch panel to switch between the two test stands.

19. Model 8212A 32 channel, 12-bit fast data logger with model 8800A memory module, LeCroy Research Systems, Spring Valley, NY.
20. Model 3457A digital multimeter, Hewlett-Packard Co., Palo Alto, CA.
21. Model 706 scanner with model 7064 low-voltage relay scanner cards, Keithley Instruments, Inc., Cleveland, OH.
22. Type AD210BN, Analog Devices, Norwood, MA.
23. Dynisco, Norwood, MA.

A HIGH RESOLUTION BARIUM FLUORIDE CRYSTAL ARRAY[1]

Ren-yuan Zhu[2]

Lauritsen Laboratory
California Institute of Technology
Pasadena, CA 91125

ABSTRACT

A high speed and highly radiation-resistant barium fluoride crystal array is under development at Caltech in collaboration with the Brookhaven National Lab, the Shanghai Institute of Ceramics in China, and the KEK in Japan. The barium fluoride array will serve as a prototype for a high precision electromagnetic calorimeter which is uniquely suitable for the Superconducting Super Collider, where high speed and radiation resistance up to 10 mega-Rads or greater doses are primary requirements. This research program aims at producing large size barium fluoride crystals with high radiation resistance, and developing a fast, effective light-sensitive readout to collect the fast scintillation components. Test results of commercially available barium fluoride and a test crystal produced at Shanghai are reported.

INTRODUCTION

The high luminosity and high center of mass energy available at the Superconducting SuperCollider (SSC) have presented a great challenge to high energy physics experimentalists faced with the problem of detector design [1]. An electromagnetic calorimeter designed for an SSC experiment has to meet the following criteria:

- Good radiation resistance to survive in a high radiation environment. Ideally, long term shifts should be 10% or less per year, at a particle flux of $10^{15}/\text{year}/r_\perp^2$ at an average luminosity of $10^{33}/\text{cm}^2/\text{sec}$, where $r_\perp$ is the distance from the beam in cm. This corresponds to doses of up to 10^7 Rads per year at small angles with respect to the beam [2].

- Fast response time of 20 nsec or less, to cope with the high interaction rate.

- Linear response over a large dynamic range, to allow precision measurement of electrons and photons (e.g. from W and Z) inclusively, down to the 10 GeV range, as well as in "special" events (e.g. decay of a second, heavy Z) to the 1 TeV range. Linearity over a large range will allow for calibrations at the GeV scale, which will be crucial for maintaining the highest possible resolution of the calorimeter.

- High granularity, to ensure that electromagnetic particles can be isolated, and precisely constructed, even when they are spatially near to hadronic jets.

[1] Work supported by U.S. Department of Energy Contracts No. DE-AC03-81-ER40050.
[2] Representing the team of H. Newman (Caltech), H. Ma, C. Woody (Brookhaven National Laboratory), Z.Y. Wei, Z.W. Yin (Shanghai Institute of Ceramics, China), T. Matsuda and F. Takasaki(KEK)

- Relatively short radiation length, to limit the lateral spread of showers. This factor must be combined with good granularity to ensure that particles in jets can be resolved, and that electromagnetic showers can be reconstructed precisely.

Barium fluoride (BaF_2) is a unique high density inorganic scintillator with three emission spectra peaking at 195 nm, 220 nm and 310 nm, with decay time constants of 0.87, 0.88 and 600 nsec respectively [3]. Because of the speed of the "short" or "fast" components and the evidence that it has high radiation resistance [4], BaF_2 has gained vast interest in recent years [5].

Armed with 5 years of experience, accumulated in the design and development of two Bismuth Germanate ($Bi_4Ge_3O_{12}$ or BGO) crystal arrays at Caltech, Caltech group proposed to develop, build and calibrate a high speed, radiation-hard BaF_2 crystal array, in collaboration with the Brookhaven National Lab (BNL), the Shanghai Institute of Ceramics (SIC) in China, and the KEK in Japan. The BaF_2 array will excel in satisfying all of the design criteria listed above. It will thus serve as a prototype for a high precision electromagnetic calorimeter for the SSC. The BaF_2 crystal array will also be an excellent design choice for other future accelerators, such as LEP II, HERA, or a TeV-range e^+e^- linear collider.

The research and development program will extend over two to three years. It will focus on the production of very large radiation hard crystals with suppressed slow component, a fast light-sensitive readout system, and an accurate calibration. Next section of this report summarizes properties of the BaF_2 crystals. Our research program is outlined in the third section. Section 4 is the preliminary result obtained in some crystal measuremnets. A brief summary is given in the last section.

BARIUM FLUORIDE CRYSTAL

Fast Scintillation Light

Table 1 lists the basic properties of barium fluoride, as compared with other commonly used scintillation crystals: NaI(Tl), pure CsI, CsI(Tl), CeF_3 and BGO. Figure 1 shows the scintillation spectrum of BaF_2 [3] and the quantum efficiencies of two photomultipliers (PMT):

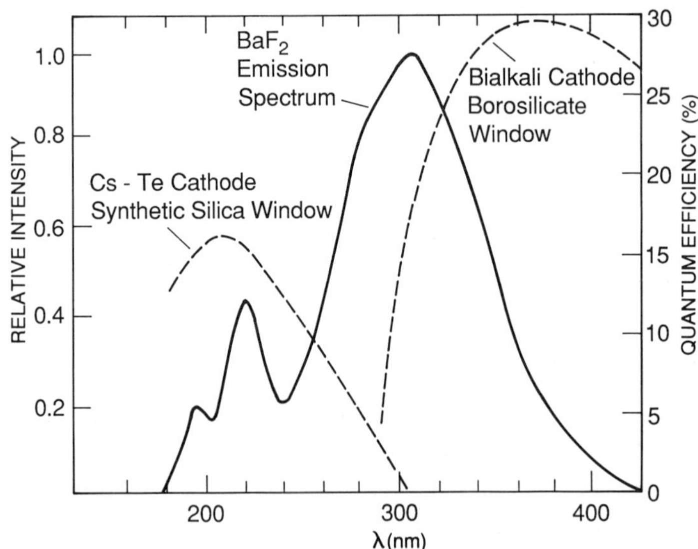

Figure 1. BaF_2 scintillation spectrum and PMT response. The scintillation spectrum is from P. Schotanus *et al.* [3].

Table 1. Properties of Scintillation Crystals.

	NaI(Tl)	Pure CsI	CsI(Tl)	BaF$_2$	CeF$_3$	BGO
Density(g/cm^3)	3.67	4.51	4.51	4.88	6.16	7.13
Radiation Length (cm)	2.59	1.85	1.85	2.10	1.7	1.12
Interaction Length (cm)	41.4	37.0	37.0	29.9	26.2	21.8
X_{rad}/X_{int}	0.063	0.051	0.051	0.068	0.065	0.051
Refractive Index	1.85	1.80	1.80	1.49	1.62	2.15
Hygroscopic	Yes	No	No	No	No	No
Luminescence(nm)	410	300	565	320	340	480
				210	310	
Decay Time(nsec)	230	10	900	630	27	300
				0.9	$\approx$2	
Light Output	100	20	45	20	3	15
				4	?	

a special UV-sensitive, solar-blind PMT with Cs-Te photocathode and synthetic silica (quartz) window (Hamamatsu R3197 or R2978) and a conventional PMT with bialkali photocathode and borosilicate glass window (Hamamatsu R1306). The fast components (195 and 220 nm) of BaF$_2$ have decay time less than 1 nsec which are the fastest scintillation light in table 1. Althogh the fast components are around 200 nm, it is spectroscopically separable from the slow component (310 nm). Figure 2 is a photo of the scintillation light pulse observed by a R3197 PMT from a BaF$_2$ crystal irradiated by a ^{137}Cs γ-ray source. The rising time of the scintillation light pulse was completely determined by the 2.3 nsec rising time of the PMT. It is thus concluded that with a proper light-sensitive readout it is feasible to build a very fast BaF$_2$ electromagnetic calorimeter for the SSC. The proposed detector array readout will select the fast scintillation components, and will have good energy resolution and radiation resistance, as detailed in the next section.

High Radiation Resistance

Early work done by S. Majewski and D. Anderson [2] showed that no color center formation is observed in BaF$_2$ up to a dose of 1.3×10^7 Rads in an 800 GeV proton beam. The crystals tested were from Harshaw. M. Marashita *et al.* [2] tested the samples made by NKK and OKEN (Ohyo - Koken in Japan) with ^{60}Co γ-rays and a 2 MeV electron beam. The measurements were done 24 hours after the end of the irradiation. The crystals from OKEN showed higher radiation resistance than crystals from NKK. While 8.8×10^7 Rads of γ-rays caused a less than 5% effect on the transmittance in the OKEN crystals, 8.8×10^8 Rads of electron dose caused a 20% effect. The crystal from NKK had intrinsic strong absorption at about 205 nm which was presumed [2] to be due to impurities (Pb) in the crystal.

Work by S. Majewski and M. Bentley [2], confirmed the early report that BaF$_2$ has exceptionally good radiation resistance, up to γ-ray doses of a few times 10^7 Rads. Figure 3 shows the transmittance curves measured before and after 2×10^7 Rads dose by Majewski and Bentley [2].

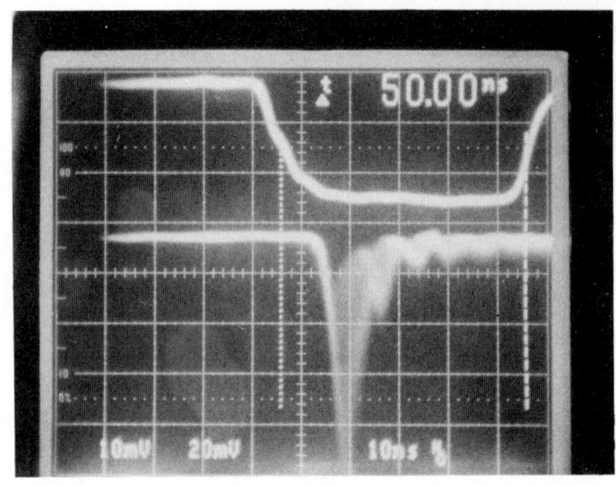

Figure 2. Scintillation Light pulse observed by a Hamamatsu R3197 PMT from a BaF$_2$ crystal irradiated by a ^{137}Cs γ-ray source.

Good Energy Resolution

M. Murashita *et al.* [2] made a systematic measurement of the electron energy between 0.25 to 5 GeV, and obtained a resolution of 4.2%/E$^{1/4}$. Another report from J. Giehl *et al.* [7] shows that the electromagnetic energy resolution obtained from BaF$_2$ by using photodiode readout with a simple wavelength shifter, for electrons, pions and photons between 60 MeV to 40 GeV, is $(1.8\pm0.1)\%/\mathrm{E}^{(0.35\pm0.03)}$.

Figure 4 shows the pulse height spectra obtained with a ^{137}Cs source from two BaF$_2$ counters which are two BaF$_2$ crystals coupled to R1306 and R3197 PMT's respectively. The FWHM resolution from the slow component (R1306) is 10.5% which is only 20–30% inferior to that measured from NaI(Tl), and is comparable with that obtainable with a high quality BGO. In summary, BaF$_2$ crystals have been demonstrated to have a very good energy resolution, comparable with BGO and NaI over a large energy range.

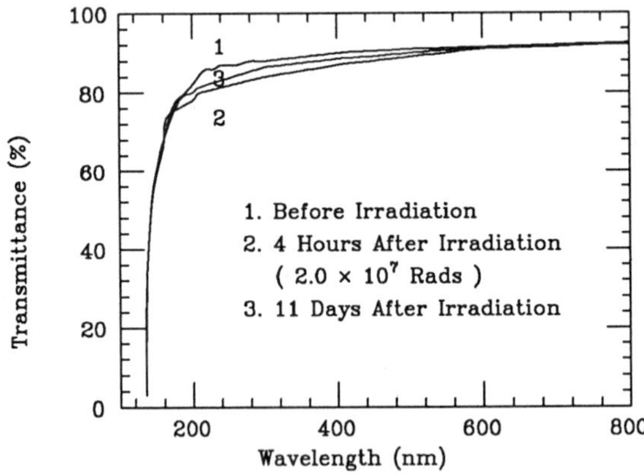

Figure 3. Transmittance curves of a BaF$_2$ sample, before and after 2×10^7 Rads dose, from Majewski and Bentley [2].

Other Properties of BaF$_2$

The properties of BaF$_2$ crystal, such as photoelectron number, linearity, temperature dependence of the scintillation light and the effect of dopant have been studied by several authors. The results are summarized as follws:

- The photoelectron number of BaF$_2$ scintillation light have been measured as 400 photoelectrons/MeV for the fast components and 2000 photoelectrons/MeV for the slow component [3,6].

- The scintillation light linearity of BaF$_2$ is known to be as good as BGO. A measurement, done in a CERN test beam by E. Lorenz *et al.* [7], showed that the linearity of light output of BaF$_2$ is good up to 50 GeV. The linearity is expected to remain good far beyond 100 GeV.

- The fast components have no temperature dependence, while the intensity of the slow component increases with decreasing temperature: –2.4%/°C [8].

- Pb^{2+} contamination in BaF$_2$, measured with Proton Induced X-ray Emission, results in an absorption band at 205 nm together with an extra scintillation emission at 257 nm with decay time of 630±40 nsec [9].

- With lanthanum doping Fast/Slow ratio improves and the crystal is radiation hard up to a level beyond 10^6 Rads [10].

- The mechanizm of BaF$_2$ scintillation has been understood in terms of the band structure of crystals. Most properties of BaF$_2$ scintillation light can be explained by the energy band mechanizm [12].

Among all these properties, the lanthanum doping is most valuable for our reaerach project.

RESEARCH PROGAM

The proposed detector consists of BaF$_2$ crystals of unprecedented size. The crystals will have a square cross-section, 2×2 cm^2 in the front, tapering to 4×4 cm^2 in the back. Each large crystal will be 50 cm long (24 radiation lengths) for a total volume of 500 cm^3, and a total weight of 2.4 Kg. This design corresponds to an SSC detector with an inner radius

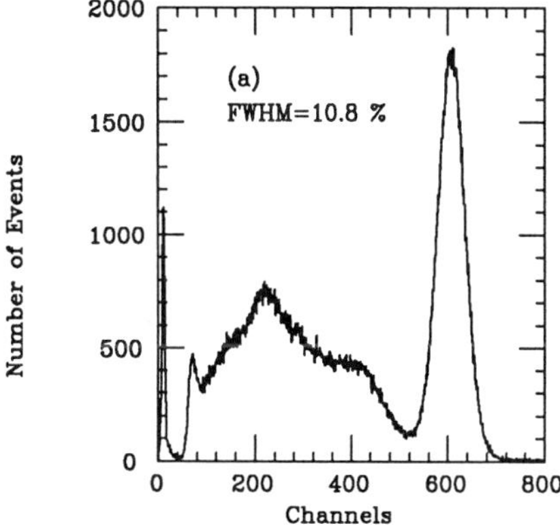

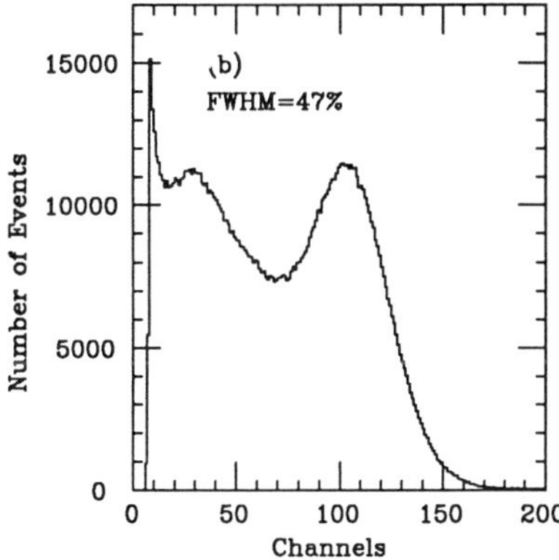

Figure 4. BaF$_2$ pulse height spectrum with ^{137}Cs γ-rays with R1306 (a) and R3197 (b) phototubes.

for a compact track detector of 50–60 cm, and a solid angle coverage of approximately 1.6 millisteradian per crystal in the calorimeter.

The research program will extend over two to three years, and will focus on following three areas:

- Development of a reliable method of growing large size, highly transparent and radiation resistant crystals, with the the slow component suppressed;

- Development of a high dynamic range, radiation resistant photodetector to readout the fast scintillation light components;

- Implementation of our recently proven new calibration technique [11], based upon radiative capture produced by a beam from a RadioFrequency Quadrupole Accelerator (RFQ).

Crystal Development and Test

Recently, P. Schotanus et al. at Delft [9] have shown that the 205 nm absorption band in BaF_2 crystals is associated with the presence of Pb impurities. F. Takasaki [13] further demonstrated a correlation between the presence of absorption band and susceptibility to radiation damage. From this, C. Woody expects that radiation damage is one-to-one correlated with the presence of lead [14]. It is well known that Pb fluoride is used to remove oxygen during the crystal production process, by some manufacturers. Hence alternative means of oxygen removal may be the key to production of radiation hard, UV transparent crystals.

These factors will be fully tested during crystal development in our project, prior to the startup of large crystal production in Shanghai. It is interesting to note that lead is also among the most harmful impurities in BGO, in terms of its effect on radiation hardness [15].

Within the past year the Delft group and C. Woody have initiated a very promising approach to the production of radiation-hard crystals with the slow component suppressed. Schotanus *et al.* [10] showed that the addition of lanthanum in 1% to 13% concentrations reduces the slow component by a factor of 3 to 30 (with increasing lanthanum concentration), and decreases the slow component's decay time to 200 nsec. Woody [10] has just completed a series of tests at BNL, demonstrating that crystals with 1% lanthanum doping are radiation hard up to a level beyond 10^6 Rads.

Following joint work by SIC and Caltech on BGO radiation damage and the role of impurities [15], SIC decided to develop large size BaF_2 crystals in collaboration with Caltech. SIC has excellent track record in providing the L3 Collaboration with 11,000 large BGO crystals of sufficient light transparency and light output to meet L3's stringent specifications, at a reasonable price [16]. Recently, SIC has successfully developed a new type of BGO crystal with high radiation resistance [17], which is especially suited for the L3 BGO end caps. A measurement of a test BaF_2 sample produced by SIC is shown in next section.

The crystals produced will be tested at Caltech. The photoelectron yield, the UV transmittance, the energy resolution and the radiation resistance will be tested. Investigations of the scintillation mechanism, and the role of impurities in radiation damage effects will be done at Caltech and BNL. We will also develop methods of surface treatment and/or coating, and of coupling to the light sensitive readout, to maximize the light output while maintaining uniformity along the entire length of the large crystals. In addition, we will perform a detailed study, using some of the methods we employed in our BGO study in 1984–1986, on the correlation between radiation damage effects and the presence of trace elements in the crystals [15]. Spectrophotmetric and radiation damage facilities available at BNL will be used during this part of the study.

Light Sensitive Readout

Without an effective readout of the fast components, the advantages of BaF$_2$ cannot be fully utilized. For BaF$_2$ there are three existing readout methods: gas-filled chambers [18], photodiodes plus wave length shifters (WLS) [20], and photomultiplier [21].

The short wavelength component, with its fast decay constant has been successfully coupled to wire chambers filled with the photosensitive gas tetrakis (dimethylamino) - ethylane (TMAE) [18]. We, however, will not pursue wire chamber readout, since hydrocarbon solvent gases in chambers suffer radiation damage well below the 10^6 Rads level [19].

The advantages of using photodiodes are obvious: they are compact and they exhibit no drift over periods of years, since they have no gain. Today's production-quality silicon photodiodes have a high quantum efficiency, of approximately 60%, down to the 190 nm range, and they can be used in magnetic fields of several kilogauss. However, the large fraction of the slow scintillation light component must be removed to avoid unnecessary saturation.

An effort has been made by E. Lorenz et al. [7] to develop a readout technique using a fluorescent flux concentration (FFC) coupled to a silicon photodiode [20]. FFC's functioning as a wave length shifter improve the signal to noise ratio by concentrating the light flux onto a small view area. A further benefit is that the light is shifted to longer wavelengths, where photodiodes are more efficient. The result of the test shows that a signal gain of 26 was obtained, where the flux concentration contributes a factor of 6 and the increase of quantum efficiency of photodiodes contributes another factor of 4. This is an attractive approach although the existing FFC cuts off the fast components and uses only the slow component.

C. Woody at BNL will look for a proper WLS or FFC which will shift the fast components to 400 nm where a conventional PMT or photodiode will function. The low cost of conventional PMT and photodiode will certainly be beneficial. However, the radiation resistance of the WLS or FFC, the effect of charged particles traversing the photodiode and WLS or FFC, and the effect of the WLS or FFC on the light uniformity of BaF$_2$ also have to be tested.

Another approach using a UV-photodiode directely together with a wide band interference filter which passes only the fast components will be tested at Caltech. Figure 5 shows the transmittance of a wide band interference filter from Twardy Technology Incorporated (TTI) together with the quantum efficiency of a Hamamatsu UV-sensitive photodiode S1723-05. One must notice that the spectral response of TTI filter matches well to the fast components of BaF$_2$, and the quantum efficiency of S1723-05 at 190 nm is around 60%.

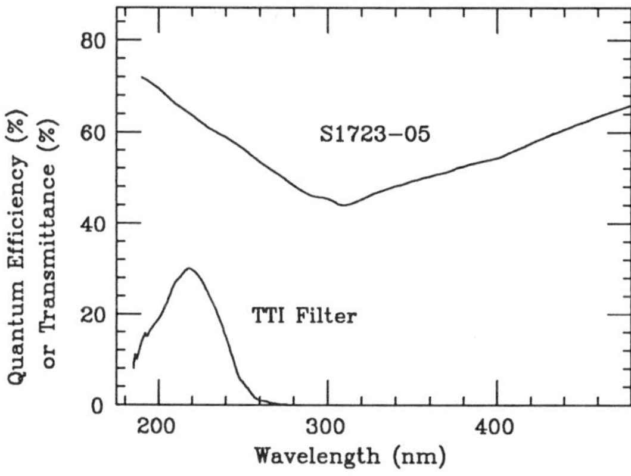

Figure 5. Spectral response of TTI wide band interference filter and Hamamatsu S1723-05 UV-sensitive photodiode.

If a photodiode turns out to be the readout element of choice, we will benefit from our experiences in the development and use of low noise preamplifiers and readout systems for the RFQ-based calibration project [11].

Unless a better readout device can be found, we will concentrate our efforts on developing a fast UV-sensitive PMT readout. A measurement [22] which used a PMT tube with a Cs-Te "solar blind" photocathode and quartz window, showed that the fast BaF_2 components can be collected with high efficiency, while the slow component is reduced by a factor of 10. As detailed in the next section, similar results were obtained in our measurements using Hammamatsu R2078 (1", 1.5 nsec rise time, 14 nsec transit time) and R3197 (2", 2.6 nsec rise time, 48 nsec transit time) PMT's. Both tubes have a Cs-Te photocathode and a synthetic silica (quartz) window and have quantum efficiencies above 10% at 200 nm (Fig. 1). We notice that the radiation resistance of the window material, synthetic silica, is excellent. Figure 6 shows the synthetic silica remains transparent after neutron irradiation of a flux up to 10^{14} n/cm^2 [23].

In this project, a high speed, low gain, high dynamic range, solar blind phototube, with maximum sensitivy in the 200 nm range will be developed by F. Takasaki (KEK) together with Hammamatsu, with final tests at Caltech. This will follow directly from our recent work, and from Takasaki's recent work on radiation damage of PMT's. It is clear that the combination of the right light-sensitive device, the new doped crystals with the slow scintillation component suppressed and a fast pulse shaping, will provide an efficient way to read out the fast scintillation light exclusively.

Multistage microcomputer-based readout systems capable of the necessary speed will also be developed for the BaF_2 array calibration. The digital, higher levels of the readout will be adapted from the VME and Motorola 68020-based, stand-alone systems developed and now used at Caltech for the RFQ calibration project [24].

Precision Calibration

Over the last 5 years, the Caltech group has developed and tested a novel calibration technique [11] based upon the radiative capture of a pulsed proton beam from an RFQ accelerator in a lithium target, $^7Li(p,\gamma)^8Be$. The resultant flux of 17.6 MeV photons can be used to calibrate the thousands of BGO crystals in the L3 electromagnetic calorimeter at once, with an absolute accuracy of better than 1% in 1–2 hours. When installed in the L3 experiment at LEP this system will help maintain the high resolution of the electromagnetic calorimeter during running. An experimental test of a 4 × 5 BGO crystal array at the RFQ facility at AccSys Technology, Inc. in Pleasanton, California, was carried out in November, 1987 [11]. An absolute calibration precision of 0.8% was achieved in the test.

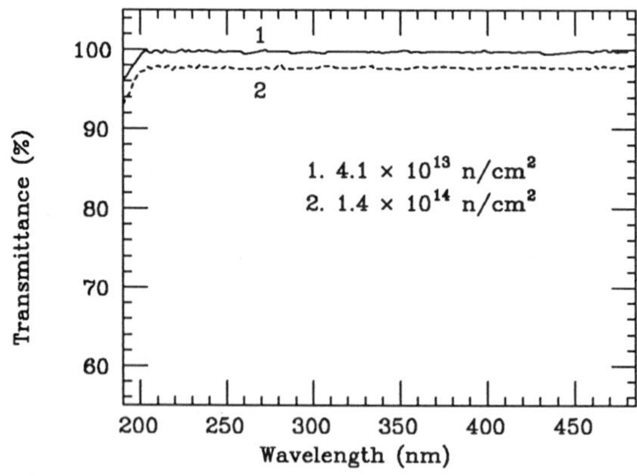

Figure 6. Transmittance of synthetic silica after neutron irradiation [23].

Following the 1987 test, we realized the possibility of a new technique, appropriate for the calibration of calorimeters at the SSC or other accelerators in the TeV range. By using the radiative capture of protons from a pulsed beam in a fluoride target, $^{19}F(p,\alpha)^{16}O^*$ [25], and the subsequent decay of the excited oxygen nucleus, $^{16}O^*$, hundreds to thousands of 6 MeV photons could be produced per millisteradian per pulse. These "equivalent high energy photons" would serve as a calibration source for electromagnetic calorimeters.

A milestone beam test was carried out in September, 1988, at AccSys under a Phase I grant from the DoE Small Business Innovative Research (SBIR) program, where 4 BaF_2 counters were set up together with a 7×7 L3 BGO crystal array as a reference counter [11]. The test demonstrated that this technique functions as a clean "pulse generator" of scintillation light with energy up to few GeV, originating several cm inside the crystals. The clean, narrow Gaussian-distributed light pulses produced with this technique were shown to provide relative calibrations with a precision of 0.5% or better in a few minutes.

There are much stronger fluorine resonances between 2.0 and 4.0 MeV [25] beyond the 1.92 MeV beam energy at AccSys. By using a 3.85 MeV RFQ and a CaF_2 target, which would have no neutron production as a by-product below 4.05 MeV [25], an equivalent photon energy per calorimeter element of 20 GeV/(0.1 μCoulomb) or more is expected [11].

During the AccSys tests we typically used a beam pulse of 0.10–0.15 μCoulomb with pulse length of a few μsec. In order to adapt this technique for future hadron colliders, such as the SSC or LHC, however, the pulse length of several μsec may not match the experimental readout electronics. Recent computer simulation studies at Saclay have shown that it should be possible to compress 4 MeV beam pulses of 0.1 μCoulomb into 100 nsec time or less, by using multiturn injection into — and single turn extraction from — a small storage ring [26]. This development will allow this technique to be used in calibrating calorimeters at the SSC and LHC. A new pulse time compressor at the output of their RFQ system will be developed by AccSys under a Phase II grand from the DoE SBIR program [26].

Caltech group will develop a precise measurement of total beam charge on the target for each pulse, to provide a direct normalization which can be cross-checked against the normalization to a standard. The design of this target has already been partially worked out. It will employ a molybdenum collimator, a precise inductive current pickup and a integrator.

After the BaF_2 array is constructed, we will carry out a series of runs at AccSys with typical beam intensity of 0.1 μCoulomb on the CaF_2 target to simulate 20 GeV pulses. These tests will span a longer time period, and will aim at proving the long term systematic stability as well.

BaF_2 CRYSTAL TEST RESULT

Begining from early 1989 we started BaF_2 crystal test at Caltech. Several BaF_2 crystals were obtained from Harshow (USA), OKEN (Ohyo - Koken in Japan) and Merck (E. Merck Darmstadt in W. Germany). Although SIC started their development late last year, we were able to obtain some test samples from SIC.

Photoelectron yield

All BaF_2 crystals were polished and wrapped with two layers of teflon tape. The scintillation light yield in terms of photoelectron numbers per MeV of energy deposited was measured by using two PMT's. One is Hamamatsu R2059 which has a bialkali photocathode and a synthetic silica window. The other is a UV-sensitive, solar-blind PMT, Hamamatsu R3197 (Fig. 1). R2059 has a wide spectral response from UV to infrared. Its quantum efficiency is around 20% at 200 nm. The PMT's were coupled to the BaF_2 crystals with Dow Corning 200 Fluid, with

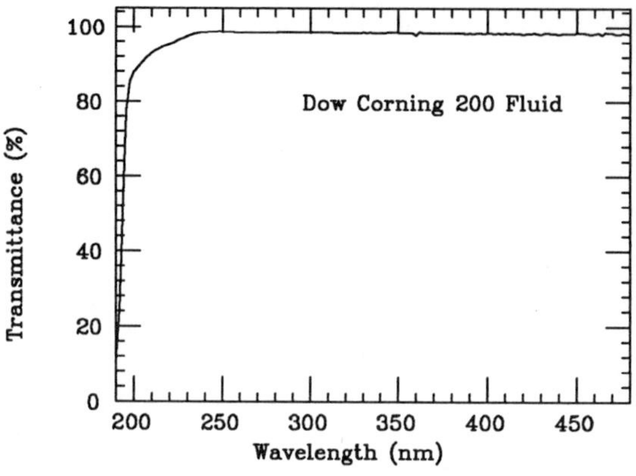

Figure 7. Transmittance of 0.5 mm thick Dow Corning 200 fluid.

a viscosity of 12,500 centipoise, which has very good UV transmittance. Figure 7 shows the transmittance of 0.5 mm thick Dow Corning 200 fluid. The transmittance at 200 nm is around 90%.

The scintillation light pulses collected by PMT's were integrated by a LeCroy 3001 QVT multichannel analyzer in the Q mode. An variable integration gate was provided by a LeCroy 2323A programmable dual gate generator. The data taking process was controled by a μVAX

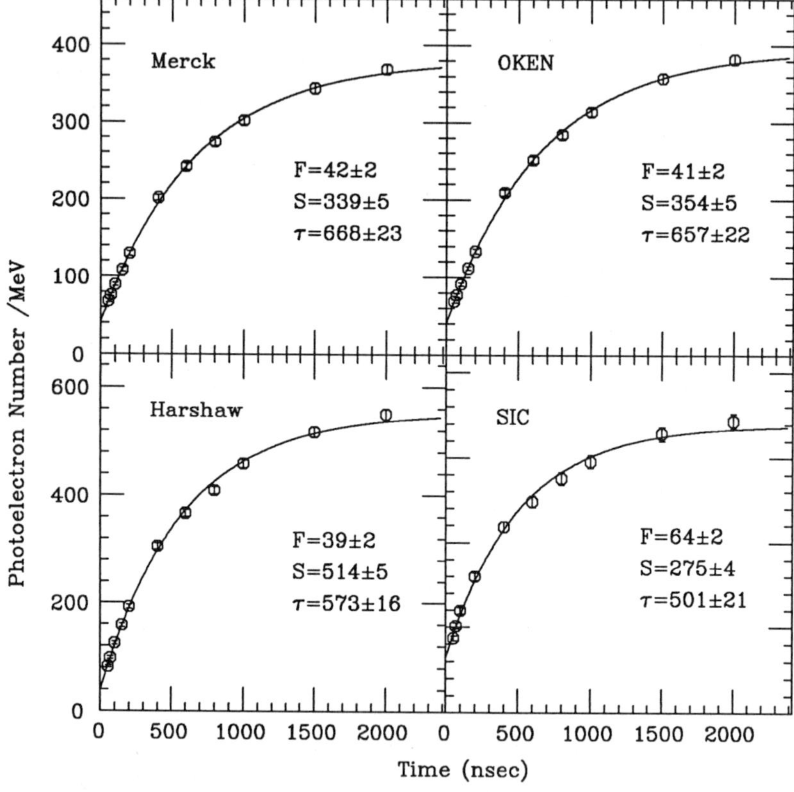

Figure 8. Photoelectron yields measured with Hamamatsu R2059 PMT are shown as a function of the integration gate width.

Table 2. Pulse Height Result Measured with ^{137}Cs and Hamamatsu R2059 Photomultiplier

Crystal Dimension(cm)	Oken $5\times5\times7$	Merck $\phi3"\times3"$	Merck H3$\times$14	Harshaw $\phi1"\times1"$	Harshaw $\phi2"\times3"$	SIC $\phi1"\times1.2"$
P.e./MeV(55ns)	69$\pm$2	68$\pm$2	62$\pm$2	82$\pm$2	90$\pm$3	87$\pm$3
P.e./MeV(2μs)	382$\pm$7	368$\pm$7	346$\pm$7	550$\pm$9	583$\pm$9	344$\pm$7
S.G./L.G.(%)	18	19	18	15	16	25
ΔE/E(%)(55ns)	45$\pm$1	50$\pm$1	45$\pm$1	42$\pm$1	35$\pm$1	43$\pm$1
(FWHM) (2μs)	19$\pm$1	22$\pm$1	18$\pm$1	16$\pm$1	14$\pm$1	27$\pm$1
Fast p.e./MeV	41$\pm$2	42$\pm$2	35$\pm$2	39$\pm$2	51$\pm$2	64$\pm$2
Slow p.e./MeV	354$\pm$5	339$\pm$5	322$\pm$4	514$\pm$5	538$\pm$6	275$\pm$4
Fast/Total(%)	10	11	9.9	6.9	8.6	19
τ_{slow} (nsec)	657$\pm$22	668$\pm$23	636$\pm$21	573$\pm$16	624$\pm$18	501$\pm$21

II station through CAMAC interface. The ^{137}Cs peak was determined by a Gaussian fit. The peak pulse height was converted to photoelectron numbers by using calibration curves for each gate width. The calibration curves were obtained from an LED test run and a subsequent Gaussian fit.

Figure 8 plots observed photoelectron numbers per MeV as a function of the integration time obtained from a R2059 PMT for four BaF$_2$ crystals. The data points in the figure were fitted to a sum of contributions from the fast components (F) and from the slow component (S) with exponential decay time constant of τ: y = F + S [(1 - exp(-t/τ)]. The numerical results of the fit are also shown in the figure.

Table 2 lists the fit results, photoelectron yield numbers, the ratios of the fast components to the total light yield and the FWHM energy resolutions measured by using a ^{137}Cs γ-ray source for 6 different crystals. Where Merck H3$\times$14 is a large crystal with a hexagon cross-section of 3 cm each edge and 14 cm long. One should note that the quantum efficiency of the PMT and the light transmittance of the Dow Corning 200 Fluid are not corrected. In summary, around 50 fast photoelectrons/MeV was observed, which is around 10% of the total light observed by R2059.

Table 3 lists the same results with R3197 PMT. Around 40 fast photoelectrons/MeV was observed, which is around 55% of the total light observed by the R3197 PMT. The lower fast photoelectron number is due to the lower quantum efficiency of R3197. The more than 1:1 fast to slow ratio promises that a readout sensitive only to the fast components can be developed.

Light Transmittance

The light transmittance of BaF$_2$ is a good measure of the crystal quality. A Hewlett Packard 8452A Diode Array Spectrophotometer was used to measure the transmittances of BaF$_2$ crystals. This spectrophotometer has wave length coverage down to 190 nm. Figure 9 shows measured transmittances curves of four BaF$_2$ crystals. While the crystals from Merck and OKEN showed very good transmittance down to 190 nm, the Harshaw crystal showed an absorption peak at 205 nm, which indicates a Pb contamination. The Pb absorption peak in

Harshow crystal also explains the lower fast photoelectron yield. The test Shanghai crystal is not bad in terms of transmittance and fast photoelectron yield. Since this is their first BaF_2 crystal ever made, we assume they will produce much better crystals in the future.

SUMMARY

A BaF_2 research research and development program now is under way at Caltech. With the existing Hamamatsu solar-blind PMT's a fast to slow ratio larger than one has be achieved. With further development in crystal manufacture and readout device a very fast and radiation-hard electromagnetic calorimeter can certainly be made with BaF_2 crystals.

We are planning to test more BaF_2 crystals, particularly BaF_2 crystals with different dopant, such as lanthanum etc. SIC is making a set of special doped BaF_2 crystals for this purpose. After testing this set of crystals an effort will be made to investigate the scintillation mechanism, and the role of impurities in radiation damage. The radiation damage tests will be carried out st Caltech and BNL.

A high speed, low gain, high dynamic range phototube sensitive in the 200 nm range will be developed by F. Takasaki (KEK) with Hamamatsu. This development will follow directly from the existing Hamamatsu UV-sensitive, solar-blind PMT's, such as R2078 or R3197. A development of grid mesh, or other fast phototubes capable of running in a magnetic field will also be pursued.

Once the production of large high quality crystals and the readout systems are established, a 49 crystal prototype array will be built and tested. The culmination of the program will be

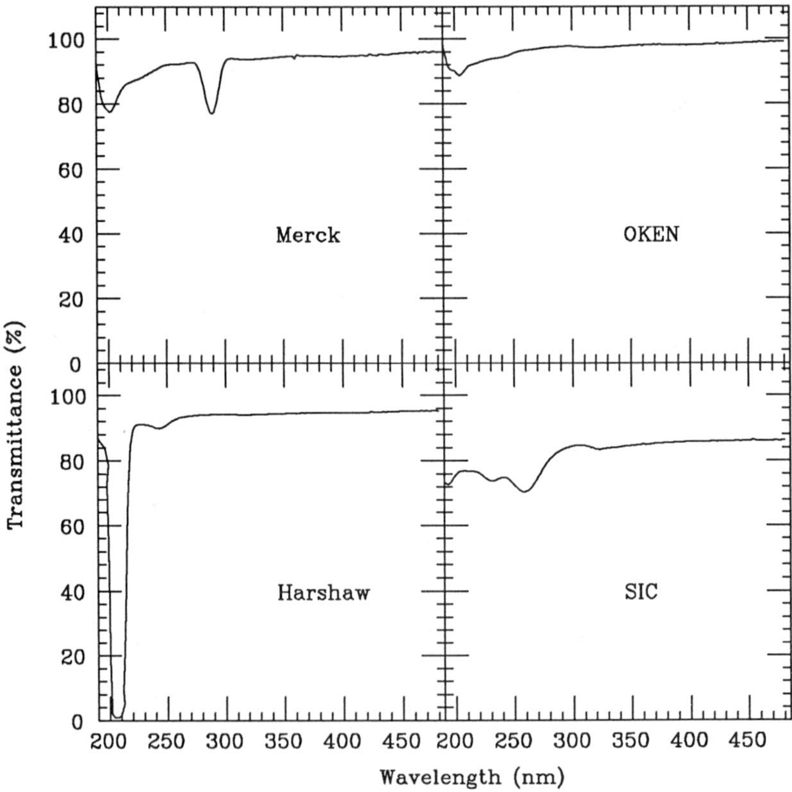

Figure 9. Transmittance measured without corrections for four BaF_2 crystals.

Table 3. Pulse Height Result Measured with ^{137}Cs and Hamamatsu R3197 Photomultiplier

Crystal Dimension (cm)	Oken 5×5×7	Merck $\phi3"\times3"$	Merck H3×14	Harshaw $\phi1"\times1"$	Harshaw $\phi2"\times3"$	SIC $\phi1"\times1.2"$
P.e./MeV(55ns)	29±1	34±1	27±1	36±1	49±1	37±1
P.e./MeV(2μsec)	53±1	56±1	48±1	81±2	91±2	51±1
S.G./L.G.(%)	55	61	56	44	54	73
ΔE/E(%)(55ns)	108±5	102±5	113±6	105±5	77±4	110±6
(FWHM) (2 μs)	75±1	82±1	71±1	55±1	49±1	78±1
Fast p.e./MeV	28±1	33±1	25±1	31±1	46±1	35±1
Slow p.e./MeV	27±1	26±1	25±1	51±1	50±2	16±1
Fast/Total (%)	51	56	50	38	48	69
τ_{slow} (nsec)	712±59	843±89	632±50	573±30	859±67	638±87

a high precision calibration, at the AccSys RFQ facility.

ACKNOWLEDGEMENTS

I am grateful to Prof. F. Takasaki and to Merck, Inc. who loaned us BaF$_2$ crystals. Dr. Z.Y. Wei helped prepare the BaF$_2$ test at Caltech. Many useful discussions with Drs. H. Newman, C. Woody, H. Ma, Z.W. Ying and T. Matsuda are greatly appreciated. The assistance of S. Sondergaard, J. Hanson, and many members of the Caltech technical staff is acknowledged.

References

[1] M.D.G. Gilchriese, in *"Proceedings of the 1984 Summer Study on the Design and Utilization of the Superconducting Super Collider"*, edited by R. Donaldson and J.G. Morfin, Snowmass, Co.(1984) 607; H.H. Williams, in *"Proceedings of the 1986 Summer Study on the Physics of the Superconducting Super Collider"*, edited by R. Donaldson and J. Marx, Snowmass, Co.(1986) 327; SSC Central Design Group, *"Report of the Task Force on Detector Research Development for the Superconducting Super Collider"*, SSC-SR-1021 (1986).

[2] D. E. Groom, in *"Proceedings of the 1988 Summer Study on the High Energy Physics in 1990's"*, Snowmass, Co. (1988).

[3] M. Lavel *et al.*, *Nucl. Instr. and Meth.* **A206** (1983) 169; and P. Schotanus *et al.*, *Nucl. Instr. and Meth.* **A259** (1987) 586; and *IEEE–NS* **34** (1987) 76.

[4] S. Majewski and D. Anderson *et al.*, *Nucl. Instr. and Meth.* **A241** (1985) 76; A. J. Caffrey *et al.*, *IEEE Trans. Nucl. Sci.* **NS-33** (1986) 230; and S. Majewski *et al.*, *Nucl. Instr. and Meth.* **A260** (1987) 373.

[5] D.F. Anderson *et al.*, *Nucl. Instr. and Meth.* **A228** (1984) 33.

[6] M. Moszynski *et al.*, *Nucl. Instr. and Meth.* **A226** (1984) 534; W.H. Wong *et al.*, *IEEE–NS* **31** (1984) 381; and Y.C. Zhu *et al.*, *Nucl. Instr. and Meth.* **A244** (1986) 577.

[7] E. Lorenz *et al.*, *Nucl. Instr. and Meth.* **A249** (1986) 235; and J. Giehl *et al.*, *Nucl. Instr. and Meth.* **A263** (1988) 392.

[8] P. Schotanus *et al.*, *Nucl. Instr. and Meth.* **A238** (1985) 564; and Kobayashi *et al.*, *Nucl. Instr. and Meth.* **A270** (1988) 106.

[9] P. Schotanus *et al.*, *Nucl. Instr. and Meth.* **A272** (1987) 917 and *Technical Univ. at Delft Preprint* **88-1**.

[10] P. Schotanus *et al.*, *IEEE–NS* **34** (1987) 272; and C. Woody *et al.*, BNL preprints.

[11] R.Y. Zhu *et al.*, *in these proceedings*; H. Ma *et al.*, *Nucl. Instr. and Meth.* **A274** (1989) 113; and *Caltech Preprint* **CALT–68–1536**, 1989, submitted to *Nucl. Instr. and Meth.*

[12] Y. Aleksandrov *et al.*, *Sov. Phys. Solid State* vol. **26** (**9**) (1984) 1734; and P. Schotanus *et al.*, *Delft preprint* (1988).

[13] F. Takasaki, *KEK Report* **84-24**, February, 1985.

[14] C. Woody, Private communication.

[15] T.Q. Zhou *et al.*, *Nucl. Instr. and Meth.* **A258** (1987) 464; and R.Y. Zhu *et al.*, *Caltech preprint* **CALT–68–1392** (1987).

[16] L3 Collaboration, *"L3 Technical Proposal"* May 1983; and J. Bakken *et al.*, *Nucl. Instr. and Meth.* **A254** (1987) 535 and **A228** (1985) 294.

[17] Z.W. Yin *et al.*, *Nucl. Instr. and Meth.* **A275** (1989) 273.

[18] D.F. Anderson *et al.*, *Nucl. Instr. and Meth.* **A217** (1983) 217; **A225** (1984) 8; **A228** (1984) 33; C.L. Woody and D.F. Anderson, *Nucl. Instr. and Meth.* **A265** (1988) 291; P. Schtanus *et al.*, *Nucl. Instr. and Meth.* **A269** (1988) 377; and P. Mine *et al.*, *Nucl. Instr. and Meth.* **A269** (1988) 385.

[19] C.L. Woody, *IEEE–NS* **35** (1988) 463 and Appendix 11 of *"Radiation Effects at the SSC* in **SSC-SR-1035**, edited by M. Gilchriese, Jun 1988.

[20] H. Dietl et al., *Nucl. Instr. and Meth.* **A235** (1985) 464; E. Blucher *et al.*, *Nucl. Instr. and Meth.* **A249** (1986) 201; H. Grassmann *et al.*, *Nucl. Instr. and Meth.* **A228** (1986) 323; and W. Klamra *et al.*, *Nucl. Instr. and Meth.* **A265** (1988) 485.

[21] S. Kubota *et al.*, *Nucl. Instr. and Meth.* **A242** (1986) 291; and T. Murakami *et al.*, *Nucl. Instr. and Meth.* **A253** (1986) 163.

[22] H. Kobayashi *et al.*, *Nucl. Instr. and Meth.* **A270** (1988) 106.

[23] S. Suzuki, private communication.

[24] R.Y. Zhu, "A portable BGO readout system for RFQ test", talk given in L3 collaboration meeting at Geneva, May 1987.

[25] F. Ajzenberg-Selove, *Nucl. Phys.* **A475** (1987) 1; and H. B. Willard *et al.*, *Phys. Rev.* **85** (1952) 849.

[26] R. Hamm, *"SSC Electromagnetic Calorimeter Calibration Source"*, Phase II proposal to the DoE SBIR program, December, 1988.

A NOVEL CALIBRATION SYSTEM
FOR SSC ELECTROMAGNETIC CALORIMETERS[1]

Ren-yuan Zhu, Hong Ma[2], and Harvey Newman

Lauritsen Laboratory
California Institute of Technology
Pasadena, CA 91125

Robert Hamm

AccSys Technology, Inc
1177A Quarry Lane
Pleasanton, CA 94566

ABSTRACT

A fast effective calibration technique has been developed for future Superconducting Super Collider (SSC) calorimeters based upon the radiative capture of protons from a pulsed RadioFrequency Quadrupole (RFQ) accelerator in a fluoride target. The intense flux of low energy photons acts as a clean "pulse generator" calibration signal equivalent to 20 GeV or more. This calibration technique has been demonstrate with a bismuth germanate (BGO) detector array, as well as with several barium fluoride detector crystals, to provide a calibration accuracy of 0.5% within two minutes. The SSC calibration has resulted from the development of this novel technique by Caltech over the past 5 years for the L3 BGO electromagnetic calorimeter, which uses a lithium target for the production of 17.6 MeV low energy photons as a calibration source. Proven techniques have also been developed to provide an *in situ* calibration for all 11,000 BGO detector crystals, with an absolute accuracy of 0.8% obtainable in 1–2 hours. The results of two experimental tests using the AccSys RFQ accelerator are reported, as is the preliminary design of a small storage ring that can be used to compress the output beam pulse from the RFQ accelerator into a target beam pulse of 100 nsec or less.

INTRODUCTION

Rapid and precise calibrations *in situ* have always been essential to maintain the design performance of high resolution calorimeters in high energy physics experiment. This will be particularly true in the case of calorimeters at the future high luminosity and high center of mass energy accelerators, such as the Superconducting Supercollider (SSC) and the large Hadronic Collider (LHC) [1], where the high radiation doses may cause rapid gain shifts in individual calorimeter elements, especially near the beam line [2].

---

[1] Work in part supported by U.S. Department of Energy Contracts No. DE-AC03-88-ER80674 and DE-AC03-81-ER40050.

[2] Present address: Department of Physics, Brookhaven Nat. Lab., Upton, NY 11973.

Many present-day calibration techniques, such as the use of radioactive sources and cosmic rays, will not work at these accelerators because the dynamic range of these calorimeter elements will, in most cases, have to extend from the GeV to the TeV range. The time to accumulate sufficient statistics for a precise calibration, for future calorimeters with 10^4 elements or more, also would be prohibitive.

Over the last 5 years, the Caltech group has developed and tested a novel calibration technique [3,4] based upon the radiative capture of a pulsed proton beam from a Radiofrequency Quadrupole (RFQ) accelerator in a lithium target, ^{7}Li(p,γ)^{8}Be. The resultant flux of 17.6 MeV photons can be used to calibrate the thousands of Bismuth Germanate (BGO) crystals in the L3 electromagnetic calorimeter at once, with an absolute accuracy of better than 1% in 1 – 2 hours. When installed in the L3 experiment at LEP (Fig. 1) this system will help maintain the high resolution of the electromagnetic calorimeter during running. An experimental test of a 4 × 5 BGO crystal array at the RFQ facility at AccSys Technology, Inc. in Pleasanton, California, was carried out in November, 1987 [4]. An absolute calibration precision of 0.8% was achieved in the test.

Following the 1987 test, we realized the possibility of a new technique, appropriate for the calibration of calorimeters at the SSC or other accelerators in the TeV range. By using the radiative capture of protons from a pulsed beam in a fluoride target, ^{19}F(p,α)^{16}O* [7], and the subsequent decay of the excited oxygen nucleus, ^{16}O*, hundreds to thousands of 6 MeV photons could be produced per millisteradian per pulse. These "equivalent high energy photons" would serve as a calibration source for electromagnetic calorimeters.

A milestone beam test was carried out in September, 1988, at AccSys under a Phase I grant from the DoE Small Business Innovative Research (SBIR) program, where 4 BaF$_2$ counters were set up together with a 7 × 7 L3 BGO crystal array as a reference counter [5]. The test demonstrated that this technique functions as a clean "pulse generator" of scintillation light, originating several cm inside the crystals. The clean, narrow Gaussian-distributed light pulses produced with this technique were shown to provide relative calibrations with a precision of 0.5% or better in a few minutes.

This report describes the RFQ-based calibration and summarizes the results of two tests.

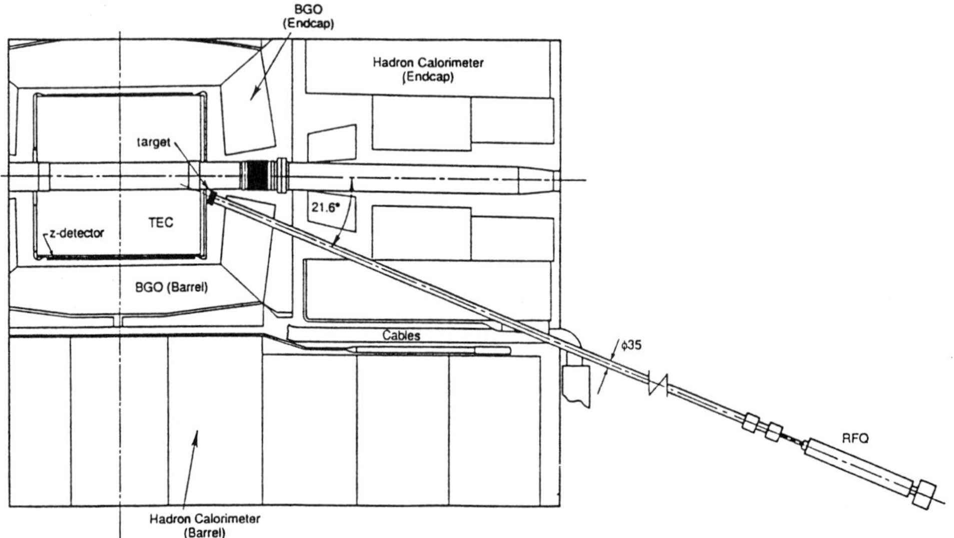

Figure 1. Plan of RFQ beam tube installation in the L3 detector.

Table 1. AccSys RFQ Specification.

Accelerated particle species	H^+
Input ion energy	30 keV
Normalized input emittance (95 %)	$<0.04\pi$ cm mrad
Nominal current limit	63 mA
Nominal phase-space acceptance	0.116π cm mrad
Final synchronous phase	30°
Nominal output energy	1.92 MeV
Operating frequency	417.7 MHz
Beam pulse width	1–50 μsec
Beam repetition rate	1–150 Hz
Input beam current (peak)	35 mA
Intervene voltage	65 kV
Maximum surface gradient	35 MV/m
Required rf power (peak)	200 KW minimum
Output beam current (peak)	up to 30 mA
Residual vacuum	$<1\times10^{-6}$ Torr
Normalized output emittance (90 %)	$<0.06\pi$ cm mrad
Out put energy spread (90%)	$< \pm20$ keV
Out put phase spread (90%)	$< \pm15°$

The next section describes the RFQ and the detector set-up. The results of the two beam tests are presented in the third and forth sections. A summary, and the outlook for SSC calibration RFQ design is given in the last two sections.

ACCELERATOR AND EXPERIMENTAL SET-UP

RFQ System

The RFQ is a quasielectrostatic accelerator, with advantages of a high beam current and a very compact size (up to 100 mA of protons). An RFQ has been the accelerator of choice (especially over last five years) whenever acceleration of intense H^+, H^- or heavy ion beams up to a fixed energy in the range from 750 keV to 4 MeV is required [6].

The AccSys RFQ uses an H^+ ion source and produces a 1.92 MeV pulsed proton beam. Figure 2 is a block diagram of the AccSys RFQ system. The principal specifications of the AccSys RFQ is listed in Table 1.

The proton beam was bent by a dipole magnet by 22.5°. A pair of small permanent quadrupole magnets was installed during the 1988 test to focus the beam spot onto the target with a diameter that could be adjusted, down to a few millimeters.

The beam pulse structure used during the 1987 (1988) test was:

- Pulse width: 3 (8) μsec full width at half height, 4 (10) μsec at the base;

- Peak current: adjustable, up to 12 (24) mA;

- Pulse repetition rate: adjustable, up to 60 (30) Hz.

Figure 3 shows a typical RFQ beam pulse together with the gate signal applied to the integrator in the 1988 test.

Target

Proton capture in a lithium or fluorine nucleus, ^{7}Li$(p,\gamma)^8$Be or ^{19}F$(p,\alpha)^{16}$O* with subsequent radiative decay of the excited oxygen nucleus, ^{16}O*, was used to produce photons. Single 17.6 MeV photons from the lithium target, and pulses consisting of many 6 MeV photons from the fluoride target, were used as the calibration source. The yield of photons was adjusted by varying the beam current [3]. For the 1987 test the photon rate was adjusted to a 0.2% hit probability per RFQ pulse for a single L3 BGO crystal with a 2 $\times$ 2 cm^2 front face situated at a distance of 50 cm from the target. For the 1988 test the photon yield was adjusted up to several GeV deposited in a typical L3 BGO crystal for each RFQ pulse.

Li and LiF were used as target materials. The target material was encapsulated in a molybdenum holder, with a thin molybdenum foil facing the beam. The foil was designed to help conduct heat away from the target and to prevent the proton beam from hitting the ^{7}Li neutron resonance which has a threshold of 1.88 MeV, since the energy of AccSys RFQ is 1.92 MeV. The foils also served as an energy degrader for the proton beam. 15 μm thick molybdenum foils were used for the lithium target allowing protons to hit the 441 keV resonance, while avoiding another undesired 1.03 MeV resonance. Table 2 lists the three LiF targets and the foils used in the 1988 test. The energy of the proton beam (E_p) hitting the crystals, after penetrating the foil, is also shown in the table.

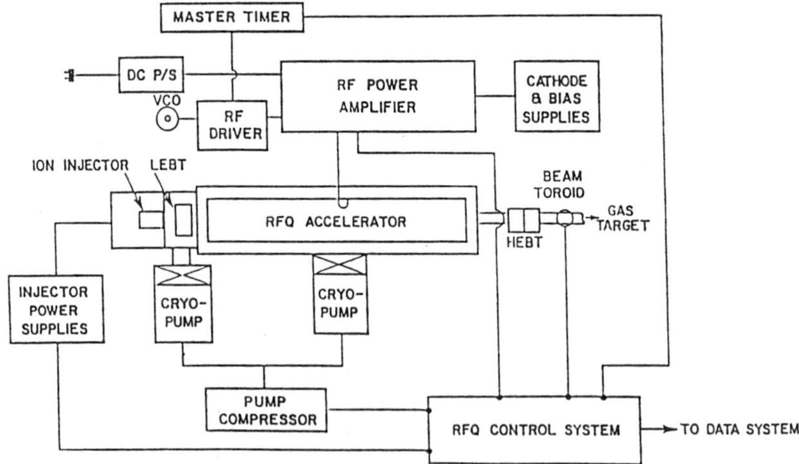

Figure 2. A block diagram of the AccSys RFQ system.

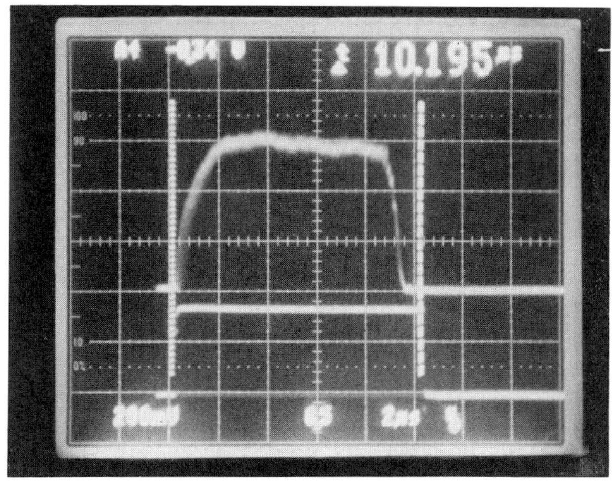

Figure 3. RFQ beam pulse and integration gate.

Table 2. LiF targets and Mo and Au foils used in the 1988 test.

Size	10mm(diameter)×1mm	10mm(diameter)×1mm	12mm(diameter)×1mm
Foil	5μm Mo	10μm Mo	5μm Au
E_p	1.6 MeV	1.2 MeV	1.5 MeV

The targets were mounted in a vacuum glass tee on movable shafts connected through an O-ring assembly to the outside. This allowed us to adjust the target position, or to pull one target out and put another target in without interrupting the RFQ operation. The target and mounting shaft were electrically connected to ground through a 100 Ω resistor. The voltage across the resistor was inverted and attenuated, then fed into a LeCroy 2249W ADC to measure the integrated beam charge per pulse.

All targets were tested up to 12 Watts average beam power at the Kellogg Van de Graaff facility at Caltech before used at AccSys. At the AccSys tests all targets survived the beam tests without apparent aging.

Detector Set-up

Figure 4 shows a schematic view of the detector setup of the 1988 test, together with the RFQ accelerator and the target at AccSys. The BGO matrix and the BaF_2 counters were placed 52cm away from the target, as shown in the figure. For the 1987 test only a 4 × 5 BGO matrix was used.

All the BGO crystals used in the tests were of the L3 standard size, approximately 2 cm × 2 cm in the front, 3 cm × 3 cm in the back and 24cm long. The crystals were produced at the Shanghai Institute of Ceramics, Shanghai, China. The crystals were polished and painted with NE560 white paint to provide a high light yield and good light collection uniformity, using methods similar to those developed by the L3 Collaboration at CERN [10].

The complete structure was installed in an aluminum box. Temperature sensors were installed inside the BGO matrix to monitor the temperature variations throughout the series

of runs, and to allow a correction of the gain shift for each crystal to sufficient accuracy off -
line. An air cooling provided by a refrigerator was installed for the 1988 test.

The BGO matrix was read out using the same Hammamatsu photodiodes, Lyon preamplifiers, and Princeton digitization boards used in the L3 experiment [10]. A Motorola 68000 ECB
CPU or 68020 - based single board computer running at 10 or 20 MHz was used to control
a token-passing ring of ADC boards, and to communicate with a VAXStation II computer
through a home-made CAMAC FIFO module for the 1987 or 1988 test, respectively.

Four BaF_2 counters were constructed for the 1988 test. The crystals were obtained from
Harshaw (USA), OKEN (Ohyo - Koken in Japan) and Merck(E. Merck Darmstadt in W.
Germany). The BaF_2 crystals were polished and wrapped with two layers of teflon tape.
Hamamatsu R2078 and R3197 PMT's were chosen to read out the fast components (195 and
220 nm) [8], while a Hamamatsu R1306 PMT was chosen to read out the slow component
(310 nm). Figure 5 shows the scintillation spectrum of BaF_2 and the quantum efficiencies of a
typical Cs-Te photocathode with a quartz window and R1306.

The R3197 and R2078 tubes both have synthetic silica (quartz) windows and Cs-Te photocathodes. The quartz window has greater than 80% transmittance for wavelengths longer
than 180 nm. The Cs-Te photocathodes are mainly sensitive only to wavelengths between 160
to 300 nm, i.e. they are UV sensitive and solar blind. The quantum efficiencies of the R2078
and R3197 at 200 nm are 18% and 13% respectively. The fast to slow ratio in the R3197 or
R2078 PMT output is approximately 1 : 1 with full integration. The fast component can then
be separated out by using a short gate integration without saturation. The R1306 PMT has a
borosilicate glass window and a conventional bialkali photocathode. As seen from Fig. 5, this
PMT is only sensitive to the slow component.

The PMT's were coupled to the crystals with Dow Corning 200 Fluid, with a viscosity of
12,500 centipoise, which has good UV transmittance. The output of the PMT's was integrated
and digitized by using a LeCroy 2249W ADC CAMAC module.

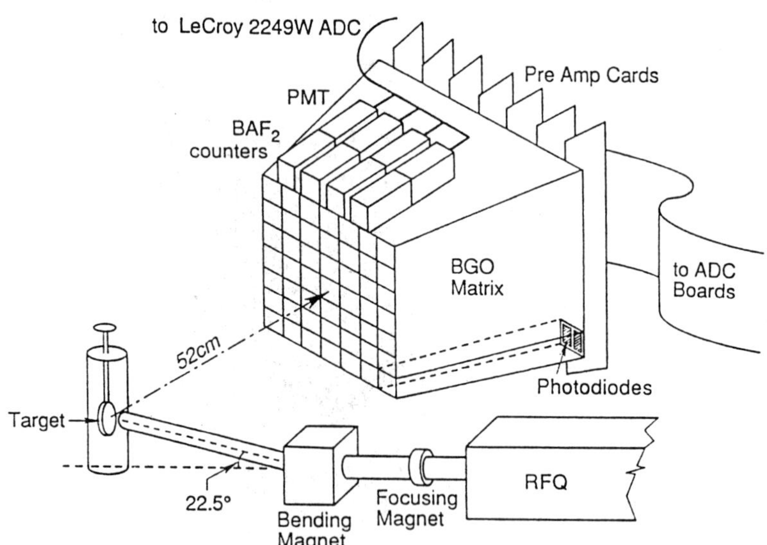

Figure 4. AccSys test experimental set-up.

RESULTS OF THE 1987 TEST

Data Analysis

The data were first filtered through a pass 1 reduction, which selects events with at least one energetic hit (>6 MeV) in any of the crystals. In a second pass, the spectrum of each crystal was first corrected to overcome the intrinsically uneven binning because of the poor differential linearity of the successive approximation ADC. The correction factors were derived by shifting the contents of the neighboring bins of typical pedestal data with high statistics to minimize the χ^2 of a Gaussian distributions. The data were then pedestal subtracted, giving the energy spectrum for each crystal. Although there was no cooling and temperature control during the 1987 test, it was found that the temperature variation within a run was less than $0.2°$ C. The temperature corrections were therefore applied as a single correction for each run.

A veto option could be selected, so that an event was accepted only if each of the 8 neighboring crystals has energy less than E_{veto}. This veto cut was used to reject two kinds of events: events with a photon whose energy was shared by two neighboring crystals and events with more than one photon hitting two neighboring crystals.

A spline fit was then applied to the upper edge of the obtained spectrum to determine a piecewise continuous cubic function which closely represents the data without bias. The function was defined in n equal intervals in the fit region with continuous derivatives at the boundaries up to the second order. The number of n was chosen as 7 to accommodate about 100 channels in the fit region, while avoiding local spurious peaks in the fit. An analytical minimization of the χ^2 uniquely determined parameters of the function [4]. The resultant function represents data very well. The bias, defined as the algebraic sum of the deviations between the data and the spline fit, was found to be zero to high precision. The calibration point HH$^+$, defined as the energy at which the upper edge of the spectrum falls to half its maximum height, was then determined from the spline function by using a numerical successive approximation method.

The statistical error on the HH$^+$ point for about 200k triggers, corresponding to 1600 events

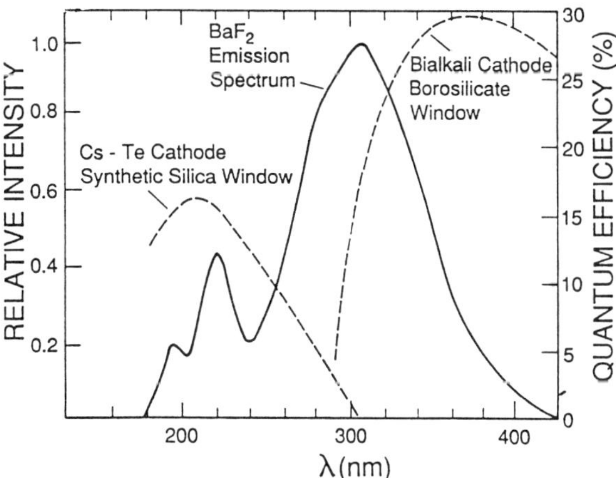

Figure 5. BaF$_2$ scintillation spectrum and PMT response. The scintillation spectrum is from P. Schotanus *et al.* [8].

above 14 MeV in each crystal's energy spectrum, was computed to be 0.8%. The computation of the error was based on: (1) the rms of the distribution of HH$^+$ points obtained from fits to a series of Monte Carlo-simulated spectra, generated according to the spline fit to the data, and taking into account the statistical fluctuation in each bin; and (2) an analytical calculation, consisting of propagating the error through the fitting procedure [4]. Both methods gave consistent results.

1987 Lithium Spectrum and Comparison with Monte Carlo

Figure 6 shows a set of typical spectra obtained from the six central crystals in the array plotted without veto option, together with the spline fit. The HH$^+$ values and the efficiency, defined as the ratio of the sum of number of events with deposited energy greater than 14 MeV to the total trigger, are also shown in the figure. It is clear from the figure that for good crystals under standard conditions the calibration point can be obtained without using the veto.

A comparison of three spectra obtained from crystal 13 for three different incident angles: normal, 30°and 50°respectively is shown in Fig. 7. The decrease of the HH$^+$ points and the efficiency versus incident angle is as predicted by the Monte Carlo simulations [11].

Table 3 compares the HH$^+$ variation of crystal 13 with the Monte Carlo simulation for different incident angles and when aluminum in front [11]. The overall agreement is very good.

Systematic Error of the Lithium Calibration

The angle effect and the aluminum effect are fixed for each crystal for a fixed geometry, so these effects will not result in a systematic calibration error. Overall, the systematics will be dominated by the long term stability. Within the limited time of several days during the 1987 test the stability of the system appeared excellent.

Table 4 lists the HH$^+$ points of the central 6 crystals without veto cut for different run

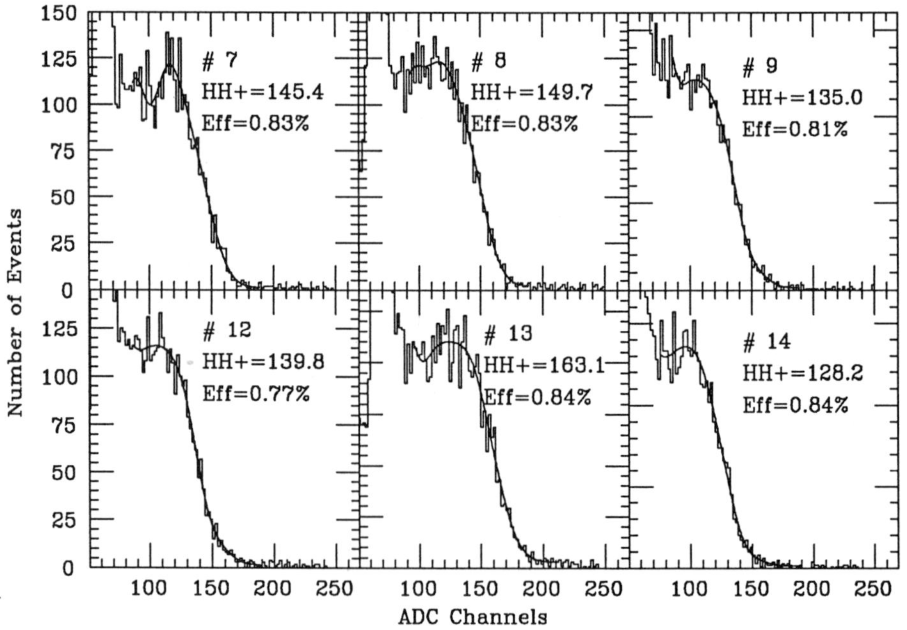

Figure 6. Photon energy spectra and HH$^+$ points for central six crystals without veto cuts.

Table 3. The averaged HH$^+$ variation (%) of the central six crystals in comparison with Monte Carlo simulation.

case	3 MeV veto		5 MeV veto		No Veto	
	M.C.	Data	M.C.	Data	M.C.	Data
30°	-0.7	-0.6	-0.8	-0.6	-0.9	-1.3
50°	-1.8	-1.4	-2.2	-1.4	-2.4	-2.5
5 cm Al	-0.5	-1.2	-0.5	-1.1	-0.8	-1.6

conditions. Also listed in the table are the statistical deviation extracted from two identical runs. The absolute average of all 24 deviations in the table is 0.8%. The same absolute average for the case with 5 MeV veto cuts is 0.7%. 1% stability has therefore been achieved. The 1987 test has thus shown that a precision of RFQ calibration of better than 1% is achievable in practice for the BGO crystals in the L3 detector.

RESULT OF THE 1988 TEST

Photon Pulse

Figure 8 is a picture taken from an oscilloscope showing typical scintillation light pulses from two BaF$_2$ counters. Figure 8a shows the slow component, as measured with an R1306 PMT. Figure 8b shows the fast component (along with a remainder of the slow component), as measured with an R2078 PMT. As shown in Fig. 8a, a pulse-generator-like scintillation light pulse is obtained, because so many photons strike the crystal within the pulse. With the fast component, many individual spikes can be seen on the scope, but the integrated pulse area is quite well defined, with the spread of the distributions dominated by statistical fluctuations in the number of incident photons.

Two Normalization Schemes

A precise normalization is needed to convert the integrated pulse read out for each crys-

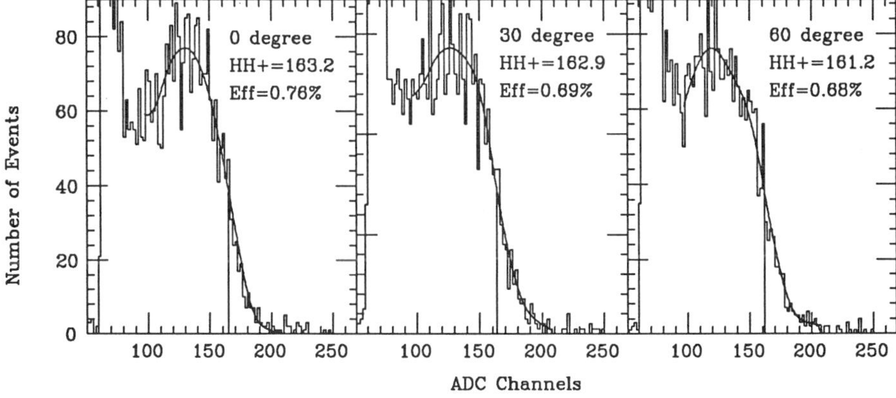

Figure 7. A comparison of photon energy spectra of crystal 13 with 5 MeV veto cut for three different incident angles: 0°, 30° and 60°.

Table 4. HH$^+$ points (channels) and deviation (δ) of central six crystals with no veto cut.

Run	Case	Crystal Number					
		7	8	9	12	13	14
2	Normal	143.4	147.8	135.0	137.0	162.0	126.3
7		145.4	149.7	135.0	139.8	163.1	128.2
	$\delta(\%)$	0.9	0.9	0.0	1.4	0.5	1.1
4	30°	145.2	148.3	135.5	135.1	161.9	126.1
9		144.6	148.2	136.9	136.1	161.4	125.5
	$\delta(\%)$	0.3	0.0	0.7	0.5	0.2	0.3
3	50°	142.8	148.6	131.8	133.7	156.9	123.8
10		147.4	148.4	135.0	135.2	159.6	125.8
	$\delta(\%)$	2.3	0.1	1.7	0.8	1.2	1.1
5	5 cm Al	142.9	146.7	128.0	136.9	160.7	127.2
13		143.9	145.2	130.4	135.6	162.4	125.7
	$\delta(\%)$	0.5	0.7	1.3	0.6	0.7	0.8

tal into a calibration constant. The energy deposited in each individual calorimeter element depends on the total number of protons striking the target on each beam pulse, the materials which the photon traverse before reaching a given calorimeter element, and the solid angle each calorimeter element subtends. The proton resonance spectrum resulting the decay of excited oxygen, ^{16}O* [7], is also complicated. It is therefore difficult to calculate precisely the absolute energy deposited in each calorimeter element directly from a knowledge of the beam conditions and the geometrical setup.

a b

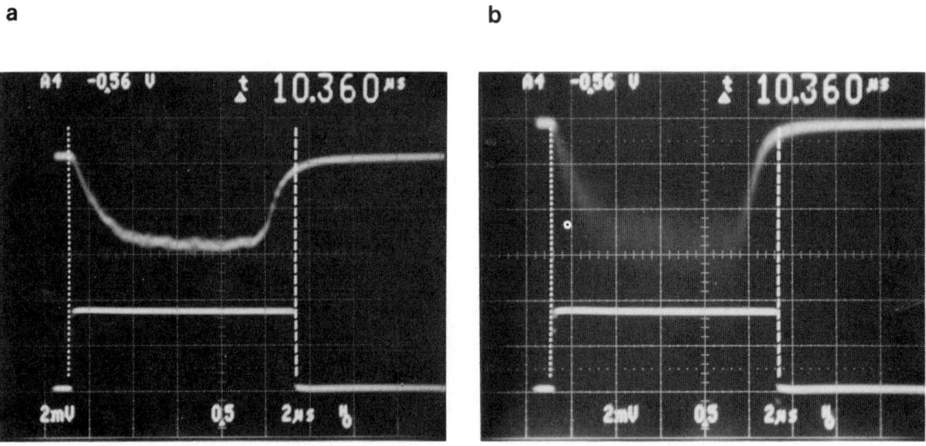

Figure 8. Scintillation light pulse: slow (a) and fast (b) components.

However, within a run or within a series of runs with an identical geometrical arrangement, it is straightforward to obtain a precise relative calibration using this technique. Apart from statistical fluctuations, the energy deposited in a calorimeter element only depends upon the number of protons hitting the target. A relative calibration constant can therefore be obtained by normalizing to (1) the total electric charge (proportional to the number of incident protons) collected from the target from each beam pulse, or (2) the response of a stable, external "standard" (i.e. a separate) counter or a set of counters with high precision ADC's. We tested these two normalization schemes during the test. The BGO array was used as our external normalization standard.

To check the consistency of the two candidate normalization methods, we measured the total "beam charge" and the total energy deposited in the BGO matrix, and took their ratio for each beam pulse. Figure 9a shows the measured beam charge distribution of run 27 without normalization. It is clear that the beam charge was not stable from pulse to pulse during the run: some deviations of up to 10% were observed. The distribution of beam charge, normalized to the total BGO energy is shown in Fig. 9b. The almost perfect Gaussian distribution, with a 1.3% standard deviation indicates that the two normalization schemes are consistent.

Normalizing the ADC distributions of the individual BGO and BaF$_2$ crystals, either to the beam charge or to the total BGO energy pulse by pulse, results in extremely clean Gaussian distributions. Figure 10 shows the ADC distribution of a BGO crystal, number 25, before and after normalization to the BGO energy. Table 5 lists the r.m.s. widths in percent of the ADC distributions of some detector crystals with and without normalizations for run 25. The BGO energy normalization appears to be slightly better than the charge normalization, although the difference is not statistically significant.

Calibration Stability

In order to check the stability of the calibration, using the two normalization schemes, the peak positions of the normalized ADC's were compared for a set of 5 runs with an identical setup but with different beam intensities.

Although the beam charge method works well within a run, it was found that the peaks of the normalized distributions for individual crystals showed significant shifts from run to run. The shifts were typically 2 – 3%, and sometimes more. This can be attributed to a change of

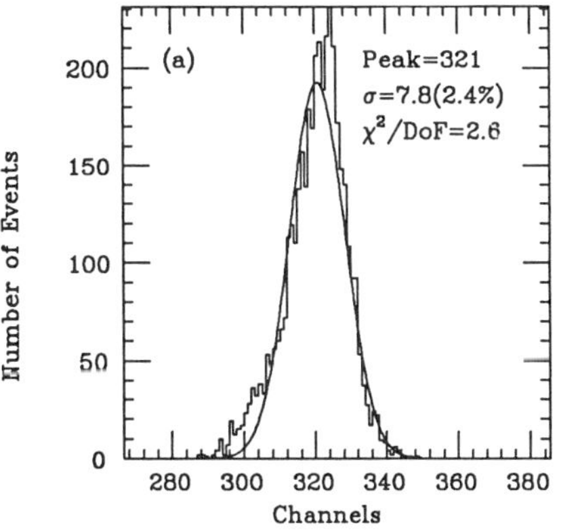

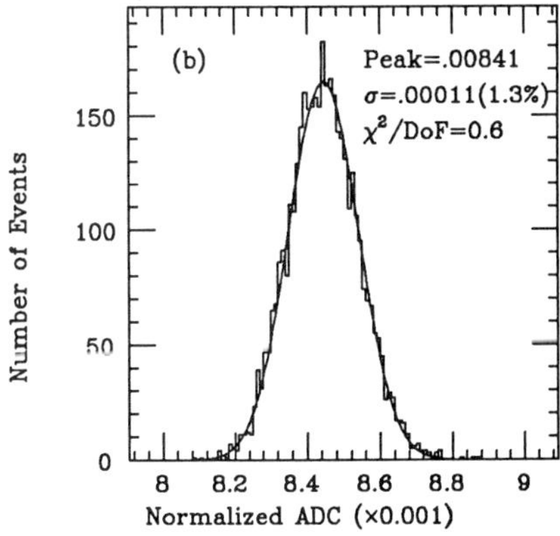

Figure 9. The Measured charge distribution of Run 27 (a) and the distribution of charge normalized to the total BGO energy (b).

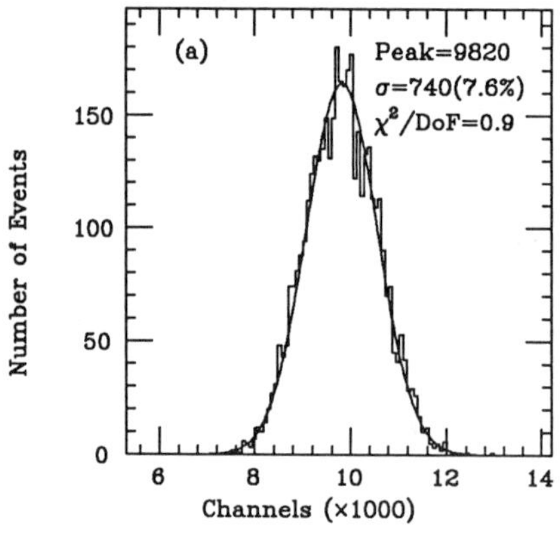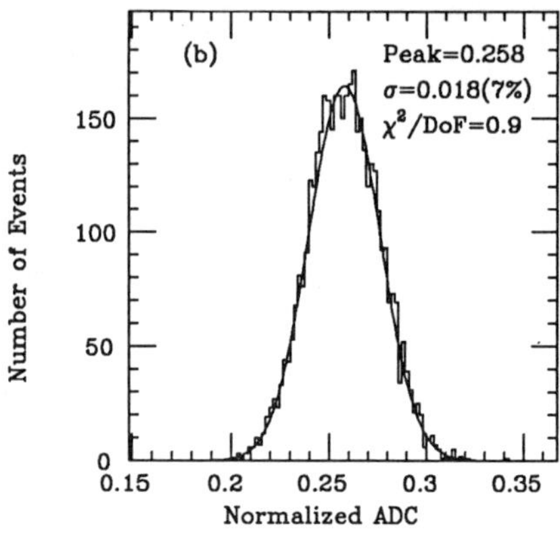

Figure 10. ADC distribution of BGO crystal 25 without (a) and with (b) normalization for Run 27.

Table 5. R.m.s. width(%) with and without normalization for Run 25.

Crystal	BGO26	BGO33	BaF1	BaF2	BaF3	BaF4	Beam
W/O Norm	6.07	6.16	5.05	4.47	3.78	8.06	3.16
Norm. BGO	5.36	5.41	4.06	3.39	2.38	7.62	1.45
Norm. Beam	5.54	5.55	4.09	3.39	2.49	7.65	N/A

the beam focus, or to slight changes in the position of the beam center relative to the active area of the target. As a result, the measured beam charge collected from the target holder is not always a precise measure of the number of protons hitting the active material in the target. Careful monitoring of beam focusing, and position, with feedback controls on the RFQ, and an improved target design are expected to help solve this problem. However, until we can demonstrate that target aging will not affect the photon yield per unit of beam charge over the long term, we have made the preliminary conclusion that the beam charge normalization method is not the best choice.

In contrast, the normalization to the BGO energy worked very well. Figure 11 shows that the relative deviation of the peaks of individual BGO distributions from run to run, after normalization to the total BGO energy in the 7 × 7 matrix was Gaussian distributed, with a standard deviation of 0.34%. Table 6 lists the relative deviation from run to run of the peaks for the BaF_2 crystals, normalized to the total BGO energy, where the numbers in bracket are the statistical errors from the fit. The r.m.s. of the relative deviation is 0.29%. These deviations are dominated by statistical errors. It is therefore evident that a stability of better than 0.5% has been achieved with this normalization scheme.

Equivalent Photon Energy

As discussed in last section, the energies of the protons hitting the fluoride crystal target were different for the three different targets. We measured the equivalent photon energy (E.P.E.), defined as the sum of the energies of photons from one beam pulse hitting one crystal detector, and were thereby able to determine the photon yield per unit of beam charge on target, at three proton energies.

Table 6. Deviation of normalized peak positions (in %) for Run 45–49.

Crystal	Run45	Run46	Run47	Run48	Run49
BAF 1	0.32(8)	-0.00(8)	0.19(7)	0.06(6)	-0.57(3)
BAF 2	-0.31(6)	-0.36(6)	0.11(5)	0.53(5)	0.04(3)
BAF 3	-0.12(4)	-0.09(4)	0.14(4)	0.24(4)	-0.17(2)
BAF 4	0.62(14)	-0.19(15)	0.04(13)	-0.12(13)	-0.36(7)
r.m.s. = 0.29%					

Table 7. Equivalent photon energy emitted into 1.6 msrad.

Target Foil	5μm Mo	10μm Mo	Punctured Au
Equiv. Photon Energy(GeV/0.1μC)	1.29	0.37	2.38

Table 7 lists the measured equivalent photon energies emitted into 1.6 millisteradian (corresponding to a single L3 BGO crystal with a 2 × 2 cm^2 front face at 50cm from the target), per 0.1 μCoulomb (μC) of beam charge on the target. The dependence of the photon energy on the proton energy is consistent with estimations from published resonance data [7]. The neutron background from the target with punctured gold did not contribute significantly to the energy measured because the proton beam energy was only 0.04MeV above the neutron production threshold, and the detectors were in the backward hemisphere.

Energy Spectrum of Individual Photons

The width of the normalized ADC distribution and the total energy in a crystal are strongly correlated. Figure 12 shows the normalized ADC distributions of BGO crystal number 32, for runs with different beam intensities. It is clear that the r.m.s. width increases as the total energy deposited in the crystal decreases.

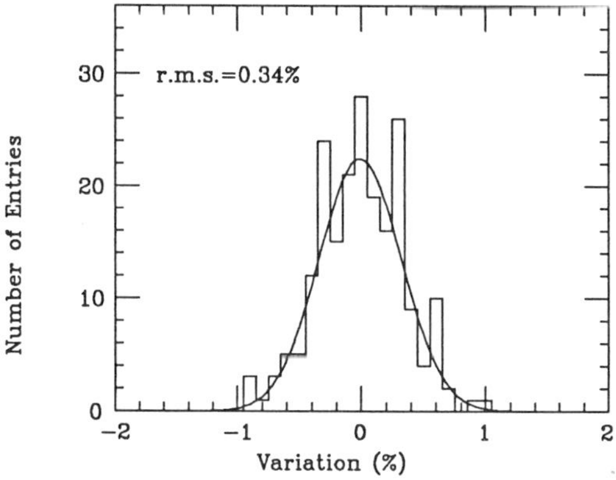

Figure 11. Variation of the normalized peak positions from BGO crystals.

Assuming that the shape of the energy spectrum for the individual photons emitted by the target does not change as a function of beam intensity, then the width of the observed energy distribution should be a measure of the inverse of the square root of the average number of photons. In order to test this, we defined the "relative r.m.s.", σ_R, as the ratio of the r.m.s. width divided by the energy corresponding to the peak (i.e. the average obtained with a Gaussian fit) of the distribution. Figure 13 is a plot of $(1/\sigma_R)^2$ versus the observed average photon energy for a BGO crystal, for a series of 9 runs where the beam intensity varied by a factor of 6. The excellent linearity shown in the figure tends to confirm that the widths of the distributions are due to statistical fluctuations. The average number of photons hitting the target per pulse should therefore be given by $1/(\sigma_R)^2$.

The average energy of the photons emitted by the target can then be obtained by dividing the equivalent photon energy by the average number of photons. Table 8 shows the equivalent photon energy in crystal 32 (E.P.E.$_{32}$), and the energy of the individual photons averaged over all the BGO crystals (E_{ave}), for 9 different runs.

Bearing in mind that the predominant photons from $^{16}O^*$ decay have an energy of 6.13 MeV, we estimated that the 4.1 MeV average energy was due to energy sharing between neighboring crystals, i.e the equivalent photon number was increased. This estimation was confirmed by a series of EGS Monte Carlo simulations [13].

A constant E_{ave} indicates that the widths of the normalized ADC distributions are dominated by the statistical fluctuations in the number of photons detected by each crystal, and that the photon energy spectrum remains the same when the beam intensity is varied.

Gaussian vs. Poisson - Distributed Spectra

When the average number of photons per pulse hitting a calorimeter cell is not very large, a Poisson distribution should describe the normalized spectrum better than a Gaussian. Figure 14 shows a semi - logarithmic plot of the normalized ADC distribution of BGO crystal 25 for Run 49, which had low intensity per pulse, and high statistics. We fitted the distribution with

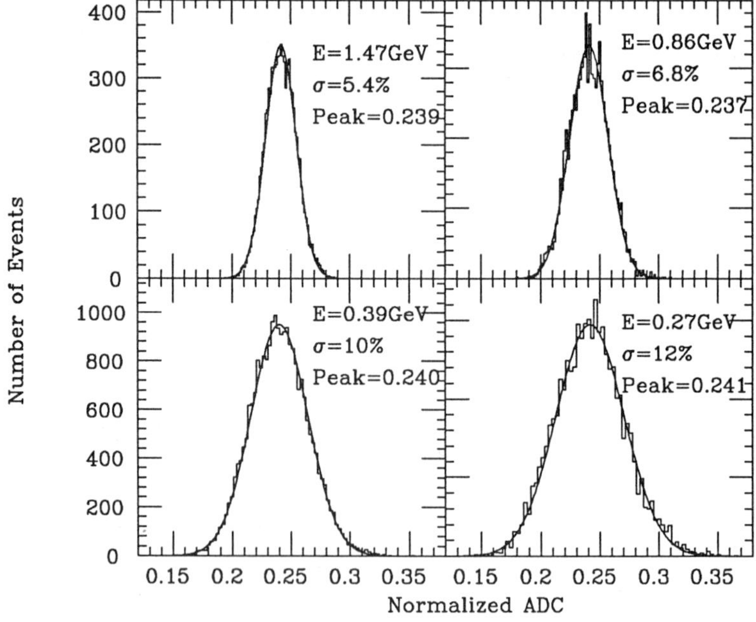

Figure 12. Normalized ADC distribution of crystal 32 at different energies.

Run	25	26	27	28	45	46	47	48	49
E.P.E.$_{.32}$(GeV)	1.47	1.49	0.86	0.52	0.42	0.27	0.35	0.44	0.39
Photon Number	359	364	209	127	105	68	88	109	98
Measured E$_{ave}$(MeV)	4.10	4.09	4.11	4.10	4.00	4.01	4.00	4.04	4.07
Average=4.06 MeV, r.m.s.=0.15 MeV									

a Gaussian, and compared the fit to the result of using a Poisson distribution with a mean energy matching the mean of the measured distribution. It should be emphasized that the Poisson distribution is not just a fit, but is completely determined by the methods summarized in the previous section. As expected, the Poisson distribution matches the data better. The quality of the match, and the fact that the distribution in Fig. 14 has no tails down to the 0.1% level, indicates that the distribution is completely free of background.

For equivalent photon energies above a few GeV, the deviation of the data from a Gaussian distribution is negligible. In this case, the data can be fit with a simple Gaussian with systematic effects from the fit which are well below the 0.1% level.

Attenuation of the Calibration Pulse

In order to check the nature of the particles which are emitted by the target, several runs were taken with the 10μm Mo target, in which shielding materials of various types and thicknesses were placed between the target and the crystals. In this way, the attenuation of the particles could be tested, and the results could be compared to the attenuation expected for 6 MeV photons. Table 9 lists the measured beam charge, the average beam charge per pulse (in ADC channels, labelled "Beam"), and the equivalent photon energy averaged over all BGO crystals, E.P.E.$_{.ave}$, for runs with 2" Al and 2" brass absorbers. A run without additional attenuation is included for comparison. The measured attenuation factors shown in the table are corrected for the beam intensity.

The measured attenuation factors were compared with the estimated attenuation factors, which are also listed in the table. The estimated attenuation factors were obtained by using the mass attenuation coefficients of 6 MeV photons. The 5% errors in the estimated attenua-

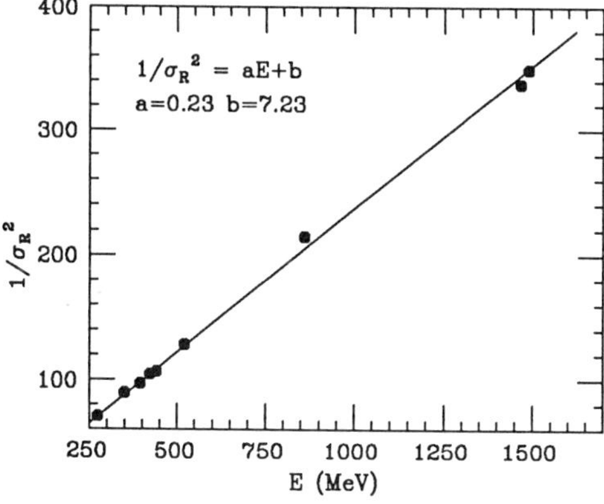

Figure 13. Correlation between the relative r.m.s. and the average energy in BGO Crystal 32.

Table 9. Attenuation of the calibration pulse by Al and Brass absorbers.

Run	Material	Beam (ch)	$E.P.E._{ave}$ (MeV)	Measured Attenuation	Estimated Attenuation
17	0	520	334		
18	2" Al	526	206	0.609 ± 0.006	0.65 ± 0.03
19	2" brass	537	100	0.290 ± 0.003	0.27 ± 0.02

tion values are dominated by the errors in the absorber thickness, by the effects of the finite transverse dimensions of the absorber, and by the uncertainty in the precise admixture of secondary metals in the aluminum and brass alloys used. The agreement between the measured and estimated attenuation factors indicates that the calibration beam pulse is indeed a photon pulse.

DISCUSSION

Short Pulse

During the AccSys tests we typically used a beam pulse of 0.10 – 0.15 μC with pulse length of few μsec. In order to adapt this technique for future hadron colliders, such as the SSC or LHC, however, the pulse length of several μsec may not match the experimental readout electronics. Recent computer simulation studies at Saclay have shown that it should be possible to compress 4 MeV beam pulses of 0.1 μC into 100 nsec time or less, by using multiturn injection into – and single turn extraction from – a small storage ring[12]. This development will allow this technique to be used in calibrating calorimeters at the SSC and LHC. A new pulse time - compressor at the output of their RFQ system will be developed by AccSys under a Phase II grant from the DoE SBIR program [12].

Equivalent Photon Energy of Short Pulse

The time structure of this kind of beam pulse is well within the range of readout speeds planned for calorimeters at the SSC and LHC. The existing RFQ (1.92 MeV) and a LiF target

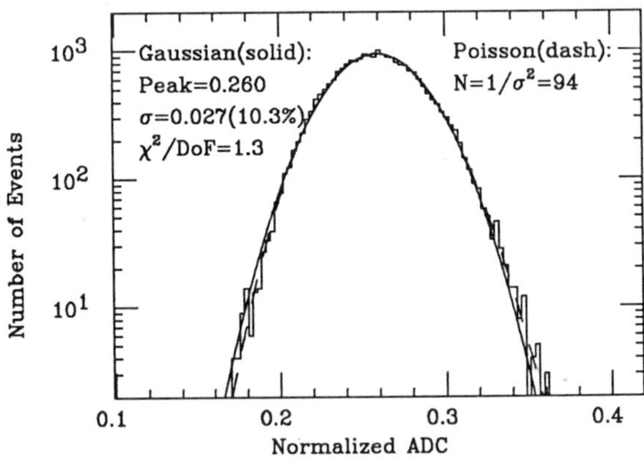

Figure 14. A Gaussian fit compared with a Poisson fit for the normalized ADC of crystal 25 from Run 49.

Table 10. Relative photon yield from a LiF target.

Proton Energy (MeV)	1.2	1.6	1.9	2.5	2.9	3.2	3.5	3.85
Photon Yield	0.4	1.0	2.0	12	22	27	29	32

could therefore be used at these accelerators to provide short pulses up to 2.3 GeV equivalent on a bare crystal target, or 1.3 GeV equivalent for a target covered with a 5 μm Mo foil (0.3 μC charge).

There are, however, much stronger fluorine resonances between 2.0 and 4.0 MeV [7]. Table 10 lists the ratio of the expected photon yield from a thick LiF target, as a function of the incident proton beam energy, to the yield from a 1.6 MeV proton beam (corresponding to a 5 μm Mo foil with the existing RFQ). A 29 - fold increase in the equivalent photon energy is expected when a 3.5 MeV proton beam hits the LiF target. Using a 3.85 MeV RFQ with the 5μm Mo foil - covered LiF target, one could thus obtain an equivalent photon energy of more than 40 GeV/(0.1 μC). Using a CaF_2 target, which is easier to handle than LiF and which would have no neutron production as a by-product below 4.05 MeV [7], an equivalent photon energy per calorimeter element of 20 GeV/(0.1 μC) or more is expected. In calorimeters where such a large equivalent energy per pulse is not required, the use of a CaF_2 can reduce the output current requirements in the RFQ design.

SUMMARY

A novel technique to produce an calibration pulse with equivalent photon energy of up to 20 GeV or more in a typical calorimeter element, using an RFQ accelerator, has been successfully tested. The photon pulse can be used to calibrate electromagnetic calorimeters at present and future high energy physics experiments. With proper normalization and an RFQ repetition rate of 30 Hz or more, an entire electromagnetic calorimeter with 10^4 elements can be calibrated in 1 – 2 minutes with an accuracy of 0.5% or better.

The technique described here could be used without modification to calibrate calorimeters at the present generation of accelerators (such as LEP and HERA) or at electron - positron colliders planned for the future in the TeV range (such as CLIC or the TLC). A precise, separately read out subarray, at a large angle to the colliding beams and with very precise ADC's, could be used as an external normalization "standard", which could be cross - checked against normalization according to the beam charge on the target, pulse by pulse.

This method can also be adapted for use at the next generation of hadron colliders, such as the SSC or LHC. In order to make the RFQ - based technique practical at these future accelerators, the external standard subarray should be made of radiation resistant material, such as a BaF_2 crystal array [14].

ACKNOWLEDGEMENTS

We are grateful to Prof. F. Takasaki and to Merck, Inc. who loaned us BaF_2 crystals. The BGO crystals, photodiodes, preamps and parts of the readout electronics were provided to us by members of the L3 Collaboration, especially the Aachen, Annecy, Lyon and Princeton groups. Mr. F. Roeber and G. Ludlam helped prepare the target and BaF_2 crystals at Caltech before the AccSys tests. Dr. R. Mount and H. Akabari participated the 1987 AccSys test. We would also like to thank Profs. C. Barnes and R. Kavanagh for many useful discussions, and for providing us with the use of the Van de Graaff facility at the Kellogg Lab at Caltech.

The assistance of S. Sondergaard, J. Hanson, M. Hamm and many members of the Caltech technical staff, and AccSys, Inc. is appreciated. The continued support and encouragement of Profs. S.C.C. Ting and A. Zichichi is also gratefully acknowledged.

References

[1] M.D.G. Gilchriese, in *"Proceedings of the 1984 Summer Study on the Design and Utilization of the Superconducting Super Collider"*, edited by R. Donaldson and J.G. Morfin, Snowmass, Co.(1984) 607; H.H. Williams, in *"Proceedings of the 1986 Summer Study on the Physics of the Superconducting Super Collider"*, edited by R. Donaldson and J. Marx, Snowmass, Co.(1986) 327; SSC Central Design Group, *"Report of the Task Force on Detector Research Development for the Superconducting Super Collider"*, SSC-SR-1021 (1986).

[2] D. E. Groom, in *"Proceedings of the 1988 Summer Study on the High Energy Physics in 1990's"*, Snowmass, Co. (1988).

[3] T.Q. Zhou *et al.*, *Nucl. Instr. and Meth.* **258** 1987 58; R.Y. Zhu, *"EGS Study of BGO Calibration Using Low Energy Photons"*, L3 Internal report, December 1985; H. Newman and R.Y. Zhu, *"Proposal for a BGO Calibration Test at AccSys"*, October, 1987.

[4] H. Ma *et al.*, *Nucl. Instr. and Meth.* **A274** (1989) 113.

[5] H. Ma *et al.*, *Caltech Preprint* **CALT–68–1536**, 1989, submited to *Nucl. Instr. and Meth.*

[6] Los Alamos National Laboratory, Collection of papers on the RFQ Linear Accelerator, presented by Accelerator Division Personnel, March 1979-May 1983.

[7] F. Ajzenberg-Selove *Nucl. Phys.* **A475** (1987) 1; and H. B. Willard *et al.*, *Phys. Rev.* **85** (1952) 849.

[8] M. Lavel *et al.*, *Nucl. Instr. and Meth.* **A206** (1983) 169; and P. Schotanus *et al.*, *Nucl. Instr. and Meth.* **A259** (1987) 586; and *IEEE Trans. Nucl. Sci.* **NS-34** (1987) 76.

[9] S. Majewski and D. Anderson *et al.*, *Nucl. Instr. and Meth.* **A241** (1985) 76; A. J. Caffrey *et al.*, *IEEE Trans. Nucl. Sci.* **NS-33** (1986) 230; and S. Majewski *et al.*, *Nucl. Instr. and Meth.* **A260** (1987) 373.

[10] L3 Collaboration, *"L3 Technical Proposal"* May 1983; and J. Bakken *et al.*, *Nucl. Instr. and Meth.* **A254** (1987) 535 and **A228** (1985) 294.

[11] R.Y. Zhu, *"EGS Study of BGO Calibration Using Low Energy Photons"*, L3 Internal report, December 1985.

[12] R. Hamm, *"SSC Electromagnetic Calorimeter Calibration Source"*, Phase II proposal to the DoE SBIR program, December, 1988.

[13] Electron Gamma Shower (EGS) program by R. Ford and W. Nelson. See Report *SLAC-210*, 1978.

[14] R.Y. Zhu, *in these proceedings*.

PIPING OF SSC SYNCHROTRON RADIATION REVISITED

Lawrence W. Jones

University of Michigan
Randall Laboratory of Physics
Ann Arbor, MI 48109

ABSTRACT

Synchrotron Radiation of 20 TeV protons at the design circulating current of the SSC dominates the refrigeration heat load of the 4.15 K cryogenic system. An earlier paper suggested that the synchrotron x-rays could be piped by total external reflection from polished metal surfaces to "warm" absorbers to relieve this heat load. Although the piping efficiency discussed earlier is now known to be unrealistic, it can be shown that significant power can be saved nonetheless on detailed consideration of a realistic lattice. Specifically, judiciously placed radiation absorbers could intercept much of the radiation. Four alternative schemes are discussed.

The synchrotron radiation of 20 TeV protons in the 6.6 T dipole magnets of the SSC carries a power of 0.142 watts per meter or about 2.35 watts per magnet at the design current of 73 milliamperes. This corresponds to 9 kW per beam, or 18 kW for the entire SSC. To carry away this heat deposited at 4.15 K requires about 500 times this power input to the refrigeration system or about 9 MW. This is the largest contributor to the refrigeration heat load, and indeed is larger than the total of all other heat loads combined. Should particular experiments choose to exploit a luminosity of 10^{34} (cm^{2}sec)$^{-1}$ (rather than the design value of 10^{33}), the synchrotron radiation from the required three-fold increase in beam would overload the designed refrigeration.

The observation that the radiation is in the soft x-ray region, corresponding to a critical energy of 284 eV (or about 44 Å) led us to suggest that this energy could be piped within the vacuum tank via the phenomenon of total external reflection and removed in a straight-section of the vacuum tank by a "warm" (77 K) liner.[1] It is well known that the index of refraction is less than unity in this range of wavelengths for many metals and for grazing incidence x-rays, so that an average reflection of over 98% over the spectrum of synchrotron x-rays is calculated.

V. Rehn[2] pointed out, however, that such a reflectivity requires a surface polish very much finer than typical of a good optical mirror, and that apparently "good" mirrors scatter the x-rays over angles of the order of a degree (for angles of incidence of milliradians). Hence piping, which might require several reflections before reaching a "warm" absorber, seemed impractical.

All of these discussions occurred early in the SSC design effort. At this time the SSC design has converged, and it seemed of interest to re-examine this issue in the light of the current design, to see if a partial absorption of the x-radiation at 77 K could be achieved with minimal effort.

The design used as a basis for these estimations is the "90° SSC Lattice."[3] Here six dipoles are grouped in a half-cell and are spaced from the subsequent dipoles by a 10.5 m straight section containing a quadrupole and other elements (Figure 1). We assume here that some or all of the 10.5-m straight section could be lined wtih a 77 K absorbing surface.

It should be noted that the characteristic divergence of the synchrotron light is $1/\gamma = 0.047$ mr. Over the 17 meter length of a bending magnet this leads to only 0.8 mm for the radius of the x-ray spot. The beam emittance is correspondingly small and contributes little more to this size.

Four specific absorber geometries are explored here:

1) A simple 77 K strip along the outer side of the straight section to absorb only x-rays directed toward it from the beam with no reflections;

2) A lining of the entire 10.5 m straight section to absorb both the x-rays from 1) and those from a single reflection in the last magnet, assuming scattering of about a degree;

3) A 77 K "finger" at the end of each magnet;

4) A 77 K tube along the entire outer radius of the beam vacuum pipe.

From the discussion of specific geometrics we assume the parameters below, and the geometry of Figure 1.

Bending angle, each dipole	1.636 mr
Vacuum tank aperture	32 mm (diam.)
Radius of curvature	10.1 km
	(magnets curved to follow orbit)

1) A 77 K strip along the outer wall of the 10.5 m straight section would intercept synchrotron radiation from the first portion of the dipole just upstream with no reflections.

As sketched in Figure 2, it is apparent that 8 meters of bending magnet produce synchrotron radiation such that a tangent to the centered proton beam trajectory strikes the outer surface of the straight section wall. A small liquid nitrogen-cooled tube would not significantly reduce the aperture and, over this 10.5-m of straight section, would absorb about 8% of the synchrotron radiation from the half cell. This saves only 700 kw power load; however, this saving can be effected at little engineering cost.

2) The entire vacuum chamber liner of the 10.5 straight section could be held at 77 K.

This would intercept not only the radiation of 1) (above), but also a significant fraction of radiation which had experienced one reflection and had scattered through an angle of one or two degrees in the upstream magnets. It is not possible to estimate this without knowledge of the reflection characteristics of the vacuum chamber, but a fair guess would be that the numbers of 1) would be at least doubled, to over 15%. From Figure 3 it is apparent that, in the limit of specular reflection, synchrotron radiation from three dipoles of the 6-dipole half-cell would experience only zero or one reflection before reaching the straight section.

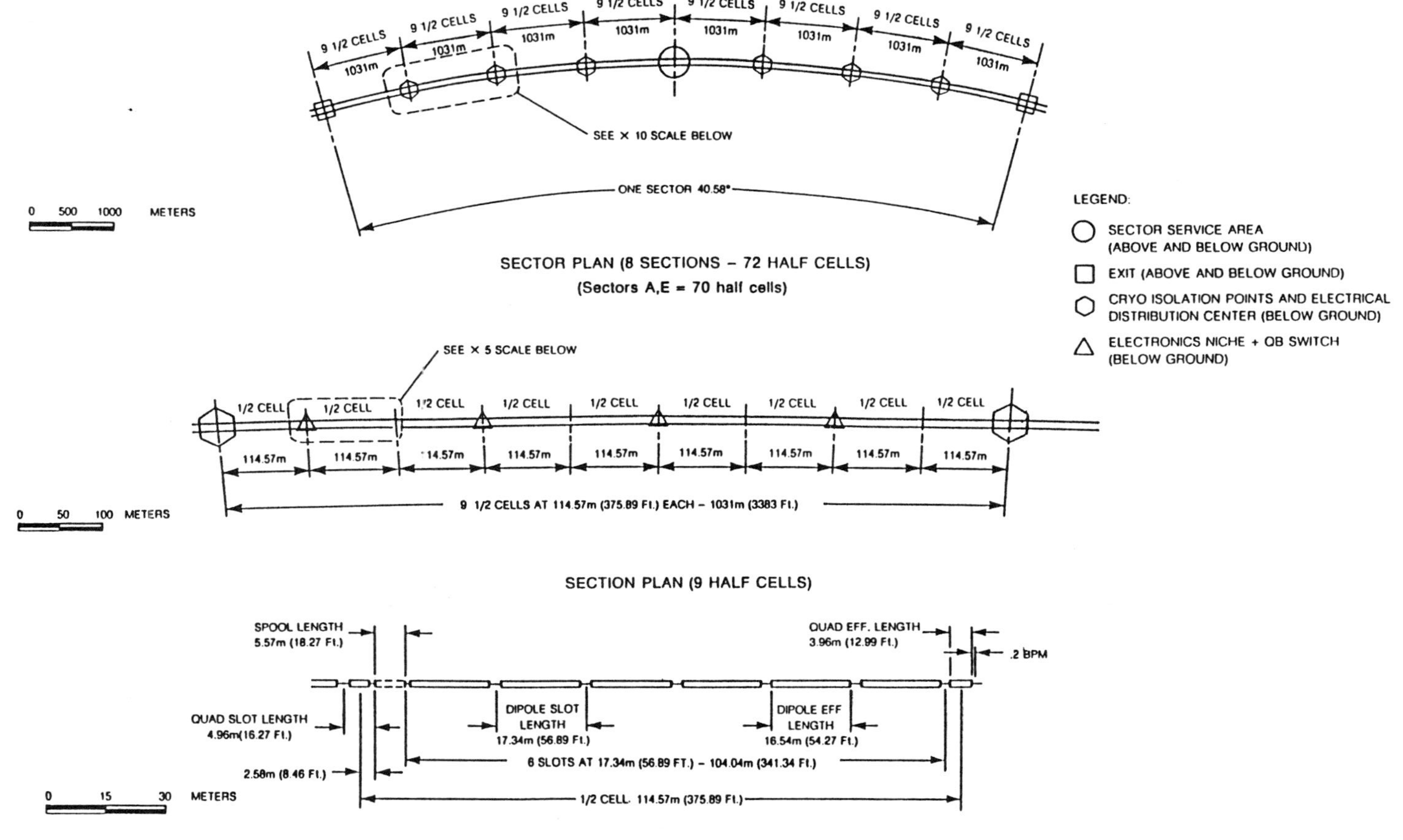

Figure 1. The dimensions of the SSC 90° lattice components.

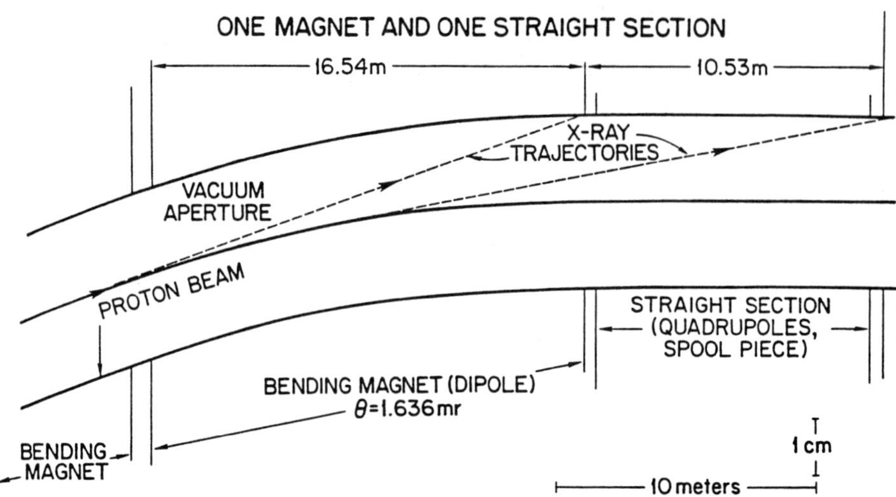

Figure 2. The geometry of one magnet and one straight section illustrating how synchrotron radiation from the upstream 8 meters of the magnet strikes the outer wall of the straight section. Note the 200:1 vertical-horizontal scale ratio.

3) A 77 K re-entrant "warm finger" could be projected into the vacuum tank at the end of each 17.5 m magnet, to absorb most of the synchrotron radiation from that magnet, as in Figure 4.

This absorber could be retracted (with a vacuum bellows) while beam was being injected and stabilized and then moved slowly toward the beam in the manner of "Roman Pots," when the beam was well established.

This scheme has the advantage of absorbing the majority (over 2/3) of the synchrotron power and requires no reflections. There are several disadvantages, however. To intercept radiation even 6 m from the probe, it is necessary to move to within 2 mm of the beam. In order to move 4000 probes towards each beam to within 2 mm without one of them scraping the beam would be an engineering *tour de force*. The effect of such fingers on the vacuum tank impedance characteristics has not been considered. Because the beam would need to be established first before the absorbers are inserted, the full refrigeration heat load would be required at first. Hence the savings in this instance would be in operating costs, not in installed capacity.

The remote deployment of this finger may not be trivial. An actuator could be frustrated by the thermal, vacuum, and magnetic environments. Perhaps a leveraged ferroelectric mechanism could be devised, to be hard wired to a beam position electrode. Of course, with a tungsten tip, this finger could also serve as a beam scraper, clearing the beam of scatter-induced halo, etc.

4) A flattened tube carrying 77 K nitrogen could be located at the outer wall of the entire vacuum system; dipoles and straight sections.

This would require no reflections, would sacrifice minimal aperture, and would reduce required installed refrigeration capacity. However, it would be quite expensive to install. If the tube were 5 mm wide it would restrict only 5% of the 4 K cold surface for purposes of cryopumping but it would also cost horizontal aperture.

The design of the absorbers for any of these schemes would require some care to assure that the absorption is nearly complete, i.e. that 30-70% of the radiation incident is not reflected. In scheme 3) this may not be a problem, as the radiation is incident on the absorber at large angles. However, in the other schemes, the radiation angle of incidence is nearly grazing.

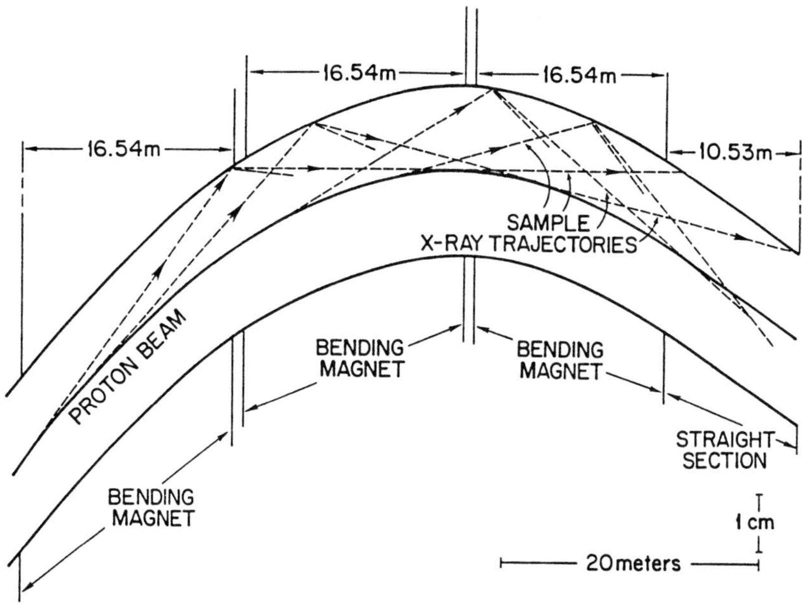

Figure 3. The geometry of three magnets and a straight section illustrating sample synchrotron radiation paths and reflections.

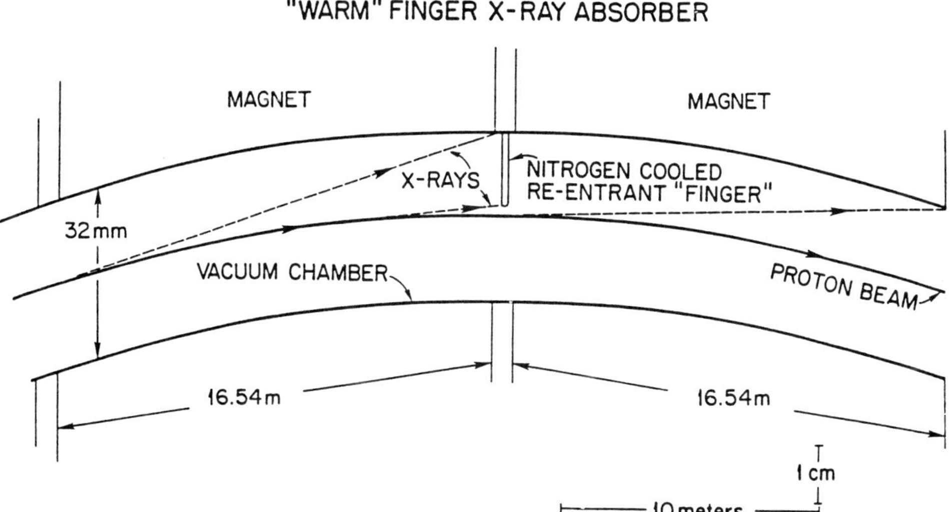

Figure 4. Two adjacent magnets with a 77K re-entrant "finger" at their juncture to absorb the synchrotron radiation from the majority of the arc.

We may conclude that it is possible but difficult to absorb much of the synchrotron radiation at 77 K, thus reducing the load on the 4 K refrigeration system. It remains an engineering cost-effectiveness decision whether any of these measures should be incorporated into the SSC design.

REFERENCES

1. L. W. Jones and T. O. Dorshem, Synchrotron Radiation from Protons in a 20 TeV, 10 Tesla SSC, p. 1138, Proc. 12th International Conference on High Energy Accelerators, Fermilab (1983).

2. V. Rehn, Synchrotron Radiation in a Superconducting Super Collider: Optical Considerations, p. 189, Accelerator Physics Issues for a Superconducting Super Collider. UM HE 84-1, Ann Arbor (1983).

3. A. A. Garren and D. E. Johnson, The 90° (September 1987) SSC Lattice. SSC 146, Berkeley (1987).

STUDY OF SCINTILLATING FIBERS FOR A HIGH RESOLUTION
TIME OF FLIGHT SYSTEM AT THE SSC

Michael Kuhlen[*]

California Institute of Technology
Charles C. Lauritsen Laboratory of High Energy Physics
Pasadena, Ca 91125, USA

ABSTRACT

We are studying properties of plastic scintillating fibers with the aim of constructing a Time of Flight system with very good time resolution (50 psec) and with high spatial segmentation. Such a system could be valuable at the SSC for two applications: 1) as a component of a first level trigger allowing the fast computation of the number of charged tracks and 2) for separating multiple events from the same bunch crossing. It would also be a valuable device for particle identification at a B-meson factory. The status of the project is being presented. Good time resolution requires a short decay time and a high light yield for the scintillator. We have measured the decay times for different samples of scintillating fibers. Photodetectors with small transit time spread, e.g. fast photomultiplier tubes or microchannel plates, will be needed.

TIME OF FLIGHT AT THE SSC: PHYSICS MOTIVATION

A system of Time of Flight (TOF) scintillation counters (see fig. 1) could provide fast information about the charged multiplicity of an event and about times of arrival of charged particles to a precision of the order of 50 psec. It may serve a variety of purposes for physics at the SSC.

---

[*] representing Chris Hawkes, M. K., Barrett Milliken, Ryszard Stroynowski and Eric Wicklund

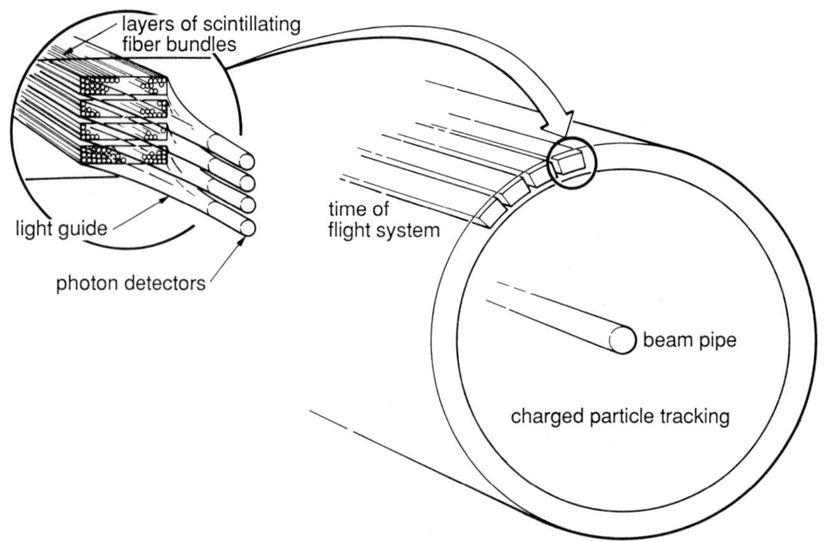

Figure 1. Schematic view of a TOF system for a SSC detector

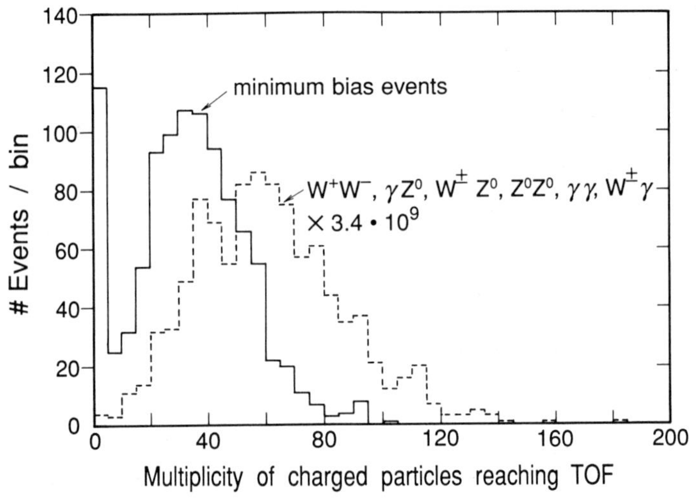

Figure 2. Charged multiplicity from the TOF. Each histogram represents 1000 generated events. The histograms are normalized to the integrated luminosity and the vector boson pair distribution is magnified by a factor $3.4\cdot10^9$ to account for the difference in cross section.

Multiplicity trigger

One of the major challenges for experiments at the SSC is the extremely high interaction rate of about 100 MHz at design luminosity. In order to be able to process the events, the trigger rate has to be reduced to about 10-100 Hz, a rate which can realistically be handled. In the trigger fast decisions have to be made to suppress "uninteresting" events by roughly seven orders of magnitude without sacrificing the events containing the physics to be studied. As ingredients for a trigger scheme total transverse energy from the calorimeter and the presence of leptons or missing transverse energy are being discussed[1] . A TOF system would provide independent and complementary information on which fast trigger decisions can be based, enhancing the ability to separate events of physics interest.

As an example let us compare QCD type "minimum bias" events, which account for almost the entire total cross section, with vector boson pair production events, which have a cross section smaller by more than a factor 10^9. We consider a typical SSC detector equipped with 5 m long TOF counters at a radius of 2.5 m surrounding the inner tracking device. The magnetic field is assumed to be 3 T. The two processes have been simulated with the PYTHIA Monte Carlo program[2] .

The TOF measures the charged multiplicity in the central rapidity region. Vector boson pair events tend to have a higher multiplicity than minimum bias events (see fig. 2). Multiplicity may be of particular use together with the transverse energy. The combined information from the calorimeter and the TOF can be used for separating the two processes (see fig. 3). For the same given transverse energy vector boson pair events have a smaller multiplicity than QCD events. It may also be possible to establish a jet trigger utilizing high multiplicity in adjacent TOF segments.

Multiple event separation

Bunch crossings at the SSC occur every 16 nsec according to the present design. Already with a moderate time resolution the TOF can distinguish particles originating from different bunch crossings. At the design luminosity one expects on average 1.5 events per bunch crossing. That means an interesting physics event will on average have one to two minimum bias events overlaid. The bunch length is 7 cm, which corresponds to a time window of 0.23 nsec (speed of light $c =30$ cm/nsec) during which multiple events may happen. With a time resolution of 50 psec it is possible to separate 47 % of the tracks in double events by more than 2σ, even if they occur in the same place and cannot be resolved by the tracking device. This can be done very quickly by the TOF, while multiple event separation from the vertex position along the beam direction must await pattern recognition in the tracking device at a later stage of the event reconstruction.

Longitudinal spatial information for charged tracks

Due to the finite effective velocity of the light pulse generated in the scintillator by a charged particle, the times recorded on both ends of the scintillator depend on the z coordinate of the intersection. A time resolution of 50 psec translates into a position resolution along the beam direction of 0.8 cm. It may be useful to aid pattern recognition in the inner tracking device and provides fast information on the rapidity of charged particles.

Particle identification

The Time of Flight system can be used to separate pions and kaons to momenta up

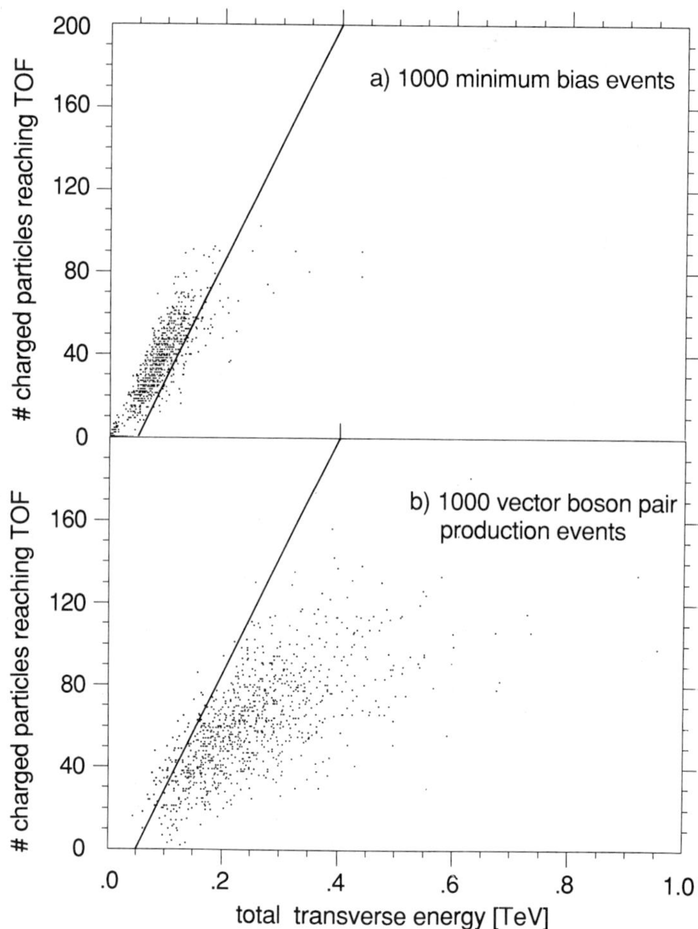

Figure 3. Charged multiplicity and transverse energy. The upper part a) shows 1000 minimum bias events and the lower b) 1000 events from vector bosom pair production. The strait lines are drawn to guide the eye when comparing the two scatter plots.

to 3 GeV/c. That accounts for roughly 70 % of all particles reaching the TOF in vector boson pair events.

PROPOSED IMPLEMENTATION OF TOF AND TECHNOLOGICAL CHALLENGES

We envision a system of approximately 200 TOF counter segments (see fig. 1) which surrounds the inner tracking device and matches the expected multiplicity. Each segment consists of about five layers of scintillation material, stacked in depth to allow for path length compensation[3]. The total thickness of the system is of the order of 10 cm. Possible advantages of using bundles of scintillating fibers instead of bulk scintillator are being evaluated. The individual layers are read out with photon detectors on both ends. For good timing resolution fast photomultiplier tubes or microchannel plates are being investigated.

Time resolution

There are different contributions to the time resolution of a TOF system[3][4] :

1. The decay time of the scintillator, which is of the order of 2.5 nsec.
2. The time variation due to different path lengths of the light travelling in the scintillator. At 1 m distance from the phototube it is 3.76 nsec for a typical counter slab and 0.24 nsec for a counter made out of a fiber bundle (see fig. 4).
3. The resolution of the photon detector. Good phototubes have a transit time spread for single photoelectrons of 0.25 nsec, microchannel plates reach 0.05 nsec.

All these contributions depend on the obtained light level. In order to get the best time resolution one triggers on the first arriving photon(s). The time resolution of the system improves as the number of photons generated (n) increases. For example the rms resolution for a single exponential decay time distribution $dN/dt = \frac{1}{\tau}e^{-t/\tau}$ is $\sigma = \tau/n$, if one triggers on the first arriving photon out of n.

To improve the time resolution, one can improve the individual contributions - we are investigating faster scintillators, photon detectors with better time resolution, and scintillating fibers, which have a smaller path length variation - and one can increase the amount of light in the system by making it thicker. A particle traversing more scintillating material will generate more light.

Apart from other restrictions however, the thickness cannot be increased indefinitely without actually degrading the resolution, because it takes a particle a finite amount of time to traverse the scintillator, which is not negligible compared with the resolution goal. One can correct for that if one knows where in depth the light one is looking at has been created. That requires radial segmentation of the system (see fig. 5).

Conventional Counters and Fiber Bundles

In a TOF counter the scintillation light produced by a charged particle is transmitted to the readout detectors at both ends by total internal reflection (see fig. 6). In order to minimize light losses the surface of the scintillator has to be very uniform and needs to be polished. That surface is usually vulnerable to damage, which poses a major problem for the production of large counters.

In contrast for scintillating fibers[5] (see fig. 7) the core-cladding interface already constitutes a nearly perfect surface for reflection. The cladding itself provides a protective layer around the fiber. Fibers may be easily arranged in arbitrary bundles and are thus well suited for application in a highly segmented system.

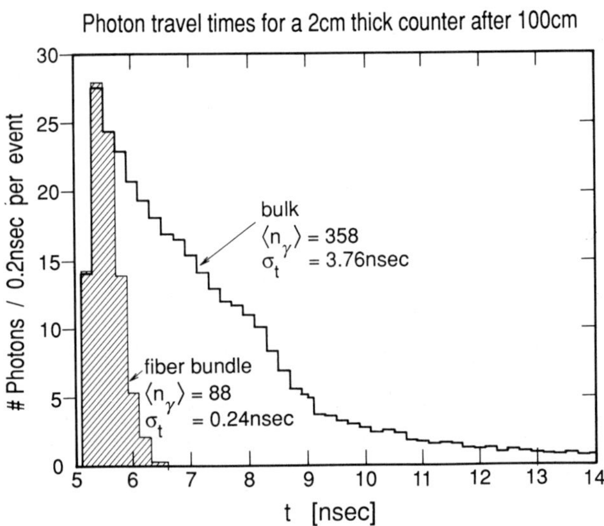

Figure 4. Photon travel times in scintillating fibers
and bulk scintillators.

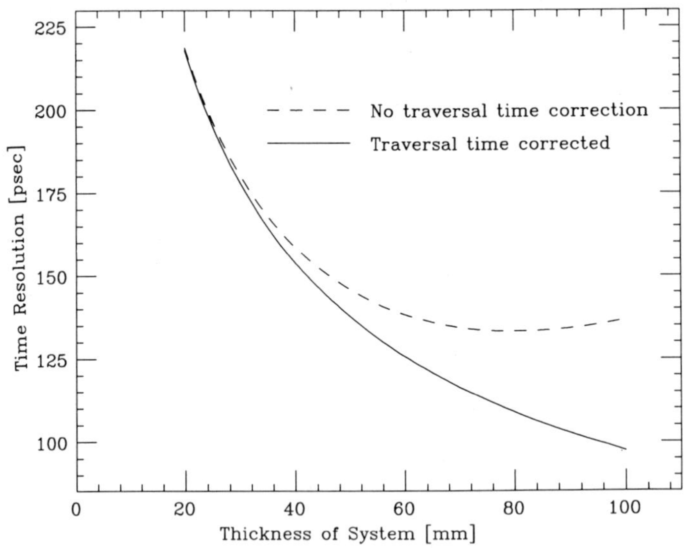

Figure 5. Time resolution with and without radial segmentation.
A dependence of the resolution on the light output
according to ref. 4 has been assumed.

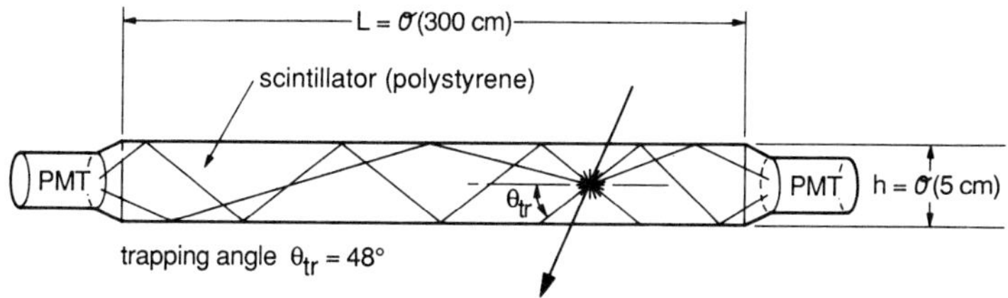

Figure 6. Conventional TOF counter.

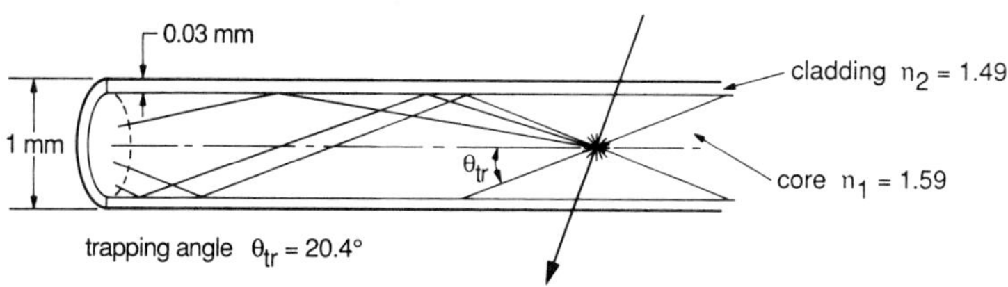

Figure 7. Scintillating fiber. The fiber consists of a core of scintillating material(index of refraction n_2). The cladding provides a smooth interface for reflections. Scintillation light emitted within the trapping angle ($\sin\theta_{tr} = n_2/n_1$)propagates along the fiber by total internal reflection.

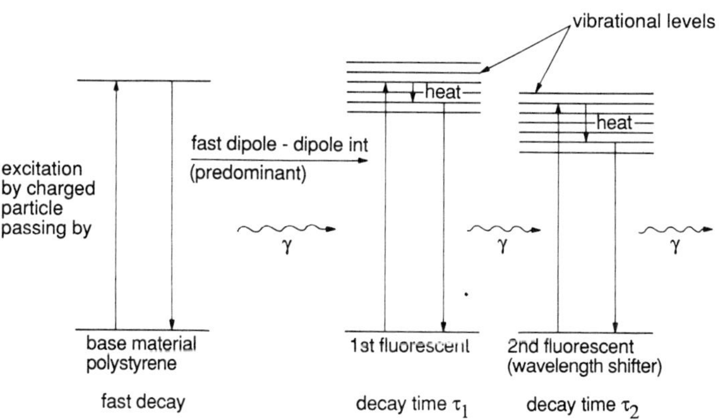

Figure 8. Model for scintillation process.

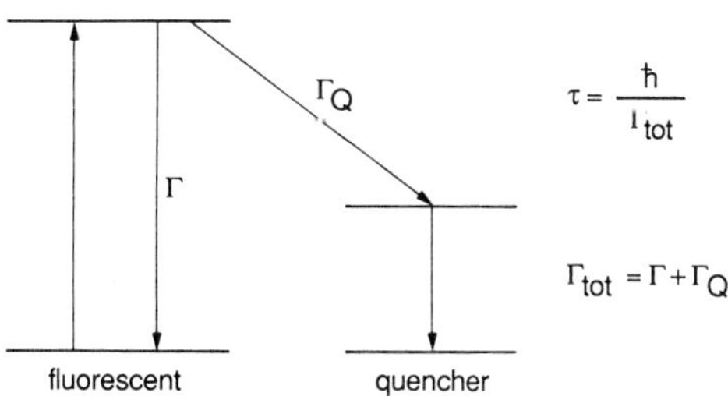

Figure 9. Effect of quenchers on decay time.

Because of the smaller trapping angle for fibers, the variation of path lengths and travel times for photons arriving at the photocathode is much smaller for fibers than for bulk material. The light outside the trapping angle does not propagate and is lost.

For optimal time resolution the earliest arriving photons are relevant. A Monte Carlo study finds the same amount of early light for a TOF system made out of fibers as for bulk scintillator, and the overall time resolution is about the same (see fig. 4). The loss of light in the fibers is compensated by the smaller path length dispersion of the remaining light.

<u>Fast Scintillators</u>

As a model for the scintillation process[6] which describes our data (see timing measurements) we adopt a three step process with two significant decay times (see fig. 8). Molecules of the basic material (polystyrene) get excited by the passing charged particle. That energy is transferred to the first fluorescent, predominantly by fast dipole-dipole transitions. The presence of the first fluorescent opens a fast decay channel from the base material to an efficient radiator. Energy transfer to the second fluorescent occurs through photon emission and absorption. The second fluorescent serves as a wavelength shifter to match the wavelength of the scintillator's light with the sensitivity range of the photocathode.

By adding a "quenching" chemical to the scintillator, one opens a new decay channel for the fluorescent and makes its decay time shorter (see fig. 9). The quencher itself decays without radiating light. One obtains a faster decay time at the cost of smaller light output. The trade off has to be investigated in terms of the effect on the resolution of the entire system. In the following chapter we present measurements of the decay times for scintillating fibers of different chemical compositions, including a quenched scintillator.

DECAY TIME MEASUREMENTS

We have measured the emission time distributions for 1 mm diameter fibers produced by Kyowa Gas Chemical Co. Ltd., Japan[7]. The core material is of the same composition as the scintillators SCSN-38, SCSN-81, and SCSN-38 with 0.5 % by weight of benzophenone added as a quencher. The quenched scintillator is referred to as SCSN-38Q.

The decay time is measured for single photoelectron events as the difference (up to a constant) in time between the excitation of the fiber and the detection of the decay photon. The experimental set up is shown in fig. 10. The fiber is irradiated with a Ru^{106} source. The excitation time is measured by a trigger scintillation counter with two photomultiplier tubes, one of which starts a TDC. The light from the fiber is recorded by a photomultiplier tube, which provides the signal to stop the TDC. A neutral density filter attenuates the light to give an average of approximately 0.05 photoelectrons per event, such that the probability for events with more than one photoelectron is negligible. An example of a measured TDC distribution is given in fig. 11.

The observed emission time distribution can be described phenomenologically by a scintillation process involving two significant steps with two different decay times, τ_1 and τ_2, plus a slow component with decay time τ_3. The short time behavior is dominated by the rise and decay time generated by the two step cascade.

The expected emission time probability distribution for this model is

$$E(t) = \frac{\dfrac{e^{-t/\tau_2} - e^{-t/\tau_1}}{\tau_2 - \tau_1} + \dfrac{R}{\tau_3}e^{-t/\tau_3}}{1+R}$$

where R is the ratio of the numbers of photons produced through the slow and the fast decay processes. This function folded with the experimental resolution (σ_t) is fitted to the data;

$$P(t) = \int_0^\infty E(t')\frac{1}{\sqrt{2\pi}\sigma_t}e^{-\frac{(t-t')^2}{2\sigma_t^2}}\,dt' = \frac{1}{1+R}\left[\frac{\tau_2 f(t,\tau_2) - \tau_1 f(t,\tau_1)}{\tau_2 - \tau_1} + Rf(t,\tau_3)\right],$$

where

$$f(t,\tau) \equiv \int_0^\infty \frac{1}{\sqrt{2\pi}\sigma_t}e^{-\frac{(t-t')^2}{2\sigma_t^2}}\frac{1}{\tau}e^{-\frac{t'}{\tau}}\,dt' = \frac{1}{2\tau}\left[1 + \mathrm{erf}(\frac{1}{\sqrt{2}}(\frac{t}{\sigma_t} - \frac{\sigma_t}{\tau}))\right]e^{-\left(\frac{t}{\tau} - \frac{\sigma_t^2}{2\tau^2}\right)}.$$

The resolution has been measured independently to be $\sigma_t = 0.280$ nsec with the same set up as above, but replacing the fiber with a piece of lucite, which yields instantaneous Čerenkov light. The function is fitted to the data with decay times up to 35 nsec. Data between 11.5 nsec and 20.8 nsec were effected by after-pulsing from the photomultiplier tube and had been excluded from the fit. The results for the fit parameters τ_1, τ_2, τ_3 and R are given in the following table:

Table 1

scintillator	rise time τ_1 [nsec]	decay time τ_2 [nsec]	slow comp. τ_3 [nsec]	ratio slow/fast R
SCSN-81	0.87 ± 0.02	2.36 ± 0.04	14.2 ± 0.4	0.27 ± 0.01
SCSN-38	0.68 ± 0.02	2.27 ± 0.05	14.9 ± 0.4	0.25 ± 0.01
SCSN-38Q	0.71 ± 0.03	2.08 ± 0.06	14.9 ± 0.5	0.21 ± 0.02

There are relatively small differences in the decay times of the scintillators under study. Adding a quencher to SCSN-38 results in slightly increased speed of the scintillation process. In order to evaluate the different scintillators for a TOF application also the light output has to be studied. These measurements are in progress.

SUMMARY

A Time of Flight system at the SSC provides fast and independent information for first level trigger decisions. Further studies on a trigger scheme involving TOF information are indicated. TOF also provides a simple method to distinguish different bunch crossings and different events in one bunch crossing. Scintillating fibers may be used for their ease of fabrication. The decay times for different samples of scintillating fibers have been measured. None of the scintillators studied however can be singled out at this time as clearly superior for this application. In order to contribute significantly to the improvement of the resolution of a TOF system scintillators with a substantially smaller ratio of decay time to light yield are needed.

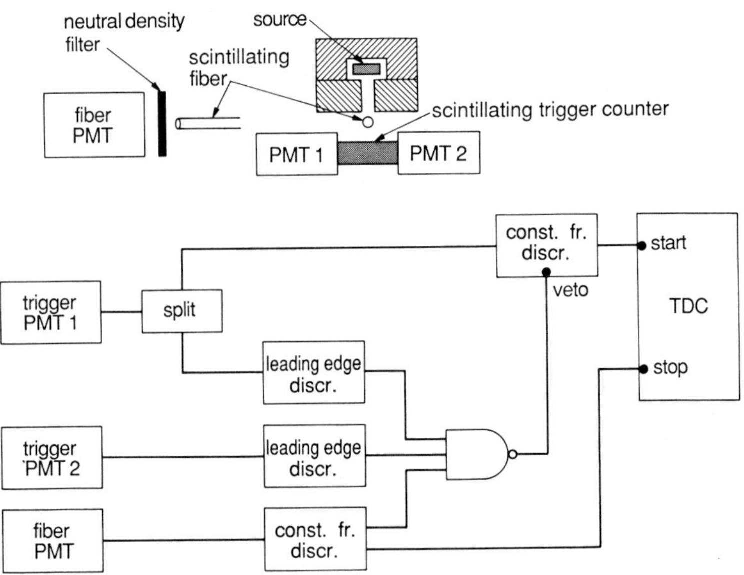

Figure 10. Block diagram of set up for decay time measurements.

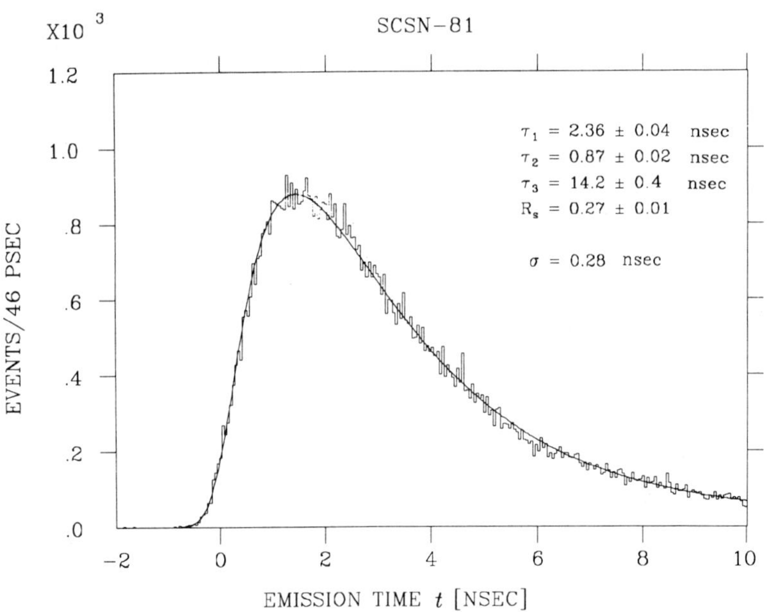

Figure 11. TDC distribution for typical decay time measurement. The smooth line represents the fit to the histogram.

REFERENCES

1. L. Price, G. Wagner, M. Abolins, Proceedings of the 1984 Summer Study on the Design and Utilization of the Superconducting Supercollider, Snowmass, p. 570

2. H. Bengtsson, T. Sjöstrand, Comp. Phys. Comm. 46 (1987), p. 43-82

3. B. Milliken, R. Stroynowski, E. Wicklund, Caltech preprint CALT-68-1538 (1989); to appear in proceedings of the Workshop on Scintillating Fibers, Fermilab, Nov. 1988

4. W. B. Atwood, Lectures at the SLAC Summer Institute on Particle Physics "The Weak Interactions", SLAC Report 239 (1981), p. 287

5. A. Grimes et al., Northeastern University preprint NUB-2943 (1988)

6. J. B. Birks, The Theory and Practice of Scintillation Counting, Oxford 1964

7. We thank Dr. T. Shimizu for providing samples of scintillating fibers for our measurements.

IMPACT OF SUPERCONDUCTIVE MAGNETS ON ACCELERATOR FACILITIES

Allen Cheng and
John Ridler

STV/Seelye Stevenson Value & Knecht

ABSTRACT

As accelerators have been pushed to higher and higher energies,
their field requirements and, therefore, their power and cooling system
requirements, have increased substantially. Superconductive magnets
have an internal electrical resistance of zero and, therefore, waste
almost no energy in the form of heat emission. As a result, their
use substantially reduces both the power and cooling system requirements
of the facility and, therefore, reduces both its capital and operating
costs. Experience at CEBAF and at Fermilab demonstrates the tremendous
savings resulting from the use of superconductive, rather than conventional magnets and RF cavities. If conventional magnets were used
to produce the 20TeV beam energies desired for the new Superconducting
Super Collider, the costs would astronomical and the project would be
economically, politically, as well as perhaps technologically,
infeasible. It is only through the use of superconductive magnets that
the desired beam energies can reasonably be achieved.

INTRODUCTION

In 1965, discussion first took place about the use of superconductive magnets in the new high-energy accelerators during the design of
the 200 BeV proton accelerator at the Lawrence Berkeley Laboratory. The
energy and cooling requirements of the increasingly powerful accelerators had become so substantial that the use of conventional magnets was
becoming prohibitively expensive. Superconductive magnets, it appeared,
could provide the means to reduce the cost and would, therefore, permit
the accelerators to become more and more powerful. While conventional
magnets were eventually chosen, the event marked the beginning of the
development of an important technology which has now made possible the
20 TeV Superconducting Super Collider.

Superconductive magnets utilize a cryogenic system to maintain a
temperature at which their electrical resistance drops to close to zero.
As a result, the power requirements of the accelerator drop substantially,
as does the associated operating cost and strain on the local utilities,
and fewer conventional cooling facilities are required. The superconductive magnets make the beam energies desired at the Superconducting
Super Collider both technologically and practically feasible.

CONVENTIONAL MAGNETS

Over the past decades, a variety of cooling systems have been proposed, studied and implemented to remove the excess heat generated by
accelerator magnets. These include the application of conventional refrigeration, the circulation of ground water with temperatures of
approximately 50°F. and, as originally utilized at Fermilab, the
circulation of warmer cooling water with temperatures in the range of
95°-115° F.

Since the intensity of a magnetic field is directly proportional to
the current, and the power loss due to heat rejection increases with the
magnet's resistance and as the square of the current, as accelerators
were pushed to provide higher and higher energies, their pulsed power
and cooling system requirements skyrocketed. When the accelerator at
Fermilab was upgraded from 200 GeV to 400 GeV in the mid-70's, for
example, its power consumption quadrupled. It has become evident that
the power and cooling requirements severely limit both the location and
potential energies of the accelerators.

SUPERCONDUCTIVE MAGNETS

Superconductive magnets operate at a very low temperature and,
therefore, have an internal resistance of almost zero. As a result,
they waste almost no power through heat emission. The consequences are
twofold: first, since almost no energy is wasted, the pulsed power
required to generate a large mangetic field is substantially less than
would be required by conventionally cooled magnets. Since the pulsed
power requirements of a large accelerator places a strain on the local
utilities, its reduction is of particular significance to the ability of
that utility to support the project. Second, since the heat rejection
is small, the magnets remain fairly cool and do not require as substantial a cooling system. Not only is water consumption reduced, but also
the cryogenic system is adaptable to an air-cooled rather than a water-
cooled condenser which provides an opportunity to heat the complex and,
therefore, reduce its total energy requirements. In sum, superconductive magnets require less power and cooling than their conventional
magnet counterparts and are thus far more economically desirable.

CASE STUDY: Fermilab

The experience at Fermilab proves the point. The accelerator was
originally designed to produce energies of 400 BeV and required 51 Megawatts, of which 45 Megawatts was pulsed power required to drive the
magnets. During the month of July, 1976, the site-wide power demand was
28.9 Gigawatt-hours, of which 15.5 Gigawatt-hours was used for the main
ring pulsed power. The costs of electric power during the month for the
accelerator alone was almost $1.5 million. As part of the Energy
Doubler Project, in which the accelerator was upgraded to provide 900
BeV particle beam energies, the conventional magnets were replaced by
superconductive magnets. In spite of the substantial increase in the
accelerator's energy, the move to cryogenically cooled magnets resulted
in an overall decrease in its power requirements. The accelerator's
total power usage dropped to 30.6 Megawatts, a decrease of over 40%, of
which only 10 Megawatts of pulsed power was required to drive the magnets. Even with the doubled energy of the accelerator and the increased
costs of electricity since 1976, Fermilab's annual power bill has
remained at about $18-$19 million.

624

Power Use for Conventional and Superconductive Magnets at Fermilab

	Conventional 400 GeV (1976)	Superconductive 900 GeV (1988)
Pulsed Power	45.0 MW	10.0 MW
Cooling Water Pumps	5.0 MW	3.5 MW
R. F. Building Power	1.3 MW	3.0 MW
Cryogenic Power	0.0 MW	11.5 MW
Additional Facilities	-	2.6 MW
Total Power	51.3 MW	30.6 MW

CASE STUDY: CEBAF

A similar reduction in power requirements occurred at CEBAF when the RF system cavities were changed from conventional to cryogenic. The LINAC AC power requirements dropped from 20 Megawatts to 6 Megawatts while the facility's overall demand dropped from 35 Megawatts to 24 Megawatts.

Power Use for Conventional and Superconductive RF Cavities at CEBAF

	Conventional RF Cavities	Superconductive RF Cavities
LINAC Power	20 MW	6 MW
Supporting Facilities	15 MW	18 MW
Total Power	35 MW	24 MW

In addition to reducing CEBAF's power requirements and, therefore, its operating costs, the move to superconducting magnets reduced the capital cost of the cooling facilities from $5.9 million to $2.3 million. This tremendous savings was accomplished without any decrease in the accelerator's capabilities.

Capital Cost Comparison of Conventional and Superconducting Systems at CEBAF

	Cost of Conventional RF Cavities	Cost of Superconductive RF Cavities
Accelerator Cool- ing Facilities	$4,600,000	$2,300,000
Cooling Ponds	$1,000,000	-
Collection Pond	$300,000	-
Total Cost	$5,900,000	$2,300,000

CONCLUSION

Were conventional magnets to be used at the Superconducting Super Collider, the facility's annual electric costs would be astronomical. With the superconductive magnets, however, the power consumption at the 20 TeV Super Collider is estimated to be approximately $30 million per year - slightly more than currently at the 1 Tev Fermilab. It is clear that the use of superconductive magnets is necessary to keep the facility's operations economically feasible.

Along with substantially diminishing the operating costs of the accelerator, the use of superconductive rather than conventional magnets significantly reduces the cost of construction. If the energy desired for the 53-mile Superconducting Super Collider were to be achieved using convenientional magnets, assuming that this was in fact technologically feasible, the ring would have to be between 300 and 400 miles in circumference. Given the estimated cost of $60 million a mile for the tunnel and related support facilities, the use of conventional magnets could add up to $20 billion to the cost of construction.

The use of superconductive magnets substantially reduces both the power and cooling system requirements of an accelerator facility and, therefore, decreases both its operating and capital costs. Without superconducting magnets, it is very unlikely that the 20 TeV beam energies envisioned for the Superconducting Super Collider could ever be reached.

A LARGE SUPERCONDUCTING DETECTOR

MAGNET WITHOUT AN IRON RETURN PATH

Michael A. Green

M/S 90-2148
Lawrence Berkeley Laboratory
Berkeley, CA 94720

ABSTRACT

This paper describes a detector magnet which returns flux between the coils rather than through an iron return path. This actively shielded, uniform field 2 T magnet can be fabricated in separate parts which can be manufactured on the SSC site. This magnet can be built so that central field is uniform enough to permit a TPC detector to be used without iron poles. The field outside of the coil can be made to fall of as R^{-N} power where N approaches 9. A major advantage of the magnet described in the paper is that there is no pole piece to block the particle jets emanating from the collision region in the forward and backward directions. Inexpensive materials such as earth and concrete can be used to provide the mass needed to analyze particles such as mu mesons. As a result, problems such as experimental hall subsidence can be reduced. Perhaps the cost of such an experiment can also be reduced. This type of magnet would require experimenters to rethink their experimental concepts.

BACKGROUND

The SSC, a large machine, will produce proton collisions of 20 TeV on 20 TeV. The collisions result in many fragments which must be detected in a large number of detectors around the collision point. Large magnetic fields can be used to bend charged particles which will permit the mass and momentum of these particles to be calculated. The volume of the magnetic field must be large (at least 5 meters in diameter and 7 or 8 meters long).

In order to generate a large volume magnetic field, superconducting magnets provide an economical way of generating the magnetic field. A large volume of magnetic field implies that there must be a large ferromagnetic return path. If the magnetic field must be uniform, the iron return path and pole pieces must be machined in order to achieve the desired field uniformity. The conventional approach to building large detector magnets would require up to 30000 tons of iron to shape the magnetic flux and return it.

A large massive iron return path has a number of disadvantages which include: 1) The iron density is high enough to cause serious subsidence problems at the site of the experiment, 2) When the desired field is uniform, the iron blocks particles which travel at low angles in the direction from which the colliding beams come. The disadvantages given above can be modified somewhat by the proper design of iron return path. The result is a larger iron return path mass and an even more complex design for the iron.

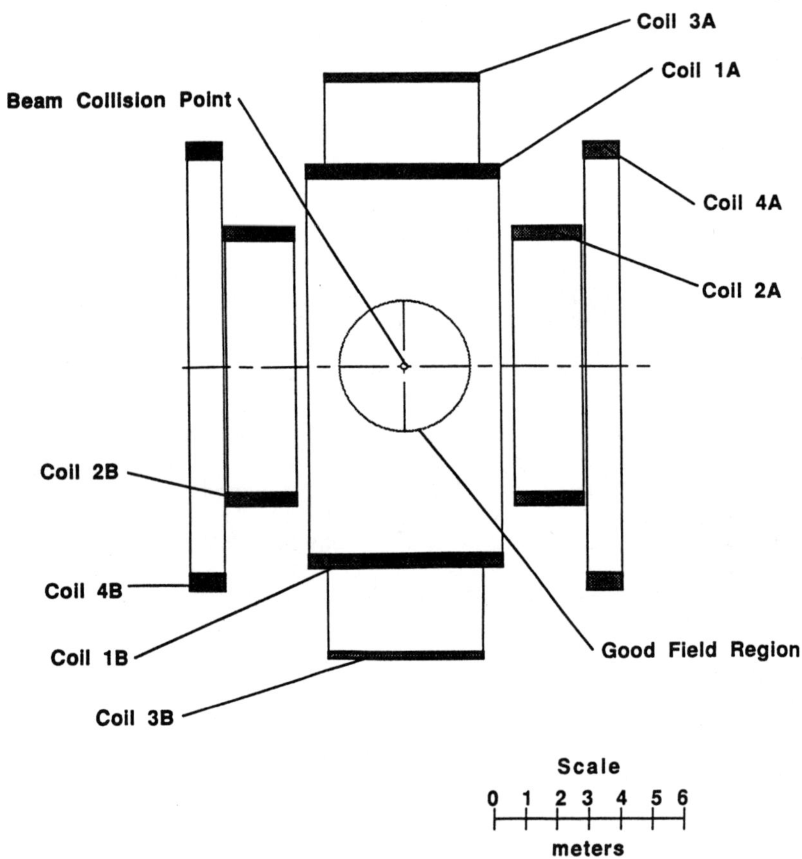

Figure 1

**A SPHERICAL DETECTOR MAGNET
WITHOUT AN IRON RETURN PATH**

This paper presents an alternative approach to detector magnets. This is a modest proposal for a large superconducting magnet which will produce a uniform field over a volume which is four meters in diameter and four meters in length. The proposed magnet has no iron return path, yet the magnetic induction 20 meters from the collision point is less than 10 Gauss. The no iron detector magnet requires that one rethink the design of SSC physics detectors.

THE IRON FREE DETECTOR MAGNET AND ITS EFFECT ON PHYSICS

For some types of SSC detectors, it may be desirable to eliminate the iron shielding and return yoke. The iron free detector magnet is in principle like the actively shielded magnetic resonance imaging magnets (MRI), which have been built by Oxford[1] and other companies.[2] Two general designs were looked at for magnets which have a free bore diameter of at least 7.5 meters.

The two designs can be described as follows:

1)   The first type of magnet is the two solenoid type of actively shielded magnet, which is shaped like a cylinder. This type of magnet is the state of the art actively shielded MRI magnet. This type of magnet, when it has an inside bore of 7.5 m would have an overall length of about 28 to 30 meters and an outside diameter of about 16 meters. Magnets of this type have been built with warm bore diameters of 1 meter. This type of magnet has a peak field which is 10 to 15 percent higher than the central field of the magnet.[3] This type of magnet is well suited for use as a high central induction magnet (>3T).

2.   The second type of actively shielded magnet is a spherical type of magnet which was developed at Stanford University[4] for MRI imaging of the heart and the circulatory system. This magnet, when it has a minimum warm bore diameter of 7.5 meters, will have an overall length of 14 meters and an outside diameter of about 19 meters. A magnet of this type has a peak induction at the conductor which is 60-80 percent higher than the central induction. This type of magnet is well suited for central inductions below 3.0T.

The advantages of the solenoidal design are: 1) The peak field in the winding is only 15 percent higher than the magnet central field. 2) The field is quite uniform over a region which is up to 6 meters in diameter over a length of about 12 meters without an iron return path. The disadvantages of this type of magnet are as follows: 1) The magnet is large, and it is difficult to fabricate in pieces on the site. The transport of magnets with a warm bore over 5 or 6 meters is difficult under the best conditions. 2) The free solid angle in the forward and backwards directions from the collision point is only about 20 degrees from the proton beam line. 3) Physics is difficult to do outside of the 3.75 m radius over the full length of the solenoid magnet. 4) The 10 Gauss induction line is 38 meters from the collision point in the radial direction and 48 meters from the collision point along the beam axis.

The advantages of the spherical design are: 1) The field is quite uniform over a diameter of 4 to 5 meters and a length of 6 to 8 meters without an iron return path. 2) The free solid angle in the forward and backwards directions from the collision point is about 35 degrees from the proton beam line. 3) At high solid angles from 60 to 90 degrees from the proton beam around the collision point, physics can be done out to a radius of 5.5 meters. 4) The 10 Gauss line is 18 meters from the collision point in the axial direction. 5) The magnet can be built in pieces on the site and it can be assembled on site. The disadvantages of this type of magnet are as follows: 1) The field rise at the conductor is 65 percent higher than the central field. This is acceptable in a 2.0 Tesla magnet, but it is not acceptable in a 4 or 5 Tesla magnet. 2) The members which carry the forces between the magnet coils must be cold. These forces are much larger than in the solenoid design.

The spherical solenoid design was selected as the candidate for an iron free detector magnet. The use of such a magnet configuration requires one to rethink how one might do the physics. If the magnet has no iron return yoke, one must use other

**PARAMETERS FOR A LARGE
SUPERCONDUCTING
DETECTOR MAGNET
WITHOUT AN IRON SHIELD**

PARAMETER	
Central Magnetic Induction (T)	2.0
Good Field Length (m)	4.0
Good Field Diameter (m)	4.0
Inside Coil Diameter (m)	7.92
Outside Coil Diameter (m)	18.12
Magnet Overall Length (m)	13.60
Magnet Overall Diameter (m)	18.50
Number of Coils	8
Number of Magnet Turns	5720
Magnet Self Inductance (H)	71.62
Magnet Design Current (A)	10126
Magnet Stored Energy* (MJ)	3672
Matrix Current Density* (A/sq cm)	1315
Winding Peak Induction (T)	3.3
Maximum 10 Gauss Distance (m)	20.5
Type of Superconductor	Nb-Ti
Matrix Material	copper
Stabilizer Material	pure Al

materials such as concrete, heavy concrete with barrites, earth and other relatively low density materials to moderate the particles generated at the collision point around which the experiment is being done. This may require one to look at different types of detectors to do the physics.

The advantages of the iron free magnet are: 1) The up to 30000 tons of iron return path can be eliminated. If iron is used in the experiment, the coil design must be altered. 2) Up to 35 degrees in the forward and backwards directions is completely free for looking at the particle jets created at the intersection point. 3) Cheap materials such as concrete and earth can be used to moderate the particles. There is a saving in foundation cost and subsidence is greatly reduced. The disadvantages are: 1) The magnet coils block particles at a radius of 4 meters from the proton beam line at solid angles from 35 to 60 degrees from the proton beam line. This radius goes out to 5.5 meters for an angle of 60 to 90 degrees from the proton beam line. 2) Some kinds of experiments require dense materials to moderate the particle produced at the collision point. Iron is one of the least expensive dense materials to use as a moderator. 3) The stray field from the magnet can have an adverse effect on some types of detectors. The colliding proton beams may have to be carried in superconducting shielded beam pipes.

A DESIGN FOR A SPHERICAL IRON FREE DETECTOR MAGNET

Figure 1 shows a schematic of an eight coil iron free detector magnet using the spherical solenoid coil concept. The good field region, which has a field uniformity of better than one part in 1000 is defined as a spherical region which is 4 meters in diameter with the center of the sphere at the SSC proton beam collision point. The magnet shown in Figure 1 has a nominal design central induction of 2T (if one makes the spherical magnet somewhat smaller, say two-thirds of the size shown in Figure 1, the central induction could be increased to 3 Tesla).

The magnet design process uses Legendre polynomials of the first kind to expand the field inside of the coil.[5] Legendre polynomial of the second kind are used for the field outside of the coils (when one looks at the problem, the use of Legendre polynomials of the second kind are not needed). The magnetic field is designed to be axially symmetric. The net magnetic Legendre dipole, sextupole and decapole moments are zero by design (for both the inner and outer fields) and the quadrupole, octupole, 12 pole and so on are zero by symmetry. Since the magnetic moments up to $N = 7$ are zero, the field outside of the magnet coils falls off rapidly (as radius to the minus nine power).

Figure 2 shows a computer drawing of the magnet coils for the actively shielded spherical axially symmetric magnet. The round dot indicates current flow into the paper; the square dot indicates current flow out of the paper. The design shown in Figure 2 produces an axially symmetric dipole of 2 Tesla, with no axially symmetric quadrupole, sextupole, octupole, decapole or 12 pole. The first higher term to appear in either the internal or external field is the axially symmetric 14 pole. A magnetic induction contour map for the magnet is shown in Figure 3 and the magnetic flux line contour map is shown in Figure 4. The flux which is inside the inner coils (coils 1 and 2) is returned between the inner coils and the outer coils (coils 3 and 4). Very little flux escapes from the magnet.

Table 1 presents parameters for the spherical no iron detector magnet shown in Figure 1 and 2. The eight coil magnet system has an overall coil length of 13.6 m and an overall diameter of 18.5 m. When a 2.0 Tesla central field is generated, the peak induction in coil 2 is 3.3 Tesla. The external induction out to a distance of 25 meters from the collision point is shown in Figure 5. The 1 gauss line is about 24 meters from the collision point in the radial direction, and it is about 28 meters from collision point in the axial direction. The shielding achieved can be controlled by thin windings mounted on the outer coils (coils 3 and 4). Table 1 assumes that the inner and outer coils are operated off of a single 12kA power supply. Quench protection can be provided by a single 0.16 ohm external dump resistor.

Table 2 presents the parameters for the four coil types. (There are two coils of each type.) A calculation of hoop stress in the four coil types indicate that the copper based superconductor within the pure aluminum stabilizer can carry the hoop forces in coils one and two. In coils three and four additional hard aluminum (6061 - T6) is required to carry

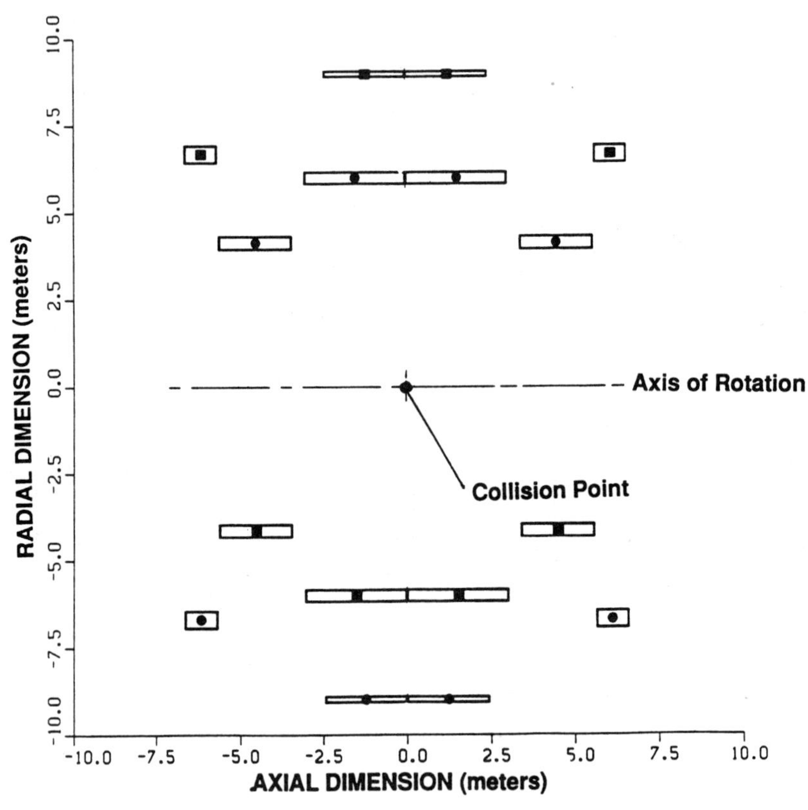

Figure 2

SUPERCONDUCTING COIL PLACEMENT
FOR A LARGE NO IRON SPHERICAL
DETECTOR MAGNET FOR THE SSC

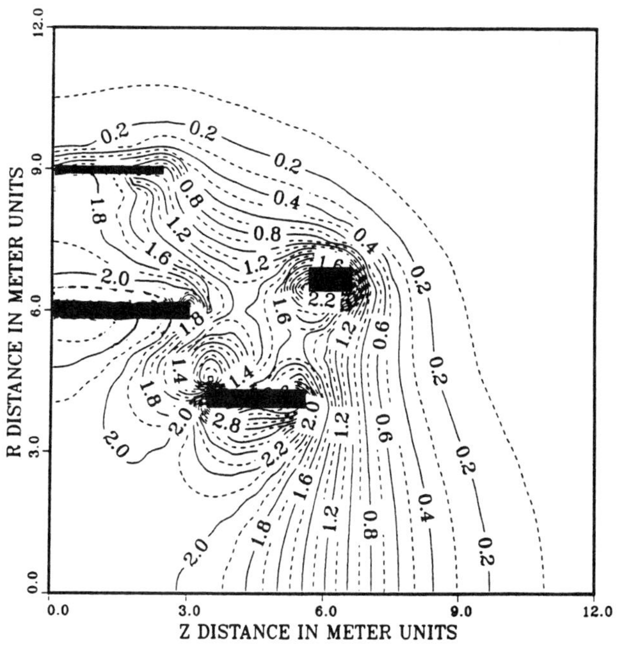

Figure 3

MAGNETIC INDUCTION CONTOUR MAP
FOR A LARGE NO IRON SPHERICAL
DETECTOR MAGNET FOR THE SSC

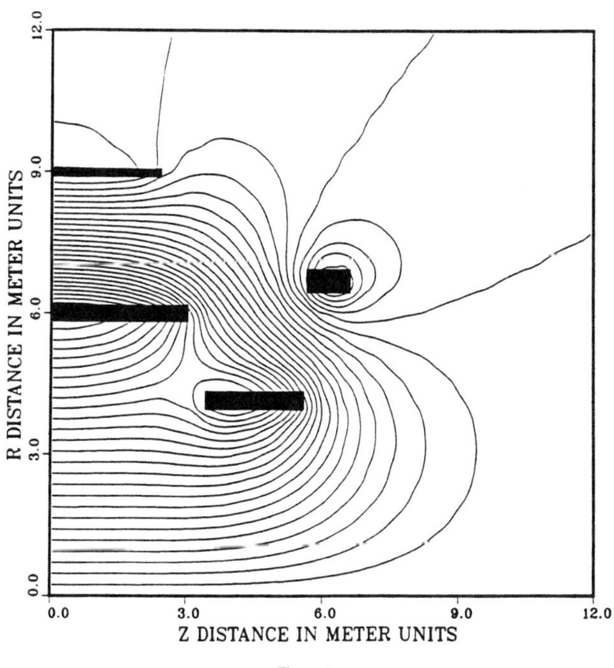

Figure 4

A MAGNETIC FLUS LINE CONTOUR MAP
FOR A LARGE NO IRON SPHERICAL
DETECTOR MAGNET FOR THE SSC

Table 2

PARAMETERS OF THE NO IRON DETECTOR
MAGNET SUPERCONDUCTING COILS

Coil Parameter	Coil 1	Coil 2	Coil 3	Coil 4
Coil Outside Diameter (m)	12.330	8.636	18.114	13.840
Coil Inside Diameter (m)	11.670	7.914	17.814	12.880
Distance from Center (m)	0.018	3.231	0.024	5.672
Coil Length (m)	3.000	2.160	2.400	0.930
Number of Layers	11	12	5	16
Number of Turns #	1100	864	400	496
Coil Design Current (A)*	10126	10126	-10126	-10126
Conductor Mass (tons)**	93	50	51	52
final Coil Mass (tons)##	145	89	120	84

\#   Conductor matrix size 2.7 by 2.8 centimeters
*   Current to generate an induction of 2.0 tesla at the magnet center
** Conductor mass for each coil in metric tons
\#\# Coil plus cryostat mass for each coil in metric tons

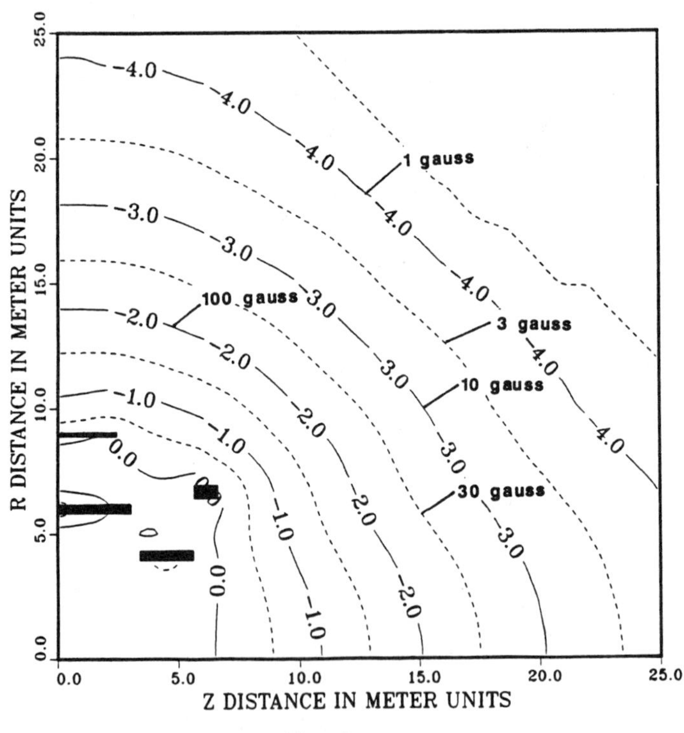

Figure 5

**MAGNETIC INDUCTION OUTSIDE OF
A LARGE NO IRON SPHERICAL
DETECTOR MAGNET FOR THE SSC**
(The Log10 of the Magnetic induction in Tesla)

the hoop forces generated in those coils. The large intercoil forces from coils 4 to coils 1 and coils 2 to coils 1 must be carried by cold members between the coils.

The total conductor mass for the eight coils is estimated to be 492 metric tons. The total magnet mass is estimated to be about 900 metric tons (the outer coils which return the flux represents about half the total mass). The total coil surface area would be around 1600 square meters. Based on experience with the TPC magnet,[6] about 500 W of refrigeration is required at 4.2 K and about 2000 W of refrigeration is required at 80 K in order to keep the magnet cold. A refrigerator equivalent to the LBL 1500 W machine[7,8] can handle the combined refrigeration liquefication load required for the no-iron detector magnet with a minimum of 7.5 diameter clear bore.

CONCLUDING COMMENTS

A large superconducting detector magnet can be built with bucking coils to return the magnetic flux. The mass of the bucking coils and the effect on the local environment is smaller than for a similar magnet with an iron return path. The induction at the beam intersection point for a magnet with a clear bore of 7.5 m can be as high as 2.0 Tesla. The field can be uniform (to 1 part in 1000) within a 4 meter diameter sphere centered at collision point. The field 25 meters from the collision point will fall off the 2.5 gauss or less.

The proposed detector magnet should be built on the site and assembled along with the experiment. A no-iron detector magnet requires one to rethink about physics options for the SSC. This type of magnet will not achieve all the physics goals for the SSC, but it will be good for analyzing particle jets up to a solid angle of 35 degrees from the proton beam line.

ACKNOWLEDGEMENTS

The author acknowledges the fruitful discussion he has had with W.M. Fairbanks of Stanford University and M.S. McAshan (now with the SSC central design group) on this magnet design concept as it pertains to MRI imaging systems.

This work was supported by the Director, Office of Energy Research, Office of High Energy and Nuclear Physics, High Energy Physics Division, U.S. Department of Energy under Contract Number DE-AC03-76F00098.

REFERENCES

1) D.G. Hawksworth, et al., "Considerations in the Design of MRI Magnets with Reduced Stray Fields", IEEE Transactions on Magnetics MAG-23, No. 2, p. 1309 (1987).
2) J.L. Carolan, W.A. Burns and M.A. Green, "A 5 Tesla Magnet for Imaging Laboratory Animals", IEEE Transactions on Magnetic MAG-25, No. 2 (1989).
3) M..A. Green and J.L. Carolan, "A 4.7 Tesla Magnet for Magnetic Resonance Imaging and Spectroscopy", IEEE Transactions on Magnetics MAG-23, No. 2, p. 1299 (1987).
4) T.L. Smith, M.S. McAshan, W.M. Fairbank, U.S. Patent 4595899 June 17, 1986.
5) M.A. Green, "The Design of Solenoid Magnets Using Legendre Functions", Proceedings of the 6th International Conference on Magnet Technology (MT-6), Bratislava, Czechoslovakia, 29 August to 2 September 1977, p. 838.
6) M.A. Green, et al., "Forced Two Phase Cooling of the TPC Superconducting Solenoid", Advances in Cryogenic Engineering 29, Plenum Press, New York (1983).
7) Fermi National Laboratory Specification Number 0418.09-ES-53497, Specification for a 1500 W Refrigerator Liquefier for FNAL and LBL (February 1975).
8) R. Byrns and M.A. Green, "The ESCAR Refrigeration System", IEEE Transactions on Nuclear Science NS-22, No. 3 (1975).

DESIGN AND ANALYSIS OF THE
SSC DIPOLE MAGNET SUSPENSION SYSTEM

T.H. Nicol, R.C. Niemann, and J.D. Gonczy

Fermi National Accelerator Laboratory
Batavia, IL

ABSTRACT

The design of the suspension system for Superconducting Super Collider
(SSC) dipole magnets has been driven by rigorous thermal and structural
requirements. The current system, designed to meet those requirements,
represents a significant departure from previous superconducting magnet
suspension system designs. This paper will present a summary of the design
and analysis of the vertical and lateral suspension as well as the axial anchor
system employed in SSC dipole magnets.

INTRODUCTION

The suspension system in a superconducting magnet performs two
functions. First it resists structural loads imposed on the cold mass assembly
ensuring stable operation over the course of the magnet's operating life.
Second it serves to insulate the cold mass from heat conducted from the
environment.

The evolution and selection of the suspension system for SSC dipole
magnets has been well documented over the course of the past several
years.[1,2] The purpose of this paper is not to reiterate the selection process,
but rather to give a detailed accounting of the current design, the analysis
used in predicting its performance, and the selection of suspension component
materials.

Figure 1 illustrates the major components of the suspension system. The
magnet assembly is supported vertically and laterally at five places along its
length. To accommodate axial shrinkage during cooldown, the magnet
assembly is free to slide axially at all but the center support. The center
serves as the anchor position. To distribute any imposed axial load to all
five supports, tie bars are used to connect the top of each post to its
neighbor(s). That is, any vertical or lateral load applied to the magnet
assembly is transmitted directly to the supports. An axial load is
transmitted to the center post and in turn to the outboard supports through
the tie bars.

Table 1 provides a summary of both the structural and thermal design
criteria for the suspension system.[3]

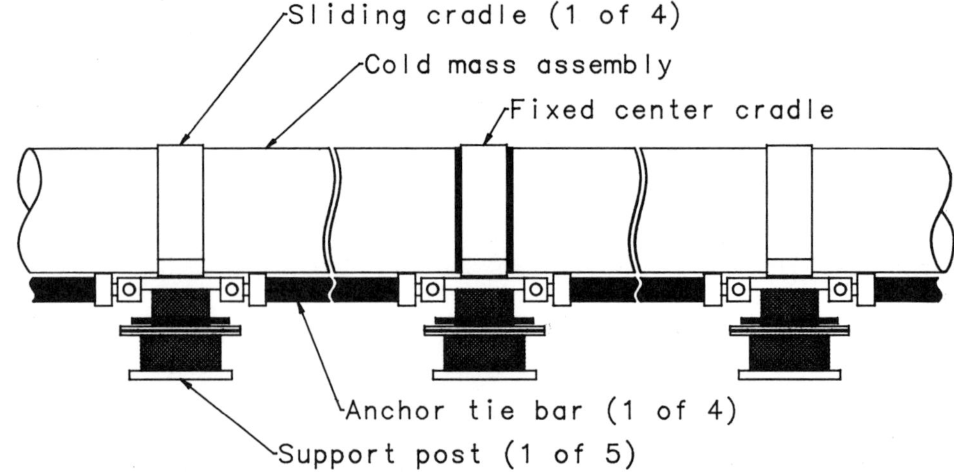

Figure 1. SSC Suspension System Components

SUPPORT POST DESIGN AND ANALYSIS

Figure 2 illustrates a cross section through one of the support posts. Each post assembly consists of inner and outer composite tubes connected by an intermediate stainless steel transition tube. Stainless steel and aluminum discs and rings serve to join the tubes and act as tie points to other cryostat components. The goal of the support post design was to select a geometry and set of materials which resulted in an assembly that satisfied both the structural and thermal design constraints referenced by Table 1.

Structural Analysis

The primary structural loads and directions are shown in Figure 3. F_a denotes an axial load applied to the top of the support post through the anchor attachment point. Axial refers to the long axis of the dipole assembly. F_l and F_v denote lateral and vertical loads respectively and are applied to the post through the cold mass cradle. The lines of action for both pass through the cold mass centerline. Shipping, handling, and seismic loads potentially act in all three directions. Quench loads act as an additional axial load. The weight of the cold mass acts as an additional vertical load although it acts opposite to the direction shown in Figure 3.

Table 1. SSC Dipole Structural and Thermal Load Summary

Shipping and handling loads:	vertical	2.0 G
	lateral	1.0 G
	axial	1.5 G
Seismic load guidelines:	Nuclear Regulatory Guide 1.61 vertical and horizontal spectra scaled by 0.3	
Maximum axial quench load:		25000 lb
Budgeted conduction heat loads per magnet:	80 K	7.20 W
	20 K	0.82 W
	4.5K	0.12 W

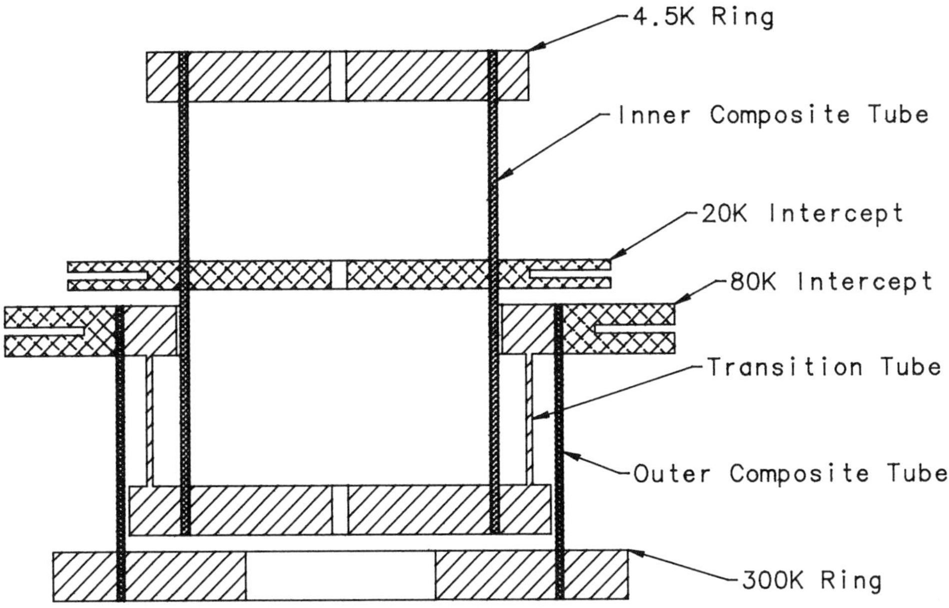

Figure 2. SSC Support Post Cross Section

Experience has shown that the bending loads resulting from the axial and lateral loads produce the highest stresses in the post assembly. Of particular interest are the membrane and shear stresses in the two composite tubes.

Using the notation in Figures 3 and 4 the maximum stresses in tubes 1 (outer) and 2 (inner) due to F_1, F_a, and F_v are

$$\sigma_{i1} = \frac{F_1(L_i)d_i}{2I_i} \quad ; \quad \sigma_{ia} = \frac{F_a(L_i-L_3)d_i}{2I_i} \quad ; \quad \sigma_{iv} = \frac{F_v}{A_i} \qquad (1);(2);(3)$$

$$\tau_{i1} = \frac{2F_1}{A_i} \quad ; \quad \tau_{ia} = \frac{2F_a}{A_i} \qquad (4);(5)$$

where σ_{i1}, σ_{ia}, σ_{iv} = bending stresses resulting from lateral, axial, and vertical loads in tube i

$\quad\quad \tau_{i1}$, τ_{ia} = shear stresses resulting from lateral and axial loads in tube i

$\quad\quad L_i$ = L_1 or L_2

The σ_i's are the stresses acting along the axis of each tube. The τ_i's are the shear stresses acting through the respective cross sections. For thin walled tubes there are three values for limiting stresses induced by F_1, F_a, and F_v. They are the ultimate tensile stress, ultimate shear stress, and the stress which causes elastic instability in the tube wall (local buckling). The ultimate tensile and shear stresses are specified to the tube manufacturer and are used to determine the fiber and resin types and the fiber orientation. The stress which causes elastic instability is determined by the composite material properties and the tube geometry. For a tube like those used in SSC supports, elastic instability will occur whenever[4]

$$\sigma_{ci} > \frac{2E_i t_i}{(1.5)\sqrt{3}\sqrt{(1-\nu_i^2)}d_i} \tag{6}$$

where σ_{ci} = critical bending stress at the onset of elastic
instablility in tube i

E_i = Young's modulus of tube i

t_i = tube thickness of tube i

ν_i = Poisson's ratio of tube i

d_i = diameter of tube i

The overall height and diameter of the support post and the ratios of
the various thermal path lengths are determined in large part by the cryostat
configuration and the conductive heat load constraints. The design
optimization of the complete assembly essentially consists of determining the
composite tube materials and wall thicknesses. Equations (1) through (5) are
set equal to the ultimate tensile and shear strengths and to the elastic
stability constraint represented by equation (6) to determine the optimum
value for the wall thickness. Note that although the wall thickness does not
appear explicit in any of equations (1) through (5), it is implicit in the
expressions for A and I.

As an example, consider the case of some lateral load, F_1 acting on a
post assembly. Using equations (1), (4), and (6) the optimum geometry
would satisfy the more stringent of the following three criteria.

$$\sigma_{i1} = \sigma_{ui}$$

$$\tau_{i1} = \tau_{ui}$$

$$\sigma_{i1} = \sigma_{ci}$$

where σ_{i1}, τ_{i1}, σ_{ci} = stresses defined above

σ_{ui} = ultimate tensile strength for tube i

τ_{ui} = ultimate shear strength for tube i

In order to satisfy all of the various load cases, a computer program
was written to calculate the tube thicknesses required to satisfy the structural
requirements given a set of input criteria. The input consists of the
structural loads, material properties, and fixed geometric parameters. The
output consists of the tube thicknesses which just satisfy the tensile, shear,
and critical stresses above, the resulting stresses, and the resulting thermal
performance.

Table 2 contains an output listing from the optimization program for an
axial load resulting from a magnet quench. Using this set of input
parameters, the resulting thicknesses are 0.109 inches for the outer tube and
0.129 inches for the inner. Both tubes are sized based on the ultimate
tensile strength (SigU). The resulting maximum stresses are 20000 psi in the
outer tube and 30000 psi in the inner. Note that these are exactly equal to
the material ultimates (SigU1 and SigU2) when the ultimates are derated by
the safety factor (SF) indicating that the solution represents an optimum
condition.

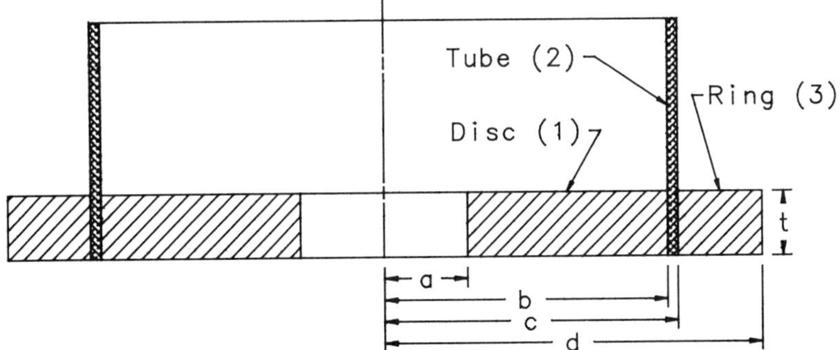

Figure 3. Structural Load and Direction Notation

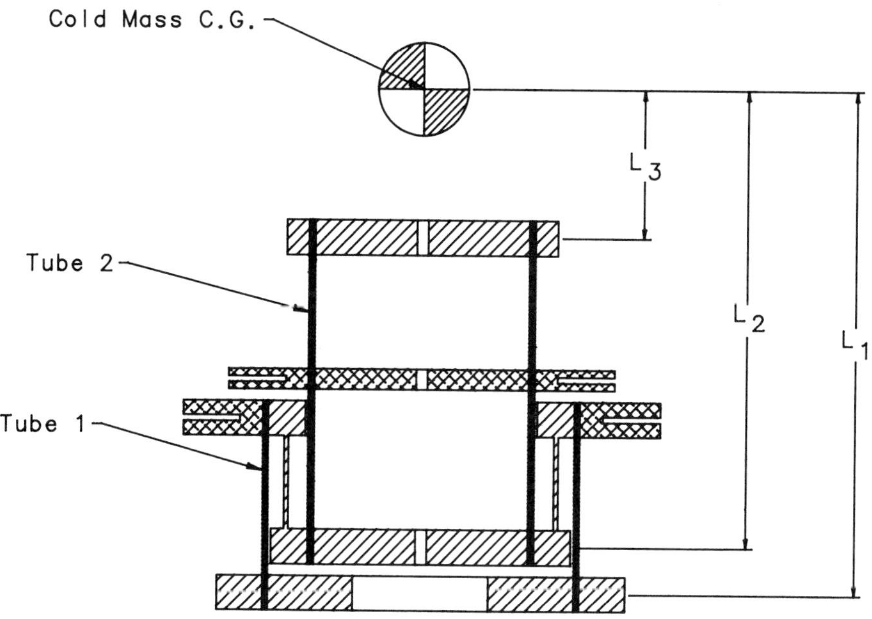

Figure 4. Structural Analysis Notation

Table 2. Summary of Structural Analysis Results

```
********************************************      *******************************
*****  Post Optimization...Input  *****          *****  Resulting Stresses  *****
********************************************      *******************************

Note...........: Design B - Axial Load           Sig1 (when sized for SigU)....(psi):   20000.0
Fg........(lb):         .0                             (when sized for SigEI)...(psi):   22099.4
Fq........(lb):     8500.0                             (when sized for TauMx)...(psi):   27598.7
W.........(lb):    -3150.0
SF............:         2.0                        Sig2 (when sized for SigU)....(psi):   30000.0
E1.......(psi):     .40E+07                            (when sized for SigEI)...(psi):   54121.6
V1............:         .200                           (when sized for TauMx)...(psi):   51044.3
G1.......(psi):     .38E+06
E2.......(psi):     .10E+08                        Tau1 (when sized for SigU)....(psi):    7186.9
V2............:         .200                            (when sized for SigEI)...(psi):    7964.3
G2.......(psi):     .38E+06                             (when sized for TauMx)...(psi):   10000.0
SigU1....(psi):    40000.0
TauU1....(psi):    20000.0                         Tau2 (when sized for SigU)....(psi):    8632.9
SigU2....(psi):    60000.0                              (when sized for SigEI)...(psi):   15930.4
TauU1....(psi):    30000.0                              (when sized for TauMx)...(psi):   15000.0
L1........(in):    14.812
L2........(in):    13.812                         ***********************************************
L3........(in):     6.000                         *****  Resulting Thermal Performance  *****
D1........(in):     7.000                         ***********************************************
D2........(in):     5.000
L300_80...(in):     3.000                         TTop (when sized for SigU).........(K):   11.2207
L80_20....(in):     3.000                              (when sized for SigEI).......(K):    9.4951
L20_Top...(in):     2.375                              (when sized for TauMx).......(K):    9.6465
LTop_45...(in):      .394
ASlide...(in2):      .062                         Q to 4.5K (when sized for SigU)....(W):     .0153
OuterMtl......:         2  G-11 (WARP)                     (when sized for SigEI)...(W):     .0097
InnerMtl......:         3  CARBON FIBER COMPOSITE (UNIAXIAL) (when sized for TauMx)..(W):   .0101
SlideMtl......:         1  304 STAINLESS STEEL
                                                  Q to 20 K (when sized for SigU)....(W):     .3198
********************************************              (when sized for SigEI)...(W):     .1719
*****  Resulting Tube Thicknesses  *****                 (when sized for TauMx)...(W):     .1827
********************************************
                                                  Q to 80 K (when sized for SigU)....(W):    2.1026
T1 (when sized for SigU)....(in):     .1093               (when sized for SigEI)...(W):    2.0182
   (when sized for SigEI)...(in):     .0984               (when sized for TauMx)...(W):    1.5591
   (when sized for TauMx)...(in):     .0782

T2 (when sized for SigU)....(in):     .1287
   (when sized for SigEI)...(in):     .0689
   (when sized for TauMx)...(in):     .0732
```

Thermal Analysis

In addition to understanding the structural performance of the support posts it is critical that an accurate prediction of the conductive heat load be made to each thermal station. Figure 5 illustrates a thermal model of an SSC support post. Q_{80}, Q_{20}, and $Q_{4.5}$ represent the heat loads to the 80, 20, and 4.5K intercepts respectively and are given by

$$Q_{4.5} = \frac{A_c}{L_c} \int_{4.5}^{T_t} \kappa_i \, dT \tag{7}$$

$$Q_{20} = \frac{A_i}{L_2} \int_{20}^{80} \kappa_o \, dT - Q_{4.5} \tag{8}$$

$$Q_{80} = \frac{A_o}{L_1} \int_{80}^{300} \kappa_o \, dT - Q_{20} - Q_{4.5} \tag{9}$$

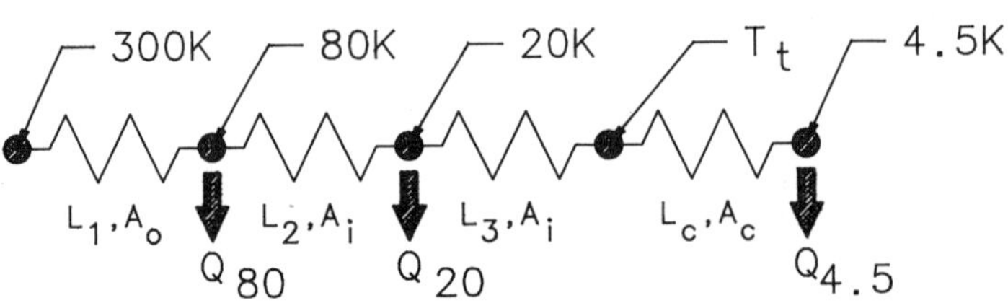

Figure 5. Thermal Analysis Notation

where A_o, A_i = outer and inner tube cross sectional areas

A_c = equivalent cold mass cradle cross sectional area

L_1 = 300K to 80K path length

L_2 = 80K to 20K path length

L_c = equivalent cold mass cradle length

κ_o, κ_i = outer and inner tube thermal conductivities

T_t = temperature at the top of the support post and is found from the steady state solution to

$$\frac{A_i}{L_3} \int_{T_t}^{20} \kappa_i \, dT = \frac{A_c}{L_c} \int_{4.5}^{T_t} \kappa_c \, dT \tag{10}$$

where κ_c = cold mass cradle thermal conductivity

L_3 = 20K intercept to top of support post length

Given a support post geometry and the thermal conductivity integrals for the composite tubes and cold mass cradle, equations (7) through (10) can be solved for the steady state heat loads and the temperature at the top of the post assembly. Again referring to Table 2, the resulting heat loads to 80K, 20K and 4.5K are 2.103, 0.320, and 0.015 W respectively. The equilibrium temperature at the top of the post is 11.2K.

Shrink Fit Joints

Connections between composite tubes and metallic end fittings have historically been made using some form of mechanical fastener or chemical bond. Mechanical fasteners typically introduce unwanted stress concentrations at the joint. Chemical bonds, e.g. epoxy joints, are susceptible to failures caused by differential thermal expansion of the joint components. To avoid these complications and to ensure long term reliability, the composite to metal joints in both the support posts and anchor tie bars are effected by shrink fitting the composite tube between an inner metal disc and an outer metal ring. A typical joint configuration is shown in Figure 6.

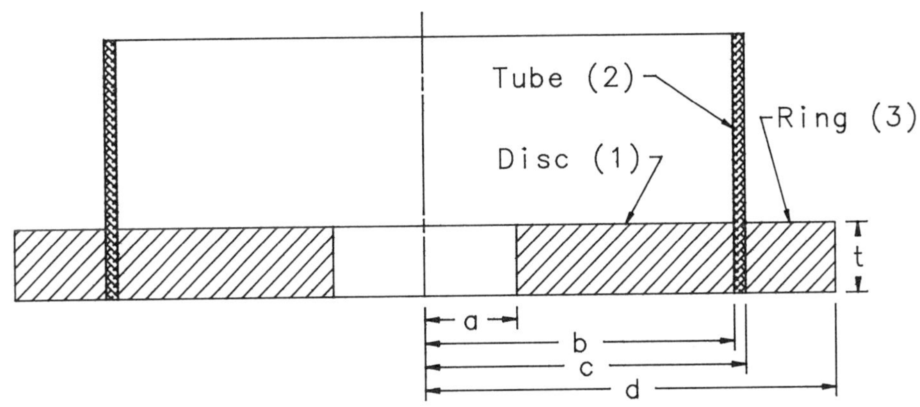

Figure 6.   Shrink Fit Joint Notation

Each of the joints in the support posts resists both axial loads and overturning moments. An axial load is one which tries to pull the joint apart. An overturning moment is one which tries to twist it apart. Using the nomenclature in Figure 6 the forces and moments required to cause the joint to fail are given by

$$F_i = P_i(2\pi bt\mu_i) \qquad ; \qquad F_o = P_o(2\pi ct\mu_o)$$

$$M_i = 4P_i\mu_i b^2 t \qquad ; \qquad M_o = 4P_o\mu_o c^2 t$$

where F_i, F_o = applied forces which induce slippage of the inner and outer interfaces respectively

M_i, M_o = applied moments which induce slippage of the inner and outer interfaces respectively

μ_i, μ_o = coefficient of friction at the inner and outer interfaces respectively

P_i, P_o = inner and outer interface pressures respectively and are given by

$$P_i = \frac{P_o(k_4+k_5)-\delta_o}{k_6} \qquad ; \qquad P_o = \frac{\delta_i k_6+\delta_o(k_1+k_2)}{(k_4+k_5)(k_1+k_2)-k_3 k_6}$$

where δ_i, δ_o = inner and outer interface radial interference fits respectively

$k_1 - k_6$ = constant parameters determined by the joint geometry and material properties and are given by

$$k_1 = \frac{b}{E_1}\left[\frac{b^2+a^2}{b^2-a^2}\right] - \nu_1 \quad ; \quad k_2 = \frac{b}{E_2}\left[\frac{c^2+b^2}{c^2-b^2}\right] + \nu_2 \quad ; \quad k_3 = \frac{b}{E_2}\left[\frac{2c^2}{c^2-b^2}\right]$$

$$k_4 = \frac{c}{E_3}\left[\frac{d^2+c^2}{d^2-c^2}\right] + \nu_3 \quad ; \quad k_5 = \frac{c}{E_2}\left[\frac{c^2+b^2}{c^2-b^2}\right] - \nu_2 \quad ; \quad k_6 = \frac{c}{E_2}\left[\frac{2b^2}{c^2-b^2}\right]$$

where E_1, E_2, E_3 = Young's modulus for the disc, tube, and ring respectively

ν_1, ν_2, ν_3 = Poisson's ratio for the disc, tube, and ring respectively

As with the structural and thermal analysis referenced above, a computer program was written which calculates the required radial interferences at the inner and outer interfaces required to produce a joint that satisfies either a maximum input axial force or overturning moment.

Table 3 contains a listing of the analysis results for a typical shrink fit joint. This particular case is for the joint at the 300K end of the support post. The input overturning moment (MRes) for this example was 90000 in-lb. The force to slip (FSlip) was input as zero. The program calculates the interference fit that satisfies the more stringent of these two parameters. The resulting radial interference is 0.0063 inches. The lowermost portion of this listing contains the resulting radial and circumferential stresses (SigR and SigC) in the disc, tube, and ring.

644

Table 3. Summary of Shrink Fit Analysis

```
*******************************************       *************************************************
*****  Shrink Fit Analysis...Input  *****         *****  Shrink Fit Analysis...Results  *****
*******************************************       *************************************************

Note............................:Design B - 300K Joint  Interference reqd at inner interface...(in):      .0189
Dimension, A..................(in):    1.5000                             outer interface...(in):     -.0127
           B..................(in):    3.3910
           C..................(in):    3.5000           Total radial interference reqd.........(in):      .0063
           D..................(in):    4.5000
           T..................(in):     .7500           Contact pressure at inner interface...(psi):     8696.5
Material property, E1......(psi):    .28E+08                             outer interface...(psi):     8163.3
                   V1...........:     .3330
                   E2......(psi):    .20E+07           Actual force to slip...................(lb):    41690.3
                   V2...........:     .2000                  resisting moment.............(in-lb):    90000.0
                   E3......(psi):    .28E+08
                   V3...........:     .3330                                           Ri         Rm         Ro
Friction coefficient, Mu.........:     .3000                                       ========   ========   ========
Force to slip, FSlip........(lb):        .0           Inner disc...SigR...(psi):        .0    -6744.3    -8696.5
Resisting Moment, MRes...(in-lb):    90000.0                      SigC...(psi):   -21624.2   -14879.9   -12927.7

                                                     Tube........SigR...(psi):    -8696.5    -8423.6    -8163.3
                                                                 SigC...(psi):     8696.5     8423.6     8163.3

                                                     Outer ring...SigR...(psi):    -8163.3    -3320.3         .0
                                                                  SigC...(psi):    33163.3    28320.3    25000.0
```

Material Selections

References have been made throughout the preceding sections to some of
the unique material property issues encountered in this design process. Of
particular interest are the composite materials used in the support post and
anchor tie bar tubes. Until recently the primary structural composite
materials found in superconducting magnets were glass reinforced composites
in an epoxy matrix. Familiar names are G-10, G-10CR, G-11, and G-11CR.
These continue to be excellent choices. They are readily available, have well
characterized structural and thermal properties, are relatively strong, have low
thermal conductivity, and are inexpensive.

Recent years have brought developments in new fibers for use in
composites, some of which offer advantages in terms of strength, some in
terms of thermal conductivity, some in both. It is well known, for example
that graphite composites can offer greater strength and stiffness than their
glass counterparts. Less known is their low thermal conductivity, particularly
at low temperature. Figure 7 is a plot of the thermal conductivity of G-
11CR and a uniaxial graphite composite. Note that at room temperature the
thermal conductivity of G-11CR is four times less than that of the GRP,
however, at approximately 40K the curves cross indicating that at low
temperature the GRP may in fact provide greater resistance to conductive
heat flow.[5]

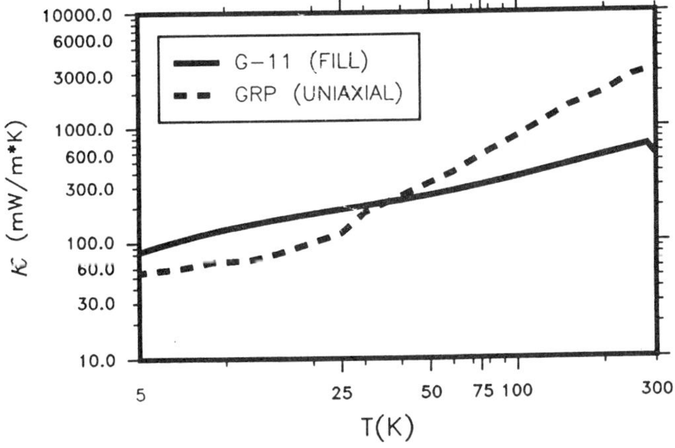

Figure 7. Thermal Conductivity of G-11CR and GRP Tube Material

Table 4.   Comparison of Two Structurally Optimized Support Posts

		FRP Outer GRP Inner	FRP Outer FRP Inner
t, outer	(inches)	0.109	0.109
t, inner	(inches)	0.129	0.201
Q to 80K	(W)	2.103	2.038
Q to 20K	(W)	0.320	0.370
Q to 4.5K	(W)	0.015	0.030

The reentrant design of the support post gave us the option of taking advantage of this behavior. For the outer tube, operating between 300K and 80K, the thermal performance of G-11CR makes it superior to GRP. For the inner tube, operating between 80K and 4.5K, GRP is better.

Table 4 contains the results of the structural optimization described previously and illustrates that the choice of GRP for the inner tube does in fact produce an assembly with lower 4.5K heat load than an assembly which uses G-11CR (FRP) for both tubes.

High strength and low thermal conductivity are not the only considerations for composite materials suitable for use in the support structure. The operating environment, dimensional stability, and fabricability must also be considered. Figure 8 illustrates a sample of a material specification given to potential manufacturers of the composite tubes needed for the suspension system. We have tried to be as specific as necessary without being overly restrictive. The hope being that each manufacturer draw from his own experience in producing a product that meets the needs of the design and which can be mass produced at a reasonable cost.

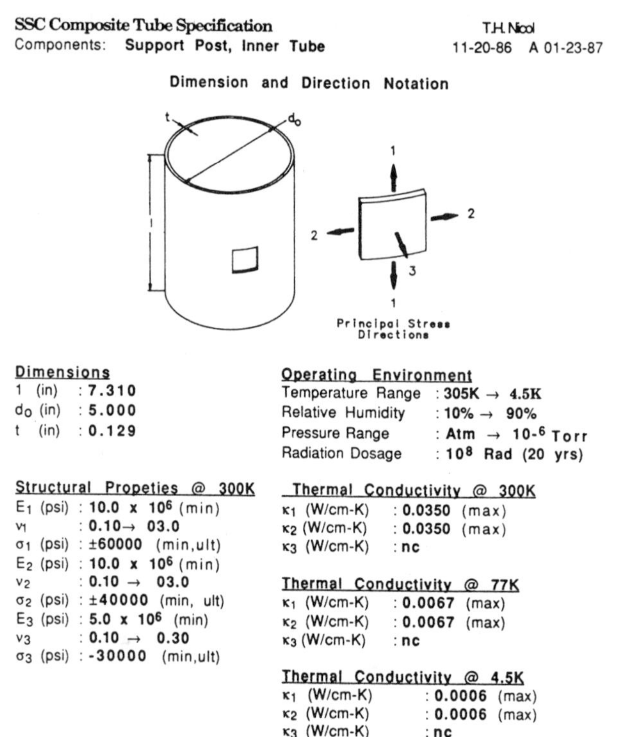

Figure 8.   Sample Composite Tube Performance Specification

ANCHOR SYSTEM

The support posts used in SSC cryostats share vertical and lateral loads induced by shipping, handling, and seismic loads. Thermal contraction of the cold mass assembly during warmup and cooldown necessitates axial sliding between the cold mass and each of the four outboard posts. The center post is attached rigidly to the cold mass assembly to ensure correct axial position within the vacuum vessel. Given no other restraint, this means that the center post would see the entire axial component of any load. A single post is incapable of handling these loads alone. Utilizing a 'strong' post at the center would impose intolerable heat loads on the cryogenic systems.

Ideally one would like an anchor system with negligible thermal impact on the cryogenic systems and which introduced no perturbations into other cryostat components. Recognizing that the bending strength of all five posts could be combined to effectively act as a single axial restraint, we have chosen to connect the 4.5K ring of each post to that of each adjacent post with axial tie bars.

To understand the effectiveness of such a scheme, we must understand the degree to which an axial load applied at the center post is shared by the remaining four posts. Specifically, we need to know the reaction forces at the top of each post, given a force applied at the center. Figure 9 is an equivalent spring diagram of the post and tie bar connections used to determine the stiffness of the total system. Points 1 through 5 represent the tie bar attachment points, point 3 being the top of the center post. The k_p's represent the bending stiffness of each support posts. The k_b's represent the axial stiffness of each tie bar. Grounded points represent the 300K attachment of each post to the cryostat vacuum vessel. The analysis consists of breaking the system into a series of equivalent k's for each post and tie bar connection. These equivalent k's are then used to calculate the total stiffness of the system as shown below.

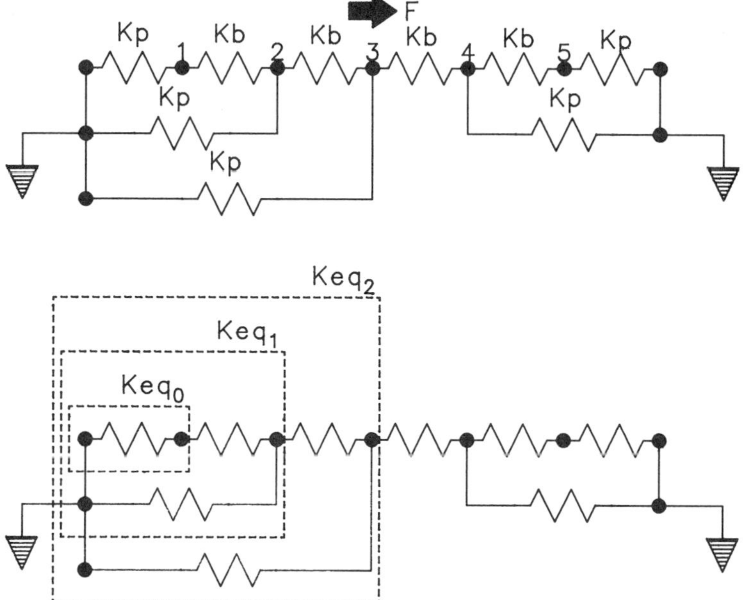

Figure 9. Equivalent Spring Diagram of the Suspension System

$$k_{eq_n} = k_p + \cfrac{1}{\cfrac{1}{k_{eq_{n-1}}} + \cfrac{1}{k_b}} \qquad n = 1..m$$

$$k_{total} = k_p + \cfrac{2}{\cfrac{1}{k_{eq_{m-1}}} + \cfrac{1}{k_b}}$$

where

k_{eq_n} = equivalent stiffnesses for each connection as shown on Figure 9

k_{eq_0} = k_p

m = (number of k_b's - 1) / 2

k_{total} = total axial stiffness

For the SSC suspension system

$$m = 2$$

$$k_p \simeq 35000 \ lb/in$$

$$k_b \simeq 200000 \ lb/in$$

resulting in

$$k_{total} \simeq 133000 \ lb/in$$

If the anchor system were 100% efficient the axial stiffness of the complete suspension system would equal 175000 lb/in (35000 lb/in x 5 posts). Given the total stiffness above, the overall anchor efficiency is 76%. Efficiency is defined as the ratio of the actual total stiffness to the sum of the stiffnesses of all five support posts.

Each discrete k_{eq} is used to determine the displacement at the top of each post resulting from the applied force, F. The product of each displacement and k_p yields the reaction force at each post. The load distribution is determined by these individual reactions. Given the above parameters, the center post in an SSC dipole carries 26.4% of the axial component of any induced load. Each of the two posts adjacent to the center carry 19.9% and the outermost two carry 16.9% each.

One final consideration remains in the design of the anchor system and relates to the material selection for the tie bars. Materials commonly selected for use in superconducting magnet cryostats; glass composites, stainless steel, and aluminum, for example, all exhibit decreases in length when cooled from 300K to 4.5K. Selecting such a material for the anchor tie bars would result either in very high tensile loads in the tie bars themselves or very high bending moments in the post assemblies because of the shrinkage which occurs during cooldown. Unlike most materials, however, graphite fibers exhibit a negative coefficient of thermal expansion meaning that they grow when cooled. In most graphite composites this effect is masked by the resin system which shrinks upon cooldown, particularly if the bulk of the fibers are oriented off-axis from the measurement direction.

By pultruding or filament winding graphite fibers with epoxy resin one can create a composite tube with an expansion coefficient from 300K to 4.5K of roughly -0.03% depending on the fiber content and on the fiber and epoxy used. Further, by attaching metal fittings to each end of the tube which shrink upon cooldown one can produce a tube assembly with no net expansion or contraction over the prescribed temperature range. For example, a composite tube 120 inches long, with an expansion coefficient of -0.03% will grow 0.036 inches when cooled from room temperature to 4.5K. Two stainless steel ends, 6 inches long, with an expansion coefficient of 0.3%, shrink 0.018 inches each resulting in a net change in length during cooldown, for the assembly, of zero.

The tie bars used in the present configuration are 120 inches long tubes with an outside diameter of 2 inches and a wall thickness of 0.25 inches. The material is filament wound using graphite fibers in an epoxy matrix. The elastic modulus of the composite is 18×10^6 psi along the axis of the tube. The measured contraction of a complete tie bar when cooled from room temperature to 77K is less than 0.001 inches.

CONCLUDING REMARKS

The suspension system for SSC magnets has evolved toward its current configuration as the result of much analysis and many design iterations, some based on established practice, others developed as complete new concepts. The system described here is a combination of old and new. Support posts in some form have been used for many years, but none, to our knowledge, have been developed into devices which afford such low heat loads and high structural stiffnesses and strengths. Further, none have been developed using shrink fit bonds at all composite to metal joints nor have they played such an integral role in the anchor system performance.

The current suspension system design meets the static structural requirements set forth for SSC magnets and exceeds the required thermal performance at 4.5K. It constitutes an assembly which requires a minimum of added perturbations to other cryostat components and lends itself well to easy fabrication and mass-production. Hopefully it is a design which serves the needs of the SSC and the needs of future applications.

REFERENCES

1. T.H. Nicol et al, A suspension system for superconducting super collider magnets, in: "Proceedings of the Eleventh International Cryogenic Engineering Conference," Butterworths, Surrey, UK (1986).
2. T.H. Nicol et al, SSC magnet cryostat suspension system design, in: "Advances in Cryogenic Engineering" Vol. 33, Plenum Press, New York (1988).
3. SSC Central Design Group, "Superconducting Super Collider Magnet System Requirements," SSC-100, October 1986.
4. R.J. Roark and W.C. Young, "Formulas for Stress and Stain," Fifth Edition, McGraw Hill, New York (1975), p. 428.
5. M. Takeno et al, Thermal and mechanical properties of advanced composite materials at low temperatures, in: "Advances in Cryogenic Engineering" Vol. 32, Plenum Press, New York (1986).

SSC SUPERCONDUCTING DIPOLE MAGNET CRYOSTAT MAGNET CRYOSTAT MODEL STYLE B CONSTRUCTION EXPERIENCE

N.H. Engler, R.C. Bossert, J.A. Carson, J.D. Gonczy
E.T. Larson, T.H. Nicol, R.C. Niemann, D.Sorenson, and R.Zink

Fermi National Accelerator Laboratory
Batavia, Illinois 60510

ABSTRACT

A program to upgrade the full scale SSC dipole magnet cryostat model function and assembly methods has resulted in a series of dipole magnets designated as style B construction. New design features and assembly techniques have produced a magnet and cryostat assembly that is the basis for Phase I of the SSC dipole magnet industrialization program. Details of the assembly program, assembly experience, and comparison to previous assembly experiences are presented. Improvements in magnet assembly techniques are also evaluated.

INTRODUCTION

The SSC dipole magnet development program includes the assembly of a premanufactured coil and cold mass sub-assembly built at Brookhaven National Laboratory into a cryostat system that was manufactured at Fermilab.[1]

The concept of different manufacturers each building a major component of the magnet which could be assembled at any site is one of the goals of the program.

A full scale dipole magnet designated model A was manufactured and assembled as a second step in the development of the final design.[2]

The complete magnet contains all of the magnet support systems, anchor system, thermal isolation shield, insulation, expansion joints, vacuum vessel, magnet interconnection devices, and a complete alignment system.

COMPLETE MAGNET

The complete magnet has been previously described in detail[3] and only its major features are presented herein. The magnet general arrangement is as shown by Fig. 1. and Fig. 2.

Cold Mass Assembly

The cold mass assembly consists of the beam tube, correction elements, collared coils, laminated iron yoke, and outer helium containment shell. The cold mass components are joined together forming a leak tight and structurally rigid assembly.

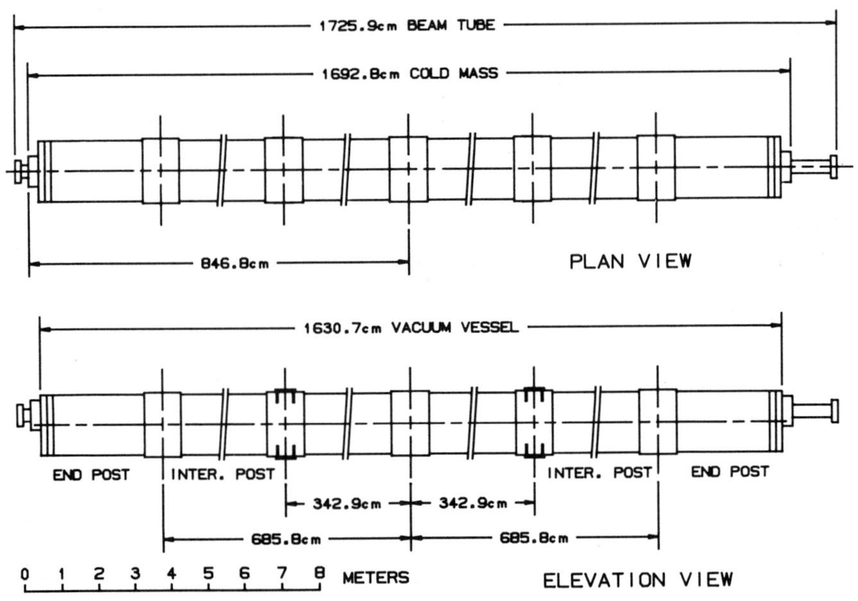

Fig. 1.   Cryostat plan and elevation views.

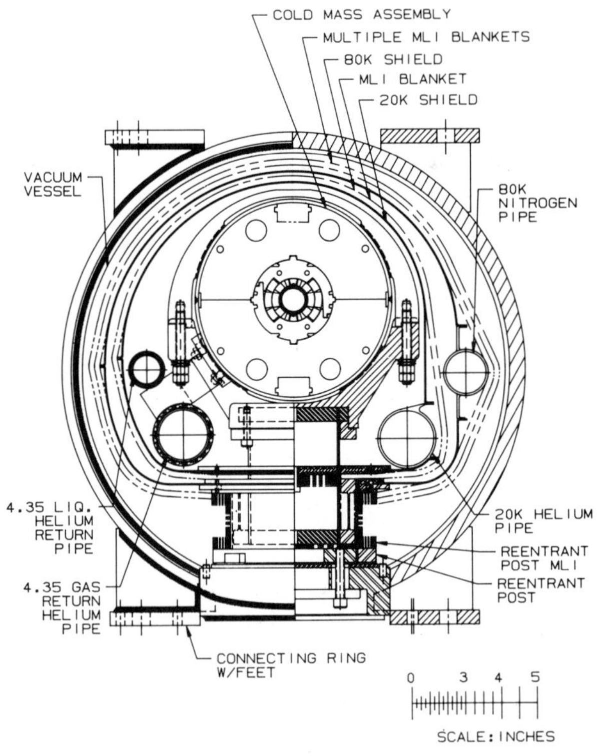

Fig. 2.   Cross Section through a support point.

<u>Cryogenic Piping</u>

The cryostat assembly contains all piping that interconnects the magnet refrigeration system throughout the circumference of the accelerator rings. A five pipe system is employed for cryogenic and safety reasons.

<u>Thermal Shields</u>

Thermal shields maintained independently at 20K and 80K surround the cold mass assembly. The shields absorb the radiant heat flux and provide heat sink stations for the suspension and anchor system. The shields are supported by and thermally connected to the cold mass assembly supports.

<u>Insulation</u>

The insulation system consists of multi-layer assemblies of aluminized polymer film with spacers fabricated and installed as blankets on the cold surfaces.[4] One blanket is installed on the cold mass circumference, two blankets are installed on the 20K shield, and four blankets are installed on the outer 80 K shield.

<u>Suspension System</u>

The cold mass and shields are supported relative to the vacuum vessel at five points by the suspension system.[5] The system incorporates a reentrant type support post. The outer tube is manufactured from a fiberglass reinforced epoxy composite. The inner tube is manufactured from a graphite reinforced epoxy composite. The support system also includes attached metalic and connections and heat intercepts that connect to the thermal shields.

<u>Anchor System</u>

The five support posts share vertical and lateral loads. The center post is rigidly attached to the cold mass to establish axial position.

<u>Vacuum Vessel</u>

The vacuum vessel provides containment for insulating vacuum and the cold mass connection to ground. The magnet assembly procedure incorporates insertion of the complete subassembly into the vacuum vessel.

<u>Interconnections</u>

Mechanical and electrical interconnections between adjacent magnets are required at the magnet ends.[6] It is essential that the connections be straight forward to assemble and disassemble, compact, and reliable. The design facilitates assembly and disassembly operations in the SSC tunnel.

ASSEMBLY

The major components of the dipole cryostat, i.e., the magnet proper in its helium containment vessel(cold mass), cryogenic piping, heat shields and insulation are located and supported by reentrant posts connected to the tow plate. This major subassembly, complete with the anchor system and heat intercepts, is inserted into an outer vacuum vessel.

Cold Mass Subassembly

As a first step in final magnet assembly, the pre-manufactured magnet cold mass is located on an assembly station, and slide cradles with their supporting subassembly, are installed. Next the five assembled support posts are installed with the cold mass supported on a locating fixture. The anchor system is installed from the center position outward in two directions interconnecting the five support posts. A single blanket of insulation material is installed around the circumference of the cold mass. After final installation of the helium return piping, the lower half of the 20K shield is installed and attached to the support system. Thermal insulation blankets are installed on the 2OK heat shield and heat conductive straps are connected to the shield. The upper half of the 20K shield is installed and the insulation is completed. The same procedure follows for the installation of the 80K shield. These steps are completed on a fixtured table that contains the cold mass slide that positions and maintains the integrety of the cold mass subassembly prior to insertion into the vacuum vessel. The arrangement is as shown in Fig. 3.

Vacuum Vessel

The vacuum vessel is fabricated from prefabricated, full section reinforcing rings connected by short sections of tubing. Two of the five support rings have external magnet mounting feet attached to them. The five support positions are mounted on a vacuum vessel assembly fixture which controls the position of the rings. Short links of tubing are used to interconnect and fix the relative ring positions. The machined surfaces of the feet are perpendicular and parallel to related surfaces of the support posts. This arrangement assures alignment of the center of the beam tube in the cold mass with respect to the outside of the vacuum vessel. End rings to support vacuum bellows and automatic welding equipment are then welded to the vessel to complete the assembly. The inspected assembly is then transferred to the final assembly station where the alignment and support tray is installed. A cross section through a support point is shown in Fig. 4.

Final Assembly

The cold mass subassembly and vacuum vessel are lined up, and a slide and pulley system draws the cold mass subassembly into the vacuum vessel. Fig. 5. Precision fixture points, together with the positive location of the support post feet, allow the cold mass to be located properly relative to the vacuum vessel. The assembly is then clamped and welded together as a unit. An optical survey is taken to assure the correct alignment of all components. The vessel is then inspected, given a final leak check, painted and prepared for protection from the weather. The assembly is now ready for shipment to the testing and measuring operations.

Factory Certification Tests

The tests required to certify each dipole for SSC acceptance, before it leaves the factory or factories, include several types and are performed at various stages of the fabrication and assembly process. These tests, conducted at room temperature, are summarized.

    (a) Electrical Measurements and Tests
        1. Resistance, inductance of each of the four windings
            and of the complete magnet.
        2. Hi-pot between windings, between windings and collars,
            and between coil and bore tubes to 2-3 kV.

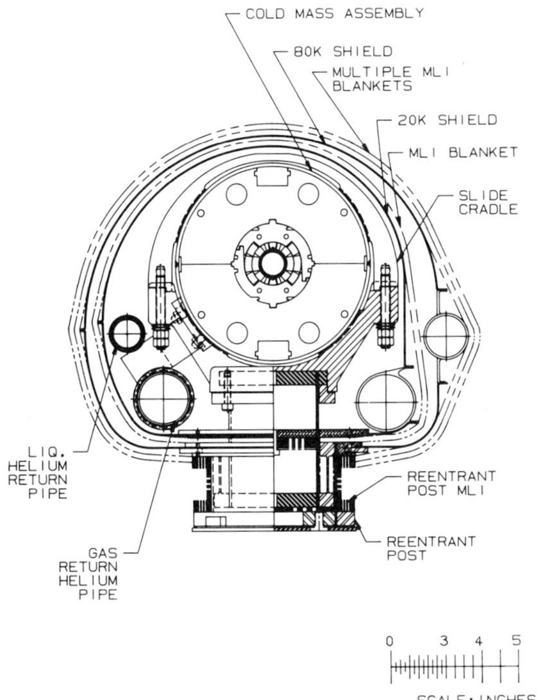

Fig. 3. Cold mass assembly.

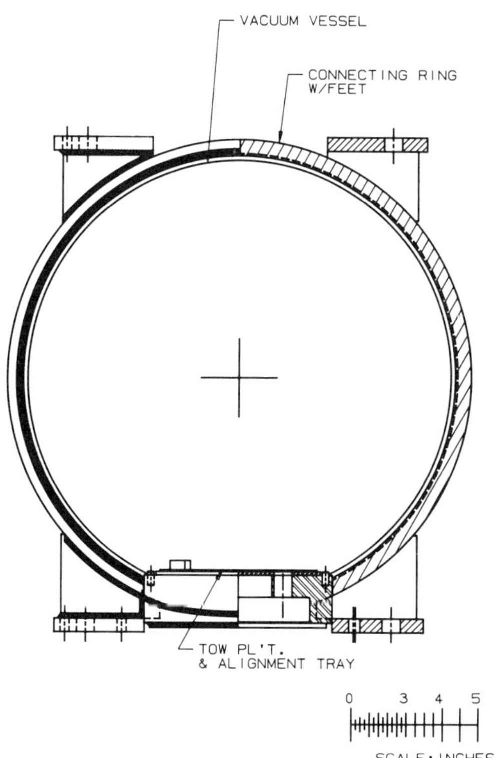

Fig. 4. Vacuum vessel subassembly.

(b) Magnetic Measurements

All magnets will have field measurements made at room temperatures. The current will be limited to +15A, and the following quantities will be measured:
B_o, a_n, and b_n $n=1$ to 6.

The field at 15 A is about 150 G.

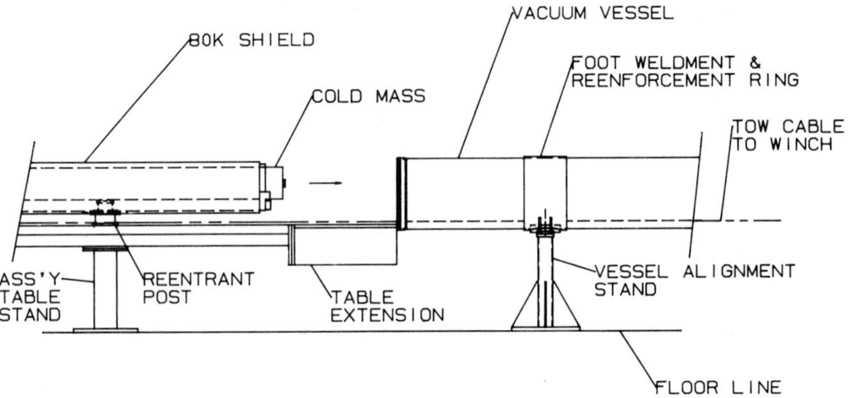

Fig. 5. Magnet final assembly fixture.

RESULTS

The completion of the Type A cryostats gave us vaulable experience to improve the assembly operation in the style B magnet assemblies. Improved assembly fistures, tooling, and measurement techniques assured a consistent magnet assembly that would duplicate itself with subsequent magnets. Specific improvements were noted in the following areas.

Cold Mass Subassembly

Cold mass assemblies were received from Brookhaven National Laboratory that had a consistent o.d. and surface finish that allowed a new slide support system to be incorporated. The new anchor system eliminated many holes cut into the thermal shields and assisted in the application of insulation on the shields. New helium piping supports were designed that permitted installation of tubing as a completed subassembly. The 20K and 80K heat shields were reconfigured to allow easier assembly and easier connection to the thermal intercepts. The ability to incorporate automatic spot welding of shields allowed for less distortion and interferences of the total assembly. The insulation blankets were made on a production style assembly fixture that gave us consistent, repeatable blankets that fit with the minimum of hand fitting on the magnet itself.

Vacuum Vessel

The new tooling that aligned the five support points and two mounting feet together with design modifications that allowed the use of automatic welding equipment greatly reduced the effort required to furnish an accurate vacuum vessel. Assembly techniques of the support rings, slide plate, and slide trays provided for the alignment of the cold mass subassembly with a minimum of special handling.

Alignment Reference

The new assembly system allowed the technicians to use the inside of the bore tube to verify the integrety of the magnet. Survey techniques were developed and improved to assure that the cold mass was in the proper relation to the mounting feet and support rings in the vacuum vessel.

IMPROVEMENTS

All of the systems will benefit from ongoing tooling and fixture upgrades. Redesign of parts for increased operating clearances, welding equipment clearances, and final assembly clearances will become part of the design C iteration and will result in an improved assembly process.

CONCLUSIONS

The design B magnet is functional and produces a satisfactory SSC dipole assembly. Automatic welding is a major time saver and must be used wherever possible. The magnet is of a satisfactory quality to allow the Phase I industrialization program to go forward.

Acknowledgements

The authors gratefully acknowledge the sincere interest, contribution, cooperation, and professional performance of the Fermilab design, production, procurement, quality assurance, and engineering test groups.

These efforts resulted in an assembled SSC dipole magnet that proved many new principles of design and assembly techniques. The work was completed on schedule, and ddelivered to the Fermilab magnet test facility for further evaluation.

The work as presented was performed at Fermilab National Accelerator Laboratory which is operated by Universities Research Association, Inc. under contract with the U. S. Department of Energy.

REFERENCES

1. SSC Central Design Group, "Conceptual Design of the Superconducting Super Collider," SSC-SR-2020, March, 1986.

2. Engler, N.H.,et al, "SSC Dipole Magnet Model Construction Experience," presented at the 1987 Cryogenic Engineering Conference, June 14-18, 1987, St. Charles, Illinois.

3. R.C. Niemann, et al, "Second Generation Superconducting Super Collider Dipole Magnet Cryostat Design," Annual Energy Souces Technology Conference, January 23-25, 1989 in Houston, Texas.

4. Ohmori, T., et al, "Thermal Performance of a Candidate SSC Magnet Thermal Insulation Systems," Adv. Cryo. Engr. Vol. 33, pp 323,(1987)

5. Nicol, T.H., et al., "SSC Magnet Cryostat Suspension System Design," Adv. Cryo. Engr., Vol. 33, pp 227, (1987).

6. R.A. Bossert, et al, "Superconducting Magnet Interconnections," International Industrial Symposium On The Super Collider Conference February 8-10, 1989, in New Orleans, Louisiana.

A FINITE ELEMENT ANALYSIS OF AN SSC DIPOLE MAGNET
(NC-9 CROSS-SECTION)

Michael S. Chapman

SSC Central Design Group*
c/o Lawrence Berkeley Laboratory
Berkeley, CA 94720

Robert H. Wands

Fermi National Accelerator Laboratory*
Batavia, IL 60510

Abstract

Finite element methods are used to calculate the mechanical behavior of
an SSC superconducting dipole magnet under different loading
conditions. A two-dimensional model of the NC-9 design (aluminum
collars) has been developed and used to calculate the transverse
deflections and stresses in the dipole after assembly of the magnet,
cooldown to 4.2 K, and energization to 6.6 T. Verification of the results
with experimental measurements and observations, and limitations of
the analysis, are also discussed.

Introduction

The Superconducting Super Collider (SSC) is a proposed high-energy synchrotron
accelerator designed to achieve proton-proton collisions in the 20-TeV range. The heart
of the machine will consist of two concentric rings, each over 50 miles in circumference,
through which counter-rotating proton beams will be accelerated to velocities approaching the speed of light and brought into collision at interaction regions where detectors
will record the energy and matter created in the collision. To direct the beam, each
particle storage ring will require approximately 4000 superconducting dipole magnets
with a central field of 6.6 T at 4.2 K. The dipoles, which are fully described in Reference 1,
consist of superconducting coils, collars, yokes, and shells (collectively referred to as the
"cold mass," Figure 1) assembled and placed in an insulating cryostat. Each dipole is to
be 17 meters long and will have a cold mass cross section 27 cm in diameter.

---

* Operated by the Universities Research Association, Inc., for the U. S. Department of
Energy.

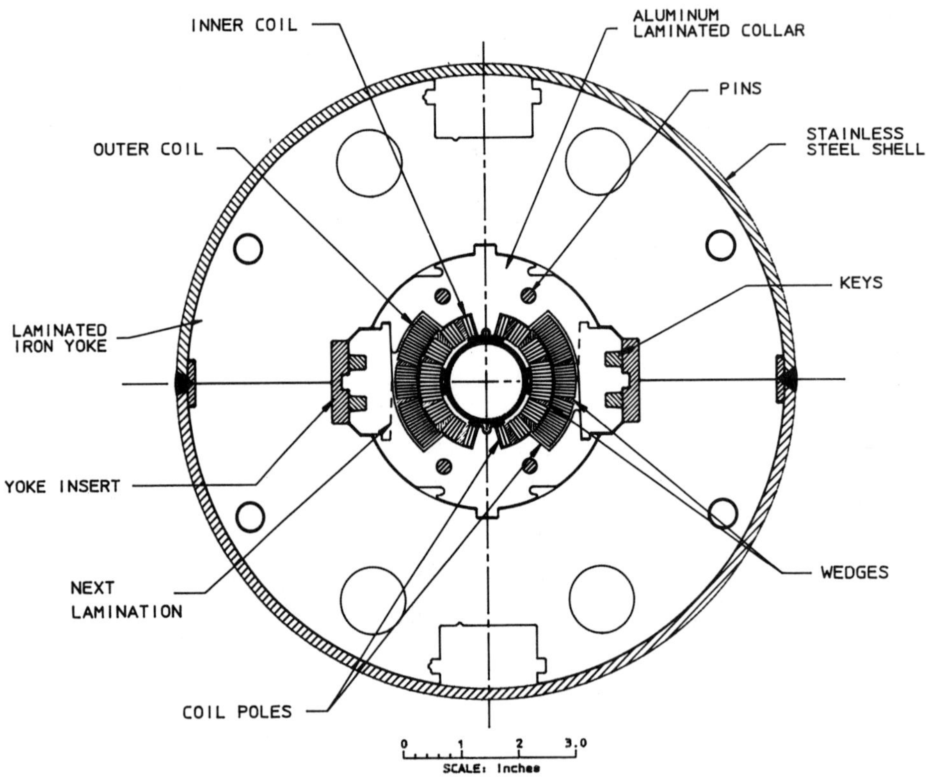

Figure 1. Cross section of the SSC dipole "cold mass" (NC-9 design).

A successful dipole must create a sufficiently uniform central magnetic field to ensure stable beam orbits and must operate at the design field (6.6 T) without quenching (a rapid loss of superconductivity in the magnet conductor). To achieve these objectives, it is necessary to minimize coil motion during operation due to the Lorentz forces acting on the conductors. Excessive coil motion can distort the magnetic field and can cause enough frictional heat generation between conductors to initiate a quench. The NC-9 dipole design, which is one of two designs currently being considered for use in the SSC, uses laminated aluminum collars tightly clamped around the coils and supported by the yoke. The collars restrain coil motion and provide azimuthal precompression to the coils, which prevents minute strand-to-strand sliding in the conductor.

Here, we use the ANSYS finite element code (a trademark of Swanson Analysis Systems) to examine the adequacy of the support provided by the collars and yoke, and, in particular, we calculate the deflections and stresses in the dipole after assembly of the magnet, cooldown to 4.2 K, and energization to 6.6 T.

The Model

The finite element model created for the analysis of the NC-9 cross section is shown in Figure 2 with some portions of the collar shown separately for clarity (see below). The model uses two-dimensional isoparametric solid quadrilateral and triangular plane stress elements to model the individual components of the dipole, as well as compression-only "gap" elements that permit separation and frictionless sliding between components. The bilinear stiffness of the gap elements necessitates the use of iterative solution methods and causes the mechanical response of the system to be nonlinear.

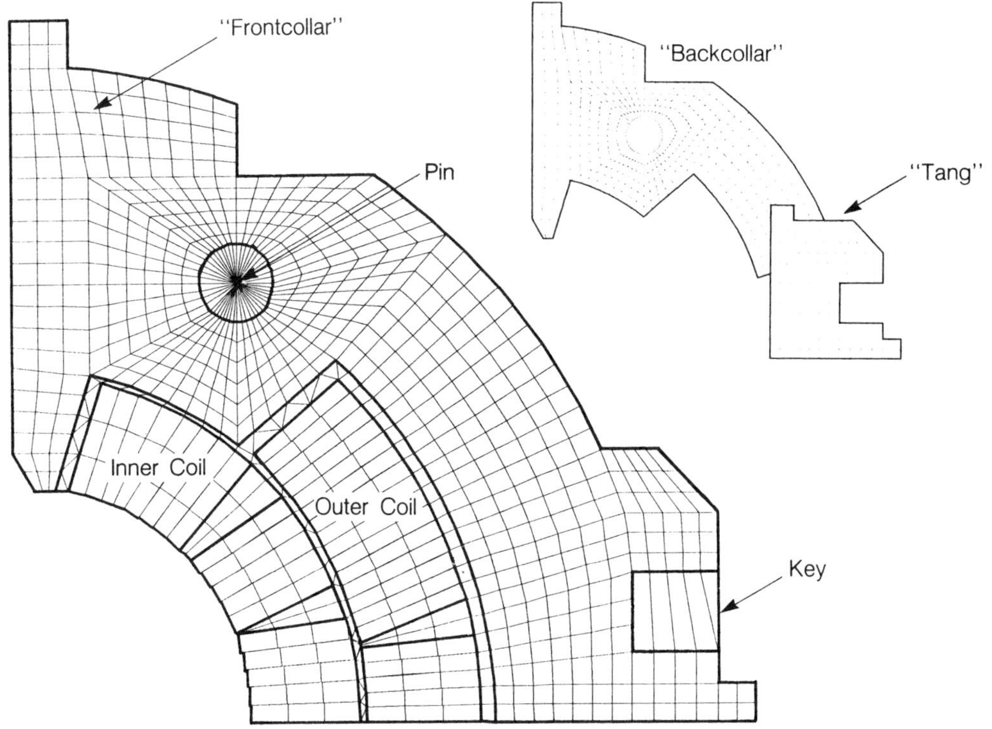

Figure 2. Finite element model of the NC-9 dipole. The "back collar" and "tang" pieces represent the next lamination longitudinally and are superimposed over the "front collar" piece.

The coils are modeled as individual turns of insulated superconducting cable and copper wedges. Each turn has three elements through the radial thickness and one element through the azimuthal thickness. For the majority of the calculations, the coils are treated as elastic, and isotropic, with a modulus of elasticity of 1.5×10^6 psi. This represents an approximation to the actual physical properties of the coil, which are very complicated due to its composite structure (multiple-strand twisted cable wrapped with Kapton insulation and epoxy-impregnated fiberglass). The calculated coil stresses, therefore, should be considered only qualitatively.

The collar assembly consists of alternating collar laminations tightly clamped together around the coils. The basic unit of the interlocking assembly consists of four collar pieces (two successive laminations), arranged as shown in Figure 1. The collars are interlocked with inserted keys and pins. In the analysis, the collars are modeled in three separated pieces: "front collar," "back collar," and "tang," which, when appropriately constrained, permit treatment of the problem in one quadrant. The models directly include the keys, and pins, as well as all specified dimensional clearances between the different components. The yoke/skin assembly has not been included in the finite element model because the yoke/collar contact point for the NC-9 design is well defined at the horizontal midplane, which permits treatment of the yoke as a boundary condition on the collars.

Loadings studied include the stresses and deflections after assembly ("assembly loading"), insertion into the yoke, cooldown to 4.2 K, and energization to 6.6 T. These loadings will be discussed separately in the following sections.

Assembly Loading

The first loading of the dipole examined is the "assembly loading" and represents the prestressed state of the collars and coils after collaring. The collaring operation, during which the collars are tightly clamped around the coils and interlocked together with inserted keys, produces compressive azimuthal stresses in the coils and counteracting tensile azimuthal stresses in the collars due to designed dimensional interferences. The amount of interference, and thus the coil prestress, can be controlled with shims and is typically chosen to produce average azimuthal stresses of 8000 psi in the inner coil and 6000 psi in the outer coil. This is modeled in the calculation by applying a vertical pressure or equivalent displacements at the coil midplane.

The assembly loading causes the collars to deflect from their original geometry. The calculated collar deflections listed in Table 1 show good agreement with measurements on actual NC-9 dipoles. The ability of the model to accurately predict the collar deflections for this type of loading provides first-order verification that collar structure is adequately modeled.

Longitudinal pins are used to align successive collar laminations and constrain them to move together. The constraint provided by the pins is not completely rigid, however, and there is significant relative motion between consecutive laminations (Figure 3). This motion is particularly significant at the outer-coil pole, where there is a

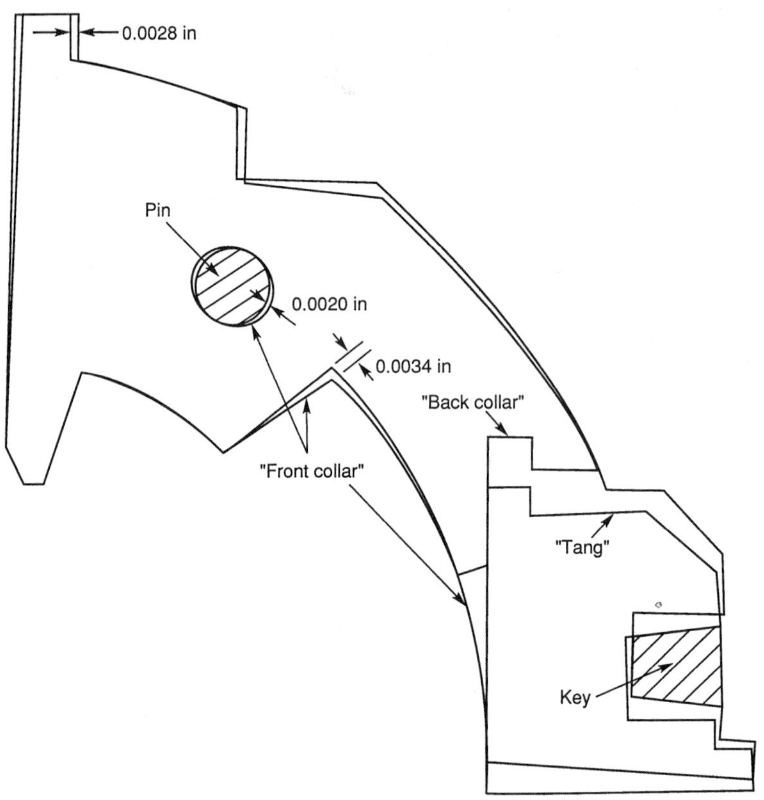

Figure 3. Relative motion between successive collar laminations due to assembly loading.

Table 1.   Calculated and Measured Collar Deflections After Assembly.

	(Change in collar diameter $\times$ 0.001 inch)	
	Vertical Deflection	Horizontal Deflection
Finite element calculation	29.0	2.0
Measured values	29.0–35.0	0.0–5.0

~0.003 inch relative motion between consecutive laminations such that the outer-coil pole is only fully contacted by every other collar lamination. Experimental confirmation of this result has been provided by measurements on magnets using pressure-sensitive paper and by observations on actual magnet cross sections. This result raises doubts about the adequacy of the outer-coil pole support and suggests a possible cause of the outer-coil training seen in the majority of NC-9 model magnets. A magnet will be tested in the near future employing a 0.060-inch thick stainless steel shim at the outer coil pole designed to "bridge" across the non-contacting lamination. It is hoped that this will make the support of the outer coil more uniform from lamination to lamination.

The calculated collar stresses for the assembly loading indicate areas of high tensile stress at the keyways and the outer pole. The areas of high stress are very localized and are caused by geometrical stress raisers. These stress raisers have been analyzed separately and have been found to be sufficiently localized to not compromise the performance of the collars.[2]

Insertion into the Yoke/Horizontal Force Loading

The NC-9 design calls for a room-temperature interference fit between the collared coil assembly and the yoke at the collar keyways on the order of 0.015 inch on the diameter. Because of this dimensional interference, it is necessary to compress the collared coil to insert it into the yoke. In the analysis, the insertion into the yoke is modeled by the application of an inward horizontal force to the collars at the keyways. The model calculates an inward collar deflection of 0.0027 inch on the diameter for an applied force of 1000 pounds. This compares very well with measured values of 0.0025 inch for the same loading, providing additional evidence that the collar stiffness is accurately modeled. The distortion of the coils produced by this loading introduces significant bending stresses into the coil. The change in azimuthal stress in the inner coil due to this loading is shown in Figure 4. Although the average stress at the coil pole remains largely unchanged, a large stress gradient is developed across the pole. For an applied force of 5000 pounds (which is closer to the actual loading applied to the magnets) the increase in stress at the inner radius of the inner coil is 6000 psi. As the inset to Figure 4 indicates, the stress gradients developed in the coil due to the horizontal force are analogous to those developed in a thick cylinder under similar loading.

Cooldown

As the magnet is cooled to 4.2 K, differential thermal contraction between various materials in the dipole produce changes in stress in the assembly. For the NC-9 design, aluminum collars are used instead of stainless steel collars in an effort to minimize coil prestress loss during cooldown.

Calculation of the change in coil stress during cooldown is very sensitive to the specified coefficient of thermal contraction and the specified change in the modulus of elasticity with temperature for the coils. There are few measurements of these properties for the SSC coils, but it is generally thought that the coefficient of thermal contraction for the coils is similar to that for aluminum and that the coil modulus either increases (perhaps up to 4.0×10^6 psi) or remains constant at 1.5×10^6 psi. The calculation uses a thermal contraction coefficient identical to that for aluminum and a coil modulus that does not change with temperature.

CHANGE IN COIL STRESS DUE TO 1000 LB HORIZONTAL FORCE

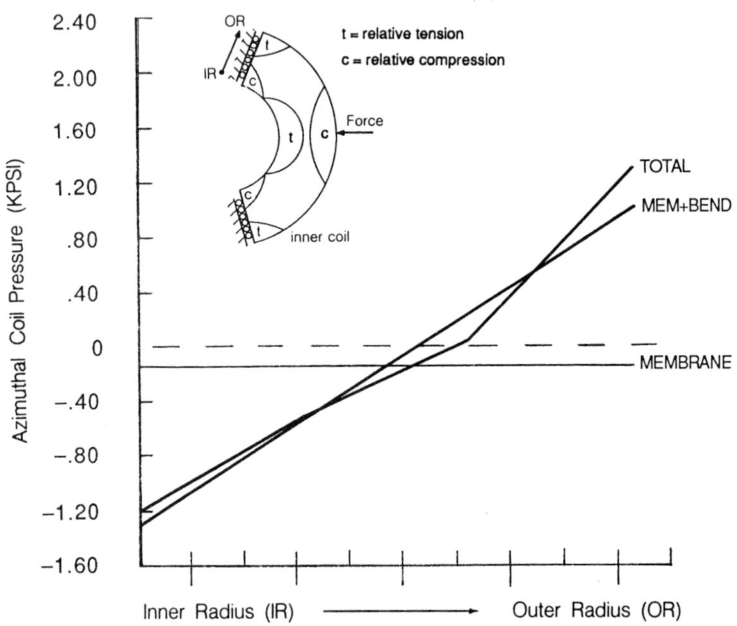

Figure 4. Change in inner coil azimuthal stress at the pole due to a 1000 pound inward force applied at the horizontal collar midplane. TOTAL refers to the actual calculated stress change; MEMBRANE represents the average stress change across the pole; MEM+BEND represents a linear fit to actual calculated stress change (A positive stress change indicates a loss in compressive stress).

As expected, the model calculates a negligible change in coil stress during cooldown using the above-specified properties. The change in coil stress is measured in magnets using pressure gauges placed at the coil poles. Two different types of gauges are used to measure coil stress: one developed at Brookhaven National Laboratory (BNL)[3] and one developed at Lawrence Berkeley Laboratory (LBL). Measurements of the change in coil stress during cooldown vary from an increase of 1500 psi for BNL-type gauges to a decrease of 1500 psi for LBL-type gauges. It is hoped that the inconsistency in the measurements will be resolved in future magnet tests.

664

Lorentz Loading

The magnetic field in the dipole produces large Lorentz forces on the current-carrying conductors. These forces have a radial outward component, which tends to expand the collars horizontally, and an azimuthal component directed toward the coil midplanes, which tends to decrease the coil prestress at the pole. Reducing collar deflections due to these forces and maintaining coil prestress at the poles are two very important elements of dipole mechanical design.

The finite element model is used to calculate the loss in prestress at the coil poles and collar deflections at different magnet currents. Energization is simulated in the models by applying independently calculated Lorentz forces directly to the conductor nodes.[4]

Two limiting cases of yoke support of the collar are examined: no yoke support and infinitely rigid yoke support. For the infinitely rigid yoke support, the collars are prevented from expanding horizontally by imposed boundary conditions at the horizontal midplane of the collars and at the keyways.

The calculated collar deflections at full field for these different cases are listed in Table 2. Although the yoke model for the NC-9 dipole has not been explicitly included for this calculation, it is possible to determine the collar deflections with the yoke support by using superposition of the collar deflections for the unsupported case and the reaction forces developed at the constraints for the rigidly supported case along with the calculated horizontal stiffness of the yoke (calculated separately).[5] Such a superposition yields a diametral deflection of 0.0026 inch of the horizontal midplane at full field. The horizontal component of the total applied Lorentz force is 3970 pounds per quadrant per longitudinal inch. For the unsupported collars, this resultant is, of course, completely reacted by the vertical collar midplanes. For the infinitely rigid yoke support, 59% of the resultant is reacted by the applied constraints, and the other 41% is reacted by the vertical collar miplanes. In the case of the actual yoke support, superposition indicates that 44% is reacted by the yoke and skin, with the remaining 56% being reacted by the vertical collar midplane.

The calculated loss in coil stress at the poles for the supported NC-9 magnet is shown in Figure 5 along with measured values obtained using BNL-type coil pressure gauges. A first-order verification of the calculated loss in stress at the inner coil is obtained using a method developed by Tollestrup[6] in which the coils are modeled by a series of springs, one per conductor, with the azimuthal Lorentz forces applied at each junction of two springs. Using this model, the stress loss at the inner pole of the NC-9 is given by

Table 2. Calculated Collar Deflections and Change in Azimuthal Coil Stress Due to Energization to 6.6 T.

	(Change in collar diameter × 0.001 inch)		(psi)	
	Vertical Deflection	Horizontal Deflection	$\Delta\sigma$ Inner Coil	$\Delta\sigma$ Outer Coil
Constrained collars	−2.4	0.0	2860	1100
Unconstrained collars	−9.5	9.5	3500	715

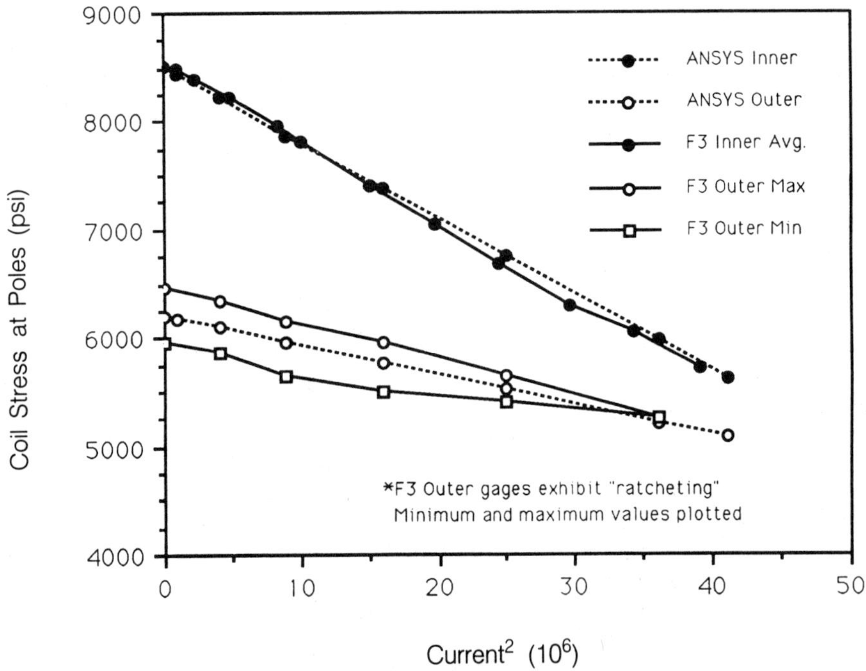

Figure 5.   Calculated and measured change in average azimuthal coil stress at the poles at different magnet currents. Measured values obtained on NC-9 prototype magnet F3 using BNL-type gauges and reflect relative stress changes rather than absolute stress.

$$\Delta\sigma_{pole} = (F_{total}/3A)(2N + 1)/(N +1) = 3200\ \text{psi},$$

where F_{total} is the total azimuthal Lorentz force for the inner coil (1850 pounds/longitudinal inch), N is the total number of turns (16), and A is the area of inner coil (0.37 square inch/longitudinal inch). The smaller stress loss measured with the BNL gauges and calculated by the finite element model (which both indicate a loss of ~2700 psi at 6400 amperes) is attributable to the elasticity of the collars, which is not included in the spring model. Although the finite element calculation of the coil prestress loss agrees well for this type of gauge, it should be noted that there is generally poor agreement with the measurements obtained using the LBL-type gauges. These gauges typically indicate a prestress loss of 5000–9000 psi at full field. The discrepancy between the two types of gauges is not currently understood. It is hoped that it will be resolved following further tests of the two types of gauges. At this point, all that can be said is that the finite element calculation and the simple spring model lend support to the BNL–type gauge results.

Although the average change in coil stress loss is not substantially different between supported and unsupported models, the peak stress loss is much greater in the unconstrained model, due to increased bending stresses in the coils caused by the increased deflection of the collars. Figure 6 shows the stress loss across the inner-coil pole face due to energization. For unsupported collars, the calculated change in stress varies from a decrease of 11000 psi at the inner radius to an increase of 3600 psi at the outer radius. While these calculated stresses should be interpreted only qualitatively, it is clear that the support offered by the yoke is essential in maintaining azimuthal prestress at the

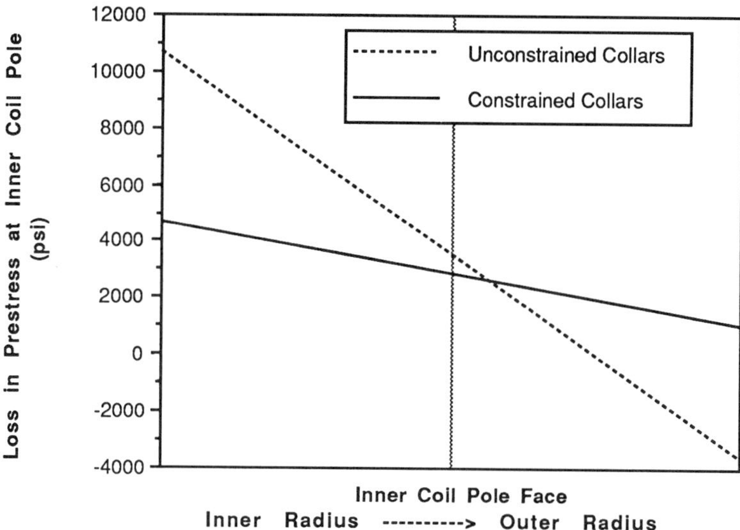

Figure 6.   Calculated stress loss at 6.6 T across the inner coil pole face.

inner radius of the inner coil. Even in the limiting case, in which the collars are rigidly constrained from horizontal expansion, bending stresses are developed in the inner coil due to the vertical deflection of the collars and horizontal compression of the outer coil and the collar itself such that the inner radius loses 5000 psi, or 1800 psi more than the average.

Summary

The finite element model used in this analysis appears to accurately capture the two-dimensional mechanical behavior of the NC-9 dipole, despite the relatively simple mechanical model utilized for the coil. As with all finite element calculations based on displacement formulations, one should have more confidence in the calculated deflections than in the calculated stresses. This is particularly true in this case because the material properties of the coil are in reality much more complex than specified in this analysis. Nonetheless, the calculations agree very well with the measurements. In addition, the analysis has focused attention on the support of the outer-coil pole provided by the laminations as well as on the difficulty of maintaining uniform stresses in the coil because of the collar elasticity.

Ongoing analysis efforts will incorporate nonlinear and orthotropic coil material properties into the model as they become available. Out-of-plane effects caused by longitudinal strains in the magnet will also be examined.

Acknowledgments

We wish to thank Bob Schermer (CDG) for his assistance and guidance in these analyses, and Julian Cortella (CDG) and Elizabeth Scholle (CDG) for their participation in this work. We would also like to thank Valerie Kelly and Annie Calinog for their skillful assistance in preparing this manuscript, as well as Kate Metropolis for her editorial assistance.

References

1. J. D. Jackson, ed., "Conceptual Design of the Superconducting Super Collider, Attachment B," SSC-SR-1020, March 1986.

2. J. M. Cortella, Areas of High Local Stress in Dipole Components, SSC/CDG Report MD-TA-113, January 1989.

3. C. L. Goodzeit, M. D. Anerella and G. L. Ganetis, "Measurement of Internal Forces in Superconducting Accelerator Magnets with Strain Gauge Transducers," IEEE Conference, Vol. 25, No. 2, March, 1989.

4. S. Caspi and M. Helm, "Lorentz Forces for LBL NC-9 Dipole Cross Section," LBL Report LBID–1304, July, 1987.

5. J. M. Cortella, and M. S. Chapman, "A Finite Element Analysis of an SSC Yoke Used for Mechanical Support of the Collared Coil," SSC/CDG Report SSC-N-568, December, 1988.

6. A. Tollestrup, "Superconducting Magnets," AIP Conference Proceedings: Physics of High Energy Particle Accelerators (Fermilab Summer School, 1981).

RECENT DEVELOPMENT OF THE Cu/Nb-Ti SUPERCONDUCTING CABLES
FOR SSC IN Hitachi Cable, Ltd.

S. Sakai, G. Iwaki, *Y. Sawada, *H. Moriai,
and Y. Ishigami

Metal Research Lab. , *Tsuchiura Works,
Hitachi Cable, Ltd.
Tsuchiura-shi, Ibaraki-ken, 300, Japan

ABSTRACTS

In this few years, Cu/Nb-Ti superconducting cables for the
dipole magnets of SSC projects have been developed in the industrial scale in Hitachi Cable, Ltd.   The features of these
developed conductors are as follows.   (1) The diameter of Nb-Ti
filaments is very small, 4-6 μm.   (2) The critical current density (J_c) is very high, 2850-3050 A/mm² at 5 T on wires,
2750-2950 A/mm² at 5 T on cables in industrial scale. The champion J_c of wires is 3460 A/mm² at 5 T in the laboratory scale.
(3) The RRR (Residual Resistivity Ratio ; $\rho_{at293K}/\rho_{at4.2K}$)
values of developed cables is very high, approximately 200, due
to the newly developed high purity Oxygen Free Copper (OFC).
(4) The conductors have been wound to the 1 m length dipole
magnet in Hitachi Ltd. , and it has generated 6.7 T in the central magnetic field at 6595 A.

The Cu/Cu-Mn/Nb-Ti composite wires which avoid the possibility of electrical coupling of the filaments have been produced in laboratory scale. The RRR of the copper stabilizer and
J_c properties have not degraded because of no metallurgical
reactions between Cu and Mn, Nb-Ti and Mn.

INTRODUCTION

Hitachi Cable, Ltd. has contributed to the completion of
many projects in the field of high energy physics. For example,
Hitachi Cable had produced the Rutherford type superconducting
cables for dipole and quadrupole magnets in KEK, the Al stabilized Nb-Ti composite cables for the colliding beam detector
magnet in FNAL. (1) These magnets have generated the specified magnetic field, and superiorities of these conductors have
been verified.

In these Rutherford type and the Al stabilized conductors,
however, Nb-Ti filaments size is not so small, 10-40 μm, and
the critical current density (J_c) is not so high too, 1600-
2000 A/mm² at 5T. In the SSC cables, they are required the

small 6μm filaments size because of the reduction in the superconducting persistent current (magnetization) at low field
level, and required the high J_c (2750 A/mm^2 at 5 T) for the
high operating field of the dipole magnets (6.6 T). Many efforts had been carried out (2), and Cables with small filament size and the high J_c have been developed in industrial
scale.

For the high thermal and electrical stability of superconductors, the high RRR (Residual Resistivity Ratio ;
$\rho_{at293K}/\rho_{at4.2K}$) of the copper stabilizer is necessary. In
this point of view, Hitachi Cable has recently developed high
purity Oxygen Free Copper (OFC) by the fully continuous casting systems.

In the superconducting composites with the small spacing
between the filaments, superconducting persistent current have
occurred because the electron mean free path of the pure copper
is long. (3) For the reduction in this superconducting persistent current, the clad Cu-Mn alloys between the filaments
have been proposed. (4 . 5) We have manufactured these
Cu/Cu-Mn/Nb-Ti composites in laboratory scale.

In this paper, the production and properties of these superconducting cable for the SSC dipole magnets, the high purity OFC, and Cu/Cu-Mn/Nb-Ti superconducting composites are
described.

DEVELOPMENT AND PROPERTIES OF SUPERCONDUCTING CABLES FOR SSC
DIPOLE MAGNETS

A flow chart of the production procedure for the SSC cables is shown·in Fig. 1. The most typical feature in these
processes is the application of the hydrostatic extrusion process. The advantages of the application of the hydrostatic extrusion to superconducting composites are the uniform deformation during extrusions because of no frictions between the billet and the container, and the high yield rate because the billet length to the billet diameter ratio is very large. (6)
Superconducting wires have been produced by double stack process in the large number of filaments composites like the SSC
wires.

Fig. 2 shows the relationships between the magnetic field
and the critical current densities on the superconducting wires
with 4-6 μm filaments size which have produced in the industrial scale and the laboratory scale.

In the laboratory scale, the flow of the manufacturing
procedure is fundamentally as same as that of the industrial
scale except the quantity. The quantity of the composites in
the laboratory scale is 1/100 of those in industrial scale.
High homogenized Nb-46.5wt%Ti content materials were used for
these composites. The J_c properties of the industrial scale
wires are 2850-3050 A/mm^2 at 5 T. This scattering is due to the
various aging heat treatment conditions. On the same heat
treatment conditions, the longitudinal J_c scatterings of the
wires are below 1.5 %. The champion J_c of wires manufactured
in the laboratory scale is 3460 A/mm^2 at 5 T.   All values are
measured at 4.2 K and determined when the resistance reached

670

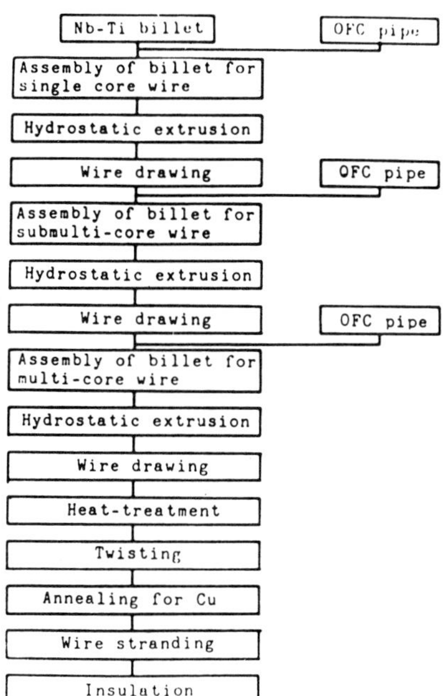

Fig. 1. A Flow Chart of the Production Procedure
for the SSC Cables.

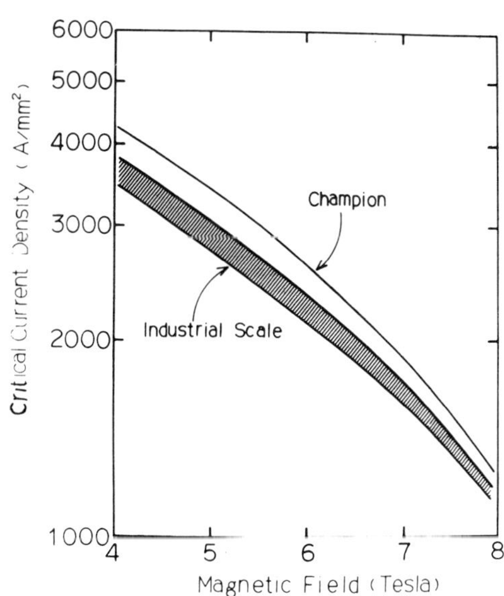

Fig. 2. Magnetic Field vs. Critical Current Density
for the Superconducting Wires with 4-6μm
Filaments Size.

1×10^{-12} Ω cm for Cu and Nb-Ti by the bifilar technique.

Fig. 3 shows the cross sectional and longitudinal photos of wires and cables in the industrial scale by a microscope and a SEM imaging. The composites have been worked very uniform in the cross sectional and longitudinal directions, and multiple necking in the filaments are not observed. The n values of wires in the V-I characteristics are the range of 35-40 at 5 T, no multiple necking in filaments are proved by this high n values. In the cables for SSC dipole magnets, the No. of wires is 23 for the Inner cables, 30 for the Outer ones.

Table 1 shows the specifications of the SSC wires and cables, the measured values of wires and cables which have produced in the industrial scale. These sufficient values in all items have been attained to the specifications. In particularly, the critical current and the RRR shows the excellent values. The critical current values leave 6-10 % margin, the RRR values are approximately three times of the specifications. High critical current properties are due to the moderate heat treatments and the cold working after the final aging heat treatment. Newly developed Oxygen Free Copper and the final annealing process after cabling provide the high RRR values of the copper stabilizer.

For the estimation of workability and flexibility of the cables in the winding magnet process, the bend tests of cables have been carried out. The bend radius is 6. 5 mm, with the loading 500 kgf for Inner cables, 200 kgf for Outer ones. The cable have not been uniform when the loads are over 600 kgf for the Inner cables, 250 kgf for the Outer ones. There were no damage in the wires of cables, in the filaments of wires after removing copper.

Table 2 shows the 1 m length dipole magnet parameter. This magnet construction has been performed in Hitachi Ltd. . The cross sectional dimensions and electro-magnetic parameters this magnet are as same as the 17 m full scale one except the magnet length. This 1 m length magnet have generated 6. 7 T in the central magnetic field at 6595 A. This obtained values is nearly equal to the magnetic field vs. critical current characteristics of short samples. High qualities and properties of the developed cables for SSC dipole magnet have been confirmed on the magnet too.

DEVELOPMENT AND PROPERTIES OF HIGH RRR OXYGEN FREE COPPER

The purity of copper greatly provides the influences in the RRR properties of the annealed copper. Investigations concerning the influences of the impurity elements on the RRR properties have been carried out (7), and high RRR properties of OFC have been achieved by the removing impurity elments using the special technique on the fully continuous casting process.

Fig. 4 shows the relationships between the magnetic field and the electric resistivity on the newly developed high purity OFC and the conventional one. At 0 T, the electric resistivity at 4. 2 K of the high purity OFC is a half of the conventional one. At 5 T, the high purity OFC is 20 % lower than the conven-

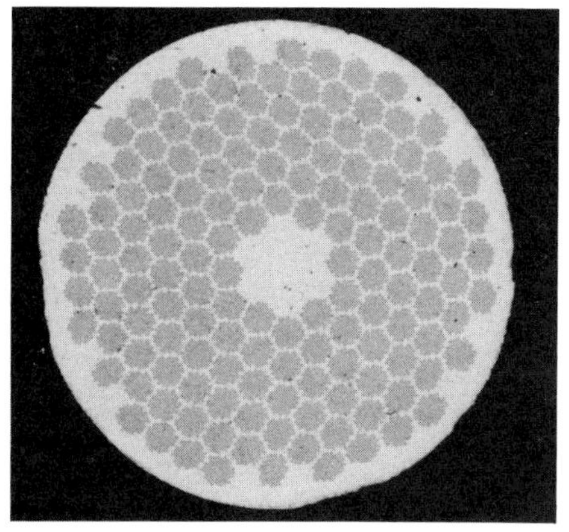

(a) Wire for Inner Cable

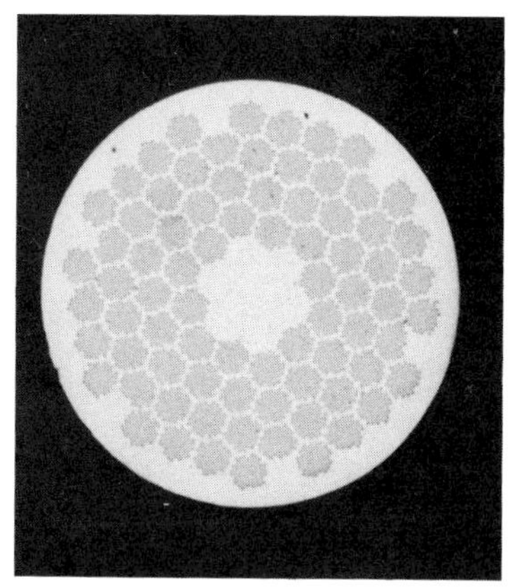

(b) Wire for Outer Cable

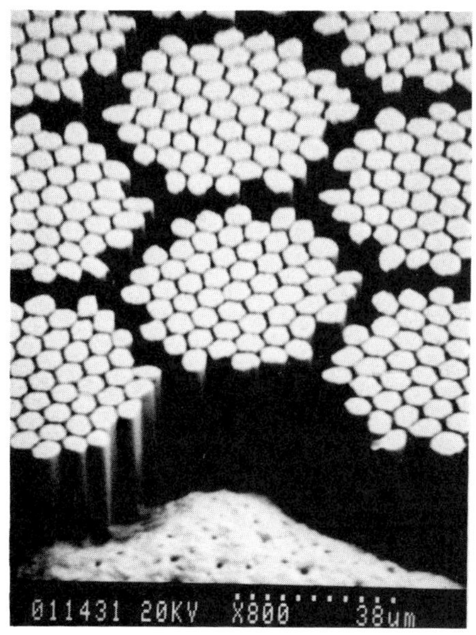

(c) SEM Imaging of Wire

(d) SEM Imaging of Filaments

Fig.3-1. Cross Sectional and Longitudinal Photos of Wires.

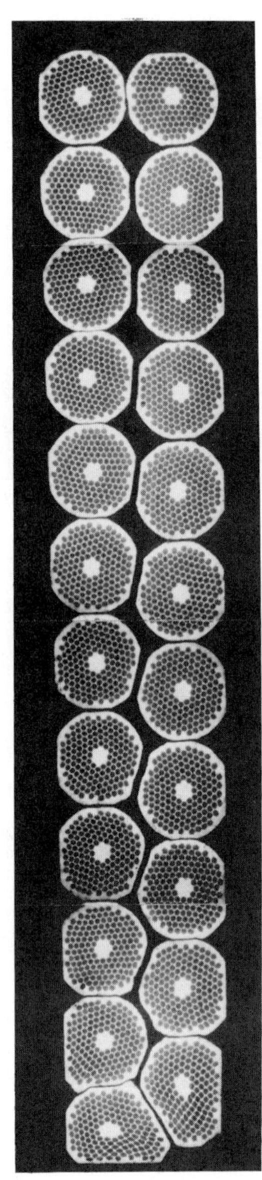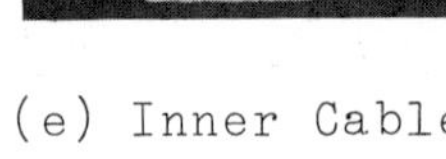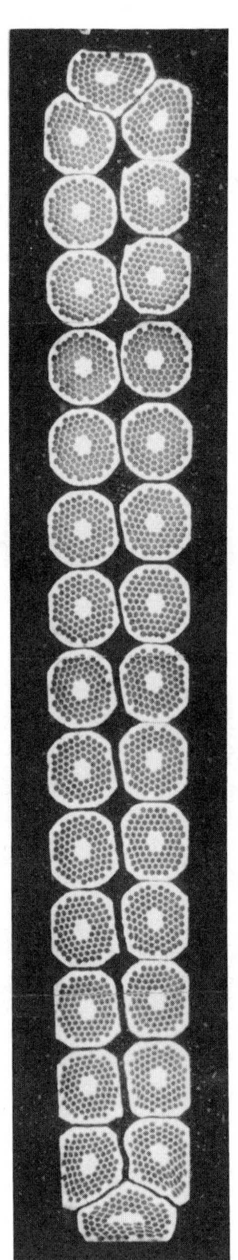

(e) Inner Cable   (f) Outer Cable

Fig.3-2. Cross Sectional Photos of Cables.

Table 1. Specifications and Measured Values of
Wires and Cables for SSC Magnets

		Specifications		Measured Values	
		Inner	Outer	Inner	Outer
Wire	Wire Diameter (mm)	0.808±0.025	0.648±0.025	0.810	0.647
	Copper/Nb-Ti Ratio	1.5±0.1	1.8±0.1	1.45	1.73
	Filament Size (μm)	6	6	5.81	5.98
	Filament Spacing (μm)	>1.0	>1.0	1.11	1.15
	Twist Pitch (mm)	12.7±1.3	12.7±1.3	13.0	12.2
	Critical Current (A)	$\geq$328 at 7T	$\geq$285 at 5.6T	337 at 7T	297 at 5.6T
	RRR	$\geq$83	$\geq$89	159	192
Cable	No. of Wire	23	30	23	30
	Cable Mid-thickness (mm)	1.458±0.006	1.166±0.006	1.463	1.170
	Cable Width (mm)	9.296±0.025	9.728±0.025	9.298	9.727
	Cable Lay Pitch (mm)	78.7±5.1	73.7±5.1	80.3	75.1
	Critical Current (A)	$\geq$7167 at 7T	$\geq$7860 at 5.6T	7640 at 7T	8670 at 5.6T
	RRR	$\geq$66	$\geq$63	172	195

Table 2. Dipole Magnet Parameters

General		
Magnetic length	1.4	m
Bore diameter	40	mm
Central field	6.6	T
Current	6504	A
Stored energy	90	kJ
Winding		
No. of turns	Inner	16
	Outer	20
Maximum field	Inner	7 T
	Outer	5.6 T
Conductors	NbTi keystoned cable	
Collars		
Material	High manganese steel	
Lamination thickness	1.52 mm	

tional one in the electric resistivity at 4.2 K. These properties have measured on the 1.0 mm in diameter samples which have annealed at 500 °Cx1 hr.. The RRR values of the cold worked pure copper is lower than that of the annealed one. And the RRR properties of the cold worked high purity OFC are almost equal to the cold worked conventional one, when the reduction in area after the final annealing is over 25 %. The maximum RRR values of the pure copper are usually attained on the samples which have annealed at 500 °C. On the superconducting Cu/Nb-Ti composites which have finally annealed at 500 °C, their J_c properties have extensively degraded. On this point of view, the final annealing treatment for Cu/Nb-Ti composites at lower temperature is necessary to obtain the excellent RRR and J_c properties.

The relationships between the annealing temperature and J_c properties are shown in Fig. 5. The J_c properties clearly degraded over 300 °Cx3 hrs.. As the RRR properties are recovered by the final annealing, the low temperature (under 300 °C) annealable OFC is required.

Fig. 6 shows the relationships between the annealing temperature and tensile strength of the developed high purity OFC and the conventional one. The full softening temperature of the high purity OFC is approximately 30 °C lower than that of the conventional one.

Fig. 7 shows the relationships between the magnetic field and electric resistivity at 10 K of the SSC cables which have clad the developed high purity OFC and the conventional one. The cables clad the high purity OFC have approximately 30 % lower electric resistivity at 5 T to the specifications (RRR=65). Newly developed high purity OFC is excellent materials for the stabilizer of Cu/Nb-Ti superconducting composites in the point of the high RRR property and the low full softening temperature.

SOME PROPERTIES OF Cu-Mn ALLOYS AND THE APPLICATION FOR SUPERCONDUCTING WIRES

Table 3 shows the electric resistivity at 293 K and 4.2 K of the various Cu-Mn alloys. These Cu-Mn alloys have been produced by the fully continuous casting systems. At 293 K, the Cu-0.5wt%Mn alloy have twice resistivity of pure copper, and the Cu-1.0wt%Mn alloy have three times resistivity of one. At 4.2 K, both the Cu-0.5wt%Mn and Cu-1.0wt%Mn alloys have a few hundred times resistivity of the pure OFC.

The electro-magnetic resistivity at 4.2 K of Cu-Mn alloys have measured, too. Fig. 8 shows the relationships between the magnetic field and the electric resistivity at 4.2 K of Cu-0.5wt%Mn and Cu-1.0wt%Mn alloys. There are slight influences when the magnetic field is applied.

By the clad Cu-Mn alloys between the filaments, the reduction in the superconducting persistent current at low field level is highly expected for the very small filaments spacing composites. The application of Cu-Mn alloys is much effective on the magnetization properties, however Cu-Mn alloys does not provide the electrical and thermal stability of superconducting cables because of their high electro-magnetic resistivities.

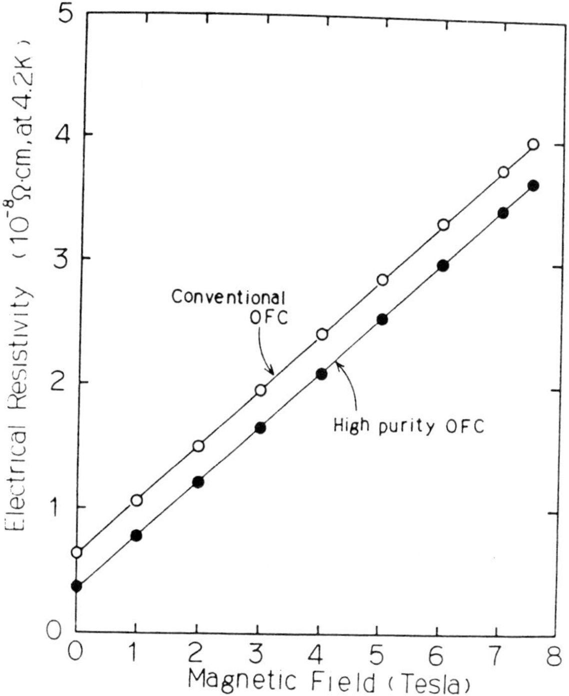

Fig. 4. Magnetic Field vs. Electric Resistivity
at 4.2 K for High Purity Oxygen Free Cu.

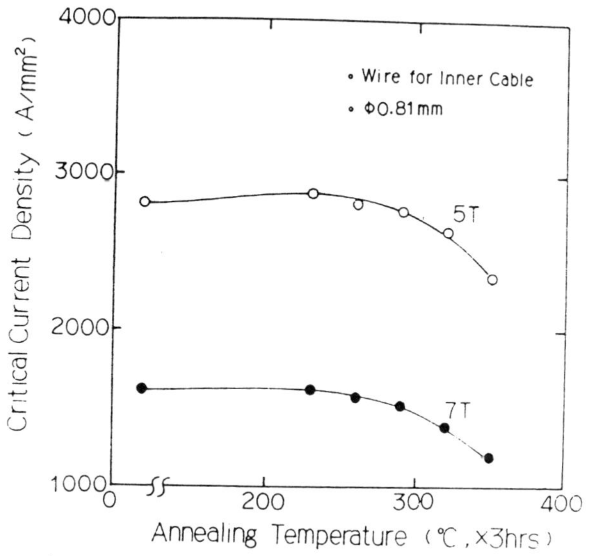

Fig. 5. Annealing Temperature vs. Critical Current
Density of wire for Inner Cable.

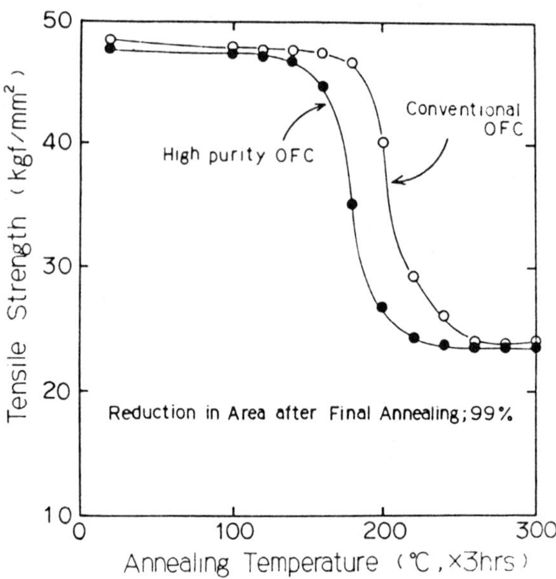

Fig. 6. Annealing Temperature vs. Tensile Strength
for High Purity Oxygen Free Copper.

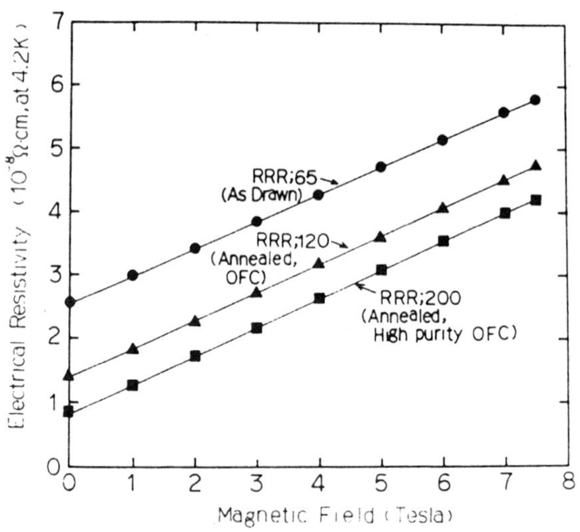

Fig. 7. Magnetic Field vs. Electric Resistivity
at 10 K for SSC Cables.

Table 3. Electric Resistivity of Cu-Mn Alloys

Mn Contents (wt%)	Temper	Resistivity($\mu\Omega\cdot$cm)	
		at 293K	at 4.2K
Cu-0.5Mn	As drawn	3.51	1.80
	Annealed (500°C×1hr)	3.50	1.77
Cu-1.0Mn	As drawn	5.16	3.43
	Annealed (500°C×1hr)	5.05	3.35

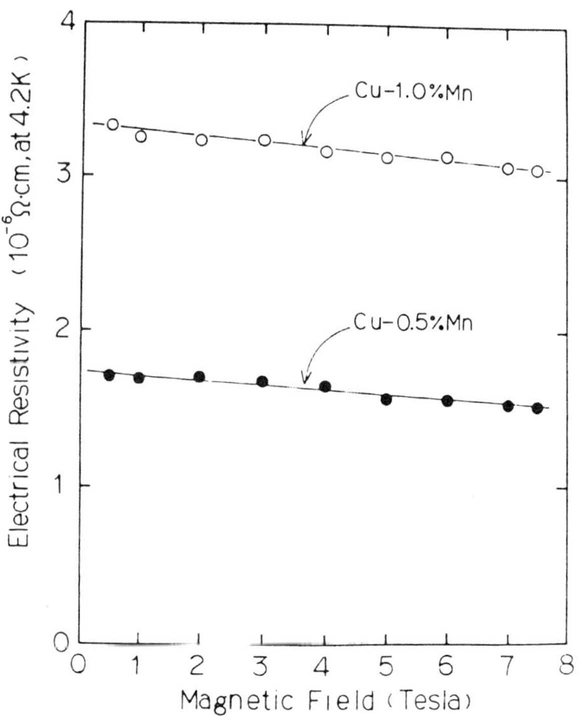

Fig. 8. Magnetic Field vs. Electric Resistivity
at 4.2 K of Cu-Mn Alloys.

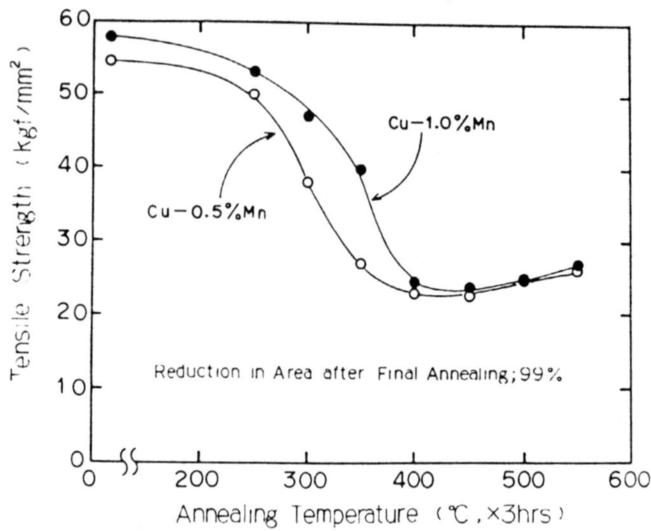

Fig. 9. Annealing Temperature vs. Tensile Strength for Cu-Mn Alloys.

Table 4. Jc and RRR Properties of Cu/Cu-Mn/Nb-Ti Superconducting Composites

Mn Contents (wt%)	Jc at 5T (A/mm²)	RRR (ρ_{295K}/ρ_{10K})
Cu-0.5Mn	2840	165
Cu-1.0Mn	2820	160

Fig. 9 shows the relationships between the annealing temperature and the tensile strength of Cu-Mn alloys. The full softening temperature of Cu-Mn alloys is 200 °C higher than that of the pure Oxygen Free Copper. These results means the Cu-Mn alloys are not softened in the range of 200-300 °C final annealing of cables.

Table 4 shows the a J_c and RRR properties of the Cu/Cu-Mn/Nb-Ti superconducting composites. On the J_c and RRR properties, there no degradations because of no metallurgical reactions between the OFC and Cu-Mn , Nb-Ti filaments and Cu-Mn alloys.

The application of the Cu-Mn alloys for the Cu/Nb-Ti composites is very useful because of the reduction in the superconducting persistent current without degradation in the stability of cables.

CONCLUSIONS

1. High critical current densities (2850-3050 A/mm^2 at 5 T in the industrial scale using the hydrostatic extrusion, 3460 A/mm^2 in the laboratory scale) were attained by the moderate processing and the high homogenized 4-6 μm filament size Nb-46.5wt%Ti wires.

2. Superconducting cables for SSC dipole magnets were developed and produced in the industrial scale. The typical features of the developed cables are their high critical current (6-10 % margin to the specifications) and the high RRR values (three times of the specifications).

3. The conductors have been wound to 1 m length dipole magnet in Hitachi Ltd. , and it has generated 6.7 T in the central field at 6595 A.

4. Newly developed high purity Oxygen Free Copper shows the excellent RRR values. The electro-magnetic resistivity at 5 T of cables is 30 % lower to the specifications when the developed high purity Oxygen Free Copper have been applied as a stabilizer of Cu/Nb-Ti superconducting composites.

5. The application of the Cu-Mn alloy for the very fine filaments Cu/Nb-Ti composites is very useful because of the reduction in superconducting persistent current at the low field level without degradations of the critical current density and the RRR values.

ACKNOWLEDGEMENTS

The author wish to express their thanks to Mr. R. Saito, Mr. T. Suzuki, and Mr. K. Asano of Hitachi Ltd. who have constructed the 1 m length dipole magnet.

REFERENCES

1. H. Hirabayashi et al., "Measurement of Propagation Velocities of Normal Zone in a 1 mϕ x1 m Superconducting Solenoid

Magnet, " Japanese Journal of Applied Physics, 20 (Nov. 1986)

2. S. Sakai et al., "Recent Development of Various Super-conducting Cables for Accelerators, " Proc. of ICFA, 95 (May 1986)

3. Gregory et al., "Importance of Spacing in the Development of High Current Densities in Multifilamentary Super-conductors, " CRYOGENICS 178 Vol. 27 (June 1987)

4. T. S. Kreilick et al., "Influence of Filament Spacing and Matrix Materials on the Attainment on the High Quality, Un-coupled NbTi Fine Filaments, " MT-10 #BC-4 (Sept. 1987)

5. Gregory, "Recent Development in Multifilamentary Nb-Ti Superconductors, " CRYOGENICS 290 Vol. 27 (June 1987)

6. S. Sakai et al., "Production of Nb_3Sn and Nb-Ti Multi-filamentary Superconducting Wires Using Warm Hydrostatic Extrusion, " Proc. of ICMC 301 (May 1982)

7. K. Noguchi et al., "The Role of Copper Products Manufacturers in the Production of Superconductors, " Proc. of Copper '83 33. 1 (Nov. 1983)

INDUSTRIAL PRODUCTION OF HERA SUPERCONDUCTING DIPOLES

P. Gagliardi, A. Laurenti, A. Martini, and R. Penco

Ansaldo Componenti
Magnet Unit
Via N. Lorenzi, 8   I-16152 Genoa, Italy

ABSTRACT

More than 40% of the total production of 242 superconducting dipoles has been carried out. The tests at room and cryogenic temperature have shown very good results, even better than expected for industrial production of such products. The well-designed tooling and plants, the high professionalism of the technicians and the operators, and the good organization have permitted us to reach these results.

INTRODUCTION

Presently Ansaldo Componenti is manufacturing 242 superconducting dipoles, including yoke and electrical connections, as shown in Figure 1. On the references 1, 2, 3 you can find a detailed description of these dipoles. Ansaldo Componenti has been the first company in the world to manufacture superconducting dipoles in an industrial way with their own plants, tooling, and technicians. The development of the design and technology made in the past by DESY laboratory and the Ansaldo experience were used as the basis for designing tooling and plants and defining the proper manufacturing methods.

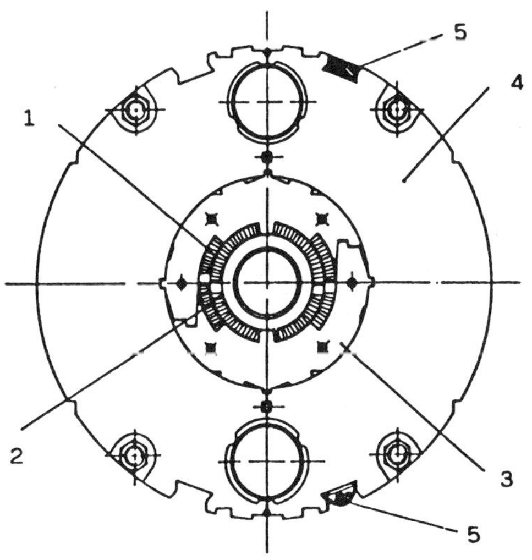

Figure 1. Cross section of HERA dipoles as supplied by Ansaldo.

MAIN PARAMETERS TO BE GUARANTEED

The main parameters requested in DESY specification are

- high magnetic field quality, without yoke, at room temperature

- current of 7400 A (depending on short-sample current of the cable) at 4.6°K and 6.0 T without yoke

- maximum total twist of 0.5 mrad on the dipole completed with yoke

- recording of each component going into each dipole and of all tests

- production rate of 20 dipoles per month.

PRODUCTION RESULTS

Up to the end of 1988, 88 dipoles have been collared and tested; coils for another 53 dipoles have been cured and tested. Only five cured layers, of more than 560, have been scrapped; two were lost for cable fault just at the beginning of production. This means that ACO had a percentage of scrap less than 0.2%. This result is very good and helpful for the cost estimation of future projects. The insulation quality control system developed by ACO in agreement with INVEX-PIRELLI gave beautiful results and avoided inconvenience. A complete summary of the results is given in Figure 2.

Scrapped cured layers

- preseries (10 dipoles)   0%

- series (130 dipoles = 520 layers)   0.9 %

Collared dipoles demounted after harmonic measurements

- preseries (10 dipoles)   20%

- series (78 dipoles)   15% a2 (out of tolerance)

                                        1.3% b3 (out of tolerance)

Collared dipoles scrapped after cryogenic tests

- preseries (5 dipoles tested)   0%

- series (11 dipoles tested)   0%

Completed dipoles scrapped after final test at Ansaldo

- preseries (10 dipoles)   0%

- series (2 dipoles)   0%

Completed dipoles scrapped after final test at DESY

- preseries (10 dipoles)   0%

Figure 2. Summary of production results (31/12/88).

EXPERIENCE ON MAGNETIC FIELD GUARANTEE

Figure 3 shows the histograms of the main multipole coefficients measured at room temperature on the first 88 dipoles. As you can see, there is a big scatter on the a2 coefficient. Some explanations were found about the difficulty to keep the skew quadrupole a2 under control. One explanation is that the coils are measured only on 9 points, one every meter, and you choose the two coils to match together based on these measurements. These measurements are made by pressing the coil at two different pressures and looking at the deviation of the middle plane of the coil from the ideal middle plane. Many dimensional uncertainties can influence the coil size and their elasticity modulus along their length. The most important are tolerance of the pressing tool for measurements, cyclic variations of cable thickness over the length, junctions of kapton tape, junctions of prepreg tape, copper fillers tolerances, tolerance of curing shims. Some extra checks were carried out on the coils to understand why the a2 calculations were different from those measured. It was found that doing measurements on the coil on 26 points, one every 360 mm along its axis, you can see in some cases a difference in the average value of 0.02 mm, in the arc length, respect to the value with 9 points. If you match two coils with this big difference with opposite sign you can find an unexpected value of about 20×10^{-5} in a2. A correlation between the measurements on 9 points and the real average size of the coil is under study in order to understand if and when it is necessary to measure the coil on 26 points. This can reduce the numbers of the collared dipoles in which a2 is out of tolerance.

Another problem met during the dimensional check was the different coil size as a function of the pressing time spent to do the measurement itself. The greater the time, the less the arc size of the coil. After some tests, it was found that 3 minutes was the best compromise, and the computer driving the pressing device was regulated accordingly. It is also very important to do a systematic check of the harmonic measuring device.

In fact, the handling can consume the supports and deform the tube of the measuring coil; so the reference line changes. Actually dipole 011 is used like a reference, and every 20 dipoles this dipole is measured; if the deviation of the value is less than $\pm 3 \times 10^{-5}$ the device is still good; otherwise deeper investigations must be carried out.

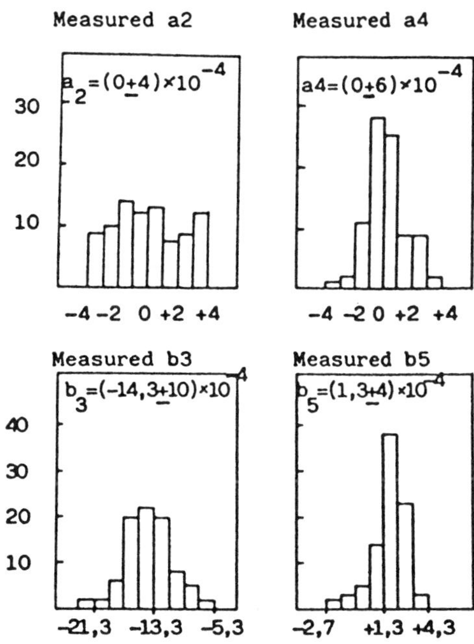

Figure 3. Distribution of the main multipole coefficients measured at room temperature.

PLANTS AND TOOLING

In order to respect the strict requests from the customer and to take into account future projects ACO has designed the plants and the tooling very precisely with a high level of automatization as well as flexibility.

The n.c. winding machine, for instance, can be used for dipoles with a length that can vary from 2 to 20 meters using the same philosophy. This machine allows winding of the coils with only one operator because it is remote controlled. The tensile force on the cable is kept constant in order to avoid any unpredictable behavior of the coil when it is removed from the mandrel, after curing. For the same reason a special tool is used for pressing the turns on the ends of the coil and the end spacers.

The two presses for curing and collaring the dipoles are also driven by computer, so you can control the pressure and the temperature on each section; all the data are recorded. The presses are also equipped with proximitors in order to be sure that each section of the coil is pressed completely. At the beginning, investigations were done for defining the best end spacer position and the shims to be used for curing. In fact, you must take into account the tensile force given to the cable during the winding and the different thermal expansion between the coil and the mandrel during the curing of the coil. Up to now no problems have been encountered with the coil length or shape.

The automatic stacking machine permits good alignment of the half yokes and presses every half-meter in order to have a constant filling factor over all the length. The automatic welding press for yoke assembly allows the insertion of the collared dipole into the half yokes on one side of the press, then the dipole moves under the pressing plates, and then two heads perform the TIG welding of the yoke simultaneously. That avoids different thermal deformation on the two sides of the yoke. This press has two pressing plates, machined very precisely, elastically pressed by a special plastic material (cellastro) that allows it to press uniformly to create a complete dipole with a very low twist (less than 0.5 mrad). Also the tooling for handling and storing of the half yokes and the complete dipoles is precisely machined to avoid introducing any additional twist.

The manufacturing area is divided into two sections. The one for collared dipoles is "clean area" with constant humidity and temperature. The second one, for yoke assembly and electrical connections, is a "clean condition." The total area is more than $3200 \, m^2$.

SOME CONSTRUCTIVE SOLUTIONS

Most of the technology developed by DESY laboratory has been used. Nevertheless some changes were necessary in order to be able to manufacture the dipoles in an industrial way with high quality. One of them is to wind and check the second layer without removing the first layer from the mandrel. That was made because of the "fish bones" problem. In fact, if you remove the first layer from the mandrel its shape changes; consequently the "fish bones," placed between the two layers, do not fit very well with all the turns of the first layer. The first one or two turns of the first coils were not covered by the "fish bones," and during the operation these turns could move. In spite of this phenomenon, the cryogenic tests on the first dipoles were good. The possibility of repairing very easily some fault of insulation on the first layer, after the second layer was already cured, has been verified. The few coils in which that happened were cryogenically tested with very good results.

Still looking to the future, alternative solutions were investigated and developed. One of them regards the brazing of the cross over. In fact the brazing joint between the two layers usually is made before the winding of the second layer starts. In one coil, made for experiment with two scrapped layers, the brazing joint was done after that the two layers were cured. The result was very good, and no problems were detected during the test of insulation. Therefore the winding and the curing of the two layers separately with two different mandrels is possible.

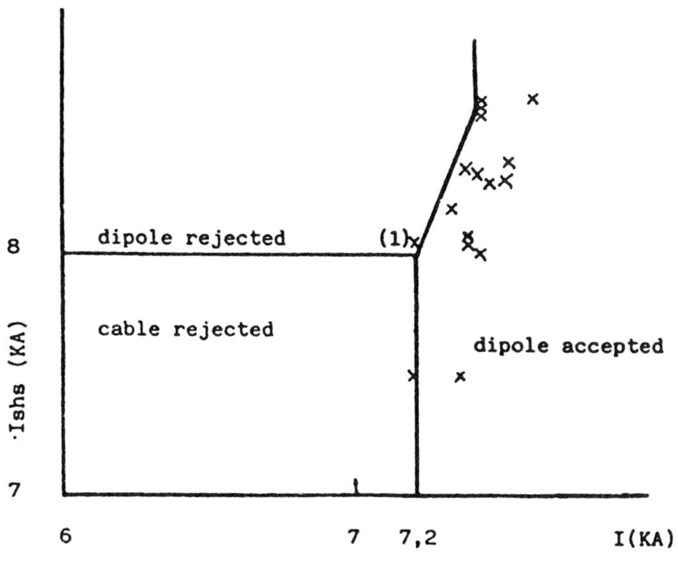

Figure 4. Results of cryogenic tests.

Other changes and improvements have been developed by ACO and will be the subjects of the next papers.

CONCLUSIONS

The results of the preseries dipoles tested at DESY laboratory, and the results of the series dipoles tested cryogenically at ACO[4-5] reported in Figure 4, have clearly shown that the industrial manufacture of high performance dipoles is possible. The production rate (actually more than one collared dipole per day with only two shifts) and the high quality of the production are satisfactory. The industry is then ready to execute the next step for new colliders with very high energy.

ACKNOWLEDGMENTS

The authors wish to thank INFN and DESY people involved in this job for their cooperation. Many thanks also to all ACO people involved and especially to the foreman, Mr. Sciutto and his operators, the production quality control team (Mr. Caserza in particular). Their high professionalism has permitted us to reach these good results.

REFERENCES

1. S. Wolff, "Superconducting Magnets for HERA." Presented at the 13th International Conference on High Energy Accelerators, Novosibirsk, USSR (1986).

2. K. Balewski, D. Degele, G. Horlitz, H. Kaiser, H. Lierl, H. Mess, S. Wolff, C. Dustmann, P. Schmuser, B. Wiik, "Cold Yoke Dipole Magnets for HERA." Presented at the 1986 ASC Baltimore, Maryland, USA (1986).

3. H. Kaiser, "Design of Superconducting Dipole for HERA." Presented at the 13th International Conference on High Energy Accelerators, Novosibirsk, USSR (1986).

4. R. Musenich, A. Bonito-Oliva, G. Gaggero, S. Parodi, S. Pepe, P. Valente, R. Penco, "Cryogenic Tests on HERA Dipoles." Presented at the ICEC 12 Southampton, England (1988).

5. R. Musenich, A. Bonito-Oliva, M. Losasso, G. Masullo, P. Valente, R. Penco "Quench Behavior of Industrial Produced HERA Dipoles." Presented at the ICEC 12, Southampton, England (1988).

DESIGN STUDY ON THE SUPERCONDUCTING DIPOLE MAGNETS WITH NON-CIRCULAR APERTURE COILS IN APPLICATION TO FUTURE COLLIDERS

Kenji Ishibashi, Akira Katase and Hiromi Hirabayashi*

Department of Nuclear Engineering, Kyushu University
Hakozaki, Fukuoka 812, Japan
*National Laboratory for High Energy Physics (KEK)
Oho, Tsukuba 305, Japan

ABSTRACT

In the near future, superconducting magnets with a small beam aperture will be required for high energy proton colliders. These magnets are usually designed to have a circular coil aperture. Their coils have many wedges for the winding to arch over the aperture as a whole, and the wedges are used to reduce the higher multipole components in the magnetic field. In this paper, the structure of a non-circular aperture is introduced to remove the wedges from the coils. Two types of windings, eccentric and elliptic coils, are studied for use of Rutherford type cables. It is shown that these coils have inherent advantages in decreasing the higher multipole components.

INTRODUCTION

Dipole magnets with two-layer coils have been developed for TEVATRON at Fermilab.[1] The Rutherford type cables keystoned in shape were wound into the coils. They were self arched over the magnet aperture without containing any wedges in the winding. The magnet coil had a coil diameter of 7.5 cm and a cable width of 0.78 cm. Then, the cable width becomes 20 % of the coil radius. This width is considered to be relatively thin in comparison to the coil radius. A high energy proton accelerator in the energy region of 10 to 20 TeV will be built in the near future.[2,3] The proton beam size decreases generally by increasing the diameter of an accelerator ring. The future high energy accelerator will have a magnet aperture of a smaller size than in the case of TEVATRON.

A model dipole magnet has been designed for the Superconducting Super Collider (SSC).[4] The magnet possesses two-layer coils. In comparison to TEVATRON, this magnet has the following characteristics. The diameter of the coil aperture is reduced by about half. The cable width is increased by about 20 %. The small coil aperture requires a considerably larger keystone angle for cables to self-arch over it. Since the Rutherford type cables can not be fabricated into such a large keystone angle, the cables with an inadequate keystone angle were used for the coils. Many wedges were therefore inserted into the magnet winding. Both the size and the position of the wedges were chosen carefully to achieve the good field quality. The use of these wedges arched the coils as a whole. Unlike that of TEVATRON, however, this arch was not made in a completely self-arched manner. This coil configuration brought about complexity in the winding procedure and in the field quality control.

It is desirable that such a magnet for the future accelerator will have a simpler coil structure. A preliminary study[5,6] has been made by a KEK group using this approach. A "braid-in-strands" structure was already deviced at KEK to fabricate the cables with the keystone angle as large as possible. It was found that the cables with the structure were useful to decrease the number of wedges in the coils.[5,6] Such magnets as TEVATRON, SSC and the above references all have a circular aperture and their coils are regarded as a kind of $\cos\theta$ winding.

In contrast, there is another type of winding which is called as intersecting circles or ellipses.[7] We are interested in introducing this feature into the TEVATRON style coils,[8] and designing the magnet with a non-circular aperture. The keystone angle for the inner cables of the SSC magnet is 1.6 degrees. The technique for cable production has advanced recently. When the conventional Rutherford type cables are made with the keystone angle twice that of the SSC cable, they show negligible degradation of the critical current.[9] The winding curvature is more relaxed in the non-circular than in the circular aperture coil design. When the magnets are designed to have the non-circular aperture coils, it may enable us to wind the Rutherford type cables of larger keystone angle without use of wedges. Eccentric and elliptic coils are considered in this work. Advantages of these coils in the field quality have not been understood yet for the SSC size magnet so far. We attempt to reveal the multipole characteristics of the magnetic field that is produced by non-circular aperture coils.

CIRCULAR APERTURE COILS WITHOUT WEDGE

Multipole characteristics of the completely self-arched coils with a circular aperture were already studied in connection with the use of the "braid-in-strands" cables with large keystone angle. Examples of coil parameters were presented[6] for a typical two-layer magnet with a small aperture. When no wedge is inserted in the coils, both 14 and 18 pole coefficients are about 10^{-4} cm^{-n} in the usual units,[6] and one order larger than the values of the SSC tolerances.[2] The reason is as follows: The angles of θ for inner and outer coils are chosen primarily to produce zero values of both 6 and 10 pole coefficients, so that there is little freedom to adjust the coil angles to make acceptably small quantities of 14 and 18 pole coefficients.

As is well known, the magnitude of the multipole component has a dependence on a coil radius in a form of a^{-n-1}, where a is the average radius of either inner or outer coil, and n the number of orders of the multipole concerned. The higher pole coefficient produced from the single layer coil decreases steeply with increasing its average radius. In the magnet size for SSC, the average radius of the inner coil is relatively small in comparison to the outer one. The higher pole component generated by the inner coil becomes much larger than that by the outer, and then is never cancelled out. Unlike this situation, the TEVATRON magnet has the inner coil of which average radius is relatively close to that of the outer coil. Hence, the size of 14 pole coefficient is successfully adjusted to a negligible magnitude by the use of no wedges in coils.

BASIC FIELD CHARACTERISTICS

Use of a current sheet configuration often helps us to understand the basic multipole characteristics of the magnetic field. The magnetic field is calculated two-dimensionally by complex numbers. A current I at a complex position $z=x+iy$ produces the field $B=B_x+iB_y$ at $z_0=x_0+iy_0$. The field is expressed as

$$iB^* = -(\mu_0/2\pi)I/(z-z_0) \quad , \tag{1}$$

where $*$ shows the complex conjugate, and μ_0 the permeability of vacuum. When the field is expanded at the origin, i.e. the magnet center, it becomes

$$iB^* = -(\mu_0/2\pi) \sum_{n=0}^{\infty} (\ I\ z^{-n-1})z_0^{\ n} \quad . \tag{2}$$

A complex multipole coefficient C_n is thus defined as

$$C_n = -(\mu_0/2\pi)\ I\ z^{-n-1} \quad . \tag{3}$$

Eccentric current sheet

The current sheet for an eccentric configuration is illustrated in Fig. 1. The current per unit length is written by λ. The field enhancement through the iron yoke is neglected for simplicity for a moment. When the current sheet of the circular arch has a curvature of r and its center deviates by d from the origin, the location of the current sheet on the median plane (x axis) becomes a=r-d.

For the current sheet configuration, C_n is generally expressed by

$$C_n = -(\mu_0/2\pi)\ \lambda \int z^{-n-1}ds \quad . \tag{4}$$

The value of ds indicates the increment along the current sheet, and it becomes ds = r dϕ. Then, a position on the eccentric current sheet is written as

$$z = -d + (a+d)e^{i\phi} \quad . \tag{5}$$

An eccentricity p=d/a is defined here. The position on the sheet is modified as

$$z = ae^{i\phi}(1+p)\ \{1 - (p/(1+p))e^{-i\phi}\} \quad . \tag{6}$$

The integration in eq. (4) is carried out in the angular range of zero to θ, and is written by Fx_n as

$$Fx_n = a^{-n-1}(1+p)^{-n-1}\{(n+1)^{-1}sin((n+1)\theta) +$$

$$\sum_{m=1}^{\infty}{}_{n+m}C_m\ (p/(1+p))^m(n+m+1)^{-1}sin((n+m+1)\theta)\} \quad . \tag{7}$$

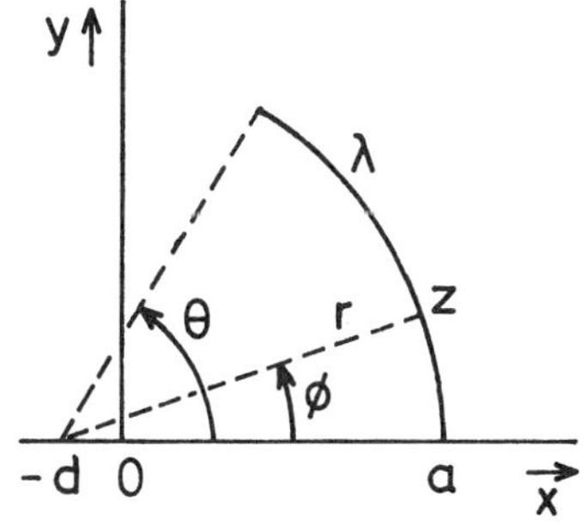

Fig. 1. Eccentric current sheet. A Coil is represented simply by the current sheet to clearly show the multipole characteristics.

The factor $(p/(1+p))^m$ in the above equation is not large because $p<1$. The sine function in the summation periodically makes the opposite sign as m increases, so that the summation procedure may produce a small value. Hence, the dominant part in the integral Fx_n may be its first term:

$$a^{-n-1}(1+p)^{-n-1}(n+1)^{-1}sin((n+1)\theta) \quad . \tag{8}$$

One can see that the conventional circular arch coil corresponds to the case of p=0. In contrast to eq. (7), the conventional arch produces the integral Fc_n as

$$Fc_n = a^{-n-1}(n+1)^{-1}sin((n+1)\theta) \quad . \tag{9}$$

From comparison between eqs. (8) and (9), $(1+p)^{-n-1}$ may be considered as a parameter which shows the effective increase of a in a sense of the average radius. When the factor has such meaning, the term of summation in eq. (7) expresses the degree of non-circularity with respect to the magnet origin.

The eccentricity p in eq. (8) is important particularly for the higher multipole components. Because the factor $(1+p)^{-n-1}$ decreases appreciably with increasing n, multipole coefficients for such poles as 14 and 18 become considerably lower for the eccentric configuration. In a rough estimate, the higher multipole characteristics for the eccentric coils may be mostly equivalent to that for the normal circular arch coils with bending radius of a+d.

For the dipole component, the factor $(1+p)^{-n-1}$ in the first term of eq. (7) is close to unity because of the least number of n=0, so that its effect on the dipole field is relatively small. The terms other than the first may overcompensate this reduction of the dipole field. Hence, Eq. (7) may give a somewhat higher value than in the case of the conventional circular aperture.

<u>Elliptic current sheet</u>

Elliptic current sheet is shown in Fig. 2. The position on the sheet is given by

$$z = acos\phi + ibsin\phi$$

$$= ae^{i\phi}\{(a+b)/2a - ((b-a)/2a)e^{-2i\phi}\} \quad , \tag{10}$$

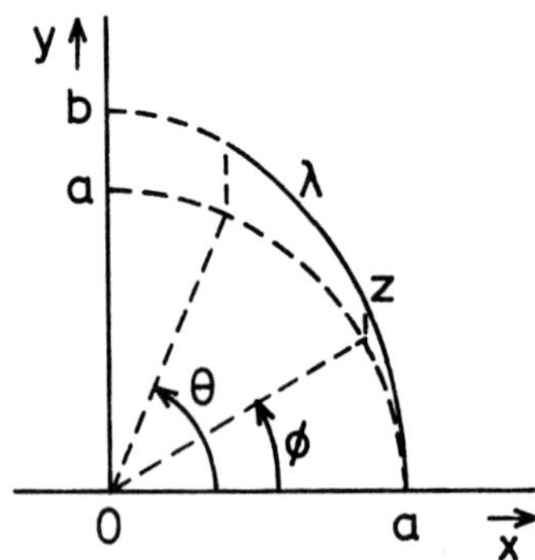

Fig. 2. Elliptic current sheet. A coil is simply represented by the current sheet to clearly show the multipole characteristics.

where ϕ is an angular parameter which indirectly determines the position as indicated in the figure. Introduction of a value of $q=(b-a)/2a$ leads to

$$z=ae^{i\phi}(1+q)\{1 - (q/(1+q))e^{-2i\phi}\} \quad . \tag{11}$$

The current sheet for the usual circular arch is represented by $q=0$. When z^{-n-1} in eq. (4) is integrated over the angel range ϕ of zero to θ, it becomes

$$Fe_n=a^{-n-1}(1+q)^{-n-1}\{(n+1)^{-1}sin((n+1)\theta) +$$

$$\sum_{m=1}^{\infty}{}_{n+m}C_m(q/(1+q))^m(n+2m+1)^{-1}sin((n+2m+1)\theta)\} \quad . \tag{12}$$

The factor $(1+q)^{-n-1}$ is again considered to represent the increase in the average radius. The term of summation in the above equation is regarded as the effect of the deviation from the circular arch. Eqs. (7) and (12) have the same form, putting aside the difference between m and 2m in the brackets. The elliptic coils are therefore considered to show multipole characteristics similar to that of the eccentric coils.

TWO-LAYER COILS WITH NON-CIRCULAR APERTURE

For the eccentric configuration, the inner and outer coils become concentric and there is no difficulty with calculation. The situation is reversed for the elliptic configuration. When the two coils are both elliptic, they can not be wound without use of highly sophisticated insulating layer between the inner and outer coils. Hence, the only innermost outline is taken as an elliptic curve whereas the others are not. The coordinates of the curves are explained in Fig. 3, and are described in the following form:

$$z=z_e+w(z_t/z_t{}^*)^{1/2} \quad , \tag{13}$$

where

$$z_e=a cos\phi +ib sin\phi \tag{14}$$

and

$$z_t=b cos\phi +ia sin\phi \quad . \tag{15}$$

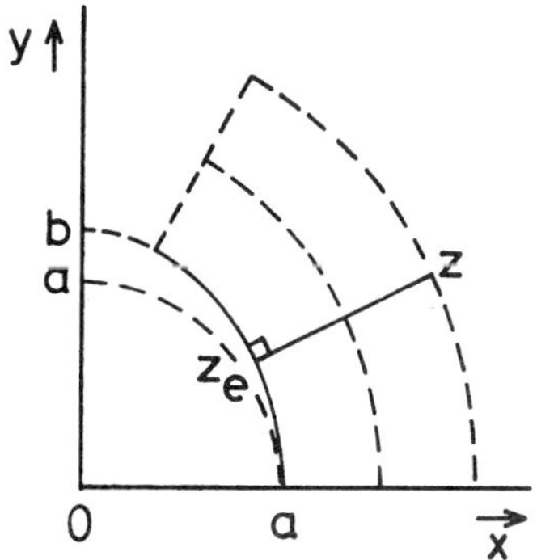

Fig. 3. Elliptic coil configuration in the field calculation.

The function $(z_t/z_t^*)^{1/2}$ indicates a unit vector, which is perpendicular to the elliptic curves shown by z_e . The value of w equals zero for the innermost curve, while for the other curves, w considers the cable width or that plus the thickness of the insulator between the inner and outer coils.

The field enhancement through the iron yoke is considered by a method of images. In the case of infinite permeability of iron, the field is written by the integral over the coil area as

$$iB^* = -(\mu_0/2\pi) \iint \{(z-z_0)^{-1} + (z-z_0)^*/R^2\}\, JdS \quad , \tag{16}$$

where J is the current density in the coil, and R the inner radius of the iron yoke. Current density is taken higher in the outer than in the inner coil, to take into account the conductor grading.

Table 1. Characteristics of eccentric coils neglecting the constraint of the integer number of turns. The figure on the right side shows the outline of the coils.

	Inner layer	Outer layer
Deviation (mm)	5.96	5.96
Radius (mm)	25.96-35.96	36.56-46.56
Max. angle (deg)	65.11	36.98
Current Density (A/mm^2)	330	400
Iron Inner radius (mm)	55.6	
Central field B_0 (T)	6.228	
Harmonics b_2 (10^{-4} @1cm)	0	
b_4	0	
b_6	0	
b_8	-0.08	
b_{10}	0.02	

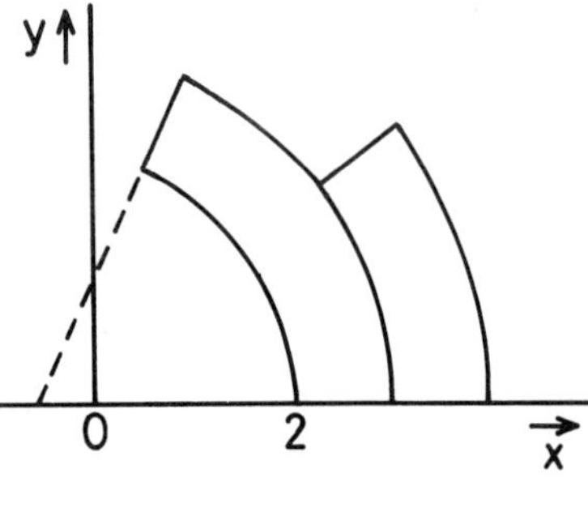

Table 2. Characteristics of elliptic coils neglecting the constraint of the integer number of turns. The figure on the right side shows the outline of the coils.

	Inner layer	Outer layer
Minor axis (mm)	40.0	
Major axis (mm)	48.44	
Position on x axis (mm)	20.0-30.0	30.06-40.06
Max. angle (deg)	74.28	43.76
Current Density (A/mm^2)	330	400
Iron Inner radius (mm)	55.6	
Central field B_0 (T)	6.376	
Harmonics b_2 (10^{-4} @1cm)	0	
b_4	0	
b_6	0	
b_8	-0.07	
b_{10}	0.01	

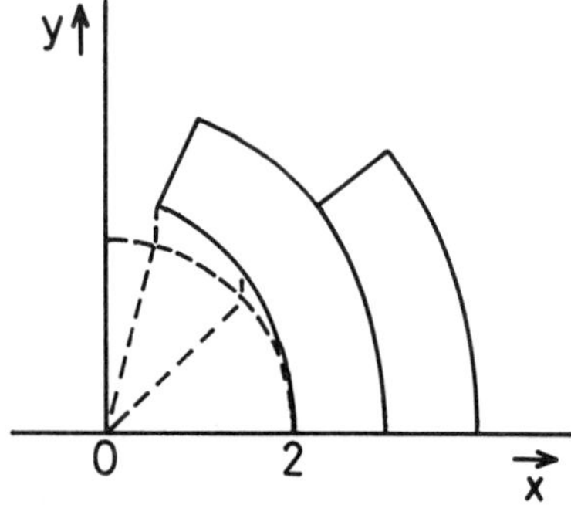

Arbitrary number of turns

A parameteric study is made at first by neglecting the constraint of
the integer number of turns of the cables. Results for the eccentric
configuration are shown in Table 1. The value of 14 pole coefficient can
be adjusted to zero. The magnitude of 18 pole component is reduced, but
still remains in a small amount. The improvement on the higher multipole
components is ascribed to the increase in the effective radius of coils. The
value of dipole component is raised by 8 % in comparison to the case of the
circular arch for the reason mentioned in the preceding section. The
transfer function then is expected to increase in comparison to the circular
arch configuration. For the elliptic coils, the results are listed in Table
2. The multipole characteristics are quite similar to that in table 1. This
similarity is understood from the current sheet consideration.

It may be interesting to adopt a coil configuration of eccentric
ellipse. As described above, however, both the eccentric and the elliptic
configuration produce mostly the same multipole field characteristics. Then,
no merit is found for such a complex coil structure of the eccentric ellipse.

Integer number of turns

The magnet coils are designed by considering the integer number of
turns. The results are shown in Tables 3 and 4 for the eccentric and
elliptic coils, respectively. The size of 18 pole coefficient becomes
slightly larger than in Tables 1 and 2. This is plausible because the more
constraint is imposed as the integer number of turns. The tolerance for 18
pole coefficient is 0.1×10^{-4} cm^{-8} for the systematic and 0.1×10^{-4} cm^{-8}
for the random errors.[2,10] As is well known, the experimental value of 18
pole coefficient usually has a very small random error and be in good
agreement with the design value. The present size of 18 pole coefficient may
be acceptable for the successful accelerator operation.

Table 3. Characteristics of eccentric coils. The figure on the right side
shows the outline of the coils, where are indicated the numbers of
turns.

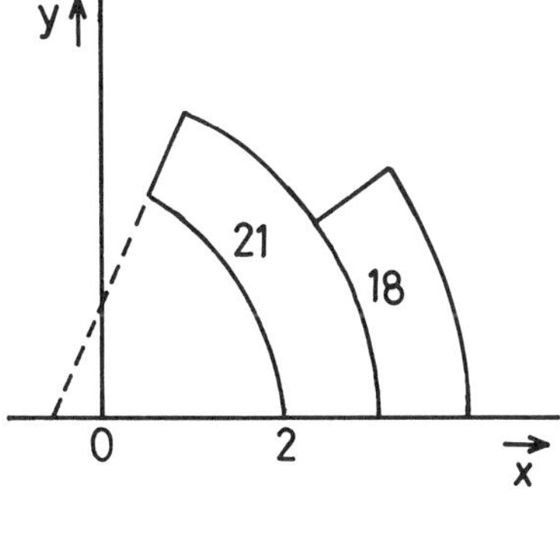

	Inner layer	Outer layer
Deviation (mm)	5.31	5.31
Radius (mm)	25.31-35.31	35.51-45.51
Max. angle (deg)	65.61	35.49
Number of turns	21	18
Cable thickn.(mm)	1.651	1.411
Keyst. angle(deg)	3.12	2.00
Current Density		
(A/mm^2) @ 2kA	121	142
Iron Inner radius(mm)	55.6	
Trans.func.(G/A) @ 2kA	10.97	
Current (A) for 6T	5.47	
Harmonics at 2kA		
b_2 (10^{-4} @1cm)	0	
b_4	0	
b_6	-0.03	
b_8	-0.12	
b_{10}	0.02	

For the eccentric configuration, the keystone angle is 3.12 degrees for the cable in the inner layer, whereas the value is 1.61 degrees for the inner layer of the SSC model magnet. The effect of the keystone angle on the critical current has been studied recently.[9] The degradation of the critical current was 1 and 4 % for the SSC size cables with keystone angles of 1.6 and 3 degrees, respectively. While the packing factor was 90 % for the cable with the angle of 1.6 degrees, it was increased to 95 % for that of 3 degrees because of the more deformation of the conductor. When the keystone angle was increased to about twice that of SSC, the decrease in the critical current was fully compensated by the enhancement of the packing factor. For the cable with the higher keystone angle, therefore, the overall critical current density keeps almost the same value.

When the number of turns in the outer layer decreases to about 15, we find the solution that also produces actually zero values of 6 to 14 pole coefficients, and gives the 18 pole value of 0.1×10^{-4} cm^{-8}. In this case, however, the transfer function is lowered by about 10 %. It may be difficult to accept this reduction of the transfer function.

In Tables 3 and 4, the calculation was made to minimize the square sum of 6 to 14 pole coefficients. This approach does not always result in the best field quality in the strict sense. The optimization to make the usable beam aperture maximum may produce the results that are slightly different from that of these tables. In fact, the 14 pole coefficient in Table 3 may be desired to have the opposite sign to the 18 pole. The value of 14 pole coefficient can be changed more easily than 18 pole, and accordingly overall coil parameters are almost the same as in these tables after this optimization in made.

Further eccentration

One may suppose that the further eccentration is effective on the

Table 4.   Characteristics of elliptic coils. The figure on the right side shows the outline of the coils, where are indicated the numbers of turns.

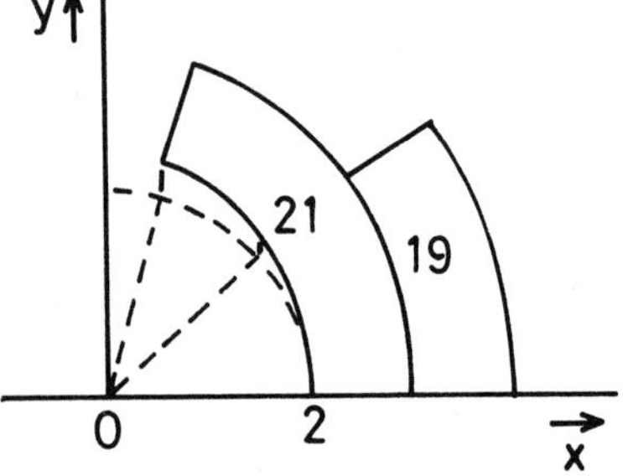

	Inner layer	Outer layer
Minor axis (mm)	40.0	
Major axis (mm)	47.01	
Position on x axis (mm)	20.0-30.0	30.02-40.02
Max. angle(deg)	73.99	41.84
Number of turns	21	19
Cable thickn.(mm)	1.655	1.400
Keyst. angle(deg)	3.40	1.96
Current Density (A/mm^2) @ 2 kA	121	143
Iron Inner radius(mm)	55.6	
Trans.func.(G/A) @ 2kA	11.20	
Current (A) for 6T	5.35	
Harmonics at 2kA		
b_2 (10^{-4} @1cm)	0	
b_4	0	
b_6	0.04	
b_8	-0.12	
b_{10}	0.02	

Table 5. Characteristics of eccentric coils with insertion of one wedge per
quadrant. The figure on the right side shows the outline of the
coils, where are indicated the numbers of turns.

	Inner layer	Outer layer
Deviation (mm)	13.0	13.0
Radius (mm)	33.0-43.0	43.2-53.2
Max. angle (deg)	53.72	37.27
Wedge angle (deg)	–	21.73-25.61
Number of turns	21	13 & 7
Cable thickn. (mm)	1.693	1.400
Keyst. angle (deg)	2.55	1.66
Current Density (A/mm^2) @ 2kA	118	143

Iron Inner radius (mm)	55.6
Trans. func. (G/A) @ 2kA	10.89
Current (kA) for 6T	5.51
Harmonics at 2kA	
b_2 (10^{-4} @1cm)	0
b_4	0
b_6	0
b_8	-0.02
b_{10}	-0.002

decrease in the 18 pole coefficient. If a single wedge is inserted in the
outer layer, the solution produces considerably small values of higher pole
coefficients. Results are shown in Table 5. The design of such larger
eccentration requires the cables of the lower keystone angle, and is
desirable from the viewpoint of cable fabrication. The current density in
the inner coils is lower by about 3 % than that of Table 3.

It has been reported that the quench has a tendency to start in the
location adjacent to a wedge in the inner coil.[11] Since the quench seldom
occurs from the outer coil, the insertion of wedge in the outer winding
alone is useful to produce the possibly higher central field. The wedge has
both an elastic modulus and friction coefficient that are different from
that of the conductor. Then, the insertion of wedge may introduce locational
errors in the magnet winding in a relatively appreciable manner. The
multipole coefficients are quite sensitive to winding accuracy of the
conductor position in the inner coil, whereas they are less influenced by
that of the outer coils. This is simply because the inner coils are closer
to the magnet center than the outer. The insertion of wedge in the outer
coil is expected to produce less variation of multipole coefficients than
in the case of that in the inner coil.

CONCLUSION

The non-circular aperture coils were studied to make the structure of
the accelerator magnet to be simpler. It was shown analytically that the
eccentric and elliptic coils have almost the same multipole characteristics.
Non circular aperture coils were useful to reduce the higher multipole
components such as 14 and 18 poles, and to increase the dipole field to some
extent. Magnet coils were designed assuming the non-circular aperture configuration without use of wedges, and they had the higher multipole coefficients of an acceptable size. When the cables were made for their keystone
angle to fit the curvature of these coils, no degradation was found in their
overall critical current density. If one wedge per quadrant was inserted in
the outer layer, the coil design led to a better field quality.

ACKNOWLEDGMENTS

The authors express their thanks to Dr. Masaru Ikeda of Furukawa
Electric Co. for his useful experiments and discussions.

REFERENCES

1. "Superconducting Accelerator," Fermilab Report (1979).
2. J. D. Jackson, ed., "Conceptual Design of the Superconducting
   Super Collider," SSC-SR-2020 (1986).
3. G. Brianti and L. Burnod, The Large Hadron Collider in the LEP
   Tunnel, "Proc. of the 1987 ICFA Seminar on Future Prospectives
   in High Energy Physics," BNL 52114: 179 (1987).
4. SSC Central Design Group, "SSC Conceptual Design - Magnet Design
   Details," SSC-SR-2020B (1986).
5. H. Hirabayashi, and K. Ishibashi, A Design Study on High-Energy
   Accelerator Magnets with Complete Arch Coils by Special Keystone
   Cable, "Proc. of the 1987 ICFA Seminar on Future Prospectives in
   High Energy Physics," BNL 52114: 327 (1987).
6. K. Ishibashi, K. Kamezawa, A. Katase, A. Terashima, Y. Sakakibara,
   Y. Funahashi, and H. Hirabayashi, Design Study on Future
   Accelerator Magnet Using Largely Keystoned Cables, Presented
   at "the 1988 Applied Superconductivity Conference,"
   San Francisco (1988).
7. H. Brechna, "Superconducting Magnet Systems", Springer-Verlag,
   Berlin(1973).
8. J. Strait, B.C. Brown, and M. Harrison, private communication,
   Fermilab people also had the same idea to improve the field
   quality of the Fermilab-style magnets, See the report: T.
   Collins, Offset Coil Design for Superconducting Magnets - a
   Logical Developement, Fermilab Report TM-1406 (1986).
9. M. Ikeda, private communication and also his presentation entitled
   "Development of SSC Cable in Furukawa" at this symposium.
10. A. W. Chao, Magnet Field Quality Requirements for the SSC, "Proc.
    of Workshop on Superconducting Magnet and Cryogenics (ICFA),"
    BNL 52006: 171 (1986).
11. P. Wanderer, J.G. Cottingham, P. Dahl, M. Garber, A. Ghosh, C.
    Goodzeit, A. Greene, J. Herrera, S. Kahm, E. Kelly, G. Morgan,
    A. Prodell, W. Sampson, W. Schneider, R. Shutt, P. Thompson,
    and E. Kelly, Test Results from Recent 1.8-m SSC Model Dipoles,
    presented at "the 1988 Applied Superconductivity Conference,"
    San Francisco (1988).

A HIGH RESOLUTION SCINTILLATING FIBER GAMMA-RAY TELESCOPE[*]

M. Atac, D. B. Cline, M. Cheng, J. Park and M. Zhou

University of California at Los Angeles

E. J. Fenyves, R. C. Chaney and H. Hammack

University of Texas at Dallas

ABSTRACT

A large variety of Compton and pair production gamma-ray telescopes using scintillator blocks coupled to vacuum photomultiplier tubes has been built earlier for space based experiments.[1,2] In these telescopes the scintillator block dimensions were large thus limiting the angular accuracies and resolutions. Here we are presenting the design and major characteristics of a large area gamma-ray telescope with high angular and energy resolution for high energy accelerator experiments, using scintillating fibers and recently developed position sensitive photomultiplier tubes.

INTRODUCTION

The University of Texas at Dallas and University of California at Los Angeles collaboration is developing a high resolution, large area gamma ray telescope for high energy accelerator experiments using scintillation fibers coupled to position sensitive photomultipliers. The general principle of the telescope design is similar to that of the High Resolution Gamma Ray Telescope developed by the same collaboration for space based experiments[3,4].

THE DESIGN CONCEPT

The high resolution, large area gamma ray telescope under development consists of a Compton-Pair Production Converter, a Gas Drift Time Projection Chamber (TPC), and a Calorimeter (Fig. 1). The Converter operates at high energies as a pair production converter measuring the position of the electron pair production with a high spatial resolution, and detecting the scattering and bremsstrahlung of the electron and positron in the Converter. The gas TPC follows the tracks of the electron pair, and the energy of the electron and positron is determined in the

---

[*]Partly supported by the Strategic Defense Initiative Organization, Office of Innovative Science and Technology.

Calorimeter with a high resolution by measuring the development of the
electromagnetic cascades generated by them in the Calorimeter. From these
data the energy and direction of the incident gamma can be precisely
determined.

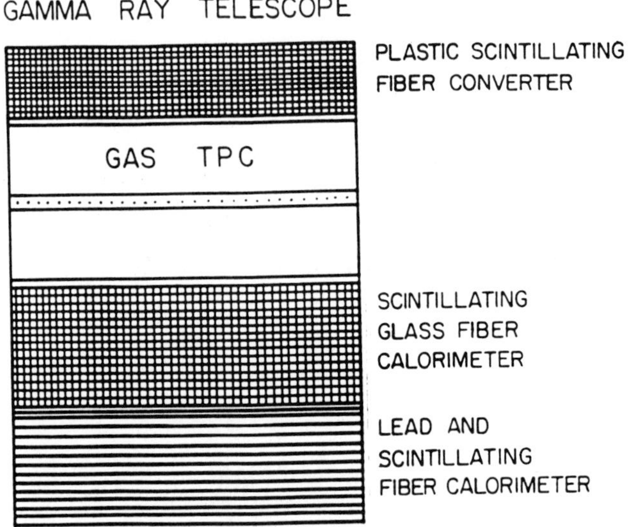

Fig. 1   Schematic drawing of the telescope.

The Converter has a surface area of 1 m^2, and the Converter and
Calorimeter are separated by 100 cm in order to obtain a high angular
resolution. The thickness of the Converter is chosen to be one radiation
length for increasing the probability of pair production. The energy loss
of the electron and positron in the Converter can be measured with a high
precision, and the secondary photons generated can be followed in most
cases into the Calorimeter.

The Converter consists of 1 mm x 1 mm cross-section scintillating
plastic fibers to provide good energy and conversion point resolution due
to the low density of the plastic fibers. The plastic scintillating
fibers are made of polystyrene doped with butyl-PBD and POPOP producing
λ = 420 nm wave length photons, and clad with PMMA (poly-
methylmetacrylate)[5] (see Fig. 2). The attenuation length of the photons
in the fibers exceeds 2 m.

A 1 MeV electron would traverse about 4 fibers in normal direction.
This way we can obtain a 3-dimensional image of the tracks of >1 MeV
Compton electrons, electron pairs or showers produced by the gammas. We
expect about 20 photoelectrons per fiber per minimum ionizing track
passing normal to the scintillating fiber using the 18% quantum efficiency
of the Hamamatsu R2487 position sensitive photomultiplier tubes[6] (Table I
and Fig. 3). The photomultiplier tubes are equipped with 17 x 19
orthogonal wire anode planes which can provide a position resolution of
σ_{rms} ~1 mm.

The scintillation fibers are stacked into 1 mm thick fiber planes U,
V and W that are rotated by 60° angle relative to each other (Fig. 4), and
directly coupled to position sensitive photomultiplier tubes (Fig. 5).

The Calorimeter of the high resolution, large area gamma ray
telescope is made of two parts. The upper part is 4 radiation lengths
deep and made of scintillating glass fibers to measure gamma energies up

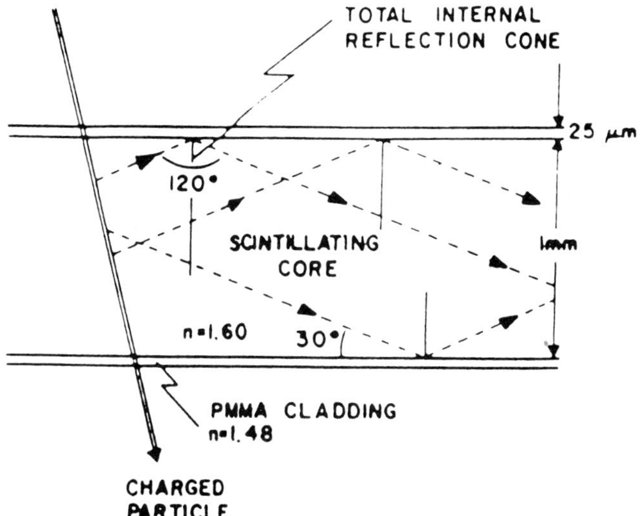

Fig. 2   Plastic Scintillating Fiber -- Polystyrene doped with Butyl-PBD
(λ = 420nm) and cladded with PMMA

to 20-30 MeV.   The lower part is made of alternating layers of 0.20
radiation length thick lead plates and 1 mm thick plastic scintillating
fiber planes. This section can cover gamma energies up to 4 GeV with its
more than 20 radiation lengths of material. The light produced in the
scintillating fibers and detected by the position sensitive
photomultiplier tubes, provide accurate position and energy information
about the electromagnetic cascade in the calorimeter.

ENERGY AND ANGULAR RESOLUTION

We obtain 3-dimensional tracking information about Compton electrons
and electron pairs in the whole telescope. This determines among others,
the conversion point of the incident gammas. The track length and the
total pulse height obtained from the position sensitive photomultipliers
provide the energy measurement of the electrons. Monte Carlo
calculations show good angular and energy resolutions, and high detection
efficiency for 10-100 MeV gammas (Figs. 6-8).

The transit time of the photomultipliers is 17 nsec with a pulse
rise time of 5 nsec (Fig. 9). Therefore, an overall time resolution of 5
nsec is achievable.

The scintillating fibers have a rate capability of ~10^8 tracks per
cm^2 of fiber cross-section area per sec. According to this very large
gamma ray fluxes or gammas in the presence of rather large backgrounds
can be detected.

The experimental testing and measuring of the energy resolution of
the scintillation fiber detectors is carried out with radioactive
sources, e.g. a Co^{60} source (Fig. 10), and accelerated particle beams at
Fermilab.

POSITION-SENSITIVE PHOTOMULTIPLIER TUBES R2487 SERIES

DIMENSIONAL OUTLINE (Unit: mm)

Figure 5: R2487

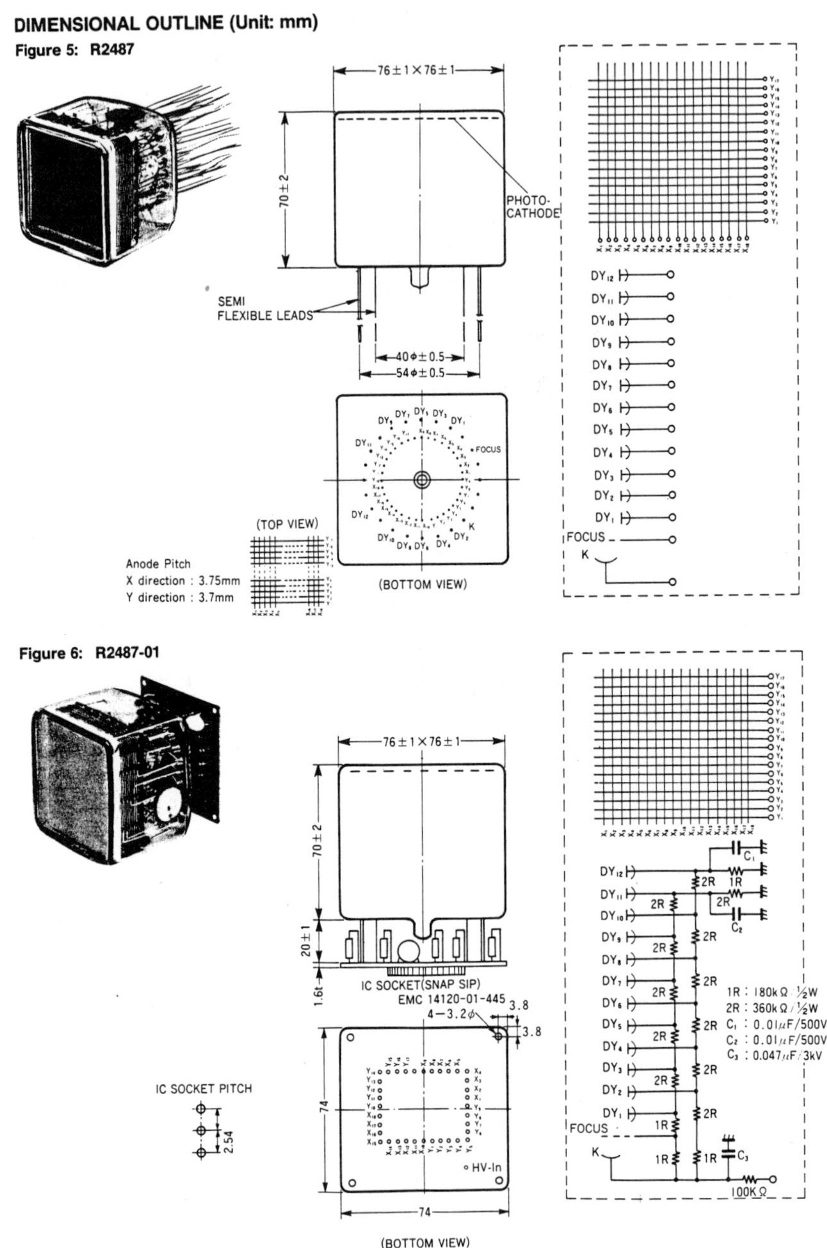

Figure 6: R2487-01

POSITION-SENSITIVE PHOTOMULTIPLIER TUBES R2487 SERIES

GENERAL

Spectral Response .... 300 to 600 nm
Wavelength of Maximum Response .... 420 nm
Photocathode Material .... Bialkali
Window
    Material .... Borosilicate glass
    Shape .... Plano – plano
Dynode
    Structure .... Proximity mesh
    Number of Stages 12
Anode
    Number of Wires .... 18 (X) + 17 (Y)
Effective Area .... 55 mm (X) x 45 mm (Y) Min.

MAXIMUM RATINGS (Absolute Maximum Values)

Supply Voltage
    Between Anode and Cathode .... 1300 Vdc
Average Anode Current[A] .... 0.1 mA
Ambient Temperature .... -80 to $+50°C$

CHARACTERISTICS (Typ. at 25°C)

Cathode Sensitivity
    Luminous[B] .... 60 μA/lm
    Radiant at 420 nm .... 60 mA/W
    Blue[C] .... 7.0 μA/lm-b
    Quantum Efficiency
    at 390 nm (peak) 18%
Anode Sensitivity[D]
    Luminous .... 6.0 A/lm
    Radiant at 420 nm .... 6.0×10^3 A/W
Current Amplification[D] .... 1.0×10^5
Anode Dark Current[E] .... 20 nA
Time Response[D]
    Rise Time[F] .... 5.5 ns
    Electron Transit Time[G] .... 17 ns

NOTES

A: Averaged over any interval of 30 seconds maximum.

B: The light source is a tungsten filament lamp operated at a distribution temperature of 2856K. Supply voltage is 150 volts between the cathode and all other electrodes connected together as an anode.

C: The value is cathode output current when a blue filter (Corning CS No. 5-58 polished to 1/2 stock thickness) is interposed between the light source and the tube under the same condition as Note B.

D: Measured with the same light source as Note B and with the anode-to-cathode supply voltage and voltage distribution ratio shown in Table 1.

E: Measured with the same supply voltage and voltage distribution ratio as Note D at 30 minutes after removal of light.

F: The rise time is the time for the output pulse to rise from 10% to 90% of the peak amplitude when the entire photocathode is illuminated by a delta function light pulse.

G: The electron transit time is the interval between the arrival of a delta function light pulse at the entrance window of the tube and the time the output pulse reaches the peak amplitude. In measurement the entire photocathode is illuminated.

Table 1: Voltage Distribution Ratio and Supply Voltage

Electrode	K	D₁	D₂	D₃	D₄	D₅	D₆	D₇	D₈	D₉	D₁₀	D₁₁	D₁₂	P
Distribution Ratio		1	1	1	1	1	1	1	1	1	1	1	1	1

Supply Voltage: 1250 Vdc
K: Cathode, D: Dynode, P: Anode

Figure 2: Example of Spatial Resolution

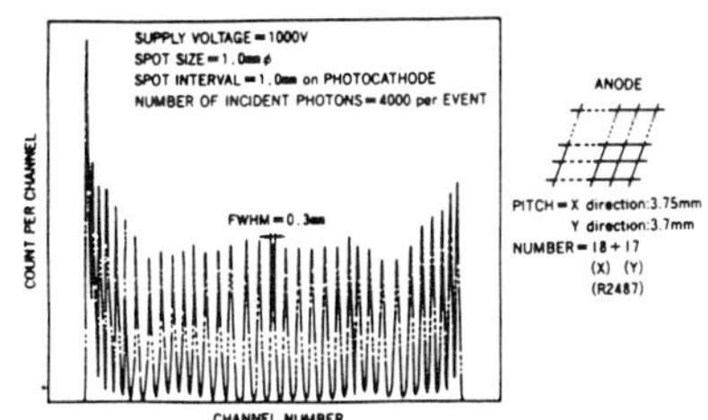

Figure 3: Example of Spatial Resolution as a Function of Incident Photons

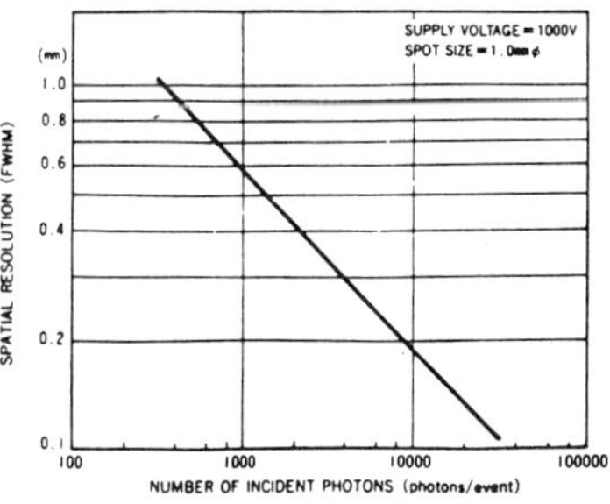

Fig. 3

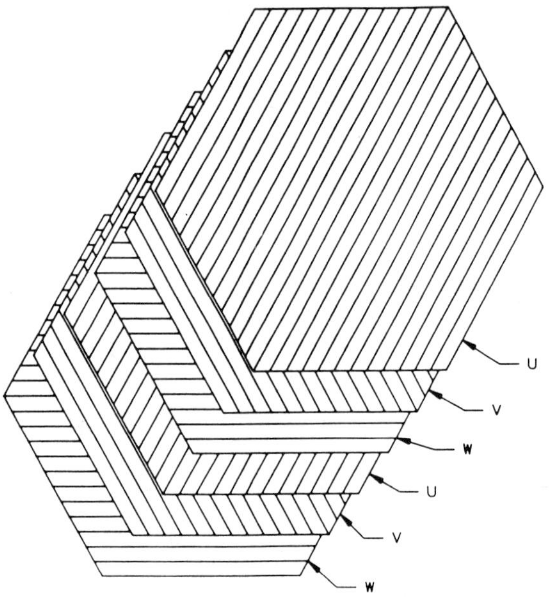

Fig. 4   Scintillating fiber plane orientation in a spaced up view

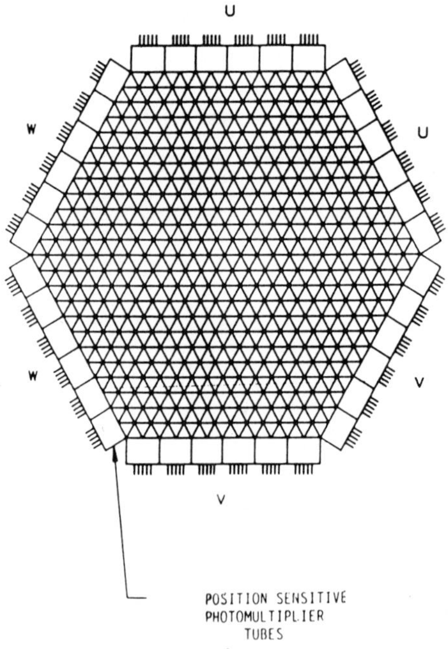

Fig. 5   Top view of the telescope showing the coupling of the fibers to the
position sensitive photomultipliers

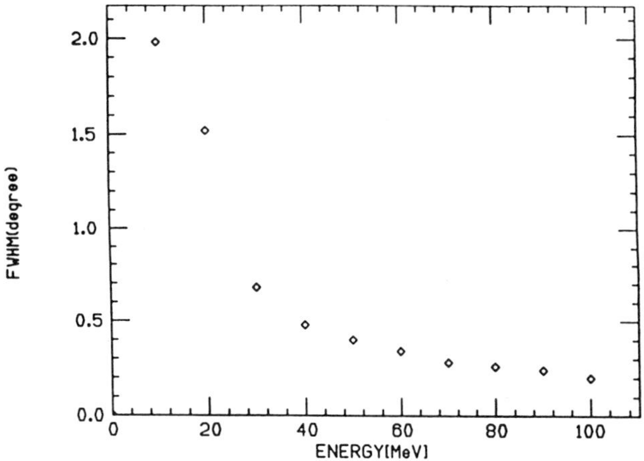

Fig. 6   Angular resolution as a function of gamma energy

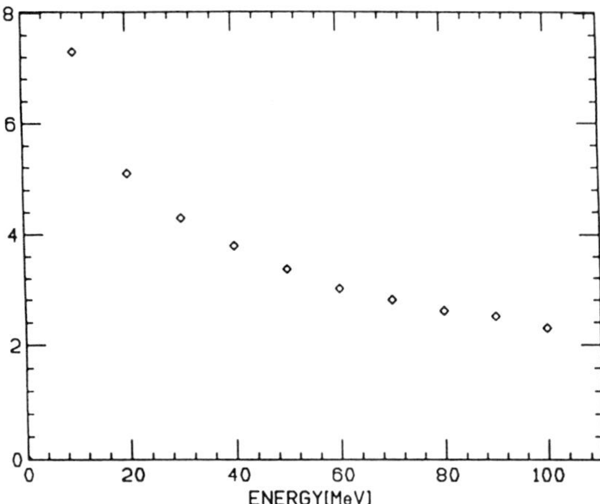

Fig. 7   Energy resolution as a function of gamma energy

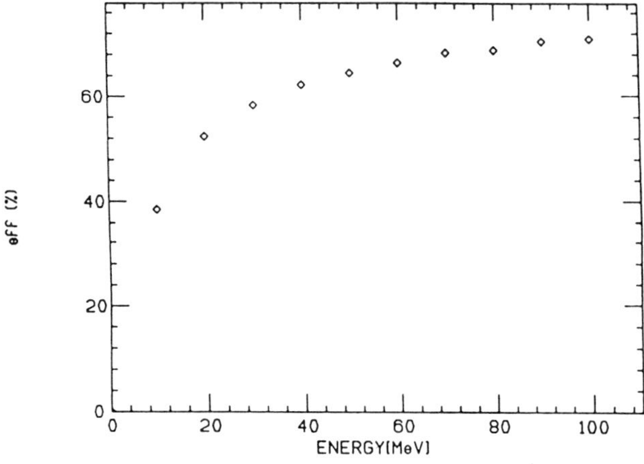

Fig. 8   Detection efficiency as a function of gamma energy

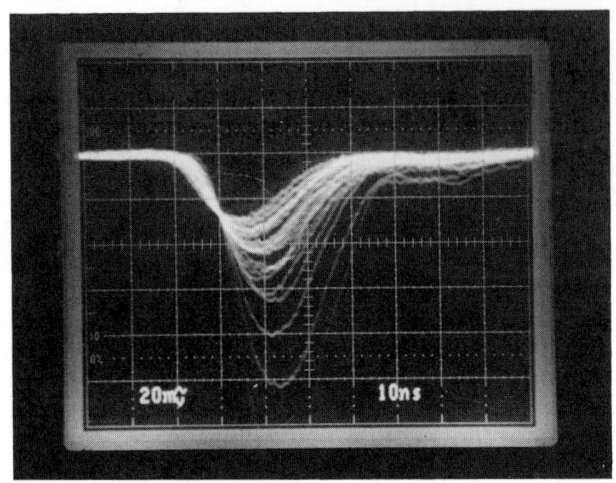

Fig. 9   Output signals on the anode of the Hamamatsu position sensitive photomultiplier tubes

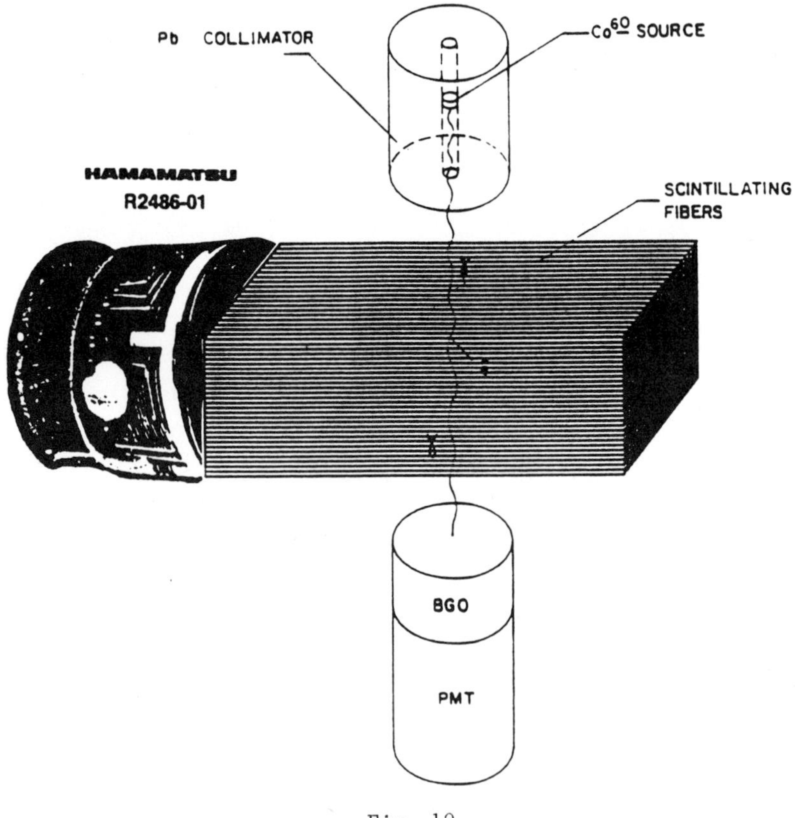

Fig. 10

REFERENCES

1. V. Schoenfelder, F. Graml, and F. P. Penningsfeld, The Astrophysical
    Journal 240 (1980) 350-362.

2. Papers presented by R. Stephen White, James M. Ryan and Robert C.
    Hartman on the Workshop on High Resolution Gamma Ray Cosmology,
    UCLA, Nov. 2-5, 1988 (to be published in Nuclear Physics B,
    Proceedings Supplement Section).

3. Ervin J. Fenyves, Proc. SPIE - The International Society for Optical
    Engineering, Vol. 879 (1988) p. 29.

4. Muzaffer Atac et al., Workshop on High Resolution Gamma Ray
    Cosmology, UCLA, Nov. 2-5, 1988 (to be published in Nuclear
    Physics B, Proceedings Supplement Section).

5. H. Blumenfeld, M. Bourdinoud and J. C. Thevenin, Nucl. Inst. and
    Meth. in Phys. Res., A 257, (1987) 603-606.

6. H. Uchida, T. Yamashita, M. Iida and S. Muramatsu, IEEE Trans. on
    Nucl. Sci. Vol. 33, No. 1, Feb. 1986.

3   Plenary Technical Session

Chairman:

R. Diebold
U.S. Department of Energy

ENGINEERING AND CONSTRUCTION EXPERIENCE

AT LEP (CERN)

G. Bachy

CERN
1211 Geneva 23
Switzerland

1. LEP IN THE EUROPEAN FRAMEWORK

CERN was born in 1953 in the centre of Europe - emerging weakened and impoverished from 6 years of war - as a cooperation between twelve Members states[1] with a goal of "nuclear research of a pure scientific and fundamental character". Since then, CERN has grown in terms of both facilities and personnel and now constitutes a community of some six thousand persons : physicists and engineers from Member and non-member states, designers, draughtsmen, electronics and computer experts etc.

The latest large facility presently under construction is LEP[2] which reflects the choice made in the 70's by the European physics community for investigating deeper the structure of matter and the forces which govern the Universe.

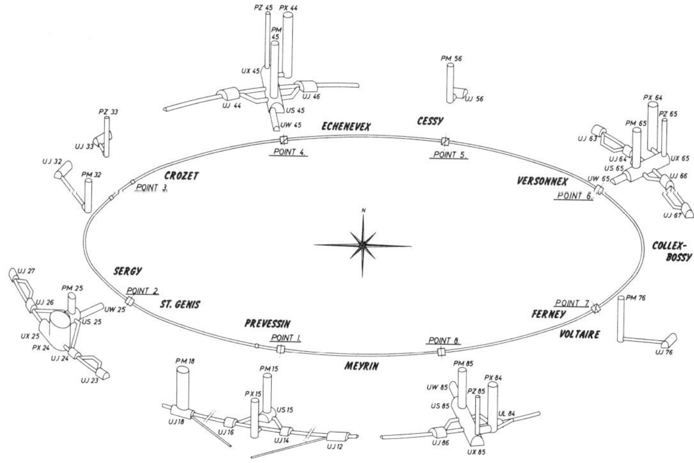

Figure 1. LEP underground work

2. LEP IN A FEW NUMBERS

LEP consists of a ring (see Figure 1), located in an underground tunnel, 26.6 km long, 3.8 m in diameter and excavated between 50 and 150 m below the surface. To minimize shaft depth, the tunnel is situated on an inclined plane (1.4%).

Three quarters of the ring lies in France and one quarter in Switzerland. Access shafts, 18 in total, with diameters between 5 and 24 m at eight points of the ring, link the LEP tunnel to the surface buildings. Four underground experimental areas are excavated and house the large detectors.
 The ring is not a perfect circle but consists of eight curved sections, each 2800 m long, equipped with 1700 dipole pairs and linked by 500 m long straight sections. The particle beams are focused by 816 quadrupole magnets and 504 sextupole magnets. The bending magnets are of the low field type (0.1 T) to keep synchrotron-radiation within acceptable limits. In two of the eight long straight sections the particles are accelerated by 128 copper cavities, equipped with energy-saving storage spherical cavities, giving a maximum energy per beam of above 50 GEV. For the next step, called LEP 200, the copper cavities will be progressively supplemented with superconducting cavities to increase the machine energy above the W mass.
As far as the vacuum system is concerned, due to the synchrotron-radiation, the aluminium alloy extruded vacuum chamber is watercooled and covered by lead radiation shielding. For the first time the main pumping system is based on a non-evaporable getter (NEG) strip installed in the pumping channel part of the extrusion.

3. LEP IN A FEW DATES

The LEP project was approved in December 1981. Civil Engineering work started in September 1983. Installation at the surface started in June 1985 and underground installation in October 1985 at a very slow rate following building delivery by the civil engineering. 50% of the underground installation was performed during 1988. The aim is to "close" the machine mid-July 1989 and to proceed with the beam tests.

4. THE DESIGN STAGE

4.1 Design thoughts and principles

The LEP project involved about 4000 man years of indoor design, coordination, labwork and construction follow-up, about one third of which is design proper.
Before setting up the necessary task force for the design of such a large project, it is necessary to decide what has to be studied in-house and what should be subcontracted. Answers to this question are not always economically optimal as several pre-existing boundary conditions need to be taken into account, in particular presence of in-house specialists. The choice is more obvious for the design of the **machine elements** proper as the know-how in these fields (magnets, RF, vacuum ..) has for years been at a superior level in our type of research laboratory. In this case usage at CERN is to design, have a prototype built and tested inside and then send out a call for tenders for the production series. The economical choice in this case is more what to subcontract : subcomponents, components or complete systems. Experience shows that when a lot of different technologies are involved, a unique firm is seldom capable of doing the complete job but will subcontract several subassemblies. Money **can** usually be saved when subcontracting the various components and by having a special contract with another firm for assembly.
At LEP a typical case is illustrated by the 1740 dipole vacuum chambers. These aluminium alloy chambers were extruded in Germany then shipped to Austria for welding the various fittings and flanges, then shipped back to Germany for lead cladding and finally to CERN for testing before assembly in the dipoles. For this unique component 5 different firms were involved
(see Figure 2).
Other reasons could lead to the same splitting principle : an example is transport as is the case for the dipole magnets of very special design (iron sheets imbedded into concrete). One dipole unit consists of a pair of 6 m cores. The sole "mechanical links" between the two cores are the vacuum chamber and the excitation bars, both having insufficient inertia for building a rigid structure out of the two cores. The assembly of the dipole subcomponents (cores, excitation bars and vacuum chamber) was thus performed at CERN, by a contractor in a hall directly located on top of an access pit to minimize the transport problems and optimize the installation sequence.

712

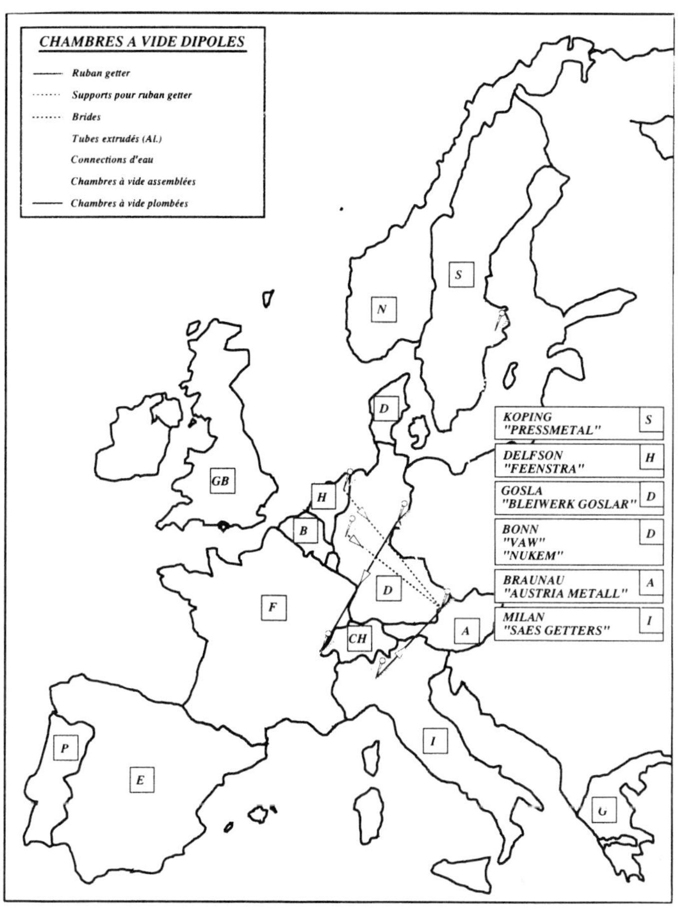

Figure 2. The round Europe trip of the dipole vaccuum chambers

It has been found that at the design stage it is very important to consider not only the component itself, but to integrate it in its environment at the final stage, and to consider the complete installation sequence. From the beginning of the project an effort was made to design all components from their functionality to installation.

A different approach can be taken in the more classical fields, the so-called **conventional facilities** : civil engineering, cooling & ventilation and electrical distribution. The main in-house role here is to define the functions, the basic requirements, the performances, the environment and the boundary conditions.

The environment and boundary conditions were found to be of prime importance.

Two other aspects during the design phase which have important economical consequences are standardization and prefabrication.

Standardization minimizes the design costs as well as the construction costs. An attempt was made, for example to keep the type of buildings as few as possible.

Figure 3. A 6m metallic module being lowered into a LEP shaft

Prefabrication is a goal in itself. Great effort was made to minimize the work on-site, especially in the underground buildings where the hour cost is higher and the risks greater. For instance the equipment in the shafts was prefabricated into modules (either 6m metallic or 3m concrete - see Figure 3).This design enabled the complete metallic framework of a 100 m shaft to be erected in 3 days!

On the machine side, the same effort was made to avoid crowding in the tunnels.

For instance subassemblies consisting of a quadrupole, sextupole, correcting dipole, beam pick-up and a vacuum chamber were assembled and aligned on a common beam at the surface, on top of a pit directly linked to the machine tunnel.

These are just a few aspects at the design stage where we tried on one hand to avoid studying each element as an isolated item, and on the other hand to prefabricate the maximum possible beforehand.

4.2 Design tools (Reference 3)

A great deal of LEP design has been achieved using CAD. The main objective in the introduction of CAD from the beginning was the complete memorization of the whole machine. The modelisation of each component is not only used to produce standard production drawings, but also in the subsequent assemblies (see Figure. 4). Following LEP experience it is thought essential to keep in-house all the so-called general layout studies which summarize all the various systems on unique documents. In conjunction with CAD, a unique, controllable source of information is vital. In the LEP project the ORACLE Relational Database Management System was the unique source of information. This information, primarily used during the design and construction phase, will be of great help during the operation of the machine.

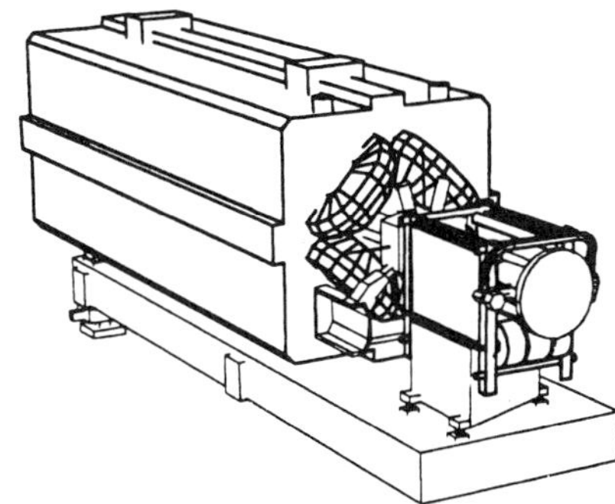

Figure 4. CAD representation of a short straight section assembly

5. THE INSTALLATION PHASE

5.1 LEP logistics (Reference 4)

The size of the LEP project, coupled with the tight construction schedule, called for organized planning, logistics and control. Considering the main tunnel alone, its 27 km's houses more than 50000 recorded different types of equipment of various dimensions and weight. In 1988, the most critical installation year for LEP, many activities were executed in parallel, spread over the 100 km² site, and performed by about 70 different contractors and subcontractors employing up to 700 people. Planning, follow-up and data feedback were essential to produce an accurate picture of the daily changes and to give a tool for immediate reaction. Conventional means were obviously out of the question for treating such an amount of information with the optimum response time. Again, all the data were managed with the Relational Database Management System, ORACLE. This powerful tool was used efficiently throughout the installation period, with the help of the various tables and application programs developed and simulated before the start of installation : POL (Planning using ORACLE for LEP), DICLEP (DICtionaries for LEP), LOI (Logistics Of Installation), Transport lists, follow-up of transport, follow-up of installation activities i.e work identified by its duration and resources, follow-up of magnetic components, feedback and statistics.

A general overview of the logistics behind LEP installation appears in Figure 5 .

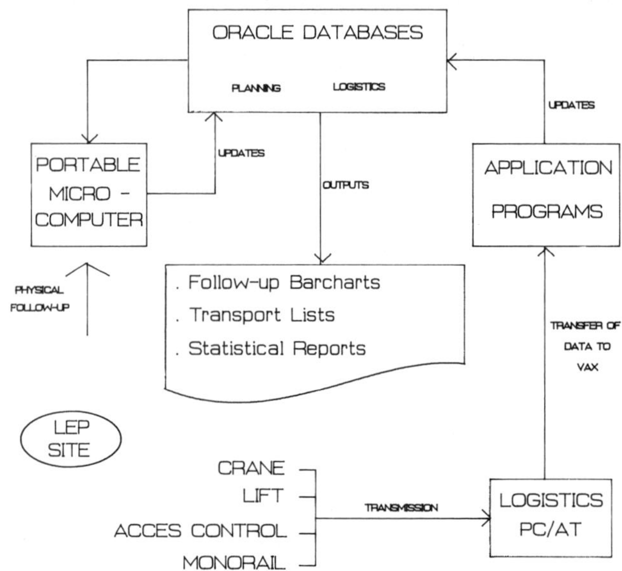

Figure 5. General overview of the logistics behind LEP installation

Figure.6 shows the number of activities followed per month since the beginning of the project.

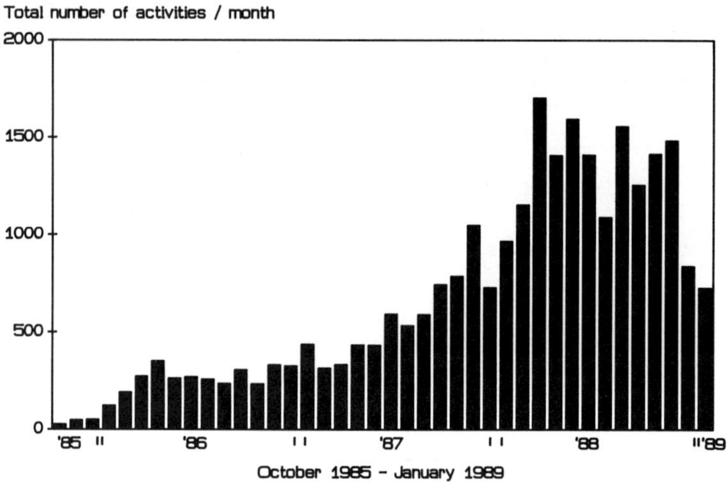

Figure 6. LEP activity follow-up.

5.2 Transport problems in underground tunnels

For the previous accelerators built at Cern (PS, ISR or even SPS) conventional ground transport means were adequate. However during the SPS installation it appeared that the speed of goods movement in the tunnel was considerably slowed down by the floor occupation of the various fitting teams. With LEP, the increase of size with the related increase of distance between pits (Reference 2) together with the very tight schedule called for an efficient means of underground transportation. The solution adopted comprises a monorail suspended from the concrete vault of the tunnel. This system has three main advantages :
- it travels off the ground which is very cluttered during the installation period

- it is self guided (a point of prime importance for safety)
- it offers an unlimited range of operation since it is powered by electric bus-bars.
The efficiency of the system was demonstrated when up to 12 Dipoles + 5 short straight section units were installed per day at distances of 18 km from the access shaft.
Another advantage of this transportation system is the standardization of the containers for handling material such as cable trays, pipes, bus-bars, cable drums and vacuum equipment. These standard containers, made available to the installation firms, implied much easier material preparation, weekly material volume evaluation and transport organization (monorail and crane loading).
In our opinion, the choice of underground transportation will be of utmost importance for the SSC installation.

6. CONCLUSION

The experience gained during the LEP project's engineering and construction / installation phase is multiple, but to relate everything in such a short paper is impossible. Nevertheless, the main features (all related to the size of the machine) appeared to us to be : a good analysis of contract splitting, minimization of costly underground work, maximization of prefabrication and standardization, use of a unique computerized source of information associated with modern CAD design tools, preparation of a management tool before the start of installation and a suitable choice of underground transport system. We certainly did not succeed in all aspects. However we tried!

<u>References</u>

1. LEP Design Report (2 Volumes), Reports CERN/LEP 84-01 (1984)
2. G. Bachy, Design, Installation problems and criteria related to large accelerators, 1987 Accelerator Conference, Washington DC
3. G. Bachy et al. Computer-Aided Engineering in High Energy Physics, 1988 European Particle Accelerator Conference, Rome
4. C. Genier et al. Logistics of LEP installation, 1988 European Particle Accelerator conference, Rome.

[1] Belgium, Denmark, France, Federal Republic of Germany, Greece, Italy, Netherlands, Norway, Sweden, Switzerland, United Kingdom. Austria became a Member in 1959, and Spain in 1983
[2] Large Electron-Positron (Collider)

INDUSTRIAL EXPERIENCE ON THE HERA ACCELERATOR AT DESY

Gustav-Adolf Voss

Deutsches Elektronen–Synchrotron DESY
Hamburg, West Germany

HERA is a unique new storage ring facility under construction at DESY,
Hamburg. A 4-mile circular tunnel houses a 1 TeV-proton-storage ring,
which will be, after completion for a number of years the world's
highest energy accelerator. In the same tunnel is a 35 GeV electron
storage ring, which, after the European LEP-machine, will be the
highest energy electron storage ring in the world. The counter-
rotating proton and electron beams can be made to collide in four
large underground halls. The project was authorized in April 1984. The
electron ring of HERA started operation in August 1988. First colliding beam experiments with protons and electrons are to start in 1990.

Besides its unique technical features, HERA is also remarkable in
other aspects: HERA is the first high energy accelerator to be built
as a national facility but with large international contributions of
equipment and labor. Contributions have been made by Canada, China,
France, Great Britain, Israel, Italy, Netherlands, Poland, USA (the
total value contributed amounts to about 200 million Deutsch Mark),
thereby significantly reducing the costs for the German government
(total costs of the project are about 1000 million Deutsch Mark). The
foreign contributions are often part of the foreign states' efforts to
develop new advanced technologies in their own countries or to gain
know-how by participating in the construction of this exciting device.

Another unusual feature is the location of this project: HERA is
within the city limits of a large metropolis. Its tunnel is built
under many private properties and public parks. This was only possible
because the laboratory enjoys trust and a very positive attitude in

its neighbourhood, a result of 25 years of a policy of open doors. The
construction of the tunnel and the huge underground halls met very
little opposition in the population and among the owners of those
private properties who had to give their consent to construction of
the tunnel under their homes. It was certainly very helpful that the
City of Hamburg was officially in charge of all the civil engineering
work for HERA. City officials could pave the way through all the
required formalities in a most efficient way. The actual supervision
of the civil engineering project was done by the private engineering
firm of WINDELS and TIMM. This firm designed tunnel, halls and other
structures in close collaboration with DESY staff, asked for tenders,
evaluated the bids and made recommendations for the award. Later they
supervised the construction. For the actual construction, four of the
most powerful construction firms in Germany and a smaller local
company (Dyckerhoff and Widmann, Hoch- und Tiefbau, Holzmann AG,
Prien GmbH, Wayss and Freytag) formed a consortium, the ARGE HERA
(ARGE = Arbeitsgemeinschaft). A tunneling machine was built, specially
suited for the sandy ground and capable of withstanding the water
pressure of up to 15m of ground water above the tunnel. The actual
tunnel was assembled out of prefabricated concrete parts with rubber
(EPDM) seals. Tunneling 10 to 20 meters under private properties was
sometimes nerve wracking for the afflicted home owners, but noise and
vibrations rarely lasted longer than one week at any given place.
Although the tunnel lies in a sloping plane and has a complicated
geometry, after each one mile trip the tunneling machine arrived at
the next underground hall with an accuracy of better than one inch.
The four underground halls were built in open pits located on public
land. Only small architect-designed entrance buildings could be built
here above ground and a considerable amount of money went into
restoring the parks and generously overcompensating the small loss of
park area with new trees, shrubs and ponds. Construction time was
4 years, only 4 months longer than the original schedule had called
for. The small delays were mostly due to some large and unexpected
boulders in the path of the tunneling machine, remnants of the last
ice age. There were no casualties and no serious injuries were
suffered during construction. The total cost of 227 million Deutsch
Mark was within 0.1% of the inflation adjusted estimate of 194 MDM in
1980. This of course is no accident but reflects the close control of
the construction project including many design changes to lower costs
and improve the lay-out.

The DESY laboratory has only a staff of slightly more than one
thousand, far too small for such a large project, if one wanted to
build it in the traditional way. This is particularly true if one
keeps in mind that much of the research program with DESY's other high
energy machines is still going on. The solution to this problem was to
increase temporarily the staff by hiring people on time limited
contracts (about 200) and by getting manpower help from universities
and laboratories outside of Germany, mostly Poland and China (100). It
also became evident that most of the work had to be farmed out to
industry. DESY know-how was used to develop crucial prototypes and to
coordinate the industrial efforts.

The most important, and as far as the SSC is concerned, the most
relevant work at HERA, is of course the construction of the
superconducting magnet system. In the HERA project, fabrication of
superconducting magnets must be done entirely by industry. DESY
developed a new type of superconducting magnet, which combines some of
the features of the Tevatron magnet with some from the ISAbelle
magnets: In the HERA dipole, aluminum clamps take up the magnetic
forces of the superconducting coil. The clamped coils are then
surrounded by cold iron yokes. The success of this new type of magnet
has already triggered a change in the design of the Russian magnets
for the UNK-project to something very similar to the HERA design. It
also had some apparent influence on the SSC design. DESY developed
this new type of magnet together with BBC (now ABB): BBC produced the
superconducting cable, DESY built the prototype coils with their
aluminum clamps, while BBC built the prototype cryostats. After the
successful tests of a few prototype magnets, tenders were invited for
the construction of the series magnets for HERA. The Italian
government offered to build half of all superconducting bending
magnets, placing orders with ANSALDO (for the coil assembly) EM-LMI
(for the fabrication of the superconducting cable) and ZANON (for the
cryostat and the final assembly). The order for the other half of the
bending magnets was placed with BBC. Preseries production of 30 magnets has been successfully concluded. All magnets built so far require
no training and easily exceeded current and field values corresponding
to energies higher than 1 TeV. Accuracy of coil fabrication was so
good, that higher field harmonics are smaller than 0.01%, well within

specification. Series production in Italy as well as in Germany has
started now.

The development of superconducting quadrupoles for the HERA-project
was done by the Saclay laboratory. The superconducting cable was
produced by the German company VACUUMSCHMELZE. 50% of all quadrupole-
magnets are being built by the French company ALSTHOM as the French
contribution to the HERA-project. The other half of the quadrupoles is
being built by the German consortium of KWU and NOELL. Series
production is in full swing. A total of 104 magnets have been
delivered. Quench currents exceed comfortably the values necessary for
1 TeV operation, field homogenuity is well within specification. Just
as with the bending magnets, training of these magnets is practically
not observed. Lack of training is an indication of proper coil
clamping at low temperatures and a proof of a good design.

Superconducting magnets need small correction fields at low excitation
to correct the effect of the so called persistent currents. These
correction coils are wound on the beampipes and are also
superconducting. The Dutch government took on the responsibility to
have most of these and other superconducting correction devices built
by Dutch industry as part of the Dutch contribution to HERA. This part
of the project has been the first one to be completed: The Dutch firm
SLE built the superconducting wire, the firm HOLEC built the
correction coils while the NIKHEF institute in Amsterdam supervised
this part of the project.

Another major part of the HERA-project is the helium refrigeration and
distribution system to cool the superconducting magnets. Three parallel refrigeration systems using screw compressors, oil separators and
turbines for adiabatic cooling each have a refrigeration power of
6.6 kW at 4.2°K and an overall efficiency of 1:285, a value which for
machines of this size is probably unmatched anywhere in the world. The
system was designed and built by SULZER and has now been in routine
operation for more than a year. Except for two compressor failures at
the running-in, the performance has been flawless over the last
$1\frac{1}{2}$ years. A contract was made with SULZER to operate and maintain the
refrigeration plant at the DESY site.

A large helium distribution system is currently being installed to
pipe liquid helium around the 4 mile ring and to collect the cold
helium gas for refrigeration in the central plant. The helium system
also carries 50°K helium for the heat shields in the magnet cryostats.
This very extended system, which also serves the cryogenic needs of
two large superconducting detector magnets, is currently being
installed by the German firms, LINDE AG and BABCOCK.

Another exciting development is the construction of superconducting
accelerating units for the electron ring. To reach an energy of
35 GeV, high gradient accelerating units are required which for high
efficiency should be superconducting. The CERN-laboratory at Geneva
and DESY worked together in the development of prototypes of helium
cooled niobium resonators. These prototypes were then successfully
tested in the DESY-storage ring PETRA. An order of 8 preseries
cryostats housing a total of 64 500 MHz accelerating cavities has been
placed with the DORNIER Comp. A second, different, approach to
superconducting resonator development is being pursued with INTERATOM.

The examples of industrial involvement in the HERA-project given so
far concerned more the advanced technology areas. A very large part of
the HERA rings of course uses very conventional technology. The
electron ring for example consists mostly of conventional steel
magnets. Here, as in earlier DESY projects, DESY did the design and
prototype work and ordered parts or subsystems from various industries
to have them assembled at DESY or outside DESY. Coordination and
responsibility for proper performance were always assumed by the
laboratory. But in contrast to earlier and smaller projects,
industrial involvement goes much further now and involves the entire
installation and commissioning of subsystems.

So far industry has been very responsive to our needs during the HERA-
construction. The fact that many components are manufactured outside
of Germany as foreign countries' contributions to the project has made
coordination slightly more complicated. But without these
contributions the project might never have been authorized. At this
moment the project is well within schedule and budget.

4 Parallel Technical Sessions II

Industrial Opportunities on the SSC

Chairman:

K. W. Chen
The University of Texas at Arlington

SSC COSTS AND PLANS FOR INDUSTRIAL INVOLVEMENT

Tom Elioff

SSC Central Design Group[*]
c/o Lawrence Berkeley Laboratory
1 Cyclotron Road
Berkeley, CA 94720

The objective of this presentation is to indicate the potential role for industry in various areas of the SSC construction program. In order to accomplish this goal, we will review and utilize the projected construction cost estimates that were developed for the SSC Conceptual Design Report (CDR) of 1986.

The CDR cost estimate was designed to be representative of SSC costs on a national average basis. Recently a specific site near Waxahachie, Texas has been selected by the Department of Energy (DOE). Most of the technical systems and components of the SSC are expected to be bid or purchased on a national basis, thus, their average costs will not be impacted significantly by the specific site. The costs of conventional systems (buildings and structures) are more likely to experience some change in cost detail due to the specifics of the geology and underground construction methods at the chosen site. These possible changes have not been fully evaluated; therefore, we will continue to use the CDR information as the best representation for relative component costs.

A Work Breakdown Structure (WBS) encompassing all aspects of the conceptual design was developed to ensure that all components of the SSC project were properly included in the design and costing process. The general philosophy of the SSC cost estimate required the inclusion of the total construction costs to bring a 20 TeV colliding beam proton accelerator to a state of operational readiness and to create a laboratory environment suitable for conducting high-energy physics experiments. The cost estimate basis thus includes all technical systems, the required conventional plant and associated laboratory support facilities, all engineering-design and inspection services for both technical-and conventional systems, and the associated management and administration services. The WBS to Level 4 is illustrated in Fig. 1. The Project Cost Summary to WBS Level 3 is provided in Table 1. In this report, all costs will be expressed in FY 88 dollars. The construction costs of Table 1 are graphically displayed in Fig. 2 in terms of the percent of total costs for each major WBS category.

In the development of costs for the CDR, a model was used to project which tasks would be accomplished by the SSC Laboratory forces and which tasks would be

---

[*]Operated by the Universities Research Association, Inc., for the Department of Energy.

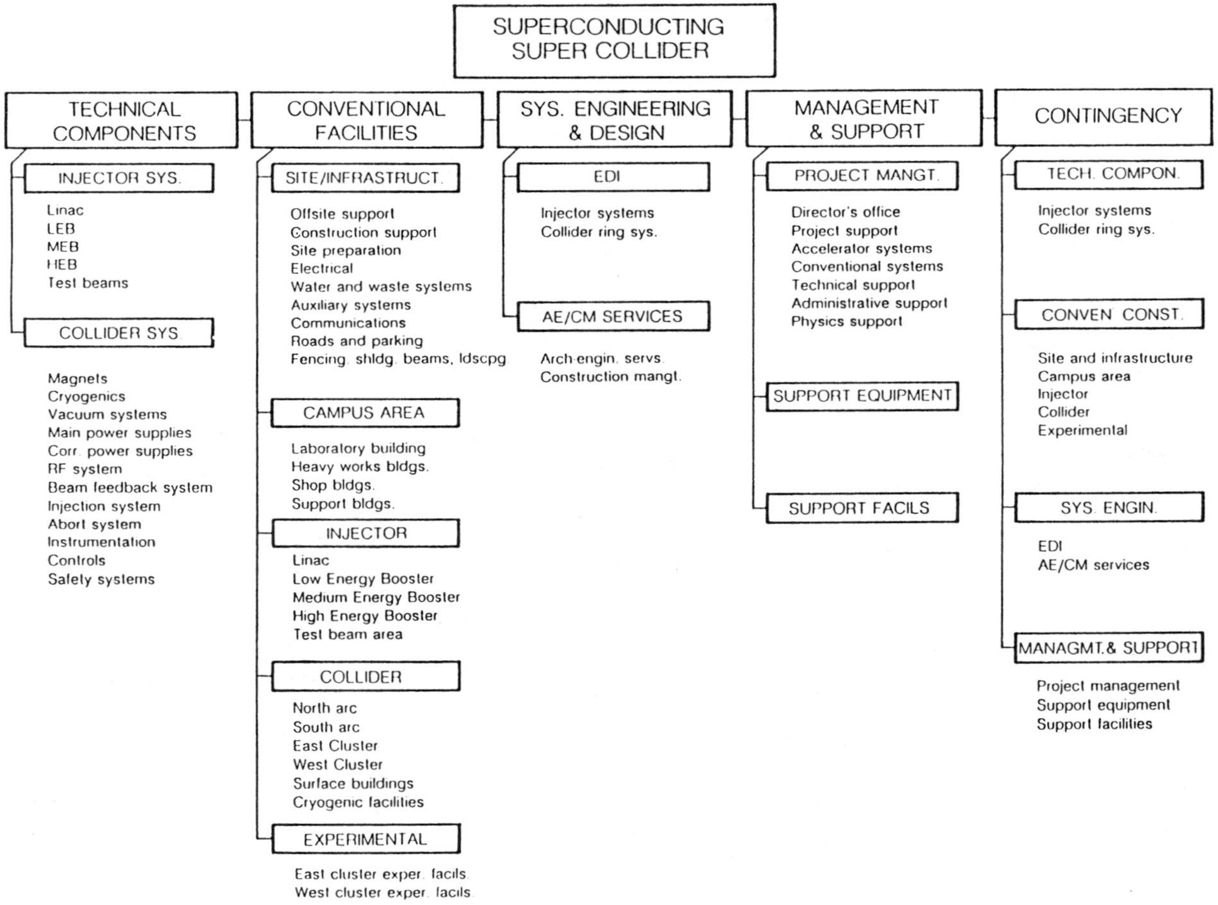

Figure 1. Work Breakdown Structure (WBS).

WBS				FY 88 M$
1.	Superconducting Super Collider			3,210
.1	Technical Components		1,518	
.1.1	Injector systems	201		
.1.2	Collider ring systems	1,317		
.2	Conventional Facilities		615	
.2.1	Site and infrastructure	91		
.2.2	Campus area	46		
.2.3	Injector facilities	42		
.2.4	Collider facilities	370		
.2.5	Experimental facilities	66		
.3	Systems Engineering and Design		307	
.3.1	EDI	209		
.3.2	AE/CM services	98		
.4	Management and Support		205	
.4.1	Project management	122		
.4.2	Support equipment	56		
.4.3	Support facilities	27		
.5	Contingency		565	

accomplished by industrial firms. The term, Industrial-Supplied Component or ISC, was utilized at the lowest WBS entry level to earmark purchased services or components from industry. The basic task assumptions were as follows:

1) Laboratory Tasks
   - Provide technical systems design or design specifications
   - Assemble and test major technical systems within the accelerator complex

2) Industry (ISC) Tasks
   - Provide component design where required
   - Provide and/or produce services, materials, components, and subsystems for both technical and conventional SSC systems

Utilizing the above model assumptions, we will interrogate the SSC cost estimate detail to summarize the ISC tasks and their projected costs in each major WBS category. We begin by reviewing the largest cost item in Table 1, collider ring systems at 1,317 M$. Table 2 provides the next WBS level of the collider ring systems and associated costs. It is seen that the superconducting magnets of the collider comprise 81% of the costs of all collider technical systems. Figure 3 provides an example of the lowest level WBS detail that is available for the dipole beam tube. For magnets, the ISC components (materials and/or labor) are identified at WBS level 9.

The Fig. 3 illustration, at WBS level 5, also identifies the various other types of collider magnets for which similar cost detail is extracted as that indicated for dipoles.

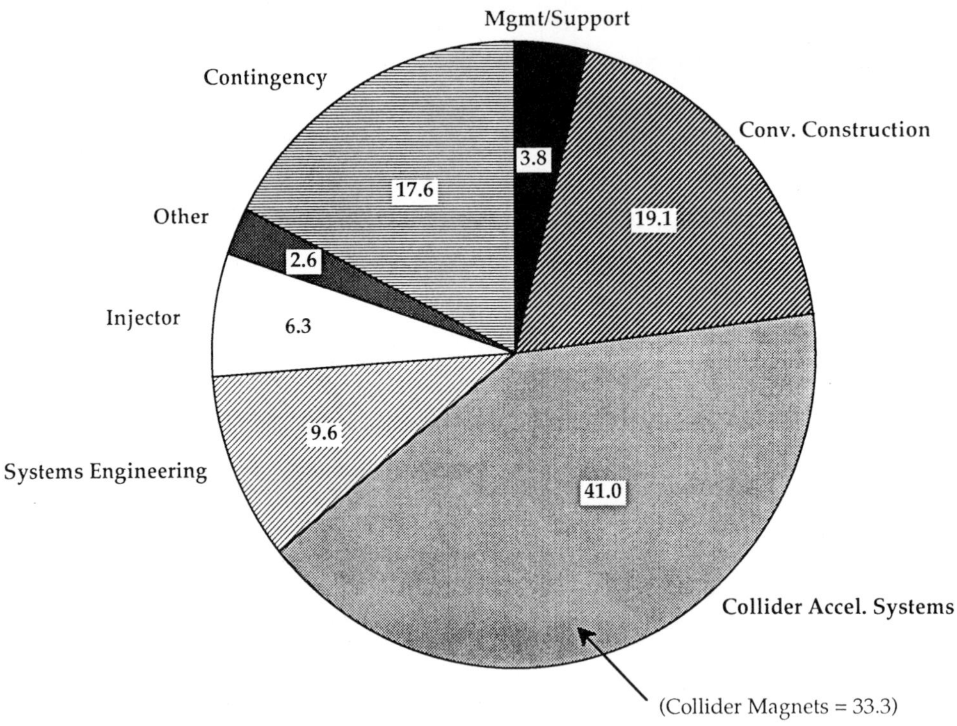

Figure 2. SSC construction costs (%).

Table 2. Collider Technical Systems Cost

WBS	System Component		FY 88 M$
.1.2	*Collider ring systems*		1,317.0
.1.2.1	Magnets	1,067.8	
.1.2.2	Cryogenics	129.2	
.1.2.3	Vacuum systems	18.5	
.1.2.4	Main power supplies	27.8	
.1.2.5	Correction power supplies	7.4	
.1.2.6	rf system	7.8	
.1.2.7	Beam feedback system	4.6	
.1.2.8	Injection system	5.6	
.1.2.9	Abort system	10.3	
.1.2.10	Instrumentation	13.7	
.1.2.11	Controls	19.2	
.1.2.12	Safety systems	5.1	

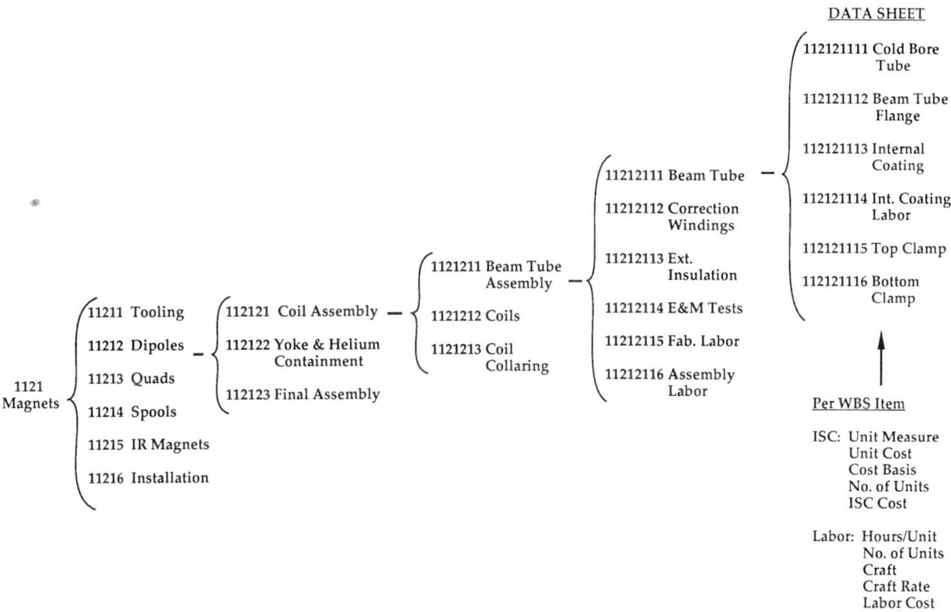

Figure 3. Example of cost estimate detail.

The quantity and variety of each magnet type are summarized in Table 3. The collider dipoles are 7,664 in number and comprise 83% of the total magnet costs. We will examine the dipole structure in the next several figures to illustrate the type of construction materials involved. Similar conceptual models of the other magnet elements can be found in the CDR. The summary tables which follow will then include the materials and labor costs for all magnets.

Figure 4 provides an overall cross section of the collider dipole magnet. The outer vacuum vessel, intermediate heat shields, superinsulation, and cold mass assembly and support structure are featured.

Figure 5 provides further details of the cold mass assembly. The stainless steel shell which forms the helium containment vessel is shown together with details of the yoke, collars, and coil windings.

Table 3. Collider Magnet Types

	Quantity	Variety
Standard Dipoles	7664	1
Standard Quadrupoles	1360	1
Spools	1580	~ 10
Interaction Region:		
Vertical dipoles	232	1
Low β quadrupoles	40	6
High β quadrupoles	40	4
Dispersion suppressor quads	256	6
Utility region quadrupoles	64	4

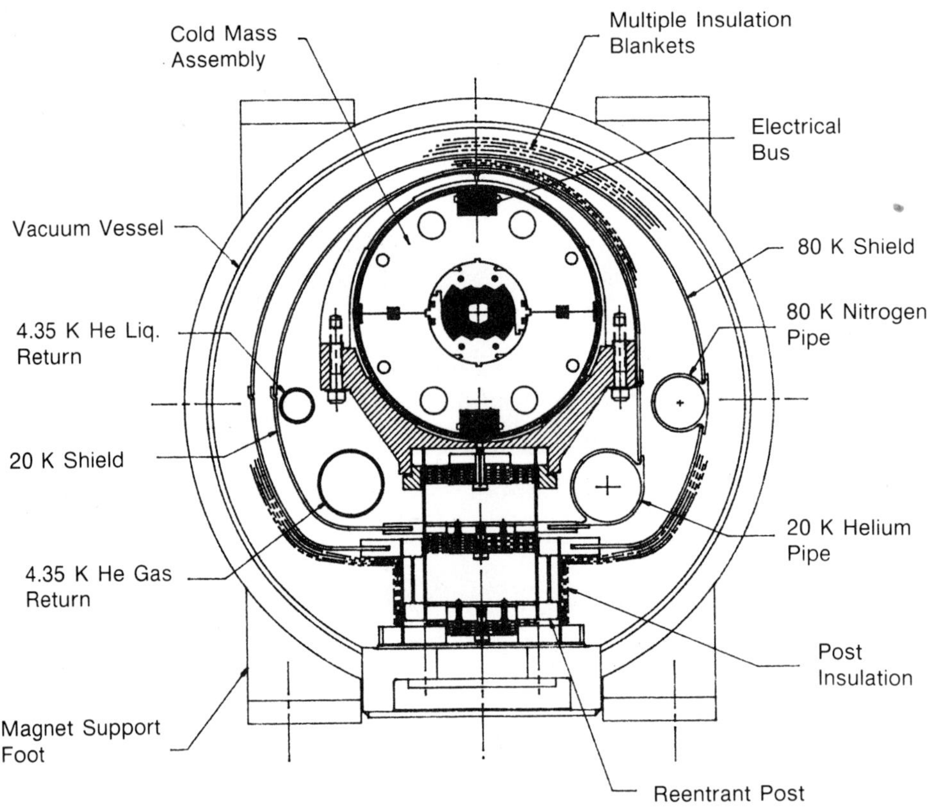

Figure 4. Dipole magnet cross section.

Figure 6 enlarges the collared coil cross section to provide details of the individual cable windings and spacers of the superconducting coil. Further detail of the cable and its insulation is provided in Fig. 7.

Figure 8 is a cross section of the central bore tube which will contain the circulating beam of the collider. A method for providing correction field windings is illustrated.

The cost of the materials (except for superconductor and tooling) for all collider magnets which was illustrated by the dipole structure provided in Figs. 4 through 8 is summarized in Table 4. The costs for the beam tube materials, coil collars, yoke laminations, cryostat components, etc., are called out in separate groupings to show the correspondence with the structures in Figs. 5 through 8. The estimated costs for other factors, such as reject allowances and transportation, are also indicated.

The total materials cost (515.8 M$) provided in Table 4 is given in the fourth column of Table 5 and distributed according to collider magnet type. The cost of superconducting cable (third column of Table 5) is called out separately because it represents a significant part (~1/3) of the total materials costs. Tooling is separately indicated mainly because it was established in the CDR as a separate Level 5 WBS item as indicated in Fig. 3. The tooling includes such items as the coil winding machine, the coil curing press, the coil collaring press, and other such devices used in the production process. Finally the last column provides the estimated costs for labor in the fabrication, assembly, test, and installation program. By the definition noted earlier, the "ISC" costs of Table 5 total 1,034.7 M$. (The only non-ISC item is the 33.1 M$ of final installation and test at the SSC site.)

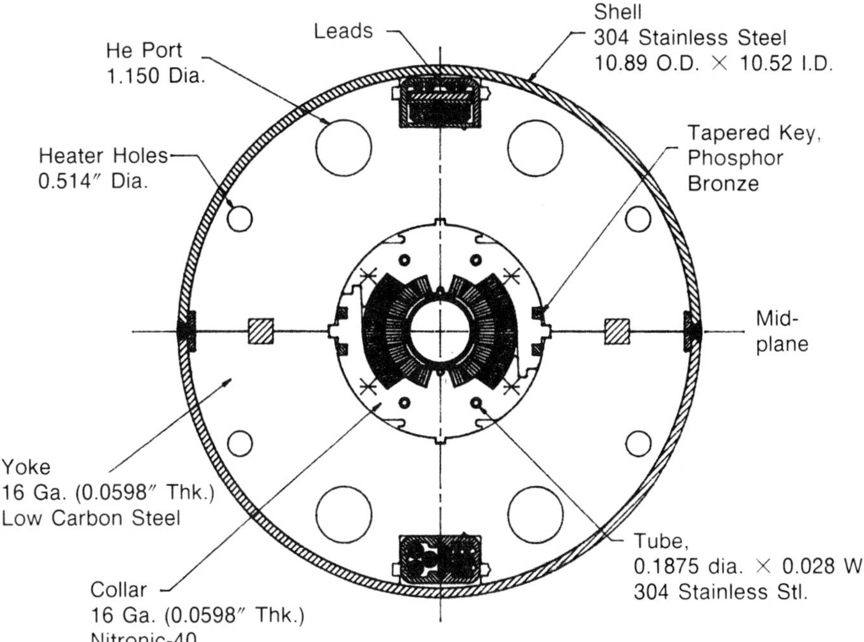

Figure 5. Dipole cold mass assembly.

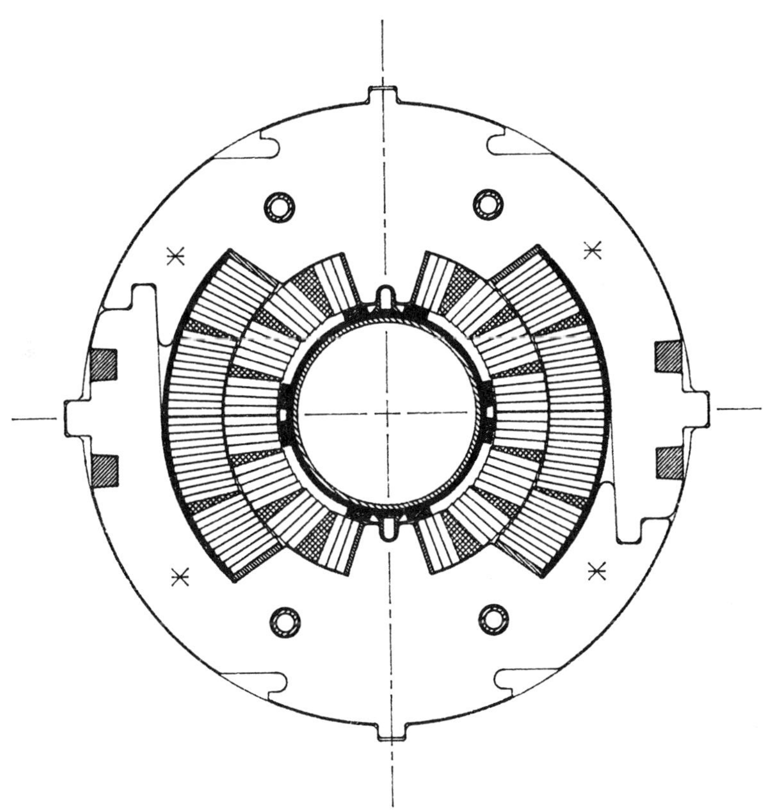

Figure 6. Collared coil cross section.

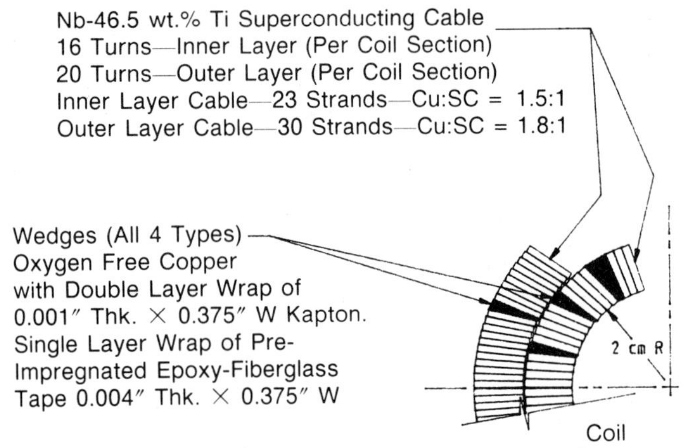

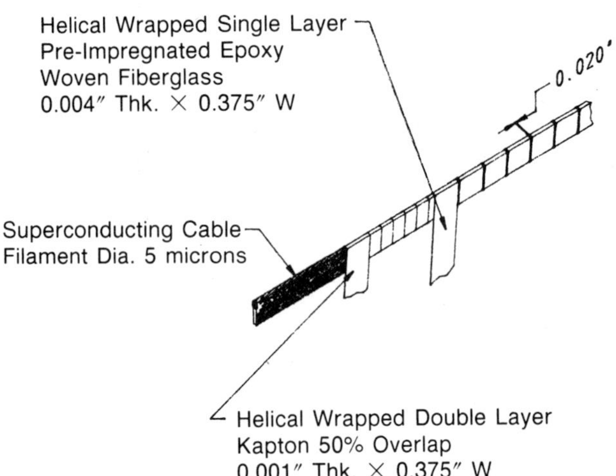

Figure 7. Cable and insulation.

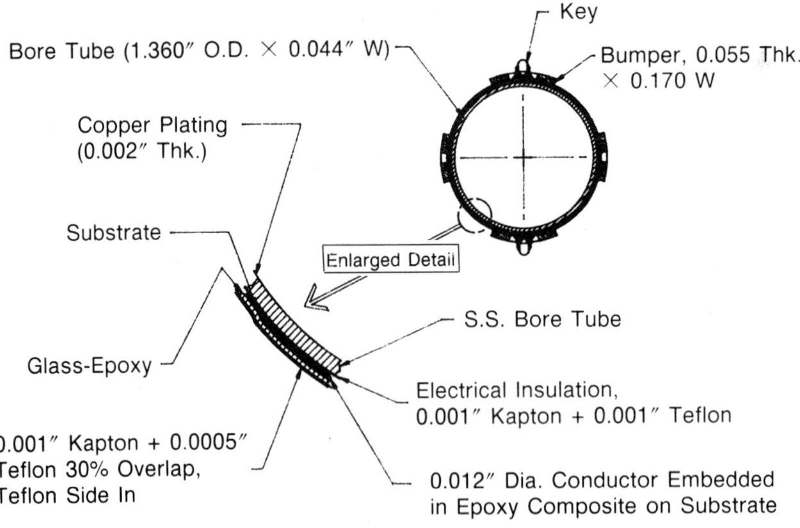

Figure 8. Dipole bore tube.

Table 4. Collider Magnet Systems – Materials

		FY 88 M$
Bore Tube Assembly (w/o SC)		36.8
Beam Tube (Nitronic 40 SS)	5.8	
Correction Winding Structure	23.7	
Insulation and Spacers	7.3	
Collared Coils (w/o SC)		79.9
Insulation (Kapton)	9.8	
Spacers (Copper)	7.7	
Misc. (End spacers, quench heaters)	7.8	
Collars (Nitronic 40 SS)	54.6	
Yoke Laminations (Low carbon steel)		93.7
Helium Containment Vessel (304 SS rolled sheet)		29.2
Cryostat		125.5
Piping (304 SS)	14.8	
Cold Mass Supports (Carbon fiber + SS)	48.3	
Heat Shields (Rolled Al sheet)	22.6	
Insulation (Mylar superinsulation)	12.2	
Vacuum Vessel (Rolled low-carbon steel)	27.6	
Internal Buss (Copper stabilized SC)		22.3
Interconnections (SS bellows)		23.2
Test and Measurements (Equipment)		3.4
Reject Allowance		11.2
Materials Use Allowance		16.9
Transportation and Handling		12.8
Procurement Expense and Industrial Fees		48.6
Installation and Survey/Alignment (Fixtures & Equipment)		12.3
	Total	515.8

The magnitude of the collider dipole system necessitates careful attention to design and production methods. A three-phase Magnet Industrialization Program (MIP) has been planned as follows:

Phase I	–	Technology Orientation
Phase II	–	Tooling Design and Preproduction
Phase III	–	Production

The Phase I program began on December 16, 1988 with DOE approval of a selected list of industrial participants. Twenty firms (eight American, seven Japanese,

Table 5.   Collider Magnet Systems Cost (FY 88 M$)

Magnet System	Total Cost	System Component Cost			
		Tooling	Supercon	Materials[a]	Labor
Dipoles	843.8	48.4	241.8	416.4	137.2
Quadrupoles	53.5	11.4	8.6	21.3	12.2
Spools	83.3	5.4	10.0	45.2	22.7
I.R. Magnets	41.8	5.8	8.0	20.6	7.4
Installation	45.4	–	–	12.3	33.1
Totals	1067.8	71.0	268.4	515.8	212.6

[a]Other ISC and materials.

and five European) will be involved. The Technology Orientation program will be completed by early March 1989, at which time the SSC Magnet Division intends to solicit proposals from industry for Phase II: Tooling Development and Magnet Preproduction. Goals in Phase II will be: (1) provision for dipole magnet tooling design and demonstration, (2) adaptation of the R&D dipole magnet design and R&D laboratory procedures to mass production, (3) proof of magnet reproducibility, and (4) demonstration of production rate capability. At the same time as these explicit program goals are being achieved, the process of developing an experienced and qualified industrial base for mass production of collider dipoles will have been put in place by the Phase II process. Up to five vendors will be selected to participate in Phase II, which is expected to be accomplished on multiyear contracts, both cost-type and fixed-price. The Request for Proposal (RFP) to begin the solicitation of industrial proposals for Phase II is anticipated to be issued in March 1989.

In order to complete the analysis of all magnet systems for this project, Table 6 shows the magnet quantities involved in each of the three injector accelerators. Table 7 provides a breakdown of the injector system magnet costs into tooling, superconductor, where applicable, other materials, and labor. This is the same breakdown that was performed for the collider magnets in Table 5. All of these costs are considered to be ISC.

Table 6. Injection System Magnet Quantities

	Dipoles	Quads
Low Energy Booster	30	40
Medium Energy Booster	216	96
High Energy Booster	528[a]	186[a]

[a]Superconducting.

Table 7.   Injection System Magnet Cost (FY 88 M$)

| | | System Component Costs | | | |
Injector	Total Costs	Tooling	Super Conductor	Materials[a]	Labor
Low Energy Booster	2.9	0.4	–	0.9	1.6
Medium Energy Booster	14.4	0.8	–	6.7	6.9
High Energy Booster	57.8	7.7	7.3	25.5	17.3

[a]Other ISC and materials.

Apart from magnets, the next largest technical system in the collider is the cryogenics. There will be 10 refrigerator stations distributed around the collider ring and one station for the HEB. Each station will consist of a compressor-cold box system connected to the magnet ring cryogenics, a helium management and storage system, and a liquid nitrogen circulator and subcooler with LN storage. Two air separation plants will supply the LN. The costs for the refrigerators, the LN plants, the associated cryo-distribution system and interconnections, and the required instrumentation and controls are summarized in Table 8. The MTF provides a test facility for magnets prior to installation. All costs for cryogenics are projected as ISC.

We have provided examples of our definition of ISC and associated costs for collider and injector magnets and for cryogenics. These items comprise 85% of all technical systems costs. We have made a similar analysis (not detailed here) for the remaining 15% of the technical systems which include vacuum, rf, power supplies, injection/extraction, and instrumentation and control systems. The overall results for all technical systems are summarized in Table 9.

The results (from Table 9) indicate that 93% of all technical systems (or 1,413 M$) are expected to be ISC. The remaining 7% of the costs (105 M$) are basically "on-site" labor costs which include, depending on the particular system, any final assembly that is required together with installation and final tests within the accelerator enclosures. While the labor identified in Part B will certainly involve SSC

Table 8.   SSC Cryogenic Costs (FY 88 M$)

	Collider	HEB[a]	MTF[b]	Totals
He Refrigerator Systems	75.4	7.3	5.9	88.6
Liquid Nitrogen Plants	9.3	–	–	9.3
Cryo-Distribution System	27.5	4.9	–	32.4
Instrumentation/Controls	10.5	2.2	0.7	<u>13.4</u>
				143.7

[a]High Energy Booster.
[b]Magnet Test Facility.

Table 9.   SSC Technical Systems Cost Summary

		FY 88 M$
A. Industrial Supplied Components & Materials		1413.4
• Collider Superconducting Magnets	1034.7	
• HEB Superconducting Magnets	57.8	
• Standard Magnets (LEB, MEB, TB)	22.3	
• Cryogenics	143.7	
• Vacuum Systems	19.9	
• RF Systems	26.9	
• Power Supplies	43.1	
• Instrumentation, Controls & Safety Systems	33.2	
• Injection, Extraction & Abort Systems	21.3	
• Injector Components (Source, RFQ, Linac Structure)	10.5	
B. Labor (Assembly, Installation & Test)		104.7
	Total	1518.1

Laboratory staff, this does not preclude that some fraction of this labor may be subcontracted to outside companies.

The conventional system costs have been analyzed in a similar manner to that of the technical systems, and the results are summarized in Table 10. Industrially supplied materials and equipment (ISC) make up 60% (371 M$) costs. The remaining 40% is basically industrially supplied labor (244 M$). This construction labor is supplied by the successful contractors who will be constructing the buildings, structures, and utilities for the SSC Laboratory.

Table 11 summarizes the system engineering and design costs which have two components: The Technical Systems EDI and the Conventional Systems AE/CM costs. The Technical Systems EDI (175.1 M$) is expected to be largely laboratory manpower. The AE/CM manpower will come from industrial organizations that are successful in the AE/CM selection process. The associated supplies and services for the design groups are listed in Table 11 and are part of the ISC group.

The last major WBS category of the project cost summary (see Table 1) is analyzed in Table 12. The labor component (93.1 M$) involves laboratory management and support services. The management includes the director, division heads, and group leaders who will plan and direct the overall design and construction effort. The support services include a wide range of activities necessary to support the management and design teams in the early construction stages as well as the functions necessary to sustain an operating laboratory in the preoperations and operations phases. Such services include, for example, budget and accounting departments, transportation, shops, library services, engineering services, etc.

The materials and supplies include general office supplies and services for the management and support group and special project equipment for the construction

Table 10.   SSC Conventional Systems Cost Summary

			FY 88 M$
A.	Industrial Supplied Components & Materials		371
	• Electrical Systems (main substations and distribution systems)	28	
	• Communications Systems	7	
	• Roads, Fencing, and Other Utilities	20	
	• Laboratory Building Structures[a]	28	
	• Accelerator (above ground) Structures[a]	35	
	• Collider Underground Structures[a]	213	
	• Experimental Halls[a]	40	
B.	Construction Labor (building contractors)		244
	Total		615

[a]Includes construction supplies, materials, and equipment.

Table 11.   SSC Systems Engineering & Design Costs.

			FY 88 M$
A.	Labor		249.4
	• Technical Engineering Design & Inspection	175.1	
	• Architectural Engineering & Construction Management	74.3	
B.	Associated Materials & Services		57.3
	• Office Supplies & Equipment	12.2	
	• Design and Drafting Equipment	6.0	
	• Computer Use/Maintenance	11.0	
	• Printing and Reproduction	10.0	
	• Reference Materials	2.8	
	• Consultant Services	3.2	
	Travel	6.0	
	• Communications and Miscellaneous	6.1	

Table 12.   SSC Management and Support Costs.

		FY 88 M$
A. Labor (Management & Staff Support Services)		93.1
B. Materials, Supply Services, & Equipment		112.0
1. Management & Support		29.2
• Office Supplies	5.5	
• Printing & Reproduction	4.5	
• Communications, Conferences	3.0	
• Travel	3.0	
• Fees and Miscellaneous	13.2	
2. Project Equipment		56.1
• Computers	5.3	
• Instrumentation & Test Equipment	10.7	
• Vehicles	11.7	
• Tunnel Transporter	9.0	
• Shop Equipment	14.9	
• Office Equipment	4.5	
3. Facility Support		26.7
• Power & Utilities	14.2	
• Rental Facilities	12.5	
Total		205.1

Table 13. SSC Construction Project Cost Summary (FY 88 M$).

	ISC[a]	"Site" Labor	Total
Technical Systems	1413	105	1518
Conventional Systems	371	244	615
Systems Engineering	57	250	307
Management & Support	112	93	205
Totals	1953	692	2645

[a]Industrial Supplied Components.

program. This latter category includes computers, transportation vehicles, and shop and office equipment. Finally, facility support includes costs for power, utilities, and rental facilities.

The results of Tables 9, 10, 11, 12 are summarized in Table 13. It is seen that the ISC make up 1,953 M$ or 74% of the projected SSC costs. The remaining 26% of the costs (classified as "site-labor") are a mixture of laboratory manpower and contract services. For example, the technical systems site labor (105 M$) and the management and support manpower (93 M$) are projected to be largely laboratory manpower while the conventional systems labor (244 M$) is entirely non-Lab manpower. The systems engineering (250 M$) will be a mixture of laboratory manpower and AE/CM manpower. Approximately half of the "site" labor would therefore be expected to be non-Laboratory or ISC manpower. With this approximation for the site labor, the total industrially supplied effort (ISC + 1/2 site labor) would be 87% of the total construction costs.

Note that the above analysis and all of the cost information has been presented in FY 88 dollars for convenience and for consistency with our reports over the last two years. The official DOE escalation factor for conversion to today's FY 89 dollars is 4.2%. Note also that we have not included the contingency estimates in our analyses. The average contingency for the total project is 21.4% (565 M$) indicated in Table 1. The CDR can be consulted for a detailed contingency analysis.

Magnet Technology II

Chairman:

P. Mantsch
Fermi National Accelerator Laboratory

MODEL SSC DIPOLE MAGNET CRYOSTAT ASSEMBLY AT FERMILAB

R.C. Niemann

Fermi National Accelerator Laboratory
P.O. Box 500
Batavia, Illinois 60510

ABSTRACT

The Superconducting Super Collider (SSC) magnet development program includes the design, fabrication and testing of full length model dipole magnets. A result of the program has been the development of a magnet cryostat design. The cryostat subsystems consist of cold mass connection-slide, suspension, thermal shields, insulation, vacuum vessel and interconnections. Design details are presented along with model magnet production experience.

INTRODUCTION

The SSC Magnet Program is developing superconducting accelerator dipole magnets in successive iterations. The initial iteration[1] is complete with six full length model magnets and a thermal model having been built and tested. This initial experience along with the evolving SSC magnet system requirements have resulted in a second generation magnet cryostat design. It is this configuration that will be employed for the near term ongoing magnetic, thermal, string and accelerated life testing and will be the design considered by the SSC Magnet Industrialization Program Phase I; i.e., Technology Orientation. Five full length second iteration model magnets have been built for magnetic testing with several more units planned for the balance of the near term cold magnetic testing program.

CRYOSTAT ASSEMBLY

The Fermilab SSC model magnet cryostat assembly area is as shown by Fig. 1.

The major components of the cryostat, which supports and provides the low temperature environment for the magnet cold mass, are the cold mass connection-slide, suspension system, thermal shields, insulation, vacuum vessel and interconnections. The magnet cross section is as shown by Fig. 2.

Fig. 1. Fermilab SSC Model Magnet Cryostat Assembly Area

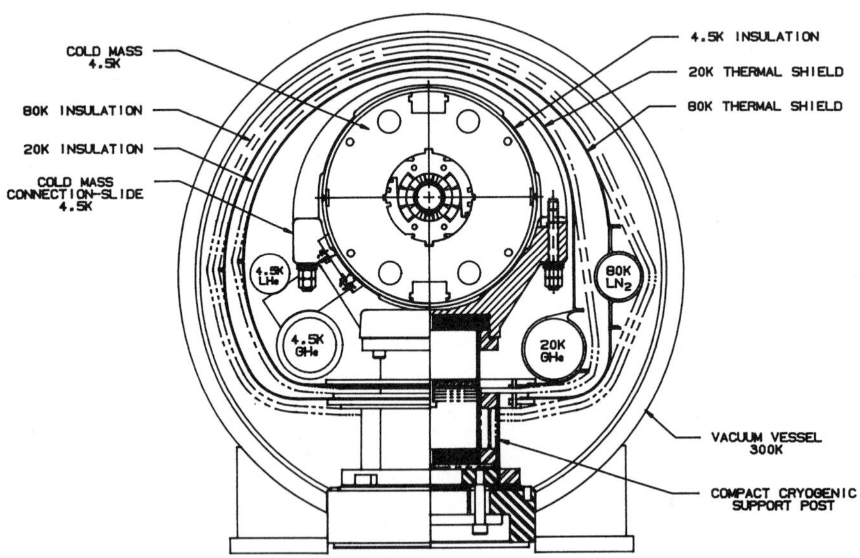

Fig. 2. Second Generation SSC Dipole Magnet Cryostat cross section

COLD MASS CONNECTION-SLIDE

The cold mass is supported by five cold mass-suspension system connections. The connections must withstand transportation and seismic loads, accommodate axial contraction of the cold mass during cooldown and warmup, accurately position the magnetic axis and fit within the compact geometry of the cryostat. Four of the connections must allow for relative axial motion, while an anchor connection fixes the cold mass axially to ground. The connections are attached to the support posts. The cold mass, connections and support posts during an initial stage of magnet assembly are as shown by Fig. 3.

The cold mass-suspension connection-slide[2] is as shown by Fig. 4. Four bearing blocks, containing removable bearing pads, contact the cold mass outer shell and establish its position. The bearing pads are supported by upper and lower cradles.

To function properly, the cold mass outer shell, or skin, must be precise in nature relative to its thickness and form and must be installed with intimate contact to the outside of the cold mass iron yoke at the bearing contact points.

The upper half of the connection retains the cold mass during transportation and seismic loading. Conical spring washers act as tensioner springs on the bolts which connect the two halves of the connection-slide. The spring action maintains uniform loading during thermal cycling and minimizes radial differential thermal contraction effects.

The connection-slides allow rotational adjustment for final alignment of the magnetic vertical plane during magnet assembly. By measuring the average vertical magnetic plane and determining its offset, the cold mass can then be rotated to compensate for this offset and fixed at the center anchor connection. The cold mass remains rotationally free at all but the center support.

Fig. 3. Cold mass connected to suspension system during early stages of assembly

SUSPENSION SYSTEM

The cryostat suspension system performs two essential functions. It resists internally and externally generated cold mass structural loads ensuring that the position of that assembly is stable over the operating life of the magnet and it insulates the cold mass from the environment.

Support Post

A compact cryogenic support post assembly[3] is employed. The outer tube is fiberglass reinforced composite. The inner tube is graphite reinforced composite. A post heat load to 4.5 K of 0.020 W has been measured.

Fig. 4   Cold mass supported in cold mass connection-slide which allows axial and rotational motion

Anchor System

The five support posts share vertical and lateral loads. Thermal contraction of the cold mass assembly during cooldown and warmup necessitates axial sliding between the cold mass and each of the four outer posts and thus they do not contribute to axial load restraint. The center post is rigidly attached to the cold mass to ensure correct axial position within the vacuum vessel. Given no other axial restraint, the center post would carry the total axial load. A single post is incapable of handling these loads alone. Utilizing a 'strong' post at the center would impose intolerable heat loads on the cryogenic system. In order that the bending strengths of all five posts be combined to effectively act as a single axial restraint,[4] the 4.5 K end of each post is connected to that of each adjacent post with anchor tie bars. The post-tie bar anchor system is as shown by Fig. 5.

The anchor tie bar is a filament wound graphite reinforced epoxy composite tube. Graphite filaments were chosen for their thermal expansion properties. When cooled from 300 to 4.5 K, the fibers tend to grow and the epoxy tends to shrink. The net effect is a tube which changes length by only a small amount over its operating temperature range.

The importance of the tie bars is their impact on the thermal performance of the anchor system. The tie bars have both ends at 4.5 K thus contribute no conductive heat load. They lie completely inside thermal shields and thus require no penetrations through the shields.

THERMAL SHIELDS

The cryostat incorporates two thermal shields to intercept radiant heat from the environment and to provide heat sinks for the suspension system conduction heat intercepts. The shields surround the cold mass and are operated at 20 and 80K.

Fig. 5. Cold mass anchor connection located at magnet midspan

The shields are aluminum and are supported by the five support posts. The post-shield interface permits relative axial motion to permit the shields to move as the cryostat is cooled down and warmed up. The shields are thermally connected to the post heat conduction intercepts by copper cables. Extruded aluminum pipes carry the 20K helium gas and the liquid nitrogen used to cool the shields. Aluminum to stainless steel transition joints are required at the end of each shield cooling pipe to permit welding to the stainless steel bellows assemblies employed in the interconnections.

The uninsulated 20K shield is as shown by Fig. 6.

INSULATION

The insulation system consists of multilayer assemblies of aluminized polymer film, fabricated and installed as blankets on the cold surfaces.[5] The blanket materials are a thermal radiation reflective layer made of double aluminized polyester film and a spacer material consisting of spunbonded polyester to separate the reflective layers. Each regular blanket is an assembly of 16 layers of reflector and 15 double layers of spacer, alternately stacked.

Fig. 6. Uninsulated 20K thermal shield

Insulation will be installed directly onto the cryostat cold mass to impede gas conduction heat transfer to the cold mass by residual gas and from desorbed gas released from cryostat surfaces during thermal upset conditions. The intent of the insulation is to act as a heat absorbing buffer surrounding the cold mass in order to slow the effects of transient heat loads by reducing the rate of heat transfer to the cold mass. A single blanket consisting of 5 reflecting layers and 4 spacer layers is used. The insulated cold mass is as shown by Fig. 7.

The 20K shield is insulated with two regular blankets.

Fig. 7. Insulated cold mass

The 80K shield is insulated with four regular blankets. The insulated
80K shield is as shown by Fig. 8.

VACUUM VESSEL

The vacuum vessel provides the insulating vacuum required for the
control of heat transfer to the internal components and provides for the
transfer of the loads of the cryostat internal structure to ground. The
vacuum vessel is circular in cross section, is equipped with bellows at its
ends for interconnection and is connected to ground by means of support foot
assemblies. The steel shell is 610 mm (24 in) in diameter and is 6.4 mm
(0.25 in) thick.

The cold mass is supported at five points relative to the vacuum vessel.
The vacuum vessel is supported at two points relative to ground by support
feet. The support feet locations coincide with the intermediate position
support posts.

Reinforcing rings are welded to the vacuum vessel at the mounting
positions of the five internal supports. These reinforcing rings reduce bending
stresses in the vacuum vessel when loads are imposed on the cold mass
assembly.

The vacuum vessel is fabricated from prefabricated, full section
reinforcing rings connected by short sections of tubing. The segmented
assembly facilitates the vertical precurving of the vacuum vessel assembly.
Precurving is employed to control; i.e., limit, the vertical sag of the cold
mass magnetic axis. The vacuum vessel loaded by a calibration cold mass
during inspection of the vertical precurve is as shown by Fig. 9.

Fig. 8. Insulated 80K thermal shield

The completed internal cryostat assembly is inserted, by sliding, into the vacuum vessel by means of a tow tray-plate assembly installed at the bottom of the vessel shell. A completed model magnet assembly being loaded for transport to a test facility is as shown by Fig. 10.

INTERCONNECTIONS

The magnet interconnection as shown by Fig. 11, consists of seven pipes, having stainless bellows which accommodate the axial motion due to cooldown and warmup, that must be connected.[5]

Fig. 9. The vertical precurve of the vacuum vessel is inspected after welding. A calibration cold mass is used to load the vessel at the five cold mass support points.

All pipe ends are stainless steel except the vacuum shell connection, which is carbon steel. All connections will be welded with automatic welding units and will be cut apart with orbital pipe cutters.

The interconnection is developed radially outward from the cold mass centerline. The first connection to be made is the beam tube. The cold mass outer helium containment shell connection is then made. The next pipes to be welded are the four small pipes surrounding the cold mass. The interconnection shield bridges are made of two pieces of aluminum. They assemble with hinge arrangement on one side and rivets on the other. Small welds attach the magnet shield to the interconnection shield on one side, primarily for thermal contraction. The other side is unattached but overlaps enough for the magnet to contract upon cooldown. The shields are insulated as they are on the body of the magnet. The final connection to be made is the vacuum vessel shell.

Typical interconnection details during magnet installation are as shown by Fig. 12.

Fig. 10. SSC full length model being loaded for transport to test facility

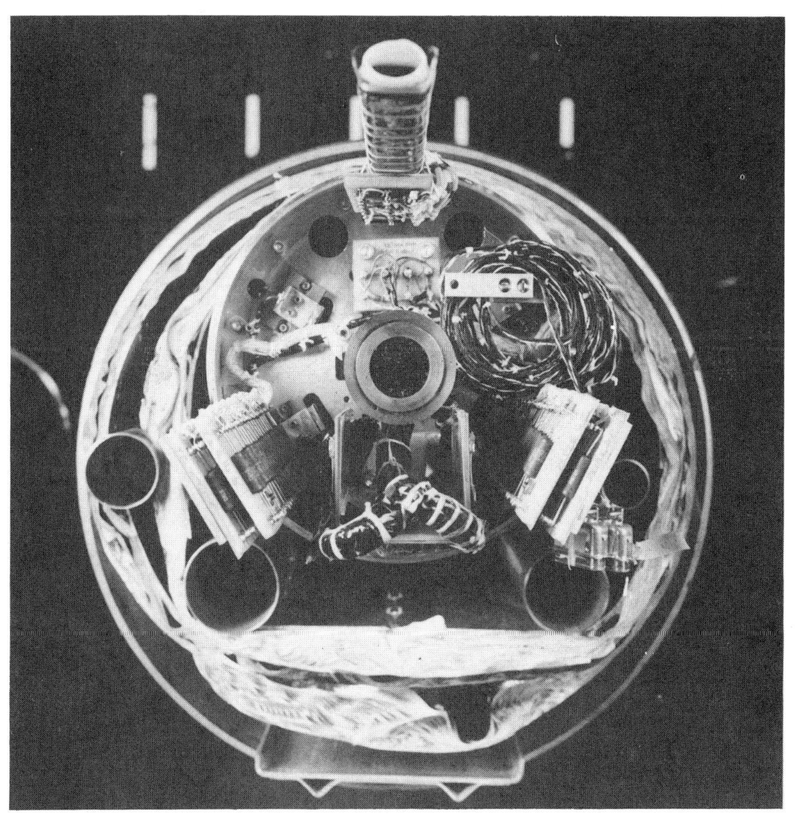

Fig. 11. End view of magnet interconnections

Fig. 12. Interconnection during assembly to cold magnetic test stand

Fig. 13. SSC model magnets being prepared for magnetic testing at Fermilab

MODEL MAGNET EXPERIENCE

Six units of the initial generation magnet design have been built and magnetic tested. Model magnets undergoing magnetic testing at the Fermilab Magnet Test Facility are as shown by Fig. 13.

Five model magnets utilizing the second generation cryostat design have been built and tested. This design will be employed for the balance of the near term SSC model magnet program.

REFERENCES

1. Niemann, R.C., et al., "Superconducting Super Collider Magnet Cryostat," Cryogenic Properties, Processes and Applications, No. 251, Vol. 82, pp. 166, (1986).

2. Larson, E.T., et al., "Status of Suspension Connection for SSC Coil Assembly," presented at the IISSC, February 8-10, 1989, New Orleans, LA.

3. Nicol, T.H., et al., "Design and Analysis of the SSC Dipole Magnet Suspension System," presented at the IISSC, February 8-10, 1989, New Orleans, LA.

4. Ibid.

5. Gonczy, J.D., et al., "Multilayer Insulation (MLI) in the Superconducting Super Collider A Practical Engineering Approach to Physical Parameters Governing MLI Thermal Performance," presented at the IISSC, February 8-10, 1989, New Orleans, LA.

6. Bossert, R.C., et al., "SSC Magnet Interconnections," presented at the IISSC, February 8-10, 1989, New Orleans, LA.

SERIES PRODUCTION OF THE FIRST 20

SUPERCONDUCTING HERA DIPOLE MAGNETS AT ABB

Dietrich Bonmann

Asea Brown Boveri (ABB)
Boveristr. 1, P.O. Box 100351
D-6800 Mannheim 31

ABSTRACT

The HERA accelerator facility at Hamburg, FRG, which is presently under construction will obtain more than 400 superconducting dipole magnets of 9 m length. ABB has completed the preseries of 20 dipole magnets as part of a contract over the production of 215 standard dipoles and the design and production of 8 vertically deflecting dipoles. More than 120 collared coils have been built and are now assembled in their cryostats. The production steps leading from the collared coil to the finished dipole magnet and some of the essential quality assurance procedures are described.

INTRODUCTION

In the early 1980's DESY* developed the first HERA dipole models, which were 'warm iron' magnets[1] (fig. 1a) similar to the dipole magnets of the Tevatron[2] at Fermi National Accelerator Laboratory. In parallel DESY had the industrial production of such magnets investigated by industry. In the course of this study ABB developed an alternative design in which the superconducting coils were installed directly into the iron yoke[3],** which is cooled to 4 K during operation (fig. 1b). This design had some advantages:

Since the large current-dependent forces between coil and yoke are supported directly by the cold yoke which also serves to apply sufficient mechanical prestress to the superconducting winding, the need for very rigid and short supports extending from ambient temperature to the cold mass becomes obsolete. The suspension for the cold mass in the cryostat can be designed with a reduced heat leak.

The cold iron yoke contributes 30 % to the useful field of the magnet. This feature allows either to save a portion of the expensive super conductor or an increase in the useful field.

---

*Deutsches Elektronen Synchrotron (DESY), Hamburg, FRG
**Superconducting cable developed and produced by ABB Baden, Switzerland

    Three 6 m prototypes of the 'cold iron' magnets were designed and
built completely by ABB. The cold iron magnets achieved a maximum field
of 5.8 T (design field 4.7 T). One disadvantage was the saturation of the
iron at magnetic fields exceeding the design value which could no longer
be compensated by holes in the yoke at locations determined by a computer.
This saturation caused a degradation of field homogeniety at high fields
which was not compatible with the requirements of beam optics.

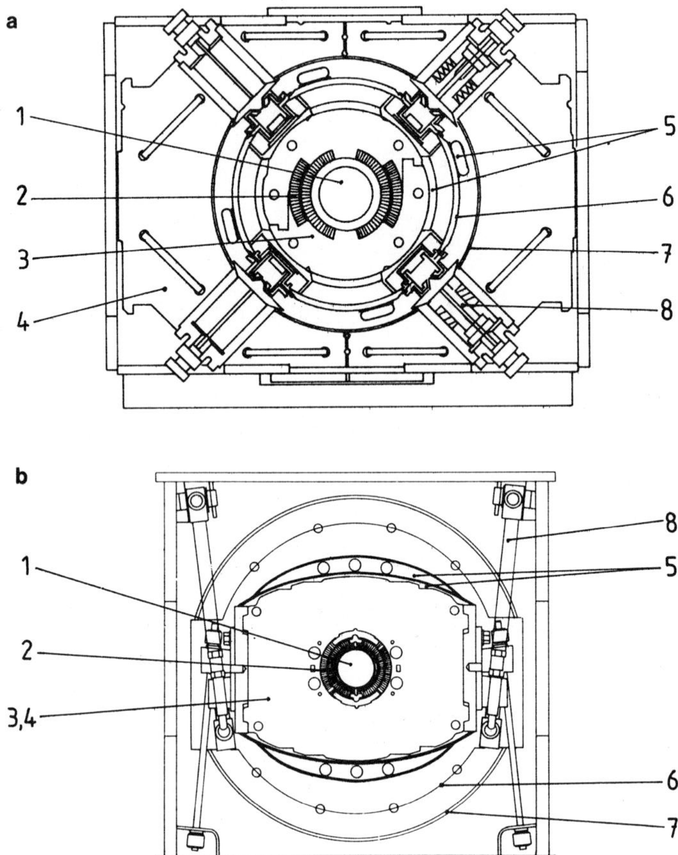

Fig. 1 Two generations of HERA prototype dipole magnets.
        (a) 'Warm iron' design by DESY. (b) 'Cold iron' design by ABB. (1)
        Beam vacuum. (2) Superconducting winding. (3) Supporting structure.
        (4) Iron yoke. (5) Single phase helium. (6) Heat shield. (7) Vacuum
        vessel. (8) Suspension.

    In a further development step in collaboration with DESY the 9 m
superconducting magnet evolved[4] (fig. 2). In this design the coil is
clamped into aluminium collars. The radial thickness of these collars is
sufficient to bear the magnetic forces in the coil without excessive

760

deflection and at the same time the distance between the winding and the
iron yoke is large enough to keep saturation effects within acceptable
limits. Still the yoke contributes 22 % to the useful field.

5 prototype magnets of this kind have been built. DESY supplied the
collared coils, ABB built the yokes and cryostats and the tooling required for the assembly of the magnets. The magnets reached short sample
current after at most one quench and with a peak field of 6.05 T significantly exceeded the specified field. During the prototype phase
various improvements of the suspension in the vacuum container and of the
multilayer insulation were introduced in order to minimize the heat leak
from ambient temperature to heat shield and cold mass.

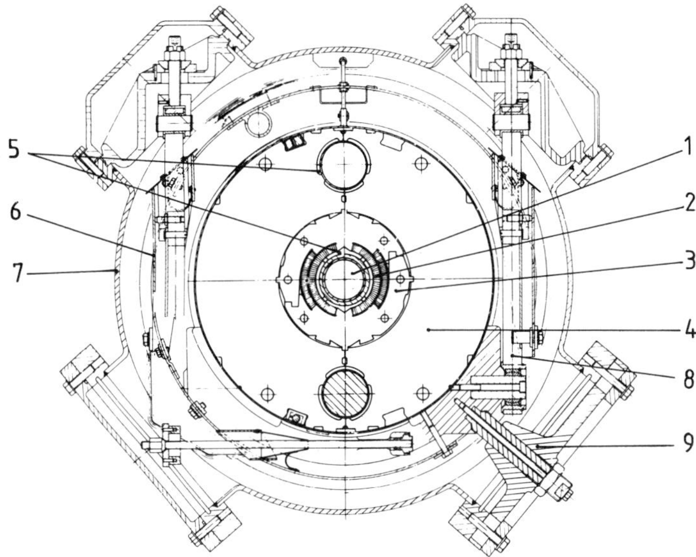

Fig. 2 Cross section of 9 m HERA dipole magnet. (1) Beam vaccum. (2)
Superconducting winding. (3) Supporting structure. (4) Iron yoke.
(5) Single phase helium. (6) Heat shield. (7) Vacuum vessel. (8)
Suspension. (9) Transportation locking device.

In 1987 ABB was awarded the contract for the production of 215 standard dipole magnets and for the design and production of 8 vertically
deflecting shorter dipole magnets. Meanwhile more than 135 collared
coils have been completed and the preseries of 20 complete magnets has
been delivered.

The production of collared coils has been described in detail in a
previous paper[5]. In the following the results of the warm field measurements of those coils will be presented. The remainder of this paper will
be dedicated to the production steps from coil to yoke assembly to the
completed magnet.

RESULTS OF WARM FIELD MEASUREMENTS

The coils are produced in two halves separated by the median plane (fig. 3). Two completed, electrically und mechanically tested and accepted half coils are paired according to their modulus of elasticity and azimuthal size, mounted around a mandrel, insulated with Kapton foils and collared in a special collaring tool[5].
Stainless steel shims at the coil angles are used to achieve the proper prestress in the collared coil and to compensate slight differences in the azimuthal sizes ($\leq$ 0.05 mm) of the two half coils.

Displacement of the separating plane of the two half coils from its ideal position results in non-zero skew quadrupole (a_2) and skew octupole (a_4) components of the magnetic field. The variation of the shim thickness causes changes in the normal sextupole (b_3) and decapole (b_5) components.
Fig. 4 shows the distribution for a_2, a_4, b_3 and b_5 of the first 124 collared coils. The multipole coefficients are the values integrated over the whole length of the coils including the ends. The measurements were conducted on equipment supplied by DESY.

All coils are within the acceptance limits. The scatter of all multipole components with the exception of the skew quadrupole a_2 fill just a part of the allowed range. The scatter of a_2 is a result of the non-ideal position of the separating plane between half coils. Since only half coils with very similar azimuthal sizes, elasticity and cable thickness were selected there must be some hidden parameter governing the mechanical behavior of the coils during the collaring process. The outer layer of the HERA coils is wound onto the inner layer. When the outer layer us cured on the previously finished inner layer they are glued together by the G10-interlayer spacer material which is covered with B-stage epoxy varnish on both sides. The azimuthal size and elasticity are measured giving an average value of both layers. Thus within certain limits the actual sizes of the individual layers remain undetermined. We suspect that this uncertainty and possibly friction within the coils and between coil and ground insulation are the reason for the fairly large scatter of a_2.

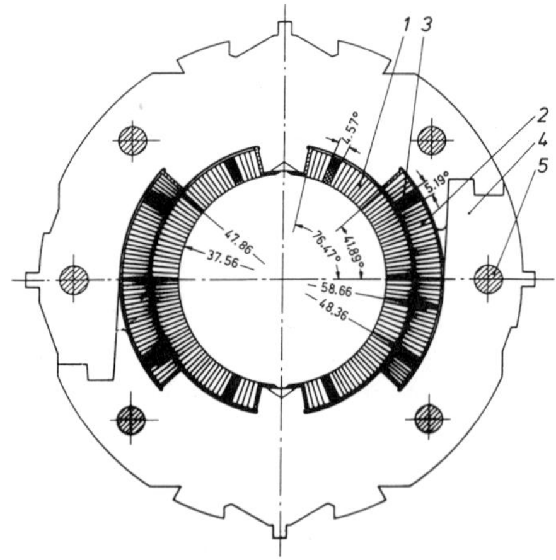

Fig. 3 Cross section of collared coil. (1) Inner coil layer. (2) Outer coil layer. (3) Copper wedges. (4) Aluminium collars. (5) 9 m stainless steel rods in the medium plane.

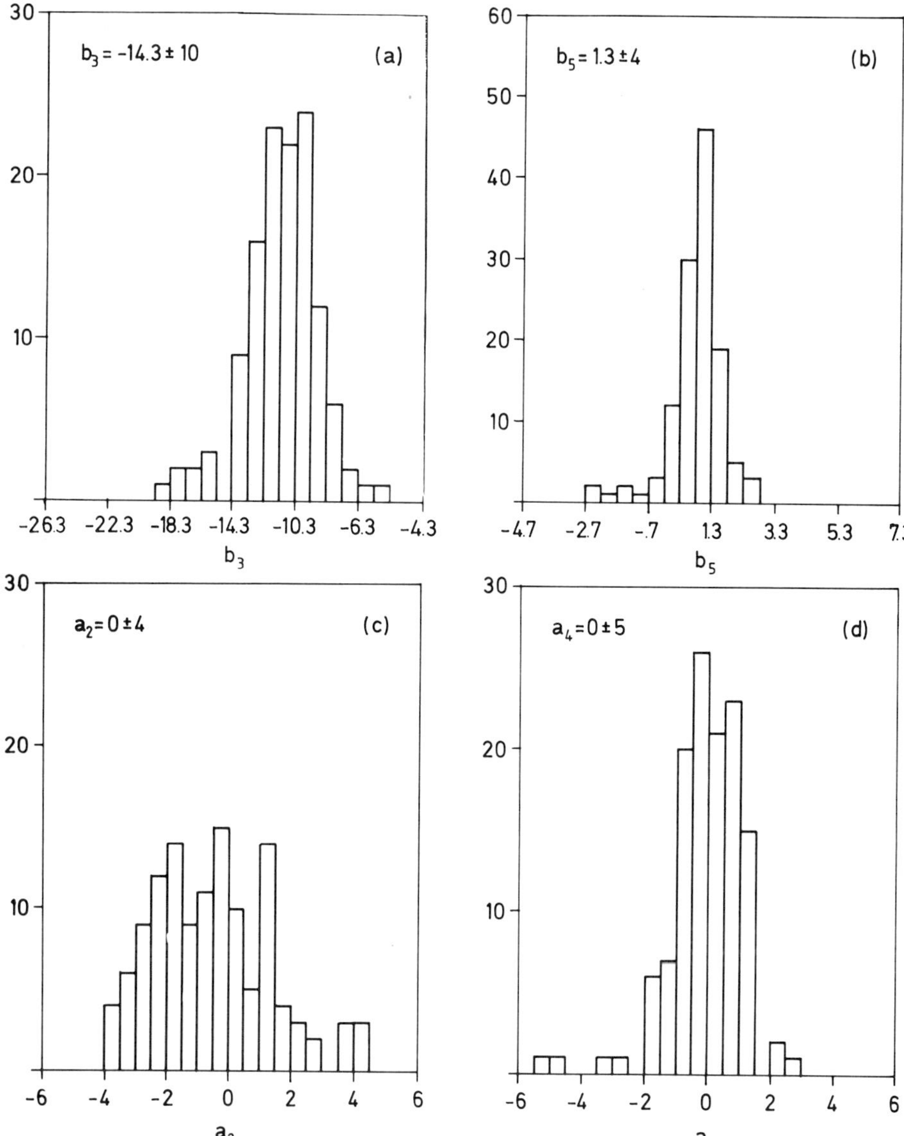

Fig. 4 Results of warm field measurements. Number of coils per interval of measured integrated multipole coefficients a_n, b_n versus measured multipole coefficients a_n, b_n. (a) Normal sextupole b_3. (b) Normal decapole b_5. (c) Skew quadrupole a_2. (d) Skew octupole a_4. Units for a_n, b_n are 10^{-4} cm $^{-(n-1)}$. Specified values and tolerances are indicated in the upper part of the graphs. Total number of coils represented is 124.

PRODUCTION OF COMPLETE HERA DIPOLE MAGNETS

Stacking of half yokes

The iron yoke of HERA-dipole magnet is split vertically in order to
allow the assembly of the collared coil, of the two phase helium tube and
a filler rod in the yoke. The half yokes are stacked on a 10 m long stacking table, which is equipped with a hydraulic press serving to adjust the
filling factor and axial compressive force. The straight edges of the laminations face the stacking table.

Due to differential thermal contraction of the stainless steel helium
vessel and the yoke during cool-down of the magnet the yoke will be compressed by 1°/oo. In this state an axial force of 4 tons per half yoke
may not be exceeded in order not to deform the yoke end plates.
By analyzing the strees-strain curve of a 1 m long yoke section the
proper axial compressive force for the stacking process is determined.

When a half yoke has been stacked to its full length and the alignment of the laminations has been checked, two 9 m long austenitic tie
bolts are inserted into corresponding holes in the laminations and the
axial compression is frozen in by tightening the nuts at both ends of
these tie bolts.

Coil and Yoke Assembly

The cold mass is assembled lying on its side. One half yoke is
turned by 180° in a turning fixture which has to assure that no permanent
twisting or bending of the half yoke occurs. The same fixture is used
when the collared coil, the two-phase-tube and the filler tube are
installed before finally the second half yoke is mounted on top of all.

The right direction of the magnetic field, the proper alignment of
the half yoke and the right positions of two-phase and filler tubes are
established by dimensional and visual checks.

The preassembled yoke is moved to the yoke welding fixture where upper
and lower halves are hydraulically clamped together. A high-pot test of
the insulation of coil and quench heaters against each other and against
the collars is performed in this state. The gap between upper and lower
halves is measured before they are TIG-welded together.

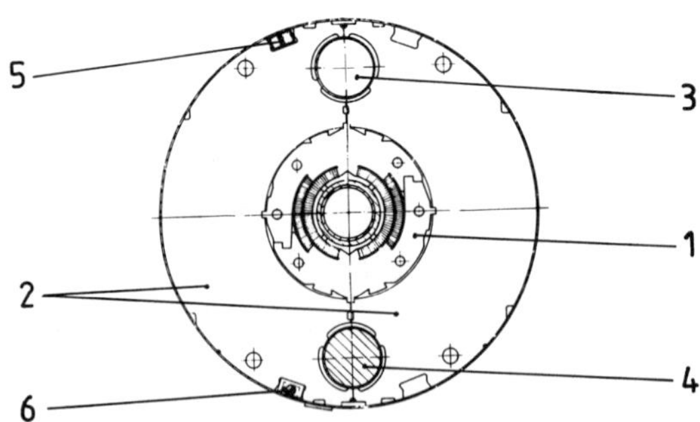

Fig. 5 Cross section of cold mass. (1) Collared coil. (2) Half yokes. (3)
Two phase helium tube. (4) Aluminium filler. (5) Main coil current
return bus. (6) Correction coil bus.

The bending radius of the HERA dipole is 588 m, which means that the magnets have to be produced with a sagitta of approximately 20 mm. The welded yoke is bent hydraulically in the yoke welding fixture and is then picked up by a very rigid transportation device (fig. 6), which keeps the yoke locked in its curved shape and also eliminates any twist in the yoke. Control of the twist of the yoke and thus of the field direction had been a problem with the five prototype magnets.

While the welded yoke is hanging in this transportation device the yoke end plates are mounted and the superconducting supply and return busses of the main coil (fig. 5) and the correction coil bus are installed in grooves at the circumference of the yoke.

Mounting and Welding of Helium Vessel Half
Shells

The completed yoke is lowered into the straight inner helium vessel half shell (fig. 6) lying in the lower part of the helium vessel welding fixture. The half shell is bent by the weight of the yoke. Special reference supports ensure a precise axial and azimuthal alignment of the yoke relative to the half shell and to the ideal plane of bending. Before removing the transportation device the yoke is rigidly clamped to the welding fixture. In this way the twist of the yoke could be kept well within the tolerance of 5 mrad maximum variation over the 9 m length of the magnet (fig. 7a). After mounting the second half shell on top of the yoke the upper part of the welding fixture is added and locked to the lower part. The half shells are pressured against the yoke by inflating fire loses between welding fixture and magnet.

The width of the welding gap between the half shells ($\leq$ 0.3 mm) is checked and recorded for the pressure vessel authorities. A high-pot test of the main return bus and of the coil at 5.6 kV is conducted now. There is no high-pot test after welding of the yoke because the risk of damaging the coil is low except at the current leads which remain accessible

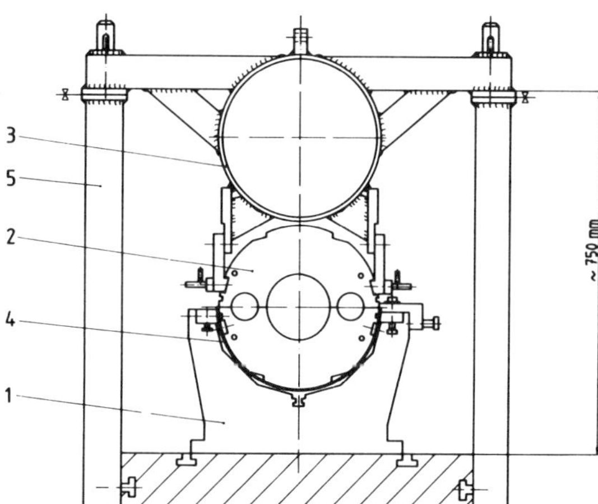

Fig. 6 Helium vessel welding fixture. (1) Lower part. (2) Assembled yoke
(3) Curved transportation device. (4) Helium vessel half shell
(5) Reference support for azimuthal and axial alignment.

at later stages. The high-pot test at 5.6 kV is made before welding the
half shells because these are expensive parts which would be destroyed if
one had to reopen the magnet.

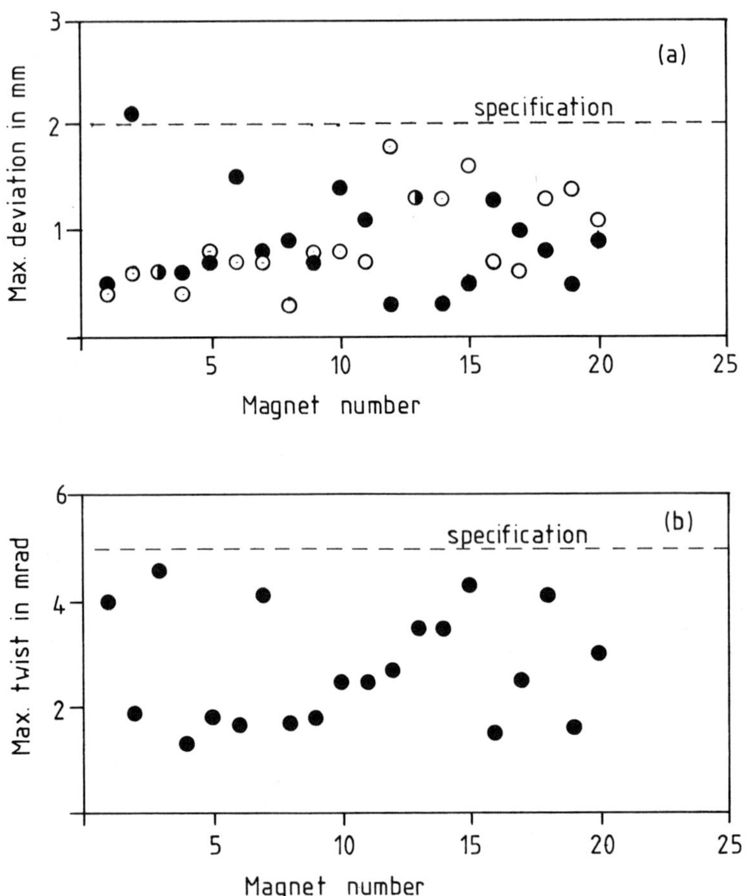

Fig. 7 Results of the measurements of twist, curvature and flatness after
welding the helium vessel half shells for the first 20 HERA magnets produced at ABB. The values shown are the maximum deviations
from the ideal shape of the magnet. (a) full circles: curvature,
open circles: flatness. (b) twist.

After simultaneously plasma welding of the two seams along the half
shells the geometry of the magnet is determined (curvature, length,
twist). The results of these measurements are presented in fig. 7.
The deviations of the magnets shape from the ideal curve in the horizontal (curvature) and vertical (flatness) plane and the twist remained
within tolerances. There is some scatter in the data between magnets 10
and 20. We expect to use about half of the tolerance width for the remaining magnets when the equipment to clamp down the magnets in the helium
vessel welding fixture will be optimized.

All pressure bearing parts of the helium vessel have to be fabricated
from certified materials by qualified suppliers. The identification numbers of these components are recorded in the quality assurance documents
accompaning each individual magnet. Since the longitudinalseams of the
helium vessel are not accessible to X-ray inspection the pressure vessel
authority demand to perform an ultrasonic test of both seams in addition
to the dye penetrant test.

Electrical Switchwork

The electrical connections of the return busses and of the diode are
now installed, the beam tube is mounted and electrically connected. Temperature sensors are installed. All connections are tested by resistance
measurements. Insulation and completeness are checked visually.

During the preseries production of the first 20 magnets several improvements to the electrical insulation of the electrical connections in
the magnet ends have been introduced by DESY and ABB. Various methods and
materials for the insulation of the flexible parts of the connections and
the details of their fixation have been tried until a safe and practical
solution was found. Industrial methods for the production of cable bundles
were introduced. The main coil return bus is now insulated with oven cured
glass-Kapton prepreg tape and glass epoxy prepreg giving a much better
protection against wear in the groove of the yoke.

Positioning and Welding of Helium Vessel
End Plates

The end plates are supplied with all flanges and bellows welded, leak
and pressure tested and approved by the pressure vessel authority.
The helium vessel end plates are positioned at the ends of the helium
vessel in a special fixture. When all electrical leads are fed through
the appropriate tubes and flanges and the end plates have been
tack welded to the helium vessel, all electrical connections are tested
again and a high-pot test at 5.3 kV is conducted. This stage has been
chosen because the insulation of the electrical connections at the
magnets ends will not be accessible later on and because the risk of
damaging the electrical leads is high while the end plates are mounted.

Final Work on the Cold Mass

After these tests the end plates are welded, beam tube, diode port,
support cradles and small lids are welded to the magnet.

The risk of damaging the electrical connections by welding is relatively low, thus only resistance measurements follow welding. The positions of the end flanges are crucial when the magnets are connected one
to the other in the accelerator tunnel, they are measured by using a
gauge. Two cylindrical references for the precise measurement of the
azimuthal field direction are installed at one end plate. A pressure
test at 28 bar and an integral helium leak test (leak $\leq 10^{-10}$ m bar
$1/s^{-1}$) complete the cold mass of the magnet.

Multilayer Insulation and Heat Shield

Now multilayer insulation (MLI) is wrapped around the cold mass,
checked visually and the heat shield is assembled around the magnet. The
shape of the heat shield is controlled before MLI is applied to it and
then visually inspected.

Insertion into Vacuum Vessel

By preassembling the suspension parts (glass fibre epoxy straps and tie rods) and diverse smaller thermal shields the magnet is prepared for insertion into the vacuum vessel. Before insertion of the magnet plus heat shield assembly into the vacuum vessel on a PTFE sled the completeness of the suspension parts is checked.
After hanging the magnet in its rough position the sled is removed from the vacuum vessel and the whole assembly is moved to one of the fine adjustment stations with specialized optical equipment. There the magnet is positioned and fixed in its final position within the vacuum vessel. The relative position of the magnet to reference screws on the outside of the vacuum vessel is recorded.

Final Work

The part of the pressure relief tube penetrating the vacuum vessel is welded to its counterpart at the magnet, the diagnostic leads are soldered to the diagnostic flange on the outside of the vacuum vessel and then the diagnostic flange is closed. A final electrical check including a high-pot test at 5 kV is conducted.

After closing of all ports of the vacuum vessel and checking all documents the transportation protection devices are installed and the magnet is ready for shipment.

REFERENCES

1. G. Horlitz, H. Kaiser, G. Knust, H. Lierl, K.-H. Mess, S. Wolff, P. Schmüser, B.H. Wiik, Warm yoke dipole prototypes for HERA, IEEE Trans. Mag. MAG-21 No. 2 (1985)
2. A Report on the Design of the Fermi National Accelerator Laboratory Superconducting Accelerator. Assembled and edited by F.T. Cole, M.R. Donaldson, D.A. Edwards, H.T. Edwards, P.F.M. Koehler, (May 1979)
3. C.-H. Dustmann, M. Förster, HERA super-microscope uses superconducting magnets from ABB, ABB Review 3/88: 17 (1988)
4. H. Kaiser, Design of superconducting dipole for HERA DESY-HERA 1986-14: (1986); 13th Int. Conf. on High Energy Accelerators, Novisibirsk, USSR, Aug. 7.-11.1986
5. C.-H. Dustmann, M. Förster, D. Bonmann, Experience during series production of the superconducting HERA dipole magnets, IEEE Trans. Mag. MAG-25: (1989)

S.C. MAGNET FABRICATION AT ANSALDO

R. Penco, A. Laurenti, A. Martini and P. Valente

Ansaldo Componenti S.p.A.
Magnet Unit
Via N. Lorenzi, 8 I-16152 Genoa, Italy

ABSTRACT

During the presentation, a videotape about the HERA DIPOLE production was shown. The workshop, the tooling and all the operation sequences for the fabrication of the dipoles were described. You can find here below the technical part of the text of that record.

Ansaldo Componenti is building 242 superconducting dipoles for the INFN (National Institute for Nuclear Physics). These dipoles have been designed to deflect the neutron beam in the 820 GeV, 6.3 km circumference "HERA" particle accelerator for protons and electrons at the DESY laboratories in Hamburg, Germany. The main technical characteristics of these dipoles are:[1]

- 9-m length

- 6.05 T magnetic field in the center

- 6,540 A rated current

- 4.6°K operating temperature

The dipoles are constructed and tested at room temperature in a 3,200 m^2 workshop divided into two areas. In the first area, which is kept at a steady temperature and humidity and subjected to strict cleanliness regulations, the collared dipoles are constructed and tested. Three winding lines ensure that more than one dipole is produced each day.

The first stage in the construction process is the winding of the first layer of a coil. The reel of superconducting cable, insulated with Kapton tape and prepreg glass tape, is positioned on the rotating head of the winding machine. The winding machine is remote controlled by the operator, and exerts a constant force on the cable to guarantee that the coil winding behaves identically when the winding mandrel is removed. The windings are laid mechanically rather than manually to ensure that the tension exerted is constant. This type of winding machine with rail mounted head allows coils of 2-m through 20-m to be produced. Fiberglass spacers are applied to the ends of the coils to obtain a magnetic field of the correct shape. As for all the dipole components, data on the spacers are recorded to allow the coil production history to be reconstructed in case of any problems with conformity.

After winding, the half-coil is mounted with its mandrel on the press curing equipment and baked in a heated press for 2 hours at 160° C. The press used can exert a force of 500 t/m and is driven by computer to ensure that coils are pressed uniformly, and that production data are recorded. The half-coil is cooled slowly and then reinstalled with its mandrel on the winding machine ready for the second layer. The fiberglass separators (fish bones) are positioned to ensure a better cooling of the two layers. Special equipment is used to solder the superconducting cable between the two layers, thus beginning the winding of the second layer. After completing the second winding phase, the coil is again baked at 160° C for further two hours.

After cooling, the coil is removed from the mold and the electrical insulation is checked under pressure in 26 different points. The elasticity of the coil along its axis is then measured at two different pressures. If the coils pass these tests, they are placed in special containers which allow 80 coils to be stored in a reduced space, protected against damage, but yet quick and easy to access.

Data on the dimensions of the coils and their modulus of elasticity are then used to pair the coils for dipole production. Next the collaring shims are computed to give the coils their correct prestress and optimize the shape of the magnetic field. The two coils are then mounted on the collaring mandrel. First to be installed, and in direct contact with the coils, are the quench heaters, followed by the 9-m long shaped kapton sheets, which provide insulation from ground. Finally the collaring shims and the aluminum alloy collar packets, which enclose the coil, are installed, providing the required prestress. These components must be precision tooled with a perfect surface finish.

The assembled dipole is then mounted on the collaring mold and placed in a computer-driven press, which records all pertinent data. The collaring mandrel, used to prevent unwanted coil movement during pressing, is removed at a pressure of about 50 bars. The pressure is then slowly increased to insert the 2-m pins, which hold the assembly together. During this phase the electrical behavior of the coils is constantly monitored to check over faults in the insulation.

The dipole is then passed to the magnetic field shape test equipment, and a test coil is inserted to allow a computer to calculate and print test data. These and other data already gathered or still to gather, along with a list of the various components, are transmitted to DESY via computer. If the dipole passes these tests it is taken to the storage area.

One out of ten dipoles is systematically tested in Ansaldo Componenti cryogenic laboratory[2,3] where it is inserted in a cryostat installed in a 10-m deep trench. The dipole is connected to a 10,000 A power supply and to a protection resistor into which all the energy stored in the dipole is discharged during quenching. The cryostat is filled with liquid helium produced by a 90 ℓ per hour liquifier/refrigerator. The helium contained in the closed circuit is recovered in two buffers (26 m^3 total capacity) and recirculated by the plant compressor. The progress of the tests, the operation of the plant and significant parameters are monitored by the control panel. The dipole is then supplied up to the critical current of about 7,500 A at a temperature of 4.6°K and with a 6 T magnetic field at the center. All the data are gathered and processed by computer, and new data are acquired every 10 msec. All 10 preseries magnets and the first series products have passed the tests. So far no dipoles have been rejected, even though the critical current to guarantee is very close to the short-sample current.

After passing this test, the dipoles are centered by means of compensation shims, and antifriction material is applied to facilitate dipole handling during cooling inside the magnetic yoke. The half-yokes are produced in the packing machine from sheared, phosphatized sheets. This is a high-precision plant, allowing the production of distortion free half-yokes, packed at a constant force and a filling coefficient greater than 97%. The dipole is inserted in the lower half-yoke, its position checked, and the upper half-yoke placed on top. After checks, the dipole is transferred to the press, where it is subjected to a constant pressure and both sides of the two half-yokes are TIG automatically and simultaneously welded together; this simultaneous welding technique avoids thermal differences which could prejudice the shape of the yoke, preloading the coil incorrectly. This technique and the precision of the pressing plans allow the production of dipoles complete, with yoke with an overall distortion of less than 0.5 mrad.

On completion, the dipoles are made, the various conductors are insulated and the sensors and beam tube installed. The dipole is now complete and is taken to the test area for final testing, according to a check list prepared in collaboration with the customer, after completion of the preseries. In addition to dimensional checks and visual inspections, this check list includes stringent high-precision electrical tests. After acceptance, the dipole is installed in the transport box, and the documentation certifying that it has passed all tests is issued. The results demonstrate that is possible to construct such a technologically advanced product on an industrial scale with a minimal reject rate.

REFERENCES

1. S. Wolff, "Superconducting Magnets for HERA." Presented at the *13th International Conference on High Energy Accelerators*, Novosibirsk, USSR (1986)

2. R. Musenich, A. Bonito-Oliva, G. Gaggero, S. Parodi, S. Pepe, P. Valente, R. Penco, "Cryogenic Tests on HERA Dipoles." *ICEC 12*, Southampton, England (1988)

3. R. Musenich, A. Bonito-Oliva, M. Losasso, G. Masullo, P. Valente, R. Penco, "Quench Behavior of Industrial Produced HERA Dipoles." *ICEC 12*, Southampton, England (1988)

PERSPECTIVES ON SSC MAGNET TECHNOLOGY TRANSFER

A.J. Jarabak, A.W. McGuigan, and E.P. Schumacher

Westinghouse Electric Corporation
The Quandrangle - 4400 Alafaya Trail
Orlando, Florida   32826-2399

ABSTRACT

The superconducting super collider provides society and it's technical community with many prime opportunities for technology transfer. It is clear that technologies such as superconductivity will impact on our nations economic future. This paper addresses technology transfer from an industrial perspective. Westinghouse experience in this area and how it relates to SSC is discussed.

INTRODUCTION

Technology transfer was introduced into this world shortly after the big bang. It has been and is being called by many names and takes many forms. An example of a very early transfer of technology is shown in Figure 1. Another form of technology transfer and probably the longest lived and perhaps the best systems is known by the term Apprendice System.

This transfer can be as simple as the older engineer talking with his younger colleague. We see that in Figure 2 this is not necessarily as simple as it sounds. As in most if not all human efforts to have success you must deal with the complications of the interactions required to make any transaction successful.

A common occurence associated with technology transfer is a collision and on the occasions when it really happens you get a big bang from which may grow new ideas, new products and possibly new industries.

For technology transfer to work one basic element is that every one must understand the meaning of technology transfer and communicate on the same level. So we will begin here by defining the words technology and transfer. By referring to various dictionaries it even appears that the people charged with defining words have a little difficulty with the word technology. Anyway for the purposes of this paper we are using the following definitions as given in Webster's Third New International Dictionary:

Technology ---- the science of the application of knowleged to practical
                purposes.

Transfer   ---- the conveyance of right, title or interest either real
                or personal property from one person to another by sale,
                gift or other process.

     This definition of transfer when looked at closely can be a very
sobering experience.

     In this paper we will outline some of Westinghouse's history in the
area of technology transfer, and what we have learned concerning technology transfer. Impediments to its success and the application of
insights we have gained concerning the SSC technology transfer program
will also be discussed.

WESTINGHOUSE'S TECHNOLOGY TRANSFER HISTORY

     Westinghouse, since its start over 100 years ago, has contineously
engaged in technology transfer. In the beginning this process was accomplished by George Westinghouse. He developed the idea, made the sketches
and went into the shop to transfer his concept into a practical product.
This appears to be a simple process because it was.

     As Westinghouse grew and technology expanded, the process became
less simple and more challanging. As shown in Figure 3, this process has
various forms and becomes more complex as organizations as well as people
become involved.

     As science and engineering technology expanded specialization occur-
red so that we now require many different expertises to design a product.
In parallel with the expansion of science and engineering, manufacturing
technology was also expanding. The result of this was the contineous
transfer of technology between design and manufacturing. This resulted
in standardization of the process to assure all parts of the organization
were applying state-of-the-art technology.

     As the process of specialization continued and technology continued
to grow we started a R&D Center. The scientist and engineers at the R&D
Center would transfer technology to manufacturing divisions requiring
their services. The divisions in turn would transfer technology back to
the laboratories. So in a sense people at Westinghouse have been on both
sides of the process.

     Westinghouse began transferring technology externally very early in
its history. One of the first examples was the transfer of technology
that convienced our society that AC should be used in place of DC for
electrical energy. As early as 1903, Westinghouse had a program that
successfully transferred steam and combustion turbine and generator
technology to Canada.

An example of a recent program was the transfer of steam turbine and generator technology to the Peoples Republic of China. This monumental program included 18 tons of documents, 500 training modules, 400 Chinese engineers and 100 Westinghouse engineers.

Technology transfer is not new nor is it easy to accomplish. You must prepare to work hard to have it succeed, and once it starts, you better be prepared to follow through to the finish.

WHAT WE HAVE LEARNED

Technology transfer, whether from laboratories to industry or from industry to laboratories like any other process has essential elements that must be considered up front. Correspondly prior to implementation of technology transfer you must ask yourself certain questions and give serious thought to the answers. We offer the following questions:

o   Do I have the technology?
o   Does someone want the technology?
o   Do I want to transfer the technology?
o   What do I want for the technology?
o   What is required?
o   Was it a success?

From the other side you need to ask:

o   Do I have the technology?
o   Do I want the technology?
o   How do I acquire the technology?
o   What is the technology worth?
o   What is required?
o   Was it a success?

On occasion the question is asked, "Should we have a formal or informal program?" The answer is formal. The reasons for this are:

o   There needs to be defined objectives. People are envolved and people can have different understandings.

o   You need to know the resources to commit.

o   There needs to be a planned and scheduled exchange of activities and accomplishments.

o   You need to know where you are with respect to the objective. Feedback is vital to the process.

o   You need an objective to verify success.

A very important consideration is that organizations sign agreements, but it is the people who make it happen. The people come together, strive toward the same objective, work as a team and and achieve the objective of a successful technology transfer.

The bottom line about what we have learned is that technology transfer is not easy. Speaking from experience when it really happens everyone gets a feeling of accomplishment.

TECHNOLOGY TRANSFER AND SSC

The Superconducting Super Collider Program, especially in the area of superconducting magnets, represents an unprecidented opportunity for the national laboratories and universities to transfer technology to industry. The drawings, specifications, technical reports and present-ations given to industry represent thousands of man-years of effort invested for and by US tax payers. The national laboratories and universities are now returning a dividend to the citizens in the form of this technology. Transferring this information from laboratories to industry is not only a vital step in assuring the success of SSC, it is also a key step toward assuring US competitiveness in a world economy in the 21st century.

The key elements to a successful program have been put in place. Some of these elements are:

o   The national laboratories have a technology that industry wants. This is obvious by the number of people wanting to participate.

o   Industry is willing to pay for this technology. This is shown by the willingness to bear the cost of Phase I.

o   There is a clearly defined formal program and schedule and all parties know their requirements to meet the common goal.

o   There is a mutual interest. This interest goes beyond the poten-tial profit to the excitement of participating in a program that will result in a new understanding of the universe and a new technology.

o   This program by getting people together from various backgrounds, capabilities and understanding will enhance the whole process of building SSC.

o   Success will be verified by the quality of the Phase II proposals.

We are on the right track and moving in the right direction and with commitment, hard work and continued cooperation, we will remain on our path to success. The SSC program will undergo changes. We must be pre-pared to assure all participants remain up to date. Open communications must be maintained as the program progresses to the point where the transfer is complete and the recipients apply their own innovations. This program will become the bench mark by which all future technology programs are measured. As shown in Figure 4, this program will be a success.

Figure 1 Without preparation technology transfer will not succeed

Figure 2 Technology transfer between individuals can have complications

I.   Individuals   Senior Person
                                             |
                                          Junior Person

II.   Specialization   R&D
                                              |
                                            Design
                                              |
                                          Manufacture

III. Standard Processes   Design
                                              |
                                          Manufacture

IV.   Joint Ventures   Company A
                                              |
                                           Company B

V.   SSC   National Labs
                                              |
                                           Industry

Figure 3 Various Forms of Technology Transfer

Figure 4 With Preparation Technology Transfer Will Succeed

Attendees

ATTENDEES

Ahuja, R. K.
Union Carbide

Albert, Gary
General Dynamics Space

Allen, Curt
SSC Central Design Group

Allgaier, Paul
Leybold Vacuum Products

Allhoff, Francis
EG&G

Altgilbers, Edward
Inland Steel

Andersen, Bruce
H & J Tool & Die Co. Inc.

Anderson, Charles E.
Air Products & Chemicals

Anderson, Cynthia
Office of Congressman
Joe Barton

Anding, Charles R.
Iron Workers Int'l Union

Ancrella, Michael
Brookhaven National
Laboratory

Anglum, Owen W.
Armco Advanced Materials

Anvin, Ulf
Outokumpu Metallverken

Appleton, William
CVI, Inc.

Arens, John
University of California
at Berkeley

Aronson, H. Harold
Anderson Consulting

Ashe, William J.
Metallized Products

Atkinson, Richard
Silvex Inc.

Aughinbaugh, John R.
Dannenbaum Engr. Corp.

Autin, Alexandre M.
SSC Central Design Group

Bachy, Gerard
C E R N

Baggett, Neil
Brookhaven National
Laboratory

Baggett, Penny
Brookhaven National
Laboratory

Baker, Richard
Congressman, Louisiana

Baker, Thomas J.
Rust International

Baker, William C.
Teledyne Hastings

Baldi, Robert W.
General Dynamics

Barnes, James L.
Votaw Precision Corp.

Barton, Joe
Congressman (R), Texas

Beatty, G. F.
LTV Steel Co.

Belanger, Chuck
Inland Steel Co.

Belding, Jeffrey S.
Belding Corp.

Belding, M. W.
Blackhawk Industrial
Service

Bell, Robert A.
SLAC

Benda, Brian J.
EQE Engineering

Bensiek, Bill
Babcock & Wilcox

Bergen, Jeffrey R.
Lake Shore Cryotronics

Berkes, Branko
SSC Central Design Group

Berley, David
University of Maryland

Berrey, Steven T.
Midlothian Nat'l Bank

Bevc, Frank
Westinghouse Electric

Bever, Daryl L.
General Dynamics Space

Bhatia, Tarlochan S.
Los Alamos National
Laboratory

Bianchi, Armand (Butch)
Fermi National Accelerator
Laboratory

Bingler, Edward C.
Texas Nat'l Research Lab
Commission

Bizilj, Charles
Howden Compressors

Black, William L.
Law Engineering

Blanar, George
LeCroy Corporation

Bland, Keith
Dalworth Quikset Company

Bogard, Ronald A.
Henningson, Durham &
Richardson, Inc.

Boivin, Michael
New England Electric Wire

Bollini, Jack
Provac Ltd.

Bonmann, Dietrick
ABB

Bonn, John W.
CVI, Inc.

Boos, Franz V.
Michigan Precision Ind.

Bordelon, J. M.
Parsons Brinckerhoff Quade
& Douglas, Inc.

Bosch, Richard A.
Fansteel Metals

Bossert, Rodger
Fermi National Accelerator
Laboratory

Both, Reinhard
Vacuumschmelze GmbH

Botwin, R.
Grumman Space Systems

Brindza, Paul
CEBAF

Britcher, L. Joseph
RMI Company

Brogan, John E.
Dawco Precision, Inc.

Brooks, Deana
C. Itoh & Co. (America)

Brothman, David J.
RTKL Associates, Inc.

Brothman, Steve
Southland Cryogenics

Brown, Bruce C.
Fermi National Accelerator
Lab

Brown, Donald P.
Brookhaven National
Laboratory

Brown, J. N.
Hughes Aircraft Co.

Brown, Rick
United Systems, Inc.

Bruno, John
J & L Specialty Products

Burke, Jack W.
Southwestern Laboratories

Burnett, Kipp
Ennis Chamber of Commerce

Burnett, Sibley
Advanced Cryo Magnetics

Burns, Cathy
Texas Nat'l Research Lab.
Commission

Butler, David
Air Products & Chemicals

Caldwell, Stan R.
Albert H. Halff Assoc., Inc.

Campbell, Glenn
Babcock & Wilcox

Campbell, R.
EBASCO Constructors, Inc.

Camper, Roddy C.
The M. W. Kellogg Co.

Carcagno, Ruben
SSC Central Design Group

Carey, Jack
Triumf

Cargo, Douglas
Texas Super Plex

Carrigan, Richard A., Jr.
Fermi National Accelerator
Laboratory

Carson, John A.
Fermi National Accelerator
Laboratory

Caspi, Shlomo
Lawrence Berkeley
Laboratory

Ceiser, Thomas A.
Trammell Crow Co.

Chao, Alex
SSC Central Design Group

Chapman, Jim
Congressman, Texas

Chapman, Michael
SSC Central Design Group

Chase, Coby
Texas Nat'l Research Lab.
Commission

Chatterjee, Nirmal
Air Products & Chemicals

Chavan, Frederic
Losinger USA Inc.

Chen, Wendell
The University of Texas at
Arlington

Cheng, Allen
S T V / S S V & K

Chester, Warren E.
Crest Products Corporation

Chrisman, B. L.
Fermi National Accelerator
Laboratory

Christopherson, Denis
Supercon

Cleburne, John Warren
I.S.D.

Clee, P.T.M
Rutherford Appleton
Laboratory

Cogger, William S.
Engineered Plastics

Coghill, Cort
Babcock & Wilcox

Colton, Eugene P.
Los Alamos National
Laboratory

Concannon, Brian
General Motors Corporation

Conroy, Peter J.
Harza Engineering Co.

Cook, Charles
Internat'l Scientific
Cooperation Subcommittee

Coombes, Roger
SSC Central Design Group

Cosmus, Thomas C.
IGC/Advanced
Superconductors

Cott, Don
MSE, Inc.

Cowart, Franz L.
Michael Baker Corp.

Cozart, John
Texas Nat'l Research Lab.
Commission

Crean, William R.
Gilbert/Commonwealth

Creutz, Jerelyn C.
General Atomics

Cross, Charles A.
Texas Superflex Devel.
Comm.

Cummins, Robert L.
ABB Electronic Components

Cunningham, Frank
Swagelok

Cunningham, Paul
PMCA

Curl, Richard L.
Morrison-Knudsen

Curtis, John E.
M-K Ferguson

Cusick, Robert L.
Stone & Webster

Dahl, Per F.
SSC Central Design Group

Dambrogi, Richard T.
Dambrogi Co.

Dann, Jeffery R.
Pollak and Skan, Inc.

Darby, Larry
Babcock & Wilcox

Davis, Michael
Martin Marietta
Astronautics

Davison, William J.
Baltimore Specialty Steels

Dawson, John W.
Argonne National Lab.

Day, Sissy
Smith-Day Associates

Dea, Al
Molectrics, Inc.

Dean, Lloyd
Westinghouse

Dean, Norm
Stanford Linear Accelerator
Center

Decker, James
U.S. Department of Energy

Deegan, Mike
Teledyne, Inc.

DeMay, Peter
Fluor Daniel, Inc.

Devred, Arnaud
SSC Central Design Group

Di Giacomo, N.
Martin Marietta
Astronautics

Dickey, Carl
Fermi National Accelerator
Lab

Diebold, Robert E.
U.S. Department of Energy

Dienst, Karl
Sargent-Welch

Diffenderfer, David
Diffenderfer Enterprises

Doddy, George
Daniel, Mann, Johnson &
Mendenhall

Dombeck, Thomas
U.S. Department of Energy

Donaldson, Rene
SSC Central Design Group

Doty, Jimmy A.
Gifford-Hill-American

Dubbeldam, R. L.
Holec

Dubroski, H. Edward
Locke Purnell Rain Harrell

Duer, Craig
GE Magnet Systems

Duncan, Alan S.
Daniel, Mann, Johnson &
Mendenhall

Durr, Charles
M. W. Kellogg

Earle, O. K.
Impell Corporation

Earsom, Deryl
Sverdrup Corporation

Eckels, P. W.
Westinghouse Electric

Edwards, Helen
Fermi National Accelerator
Laboratory

Egilsrud, Philip
Sverdrup Corporation

Ehasz, Joseph L.
EBASCO Services, Inc.

Elioff, Tom
SSC Central Design Group

Ely, Lowell
U.S. Department of Energy

Emi, T.
Kawasaki Steel America

Engler, Norbert
Fermi National Accelerator
Laboratory

Erdman, Karl
EBCO Industries Ltd.

Erkolahti, Timo
Outokumpu Copper

Eskola, Antti
Ministry of Trade and
Industry/Finland

Espy, Mike
Congressman (D),
Mississippi

Eudy, Bob
CRS Sirrine, Inc.

Ewoldsen, Hans
Woodward Clyde

Fagan, Thomas J.
Westinghouse Electric

Farrar, Edwin
Waxahachie C of C

Farrell, Roger A.
Intermagnetics General Corp.

Faulkner, J. Ross
Varian Continental
Electronics Divison

Favale, A.
Grumman Space Systems

Fenyves, Ervin J.
University of Texas

Fieseler, Robert W.
Neuman, Williams,
Anderson & Olson

Fietz, W. A.
U.S. Department of Energy

Fincher, Ron
AT&T

Finks, James E., Jr.
Fermi National Accelerator
Laboratory

Fleener, Terry N.
Ball-Electro Optics Cryo
Division

Fluehmann, Werner
Werner Fluehmann, Inc.

Forsen, H. K.
Bechtel National, Inc.

Fox, Gilbert
Plainfield Stamping

Fraivillig, Jim
E. I. DuPont Company

Francoline, Rocco J.
Tesla Engineering

Fricken, Raymond L.
U.S. Department of Energy

Frobenius, Peter
Bechtel National, Inc.

Fromont, Richard
Dour Metal

Fujino, Haruyuki
Fuji Electric Co. Ltd.

Fukuda, Hideki
Shinko Wire America, Inc.

Funk, L. Warren
Chalk River Nuclear Labs

Fusco, Jim
Iten Industries

Gaalema, Steve
Hughes Aircraft

Gandy, Henry
State of Texas

Garber, M.
Brookhaven National
Laboratory

Garvey, John R.
National Metal Products

Gauhar, Mohammad A.
Mycom Corp.

Gaytan, Eduardo
Texas Higher Ed. Coord. Bd.

Geannaris, George S.
VAT Inc.

Gensamer, Daniel
Fansteel Metals

Gezelle, Donna
U.S. Department of Energy

Gibson, Chuck
General Atomics

Gibson, Sheila
Parsons Brinckerhoff Quade
& Douglas, Inc.

Gifford, Peter E.
Cryomech, Inc.

Gilbert, Paul
Parsons Brinckerhoff Quade
& Douglas, Inc.

Gilbert, William
Lawrence Berkeley
Laboratory

Gilchriese, M.G.D.
SSC Central Design Group

Gilmore, Greg
General Dynamics

Gistau, Guy
L'Air Liquide

Glover, Paul
Brandt Engineering

Gnilsen, Reinhard
Geoconsult, Inc.

Gober-Raynor, Donna
Martin Marietta

Golden, David E.
University of North Texas

Goldman, Leonard M.
Bechtel National, Inc.

Goldwasser, Edwin L.
SSC Central Design Group

Gonczy, J. D.
Fermi National Accelerator
Laboratory

Goodenow, R. H.
LTV Steel Co.

Goodzeit, Carl L.
Brookhaven National
Laboratory

Gravallese, Albert
Daniel, Mann, Johnson &
Mendenhall

Green, David
SAES Getters

Green, Michael A.
Lawrence Berkeley
Laboratory

Greene, Arthur F.
Brookhaven National
Laboratory

Greenfield, Steve
Parsons Brinckerhoff Quade
& Douglas, Inc.

Gregory, Eric
Supercon

Grenfell, J.
Tesla Engineering Ltd.

Grimm, Rodman D.
Masson Grimm & Burgum
Ltd.

Gruver, John A.
Westinghouse Electric

Guertin, Janet R.
Spaulding Composites Co.

Guglielmo, Donald
Europa Metalli - LMI

Gursoy, Ahmet
Parsons Brinckerhoff Quade
& Douglas, Inc.

Gutzler, Marc H.
Standard Mfg. Co., Inc.

Gyssler, George
Asea Brown Boveri
Technology Company

Hagopian, Vasken
Florida State University

Haikonen, Timo A.
Finnish Trade Commission

Hakushi, T.
Kawasaki Steel America

Hampl, Edward F., Jr.
3M Company

Harlan, Thomas S.
Harlan Associates, Inc.

Harrison, J. A.
Parsons Brinckerhoff Quade
& Douglas, Inc.

Harrison, Michael
Fermi National Accelerator
Laboratory

Harrison, William E.
Brookhaven National
Laboratory

Hayashida, E.
Kawasaki Steel America

Hayden, Pat
Inland Steel Co.

Hayes, Jimmy
Congressman (D), Louisiana

Hazlewood, Keith H.
MKS Instruments, Inc.

Heimann, Kermit
Johnson County

Heimdahl, Bob
Sulzer Bingham

Heitowit, Ezra
Universities Research
Association

Henley, Aubrey D.
Mason-Johnston & Assoc.

Hensely, Steve
CVI, Inc.

Hervey, Frank, Jr.,
Hyspan Precision

Hess, Wilmot N.
U.S. Department of Energy

Hester, John C.
National Projects, Inc.

Hicks, Steve
IBM

Hilal, Mohamed
ASC, University of
Wisconsin

Hildebrandt, Don C.
Gundle Lining Systems Inc

Hill, Gale
MK-Ferguson Company

Hill, Norman
Argonne National
Laboratory

Hindle, Jim
Inland Steel Co.

Hirano, K.
Kawasaki Steel America

Hisada, M.
Hitachi America Ltd.

Hoag, Kim K.
Ralph M. Parsons Co.

Hoang, Brian
Mitsui & Co.

Hong, Seung
Oxford Superconducting

Hood, Charles B., Jr.
CVI, Inc.

Hoshino, Hiroyuki
Hitachi Cable, Ltd.

Howerton, Steve
City of Ennis

Hughes, Robert E.
Associated Universities, Inc.

Hulm, John K.
(Retired)

Hunter, Robert, Jr.
U.S. Department of Energy

Hurlbut, Chuck
Bicron Corporation

Huson, F. R.
Texas Accelerator Center

Hutton, Ralph
E. I. DuPont Company

Hyatt, Delton
HPS Division of MKS Inst.

Hyde, Jerry M.
U.S. Department of Energy

Ichimura, K.
Nissho Iwai American Corp.

Ikeda, Junji
Panasonic

Ikeda, Masaru
Furukawa Electric

Irish, Ellwood
UNISTRUT Corporation

Irons, Jeff
GD Space Systems Division

Ishibashi, Kenji
Kyushu University

Isobe, Tsutomu
The Furukawa Electric Co.,
Ltd.

Israel, H. G.
Holec

Jacobsen, Robert
Maxwell Laboratories

Jacyno, Gerald
REA Engineered Wire
Products

Jake, Robert A.
American Magnetics, Inc.

James, David W.
Stone & Webster Engrg.
Corp.

Jarabak, Andrew J.
Westinghouse Electric

Jayakumar, Raghavan
GE Magnet Systems

Jodoin, Peter Paul
U.S. Department of Energy

Joensuu, Jorma
Nokia TK

Johnson, David
Howden Compressors

Johnson, Evan
General Atomics

Johnson, Robert A.
General Dynamics Space

Johnson, T. Ed
General Dynamics Space

Jones, Lawrence W.
University of Michigan

Jones, Wayne
AT&T

Jordan, BN. B. Buck
Waxahachie C of C

Jordan, Donald
Spaulding Composites Co.

Jumonville, Wade
John E. Chance & Assoc.

Kamiya, Shouji
Kawasaki Heavy Industries,
Ltd.

Karesto, Jarmo
Finnish Foreign Trade Assn

Karhumaki, Pekka
Outokumpu Metallverken

Kaugerts, Juris
SSC Central Design Group

Kavlick, Vincent J.
Fluor Daniel

Kawai, Tsuguyuki
Nippon Steel Corp.

Kawamura, Shigeto
Hitachi Ltd

Kays, David D.
The Ralph M. Parson
Company

Kazimierzak, B.
Dour Metal

Kelly, Eugene R.
Brookhaven National
Laboratory

Kelly, Robert
Minnesota Supercomputer
Center

Kephart, Robert
Fermi National Accelerator
Laboratory

Kernan, Timothy C.
Fluor Corporation

Kimmy, Mike
GD Space Systems Division

King, R. Ed
Air Products & Chemicals

Kirk, Thomas
SSC Central Design Group

Klein, Stephen
GD Space Systems Division

Klemm, Stefan
Minnesota Supercomputer
Center

Knafelc, J.
Martin Marietta
Astronautics

Kobayashi, Hitoshi
Mitsubishi Heavy Industries
Ltd., Mihara Mach Works

Koch-Mathian, M. C.
ALSTHOM Intermagnetics

Koizumi, Ted
Furukawa Electric

Komuro, Hiroshi
Mitsui & Co., Inc. (USA)

Kramer, Gordon
Hughes Aircraft Co.

Krause, Robert
Inland Steel Co.

Krauth, Helmut
Vacuumschmelze GmbH

Kreilick, T. Scott
Supercon

Kreinbrink, Kenneth
CVI, Inc.

Krischel, Detlef
Interatom GmbH

Krueger, Thomas A.
Johnson Controls, Inc.

Kuhlen, Michael
Caltech

Kurauchi, Makoto
Shinko Wire America, Inc.

Kurowski, James Paul
Standard Mfg. Co., Inc.

La Fleur, David S.
MKS Instruments, Inc.

Laintz, D.
Martin Marietta
Astronautics

Lamm, Michael
Fermi National Accelerator
Laboratory

Lander, Horace N.
Nippon Steel USA

Larbalestier, David
University of Wisconsin

Larson, E. T.
Fermi National Accelerator
Laboratory

Laughton, Chris
SSC Central Design Group

Laurenti, Adamo
Ansaldo N.A., Inc.

Laverick, Charles
Consultant

Ledbetter, Donald L.
Ralph M. Parsons Co.

Lederman, Leon
Fermi National Accelerator
Laboratory

Leibold, Phil
EG&G

Lemons, Richard M.
Lester B. Knight & Assoc.

Lesage, Herbert
Klockner Wilhelmsburger
GmbH

Leung, Eddie M.
General Dynamics Space

Lewis, Joyce
U.S. Department of Energy

Lien, Neil C.
Baker Manufacturing Co.

Lindholm, Goran
Consulate Gen'l of Finland

Linteau, Jos
Climax Specialty Metals

Littlejohn, John P.
CRS Sirrine, Inc.

Lombard, Harry
Parsons Brinckerhoff Quade
& Douglas, Inc.

Longfellow, Conrad
Caldwell Culvert Co.

Longsworth, Ralph
APD Cryogenics, Inc.

Lowe, Anna
U.S. Department of Energy

Luce, Thomas W., III
Hughes & Luce

Malnar, Robert J.
General Electric Co.

Mantsch, Paul
Fermi National Accelerator
Laboratory

Mara, Gerhard
Elin Union

Marancik, William
Oxford Superconducting

Marcum, Doug
Huntington Labs

Marine, Gary R.
Chicago Bridge & Iron Co.

Marshall, Paul E.
Brown & Root

Marston, Peter G.
MIT/Plasma Fusion Center

Martin, James
Iron Workers International

Matlach, William
Grumman Aerospace

Matsumura, Kazuyoshi
Kawasaki Steel America

Matuszak, Thomas J.
Trent Tube Division

Mayes, Jim
Walk, Haydel & Associates

Mazur, Peter O.
Fermi National Accelerator
Laboratory

Mc Allister, G. L.
Bechtel National, Inc.

McAshan, Mike
SSC Central Design Group

McAuley, William J.
Koch Process Systems

McCabe, Jack
TU Electric

McCarthy, Jim
Pagonis & Donnelly

McDonald, J. R.
Parsons Brinckerhoff Quade
& Douglas, Inc.

McGuigan, Art W.
Westinghouse Electric

McHale, Frank
Pennsylvania Dept of
Commerce

McIntosh, Glen E.
Cryogenic Technical Services

Meade, Anthony
Brookhaven National
Laboratory

Meinke, Rainer
DESY

Merrigan, John
Verner Liipfert, et al

Mesch, Vincent A., Jr.
Spaulding Composites Co.

Meserve, Robert F.
New England Electric Wire

Metzler, John
U.S. Department of Energy

Meyers, Jim
Mitsui & Co., Inc. USA

Michaels, Richard
Laborers International

Miecyjak, Richard C.
Edwards High Vacuum

Milder, Nelson
Staff, House Energy R&D
Subcommittee

Miller, James R.
Sheet Metal Workers
International Assn.

Miller, Peter W.
MK-Ferguson Company

Missig, James R.
Meyer Tool & Mfg., Inc.

Mizuno, Chiaki (Chris)
Mitsui & Co. Inc., USA

Moe, James A.
Bechtel National, Inc.

Moger, Jack B.
Brown & Root

Mokhov, Nikolai
Fermi National Accelerator
Laboratory

Monsees, J. E.
Parsons Brinckerhoff Quade
& Douglas, Inc.

Mookerjee, Suhas
ABB Technology Company

Moore, Terry
Engelhard

Morce, John
Internat'l Union of
Operating Engineers

Moretti, Stefano
Ansaldo N.A., Inc.

Morgan, Gerry H.
Brookhaven National
Laboratory

Morris, Edgar E.
Structural Composites
Industries (SCI)

Morris, Vicki Lynn
Structural Composites
Industries (SCI)

Moss, John E.
Mott Hay, Inc.

Moss, Lloyd
Johnson County

Mundo, Pam
Midlothian Chamber of
Commerce

Murayama, Hiroshi
Nippon Steel USA

Murphy, John C.
Pitt-Des Moines, Inc.

Murphy, Michael
Oxford Superconducting

Naftzger, Donald E.
RMI Company

Nahmias, David
Air Products & Chemicals

Naruse, Koichi
Mitsui & Co., Inc.

Nash, Thomas
Fermi National Accelerator
Laboratory

Nelson, Priscilla
The University of Texas

Newbold, Bill
Daniel, Mann, Johnson &
Mendenhall

Newman, Harvey B.
Caltech

Nichols, Ben
Furnas Electric Co.

Nichols, Christopher
EG&G Sealol

Nicol, T. H.
Fermi National Accelerator
Laboratory

Niemann, Ralph
Fermi National Accelerator
Laboratory

Nishikawa, H.
Nissho Iwai American Corp.

Nohara, K.
Kawasaki Steel America

Nordberg, Markus
LEP-Group

Norman, Pam
Midlothian Independent
School District

Norris, Barry
Fermi National Accelerator
Laboratory

Nukata, M.
Nissho Iwai American Corp.

Nunnally, W. C.
University of Texas at
Arlington

O'Mahony, Mark
Armco Advanced Materials

Obman, Allan R.
Armco Advanced Materials

Ochsner, Jim
E. I. DuPont Company

Ohno, Isamu
Ishikawajima - Harima
Heavy Industries Co., Ltd.

Oldefest, Bob
Advanced Plastics

Olsen, Kenneth O.
Martin Marietta Corporation

Olson, Eric N.
R. Olson Mfg. Co., Inc.

Olson, R. R.
Union Carbide Industrial

Opbroek, Edward G.
Armco Advanced Materials

Orava, Risto
LEP-Group

Orris, Darryl
Fermi National Accelerator
Laboratory

Osborn, Maurice
City of Midlothian

Oster, Gene
Oster Engineering

Owen, Janet L.
Bechtel National, Inc.

Paajanen, Esko
VTT Tech

Pahade, R. F.
Union Carbide Industrial

Paine, Robert G., III
Stone & Webster

Panfil, Peter A.
Lake Shore Cryotronics

Parish, Joe
E. I. DuPont Company

Parnham, K. B.
N. E. America

Patterson, Lee R.
General Dynamics Space

Peeples, James B.
Air Products & Chemicals

Pendo, Roberto
Ansaldo N.A., Inc.

Petersen, Henning
SLAC

Peterson, Jack M.
SSC Central Design Group

Peterson, Thomas J.
Fermi National Accelerator
Laboratory

Petito, Fred
Union Carbide Industrial
Gas

Pfister, Karl A.
Michigan Precision Ind., Inc.

Phelps, J.
Martin Marietta
Astronautics

Pierce, James G.
CVI, Inc.

Pilkington, Richard
EG&G Sealol

Pinto, Charles W.
City of Midlothian

Piquette, Thomas
EG&G Pressure Science

Pollard, Barry
Armco Advanced Materials

Poole, Charles R., Jr.
Mason & Hanger-Silas
Mason

Powell, Jim
Giffels Association, Inc.

Price, Peter E.
Industrial Materials
Technology

Prichard, Ben, Jr.
Science Applications
International Corporation

Prieto, Robert
Parsons Brinckerhoff Quade
& Douglas, Inc.

Puippe, J. C.
Werner Fluehmann, Inc.

Pursell, Carl
Congressman (R), Michigan

Quack, Hans H.
Sulzer Brothers, Ltd.

Rackley, Patricia
Lawrence Berkeley
Laboratory

Rae, Jim
Triumf

Ratz, George A.
Niobium Products Company

Rawls, John M.
General Atomics

Reardon, Paul
Science Applications
International Corporation

Reeves, Gary
Freese and Nichols

Reichard, Jo Anne
Hitachi America, Ltd.

Reidy, James J.
University of Mississippi

Reinarts, Thomas M.
Keller & Gannon

Remsbottom, Robert
SSC Central Design Group

Riddle, G. Mack
Kaiser Engineers, Inc.

Rider, M.
Martin Marietta
Astronautics

Ridley, Philip
Silvex, Inc.

Rifflet, Jean-Michel
CEN/SACLAY

Robbins, Robert L.
Sverdrup Corporation

Rodgers, Gene
First National Bank
Of Midlothian

Rohleder, Stephen J.
Andersen Consulting

Rohrer, E. Parke
Brookhaven National
Laboratory

Root, W.
EBASCO Constructors, Inc.

Rosenblad, Lyndon
Stone & Webster

Rosner, Carl H.
Intermagnetics General Corp.

Rosson, Lee H.
S&W Tech Services

Royet, John
Lawrence Berkeley
Laboratory

Rueb, William G.
MK-Ferguson Company

Rusche, Ben C.
Law Engineering

Russell, Linda W.
Price Waterhouse

Ryn, Vincent B.
Iron Workers International

Sabado, Maurice M.
Science Applications
International Corporation

Saeki, Hiroshi
Panasonic

Sah, Richard
SSC Central Design Group

Saiki, Yukio
Mitsubishi Heavy
Industries, Ltd.

Saito, Ryusei
Hitachi Ltd./Hitachi
Works

Sakai, Shuji
Hitachi Cable, Ltd

Salpaka, Glenn L.
General Motors Corporation

Samuel, Arnold W.
U.S. Department of Energy

Sanchez, Ruben A.
U.S. Department of Energy

Sanders, Whitney
STV/SSV & K

Sarao, Tom
Day Association

Scalabrin, Joseph J.
RTKL Associates, Inc.

Scango, Guy J.
U.S. Department of Energy

Scanlan, Ron
Lawrence Berkeley
Laboratory

Scarfi, Giuseppe
Ansaldo N.A., Inc.

Schad, Doug
H. B. Zachry

Schellmann, Wolfgang
Michigan Precision Ind.

Schermer, Robert
SSC Central Design Group

Schieber, Len
PCK Technology

Schiesser, W. E.
Lehigh University

Schlosser, Cecil
Remmele Engineering

Schuermann, Stephen F.
Morrison-Knudsen

Schumacher, Pete
Westinghouse

Schumacher, William
Armco Advanced Materials

Schwitters, Roy F.
Harvard University

Selleck, Clyde A.
Stone & Webster

Shapiro, Stephen
SLAC

Shearer, Ken
Martin Marietta
Astronautics

Shih, Hsiuan-Jeng
Lehigh University

Shultz, Jack
Tempel Steel Co.

Siebs, Carl V.
Cray Research

Simmers, Robert
Westinghouse - PSD

Singletary, Lil
Oster Engineering

Sintchak, George
Brookhaven National
Laboratory

Sipila, Heikki
Outokumpu Electronics

Sixsmith, Herbert
Creare, Inc.

Sjoblom, Bjarne
Consulate of Finland

Skaritka, John
Brookhaven National
Laboratory

Skubic, Patrick
University of Oklahoma

Smathers, David
T.W.C.A.

Smith, Craig
Trent Tube Division

Smith, Jan
Price Waterhouse

Sokoll, Bob
City of Waxahachie

Sparrow, Rob
Optovac

Stanko, Michael J.
Westinghouse Electric

Staub, Gerald
General Dynamics

Stekley, Z.J.J.
Intermagnetics General Corp.

Stever, M. F.
MIT-LNS

Steining, Rae
SSC Central Design Group

Story, E. Jack
EG&G

Strait, James
Fermi National Accelerator
Laboratory

Strauss, Bruce P.
Powers Associates, Inc.

Stringer, Robert T.
Sheet Metal Workers
International Assn.

Stropeni, Nancy
Control Data Corp.

Stuard, Astrid
S&W Technical Services

Stuber, Wayne G.
Air Products & Chemicals

Sullivan, Gael M.
The LTV Corp.

Sunderman, Wallace
Westinghouse Electric

Susko, Larry
The DOSCO Corp.

Suzaki, Takeshi
Kawasaki Heavy Industries,
Ltd.

Sweeney, Edward A.
Stone & Webster Engrg.
Corp.

Swenson, Bernie
Personal Computer Support
Group

Swift, Norman
U.S. Department of Energy

Swift, Walter L.
Creare, Inc.

Swoboda, John, Jr.
VMW Industries

Szurek, Peter
Sentek

Tanaka, Jiro
Mitsubishi Heavy Industries
Ltd., Technical HQ

Taylor, Clyde E.
Lawrence Berkeley
Laboratory

Taylor, David W.
Armco Advanced Materials

Temple, L. Edward
U.S. Department of Energy

ten Kate, H. H. J.
University of Twente

Tener, Robert K.
North Texas Commission

Teuho, Juhani
Outokumpu Copper

Thomas, Floyd
Grumman Corp.

Thomas, Kenneth M.
Maxwell Laboratories

Thompson, Randell
U.S. Department of Energy

Thorne, John C.
GE Magnet Systems

Tigner, Maury
SSC Central Design Group

Tilford, Norm
Texas A&M

Tkaczyk, Steve
U.S. Department of Energy

Toga, Kuniyasu
Hitachi, Ltd.

Tokunaga, Yoshi
Nippon Steel USA

Tomita, Nobuyoshi
Mitsubishi Heavy
Industries, Ltd.

Tompkins, John
SSC Central Design Group

Toyoda, Katsuyoshi
Kobe Works of Mitsubishi
Electric Corp.

Trendler, Robert
Fermi National Accelerator
Lab

Trivelpiece, Alvin
Oak Ridge Nat'l Lab.

Tsavalas, Yannis P.
GE Medical Systems

Ueda, Keisuke
Kawasaki Heavy Industries,
Ltd.

Ulrich, R. C.
Parsons Brinckerhoff Quade
& Douglas, Inc.

Utsonomiya, T.
Furukawa Electric Co.

Vacca, Joseph G.
Intermagnetics General Corp.

Valaris, Peter
Supercon, Inc.

Vanpeteghem, Peter M.
Texas A&M University

Virts, Judy
U.S. Department of Energy

Voss, G.
DESY

Waddell, John
Mitsui & Co.

Wagner, Rodger L.
Spaulding Composites Co.

Walker, Ian J.
GMW Associates

Wallace, Roy
Remmele Engineering

Conference Program

Wednesday, 8 February 1989

4:00 p.m.	Registration
6:00 p.m.	Reception
7:30 p.m.	Adjourn

Thursday, 9 February 1989

7:30 a.m.   Registration

8:45 a.m.   *Introduction*

R. W. Baldi, Conference Chairman
General Dynamics Space Systems Division

Session I: *Overview*

L. Goldman, Chairman
Bechtel National, Inc.

9:00 a.m.   I-1   *The Genesis of the SSC Project*

Alvin Trivelpiece, Director
Oak Ridge National Laboratory

9:30 a.m.   I-2   *The Impact of the SSC on Science and Technology*

Leon Lederman, Director
Fermi National Accelerator Laboratory

10:00 a.m.   Break

Session II: *Growing Role of Industry in Accelerator Technology*

P. Reardon, Moderator
Science Applications International Corporation

10:20 a.m.   *Panelists:*

RF Systems for Accelerators
J. R. Faulkner
Varian-Continental Electronics Division

High-Performance Proton Injectors
A. J. Favale
Grumman Space Systems

LSU Turnkey Synchrotron Storage Ring

R. Jacobsen
Maxwell Labs, Inc.

Loma Linda Proton Therapy Accelerator System

B. Prichard, Jr.
Science Applications International Corporation

High-Power Neutral Beam Systems

J. M. Rawls
General Atomics

11:45 a.m.   I-3   *The SSC Program*

Roy F. Schwitters, Director
SSC Laboratory

12:30 p.m.   Luncheon

Address

James F. Decker
Deputy Director, Office of Energy Research
U. S. Department of Energy

Session III: Parallel Technical Sessions

2:00 p.m.   III-A   Magnet Technology

R. Coombes, Chairman
SSC Central Design Group

III-A-1   *Superferric Collider Magnets*

F. R. Huson, J. Colvin, W. W. MacKay,
S. Pissanetzky, R. Rocha, W. Schmidt,
G. Shotzman, and J. Zeigler
Texas Accelerator Center

III-A-2   *Performance of Full-Length SSC Model Dipoles: Results from 1988 Tests*

J. C. Tompkins
for the teams at SSC Central Design Group,
Brookhaven National Laboratory, Fermi National
Accelerator Laboratory, and Lawrence Berkeley
Laboratory

III-A-3   *SSC Dipole Coil Production Tooling*

J. A. Carson, E. J. Barczak, R. C. Bossert,
J. S. Brandt, and G. A. Smith
Fermi National Accelerator Laboratory

III-A-4   *Development and Production of the Beam Tube Assembly of the SSC Superconducting Magnet*

J. R. Skaritka
Brookhaven National Laboratory

<table>
<tr><td></td><td>III-B-6</td><td>Systems Engineering and Integration for the SSC</td></tr>
<tr><td></td><td></td><td>D. J. Laintz
Martin Marietta Astronautics</td></tr>
<tr><td>2:00 p.m.</td><td>III-C</td><td>Advanced Particle Detectors</td></tr>
<tr><td></td><td></td><td>M. G. D. Gilchriese, Chairman
SSC Central Design Group</td></tr>
<tr><td></td><td>III-C-1</td><td>Advanced Detector Technology for Experiments at the SSC</td></tr>
<tr><td></td><td></td><td>H. H. Williams
University of Pennsylvania</td></tr>
<tr><td></td><td>III-C-2</td><td>Engineering Issues in the Design and Construction of Large Experiments</td></tr>
<tr><td></td><td></td><td>R. Bell
Stanford Linear Accelerator Center</td></tr>
<tr><td></td><td>III-C-3</td><td>An Integrated 3D Design, Modeling and Analysis Resource for SSC Detector Systems</td></tr>
<tr><td></td><td></td><td>N. J. DiGiacomo
for the team at Martin Marietta Astronautics</td></tr>
<tr><td></td><td>III-C-4</td><td>Silicon Pin Diode Hybrid Arrays for Charged Particle Detection: Building Blocks for Vertex Detectors at the SSC</td></tr>
<tr><td></td><td></td><td>S. Gaalema
Hughes Aircraft Company</td></tr>
<tr><td></td><td>III-C-5</td><td>Development of New Scintillating Materials for the SSC</td></tr>
<tr><td></td><td></td><td>C. R. Hurlbut
Bicron Corporation</td></tr>
<tr><td></td><td>III-C-6</td><td>Development of Radhard VLSI Electronics for SSC Calorimeters</td></tr>
<tr><td></td><td></td><td>J. W. Dawson and L. J. Nodulman
Argonne National Laboratory</td></tr>
<tr><td></td><td>III-C-7</td><td>Frontiers in Computing in the 1990's and SSC Experiments</td></tr>
<tr><td></td><td></td><td>T. Nash
Fermi National Accelerator Laboratory</td></tr>
<tr><td>2:00 p.m.</td><td>III-D</td><td>Superconducting Wire and Cable</td></tr>
<tr><td></td><td></td><td>C. Taylor, Chairman
Lawrence Berkeley Laboratory</td></tr>
<tr><td></td><td>III-D-1</td><td>Development of Superconducting Strand and Cable with Improved Properties for Use in SSC Magnets</td></tr>
<tr><td></td><td></td><td>R. M. Scanlan
Lawrence Berkeley Laboratory</td></tr>
</table>

III-F-44 *Manufacturing of the HERA Quadrupoles: Coil and Cryostat*

M. C. Koch-Mathian
ALSTHOM

5:00 p.m.   Adjourn

7:00 p.m.   Dinner

Address

Edward C. Bingler, Executive Director
Texas National Research Laboratory Commission

Friday, 10 February 1989

7:30 a.m.   Registration

Session IV: Plenary Technical Session

R. Diebold, Chairman
U. S. Department of Energy

8:15 a.m.   *Introduction*

Robert O. Hunter
Director, Office of Energy Research
U. S. Department of Energy

8:30 a.m.   IV-1   *The SSC Accelerator Systems*

H. Edwards
Fermi National Accelerator Laboratory

9:00 a.m.   IV-2   *The SSC Detector Systems*

M. G. D. Gilchriese
SSC Central Design Group

9:30 a.m.   Break

9:45 a.m.   IV-3   *Engineering and Construction Experience on LEP*

G. Bachy
CERN

10:15 a.m.   IV-4   *Industrial Experience on the HERA Accelerator at DESY*

G. Voss
DESY

Session V: *Congressional Perspective on the SSC*

P. H. Gilbert, Moderator
Parsons, Brinckerhoff, Quade & Douglas, Inc.

11:00 a.m.   *Panelists:*

The Honorable Mike Espy (D–Mississippi)
House Budget Committee

The Honorable Jimmy Hayes (D–Louisiana)
House Public Works and Transportation Committee
House Science, Space and Technology Committee

The Honorable Carl D. Pursell (R–Michigan)
House Appropriations Committee

The Honorable Charlie Wilson (D–Texas)
House Appropriations Committee

12:30 p.m.   Luncheon

Address

The Honorable Joe Barton (R–Texas)
House Energy and Commerce Committee

Session VI: *Parallel Technical Sessions II*

2:00 p.m.   VI-A   *Industrial Opportunities on the SSC*

K. W. Chen, Chairman
The University of Texas at Arlington

VI-A-1   *Technological Challenges and Opportunities*

S. G. Wojcicki
SSC Central Design Group

VI-A-2   *SSC Costs and Plans for Industrial Involvement*

T. Elioff
SSC Central Design Group

VI-A-3   *Doing Business With DOE/SSC*

B. L. Chrisman
Fermi National Accelerator Laboratory

2:00 p.m.   VI-B   Magnet Technology II

P. Mantsch, Chairman
Fermi National Accelerator Laboratory

VI-B-1   *SSC Model Dipole Cold Mass Production at BNL*

E. Kelly
Brookhaven National Laboratory

VI-B-2   *SSC Model Dipole Cryostat Production at Fermilab*

R. C. Niemann
Fermi National Accelerator Laboratory

VI-B-3   *Series Production of the First 20 Superconducting
HERA Dipole Magnets at ABB*

D. Bonmann
ASEA Brown Boveri

VI-B-4   *S.C. Magnet Fabrication at Ansaldo*

A. Laurenti and R. Penco
Ansaldo Componenti, S.p.A.

VI-B-5   *Perspectives on SSC Magnet Technology Transfer*

A. J. Jarabak and A. W. McGuigan
Westinghouse Electric Corporation

4:00 p.m.   Adjourn

recovery from, see Quench
        recovery
study of, 561, 565
suspension systems and, 640
training, 75
Quench loads, 638
Quench origins, 75, 76
Quench performance, 39-46
Quench propagation velocity,
        76, 81-82
Quench protection, 321, 322
    in Fermilab Magnet Test
        Facility, 566-568
    in iron-free detector
        magnets, 631
    in superconducting
        solenoids, 86
Quench recovery, 297, 321,
        327
Quench start localization,
        73-82
Quench tolerance, 300

                R

Radhard process, 214-215
Radhard scintillators, 187,
        198
Radhard VLSI electronics,
        203-216
Radiation, see also
        Radioactivity;
        specific types
    barium fluoride crystal and,
        575-576
    baseline, 151
    calorimeters and, 204,
        207-211
    damage from, see Radiation
        damage
    detection devices and,
        191-193
    at liquid argon temperature,
        211-213
    prompt, 138
    scintillators and, 187,
        198-200, 537-549
    shielding requirements for,
        see Radiation
        shielding
        requirements
    synchrotron, 139, 605-610
    thermal, 483
    transient, 215
    types of, 138
Radiation damage
    calorimeters and, 203,
        204-205, 206
    to scintillators, 198-200
Radiation design goals,
        141-151

Radiation hardness, 203-216,
        see also Radhard
Radiation heat leak, 485-486
Radiation resistance
    of barium fluoride crystal,
        575-576
    of scintillators, 187,
        537-549
    testing for, 518
Radiation shielding
        requirements, 137-152
    baseline radiation and, 151
    design goals and, 141-151
    types of radiation and,
        138-140
Radioactivity, see also
        Radiation
    induced, 138, 139
    residual, 139
RadioFrequency Quadrupole
        (RFQ) accelerator, 3,
        4, 5, 6, 9, 11, 578
    calorimeter calibration and,
        587, 588, 589-590,
        591, 603
Ramp-splice quenches, 47-48
Rapid transit systems, 93
Raster control power supplies,
        18
Ratiometer tests, 67, 71
Recovery phenomenon, 198
Refractive index solvents,
        197
Refrigeration, 330, 341-348,
        see also Cryogenics
    capacity requirements of,
        291-293
    centrifugal pumps for, see
        Centrifugal pumps
    cold box in, 347-348
    computer simulation of,
        301-319
    in HERA, 722, 723
    hydrogen removal in, 343
    liquefaction in, 344,
        346-348
    liquid storage in, 346
    loss of capacity in
        compressor, 313-314
    oil removal in, 347
    plant operating requirements
        of, 297-299
    purification in, 343-344
    purity upgrading in,
        342-343
    superconducting solenoids
        and, 88-89
    system design of, 287-300
Residual gas conduction, 486
Residual radioactivity, 139

820